Lecture Notes on Data Engineering and Communications Technologies

Volume 116

Series Editor

Fatos Xhafa, Technical University of Catalonia, Barcelona, Spain

The aim of the book series is to present cutting edge engineering approaches to data technologies and communications. It will publish latest advances on the engineering task of building and deploying distributed, scalable and reliable data infrastructures and communication systems.

The series will have a prominent applied focus on data technologies and communications with aim to promote the bridging from fundamental research on data science and networking to data engineering and communications that lead to industry products, business knowledge and standardisation.

Indexed by SCOPUS, INSPEC, EI Compendex.

All books published in the series are submitted for consideration in Web of Science.

More information about this series at https://link.springer.com/bookseries/15362

V. Suma · Xavier Fernando · Ke-Lin Du ·
Haoxiang Wang
Editors

Evolutionary Computing and Mobile Sustainable Networks

Proceedings of ICECMSN 2021

 Springer

Editors
V. Suma
Department of Information Science
and Engineering
Research and Industry Incubation Center
Dayananda Sagar College of Engineering
Bengaluru, Karnataka, India

Ke-Lin Du
Department of Electrical and Computer
Engineering
Concordia University
Montreal, QC, Canada

Xavier Fernando
Ryerson Communications Lab
Toronto, ON, Canada

Haoxiang Wang
Go Perception Laboratory
Cornell University
Ithaca, NY, USA

ISSN 2367-4512   ISSN 2367-4520 (electronic)
Lecture Notes on Data Engineering and Communications Technologies
ISBN 978-981-16-9607-7   ISBN 978-981-16-9605-3 (eBook)
https://doi.org/10.1007/978-981-16-9605-3

This Springer imprint is published by the registered company Springer Nature Singapore Pte Ltd.
The registered company address is: 152 Beach Road, #21-01/04 Gateway East, Singapore 189721,
Singapore

The ICECMSN 2021 is solely dedicated to all the editors, reviewers, and authors of the conference event.

Foreword

I extend my warm welcome in inviting you all to the proceedings of the International Conference on Evolutionary Computing and Mobile Sustainable Networks (ICECMSN 2021) organized at the Department of CSE/ISE and ECE, RV Institute of Technology and Management, Bengaluru, India, on September 28–29, 2021.

The theme of the conference event is "Emerging Advances in Sustainable Mobile Networks and Computational Intelligence," topics that are quickly gaining research attention from both academia and industries due to the relevance of maintaining sustainability and enhancing intelligence in smart mobile networks. The already established track record of computational intelligence models and sustainable mobile networks seems to be very functional and reliable, where it mandates the need for further exploration in this research area. This makes the ICECMSN 2021 an excellent forum for exploring innovative research ideas in the smart and intelligent networks domain.

The entire success of the ICECMSN 2021 event depends on the research talents and efforts of the authors in the intelligent mobile networks and computer science domains, who have contributed their submissions on almost all the facets of the conference theme. An extensive appreciation is also deserved for all the conference program and review committee members who have invested their valuable time and professional expertise in assessing research papers from multiple domains by maintaining the quality standards for this conference. We extensively thank Springer for their guidance before and after the conference event.

Conference Chair
Dr. J. Anitha
HOD and Professor
Department of Computer Science and Engineering
RV Institute of Technology and Management
Bengaluru, India

Preface

It is our pleasure to welcome you to the International Conference on Evolutionary Computing and Mobile Sustainable Networks (ICECMSN 2021) at RV Institute of Technology and Management, Bengaluru, India. The main goal of this conference is to bring academicians, researchers, and industrialists together under one platform to share and exchange research experience and results on various aspects of mobile sustainable networks and computational intelligence research, as well as to discuss real-time challenges and solutions adopted for it.

ICECMSN 2021 has received ample submissions of about 382 papers from both academia and industrial tracks, and based on the selection of conference review committee and advisory committee members, a total of 74 papers appear in the proceedings of ICECMSN 2021. It is to be noted that all the papers regardless of their allotted tracks have extensively received at least three reviews from the research experts. We extend our sincere thanks to our keynote speakers "Dr. Manu Malek, Alcatel-Lucent Bell Labs and Stevens Institute of Technology (ret.), New Jersey, USA, and Dr. R. Kanthavel, Professor, Information and Computing Technology, University of Bisha, Saudi Arabia."

We hope the readers will have a productive, satisfying, and informative experience from the research works gathered from all over the world. Nevertheless, this proceeding will provide a written record of a synergy of research works that exists in communication networks communities and provides a significant framework for a new and futuristic research interactions. Moreover, this proceeding will pave way for the applications of computational intelligence in mobile sustainable networks (MSN).

Bengaluru, India   V. Suma
Toronto, Canada   Xavier Fernando
Montreal, Canada   Ke-Lin Du
Ithaca, USA   Haoxiang Wang

Acknowledgments

We would like extend our sincere thanks to all who have helped in making this conference event a great success. We are much pleased in thanking our educational institution RV Institute of Technology and Management, Bengaluru, India, for their pervasive support and effective help during the conference.

The extended support of the conference committee members before and during the conference event has helped to tackle many challenging tasks in a smooth and efficient way, where it has significantly contributed to excel the quality of the conference. Our special thanks belong to all the conference reviewers, who played an indispensable role in providing technical and semantic reviewing assistance to all the research manuscripts received for the conference. We are thankful for their help in guiding us to select the state-of-the-art high-quality papers that deserve the publication under this conference. We also wish to thank all our faculty members and staffs for their technical and non-technical contribution for maintaining the conference participants' contentment.

Our very special thanks will go exceptionally to all the conference delegates for their active participation in the conference event.

At last, the editors would like to gladly acknowledge the local organizing committee and conference organizers, who ensured that all the formal steps of the conference event have been completed in an effortless way.

Contents

About the Editors

Dr. V. Suma has obtained her B.E. in Information Science and Technology, M.S. in Software Systems, and her Ph.D. in Computer Science and Engineering. She has a vast experience of more than 17 years of teaching. She has published more than 183 international publications which include her research articles published in world class international journals such as ACM, ASQ, Crosstalk, IET Software, and International journals from Inderscience publishers, from journals released in MIT, Darmout, USA, etc. Her research results are published in NASA, UNI trier, Microsoft, CERN, IEEE, ACM portals, Springer, and so on.

Dr. Xavier Fernando is Director of Ryerson Communications Lab that has received total research funding of $3,185,000.00 since 2008 from industry and government (including joint grants). He has (co-)authored over 200 research articles, three books (one translated to Mandarin), and several chapters and holds few patents. The present and past members and affiliates of this laboratory can be found at this LinkedIn group and Facebook page. He was IEEE Communications Society Distinguished Lecturer and has delivered over 50 invited talks and keynote presentations all over the world. He was Member in the IEEE Communications Society (COMSOC) Education Board Working Group on Wireless Communications. He was Chair of IEEE Canada Humanitarian Initiatives Committee 2017–2018. He was also Chair of the IEEE Toronto Section and IEEE Canada Central Area.

Dr. Ke-Lin Du received a Ph.D. degree in electrical engineering from Huazhong University of Science and Technology (HUST) with a thesis on artificial intelligence and robotics in 1998. Currently, Ke-Lin Du is Member of editorial board of IEEE Spectrum Chinese Edition and is Associate Editor of IET Signal Processing and Circuits, Systems & Signal Processing. Ke-Lin Du also served on program committees of dozens of international conferences. His current research interest is in signal processing for wireless communications, machine learning, and intelligent systems. On February 14, 2018, Ke-Lin Du was included in Marquis Who's Who for professional excellence.

Dr. Haoxiang Wang is currently Director and Lead Executive Faculty Member of GoPerception Laboratory, NY, USA. His research interests include multimedia information processing, pattern recognition and machine learning, remote sensing image processing and data-driven business intelligence. He has co-authored over 60 journal and conference papers in these fields on journals such as *Springer MTAP, Cluster Computing, SIVP*; *IEEE TII, Communications Magazine*; *Elsevier Computers & Electrical Engineering, Computers, Environment and Urban Systems, Optik, Sustainable Computing: Informatics and Systems, Journal of Computational Science, Pattern Recognition Letters, Information Sciences, Computers in Industry, Future Generation Computer Systems*; Taylor & Francis *International Journal of Computers and Applications* and conference such as IEEE SMC, ICPR, ICTAI, ICICI, CCIS, and ICACI. He is Guest Editor for *IEEE Transactions on Industrial Informatics, IEEE Consumer Electronics Magazine, Multimedia Tools and Applications, MDPI Sustainability, International Journal of Information and Computer Security, Journal of Medical Imaging and Health Informatics*, and *Concurrency and Computation: Practice and Experience*.

Improved Grey Wolf Optimization-Based Feature Selection and Classification Using CNN for Diabetic Retinopathy Detection

Anas Bilal, Guangmin Sun, Sarah Mazhar, and Azhar Imran

Abstract This research offers a new prediction structure coupling improved grey wolf optimization (IGWO) and convolutional neural network (CNN), called IGWO-CNN, to diagnose diabetic retinopathy. Grey wolf optimizer (GWO) is achieving success among other swarm intelligence procedures due to its broad tuning features, scalability, simplicity, ease of use and, most importantly, its ability to ensure convergence speed by providing suitable exploration and exploitation throughout a search. The suggested methodology used a genetic algorithm (GA) to build diversified initial positions. GWO was subsequently applied to adjust existing population positions in the discrete search procedure, getting the optimal feature subset for a higher CNN-based classification challenge. The presented technique contrasts with GA, GWO and numerous existing state-of-the-art diabetic retinopathy classification approaches. The suggested strategy outperforms all other methods by increasing classification accuracy to 98.33%, indicating its efficacy in detecting the DR. The simulation outcomes have shown that the proposed approach outperforms the other two competing methods.

Keywords Grey wolf optimization · Genetic algorithm · Convolutional neural network · Diabetic retinopathy · Feature extraction · Classification

A. Bilal · G. Sun (✉) · S. Mazhar
Faculty of Information Technology, Beijing University of Technology, Chaoyang District, Beijing, China
e-mail: gmsun@bjut.edu.cn

S. Mazhar
e-mail: sarah.mazhar@emails.bjut.edu.cn

A. Imran
School of Software Engineering, Beijing University of Technology, Chaoyang District, Beijing, China

1   Introduction

Diabetic retinopathy (DR) is a syndrome in which the retina is damaged as a result of diabetic retinopathy complications, resulting in irreversible eye impairment and, in some cases, blindness. The possibility for untreated patient blindness highlights this type of problem among patients. Figures from the International Diabetes Federation (IDF) suggest that over 425 million diabetic adults have an expected 629 million by 2045. Most people are aged 40 to 59, with 1 out of 2 among 212 million people ignorant of their disease [1].

Previous research has shown that diabetic retinopathy is the most common chronic issue among early-stage health professionals. Therefore, several techniques based on feature selection, such as lesion shape uniformity and lesion size, were employed to diagnose diabetic retinopathy.

Various classification systems have been established on several generally accessible datasets. Mateen et al. [2] offered a methodology for classifying fundus images based on a deep convolution network. A Gaussian mixture model was employed for region segmentation, SVM and PCA for feature selection, VGGNet for feature extraction and softmax to classify retinal fundus images. The studies used the Kaggle datasets, AlexNet and the spatial invariant feature transformation offered better results. The authors in [3] performed a retinal segmentation and acquired numerous critical metrics for better retinal segmentation. For effective segmentation of retinal fundus image, the authors adopted PixelBNN architecture. The suggested system was tested using CHASE DB1, STARE and DRIVE retinal datasets. Using F1-score, the efficacy and computing speed of the technique were examined.

The authors in [4] described a new methodology that utilizes a random forest classifier (RFC) to assist ophthalmologists in effectively identifying anomalies in retinal images. To determine the retinal images, the authors used a K-means clustering-based method and machine learning approaches. Experiments indicated that the random forest classifier correctly diagnoses 93.58% retinal disease accuracy. This is better than the 83.63% accuracy of the naive Bayes classifier. The authors proposed in [5] a computational diagnostic system that relies on generic type 2 fuzzy logic and exhibited applications that depend on a type 2 fuzzy inference system. They lowered computing costs to address data uncertainty. In [6], the authors provide a detailed survey on EIT by variable current pattern techniques, and in [7], a detailed overview of deep learning image preprocessing is provided. In [8], modified blue whale algorithm based on the bat and local search algorithm is proposed.

Evolutionary computing (EC), an essential member of the metaheuristic class, has also got a lot of attention. Several EC-based feature selection algorithms have been offered. The authors of [9] use a genetic approach (GA) to select features. The authors in [10] recommended that binary particle swarm optimization (PSO) be used to determine features. The authors investigated the use of tabu search in the selection of features [11]. Compared to the outlined in the previous EC procedures, grey wolf optimization (GWO) is a novel EC methodology newly developed [12]. GWO is based on the social structure and hunting grey wolves' behaviour in the natural environment. Its superior search capabilities successfully tackle a wide range of real-world concerns, including the optimal reactive power dispatching problems [13], surface acoustic wave parameter approximation [14] and static VAR compensator design [15].

It should be emphasized that the original GWO's initial population is created at random. As a consequence, there may be a lack of diversity among wolf flocks throughout the search area. Numerous studies have shown that the starting population's performance significantly impacts the global convergence rate and the desired outcome for flock intelligence optimization methodologies. Therefore, the initial population size with high diversity is advantageous in enhancing the efficiency of initial population optimization techniques. Encouraged by this underlying principle, we sought to employ GA to create a better initial population. Then, using the diversified population, a binary form of GWO was offered to execute selecting features.

Conversely, the IGWO approach, in cooperation with CNN, has been recommended to increase the system's classification efficiency further. The recommended IGWO-CNN approach would assess the optimal combination of features to enhance the classification effectiveness of the CNN classifier. IGWO will train the CNN classifier in an optimum subset of features.

1.1 Problem Statements and Contributions

Detection of retinal image lesions is challenging and has low accuracy owing to the complexity of eye structure. Therefore, there is a need to develop a new technique to extract features from images and guaranty the correct classification of diabetic retinopathy. In this study, we proposed the following main objectives:

- A new prediction mechanism based on GWO and CNN is demonstrated.
- GA is implemented in the IGWO to provide GWO with the most suitable initial positions.
- To classify diabetic retinopathy by using improved CNN integrated with IGWO.
- To improve the accuracy of diabetic retinopathy detection based on novel combo technique based on IGWO-CNN as IGWO-CNN methodology has not been tested and developed till now for the diabetic retinopathy detection and classification.
- The established structure, IGWO-CNN, is effectively analysed in terms of diabetic retinopathy detection problems and has achieved superior classification.

The proposed method's classification output has been validated and contrasted to various current, state-of-the-art and newly reported approaches to diabetic retinopathy detection and classification. Therefore, the rest of the article is structured as follows: Sect. 1 provides a concise outline of the proposed method. Sect. 2 explains the GWO, IGWO, GA and CNN. The proposed methodology is discussed in Sect. 3. Performance, discussions and contrasts with the current study are provided in Sect. 4, and the work is concluded in Sect. 5.

2 Background

2.1 Grey Wolf Optimization

Mirjalili et al. [9] developed the grey wolf optimizer (GWO) in response to the grey wolves' leadership and hunting behaviour framework.

The initial wolf populations are randomly identified in GWO. And, α, β and δ wolves are the first, second, and third finest solutions. Each grey wolf's position may be expressed in the form of a vector as follows $\vec{X} = \vec{x_1} + \vec{x_2} + \cdots + \vec{x_n}$ where n indicates the search space dimension. α, β and δ wolves move towards prey, and ω wolves approach them. Wolves hunting technique is characterized as follows: (i) encircling prey mechanism, (ii) hunting prey mechanism and (iii) attacking prey.

2.1.1 Encircling Prey Mechanism

In this mechanism, wolves modify the positions during the optimization by representing the positions of α, β and δ wolves. The following Eqs. 1 and 2 express this mathematically.

$$\vec{D} = \left| \vec{C} \cdot \vec{X}_p(t) - \vec{X}(t) \right| \tag{1}$$

$$\vec{X}(t+1) = \vec{X}_p(t) - (\vec{A}\vec{D}) \tag{2}$$

where t indicates the current iteration, $\vec{X}_p$ signifies the position of prey, $\vec{X}$ represents the wolfs' position, $\vec{A}$ and $\vec{C}$ show the coefficient vectors, and $\vec{a}$ is reduced linearly from 2 to 0 throughout the iteration courses. It is used to approach the solution range. $\vec{r}_1$ and $\vec{r}_2$ represent the random vectors in the range of 0 and 1 as displayed below in Eqs. 3–5, respectively.

$$\vec{A} = 2\vec{a} \cdot \vec{r} - a \tag{3}$$

$$\vec{C} = 2\vec{r}_2 \tag{4}$$

$$\vec{a} = 2\left(1 - \frac{y}{T}\right) \tag{5}$$

2.1.2 Hunting Prey Mechanism

Hunting activity is performed by α, β and δ wolves; they have a superior understanding of prey position. The three finest solutions, as described in Eq. 6, are used by ω wolves to update their position.

$$\vec{D}_\alpha = \left| \vec{C}_1 \cdot \vec{X}_\alpha - \vec{X} \right|, \quad \vec{D}_\beta = \left| \vec{C}_2 \cdot \vec{X}_\beta - \vec{X} \right| \quad \vec{D}_\delta = \left| \vec{C}_3 \cdot \vec{X}_\delta - \vec{X} \right| \qquad (6)$$

$\vec{X}$ represents the wolfs' position. $\vec{X}_\alpha$ and $\vec{D}_\alpha$ indicate the current and updated α wolves position. $\vec{X}_\beta$ and $\vec{D}_\beta$ indicate the current and modified β wolves position. $\vec{X}_\delta$ and $\vec{D}_\delta$ indicate the current and modified δ wolves position. After computing distances, the absolute position of current wolves $\vec{X}_1$, $\vec{X}_2$ and $\vec{X}_3$ is estimated as representing in Eqs. 7 and 8.

$$\vec{X}_1 = \vec{X}_\alpha - \vec{A}_1 \cdot \vec{D}_\alpha, \quad \vec{X}_2 = \vec{X}_\beta - \vec{A}_2 \cdot \vec{D}_\beta, \quad \vec{X}_3 = \vec{X}_\delta - \vec{A}_3 \cdot \vec{D}_\delta \qquad (7)$$

$$\vec{X}(t+1) = \frac{\vec{X}_1(t+1) + \vec{X}_2(t+1) + \vec{X}_3(t+1)}{3} \qquad (8)$$

where $\vec{A}_1$, $\vec{A}_2$ and $\vec{A}_3$ represent the random vectors and t signifies the current number of iteration.

2.1.3 Attacking Prey

While seeking prey (exploration), grey wolves diverge from each other; when attacking prey (exploitation), they converge.

2.2 Improved Grey Wolf Optimization

The GWO core is gaining favour among other approaches because of its extensive tuning parameters, scalability, simplicity, ease of use and, most significantly, its ability to guarantee convergence speed by assuring appropriate exploration and exploitation throughout a search. Why would the dominant alpha-searching agent employ low-beta and delta-searching agents to update their position or use a lower-delta-searching agent to update their location? This is the core issue with GWO. This is a significant contributor to the pack's inability to achieve the global optimal performance level.

As a result, selecting the three most crucial primary search agents is critical in each iteration. In this part, initially, utilize GA to construct GWO's initial position. Second, we offer an improved GWO (IGWO) method that enhances the leader selection procedure and prevents premature convergence owing to local optimum stagnation. α, β and δ search agents are created to indicate diverse solutions to explore the globally best solution (to encircle the prey). Finally, we established the fitness sharing principle to increase the diversity of GWO's solutions. Fitness sharing is a mechanism where a search agent's fitness is combined with other search agents circling a comparable solution (or a peak). Merging the fitness sharing mechanism with the GWO core, the suggested IGWO method can effectively identify all global objective function solutions and prevent convergence to a local solution.

2.3 Genetic Algorithm (GA)

GA is an adaptive optimization search approach based on the biological system genetics and the analogy of Darwinian natural selection, which was initially presented by Holland [13]. A population in GA is made up of chromosomes, which are candidate solutions. Several genes containing binary values of 0 and 1 are found on each chromosome. GA was utilized to establish the GWO initial positions for this investigation.

The phases of GA's initialization positions are explained below.

- Initialization: chromosomes are created randomly.
- Selection: selecting parent chromosomes uses a roulette selection strategy.
- Crossover: construct offspring chromosomes using a single-point crossover approach.
- Mutation: it adopts uniform mutation.
- Decode: decode mutated chromosomes as the population's initial positions.

2.4 Convolutional Neural Network

The importance of the proposed CNN for diabetic retinopathy detection is provided in this section. CNNs are helpful in situations in which we handle enormous and complex dataset structures such as images. CNNs have fewer connections and fewer parameters to learn than simple feedforward networks with the same number of layers [16]. Therefore, CNNs are a popular selection for object detection and recognition. There are several applications in which CNNs are beneficial, such as person recognition using the face, finger veins, palm veins and traffic sign recognition. The structures of CNNs can be varied to improve their accuracy. Owing to the abstruse structure of images and other similar data, manual feature extraction becomes difficult, and thus, CNN is a better option for achieving a holistic solution. CNN automatically extracts and learns features and abates specific field knowledge impediments to obtaining a probable solution.

3   Methodology

3.1   Proposed IGWO-CNN Approach

This research offers a novel procedure for diagnosing diabetic retinopathy using retinal imaging. This paper provides a new computational diagnostic framework called IGWO-CNN. The design suggested contains two main parts. In the first step, IGWO eliminates irrelevant and redundant information by adaptively examining the optimal combination of medical data features. Firstly, the designed IGWO uses GA to create the population's initial position. Then, GWO is used to modify current population positions in search space. The efficient and effective CNN classifier is applied in the second phase, depending on the best feature subset acquired in the first phase. The designed IGWO-CNN mechanism is shown in Fig. 1. The IGWO is used chiefly for adaptively exploring feature space in search of the optimal feature combinations. The optimum diversity of features is one with the most significant classification precision and a minimal number of the selected features.

To estimate the selected features, the fitness function employed in the IGWO is expressed as the following expression:

$$\text{Fitness} = \alpha P = \beta \frac{N - L}{N} \tag{9}$$

where P represents the classification model precision, N indicates the total quantity of features in the dataset, L means the dimension of the selected feature subset, and α and β are two factors of classification accuracy weight and selection feature, $\alpha \in [0, 1]$ and $\beta = 1 - \alpha$. Figure 2 shows a feature selection flag vector.

A subset of features, real feature vectors, is represented by vectors containing a sequence of binary numbers 0 and 1, which has been standardized [17]. There are n bit pieces in the vector for challenges with dimension n. If the bit value equals one, the feature will be selected; otherwise, it is not collected ($i = 1, 2 \ldots, n$). Thus, a feature subset's size is the number of bits, where its values equal one in the vector.

3.2   Classification

For classification, we make a CNN-based classifier that contains 13 layers, including the classification layer. Before passing into CNN, we resized the data from 512×512 to $227 \times 227 \times 3$ in 3D dimensions. The first layer of the image input layer is getting resized images as input with zero centre normalization. Second layer is the convolutional layer, which filters 20 filters of 5×5 area in a single image. In the next layer, we perform activation using the ReLU function. In the fourth layer, we performed max-pooling by setting stride equal to 2. We again applied 20 filters of the 5×5 area in the next convolutional layer and then passed it to the next layer of ReLU

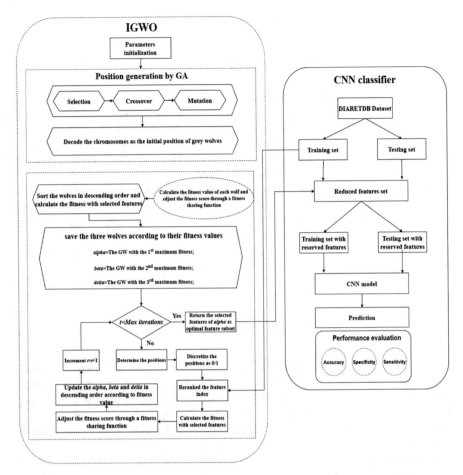

Fig. 1 IGWO-CNN flowchart

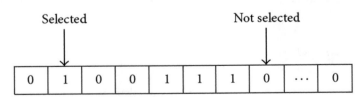

Fig. 2 Flag vector for feature selection

activation. In the seventh layer, we again applied max-pooling of 2D by setting stride as 2. We again used the same three layers to get a more depth max-pooled image to pass the fully connected layer, making the two types of classes.

By starting the training of CNN, we set some parameters and we set random epochs to check how many iterations are necessary to get a good classification accuracy. After applying different training options, we get 98.33% accuracy with a mini-batch loss value of 0.0279. We can see the training time, validation accuracy in Fig. 3a. Accuracy and validation loss in Fig. 3b. Loss. The segmented and preprocessing process was so strong. When we start training, we can see that validation accuracy reaches 90% when training reaches five epochs.

Figure 4 represents the exudate extraction, and Fig. 5a, b represent proliferative and non-proliferative types of hard exudates.

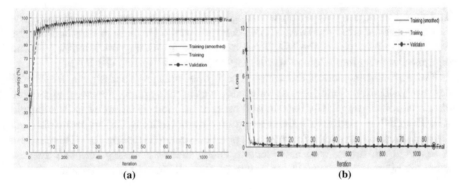

Fig. 3 **a** Accuracy, **b** loss

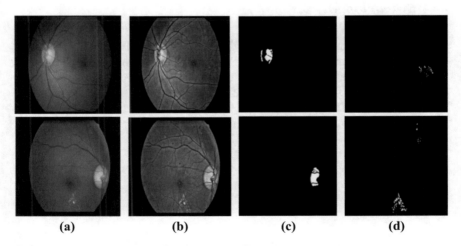

Fig. 4 Edudate extraction (**a, b**) Input image and contrast enhancement image (**c, d**) Optic disc and hard exudates detection

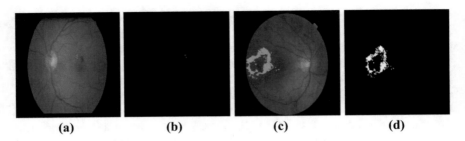

(a)   (b)   (c)   (d)

Fig. 5 **a, b** NPDR, **c, d** PDR

3.3 Experimental Setup

The experimentation was performed using the diabetic retinopathy Stare dataset. The feature retrieved from the DIARETDB dataset was the attribute in this dataset. The tests were carried out in the MATLAB environment, running on Windows 10 and equipped with an Intel Core i5 4210H CPU (2.90 GHz) and 32 GB RAM.

Before feature selection, this analysis performed data preprocessing in the form of data normalization. Tenfold cross-validation was used to attain unbiased and reliable classification findings. The dataset will be split into ten equal sections in a tenfold CV. The CNN classifier is trained using nine parts (training set) in each fold, and the classification precision for the tenth part is calculated (testing set). This process is repeated ten times, with each section designated as a testing set. The average accuracy of each fold is used to calculate the overall classification accuracy.

3.4 Evaluation Metrics

Specificity (spc), accuracy (acc), sensitivity (sen), precision (prc) and $F1$-score (f) are the performance measures evaluated. Suppose, TP is the number of appropriately classified DR samples in a certain cross-validation run.

In that example, TN is the number of adequately classified normal cases. $N = TN + FN$ and $P = TP + FP$ are the total numbers of normal and DR and normal in the testing data. These performances are determined as:

$$acc = \frac{TP + TN}{TP + FP + FN + TN} \times 100\% \tag{10}$$

$$spc = \frac{TN}{TN + FP} \times 100\% \tag{11}$$

$$Sen = \frac{TP}{TP + FN} \times 100\% \tag{12}$$

4 Results and Discussions

4.1 Dataset

Using EyePACS, California Healthcare Foundation's most enormous publicly accessible dataset for diabetic retinopathy, collected via the Kaggle Diabetic Retinopathy Detection competition [18], EyePACS has 88,702 retinal fundus images combined under diverse imaging circumstances. Due to 75% of Level 0 (no DR) images, data collection is substantially imbalanced. The dataset was split into 35 parts: the training set contains 126 images, and the test set has 576 images as specified by the setting of the Kaggle competition [18]. We also utilized 96% of each class' images in the training phase and 4% for validation.

4.2 Results and Analysis

The classification efficacy of the proposed IGWO-CNN methodology on the Kaggle dataset is provided in this section. It is compared to numerous state-of-the-art and newly published breast cancer prediction models.

To assess the performance of the suggested technique on the DR prediction issue, tests were conducted among IGWO-CNN and two other competing approaches (namely GWO-CNN and GA-CNN). The classification precisions of each strategy were estimated using a tenfold CV, and the mean values across ten times of the tenfold CV were employed as the final test findings. The suggested model is equated to numerous state-of-the-art approaches, including the VGG-19 architecture with SVD and PCA [2], random forest classifier [4], PixelBNN deep technique [3], adaptive histogram equalization, Gabor and top hat transformation and iterative thresholding technique [17]. Performance measures such as sensitivity, accuracy and specificity are used to assess various techniques. Table 1 shows the results of the comparison. In all assessments, the suggested approach has outperformed others.

Figure 6 outlined the iteration course to analyse algorithm optimization methodology, namely GA, GWO and IGWO. Figure 6 shows that IGWO's fitness curve finally converges after the 17th iterations. On the other hand, GA and GWO's fitness curves eventually intersect after the thirtieth iterations and forty-fifth iterations, respectively. It indicates that the proposed IGWO is much better than the two previous techniques and can quickly find the most satisfactory solution in the search area. In addition, we can observe that the fitness value of IGWO is always better than GWO and GA throughout the iteration process.

Table 1 Comparison of IGWO with the state-of-the-art methods

Author	Methods	Accuracy (%)	Specificity (%)	Sensitivity (%)
Mateen et al. [2]	VGG-19 architecture with PCA and SVD	96.34	–	–
Leopold et al. [3]	PixelBNN deep method	91	–	–
Chowdhury et al. [4]	Morphological operations and Gaussian mixture model	95.20		
Adem et al. [19]	AHE, Gabor and top-hat tranformations, iterative thresholding approach	94.10	91.40	96.70
Proposed	IGWO-CNN	98.33	93.37	100
	GWO-CNN	95.90	92.69	96.34
	GA-CNN	92.90	92.50	95.67

Fig. 6 Comparison of fitness among three approaches using the Kaggle dataset

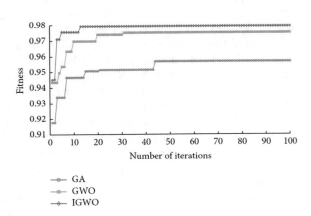

5 Conclusion

This work enables automated retinopathy identification and classification based on IGWO-CNN. The suggested methodology has two parts: features selection and classification. First, the IGWO was utilized to select the most significant features, particularly for medical data. Secondly, the very effective CNN classifier was used to anticipate the first-step representative feature subset generated. The suggested technique on the Kaggle dataset is associated with famous approaches like GA and GWO for selecting the features, which use a set of standards to analyse essential aspects of the proposed model. Simulation findings show that the suggested IGWO methodology converges faster, providing much more excellent solution quality and collecting fewer selected features while achieving satisfactory classification results.

We will develop the recommended methodology for different realistic situations and strive to incorporate our methods in future work alongside high-performance tools.

References

1. Global report on diabetes. World Health Organization (2016). https://www.who.int/
2. Mateen, M., Wen, J., Nasrullah, Song, S., Huang, Z.: Fundus image classification using VGG-19 architecture with PCA and SVD. Symmetry (Basel) **11**(1) (2019). https://doi.org/10.3390/sym11010001
3. Leopold, H.A., Orchard, J., Zelek, J.S., Lakshminarayanan, V.: PixelBNN: augmenting the Pixelcnn with batch normalization and the presentation of a fast architecture for retinal vessel segmentation. J Imag **5**(2) (2019). https://doi.org/10.3390/jimaging5020026
4. Chowdhury, A.R., Chatterjee, T., Banerjee, S.: A Random Forest classifier-based approach in the detection of abnormalities in the retina. Med. Biol. Eng. Comput. **57**(1) (2019). https://doi.org/10.1007/s11517-018-1878-0
5. Ontiveros-Robles, E., Melin, P.: Toward a development of general type-2 fuzzy classifiers applied in diagnosis problems through embedded type-1 fuzzy classifiers. Soft Comput. **24**(1) (2020). https://doi.org/10.1007/s00500-019-04157-2
6. Babikir Adam, E.E., Sathesh, Survey on medical imaging of electrical impedance tomography (EIT) by variable current pattern methods. J. ISMAC **2**(2) (2021). https://doi.org/10.36548/jismac.2021.2.002
7. Ranganathan, G., A study to find facts behind preprocessing on Deep Learning algorithms. J. Innov. Image Process. **3**(1) (2021). https://doi.org/10.36548/jiip.2021.1.006
8. Dutta, S., Banerjee, A.: Highly precise modified Blue Whale Method framed by blending Bat and Local Search Algorithm for the optimality of Image Fusion Algorithm. J. Soft Comput. Paradig. **2**(4) (2020). https://doi.org/10.36548/jscp.2020.4.001
9. Raymer, M.L., Punch, W.F., Goodman, E.D., Kuhn, L.A., Jain, A.K.: Dimensionality reduction using genetic algorithms. IEEE Trans. Evol. Comput. **4**(2) (2000). https://doi.org/10.1109/4235.850656
10. Tanaka, K., Kurita, T., Kawabe, T.: Selection of import vectors via binary particle swarm optimization and cross-validation for kernel logistic regression (2007). https://doi.org/10.1109/IJCNN.2007.4371101
11. Zhang, H., Sun, G.: Feature selection using Tabu search method. Pattern Recognit. **35**(3) (2002). https://doi.org/10.1016/S0031-3203(01)00046-2
12. Mirjalili, S., Mirjalili, S.M., Lewis, A.: Grey wolf optimizer. Adv. Eng. Softw. **69**:46–61 (2014)
13. Sulaiman, M.H., Mustaffa, Z., Mohamed, M.R., Aliman, O.: Using the gray wolf optimizer for solving optimal reactive power dispatch problem. Appl. Soft Comput. J. **32** (2015). https://doi.org/10.1016/j.asoc.2015.03.041
14. Song, X., et al.: Grey Wolf Optimizer for parameter estimation in surface waves. Soil Dyn. Earthq. Eng. **75** (2015). https://doi.org/10.1016/j.soildyn.2015.04.004
15. Mohamed, A.A.A., El-Gaafary, A.A.M., Mohamed, Y.S., Hemeida, A.M.: Design static VAR compensator controller using artificial neural network optimized by modify Grey Wolf Optimization. In: Proceedings of the International Joint Conference on Neural Networks, vol. 2015, Sept 2015. https://doi.org/10.1109/IJCNN.2015.7280704
16. Krizhevsky, A., Sutskever, I., Hinton, G.E.: ImageNet classification with deep convolutional neural networks. Commun. ACM **60**(6) (2017). https://doi.org/10.1145/3065386

17. Huang, D.S., Yu, H.J.: Normalized feature vectors: a novel alignment-free sequence comparison method based on the numbers of adjacent amino acids. IEEE/ACM Trans. Comput. Biol. Bioinforma **10**(2) (2013). https://doi.org/10.1109/TCBB.2013.10
18. Kaggle: Diabetic Retinopathy Detection (2015) [Online]. Available: https://www.kaggle.com/c/diabetic-%0Aretinopathy-detection
19. Adem, K., Hekim, M., Demir, S.: Detection of hemorrhage in retinal images using linear classifiers and iterative thresholding approaches based on firefly and particle swarm optimization algorithms. Turkish J. Electr. Eng. Comput. Sci. **27**(1) (2019). https://doi.org/10.3906/elk-180 4-147

Feature Selection Using Modified Sine Cosine Algorithm with COVID-19 Dataset

Miodrag Zivkovic⑩, Luka Jovanovic⑩, Milica Ivanovic⑩, Aleksa Krdzic⑩, Nebojsa Bacanin⑩, and Ivana Strumberger⑩

Abstract The research proposed in this paper shows application of the sine cosine swarm intelligence algorithm for feature selection problem in the machine learning domain. Feature selection is a process that is responsible for selecting datasets' features that have the biggest effect on the performances and the accuracy of the system. The feature selection task performs the search for the optimal set of features through a enormous search space, and since the swarm intelligence metaheuristics have already proven their performances and established themselves as good optimizers, their application can drastically enhance the feature selection process. This paper introduces the improved version of the sine cosine algorithm that was utilized to address the feature selection problem. The proposed algorithm was tested on ten standard UCL repository datasets and compared to other modern algorithms that have been validated on the same test instances. Finally, the proposed algorithm was tested against the COVID-19 dataset. The obtained results indicate that the method proposed in this manuscript outperforms other state-of-the-art metaheuristics in terms of features number and classification accuracy.

Keywords Sine cosine algorithm · Swarm intelligence · Feature selection · Classification · COVID-19

M. Zivkovic (✉) · L. Jovanovic · M. Ivanovic · A. Krdzic · N. Bacanin · I. Strumberger
Singidunum University, Danijelova 32, 11000 Belgrade, Serbia
e-mail: mzivkovic@singidunum.ac.rs

L. Jovanovic
e-mail: luka.jovanovic.191@singimail.rs

M. Ivanovic
e-mail: milica.ivanovic.17@singimail.rs

A. Krdzic
e-mail: aleksa.krdzic.19@singimail.rs

N. Bacanin
e-mail: nbacanin@singidunum.ac.rs

I. Strumberger
e-mail: istrumberger@singidunum.ac.rs

1   Introduction

Rapid developments in information science have resulted in a dramatic increase in dataset dimensions over the past decade. Potential dimension reduction algorithms are needed to remove redundant or irrelevant information from these datasets, since these features can lead to reduced performance of learning algorithms [22].

Typically considered a mechanism for preprocessing, feature selections are used for decreasing the total number of input variables, as well as finding the most relevant subset from a complete features set. Feature selection reduces the dimensionality of data by removing the noise and irrelevant attributes. This challenge is very important, especially when the real-time classification is needed by finding optimal or near-optimal subset of features, the training process can be shortened and classification accuracy can be improved. It is applied so as to increase the precision of prediction results given by the machine learning model, by reducing complexity, and diminishing redundant and irrelevant features in the dataset. This can be crucial in case of some critical applications, such as medical diagnostic [10]. Feature subset evaluation and search strategy are the two primary stages of preprocessing. Search strategy uses techniques for subset feature selection, while feature subset evaluation utilizes a classifier for evaluating the quality for the selected feature subset. All methods for feature selection, according to reviewed literature, are defined as either filter based or wrapper based.

Metaheuristic algorithms are considered the most reliable and efficient techniques for optimization and show great results when applied to problems considered more challenging or with higher-dimensional datasets. As a result, these algorithms show great promise and have been applied to many real-world problem that require optimization and performance improvements [3, 4, 25, 32, 34, 36]. Although these algorithms are often nature inspired, this is not necessarily always the case as shown in the sine cosine algorithm (SCA) [20].

Because of the high accuracy results achieved, as well as the reduced computational times when compared to traditional discrete methods, the metaheuristic approach has been employed by researchers, in wrapper-based methods, when solving the problem of feature selection. A Gaussian mutational chaotic fruit fly optimization algorithm's [31] application has been suggested for tackling the problem of feature selection, specifically to classification tasks. An augmented model of the dragonfly algorithm (DA), the hyper-learning binary dragonfly algorithm (HLBDA), has been implemented for feature evaluation and utilized on coronavirus (COVID-19) datasets [28].

SCA is a population-based algorithm, named after its use of the sine and cosine functions in its formulation, originally intended for use in solving optimization problems [20]. The algorithm initially creates a collection of multiple randomized solutions requiring them to fluctuate toward the best solution during the exploitation phase or outwards to encourage exploration employing a mathematical model formed from the sine cosine functions.

Some deficiencies were observed in the original SCA while performing practical empirical simulations with standard unconstrained benchmarks. Because of this, we have attempted to improve the basic SCA by performing hybridization with the well-known ABC algorithm. The mSCA is benchmarked using ten datasets form the University of California, Irvine (UCI) repository, and Arizona State University, as well as a single dataset of the coronavirus disease (COVID-19).

The main contribution of this conducted research can be outlined in the following:

- Proposal of a mSCA applied to the problem of feature selection elements of the ABC algorithm is integrated into the SCA to improve exploratory behavior.
- Testing the mSCA on ten standard benchmark datasets with low medium and high dimensions sets represented.
- Comparing the mSCA to other advanced feature selection algorithms and demonstrating the improvements made.
- Applying the proposed mSCA to solving a case study of COVID-19.

The remainder of this article is organized according to the following order: Sect. 2 shows a summary of the reviewed literature. Section 3 consists of a description of the original SCA. Section 4 shows experimental results and discusses the findings based on said results. Section 5 summarizes the findings and presents proposals for the direction of further work in this field.

2 Literature Review

When we have large datasets that are too difficult to classify, the use of swarm intelligence-based algorithm is suggested. Each large dataset contains features that are insignificant and irrelevant which can prove to be difficult when trying to analyze and interpret data. Swarm intelligence algorithm's purpose is to reduce dimensionality (feature selection) by keeping only useful features and those containing rich information. As a result of using dimensionality reduction technique, we have better understanding and interpretation of data, as well as higher accuracy of the results. There are two main steps in dimensionality reduction process, extracting features and selecting features. Before any further explanation of those features, we should give a short overview of swarm intelligence algorithms.

Swarm intelligence algorithms are part of the artificial intelligence (AI) field, and they are so-called nature-inspired metaheuristics [29]. Many groups of animals form collective intelligence which means that every member acts independently, but they mutually exchange information. That information eventually takes the group toward the optimal solution of their problem. Such animal colonies are ants, birds, hawks, fish, and more [16, 21, 29]. Nature-inspired metaheuristics are not that efficient at finding the most optimal solutions inside the search area, but they are efficient at finding the candidate solutions. Furthermore, they are especially good at finding possible solutions inside very large search areas. Because they take unreasonable amount of time to find the most optimal solution, swarm intelligence algorithms are also classified as NP-hard problems [15]. Many diverse problems can be solved with swarm intelligence algorithms such as wireless sensor network optimization [4, 32], cloud computing [6, 8, 35] and optimization of neural networks [2, 5, 12, 24], machine learning, and COVID-19 prediction [33], all the way to solving complicated problems in the field of medicine [7].

In order to prepare raw and unprocessed data, feature extraction is used [17]. A new dataset is formed by keeping some of the core features after which new features can be derived. Eventually, we have a new dataset that is cleaner, containing only features relevant to the specific problem and with fewer dimensions compared to the original dataset.

Since we have our most relevant and important data after feature extraction, the next step is feature selection. With feature selection, we select attributes previously defined in original dataset. This step is extremely important since the combination of the right attributes can improve the model's performance and accuracy. A common example of feature selection, alongside feature extraction, is image processing and analysis. Large amount of statistical features can be retrieved from the image, but a combination of only a few gives satisfactory results.

A side effect of feature selection is a possible loss of a certain amount of information, but, due to achieving simplicity of the model and significant performance improvement, it is well worth it. There are three distinct categories of techniques for selecting features, the wrapper, filter, and embedded technique [9].

Filter techniques choose the features that should contain the most information, without taking into consideration whether there are any relationships between the features or not. Wrapper techniques choose features that are most accurate to our machine learning model by going through all feature combinations. As for the embedded technique, the features are chosen while the model is still being constructed [9]. With these techniques, a decent performance can be achieved on relatively small datasets, but, for larger datasets, because of the decline in performance, a different method should be used such as swarm intelligence algorithm. In a reasonable amount of computational time, satisfactory results on large datasets are provided by the algorithm.

3   Original and Proposed Modified Sine Cosine Algorithm

SCA originally designed with the purpose of solving optimization problems, and first introduced by Seyadali Mirjalili [20], is a generally new population-based algorithm. The algorithm stochastically looks for the most optimum solution to our problems. At the very beginning, it starts with a randomized set of solutions, then repeatedly evaluates this set against an objective function, and follows a given ruleset that forms the core of the given optimization technique. As such, finding the most optimal solution in the first iteration is not guaranteed; however, given enough iterations and a large enough collection of randomized solutions, the probability of the global optimal solution being found increases.

Fig. 1 Sine and cosine effects on the upcoming position from Eq. 1

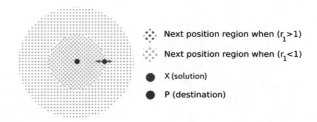

The process of optimization in the stochastic population-based approach, regardless of the algorithm being applied, can be split across two distinct phases: exploration phase and exploitation phase. In the exploration phase, the algorithm quickly, in a very random manner, combines solutions from a given random set, looking through the search space for the most favorable regions. With the exploitation phase, the changes are gradually made, however, noticeably less severe than those from the exploitation phase.

The original SCA proposes the use of the following equations for position updating in both phases Eq (1):

$$X_i^{t+1} = X_i^t + r_1 \times \sin(r_2) \times |r_3 P_i^t - X_i^t|$$
$$X_i^{t+1} = X_i^t + r_1 \times \cos(r_2) \times |r_3 P_i^t - X_{i_i}^t| \tag{1}$$

where X represents the current solution's position in the i-th dimension after the t-th iteration, P_i represents the point of destination in the i-th dimension, r_1, r_2 and r_3 are random numbers, and $||$ indicates an absolute value.

In Eq. (2), a combination of these two Eq. (1) can be seen:

$$X_i^{t+1} = \begin{cases} X_i^{t+1} = X_i^t + r_1 \times \sin(r_2) \times |r_3 P_i^t - X_i^t|, & r_4 < 0.5 \\ X_i^{t+1} = X_i^t + r_1 \times \cos(r_2) \times |r_3 P_i^t - X_{i_i}^t|, & r_4 \geq 0.5 \end{cases} \tag{2}$$

where r_4 represents a random value in $[0,1]$.

The four major parameters of the SCA are r_1, r_2, r_3, and r_4, as shown in the equations above. Parameter r_1 defines region of the following position. Said position signifies one of the two possible spaces: the space between the solution and destination or the space outside of the two. Parameter r_2 dictates the movement away from or toward the destination, or more precisely, how distant the movement is. The role of parameter r_3 is to stochastically diminish ($r_3 < 1$) or emphasize ($r_3 > 1$) the distribution effects on distance definition. Lastly, parameter r_4 plays the part of switching between the sine and cosine components in Eq. (2).

The effects of the sine and cosine functions on Eqs. (1) and (2) are depicted in Fig. 1. The search space in between the two solutions is dictated by these two equations as depicted in said figure. These two equations can also be expanded to include higher dimensions; however, Fig. 1 depicts a two-dimensional model.

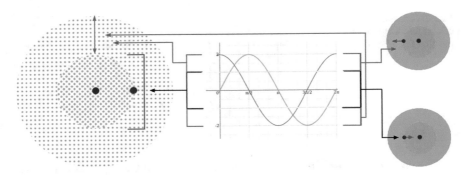

Fig. 2 Sine and cosine with the range in $[-2, 2]$ allow a solution to go around (inside the space between them) or beyond (outside the space between them) the destination

The sine and cosine functions cyclic pattern allows for solution repositioning around a different solution. This can provide a guarantee of exploitation in the defined space enclosed by the two calculated solutions. Altering the range of sine and cosine function enables the solutions to search outside the space that is defined by the corresponding destinations, and this is done so as to ensure exploration.

While changing the function range, as shown in Fig. 2, it is necessary to update the new position of the solution taking into account positions of the existing solutions. The updated position is attained by choosing a random value in range $[0, 2\pi]$ for r_2 from Eq. 2, and it can be either on the outside or on the inside. This mechanism ensures both exploitation and exploration of the search space.

The algorithm needs to have the ability to balance both exploration and exploitation when searching for promising regions inside a given search space. This is done to eventually converge on a global optimum. The SCA does this by changing range of the sine and cosine adaptively in Eq. 2 according to Eq. 3:

$$r_1 = a - t\frac{a}{T} \tag{3}$$

where a represents a constant, T represents the maximum amount of allowed repetitions, and finally t represents the active iteration.

Through many repetitions of Eq. 2, we get a decreasing range of sine and cosine as shown in Fig. 3.

By taking into consideration both Figs. 2 and 3, it can be deduced that the SCA focuses on exploitation when the given ranges are in $[-1, 1]$, and on exploration when the ranges are in between $(1, 2]$ and $[-2, -1)$.

Finally, the pseudocode for the SCA are shown in Fig. 4. As depicted, the algorithm begins the process of optimization with randomized set of solutions. Every time the algorithm encounters a solution, it considers the most optimal so far, and it assigns it as a target point. The algorithm then, in regard to the most optimal solution, updates other solutions. During this process, the iteration counter is increased, and

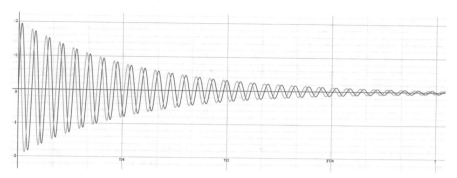

Fig. 3 Decreasing range of sine and cosine ($a = 3$)

```
initialize a randomized of search agents (X)
do
    evaluate each of the search agents by the objective function
    update the best solution obtained so far (P = X*)
    update r1, r2, r3, r4
    update the positions of search agents
while t < maximumnumberofiterations
return the best solution obtained so far as the global optimum
```

Fig. 4 General steps of the original SCA algorithm

the ranges of sine and cosine function are, after every iteration, updated emphasizing exploitation of the defined search space. When the counter reaches the maximum allowed amount of iterations, the optimization process of the original SCA stops. Other conditions for termination can be implemented as well, including the total number of functional evaluations or reaching a desired global optimum accuracy.

3.1 Proposed Modified SCA Approach

Notwithstanding the fact that the basic SCA metaheuristics establish excellent results for standard benchmark instances [20], based on additional conducted experiments with basic congress on evolutionary computation (CEC) benchmark suites, it was concluded that the basic SCA can be further improved.

As many other swarm intelligence approaches, original SCA may be stuck in non-optimal regions of the search domain in early iterations of execution. In this early phase, due to the lack of exploration power, if the search process is not "lucky" and if does not register optimal domain of the search space, algorithm may stuck in sub-optimal domain for many iterations. As a consequence, worse mean values are generated, and performance of the metaheuristic is seriously degraded.

Without adding complexity to the algorithm, abovementioned drawback of original SCA can overcome by introducing simple mechanism in the search process as follows: after every iteration, 5% of worst solutions from the population are replaced with the randomly generated individuals within the boundaries of the search space in the first 50% of iterations:

$$X_{rnd}^{j} = L^{j} + \phi \cdot (U^{j} - L^{j}), \tag{4}$$

where X_{rnd}^{j} is j-th component of the newly generated random solution, phi is the value derived from the uniform distribution, and U^{J} and L^{j} are upper and lower boundaries of j-th parameter, respectively.

Based on conducted simulations, it was concluded that in approximately first 50% of iterations described exploration mechanism should be triggered. However, in later iterations, this mechanism is not needed, and it would only represent an obstacle in performing a fine-tuned search around the promising domain of the search region. Proposed method is named modified SCA (mSCA), and its pseudocode is shown in Algorithm 1.

Algorithm 1 Pseudocode of proposed mSCA

Initialization. Generation of the starting random population of N individuals X within the boundaries of the search space and calculation of its fitness.
Initialize the maximal number of iterations T.
do
    for all X in the generated population **do**
        Evaluate utilizing the fitness function.
        if $f(X)$ better than $f(P)$ **then**
            Update the position of the best solution so far ($P = X^{*}$).
        end if
    end for
    Update r_1 parameter
    Update r_2, r_3, and r_4 parameters.
    Update the positions of search agents
    if $t < T \cdot 0.5$ **then**
        Replace 5% worst solution with random one using Eq. (4)
    end if
while $(t < T)$
return P the best solution found.

---

4 Experiments and Discussion

In the research presented in this manuscript, the proposed mSCA algorithm was tested on ten basic datasets and one additional COVID-19 dataset. The experimental simulations in this research were executed through 20 independent runs, while each run consisted of 70 iterations. The size of the population was set to 8, and a mixed initializer was utilized to randomly select 2/3 from the available amount of features. The suggested improved optimization method's performance has been tested on ten UCI datasets that are very popular among researchers and used as a benchmark in Table 1.

Table 1 List of experimental simulation datasets

No.	Name	Features	Samples
1	Glass	10	214
2	Hepatitis	19	155
3	Lymphography	18	148
4	Primary tumor	17	339
5	Soybean	35	307
6	Horse colic	27	368
7	Ionosphere	34	351
8	Zoo	16	101
9	Musk 1	166	476
10	Arrhythmia	279	452
11	COVID-19	15	TBD

The performance of mSCA was evaluated on a computer with a central processing unit (CPU) with a clock frequency of 2.90 GHz, additionally with 16.0G of available random access memory (RAM) and programmed in the language of Python with Anaconda framework using machine learning libraries including NumPy, SciPy and scikit-learn. The performance is judged based on five calculated evaluation metrics. The evaluation metrics include optimal fitness value, average fitness value, fitness value normal divination, precision of classification, and the ratio of feature selection with each method executed and evaluated 20 times. The repetition is performed to better represent results and avoid bias caused by optimization algorithms stochastic nature. The result averages are logged and presented after the last iteration of the 20 individual runs.

The mSCA in tested against ten standard datasets and COVID-19 dataset. And its performance is then evaluated. The datasets are acquired from the UCI repository [11] and Arizona State University [18]. Table 2 represents best overall fitness while Table 3 represents the mean fitness metric. Tables 4 and 5 each represent standard deviation, average classification accuracy and feature selection of already referenced ten datasets. The best results are marked in bold in each table, except in the case of tie, where none of the results are marked. Tests of the proposed mSCA have been conducted on different structures, so as to provide evidence of the algorithms efficiency and performance in differing dimension.

The obtained results from Tables 2, 3, 4 and 5 from conducted experiments proved the efficiency and efficacy of mSCA proposed algorithm. Based on the empirical analysis, a deduction can be made that the proposed mSCA can yield higher-quality results than the algorithms it has been tested against. The eight algorithms tested in this paper are (BDA) [19], binary artificial bee colony (BABC) [14], binary multiverse optimizer (BMVO) [1], binary particle swarm optimization (BPSO) [30], chaotic crow search algorithm (CCSA) [23], binary coyote optimization algorithm (BCOA) [27], evolution strategy with covariance matrix adaptation (CMAES) [13]

Table 2 Best fitness metric over ten UCI datasets for the compared approaches

No.	Dataset	HLBDA	BDA	BABC	BMVO	BPSO	CCSA	BCOA	CMAES	LSHADE	SCA	mSCA
1	Glass	0.0067	0.0067	0.0067	0.0067	0.0067	0.0067	0.0067	0.0067	0.0067	0.0067	0.0067
2	Hepatitis	0.1154	0.1245	0.1305	0.1226	0.1235	0.1309	0.1220	0.1229	0.1235	0.1223	**0.1151**
3	Lymphography	0.1116	0.1181	0.1122	0.1297	0.1177	0.1309	0.1257	0.1171	0.1226	0.1288	**0.1109**
4	Primary tumor	0.5646	0.5731	0.5673	0.5887	0.5623	0.5756	0.5642	0.5623	0.5883	0.5685	0.5623
5	Soybean	**0.2009**	0.2073	0.2035	0.2421	0.2189	0.2294	0.2035	0.2010	0.2037	0.2112	0.2010
6	Horse colic	0.1298	0.1330	0.1350	0.1440	0.1309	0.1418	0.1304	0.1298	0.1327	0.1429	**0.1297**
7	Ionosphere	**0.0694**	0.0731	0.0830	0.0980	0.0814	0.0906	0.0715	0.0746	0.0719	0.0719	0.0697
8	Zoo	0.0332	0.0325	0.0332	0.0332	0.0325	0.0338	0.0325	0.0334	0.0325	0.0337	0.0325
9	Musk 1	0.0608	0.0626	0.0879	0.0942	0.0783	0.0836	0.0662	0.0739	0.0633	0.0792	**0.0606**
10	Arrhythmia	0.2926	0.3179	0.3330	0.3352	0.3282	0.3399	0.3106	0.3269	0.2999	0.3215	**0.2923**

Table 3 Statistical mean fitness metric over ten datasets for the compared approaches

No.	Dataset	HLBDA	BDA	BABC	BMVO	BPSO	CCSA	BCOA	CMAES	LSHADE	SCA	mSCA
1	Glass	0.0112	0.0112	0.0111	0.0116	0.0111	0.0116	0.0111	0.0119	0.0116	0.0116	0.0112
2	Hepatitis	0.1311	0.1368	0.1386	0.1454	0.1334	0.1457	0.1425	0.1430	0.1399	0.1382	**0.1310**
3	Lymphography	0.1311	0.1359	0.1350	0.1527	0.1342	0.1515	0.1416	0.1392	0.1474	0.1305	**0.1309**
4	Primary tumor	0.5850	0.5932	0.5845	0.6088	0.5850	0.5996	0.5944	0.5935	0.5989	0.5949	0.5852
5	Soybean	0.2124	0.2212	0.2254	0.2594	0.2245	0.2479	0.2168	0.2177	0.2211	0.2238	**0.2121**
6	Horse colic	0.1358	0.1427	0.1480	0.1674	0.1409	0.1699	0.1434	0.1419	0.1479	0.1483	**0.1355**
7	Ionosphere	0.0842	0.0928	0.1016	0.1106	0.0958	0.1114	0.0868	0.0882	0.0909	0.0913	0.0842
8	Zoo	0.0401	0.0409	0.0397	0.0482	0.0369	0.0493	0.0437	0.0470	0.0503	0.0425	**0.0398**
9	Musk 1	0.0673	0.0832	0.0961	0.1082	0.0929	0.1017	0.0792	0.0842	0.0793	0.0903	**0.0669**
10	Arrhythmia	0.3160	0.3341	0.3450	0.3538	0.3413	0.3529	0.3290	0.3352	0.3270	0.3275	0.3160

Table 4 Standard deviation results for ten datasets included in the comparative analysis

No.	Dataset	HLBDA	BDA	BABC	BMVO	BPSO	CCSA	BCOA	CMAES	LSHADE	SCA	mSCA
1	Glass	0.0033	0.0033	0.0033	0.0033	0.0033	0.0032	0.0033	0.0036	0.0033	0.0034	0.0033
2	Hepatitis	0.0093	0.0062	**0.0057**	0.0099	0.0080	0.0065	0.0139	0.0132	0.0091	0.0069	0.0062
3	Lymphography	0.0130	0.0121	0.0125	0.0104	0.0138	0.0117	0.0134	0.0151	0.0147	0.0121	**0.0111**
4	Primary tumor	0.0104	0.0099	0.0085	0.0080	0.0116	0.0120	0.0135	0.0136	0.0134	0.0134	**0.0078**
5	Soybean	0.0087	0.0094	0.0089	0.0118	**0.0041**	0.0108	0.0103	0.0116	0.0109	0.0104	0.0088
6	Horse colic	**0.0039**	0.0088	0.0073	0.0164	0.0069	0.0140	0.0110	0.0097	0.0158	0.0087	0.0041
7	Ionosphere	0.0086	0.0105	0.0101	0.0053	0.0056	0.0090	0.0104	0.0070	0.0099	0.0098	**0.0052**
8	Zoo	0.0079	0.0080	0.0073	0.0101	0.0065	0.0097	0.0092	0.0101	0.0116	0.0085	0.0073
9	Musk 1	**0.0062**	0.0099	0.0063	0.0076	0.0079	0.0079	0.0077	0.0075	0.0097	0.0081	0.0063
10	Arrhythmia	0.0094	0.0088	0.0042	0.0077	0.0073	0.0065	0.0100	0.0057	0.0130	0.0118	**0.0039**

Table 5 Percentage of selected feature for ten datasets included in comparative analysis

No.	Dataset	HLBDA	BDA	BABC	BMVO	BPSO	CCSA	BCOA	CMAES	LSHADE	SCA	mSCA
1	Glass	0.2900	0.2900	0.3050	0.3250	0.3050	0.3300	0.3050	0.2900	0.3050	0.3050	0.2900
2	Hepatitis	0.3184	0.3658	0.3158	0.3605	0.3263	0.3526	0.3053	0.3500	0.3368	0.3368	**0.3043**
3	Lymphography	0.4500	0.4806	0.5083	0.4889	0.5000	0.4972	0.4472	0.5083	**0.4333**	0.4806	0.4500
4	Primary tumor	0.6676	0.6118	0.6706	0.5941	0.6676	0.6441	0.6059	0.6265	0.6412	0.6402	**0.5923**
5	Soybean	0.6529	0.6229	0.6429	0.5743	0.6414	0.5971	0.6371	0.6286	0.6486	0.6362	**0.5723**
6	Horse colic	**0.0870**	0.1370	0.2426	0.2574	0.1963	0.2926	0.1056	0.1130	0.1500	0.0945	0.0892
7	Ionosphere	0.2191	0.2676	0.2897	0.2882	0.2441	0.3426	0.2265	0.2456	0.2824	0.2741	**0.2183**
8	Zoo	0.4563	0.4469	0.4969	0.5188	0.4250	0.4938	0.4531	**0.4500**	0.4938	0.5091	0.4523
9	Musk 1	0.4687	0.4783	0.4946	0.4602	0.4964	0.5033	0.4458	0.4946	0.4849	0.5012	**0.4451**
10	Arrhythmia	0.4050	0.4699	0.4803	0.4303	0.4706	0.4787	0.4048	0.4627	0.4495	0.4152	0.4048

and success history-based adaptive differential evolution with linear population size reduction (LSHADE) [26] algorithms.

Based on the presented results, it can be concluded that the proposed mSCA metaheuristics clearly outperformed the original SCA approach for all observed metrics. In general, when compared to other approaches included in the simulations, mSCA obtained the best performances. Based on the results from Table 2, the proposed mSCA approach obtained the best results for best fitness metrics on five out of the ten UCI datasets. When the statistical mean fitness metric is observed, from Table 3, it can be concluded that the mSCA obtained the best results on six out of ten UCI datasets. In case of the standard deviation, Table 4 shows that the mSCA obtained the best results on four datasets and tied the best results on the Glass dataset. In Table 5, comparative analysis between proposed mSCA and other approaches in terms of selected features (expressed as ratios of total number of features in the datasets) is presented. From results, it can be seen that proposed mSCA in average utilizes a smaller number of features than other methods which means that it managed to substantially reduce the problem dimensions, which makes the training process of a classifier much faster (Figs. 5 and 6).

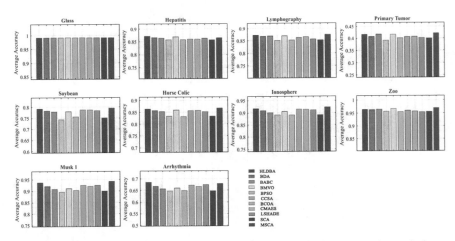

Fig. 5 Average classification accuracy over ten datasets included in the comparative analysis

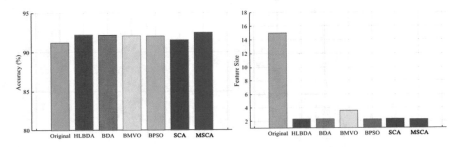

Fig. 6 Accuracy and feature size of the proposed SCA and mSCA on the COVID-19 dataset

5 Conclusion

The conducted research that is presented in this manuscript proposes a novel feature selection method. The implemented mSCA metaheuristics address the drawbacks of the original SCA method that are observed from the results of the conducted experiments. The proposed mSCA approach was later used to help find the crucial features for the classification process. The presented algorithmic method of optimization was validated on ten benchmark datasets, and the results are represented in comparison with other swarm intelligence metaheuristics. Finally, the mSCA method was used on COVID-19 dataset. The conducted experiments results indicate that the mSCA approach outperformed other methods included in the comparative analysis. Based on defined research contributions, the novelty of proposed research can be summed as follows: more efficient SCA metaheuristics are devised, solving feature selection challenge was improved in terms of classification accuracy, and the number of employed features and classification for the most recent and important COVID-19 dataset was performed.

The future research in this area will be focused on including additional datasets to the experimental simulations. Also, the future work will deal with adaptation of other swarm intelligence metaheuristics, with a goal to further enhance the classification accuracy.

References

1. Al-Madi, N., Faris, H., Mirjalili, S.: Binary multi-verse optimization algorithm for global optimization and discrete problems. Int. J. Mach. Learn. Cybern. **10**(12), 3445–3465 (2019). https://doi.org/10.1007/s13042-019-00931-8
2. Bacanin, N., Bezdan, T., Tuba, E., Strumberger, I., Tuba, M.: Monarch butterfly optimization based convolutional neural network design. Mathematics **8**(6), 936 (2020)
3. Bacanin, N., Bezdan, T., Tuba, E., Strumberger, I., Tuba, M., Zivkovic, M.: Task scheduling in cloud computing environment by grey wolf optimizer. In: 2019 27th Telecommunications Forum (TELFOR). pp. 1–4. IEEE (2019)
4. Bacanin, N., Tuba, E., Zivkovic, M., Strumberger, I., Tuba, M.: Whale optimization algorithm with exploratory move for wireless sensor networks localization. In: International Conference on Hybrid Intelligent Systems. pp. 328–338. Springer (2019)
5. Bezdan, T., Tuba, E., Strumberger, I., Bacanin, N., Tuba, M.: Automatically designing convolutional neural network architecture with artificial flora algorithm. In: Tuba, M., Akashe, S., Joshi, A. (eds.) ICT Systems and Sustainability, pp. 371–378. Springer Singapore, Singapore (2020)
6. Bezdan, T., Zivkovic, M., Antonijevic, M., Zivkovic, T., Bacanin, N.: Enhanced flower pollination algorithm for task scheduling in cloud computing environment. In: Machine Learning for Predictive Analysis, pp. 163–171. Springer (2020)
7. Bezdan, T., Zivkovic, M., Tuba, E., Strumberger, I., Bacanin, N., Tuba, M.: Glioma brain tumor grade classification from mri using convolutional neural networks designed by modified fa. In: International Conference on Intelligent and Fuzzy Systems. pp. 955–963. Springer (2020)
8. Bezdan, T., Zivkovic, M., Tuba, E., Strumberger, I., Bacanin, N., Tuba, M.: Multi-objective task scheduling in cloud computing environment by hybridized bat algorithm. In: International Conference on Intelligent and Fuzzy Systems. pp. 718–725. Springer (2020)

9. Chandrashekar, G., Sahin, F.: A survey on feature selection methods. Computers and Electrical Engineering **40**(1), 16–28 (2014). https://doi.org/10.1016/j.compeleceng.2013.11.024. https://www.sciencedirect.com/science/article/pii/S0045790613003066, 40th- year commemorative issue

10. Chen, J.I.Z., Hengjinda, P.: Early prediction of coronary artery disease (cad) by machine learning method-a comparative study. J. Artif. Intell. **3**(01), 17–33 (2021)

11. Dua, D., Graff, C.: UCI machine learning repository (2017). http://archive.ics.uci.edu/ml

12. Gajic, L., Cvetnic, D., Zivkovic, M., Bezdan, T., Bacanin, N., Milosevic, S.: Multi-layer perceptron training using hybridized bat algorithm. In: Computational Vision and Bio-Inspired Computing, pp. 689–705. Springer (2021)

13. Hansen, N., Kern, S.: Evaluating the CMA Evolution Strategy on Multimodal Test Functions, pp. 282–291. Lecture Notes in Computer Science. Springer (2004)

14. He, Y., Xie, H., Wong, T.L., Wang, X.: A novel binary artificial bee colony algorithm for the set-union knapsack problem. Future Gener. Comput. Syst. **78**, 77–86 (2018). https://doi.org/10.1016/j.future.2017.05.044. https://www.sciencedirect.com/science/article/pii/S0167739X17310415

15. Johnson, D.S.: The np-completeness column: an ongoing guide. J. Algorithms **6**(3), 434–451 (1985). https://doi.org/10.1016/0196-6774(85)90012-4. https://www.sciencedirect.com/science/article/pii/0196677485900124

16. Karaboga, D.: Artificial bee colony algorithm. Scholarpedia **5**(3), 6915 (2010). https://doi.org/10.4249/scholarpedia.6915, revision #91003

17. Levine, M.: Feature extraction: a survey. Proc. IEEE **57**(8), 1391–1407 (1969). https://doi.org/10.1109/PROC.1969.7277

18. Li, J., Cheng, K., Wang, S., Morstatter, F., Trevino, R.P., Tang, J., Liu, H.: Feature selection: a data perspective. ACM Comput. Surv. (CSUR) **50**(6), 94 (2018)

19. Mirjalili, S.: Dragonfly algorithm: a new meta-heuristic optimization technique for solving single-objective, discrete, and multi-objective problems. Neural Comput. Appl. **27**(4), 1053–1073 (2016). https://doi.org/10.1007/s00521-015-1920-1

20. Mirjalili, S.: Sca: a sine cosine algorithm for solving optimization problems. Knowl Based Syst **96** (2016). DOI 10.1016/j.knosys.2015.12.022

21. Mirjalili, S., Mirjalili, S.M., Lewis, A.: Grey wolf optimizer. Adv. Eng. Softw. **69**, 46–61 (2014). https://doi.org/10.1016/j.advengsoft.2013.12.007. https://www.sciencedirect.com/science/article/pii/S0965997813001853

22. Ranganathan, G.: A study to find facts behind preprocessing on deep learning algorithms. J. Innov. Image Proces. (JIIP) **3**(01), 66–74 (2021)

23. Sayed, G.I., Hassanien, A.E., Azar, A.T.: Feature selection via a novel chaotic crow search algorithm. Neural Computing and Applications **31**(1), 171–188 (2019). https://doi.org/10.1007/s00521-017-2988-6

24. Strumberger, I., Tuba, E., Bacanin, N., Zivkovic, M., Beko, M., Tuba, M.: Designing convolutional neural network architecture by the firefly algorithm. In: 2019 International Young Engineers Forum (YEF-ECE). pp. 59–65. IEEE (2019)

25. Strumberger, I., Tuba, E., Zivkovic, M., Bacanin, N., Beko, M., Tuba, M.: Dynamic search tree growth algorithm for global optimization. In: Doctoral Conference on Computing, Electrical and Industrial Systems. pp. 143–153. Springer (2019)

26. Tanabe, R., Fukunaga, A.: Improving the search performance of shade using linear population size reduction. 2014 IEEE Congress on Evolutionary Computation (CEC) pp. 1658–1665 (2014)

27. Thom de Souza, R.C., de Macedo, C.A., dos Santos Coelho, L., Pierezan, J., Mariani, V.C.: Binary coyote optimization algorithm for feature selection. Pattern Recognition **107**, 107470 (2020). https://doi.org/10.1016/j.patcog.2020.107470. https://www.sciencedirect.com/science/article/pii/S0031320320302739

28. Too, J., Mirjalili, S.: A hyper learning binary dragonfly algorithm for feature selection: A covid-19 case study. Knowledge-Based Systems **212**,(2021). https://doi.org/10.1016/j.knosys.2020.106553. https://www.sciencedirect.com/science/article/pii/S0950705120306821

29. Yang, X.S.: A new metaheuristic bat-inspired algorithm. In: Nature inspired cooperative strategies for optimization (NICSO 2010), pp. 65–74. Springer (2010)
30. Yin, P.Y.: A discrete particle swarm algorithm for optimal polygonal approximation of digital curves. Journal of Visual Communication and Image Representation **15**(2), 241–260 (2004). https://doi.org/10.1016/j.jvcir.2003.12.001. https://www.sciencedirect.com/science/article/pii/S1047320303000981
31. Zhang, X., Xu, Y., Yu, C., Heidari, A.A., Li, S., Chen, H., Li, C.: Gaussian mutational chaotic fruit fly-built optimization and feature selection. Expert Systems with Applications **141**,(2020). https://doi.org/10.1016/j.eswa.2019.112976. https://www.sciencedirect.com/science/article/pii/S0957417419306943
32. Zivkovic, M., Bacanin, N., Tuba, E., Strumberger, I., Bezdan, T., Tuba, M.: Wireless sensor networks life time optimization based on the improved firefly algorithm. In: 2020 International Wireless Communications and Mobile Computing (IWCMC). pp. 1176–1181. IEEE (2020)
33. Zivkovic, M., Bacanin, N., Venkatachalam, K., Nayyar, A., Djordjevic, A., Strumberger, I., Al-Turjman, F.: Covid-19 cases prediction by using hybrid machine learning and beetle antennae search approach. Sustain. Cities Soc. **66**, 102669 (2021)
34. Zivkovic, M., Bacanin, N., Zivkovic, T., Strumberger, I., Tuba, E., Tuba, M.: Enhanced grey wolf algorithm for energy efficient wireless sensor networks. In: 2020 Zooming Innovation in Consumer Technologies Conference (ZINC). pp. 87–92. IEEE (2020)
35. Zivkovic, M., Bezdan, T., Strumberger, I., Bacanin, N., Venkatachalam, K.: Improved harris hawks optimization algorithm for workflow scheduling challenge in cloud–edge environment. In: Computer Networks, Big Data and IoT, pp. 87–102. Springer (2021)
36. Zivkovic, M., Zivkovic, T., Venkatachalam, K., Bacanin, N.: Enhanced dragonfly algorithm adapted for wireless sensor network lifetime optimization. In: Data Intelligence and Cognitive Informatics, pp. 803–817. Springer (2021)

Blood Cell Image Denoising Based on Tunicate Rat Swarm Optimization with Median Filter

M. Mohana Dhas and N. Suresh Singh

Abstract The significant challenge that occurs due to medical image processing is to acquire the image without the loss of any crucial data. The image data can be degraded by noise or other factors while acquiring or processing the image. This noise affects the image quality since the contrast of medical images is already very low, and it is hard for the experts to identify the infections from the images. Henceforth, image denoising is an essential process in the medical imaging systems. In this paper, a hybrid tunicate rat swarm optimization with median filter (TRSWOAMF) has been proposed to remove or reduce the noise from the medical blood cell image, thus to restore a high-quality image. The proposed TRSWOAMF method uses median filter to remove the noise from the blood cell images, which then optimizes the parameters by using the tunicate rat swarm optimization algorithm. This TRSWOAMF method detects the noise in the blood cell image, wherein the median filter computes the median value for every pixel and best value is replaced. Then the parameters in the image are optimized by the tunicate rat swarm optimization algorithm to retain the original quality of the image. The result shows that the proposed TRSWOAMF method produces high-quality denoised image with a significantly reduced error rate.

Keywords Blood cell · Image denoising · Optimization

1 Introduction

In many facets of our everyday lives, monitoring the traffic, check recognition, approval of signature, satellite television, medical images [2] and so on utilizes the digital images [1]. While transferring the images, the noise and other factors affect the images. Image acquisition, compression and transmission are the processes that

M. Mohana Dhas (✉)
Department of Computer Science, Annai Velankanni College, Tholayavattam, India

N. Suresh Singh
Department of Computer Applications, Malankara Catholic College,
Mariagri, Kaliyakkavilai, India

© The Author(s), under exclusive license to Springer Nature Singapore Pte Ltd. 2022
V. Suma et al. (eds.), *Evolutionary Computing and Mobile Sustainable Networks*,
Lecture Notes on Data Engineering and Communications Technologies 116,
https://doi.org/10.1007/978-981-16-9605-3_3

take place during the transferring and receiving of the images. Image denoising is the process of removing the noise and other factors that affects the original image and recovers the image like original image. The high-frequency component that affects the image is noise, other high-frequency components are edge and texture, so it is tremendously difficult to remove only the noise from the images. There are various denoising approaches, which can be chosen based on the type of the noise to be removed from the images.

The main purpose of image denoising is to eliminate or reduce the noise from the original image. Original pixel values of the images are changed randomly, and this is called as noise. This noise affects the quality of the image during edge detection, image segmentation, feature extraction, etc. Henceforth, removing these noises is very important. Spatial domain or frequency domain based is the category of image denoising approaches [3]. The spatial domain-based approach works on the intensity of every pixel of the image thus restoring the original image. However, frequency domain-based approach works on after decomposition of image in the frequency domain, and the coefficients of multiple frequencies are modified. By using these modified coefficients, the reverse transformation is applied to restore the image [4].

One of the main challenges in recent days is removing the noise from the medical images. This happens because all the medical images are not similar. There are various filters that help to remove the noise from the medical images such as bilateral filter and Gaussian filter, while removing the noise using bilateral filter [13], the texture of the image is changed, flat intensity areas are created in the image, and new contours are also created that affect the image quality. The Gaussian filter [14] removes the noise effectively, but the image content also reduced which affects the quality of the image and my leads to false diagnosis in case of medical images. To overcome these limitations in this article, a novel hybridization of tunicate rat swarm optimization algorithm (TRSWOAMF) is proposed based on median filter for denoising the medical blood cell images. The information about the image is protected, and the unwanted noises in the images are removed by the median filter. After that, the proposed tunicate rat swarm optimization algorithm is utilized to optimize median filter parameters of the blood cell images. The TRSWOAMF with median filter computed random population for each and every solution in the objective function. And then by transferring between the TSA and RSO, the present solution is updated, and the best solution is obtained. The tunicate rat swarm optimization algorithm with median filter (TRSWOAMF) is compared with other filters like bilateral filters and median filters. In this method, the swarm is the victim, and the tunicate and rat are the predator. If the random value (rand) is greater than 0.5, then the rat swarm optimization algorithm alters the pixel location; otherwise, the tunicate swarm algorithm alters the pixel location until the optimum solution is obtained. The optimum pixel position is selected according to the position of the present filtering window values. The results show that the proposed optimization algorithm is faster and have less complexity compared to other optimization algorithms.

The remaining sections of the article are organized as follows: Sect. 2 describes the literature review. The proposed methodology and the algorithms used are explained in Sect. 3. Section 4 explains the implementation and the experimental results of the denoised images. At last, conclusion of the article along with future works is described in Sect. 5.

2 Literature Survey

Pan et al. [5] introduced a novel pre-trained convolutional neural network (CNN) for image denoising. The proposed work uses synthetic aperture radar (SAR) images that were decomposed into low-frequency bands and high-frequency bands. The result of these two bands is combined together to enhance the image. Then a pre-trained fast and flexible denoising convolutional neural network (FFDNet) with additive white Gaussian noise (AWGN) was included in multi-channel logarithm with Gaussian denoising (MuLoG) framework. This system removes the noise from both the single channel and multiple channels SAR images. The limitation of this system was it does not remove the halo artifacts from the images. Yang et al. [6] suggest a pre-trained convolutional neural network approach for additive white Gaussian noise to solve the problem of maximum a posteriori (MAP) optimization with an alternating direction method of multipliers (ADMM) algorithm for denoising the SAR images. The pre-trained denoiser deep denoising convolutional neural network (DnCNN) was incorporated within the MuLoG framework to denoise the natural images. The MuLoG-DnCNN sometimes may over smooth the images which leads to loss of image information, and also the images have some spots that are not removed.

Dutta et al. [11] designed a modified whale optimization algorithm to optimize the image fusion. Catching the local minima and premature convergence are the drawbacks occurred during image fusion by whale optimization algorithm (WOA). So the WOA was modified by integrating bat algorithm (BA) and local search algorithm (LSA), thus the efficiency of searching is improved. The LSA was applied to catch the prey within two unit diameters. If the prey was exterior to two units, then BA algorithm was used to catch the prey. Thus, the accuracy of the image fusion is enhanced by the combination of BA and LSA with WOA. By the wise planning approach, the drawback of WOA was overcome by MWOA. Samuel et al. [12] introduces a spatio-frequency domain anisotropic filtering joined with discrete wavelet transform (DWT). The approach is motivated by the fact that the filtering procedure was carried out in the frequency domain that reflects any changes in the time domain representation the least. The primary dataset for examination in this study includes histopathological pictures of platelets, white blood cells (WBC) and red blood cells (RBC) whose analysis is of significant importance in the medical sector for the diagnosis of various disorders. Gaussian noise, speckle noise and impulse noise were included in the medical images, and these images were put into the spatio-frequency domain anisotropic-based DWT filtering to increase the image quality by filtering the noise. In terms of structural cost and reduced complexity, the performance of this approach was good.

Mandić et al. [9] introduce a hybrid adaptive approach data-adaptive 2D spatial filters with the local polynomial approximation (LPA) estimators linked by a relative intersection of confidence intervals (RICI) rule (2D LPA-RICI) for denoising the medical X-ray images. The 2D shape for every pixels estimator size was computed by the RICI rule. LPA weighted averaging was applied to each and every pixel in the image to remove the noise from the X-ray image. This method easily parallelizes the images, thus it reduces the computational costs. The 2D LPA-RICI does not apply on quadrilateral, octagonal and hexadecagonal portions of the images, and the polygonal angles in these portions were not denoised accurately. Sam and Lenin Fred [10] suggest a hybrid filter with a firefly algorithm to eradicate the noise from the medical images. The hybrid filter was a combined Kirsch filter, anisotropic filter and kalman filter. The resultant images of the filter were applied by a firefly neural network algorithm to acquire a high-quality original image. Backpropagation neural network (BPNN) was included in the firefly algorithm to produce the fast convergence with less training error. Due to the use of various filters and algorithms, there may be a loss of information in the images. The above discussed all the denoising techniques have their own limitations. In the proposed method, tunicate rat swarm optimization algorithm with median filter is utilized to denoise the blood cell images. This method removes the noise from the blood cells and produces a high-quality image with less complexity.

3   Proposed Method

In the proposed method, tunicate rat swarm optimization algorithm-based median filter (TRSWOAMF) for denoising of medical blood cell images is proposed. The median filter is utilized to remove the salt and pepper noise from the blood cell images. Figure 1 shows the block diagram of the proposed method. The medical blood cell image from the BCCD dataset is gathered, and then the salt and pepper noise is introduced into the blood cell image. The noise is filtered from the image by median filter with the parameters σ_n, defines the range of the noise variance such as 0.01, 0.05 and 0.1 and s defines the optimization value that is optimized by the optimization algorithm tunicate rat swarm optimization algorithm-based median filter (TRSWOAMF). Thus, the denoised image is obtained. The performance of the denoised image is analyzed using CPU time, mean-squared error (MSE), peak signal-to-noise ratio (PSNR) and feature similarity index (FSIM) entropy. The other filters such as Weiner filter and median filters are compared with the proposed method. From the comparison, the proposed TRSOA method produces faster result with minimum complexity.

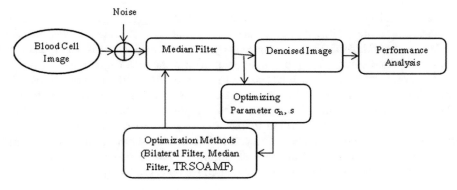

Fig. 1 Block diagram of the proposed method

3.1 Median Filter

Median filter (MF) is one of the most popular nonlinear filters that remove the salt and pepper noise from the images. The blood cell images are taken from the small-scale dataset for blood cells detection BCCD dataset. These blood cell images are then filtered to remove the noise by a median filter. At first, the shape and size of the filter window are set. The shape can be a line, a rectangle, a circle or can change based on the situations, but here the shape square is used to filter the blood cell images. The size of the filter window cannot be changed once it is assigned. Once the MF is applied to the blood cell image, the pixel values in the present filtering window are arranged in ascending order. Based on the median in the filtering window, the MF changes the pixel values. The sequence of pixel values is $Px_1, Px_2, Px_3, \ldots, Px_m$. The median filter is mathematically represented as follows.

$$\text{MF} = \text{Median}(px_k) = \frac{\sum_{k=1}^{m} px_k (m+1)}{2} \tag{1}$$

$$\text{MF} = \text{Median}(px_k) = \sum_{k=1}^{m} Px\left(\frac{m}{2}\right) + 1 \tag{2}$$

Here, $k = 1, 2, 3, \ldots, m$, Px represents the pixel value and m represents the number of blood cell images.

3.2 Tunicate Rat Swarm Optimization Algorithm

The proposed tunicate rat swarm optimization algorithm with median filter (TRSWOAMF) is a bio-inspired hybrid metaheuristic algorithm that optimizes the median pixel values of the blood cell images. Manually tuning the optimization parameters is quite difficult, and it takes more time. So, tunicate rat swarm optimization algorithm (TRSWOA) is used to optimize the median pixels and to get the best optimized value. This process also diminishes the mean square error (MSE) of the proposed method when compared to other similar methods. The TRSWOA describes the public activities of tunicate, rat and swarm. Tunicate swarm algorithm [8] describes the swarm behavior and jet propulsion of tunicates (sea squirts). Rat swarm optimization [7] algorithm defines the way the rat catches the swarm for its food. The proposed method TRSWOAMF combines both the tunicate swarm algorithm and rat swarm optimization algorithm. Figure 2 shows the structure of the proposed method. A fitness value is computed for each search agent, and the value

Fig. 2 Block diagram of the proposed method

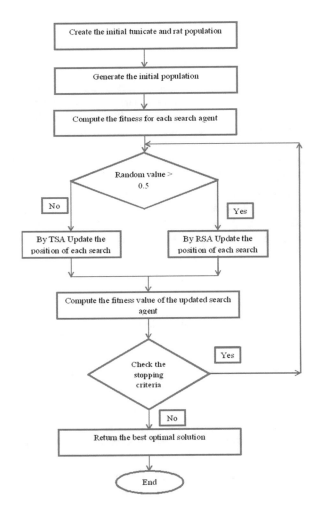

is in a range of [0,1] called as random value. The tunicate swarm algorithm (TSA) detects the best search agent when the random value is less than 0.5; otherwise, rat swarm algorithm (RSA) detects the best search agent.

In this method, the swarm is the victim, and the tunicate and rat are the predators. If the random value ($r_a nd$) is greater, then the rat swarm optimization algorithm alters the pixel location; otherwise, the tunicate swarm algorithm alters the pixel location until the optimum solution is obtained. The optimum pixel position is selected according to the position of the present filtering window values $\overrightarrow{PV}$.

Tunicate Swarm Algorithm (TSA) The mathematical model to avoid the conflicts between the pixel values is defined as in Eqs. (3)–(6).

$$\overrightarrow{A} = \frac{\overrightarrow{GF}}{\overrightarrow{M}} \tag{3}$$

$$\overrightarrow{GF} = r_2 + r_3 - \overrightarrow{S} \tag{4}$$

$$\overrightarrow{S} = 2.r_1 \tag{5}$$

$$\overrightarrow{M} = \lfloor Px_{\min} + r_1 \cdot Px_{\max} - Px_{\min} \rfloor \tag{6}$$

where force of gravity is defined as $\overrightarrow{GF}$ and the advection of water flow in deep ocean is represented as $\overrightarrow{S}$. r_1, r_2 and r_3 are random number variables between the range [0, 1]. $(Px)_{\min}$ and $(Px)_{\max}$ show the present and selected speeds to create public interaction. $(Px)_{\min}$ and $(Px)_{\max}$ values are set as 1 and 4, respectively. $\overrightarrow{PD}$ is defined as the distance between the denoising feature and the position of best denoising feature. It can be computed using Equation (7).

$$\overrightarrow{PD} = \left| \overrightarrow{GF} - r_{and} \cdot \overrightarrow{P_i(p)} \right| \tag{7}$$

The tunicate can move near the best optimum solution that is nearer to the food. This is computed in Equation (8).

$$\overrightarrow{P_i(p)} = \overrightarrow{GF} + \overrightarrow{A} \cdot \overrightarrow{PD} \tag{8}$$

Rat Swarm Optimization (RSO) The position of the swarm is known by the best search agent. Based on the position of the best search agent, the other rats in the group change their positions. This is mathematically modeled in Eq. (9).

$$\overrightarrow{L_u(x)} = \overrightarrow{A} \cdot \overrightarrow{L_i(x)} + \overrightarrow{S} \cdot (\overrightarrow{L_b(x)} - \overrightarrow{L_i(x)}) \tag{9}$$

where the location of the rat is represented as $\overrightarrow{L_u(x)}$. The best solution is represented as $\overrightarrow{L_i(x)}$. The aggressive fighting of the rat with the swarm to kill him is mathematically computed as follows:

$$\overrightarrow{L_i}(x+1) = \left| \overrightarrow{L_r}(x) - \overrightarrow{L} \right| \tag{10}$$

Here, the next altered position of the rat is represented by $\overrightarrow{L_i}(x+1)$. When the position of the rat is altered, the best ideal solution is stored in $\overrightarrow{L_i}(x+1)$. By modifying the parameter values of A and S, the optimum solution can be obtained.

4 Experimental Results and Discussion

4.1 Dataset Description

The dataset tested comprises blood cell images which are be taken from the BCCD dataset [15]. The BCCD dataset is an open dataset that can be accessed by everyone for the research purpose (https://www.kaggle.com/paultimothymooney/blood-cells/data). It consists of enlarged 12,444 blood cell images in .jpeg format, in which 9957 images are used for training, and the dataset and 2487 images are used for testing the dataset. Neutrophil, lymphocyte, eosinophil and monocyte are the categories of the blood cell images. Neutrophil 2499, lymphocyte 2483, eosinophil 2497 and monocyte 2478 blood cell images are taken for training the dataset. Neutrophil 624, lymphocyte 623, eosinophil 623 and monocyte 620 blood cell images are taken for testing the dataset. The size of the stored blood cell images is $320 \times 240 \times 3$ pixels.

4.2 Simulation Results

The proposed method is simulated by MATLAB2018a software running on a Windows 8.1 operating system. The salt and pepper noise is included in the blood cell images with different variances of σ_n set to $0.01, 0.05$ and 0.1. The proposed method TRSOA in median filter is compared with various filters like bilateral filter and median filter. The metrics used to compare the various denoising methods are peak signal-to-noise ratio (PSNR), mean square error (MSE), SMAPE, NC, image quality assessment (SSIM), absolute value of the error (MAE), PC and root mean square error (RMSE).

The experiments are done by applying three levels of salt and pepper noise variance $0.01, 0.05$ and 0.1. Thus, the proposed method median filter with TRSWOA's efficiency is validated. The filtering window size is set to 3×3 which cannot be changed throughout the experiment once it is set. The sum of the elements in the

Table 1 PSNR, MSE, NC, SSIM, MAE, SMAPE, PC and RMSE for neutrophil blood cell image at noise 0.01

Filters	PSNR	MSE	NC	SSIM	MAE	SMAPE	PC	RMSE
Bilateral	24.9422	208.38 42	0.99 734	0.73116	1.3975	0.005 0405	162777. 5129	1.4877
Median	28.5116	91.60 47	0.99 88	0.96519	1.5027	7.629 4e−05	179244. 9098	2.2543
Proposed	41.8855	4.2124	0.99 995	0.98853	0.35687	7.120 8e−05	195738. 1465	0.98209

filter coefficients for the median filter dimensionality is defined as d. The $n \times n$ filter window size is set to 3×3, so that the d is set to $n2$, i.e., 9. The values of other parameters are set from their corresponding references. Tables 1, 2 and 3 illustrate the values of PSNR, MSE, NC, SSIM, MAE, SMAPE, PC and RMSE for neutrophil blood cell image at various noise levels such as 0.01, 0.05 and 0.1. These values of the proposed TRSWOAMF are compared with the median filter and bilateral filter. The results of PSNR, MSE, NC, SSIM, MAE, SMAPE, PC and RMSE show that the performance of the proposed TRSWOAMF is superiorly high compared to others.

Figure 3 shows the noise removal of salt and pepper noise from the neutrophil blood cell image using various filters such as bilateral filter and median filter. From the figure, it is evident that the proposed TRSWOAMF approach removes the noise better than other filters, and it also preserves the quality of the image.

Figure 4 represents the performance of PSNR with various denoising techniques for same noise levels. From that, it is proved that the proposed system denoises the image degraded by salt and pepper noise with high noise ratio. Figure 5 represents the performance of RMSE with various denoising techniques for same noise levels 0.01, 0.05 and 0.1. The proposed method produces significantly low error rate compared to other denoising technique.

Figure 6 represents the convergence rate analysis between various noise variance levels of 0.01, 0.05 and 0.1 for the proposed TRSWOAMF method. Different denoising techniques are used to test the blood cell images by setting the number of iteration to 100. Least possible fitness MSE value is obtained by the proposed method. The high convergence rate allows the TRSWOAMF to design the real-time applications. So, the proposed approach can also be used to denoise other medical images like computed tomography (CT) images, magnetic resonance imaging (MRI) and X-ray images.

5 Conclusion

In this article, a tunicate rat swarm optimization algorithm based on median filter is proposed for denoising the blood cell images which is corrupted by the salt and pepper

M. Mohana Dhas and N. Suresh Singh

Table 2 PSNR, MSE, NC, SSIM, MAE, SMAPE, PC and RMSE for Neutrophil blood cell image at noise 0.05

Filters	PSNR	MSE	NC	SSIM	MAE	SMAPE	PC	RMSE
Bilateral	17.98450,0.69,0.941	1034.2 5450, 0.69, 0.941	0.986 860, 0.69, 0.941	0.26 5790, 0.69, 0.941	5.18230,0.69,0.941	0.0249 790, 0.69, 0.941	105407. 20060, 0.69, 0.941	2.69160,0.69,0.941
Median	26.6822	139.5 931	0.998 23	0.87 446	1.8734	0.0004 4759	171604. 8729	2.4686
Proposed	36.167	15.71 73	0.999 82	0.96 209	0.874 78	0.0002 5431	193354. 2379	1.9333

Table 3 PSNR, MSE, NC, SSIM, MAE, SMAPE, PC and RMSE for neutrophil blood cell image at noise 0.1

Filters	PSNR	MSE	NC	SSIM	MAE	SMAPE	PC	RMSE
Bilateral	15.0075	2052.7 348	0.97 275	0.109 93	9.8631	0.0496 06	77399. 5582	3.6768
Median	23.7431	274.6 429	0.99 629	0.680 83	2.9526	0.0012 817	154101. 1662	2.9716
Proposed	32.2279	38.9 303	0.99 947	0.959 89	1.0256	0.0008 3923	188719. 451	2.0403

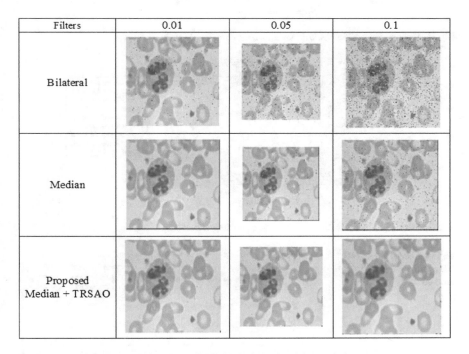

Fig. 3 Neutrophil blood cell image applied with various denoising techniques

noise. Various performance measures like PSNR, MSE, NC, SSIM, MAE, SMAPE, PC and RMSE are applied to find the performance of TRSWOAMF. The simulation is done for denoising the blood cell images in three different salt and pepper noise levels such as 0.01, 0.05 and 0.1 to evaluate the performance of the proposed filter. In these evaluations, it is proved that the proposed TRSWOAMF produces high PSNR, low SSIM, MAE and MSE. The TRSWOAMF method removes the salt and pepper noise from the blood cell images and produces a high-quality image which looks similar to the original image. The resultant image has low error rate and high convergence rate that helps to implement the real-time applications.

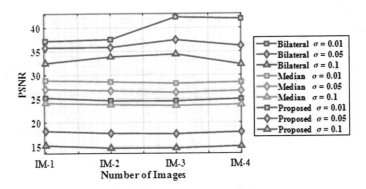

Fig. 4 Performance of PSNR with various denoising techniques

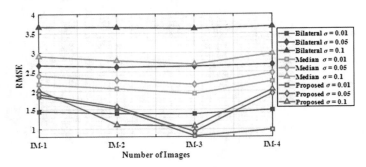

Fig. 5 Performance of RMSE with various denoising techniques

Fig. 6 Convergence rate analysis of the proposed method

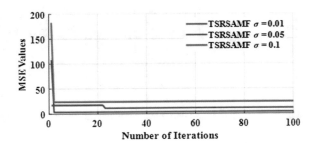

References

1. Ghose, S., Singh, N., and Singh, P.: Image denoising using Deep Learning: convolutional neural network. In: 2020 10th International Conference on Cloud Computing, Data Science and Engineering (Confluence) (2020). https://doi.org/10.1109/confluence47617.2020.9057895
2. Gondara, L.: Medical image denoising using convolutional denoising autoencoders. In: 2016 IEEE 16th International Conference on Data Mining Workshops (ICDMW) (2016). https://doi.org/10.1109/icdmw.2016.0041

3. Liu, Z., Yan, W.Q., and Yang, M.L.: Image denoising based on a CNN model. In: 2018 4th International Conference on Control, Automation and Robotics (ICCAR) (2018). https://doi.org/10.1109/iccar.2018.8384706

4. Milanfar, P.: A tour of modern image filtering: new insights and methods, both practical and theoretical. IEEE Signal Process.(Magazine) 30(1), 106–128 (2013)

5. Pan, T., Peng, D., Yang, W., Li, H.-C.: A filter for SAR image despeckling using pre-trained convolutional neural network model. Remote Sens. 11(20), 2379 (2019)

6. Yang, X., Denis, L., Tupin, F., and Yang, W.: SAR image despeckling using pre-trained convolutional neural network models. In: 2019 Joint Urban Remote Sensing Event (JURSE) (2019). https://doi.org/10.1109/jurse.2019.8809023

7. Dhiman, G., Garg, M., Nagar, A., Kumar, V., Dehghani, M.: A novel algorithm for global optimization: Rat Swarm Optimizer. J. Ambient Intell. Human. Comput. (2020)

8. Kaur, S., Awasthi, L.K., Sangal, A.L., Dhiman, G.: Tunicate Swarm Algorithm: a new bio-inspired based metaheuristic paradigm for global optimization. Eng. Appl. Artif. Intell. 90, 103541 (2020)

9. Mandić, I., Peić, H., Lerga, J., Štajduhar, I.: Denoising of X-ray images using the adaptive algorithm based on the LPA-RICI Algorithm. J. Imaging 4(2), 34 (2018)

10. Sam, B.B., Lenin Fred, A.: Denoising medical images using hybrid filter with Firefly Algorithm. In: 2019 International Conference on Recent Advances in Energy-Efficient Computing and Communication (ICRAECC) (2019). https://doi.org/10.1109/icraecc43874.2019.8995015

11. Dutta, S., Banerjee, A.: Highly precise modified Blue Whale method framed by blending Bat and Local Search Algorithm for the Optimality of Image Fusion Algorithm. J. Soft Comput. Paradigm (JSCP) 2(4), 195–208 (2020). https://doi.org/10.36548/jscp.2020.4.001

12. Samuel Manoharan, J., Parthasaradi, V., Suganya, K.: A spatio-frequency domain anisotropic filtering for contrast enhancement of histopathological images. Ann. Romanian Soci. Cell Biol. 29(5), 4945–4958 (2021)

13. Salehi, H., Vahidi, J.: A novel hybrid filter for image despeckling based On improved adaptive Wiener Filter, Bilateral Filter and Wavelet Filter. Int. J. Image Graph. 21(3), 2150036 (2021). https://doi.org/10.1142/S0219467821500364

14. Chen, Z., Zhou, Z., Adnan, S.: Joint low-rank prior and difference of Gaussian filter for magnetic resonance image denoising. Med. Biol. Eng. Comput. 59(3), 607–620 (2021). https://doi.org/10.1007/s11517-020-02312-8

15. Khan, M.B., Islam, T., Ahmad, M., Shahrior, R., Riya, Z.N.: A CNN based Deep Learning Approach for leukocytes classification in peripheral blood from microscopic smear blood images. In: Proceedings of International Joint Conference on Advances in Computational Intelligence, pp 67–76 (2021). https://doi.org/10.1007/978-981-16-0586-4_6

EVGAN: Optimization of Generative Adversarial Networks Using Wasserstein Distance and Neuroevolution

Vivek K. Nair and C. Shunmuga Velayutham

Abstract Generative Adversarial Networks (or called GANs) is a generative type of model which can be used to generate new data points from the given initial dataset. In this paper, the training problems of GANs such as 'vanishing gradient' and 'mode collapse' are reduced using the proposed EVGAN which stands for 'evolving GAN'. An improvement was done with the help of using Wasserstein loss function instead of the Minimax loss in the traditional GAN. Also, coevolution ensures that the best models from both the generator and discriminator pool are selected for further evolution thereby making the training of GAN more stable. Speciation is also included with a threshold of min. no of species thereby helping in increasing the diversity of the generated image. The model is evaluated in the MNIST dataset and shows better performance in accuracy as well as convergence when compared to the traditional GAN and WGAN.

Keywords GAN · Neuroevolution

1 Introduction

Generative Adversarial Networks (or called GANs) is a generative type of model which can be used to generate new data points from the given initial dataset. This is particularly useful in instances where there is a small amount of data available for training. A basic GAN model is comprised of two sub-models namely:

- **Generator**: It is the function of this model to generate realistic data from the given random input values and noises and feed it to the discriminator for evaluation.

V. K. Nair (✉) · C. Shunmuga Velayutham
Department of Computer Science Engineering, Amrita School of Engineering, Amrita Vishwa Vidyapeetham, Coimbatore, India

C. Shunmuga Velayutham
e-mail: cs_velayutham@cb.amrita.edu

© The Author(s), under exclusive license to Springer Nature Singapore Pte Ltd. 2022   47
V. Suma et al. (eds.), *Evolutionary Computing and Mobile Sustainable Networks*,
Lecture Notes on Data Engineering and Communications Technologies 116,
https://doi.org/10.1007/978-981-16-9605-3_4

- **Discriminator**: The discriminator basically acts as a classifier and evaluates whether the data given by the generator is real or fake based on the existing dataset points.

GANs have a wide variety of applications such as image synthesis, generating photo-realistic images, image-to-image translation, image inpainting, etc. The training of the GANs is very difficult as it requires a balance between the discriminator and generator model's learning pace. If one of the models becomes too good, then the other one will not be getting any useful feedback to improve and thus the overall learning will stagnate. This is especially true in cases where the GANs are required to work on complex tasks. One other issue is the overall time taken for training itself as it will make models not viable for implementation in time-constrained use cases. Thus, there is a need for optimization to make the training of GANs more efficient.

2   Related Works

GANs have been very popular ever since its inception in 2014 and there have been a lot of papers on them. The survey work was started off by looking into the paper that started it all, i.e., the discovery of Generative Adversarial Networks [1]. Here, they are proposing a novel generative framework using an adversarial approach in which there will be a generator (G) as well as a discriminator (D). The generator will be trying its best to replicate the given data distribution while the discriminator will be classifying whether the given data is a real one or generated. Thus, the generator will be becoming better by maximizing the probability of discriminator making a mistake. Overall, the ideal solution was found out to be with the generator replicating the data distribution while the discriminator has a 50% chance of correctly identifying the data. In this paper, they are using the minimax loss function [2].

Also, it was discovered that GAN models do not follow the traditional way of minimizing a single criterion (or cost function) to achieve convergence. Here, there is a constant competition going on between the generator and discriminator and as it is adversarial in nature, one side cannot reduce its cost function without affecting the parameters of the other. Thus, it was concluded that the GANs need not try to decrease the divergence at each and every epoch and instead the model could achieve a good endpoint when it reaches Nash equilibrium [3]. Convolutional Neural Networks (CNNs) were also introduced into the GANs [4] resulting in a new type called DCGANs. This was a very important step in bringing in CNNs into the field of unsupervised learning. Also, it helped pave the way for using GANs in the many areas requiring computer vision tasks.

Among the many training issues involved in the GAN model, one of the prominent issues is the 'vanishing gradients problem'. This happens because of the Binary Cross Entropy (BCE) loss function used in the traditional GANs. But this issue was overcome by replacing the discriminator's BCE loss with the 'Least Squares' loss function. The resultant model was proven to be more stable in the training phase and also lead to producing higher quality images [5].

Also, it was found out that the task assigned to the discriminator which is to classify whether the input was fake or real is easier when compared to generator which has to produce data that would fool the discriminator. Thus, more often, it was seen that the discriminator becomes better quickly thereby disrupting the balance between the two models learning. This leads to training instability. A novel solution for this was proposed by restricting the discriminator by putting on weight normalization into the corresponding value parts [6].

Another alternative to mitigate some of the training problems of GANs came to light with the addition of a new loss function called the Wasserstein loss (or called Earth Mover's Distance) [7]. The corresponding model was good at getting a stable training phase and achieve more relatively better outputs. It is achieved here because the discriminator is not restricted between a boundary of values as seen in the case BCE loss. Instead, its loss function can increase as needed thereby overcoming the issue of vanishing gradient [8]. But on the other hand, it requires the discriminator (or called critic in this case) to follow the Lipschitz constraint thereby making it not suitable for every objective function.

This problem was addressed by using some weight clipping techniques thereby ensuring that the gradient of the critic never overshoots 1 but resulted in poorer quality images and also convergence issues [9, 10].

Now, coming onto the evolutionary part of the project, there are some foundational works done by researchers the starting one being the evolutionary GANs [11] where instead of the traditional 1 generator—1 discriminator pair, a pool of generators is introduced into the equation. Here, the best out of the pool is chosen in each epoch thereby ensuring that the quality of generators is increased as training goes on. But the discriminator is kept to a single one in this paper. Now, another approach on top of this is the coevolutionary one where both discriminators and generators are present in the pool [12, 13]. Here, they do not variate the adversarial objective function but the discriminator training becomes better as they will also be improved faster. Thus, they ensure a balance in the stability of training. In both of these cases, mutations are performed on the samples in the pool thereby upgrading their quality based on a fitness criterion and then they are ranked accordingly before putting back into the pool for the next epoch. In image processing applications also, the evolutionary approaches can be used for getting better results [14]. Also, care needs to be done so that the model will not have premature convergence [15] as this will cause the overall training to be a failure.

The above part described the quality improvement in terms of fitness criterion. But another equally important thing to make sure of is preservation of the diversity of samples. Exploring the Evolution of GANs through Quality Diversity [16] makes use of a Novelty search algorithm to ensure that there is enough variety in the samples of the training pool at all times. Another important factor to consider is to ensure a progressive growth in the GAN model which includes ensuring proper diversity [17] and also gives more insight into having a stable training for the generator—discriminator pair. But here, they are sticking with a single pair similar to the

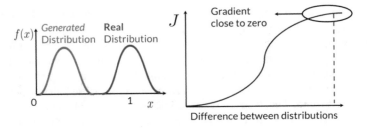

Fig. 1 Vanishing gradient problem

traditional GAN approach. Adapting strategies for the variation operators such as mutation and crossover rates can also ensure that there will be a variety of samples in the population pool at all times [18].

3 Problem Formulation

Some of the training issues that may arise in a GAN model are as follows:

Mode Collapse: Ideally, the generator should be capable of producing all the different kinds of data types present. But in some cases, when a generator produces a very plausible output, it will then focus on producing only those types of input which can be used to fool the discriminator. If the discriminator is failing to identify this (maybe because of hitting local minima), then the generator will over-optimize for that type of data only thereby failing to produce other data types.

Vanishing Gradients: This kind of problem occurs when the discriminator becomes too good and as a result, the training of the generator will stall as there is no useful feedback to improve (i.e., the gradient becomes flat) (Fig. 1).

The objective is to overcome some of the above-mentioned training problems such as 'vanishing gradient' and 'mode collapse' using the proposed model. Meanwhile, the convergence problem of GAN is another area where improvement can be made by adding noise vector along with the input to the generator network. This helps in increasing the diversity of the generated images as well as adding some element of stochasticity to the model. Thus, it will help the model if it gets stuck in a local minimum during the training phase by making a leap across it.

4 Dataset

The MNIST dataset of handwritten digits is used to measure the performance of our proposed model against the traditional GAN structure. This dataset is a subset of a larger NIST data and comprises a training set of 60,000 examples and a test set of

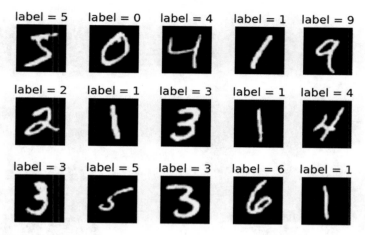

Fig. 2 MNIST dataset

10,000 examples. It is known as one of the benchmark datasets when it comes to Computer Vision and Digital Image Processing.

Each data point is a '28 × 28' grayscale image, with an attached label from 10 classes (namely from 0 to 9). As a result of such a small-scale image, the training period will require smaller computation power when compared to color images and high-resolution ones. Given below are some of the samples from the MNIST dataset (Fig. 2):

5 Proposed Work

5.1 Architecture of System

One of the basic changes in the model is the addition of the Wasserstein-based GAN model. Here, instead of using the 'minimax' loss function, 'Wasserstein loss' function will be used for the proposed GAN architecture. One of the main problems of GAN training namely 'vanishing gradient problem' can be overcome with the use of this loss function. This is because, in case of Wasserstein loss, it measures not only the distance between the generated and real distribution but also the amount of effort needed to move the generated distribution to make it look similar to the real one. It can be represented as follows:

$$\nabla_w \frac{1}{m} \sum_{i=1}^{m} \left[f\left(x^{(i)}\right) - f\left(G\left(z^{(i)}\right)\right) \right] \tag{1}$$

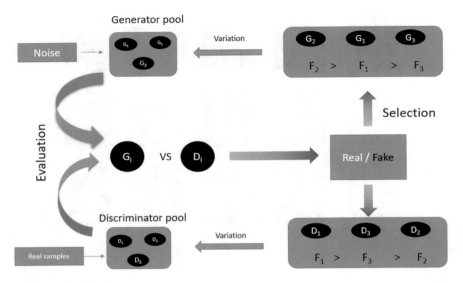

Fig. 3 Proposed architecture

where $G(z)$ is the sample produced by the generator from the noise vector 'z' while the first part of the formula with $f(x)$ represents the real sample taken from the dataset during training. Also, the average of the batch is taken as seen where 'm' is the number of samples in each batch. The problem with minimax loss function is that after the discriminator improves, it will start giving extreme values close to 0 and 1 thereby making the learning of generator stop. This will lead to gradient of the discriminator flattening thus causing vanishing gradient issue. But this will not happen with Wasserstein loss function as here, the cost will increase along with the increase in distance between the two distributions (Fig. 3).

The workflow of the architecture is as follows:

1. Create an initial pool of generators and discriminators of different characteristics.
2. Pit each generator against the given discriminator and record their fitness value: All versus Best evaluation used.
3. Ranking based selection scheme to select survivors:

   3.1   Discriminator—loss function.
   3.2   Generator—FID score.

4. Variational operators used:

   4.1   Adding, removing layers—mutation operations.
   4.2   Uniform crossover—crossover rate.

5. Top 10 samples selected are put back to the next epoch.

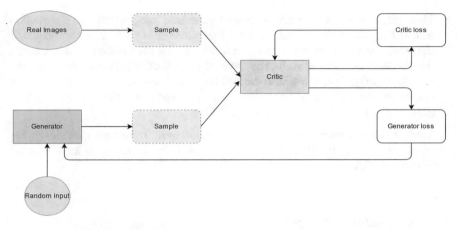

Fig. 4 Architecture for each sample

Here, each and every sample in their respective pools represent an individual generator or discriminator. The architecture of each sample is similar to the traditional GAN with having both a Generator part and a Discriminator. But in this case, the discriminator is called as 'Critic' instead since there will not be any discrimination between two values here. Thus, with the usage of critic and it is accompanying Wasserstein loss function, the model is able to overcome the issue of vanishing gradients. The training is performed by alternating between the generator and critic models meanwhile keeping the other one constant (Fig. 4).

5.2 Fitness

Here, since there are both generators and discriminators in the model, both of them also have to be ranked according to their own fitness criteria. In case of discriminators, the basic loss value is taken into consideration as their fitness which can be the minimax loss in case of Vanilla GAN while the Wasserstein loss function for WGANs.

But in case of generators, it was found that the loss function was not the best metric to evaluate the generators. Its reasons are as given below:

- The loss for generators is not stable enough during the training, making it not suitable to be used as an evaluation metric. Thus, the Fréchet Inception Distance (FID) is used as the fitness metric.

$$\text{FID} = |\mu - \mu_r|^2 + tr\left(\sum + \sum_r - 2\left(\sum \sum_r\right)\right)^{1/2} \qquad (2)$$

- This metric compares the distribution of generated images with the distribution of real images that were used to train the generator. In Eq. 2, it compares the difference between the mean and covariance of both distributions and calculates the resultant FID score. Here, the lower the value of FID score, the better the generated image is equivalent to the real images. It was found to be giving better results for the outputs of generators when compared to other metrics such as generic loss functions, Inception Score (IS), etc.

Thus, the model uses Wasserstein loss function to rank the survivors among the discriminators while the FID score is used to rank the survivors in the evaluated generator samples.

5.3 Evaluation Strategy

In the proposed architecture of Fig. 3, it can be seen that the discriminator & generator each is taken from their respective population pools and then first evaluated against each other. Here, the planned strategy is of taking the best two samples from each of the pools and then pitting them against all the samples in the other pool. That is, for example, the best generator is taken first and then it will be evaluated with all the discriminators and then ranked accordingly. This evaluation criterion was found to be the best as it is less computationally expensive while at the same time giving good results.

5.4 Variation Operator

Here, the mutation and crossover operations are used to bring variation into both the generator's and discriminator's population pool. Mutation operations can be as follows:

- Addition of layer where a new layer is added to the existing structure and this can be either a linear layer or a convolution layer.
- Removing a layer at any random level from the structure.
- Changing of a layer where the changes can be either in the activation function, no. of nodes present there, etc.

In addition to this, crossover operations are also performed where two parents of either discriminators or generators will be taken at random and then a uniform crossover is performed where each gene of the parent is taken one at a time and then passed on to either of the two children based on the crossover probability factor.

Another important factor to be taken into consideration here is ensuring a minimum amount of variety in the population pool. Preserving diversity is very essential as otherwise, it will result in 'modal collapse' where the generator captures only a small fraction of the input distribution, limiting the type of produced samples. Some solutions are:

- Speciation: By making sure that there should be a min. no of species at all time, diversity is ensured ... min_ species_ threshold.
- Novelty search: It creates a gradient of behavioral differences rewarding samples for being different thereby making sure that promising fitness candidates do not fall into local maxima.
- Non-dominated sorting genetic algorithm (NSGA2): The advantage of NSGA2 is that it uses an elitist principle where the best among the pool will be carried over to the next generation. Meanwhile, even if there is a clash of equally fit individuals, it will select one among them based on the crowding distance which essentially picks the one with lesser crowd around them. Thus, it ensures that diversity is also preserved in the overall population.

6 Experimental Setup

To validate the performance of our model, experimentation is done on the MNIST dataset mentioned above. Some of the standard techniques for comparison are as follows:

GAN (with BCE): Traditional GAN network with a generator—discriminator pair.

- Min–max optimization formulation.
- Suffers from training problems such as 'vanishing gradient' & 'modal collapse'.

Wasserstein GAN (with GP): This provides an alternate way of training the generator model to better approximate the distribution of data observed.

- Wasserstein loss function used.
- Critic can have more value range than discriminator.
- Basic training problems of Vanilla GAN are somewhat improved.
- Gradient penalty (GP) is added so as to keep a threshold on the critic's range.

Parameters settings:
The following are the various parameters set during the course of the experimentations in the GAN part:

1. No. of epochs = 100
2. Noise vector dimension (z_dim) = 64
3. Display step = 500
4. Batch size = 128
5. Learning rate = 0.0002
6. Discriminator loss function = Wasserstein loss.
7. Weight of the gradient penalty = 10
8. Number of times to update the critic per generator update = 5

Parameters for the evolutionary part are as follows:

1. Number of generations = 50
2. Population size (generators) = 10
3. Population size (discriminators) = 10
4. Elitism = 0.2,
5. Tournament size = 2,
6. Max. layers = 4,
7. Adding layer prob = 0.3,
8. Removing layer prob = 0.1,
9. Gene mutation prob = 0.1,
10. Speciation size = 4
11. Speciation min. threshold = 2
12. Optimizer = Adam
13. FID samples = 256

7   Results

Metrics comparison:

Here, the first part consists of comparing the different objective functions of GAN and WGAN-GP and their corresponding performance based on the loss of both generator and discriminator as represented below:

It can be seen from Fig. 5. that in case of GAN, the 'Discriminator loss' function flats out after a few epochs and as a result, the gradient or feedback given back to the generator becomes bad. This results in the vanishing gradient problem and affects the overall performance of GAN model.

Meanwhile, in the case of WGAN-GP, the 'Critic' is not limited by the boundary between 0 and 1. Thus, it can take any values as the model training progresses

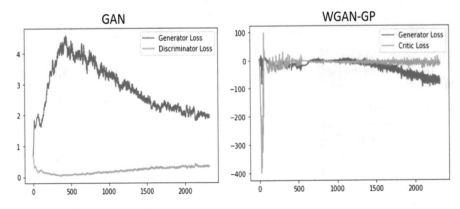

Fig. 5 Loss function comparison

Table 1 Metrics comparison

	Generator (FID score)	Discriminator (loss)
GAN (with BCE)	198.34	0.3296
WGAN (with GP)	70.12	−8.575
EVGAN (our model)	36.45	0.2811

and it can be seen that it is always producing gradients and thereby eliminating the vanishing gradient issue. Here, to somewhat regulate the critic so it is following the Lipschitz continuity principle, a 'Gradient Penalty (GP)' is used to ensure the gradient magnitude never crosses above 1.

After that, some of the other metrics such as Inception Score (IS) was looked into but as the loss function was not stable enough for the generator evaluation, it was finally settled on FID score as loss function. The various model's best performance is as shown in Table 1.

Mode collapse comparison

The mode collapse problem can be analyzed qualitatively by comparing the sample outputs produced by the two models. Here, since the MNIST dataset is used, there will be 10 modes present (namely one each for the digits 0–9). Thus, the ideal model will be the one that retains all 10 modes and is capable of reproducing each of them at the end of the training phase.

From Fig. 6, it is clear that the model is focusing only on some digits such as '7, 8 and 9' while modes such '3, 5, 6' are lost and thus showing that the model is suffering from mode collapse. On the other hand, with WGAN-GP and EVGAN, it is retaining all the digits thereby overcoming the mode collapse issue.

Furthermore, the EVGAN is quicker to converge than WGAN-GP as well takes lesser time per epoch thereby making it more computationally efficient. Even after

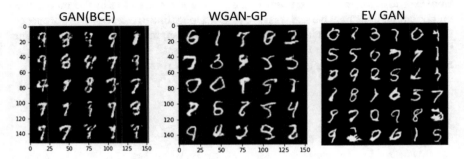

Fig. 6 Mode collapse comparison

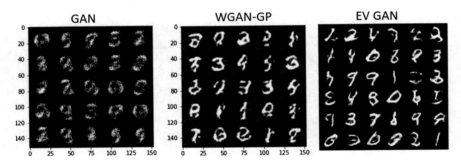

Fig. 7 Sample output after epoch 10

a few epochs, our model will start generating recognizable digits. This can be seen in Fig. 7 sample output images.

8 Conclusion

In this paper, a basic GAN model is first built to generate realistic images from the given dataset. Then, the various challenges faced during the training phase of the model were analyzed and how they will affect its performance. Here, two main problems were found out. The first is the vanishing gradients which occur when the discriminator becomes too good and as a result, the training of the generator will stall as there is no useful feedback to improve (i.e., the gradient becomes flat). The second problem is the mode collapse which occurs when a generator produces a very plausible output, it will then focus on producing only those types of input which can be used to fool the discriminator. If the discriminator is failing to identify this (maybe because of hitting local minima), then the generator will over-optimize for that type of data only thereby failing to produce other data types.

With the interim model where Wasserstein's loss function is used instead of the Minimax loss, the above-mentioned issues were solved. But still, the model was computationally expensive and was also taking a long time to converge. The proposed model involving the introduction of coevolutionary approach was able to overcome these issues all the while maintaining the performance showcased by the interim model and also somewhat getting a better qualitative result.

References

1. Goodfellow, I.J., Pouget-Abadie, J., Mirza, M., Xu, B., Warde-Farley, D., Ozair, S., Courville, A., Bengio, Y.: Generative adversarial nets. In: Proceedings of NIPS, vol. 27, pp. 2672–2680 (2014)

2. Goodfellow, I.: NIPS 2016 tutorial: generative adversarial networks (2016). arXiv preprint arXiv:1701.00160
3. Fedus, W., Rosca, M., Lakshminarayanan, B., Dai, A.M., Mohamed, S., Goodfellow, I.: Many paths to equilibrium: GANs do not need to decrease a divergence at every step. In: Proceedings of ICLR 2018 (2018)
4. Radford, A., Metz, L., Chintala, S.: Unsupervised representation learning with deep convolutional generative adversarial networks. In: Conference paper at ICLR 2016 (2016)
5. Mao, X., Li, Q., Xie, H., Lau, R.Y.K., Wang, Z., Smolley, S.P.: Least squares generative adversarial networks. In: Proceedings of CVPR (2017)
6. Miyato, T., Kataoka, T., Koyama, M., Yoshida, Y.: Spectral normalization for generative adversarial networks. Published as a conference paper at ICLR (2018)
7. Arjovsky, M., Chintala, S., Bottou, L.: Wasserstein GAN. In: Proceedings of MLR 2017
8. Gulrajani, I., Ahmed, F., Arjovsky, M., Dumoulin, V., Courville, A.: improved training of Wasserstein GANs. In: NIPS proceedings, pp. 1767–1777 (2017)
9. Salimans, T., Goodfellow, I., Zaremba, W., Cheung, V., Radford, A., Chen, X.: Improved techniques for training Gans. In: Advances in Neural Information Processing Systems, pp. 2226–2234 (2016)
10. Liu, S., Bousquet, O., Chaudhuri, K.:. Approximation and convergence properties of generative adversarial learning (2017). arXiv preprint arXiv:1705.08991
11. Wang, C., Xu, C., Yao, X., Tao, D.: Evolutionary generative adversarial networks. Proc. IEEE **23**(6), 921–934 (2019)
12. Costa, V., Lourenço, N., Machado, P.: Coevolution of generative adversarial networks. In: International Conference on the Applications of Evolutionary Computation, pp. 473–487. Springer (2019)
13. Nambiar, S., Jeyakumar, G.: Co-operative Co-evolution based hybridization of differential evolution and particle swarm optimization algorithms in distributed environment. In: Emerging Research in Computing, Information, Communication and Applications: ERCICA 2015, vol. 2. Springer India, New Delhi, pp. 175–187 (2015)
14. Sree, K., Jeyakumar, G.: An evolutionary computing approach to solve object identification problem in image processing applications. J. Comput. Theor. Nanosci. **17**, 439–444 (2020)
15. Reddy, R.R., Jeyakumar, G.: Differential evolution with added components for early detection and avoidance of premature convergence in solving unconstrained global optimization problems. Int. J. Appl. Eng. Res. 10(5), 13579–13594 (2015)
16. Costa, V., Lourenco, N., Machado, P.: Exploring the evolution of GANs through quality diversity. In: GECCO (2020)
17. Karras, T., Aila, T., Laine, S., Lehtinen, J.: Progressive growing of GANs for improved quality stability and variation. In: Proceedings of International Conference on Learning Representations (ICLR), pp. 1–26 (2018)
18. Dhanalakshmy, D.M., Pranav, P., Jeyakumar, G.: A survey on adaptation strategies for mutation and crossover rates of differential evolution algorithm. Int. J. Adv. Sci. Eng. Inf. Technol. (IJASEIT) (Scopus) **6**(5), 613–623 (2016)

A Hybrid Approach for Deep Noise Suppression Using Deep Neural Networks

Mohit Bansal, Arnold Sachith A. Hans, Smitha Rao, and Vikram Lakkavalli

Abstract Reducing noise to generate a clean speech in stationary and non-stationary noise conditions, or denoising, is one of the challenging tasks in the areas of speech enhancement for single channel data. Traditional methods depend upon first-order statistics, deep learning models, through their power of multiple nonlinear transformations can yield better results compared to traditional approaches for reducing stationary and non-stationary noise in speech. To denoise a speech signal, we propose a deep learning approach called UNet with BiLSTM network (bi directional long short-term memory) to enhance speech. A subset of LibriSpeech speech dataset is used to create training set by using both stationary noise and non-stationary noise with different SNR ratios. The results were evaluated using PESQ (perceptual evaluation of speech quality) and STOI (short-term objective intelligibility) speech evaluation metrics. We show through experiments that the proposed method shows better denoising metrics for both stationary and non-stationary conditions.

Keywords Convolution neural network (CNN) · Long short-term memory (LSTM) · UNet · Perceptual evaluation of speech quality (PESQ) · Short time objective intelligibility (STOI) · White noise · Urban noise · Stationary noise · Non-stationary noise

M. Bansal (✉) · A. S. A. Hans
Department of CSE, Presidency University, Bangalore, India

S. Rao
Vidyashilp University, Bangalore, India
e-mail: smitha.rao@vidyashilpuniversity.edu.in

V. Lakkavalli
Kaizen Secure Voiz Pvt Ltd, Chennai, India

© The Author(s), under exclusive license to Springer Nature Singapore Pte Ltd. 2022
V. Suma et al. (eds.), *Evolutionary Computing and Mobile Sustainable Networks*,
Lecture Notes on Data Engineering and Communications Technologies 116,
https://doi.org/10.1007/978-981-16-9605-3_5

1 Introduction

The most important aspect of speech enhancement is noise reduction, which finds its use in denoising telephone speech, conferencing systems, automatic speech recognition, and in the hearing aids among others. Speech enhancement [1–5] seeks to improve speech quality, thereby promising better results for automatic speech recognition and speaker recognition systems. Speech systems are particularly vulnerable to performance degradation due to variability of acoustic conditions in which the signal is acquired. To overcome data variability, deep learning approach-based system design requires large diverse data to compensate for the variability. Thus, reduction in the level of noise in input speech will help in the reduction of design complexity and data requirement of speech systems for having to cater less toward data variability contributed.

Noise reduction has been a topic of discussion since 1970s. A noise spectral estimator, which is driven by a voice activity detector (VAD), is used in spectral estimation of noise to be used in spectral subtraction. Spectral subtraction [1] technique involves identifying noise regions in the signal using a voice activity detector (VAD), and then use the spectral information of the noise to remove it from the noisy speech. An estimate of the noise [2, 3] is made from the identified noisy regions in the signal is subtracted from the speech signal. Spectral subtraction is simple and effective solution to denoise a signal where the noise characteristics do not vary with time. The technique has further been made robust by adapting the spectral estimate of the noise over time; however, its performance remains poor for non-stationary noise.

Deep learning techniques have been explored to denoise single-channel speech in recent times with success. The deep noise suppression (DNS) model was developed using deep learning-based technique, which takes the noisy version of speech as input and the clean version of the speech as the reference or target. A transformation of the speech representation such as short-time Fourier transform (STFT) is used as input and output of the network. These representations of speech are then converted back to speech using inverse transforms. The network acts as a mapping function which learns to denoise speech data in transform domain through many layers of nonlinear transformations.

The UNet architecture [6] was first built for the analysis of biomedical images. A segmented output map is generated from the input images. The most notable feature of the architecture is in the second half that the network does not provide a fully connected layer at the last; instead of that, it provides a convolutional layer followed by the ReLU activation function. Same UNet architecture [6] with some enhancements is used in this paper where noisy speech spectrogram is given as input and noise estimate is generated as the output.

Libri Speech data (clean speech mixed with stationary and non-stationary noises) is used in this paper to report the results on denoising. LibriSpeech [7] is a corpus of approximately 1000 h of 16 kHz read English speech. White noise was collected from free sound, which has stationary noise. Urban noise [8] samples were collected from the urban sound 8 k website [8], which has 10 different environmental noises such as dog barking, children playing, gunshots, and other kinds of non-stationary noise.

In this paper, literature survey is presented in Sect. 2, Sect. 3 will explain proposed system, and the relevant algorithms. Section 4 will conclude with the result and its analysis obtained from the UNet [6] BiLSTM models, and the paper ends with Sect. 5 along with the conclusions.

2 Literature Review

Spectral subtraction [1–3] method suppresses stationary noise from a speech by deducting the spectral distortion bias calculated during non-speech activity, which requires roughly the same processing as high-speed convolution. Secondary methods were being used to reduce the residual noise that remains after subtraction.

In audio processing chains aimed at communication and speech recognition, statistical techniques for speech enhancement are being popularly used. Recent developments in deep learning methods [4, 5, 9–16] have enabled the development of speech enhancement methodologies that typically outperform a traditional noise suppressor [1–3]. The proposed technique in [10] is a hybrid system for single-channel speech enhancement that combines conventional and deep learning methods to enhance speech more effectively. A regression long short-term memory recurrent neural network (LSTM-RNN) [9, 12, 14, 17] was trained on bilateral learning, with one output variable estimating clean speech features and the other calculating the suppression rule by combining the predicted suppression rule with the one estimated by a traditional speech enhancement algorithm [1–3].

In [11], the proposed new learning algorithm employs a fully convolutional neural network (CNN) [11, 13] to solve time-domain speech enhancement. The technique proposed in [11] takes noisy utterance periods as the input to CNN, and their outputs improved utterance time frames. To train the CNN, employ mean absolute error loss between the enhanced short-time Fourier transform (STFT) magnitude and the clean STFT magnitude.

3 Proposed System

This section provides an overview of the deep noise suppression framework which performs the denoising of audio data in both stationary and non-stationary conditions (Fig. 1).

The data augmentation system takes into account the noise level to be added, or signal to noise ratio (SNR) consideration of noisy speech, and other parameters such as duration of speech to be considered for processing, etc.

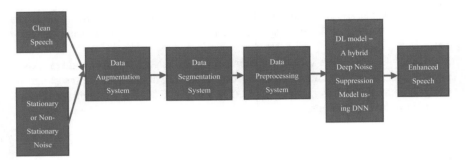

Fig. 1 Noise suppression system block diagram

The output from the data augmentation system is then forwarded to the data segmentation system. To train the Denoiser model, the proposed system requires two inputs, clean reference speech, and its noisy version. To create noisy version of input speech, the clean input signal is augmented with noise data. Here, input audio (noisy speech) of any length is segmented into chunks of a constant duration (here, it is 2 s duration is chosen) for further processing. The clean reference speech is also divided into chunks corresponding to its noisy version of speech, such that each noisy speech chunk has its corresponding clean chunk of speech for training. Division of a long input speech file into smaller chunks helps to parallelize the operation. The data preprocessing system's output is then used to train a Deep Learning model.

In the modified UNet architecture, we use a BiLSTM layer along with the UNet architecture. The architecture predicts noise in the signal and provides an estimated noise spectrum. This noise spectrum is subtracted from the noisy speech to obtain a denoised signal. Here, the UNet architecture acts a better predictor for noise from the input speech signal. The prediction is learnt through data intensive /statistical approach. The architecture is described in Fig. 2.

3.1 Data Collection

In this paper, the dataset is divided into three parts, clean speech from LibriSpeech corpus [7], stationary white noise from free sound website [18], and non-stationary urban noise from Urban Sound 8 k website [8]. Vassil Panayotov created LibriSpeech Corpus of 1000 h of English speech with the help of Daniel Povey. The data was obtained from audiobooks read through the LibriVox project and was meticulously segregated and aligned. The audio data is in.flac format, with a sample rate of 16,000 Hz and audio lengths ranging from 3 to 12 s. In this work, only 2000 audio files are chosen for training and testing the deep noise suppression model.

Different types of noise were used to augment the data to train the model, stationary and non-stationary noise. Stationary noise has predictable spectrum over long durations of time, while, non-stationary noise has time varying spectral characteristics.

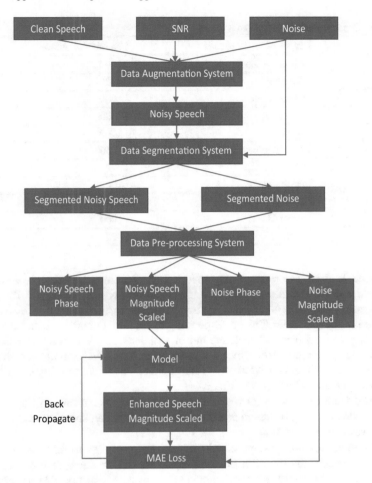

Fig. 2 Process of model training

It is difficult to handle non-stationary noise because of their characteristics. Non-stationary urban noise was collected from the urban sound 8 K website which consists of 10 different classes as mentioned in Table 1. The urban noise dataset consists of 8732 labeled sounds with each sound ≤4 s. 8732 labeled audio files are in RIFF ".wav" format. Table 1 shows the number of data available in each class.

3.2 Data Augmentation System

2000 audios belonging to the clean speech data was used from the LibriSpeech dataset and white noise and urban noises from online sources as described in data

Table 1 Urban sound data description

ID	Types of urban sounds	Number of audios
1	Air conditioner	1000
2	Car horn	429
3	Children playing	1000
4	Dog bark	1000
5	Drilling	1000
6	Engine Idling	1000
7	Gun shot	374
8	Jackhammer	1000
9	Siren	929
10	Street Music	1000
	Total	8732

collection section. To prepare noisy speech data, the clean speech data was mixed with white noise at different signal to noise ratios (SNRs), 10 dB, 15 dB, 20 dB, 25 dB, using a data augmentation system. The output generated by the system is 2000 noisy speech audio files corresponding to the input clean speech. 2000 audio files were selected from 40 speakers out of which 30 speaker data was considered to train the system consisting of 1329 audio files. Ten speaker data was chosen to test the system which consisted of 671 audio files. Out of the training set, 10% was earmarked as the validation set (Fig. 3).

Figure 4 shows the waveform of clean speech, noisy speech, and mix speech with 20 dB SNR. We can see the clear difference between clean speech and mix speech which was used for training.

Data augmentation systems consist of various processes to mix the noise with the clean speech, the duration of noise audio should be equal to the duration of clean speech, otherwise the data augmentation system will not able to mix noise with clean speech. Therefore length of noise (L_N) should be equal to length of clean speech ($L_{CS)}$.

$$L_N = L_{CS} \tag{1}$$

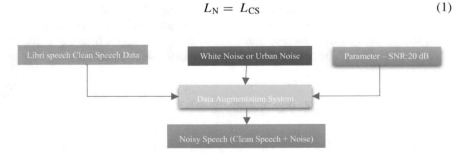

Fig. 3 Data augmentation system

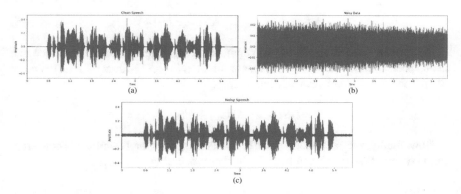

Fig. 4 Audio waveform of **a** clean speech (LibriSpeech Data) **b** stationary noise (white noise with 20 **dB**) **c** noisy speech

Care is taken to make sure that the audio length and the noise file have corresponding lengths. If the noise file has a smaller duration compared to the clean speech, then a part of the noisy file is used to make up for the duration shortfall. Before noise is added to clean speech, SNR is a parameter to be considered. Higher the SNR, better is the quality of speech, and vice versa. Speech amplitude is a pitch of the voice which is perceived as the frequency of vibration of the vocal folds. Root mean square (RMS) is just calculated as square root of the mean square of their amplitude, RMS of clean speech (CS_{RMS}) is the square root of the mean square of the clean speech amplitude (CS_{amp}). Similarly, noise RMS (N_{RMS}) is calculated using square root of the mean square of noise amplitude (N_{amp}).

$$CSRMS = \sqrt{mean\left(\left(CS_{amp}\right)^2\right)} \tag{2}$$

$$NRMS = \sqrt{mean\left(\left(N_{amp}\right)^2\right)} \tag{3}$$

Now, we need to calculate adjusted noise RMS which means the noise is manipulated by increasing or decreasing the noise level using signal to noise ratio (SNR) to mix clean speech audio with the noise. Adjusted noise RMS was calculated on 10 dB, 15 dB, 20 dB, and 25 dB noise level. The adjusted noise RMS was calculated using the below formula. Adjusted noise RMS (AN_{RMS}) is equal to clean speech RMS (CS_{RMS}) to the ten to the power of SNR ration in decibels divided by 20.

$$ANRMS = CSRMS / 10^{SNR(dB)/20} \tag{4}$$

Now, using adjusted noise RMS, the adjusted noise amplitude is calculated. Adjusted noise amplitude (AN_{amp}) is the product of noise amplitude (N_{amp}) and adjusted noise RMS (AN_{RMS}) to the noise RMS (N_{RMS}).

$$AN_{amp} = N_{amp} * (AN_{RMS}/N_{RMS}) \tag{5}$$

Now, noisy speech amplitude is calculated using adjusted noise amplitude and clean speech amplitude. Noisy speech amplitude (NS_{amp}) is a sum of adjusted noise amplitude (AN_{amp}) and clean speech (CS_{amp}).

$$NS_{amp} = AN_{amp} + CS_{amp} \tag{6}$$

Now, finally, we have added noise to the clean speech, and finally, it is converted into.wav format so that it can used for further processing.

3.3 Data Segmentation System

The long input speech file is divided into smaller chunks for parallel processing of data; also, it has been found that the smaller the duration of the data, can help in global and local modeling [19].

Since every audio data considered from LibriSpeech dataset that has a different length ranging from 2 to 13 s, each audio was cut into 2 s length and saved in as a separate file. The audios that were less than 2 s in length were adjusted for 2 s by adding silence at the end of the speech. For example, if audio has a length of 6.5 s, the audio was cut into four segments, the first three of which are 2 s in length, and the last segment is only 0.5 s in length, so 1.5 s of silence was added to the last segment to make it 2 s as mentioned in Fig. 5. After this data preparation activity, we obtained 11,085 audio files, out of which 5000 audios were selected for training, and the rest 500 audios were used as validation set.

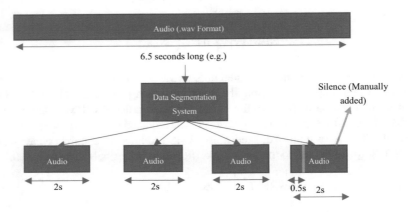

Fig. 5 Data segmentation system

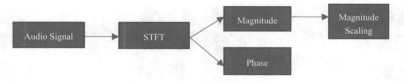

Fig. 6 Data preprocessing system

3.4 Data Preprocessing System

In the preprocessing stage, the audio files obtained from the data preparation stage were converted to the feature representation. Here, short-time Fourier transform (STFT) is used to represent speech. STFTs are obtained by dividing the audio into overlapping frames of a fixed duration (usually, 20–30 ms) with shift of 10 ms. Each of these frames is then transformed into a magnitude Fourier representation, and the stack of all such transformations constitutes a STFT. The STFT representation is a representation of the continuity or evolution of frequency contents in speech which is a very important factor determining the naturalness. The phase of the Fourier transform is also stored to be used for the reconstruction of the denoised speech (Fig. 6).

Data preprocessing was done with 5000 audios. Each audio has a signal length of 32,000, with a sample rate of 16 kHz. STFT was applied for chunks of audio with DFT length of 1024 samples, hop length 251 samples and the signal is windowed with a Hanning window. The magnitude of STFT was normalized between 0 and 1.

3.5 Proposed Architecture

After preprocessing, the data dimension was (5000,512,128,1), where 5000 is the number of audios, 512 is half the DFT length, and 128 represents the number feature vectors extracted from 2 s of speech data chunks. The architecture received noisy speech as input, and the output generated by the architecture is noise. The architecture consists of four layers of down sampling convolutional blocks and four layers of up sampling convolutional blocks, with a 2D convolutional block with an LSTM layer sandwiched between the two. The architecture's output has the same dimension as the input.

From Fig. 8. we can see that the convolution block consists of two convolutions 2D layer followed by Batch Normalization and ReLu activation function, and up convolution, a block consists of Upsample layer followed by Convolutional 2D layer with Batch Normalization and ReLu activation function (Fig. 7).

Figure 7a shows the UNet [6] Architecture and Fig. 7b shows the proposed UNet Architecture with BiLSTM. In downsampling part, 4 convolution blocks were used

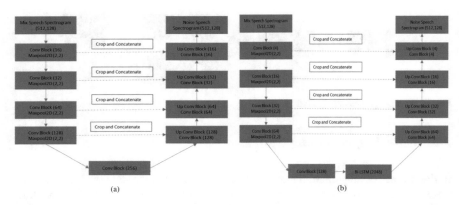

(a) (b)

Fig. 7 **a** UNet architecture, **b** UNet BiLSTM architecture (proposed)

Fig. 8 Convolution block and up sampled convolution block

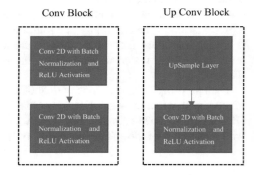

followed by a max pool layer, and have a filter size of 4,16,32,64. After that in-between downsample and up to sample layers there is one more convolution block was added with a filter size of 128 and followed by the LSTM layer with 2048 units. Now the architecture consists of upsample layers in which the Up convolution block is concatenated with downsample convolution blocks and now the filter size in the Unsampled layer is 64,32,16,4. The model was trained on various parameters like Adam optimizer with a learning rate of 0.0001, trained for 50 number of epochs with a batch size 8, the loss was L1 loss (mean absolute error loss) which performs the sum of absolute differences between predicted output and the target output. The model was trained with different SNR ratios namely 10 dB, 15 dB, 20 dB, 25 dB, and with stationary and non-stationary noises separately. So, a total of eight models were created four models for stationary and four models for the non-stationary, time taken to train the model 1 h per model 16 GB Tesla T4 GPU.

3.6 Model Testing and Prediction

For testing the model, test audios were chosen from noisy speech data, e.g., consider an audio with a duration of 5.5 s. The data segmentation system will segment the 5.5 s of noisy speech audio into three segments; the first two segments have 2 s of audio length, and the last segment has 1.5 s of audio length. To make the last segment 2 s long, 0.5 s of silence were added. All three segments are now of the same length. The segmented chunks of speech are then transformed to STFT representation with DFT 1024 length and hop length 250 samples to obtain magnitude and phase outputs. The magnitude of each of the three segments was scaled from 0 to 1. The model denoises each chunk to provide a denoised output as separate chunks. Each successive chunk consisting of one enhanced speech magnitude and noisy speech phase is taken back into the time domain using the ISTFT and added to the result of the previous shift with the frame length of 32,000. By discarding the appended data during data preparation, the test audio is reconstructed with an original length of 5.5 s that contains the enhanced speech signal. The diametric representation of overall noise suppression model testing is shown in Fig. 9.

4 Result Analysis

Perceptual evaluation of speech quality is a set of principles that includes a test method for auto evaluation of speech quality as perceived by a telephony system user. The test compares an audio output to the original voice file to produce a completely objective and unbiased indicator of the actual audio being heard. This technique is more effective than other techniques of measuring sound quality, which frequently relies on audio quality predictions based on network performance. PESQ assigns a score between -0.5 and 4.5, with higher scores indicating higher quality. Almost all mobile devices' analog circuits can process wideband or fullband speech. The ability of the phone, network, and call setup to transmit narrowband or wideband speech determines whether a call transmits narrowband or wideband speech. The subscriber has no control over whether or not the phone connects in narrowband or wideband. PESQ narrow band is calculated between the pass band of 300–3400 Hz and PESQ wide band is calculated between the 100 and 7000 Hz.

Existing objective speech-intelligibility measures are appropriate for several types of decay; however, they are less relevant for techniques that analyze noisy speech using a time–frequency weighting, such as speech enhancement and source separation. So, researcher have introduced new intelligibility measures called STOI.

There was a total of eight models built, four with LibriSpeech + Urban Noise (LU) and four with LibriSpeech + White Noise (LW), each with a different SNR ratio, primarily 10 dB, 15 dB, 20 dB, and 25 dB. PESQ and STOI, two speech quality parameters, were used to evaluate these eight models. The results obtained from all eight models are shown in the Tables 2, 3 and 4. The lower the SNR dB, the higher

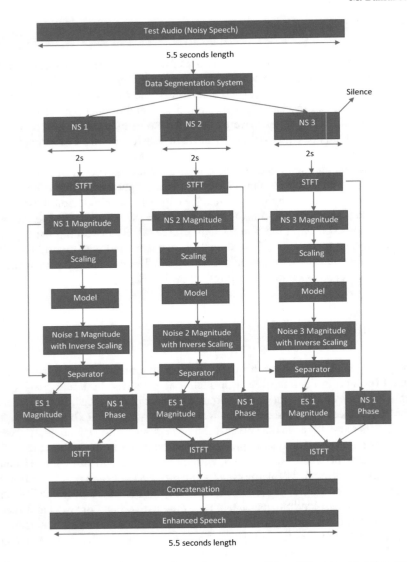

Fig. 9 Overall noise suppression system for testing purposes. Where NS represents noisy speech, ES represents enhanced speech

the noise level. For UNet BiLSTM (LibriSpeech + UrbanNoise), the PESQ wide band score was highest for 25 dB Noise level with 2.6847, the PESQ Narrow Band score was highest for 25 dB noise level with 3.368, and the STOI score was highest for 25 dB noise level with 0.9722. For UNet BiLSTM (LibriSpeech + WhiteNoise), the PESQ wide band score was highest for 25 dB Overall, stationary noise produced

Table 2 Results of PESQ wide band evaluation

SNR	PESQ wide band			
	10 dB	15 dB	20 dB	25 dB
UNET (LW)	2.0999	2.4047	2.7206	**2.9757**
UNET-BiLSTM (LW)	**2.1257**	**2.4050**	**2.8282**	2.9737
UNET (LU)	1.7656	2.1397	2.4409	2.7563
UNET-BiLSTM (LU)	1.7925	2.1569	2.4567	2.8943

Table 3 Results of PESQ narrow band evaluation

SNR	PESQ narrow band			
	10 dB	15 dB	20 dB	25 dB
UNET (LW)	2.7291	3.0741	3.39226	**3.6247**
UNET-BiLSTM (LW)	**2.7323**	**3.1111**	**3.4020**	3.6245
UNET (LU)	2.4678	2.8949	3.2276	3.5272
UNET-BiLSTM (LU)	2.4525	2.9033	3.2651	3.5335

Table 4 Results of STOI evaluation

SNR	STOI			
	10 dB	15 dB	20 dB	25 dB
UNET (LW)	0.9208	**0.9487**	0.9660	0.9735
UNET-BiLSTM (LW)	**0.9252**	0.9418	**0.9686**	**0.9757**
UNET (LU)	0.9025	0.9436	0.9608	0.9742
UNET-BiLSTM (LU)	0.9112	0.9457	0.9699	0.9722

better results than non-stationary noise, but non-stationary noise results were close to stationary noise results.

A spectrogram is defined as frequency versus time plot as shown in Fig. 10, Noisy speech in Fig. 10a is the input speech given to the model by converting speech signal into a spectrogram, once the spectrogram of noisy speech is denoised by the model, it will generate a spectrogram of noise without inverse scaling as shown in Fig. 10b. To separate noise from a noisy speech the noise should be inverse scaled as shown in Fig. 10c. Now, noise after inverse scaling is separated from noisy speech spectrogram to generate enhance speech as shown in Fig. 10d.

Table 5 shows the comparison of the PESQ Score between various techniques with our approach; it has been noted that [10] have got PESQ score of 3.41 on the LSTM-LPS approach which was trained on the synthetically generated dataset, [11] have got PESQ to score 3.02 on AECNN-SM2 approach which was trained on TIMIT clean speech with a mixture of NOISEX Data which consist of noisy speech, [13] obtained PESQ Score 3.21 on DCCRN-E-Aug approach which was trained on DNS Challenge Dataset, our proposed Architecture got PESQ to score 3.26 which

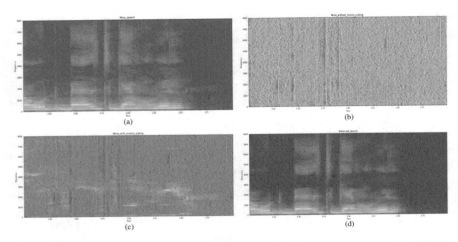

Fig. 10 Audio Spectrogram of 2 s chunk **a** noisy speech, **b** noise without inverse scaling (10 dB), **c** Noisy with inverse scaling (10 dB), **d** enhanced speech

Table 5 Comparison of PESQ scores with existing research

Research paper	PESQ	Year	Dataset	Architecture
Tu et al. [10]	**3.41**	2018	Synthetically generated dataset	LSTM-LPS
Pandey et al. [11]	3.02	2019	TIMIT + NOISEX	AECNN-SM2
Hu et al. [13]	3.21	2020	DNS Challenge Dataset	DCCRN-E-Aug
Proposed architecture	3.26	2021	LibriSpeech + Urban Noise	UNet-BiLSTM

was second-highest score in the table, the approach we used was UNet–BiLSTM approach which was trained on LibriSpeech as a clean speech with a mixture of 10 different Urban Noise from Urbansound 8 k dataset.

Table 6 shows the comparison of STOI Score between various techniques with our approach, Pandey and Wang [11] have obtained STOI score of 92.8% on AECNN-SM2 Approach which was trained on TIMIT clean speech data with a mixture of NOISEX data which consist different noise environment, Tu et al. [10] have obtained 94.76% STOI Score on DTLN approach on DNS Challenge Dataset, our proposed approach obtained 94.87% STOI score on UNet BiLSTM approach on Librispeech as clean speech with a mixture of different urban noises from urbansound 8 k dataset.

Table 6 Comparison of STOI scores with existing research

Research Paper	STOI (%)	Year	Dataset	Architecture
Pandey et al. [11]	92.8	2019	TIMIT + NOISEX	AECNN-SM2
Westhausen et al. [10]	94.76	2020	DNS challenge Dataset	DTLN
Proposed architecture	**94.87**	2021	LibriSpeech + Urban Noise	UNet-BiLSTM

5 Conclusion

Reducing noise in a non-stationary environment is not an easy task; most researchers developed various deep learning approaches and worked in a variety of noise environments. We proposed an architecture called UNet BiLSTM, which takes the shape of a U and convolution layer along with BiLSTM layer is sandwiched between downsampling and upsampling layers. The proposed architecture was used to address noise reduction issues in both stationary and non-stationary environments. The proposed architecture was trained on the Librispeech dataset with white noise as the stationary environment and urban noise as the non-stationary environment; in both cases, the proposed architecture performed well in noise reduction and obtained higher PESQ and STOI scores. For stationary conditions, the PESQ wide band score 2.1257 of 10 dB, 2.4050 of 15 dB, and 2.8282 of 20 dB were highest score when trained on UNet BiLSTM Architecture and PESQ narrow band score 2.7323 of 10 dB, 3.1111 of 15 dB, and 3.040 of 20 dB were highest score when trained on UNet Architecture, while for non-stationary conditions, the PESQ wide band score 1.7925 of 10 dB, 2.1569 of 15 dB, 2.4567 of 20 dB and 2.8943 of 25 dB and PESQ narrow band score 2.4525 of 10 dB, 2.0933 of 15 dB, 3.2651 on 20 dB, and 3.5335 on 25 dB were the highest score when trained UNet Architecture. Overall, the proposed architecture produced better results for stationary noise when compared with non-stationary noise on UNet BiLSTM Architecture. The proposed architecture was able to reduce noise in both stationary and non-stationary conditions.

References

1. Boll, S.: Suppression of acoustic noise in speech using spectral subtraction. IEEE Trans. Acoust. Speech Signal Process. **27**, 113–120 (1979)
2. Gerkmann, T., Hendriks, R.C.: Unbiased MMSE-based noise power estimation with low complexity and low tracking delay. IEEE Trans. Audio, Speech Lang. Process. **20**(4), 1383–1393 (2012)
3. Ephraim, Y., Malah, D.: Speech enhancement using a minimum mean-square error log-spectral amplitude estimator. IEEE Trans. Acoust. Speech Signal Process. **33**(2), 443–445 (1985)
4. Valin, J.: A hybrid DSP/deep learning approach to real-time full-band speech enhancement. In: 2018 IEEE 20th international workshop on multimedia signal processing (MMSP), pp. 1–5. https://doi.org/10.1109/MMSP.2018.8547084 (2018)
5. Nossier, S.A., Wall, J., Moniri, M., Glackin, C., Cannings, N.: An experimental analysis of deep learning architectures for supervised speech enhancement. Electronics **10**, 17. https://doi.org/10.3390/electronics10010017 (2021)
6. Ronneberger O., Fischer P., Brox T.: U-Net: convolutional networks for biomedical image segmentation. In: Navab N., Hornegger J., Wells W., Frangi A. (eds) Medical image computing and computer-assisted intervention—MICCAI 2015. MICCAI 2015. Lecture Notes in computer science, vol 9351. Springer, Cham (2015). https://doi.org/10.1007/978-3-319-24574-4_28

7. Panayotov, V., Chen, G., Povey, D., and Khudanpur, S.: Librispeech: an ASR corpus based on public domain audio books. In: 2015 IEEE International Conference on Acoustics, Speech and Signal Processing (ICASSP), pp. 5206–5210 (2015). https://doi.org/10.1109/ICASSP.2015.7178964. https://www.openslr.org/12

8. Justin, S., Christopher, J., Juan Pablo, B.: A dataset and taxonomy for urban sound research. In: Proceedings of the 22nd ACM international conference on Multimedia (MM '14). Association for Computing Machinery, New York, pp. 1041–1044 (2014). https://doi.org/10.1145/2647868.2655045. https://urbansounddataset.weebly.com/urbansound8k.html

9. Strake, M., Defraene, B., Fluyt, K. et al.: Speech enhancement by LSTM-based noise suppression followed by CNN-based speech restoration. EURASIP J. Adv. Signal Process. (2020). https://doi.org/10.1186/s13634-020-00707-1

10. Tu, Y., Tashev, I., Zarar, S., Lee, C.: A hybrid approach to combining conventional and deep learning techniques for single-channel speech enhancement and recognition. In: 2018 IEEE International Conference on Acoustics, Speech and Signal Processing (ICASSP), pp. 2531–2535 (2018). https://doi.org/10.1109/ICASSP.2018.8461944

11. Pandey, A., Wang, D.: A new framework for cnn-based speech enhancement in the time domain. IEEE/ACM Trans. Audio, Speech, Lang. Process. 27(7), 1179–1188 (2019). https://doi.org/10.1109/TASLP.2019.2913512

12. Bando, Y., Sekiguchi, K., Yoshii, K.: Adaptive neural speech enhancement with a denoising variational autoencoder. In: ISCA, INTERSPEECH (2020)

13. Hu, Y., Liu, Y., Lv, S., Xing, M., Zhang, S., Fu, Y., Wu, J., Zhang, B., Xie, L.: DCCRN: deep complex convolution recurrent network for phase-aware speech enhancement. In: ISCA, INTERSPEECH (2020)

14. Westhausen, N.L., Meyer, B.: Dual-signal transformation LSTM network for real-time noise suppression. In: ISCA, INTERSPEECH (2020)

15. Manoharan, J.S., Sathesh, A.: Super-resolution reconstruction model using Compressive Sensing and Deep Learning. Int. J. Res. Dev. Technol. 7(4), 884–889 (2017)

16. Samuel, M., Ponraj, N.: Analysis of complex non-linear environment exploration in speech recognition by hybrid learning technique. J. Innov. Image Process. (JIIP) 2(04) (2020)

17. Chi-Chang, L., Yu-Chen, L., Hsuan-Tien, L., Hsin-min, W., Yu, T.: SERIL: noise adaptive speech enhancement using regularization-based incremental learning. In: ISCA, Interspeech (2020)

18. Free Sound, https://freesound.org/

19. Luo, Y., Chen, Z., Yoshioka, T.: Dual-Path RNN—efficient long sequence modeling for time-domain single-channel speech separation, pp. 46–50 (2020). https://doi.org/10.1109/ICASSP40776.2020.9054266

A New Hybrid Approach of NaFA and PSO for a Spherical Interacting System

S. Meena⑩, M. Mercy Theresa⑩, A. Jesudoss⑩, and M. Nivethitha Devi⑩

Abstract Spherical tanks are widely used as storage system in Petroleum and Chemical industries. The objective of this work is to develop an optimal controller for the liquid level process of the spherical interacting system which is highly non-linear in nature. A hybrid approach called NaFA-PSO has been developed with PSO and Firefly algorithm with neighbourhood attraction model. The developed approach is validated through simulations in MATLAB and it outperformed both the conventional approach (ZN) and the optimization approach, Grey Wolf Optimization (GWO).

Keywords Spherical interacting system · Firefly Algorithm · Particle swarm optimization · FA with neighbourhood attraction · TTSIS

1 Introduction

Spherical tanks are particularly used as storage systems as they are subjected to uniform tensile stress is all directions [1]. They are particularly used in petroleum and chemical process industries as pressure vessels as these are the structures that have null weak points [1, 3, 4]. These are also used to store various type of liquid ranging from non-flammble to highly flammable liquids or toxic chemicals that are explosive in nature. These type of tanks are almost in chemical, nuclear industries [2].

Controlling such liquid levels is paramount as an unnoticeable and uncontrollable increase in liquid level may lead to spillage and sometimes, may be an explosion. It is also obvious that an unnoticeable and uncontrollable reduction in liquid level

S. Meena (✉) · M. Nivethitha Devi
Department of EIE, St. Joseph's College of Engineering, OMR, Chennai, India

M. Mercy Theresa
Department of ECE, Prince Shri Venkateswara Padamvathy Engineering College, Chennai, India

A. Jesudoss
Department of CSE, Sathyabama Institute of Science and Technology, Chennai, India

© The Author(s), under exclusive license to Springer Nature Singapore Pte Ltd. 2022   77
V. Suma et al. (eds.), *Evolutionary Computing and Mobile Sustainable Networks*,
Lecture Notes on Data Engineering and Communications Technologies 116,
https://doi.org/10.1007/978-981-16-9605-3_6

may imbalance the reaction and subsequently affects the end product. To effectively control any process a modelling is required and the mathematical analysis and modelling was done by Ravi et al. [5].

Controller tuning is being redefined day by day by researchers across the globe through their innovative dynamic strategies. There are various tuning methods and they are Zeigler Nichols [6], Cohen coon [7], Frequency domain-based PID tuning [8], adaptive and IMC. All the methods adapted so far are for reaching an optimal solution. The best approach for optimal solutions is trial and error approach and all the metaheuristic algorithms are trial and error in nature. There are various optimization algorithms such as GA [9], simulated annealing [10], Particle swarm optimization [11], Ant colony optimization [12], Cuckoo search [13, 14] etc., Soft computing techniques and soft computing with SVM are used in various applications like wireless networks [15] and in wind turbine monitoring mechanism [16].

After taking into account of various optimization algorithms, FA has been considered for finding the optimal solutions as the solutions are obtained in a relatively low time by means of trial-and-error approach [17]. Zouache et al. [18] confirmed that FA has been used extensively in continuous optimization problems. Fister et al. [19] in his review confirmed that most of the problems solved through optimization were FA based. But in all the metaheuristic algorithms there are some randomness in the population. The randomness in the population leads to a greater delay in finding the appropriate solution. Though it can't be eliminated, it can be reduced to some extent. In [20] Hsaio, created a sequence of arrays to improve bio sequence alignment in a biological application. He sequenced the population before PSO optimization. Similarly, in the proposed work the randomness in population is reduced by creating pre-optimized solution space and thus the so-called solution space is created by PSO.

One of the promising areas in optimization is always hybridization and the proposed work deals with the hybridization of FA and PSO. But Hybridization leads to increase in the computation time and to reduce the computation time FA with neighbourhood attraction (NaFA) is adapted. Because in NaFA [21] each firefly is attracted to the predefined neighbour rather than the entire population.

Thus, the proposed work satisfies the following criteria such as continuous optimization in a relatively low time by adapting FA, reduction in randomness by creating a pre-solution space by PSO and less computation time in the hybrid approach by adapting NaFA. The algorithm is framed and applied in an interacting non-linear system modelled by Ravi et al. [5].

The paper is organized as follows: In Sect. 2, the process and its mathematical model has been discussed. Section 3 has three subdivisions. In Sects. 3.1 and 3.2, the Firefly algorithm (FA) and PSO has been discussed respectively. In the Sect. 3.3, the development of a new hybrid algorithm NaFA-PSO has been discussed. In Sect. 4, results have been discussed in both qualitative and quantitative way. It is finally followed by conclusions in Sect. 5.

2 Process Description

The process that is considered into account is a two-tank spherical interacting system (TTSIS). The tank consists of two inlets called F_{in1} and F_{in2} controlled by two independent pumps and are interconnected at the bottom by a manually operated valve (MV_{12}). The spherical system is highly non-linear as its radius is varying from top to bottom and thereby the cross-sectional area. The non-linearity gets enhanced by its interacting fashion. The input considered for analysis is F_{in1}, the control variable is the height of the second tank (H_2) here and thus the process that has been analyzing here is a SISO process.

It is well known that the mass balance equation for the spherical tank system that has its area as varying with respect to height.

$$F_{in} - F_{out} = \frac{1}{3}\left[A\frac{dh}{dt} + h\frac{dA}{dt}\right] \tag{1}$$

As the tank is a SISO process, at tank1 the balance equation can be written as

$$F_{in1} - \beta_{12}\sqrt{(h_1 - h_2)} = \frac{4}{3}\left(A_1\frac{dh_1}{dt} + h_1\frac{dA_1}{dt}\right) \tag{2}$$

where
A_1 is area of a circular cross section of the spherical tank.
H_1 is the level of tank 1.
F_{in1} is the input flow rate.
β_{12} is the valve coefficient of interacting valve.
As the tank is a SISO process, at tank 1 the balance equation can be written as

$$\beta_{12}\sqrt{(h_1 - h_2)} - \beta_2\sqrt{h_2} = \frac{4}{3}\left(A_2\frac{dh_2}{dt} + h_2\frac{dA_2}{dt}\right) \tag{3}$$

where
A_2 is area of a circular cross section of the spherical tank.
H_2 is the level of tank 1.
β_2 is the valve coefficient of outlet valve.
β_{12} is the valve coefficient of interacting valve.
From Eq. 2 it can be written as

$$\frac{dh_1}{dt} = \frac{F_{in1} - \beta_{12}\sqrt{(h_1 - h_2)} - \frac{4}{3}h_1\frac{dA_1}{dt}}{\frac{4}{3}A_1} \tag{4}$$

Similarly, Eq. 3 can be written as

Table 1 Transfer function of TTSIS system for different regions

Region	Flow rate in cm³/s	H_1 (cm)	H_2 (cm)	Transfer function
I	0–25	1.714	1.612	$\frac{0.1317}{68.39s^2+68.06s+1}$
II	26–50	6.856	6.448	$\frac{0.2588}{3452.73s^2+483.7s+1}$
III	31–75	14.97	14.06	$\frac{0.3823}{24269s^2+1276.09s+1}$
IV	76–107.85	31.9	30	$\frac{0.558}{67123.71s^2+2126.7s+1}$

$$\frac{dh_2}{dt} = \frac{\beta_{12}\sqrt{(h_1-h_2)} - \beta_2\sqrt{h_2} - \frac{4}{3}h_1\frac{dA_1}{dt}}{\frac{4}{3}A_1} \tag{5}$$

The mathematical model of the TTSIS is given through Eqs. 4 and 5. By applying Laplace transform and by setting down the values of time constant and radius, given by Ravi et al. [5] the transfer function of four different regions for the said system had been found out. The transfer function of the four different regions is given in the following tabular column Table 1.

3 Methodology

3.1 NaFA

In 2008,Yang, on seeing the behaviour of fireflies, developed an algorithm called Firefly Algorithm [22]. In FA the solutions are considered as fireflies and each firefly (solution) is compared with the other one. If the brightness of the firefly (X_i) is less than the other (X_j) then it would take other's brightness. By doing so the optimal solution is identified.

From the year 2008 to till date a lot of new variants of FA has been proposed by the researchers across the globe. Fister et al. [19] have said that most of the industrial optimization problems have been solved by using FA as a tool. In the list given in Fister et al. [19] image processing is the second one which has been solved by FA.

Wang in the year 2017 [21] proposed a new FA model, NaFA as neighbourhood attraction model. In FA full attraction model had been followed. That is each firefly had been compared with all the other fireflies. Whereas in NaFA, the fireflies have been compared with a group of fireflies that are nearer to its location. That is the brightness of ith firefly gets compared with $i - k$, ..., i, ... $i + k$ fireflies where $k = (N - 1)/2$. Thus too much of comparisons, too much of oscillations and too much of computation time can be avoided. The algorithm steps are as follows:

1. Initially random solutions are considered as fireflies.
2. Forthesefireflies, the brightness is calculated with respect to the objective functions and they are compared. Let us consider the brightness of X_i and X_j is

compared and if $X_i < X_j$ the brightness is modified as per the Eq. (6). Here the range of j starts from $i - k$ and ends with $i + k$ where $k = (N - 1)/2$. It must be noted down that if $j < 1$ or $j > N$ then j would be $(j + N) \% N$

$$X_{it}(n + 1) = X_{it}(n) + \beta_0 e^{-\gamma r_{ij}^2}(X_{jt} - X_{it}) + \alpha \epsilon_i \qquad (6)$$

where

β_0 is the maximumm brightness of the firefly.
α is the step size of the firefly.
γ is the adsorption co-efficient.

3. The brightness is updated till the stopping criteria is satisfied.

3.2 PSO

The social behaviour that revolves around fish schooling was elaborated by Ebenhart and Kenedy [11] in the year 1995 as PSO. Initially random solutions are considered as the population. The optimal solution is obtained by updating the population with new velocity and position. The balance between the exploration and exploitation was done through the inertia weight W and constraint factors C_1 and C_2. The updation is through Eq. 7.

$$v_{new} = w * v_{old} + C_1(p_{best} - p_{current}) + C_2(g_{best} - p_{current}) \qquad (7)$$

where

v_{new} is the updated velocity; p_{best} is the local best position;
g_{best} is the global best position; $p_{current}$ is the current position.
W is the inertia weight; C_1 and C_2 are constants.

3.3 NaFA-PSO

The hybrid of FA and PSO approach is an upcoming technique as there would be a balance between exploration and exploitation [23]. The main constraints in any optimization are its randomness, reliable optimal solutions and computation time to find the solution. In [20] the randomness is reduced by sequencing the initial population in a biological application and then it was made to run through PSO. In the proposed hybrid approach the randomness is reduced by creating an initial solution space through PSO. For the considered spherical interacting systems, the reliable solutions were found through various optimization approaches like PSO, BFO and ACO by Kumar et al. [24] in which it was concluded that PSO had yielded very less error values. Therefore, PSO is considered for initial population determination. Chen in 2011 [25] proposed a sigmoidal function to have a balance between exploration and exploitation. In [25] it was said that the small step must be adapted when the solution is nearer the global optimum and a large step is required when it is far away the global optimum and therefore a sigmoidal function is proposed for W. That is the step size must be dynamically varying. To keep a dynamic varying step size the PSO is made to be in hybrid with FA. Because in FA, the brightness is updated through $\beta_0 e^{-\gamma r_{ij}^2}$ which indirectly influences the population given by PSO. As per [17] FA gives reliable solutions in the stipulated time and as per [19] FA has been used in most of the industrial optimization problems. However, in FA there are too many comparisons and too many oscillations because of the full attraction model. Both the hybrid and full attraction model leads to an increase in computation time. The time can be minimized by doing comparisons with the predefined neighbourhood models instead of the entire model. Therefore, to reduce computation NaFA is adapted. The hybridization of FA and PSO is done in various forms for different applications [25–28]. In the hybrid approach the local solutions are obtained through PSO. Then the local solution is optimized through NaFA and thereby gaining a global optimum. The steps for the proposed algorithm are

1. Initially solutions are considered from PSO and thus a group of local optimal solutions are obtained.
2. The solutions are considered as fireflies. The brightness is calculated with respect to the objective functions and they are compared. Let us consider the brightness of X_i and X_j is comapred and if $X_i < X_j$ the brightness is modifed as per the Eq. (6). Here the range of j starts from $i - k$ and ends with $i + k$ where $k = (N - 1)/2$. It must be noted down that if $j < 1$ or $j > N$ then j would be $(j + N)\%N$.
3. The brightness is updated till the stopping criteria is satisfied.

4   Results and Discussions

The algorihtm NaFA-PSO is made to run in MATLAB-SIMULINK environment for all the four regions of TTSIS. For PSO the algorithm parameters are $C_1 = 1.2$, $C_2 = 0.12$ and $W = 0.75$. The step size, minimum brightness and adsorption co-efficient for firefly is $\alpha = 0.8$, $\beta_{min} = 0.4$ and $\gamma = 1$ respectively. The number of birds and the number of fireflies that are scrutinized for the 3D problem is 50. It is noted down that the sampling time instant is 0.01 and the ser point is 1.0 for all regions. The stopping time criteria for region I, II, III and IV is 100, 200, 500 and 700 respectively as the natural frequency and thereby the time constants are different for these regions. In addition to that the load disturbances are given at diffrent instants for these different regions such as a disturbance of 0.1 at 70th instant for region I and an interfernce of 0.1 at 150th instant for region II. Similarly, interference of magnitude 0.1 at 300 th instant and 500 th instant for the regions III and IV respectively (Fig. 1).

The proposed algorithm is compared with the conventional technique ZN [5] and the optimized algorithm GWO [29] in terms of performance indices and time domain specifications. The tuning parameters of all three methods for all the four regions

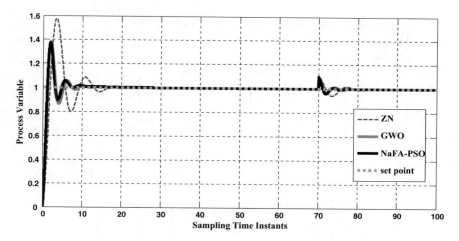

Fig. 1 Comparison response of NaFA-PSO for Region I

have been listed in the Table 2. The corresponding performance indices such as IAE, ISE and ITAE and the corresponding time domain specifications such as tr, tp, ts and PO have also been tabulated in Table 2.

4.1 Results and Discussions for Region I

For region I, the objective functions, PO and ITAE are reduced to 37.6% and 15.596 respectively. Thus, the PO is reduced by almost 1% and the ITAE is reduced by almost 10% when it is compared to GWO. The results are shown in the first three rows of Table 2. The corresponding response is shown in Fig. 2 and the comparison of time domain specifications and performance indices are depicted in Fig. 5 and 6 respectively. The comparison bar chart reveals that not only the objective functions but also all the time domain specifications and performance indices have been reduced significantly.

4.2 Results and Discussions for Region II

PO and ITAE are the objective functions here and they have been reduced to 37% and 370 from 40% and 660, on comparing with GWO. The objective functions are diminished by 3% and by 43.9%. From Fig. 2 it is shown that proposed approach excels in set point tracking and disturbance rejection. All the time domain specifications and performance indices are compared through the bar graph shown in Figs. 7

Table 2 Comparison of NaFA-PSO with ZN and GWO for Region I

Region	Tuning methods	Tuning parameters	IAE	ISE	ITAE	tr	tp	ts	PO in %
I	ZN	$K_p = 539.7$ $K_i = 207.6$	3.599	1.577	37.596	124	345	7732	57%
	GWO	$K_p = 1350$ $K_i = 36$	1.729	0.721	16.955	82	206	7276	37.8%
	NaFA-PSO	$K_p = 842.15$ $K_i = 113.74$ $K_d = 137.24$	**1.628**	**0.661**	**15.396**	**76**	**192**	**7251**	**37.6%**
II	ZN	$K_p = 262.3$ $K_i = 14.57$	25.86	11.57	1020	912	2507	19,106	58
	GWO	$K_p = 531$ $K_i = 8$	13.85	5.97	433.7	660	1692	17,174	40
	NaFA-PSO	$K_p = 785.384$ $K_i =$ $K_d = 462.535$	**11.46**	**4.71**	**370**	**540**	**1351**	**16,824**	**37**
III	ZN	$K_p = 180$ $K_i = 3.673$	66.54	30	6317	2353	6596	45,609	57
	GWO	$K_p = 600$ $K_i =. 0.5$	29.5	12.5	1875	1316	3353	34,610	42
	NaFA-PSO	$K_p = 1275$ $K_i =$ $K_d = 1361$	**27.2**	**10.5**	**1824**	**842**	**2257**	**34,129**	**53**
IV	ZN	$K_p = 123$ $K_i = 1.522$	110	50	16,771	32	10,985	67,312	57
	GWO	$K_p = 430$ $K_i = 1.5$	55	23	6266	2072	5470	6144	49

(continued)

Table 2 (continued)

Region	Tuning methods	Tuning parameters	IAE	ISE	ITAE	tr	tp	ts	PO in %
	NaFA-PSO	$K_p =$ 630.2 $K_i =$ $K_d =$ 70.95	**48**	**19**	**5733**	**1687**	**4437**	**59,704**	**47**

Bold indicates the specifications of all regions, from I to IV, corresponding to NaFA-PSO are considerably diminished than that of its counterpart

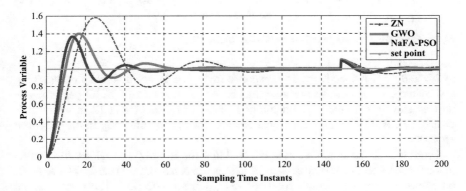

Fig. 2 Comparison response of NaFA-PSO for Region II

and 8. From the bar graphs it can be said that the proposed approach outperforms the other approaches.

4.3 Results and Discussions for Region III

The objective function ITAE is minimized by the proposed approach to the extent of 1824 from 1875 for the region III. The reduction is almost 2.72%. All the performance indices and the time domain specifications have been reduced significantly and they are depicted as bar graph in Figs. 9 and 10. From the response Fig. 3 and from the bar graph 9 and 10 it can be said that the proposed approach exceeds the expectations.

4.4 Results and Discussions for Region IV

The objective function PO is narrowed down to 47% from 49% and ITAE is lowered to 5733 from 6266. Thus, on comparing with GWO, PO and ITAE has been diminished

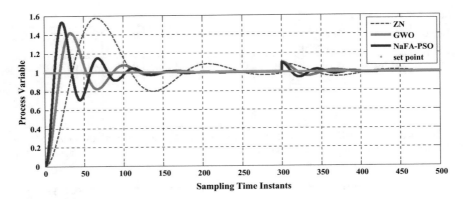

Fig. 3 Comparison response of NaFA-PSO for Region III

to 2% and 8.5% respectively. The response for region IV for both set point tracking and disturbance rejections is shown as Fig. 4. The comparison of all performance indices and time domain specifications is shown in Fig. 11 and 12 in which it is clearly revealed that all the specifications are lowered significantly in the proposed approach.

From the tabular column 2 it is inferred that for all the four regions, the performance indices and time domain specifications are considerably less for NaFA-PSO, on comparing with the counterparts of ZN [5] and GWO [29]. The responses are shown as Fig. 2, 3, 4 and 5 for regions I, II, III and IV respectively. From the responses it is obvious that the proposed NaFA-PSO outperforms the fellow methods such as ZN and GWO [29] in both set point tracking and load disturbances.

The hybrid algorithm provides a balance between exploration and exploitation and thereby optimal solutions are obtained at the earlier iterations itself (less than 5 for all the 4 regions).The computation time for the algorithm gets reduced by 20–30%

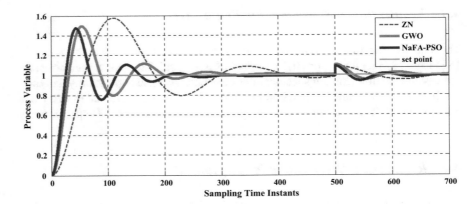

Fig. 4 Comparison response of NaFA-PSO for Region IV

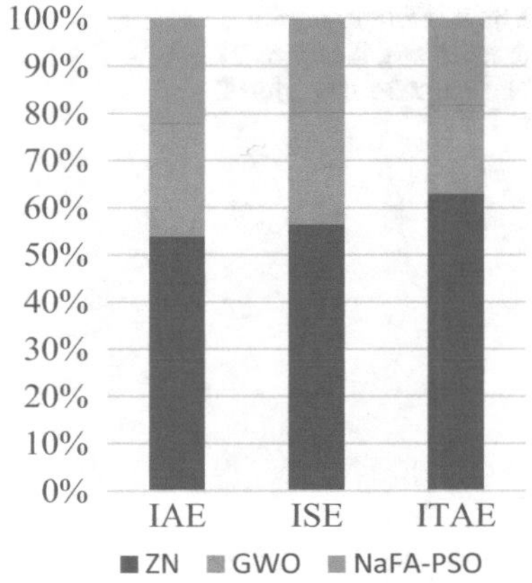

Fig. 5 Comparison of
performance indices of
NaFA-PSO for Region I

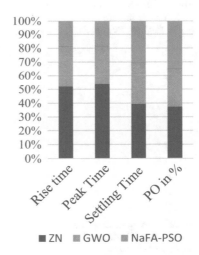

Fig. 6 Comparison of time
domain specifications of
NaFA-PSO for Region I

Fig. 7 Comparison of
performance indices of
NaFA-PSO for Region II

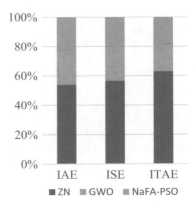

Fig. 8 Comparison of time domain specifications of NaFA-PSO for Region II

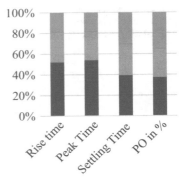

Fig. 9 Comparison of performance indices of NaFA-PSO for Region III

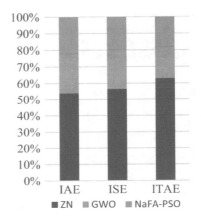

Fig. 10 Comparison of time domain specifications of NaFA-PSO for Region III

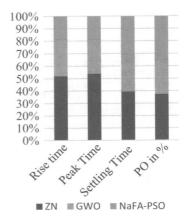

Fig. 11 Comparison of performance indices of NaFA-PSO for Region IV

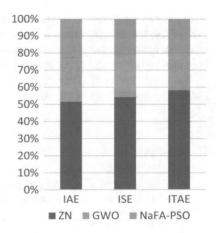

Fig. 12 Comparison of time domain specifications of NaFA-PSO for Region IV

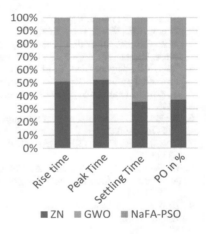

for the neighbourhood hybrid model (NaFA-PSO) when compared with FA-PSO with same number of birds and fireflies. The performance indices and time domain specifications are represented through stacked columns in Figs. 6, 7, 8, 9, 10, 11 and 12. In all these graphical representation the metrics are shown in percentage and it is quite obvious that the performance metrics are reduced in an appreciable manner.

5 Conclusions

A hybrid algorithm NaFA-PSO has been proposed for TTSIS for four different regions. Subsequently it has been concluded that the perfomance metrics such as error indices and time domain metrics have come down for the proposed hybrid algorithm. The computation time of the algorithm is also appreciably reduced because of the neighbourhood attraction model. Finally, exploration and exploitation gets balanced in the hybrid approach and thus the optimal solutions are obtained in the earlier iterations itself. The scope of the proposal is that the algorithm's parameter tuning and application of problems that has the dimension more than three.

References

1. Oyelami, A.T., Olusunle, S.O.O.: Spherical storage tank development through mathematical modeling of constituent sections. Math. Model. Eng. Probl. **6**(3), 467–473 (2019) https://doi.org/10.18280/mmep.060320
2. Hutchinson, J.W.: Buckling of spherical shells revisited. Proc. R. Soc. A **472**(2195), 1–25 (2016). https://doi.org/10.1098/rspa.2016.0577
3. Afkar, A., Camari, M.N., Paykani, A.: Design and analysis of a spherical pressure vessel using finite element method. World J. Model. Simul. **10**(2), 126–135 (2014)
4. Zhang, S.H., Wang, B.L., Shang, Y.L., Kong, X.R., Hu, J.D., Wang, Z.R.: Three-dimensional finite element simulation of the integral hydrobulge forming of a spherical LPG tank. Int. J. Press. Vessels Pip. **65**, 47–52 (1994). https://doi.org/10.1016/0308-0161(94)00158-F
5. Ravi, V.R., Thyagarajan, T., Monika Darshini, M.: A multiple model adaptive control strategy for model predictive controller for interacting nonlinear systems. In: International conference on process automation, control and computing (2011). https://doi.org/10.1109/pacc.2011.5978896
6. Ziegler, J.G., Nichols, N.B.: Optimum settings for automatic controllers. Trans. ASME **64**, 759–768 (1942)
7. Cohen, G.H., Coon, G.A.: Theoretical considerations of retarded control. Trans. ASME 827–834 (1953)
8. Ho, W.K., Hang, C.C., Cao, L.S.: Tuning of PID controllers based on gain and phase margin specifications. Automatica **31**(3), 497–502. https://doi.org/10.1016/B978-0-7506-2255-4.50009-X (1995)
9. Holland, J.: Adaptation in natural and artificial systems. University of Michigan Press, Ann Anbor (1975)
10. Kirkpatrick, S., Gellat, C.D., Vecchi, M.P.: Optimization by simulated annealing. Science **220**, 671–680 (1983)
11. J. Kennedy, R. Eberhart: Particle swarm optimization. In: Proceedings of the IEEE International Conference on Neural Networks, Piscataway, NJ, pp. 1942–1948 (1995)
12. M. Dorigo: Optimization, learning and natural algorithms. Ph.D. thesis, Politecnico di Milano, Italy (1992)
13. X. S. Yang, S. Deb, Cuckoo search via L'evy flights. In: Proceedings of World Congress on Nature & Biologically Inspired Computing. IEEE Publications, USA, pp. 210–214 (2009)
14. Yang, X.S., Deb, S.: Engineering optimization by cuckoo search. Int. J. Math. Model. Num. Optim. **1**, 330–343 (2010)
15. Haoxiang, W., Smys, S.: Soft computing strategies for optimized route selection in wireless sensor network. J. Soft Comput. Paradigm (JSCP) **2**(01), 1–12 (2020)
16. Shakya, S.: Performance analysis of wind turbine monitoring mechanism using integrated classification and optimization techniques. J. Artif. Intell. **2**(01), 31–41 (2020)
17. Patel, B., Patle, B.: Analysis of firefly–fuzzy hybrid algorithm for navigation of quad-rotor unmanned aerial vehicle. Inventions **5**, 48 (2020). https://doi.org/10.3390/inventions5030048
18. Zouache, D., Nouioua, F., Moussaoui, A.: Quantum-inspired firefly algorithm with particle swarm optimization for discrete optimization problems. Soft Comput. **20**, 2781–2799. (2016). https://doi.org/10.1007/s00500-015-1681-x

19. Fister, I., Fister, I., Yang, X.-S., Brest, J.: A comprehensive review of firefly algorithms. Swarm Evolut. Comput. **13**, 34–46. ISSN 2210-6502 (2013). https://doi.org/10.1016/j.swevo.2013.06.001
20. Hsiao, Y.T., Chuang, C.L., Jiang, J.A.: Particle swarm optimization approach for multiple biosequence alignment. In: Proceedings of the IEEE international workshop on genomic signal processing and statistics. Rhode Island, USA (2005)
21. Wang, H., Wang, W., Zhou, X., Sun, H., Zhao, J., Yu, X., Cui, Z.: Firefly algorithm with neighborhood attraction. Inf. Sci. **382–383**, 374–387 (2017). ISSN 0020-0255. https://doi.org/10.1016/j.ins.2016.12.024
22. Yang, X.S.: Firefly algorithm. Nature-Insp. Metaheur. Algorithms **20**, 79–90 (2008)
23. Meena, S., Chitra, K.: An approach of firefly algorithm with modified brightness for PID and I-PD controllers of SISO systems. J. Ambient Intell. Human. Comput. (2018). https://doi.org/10.1007/s12652-018-0747-x
24. Kumar, D., Meenakshipriya, B., Ram, S: Design of PSO, BFO, ACO Based PID Controller For Two Tank Spherical Interacting System. Int. J. Power Control Signal Comput. **8**, 129–133 (2016)
25. Windarto, W., Eridani, E. :Comparison of particle swarm optimization and firefly algorithm in parameter estimation of Lotka-Volterra. In: The 4th Indoms International Conference on Mathematics and Its Application (IICMA 2019) (2020). https://doi.org/10.1063/5.0017245
26. Gebremedhen, H.S., Woldemichael, D.E., Hashim, F.M.: A firefly algorithm-based hybrid method for structural topology optimization. Adv. Model. Simul. Eng. Sci. **7**, 44 (2020). https://doi.org/10.1186/s40323-020-00183-0
27. Zhao, C., Jiang, L., Teo, K.L.: A hybrid chaos firefly algorithm for three-dimensional irregular packing problem. J. Ind. Manag. Optim. **16**(1), 409–429 (2020). https://doi.org/10.3934/jimo.2018160
28. Ab Talib, M.H., Mat Darus, I.Z., Mohd Samin, P., et al: Vibration control of semi-active suspension system using PID controller with advanced firefly algorithm and particle swarm optimization. J. Amb. Intell. Human. Comput. **12**, 1119–1137 (2021). https://doi.org/10.1007/s12652-020-02158-w
29. Singh, R.K., Yadav, S.: Optimized PI controller for an interacting spherical tank system. In: 1st International conference on electronics, materials engineering and nano-technology (IEMENTech) (2017), pp. 1–6. https://doi.org/10.1109/IEMENTECH.2017.8076977

BitMedi: An Application to Store Medical Records Efficiently and Securely

Rahul Sunil⬤, Kesia Mary Joies⬤, Abhijeet Cherungottil, T. U. Bharath, and Shini Renjith⬤

Abstract Accurate medical history plays a crucial role in the diagnosis by providing physicians relevant data regarding the health of the patient. It helps in preventing prescription errors and the consequent risks to patients. Securing and storing medical records are essential for attaining proper treatment. This paper details the solution BitMedi, a mobile application that helps in uploading medical records and storing them in a well-organized and secure manner. The struggle of all in finding their previous records and risking their lives on giving unsure details of their health to the doctors can hence be eliminated with the solution which helps users save their records in their mobile phones in a user-friendly manner while not compromising on the security of the records. The solution extracts data from the uploaded records while maintaining the user's privacy and provides graphical representation that helps guide patients to maintain their health.

Keywords IOTA tangle · Electronic health record · Smart health care · Tesseract OCR · Flutter mobile app · Tesseract · Django web app

1 Introduction

Health records empower doctors, nurses, or other healthcare professionals to treat patients to the best of their ability. It also helps in eliminating any potential drug interactions. Access to health records has become a primary concern for many individuals. Traditional methods of storing records were in control with clinicians of a particular health center and restricting others, including patients or doctors from other hospitals to access it. This has caused several issues, including a wastage of time in analyzing and providing the right treatment to the patients.

Over the past years, globalization and rapid digitalization have led to the equity of only resources and services to every human being, regardless of geographical, economical, and cultural barriers. This has created an immense opportunity as well

R. Sunil (✉) · K. M. Joies · A. Cherungottil · T. U. Bharath · S. Renjith
Department of Computer Science and Engineering, Mar Baselios College of Engineering and Technology, Thiruvananthapuram, Kerala 695015, India

© The Author(s), under exclusive license to Springer Nature Singapore Pte Ltd. 2022
V. Suma et al. (eds.), *Evolutionary Computing and Mobile Sustainable Networks*,
Lecture Notes on Data Engineering and Communications Technologies 116,
https://doi.org/10.1007/978-981-16-9605-3_7

as a threat in the storage of health records as well. The electronic health records system developed by time has helped in solving the issue to some extent by making it accessible to different healthcare settings. But, the need for a system for the patients to store and manage their health records was considered highly important. Mobile phones have become a necessity and are used by almost all individuals. For example, taking the instance of the ongoing Covid-19 lockdown in India, immediate access to health records has become vital. According to a cross-sectional hospital-based study, it was included that 7008 teleconsultations were conducted between March 23rd, and April 19th, 2020 [1]. Providing accurate recent details are highly important for receiving the right consultation. The idea of storing medical records, prescriptions, and other important information about an individual on a mobile application is considered and proposed in this paper.

To track one's health records, store them in an organized manner, and extract the necessary information, a software solution BitMedi is proposed. Consider a scenario where a person is in a remote place, and his/her health records are not immediately accessible. And suppose the person requires immediate treatment for an injury/accident, then getting his/her blood group or important previous medical history will be difficult. For such a situation, an all-in-one mobile app with all the important medical data of the user will be really helpful. The BitMedi application will store the allergies, blood group, and other relevant details. This thereby prevents the risk of death caused by intaking prescribed medications that are allergic to the patient. Mismatched blood transfusion can cause several health hazards to the patient [2]. BitMedi helps in solving the issue by storing the blood group of the patient and important contact details. The records are made easily accessible to users and reduce their time and effort in reading the record data values. Important health data from records are extracted and displayed alongside the record helping users track their health easily. The proposed solution, BitMedi, stores the records securely by encrypting the username details and storing them in a secure distributed ledger network.

2 Existing Works

Before we started working on this problem, we needed to have an understanding of all products and solutions that were already available on the market. So, we started our literature review by going through papers that are available on platforms like IEEE and Springer, and we came across several papers which discussed the several ways by which the medical records are converted to digital format, stored, and used by the patients. When we started looking for solutions that used mobile as a medium for scanning the records, we came across papers on mobile applications for storing medical records [3–5]. But, some limitations that these papers posed were that they do not allow the users to view their health reports at times of emergency or other situations, and also, there is neither an option for the doctors to view the medical history of a patient nor for the patient to share it with the doctors efficiently. One of the main challenges we had to go through was to ensure the security of the data that we collect from the user, and to solve this problem, we went through several papers discussing the security aspect of cloud-based systems for storing information [6], the feasibility and efficiency of implementing it using an IoT system [7], and the viability of storing user's sensitive information securely on cloud [8]. We thought about implementing blockchain [9] in our system after analyzing its viability through some papers [10], but the main issue about blockchain was that the entire system will become heavy and slow as we add more data to it, and we also have to incentivize the miners for adding the data to the chain. So, we had to think of a system that was as robust as blockchain, but yet, didn't have the limitations posed by blockchain. We also went through some other papers which provide details [11, 12] to further improve the quality of our system. One of the papers proposes a framework for the automated assessment of medical history taken within a VP software and then tests this framework by comparing VP scores with the judgment of 10 clinician educators. There was a study conducted in an article [13] on similar implementations from Germany studying how and when the EHR is needed to be accessed and how the patient feels about sharing their records, and it shows it is viable to implement the proposed solution.

Then, to know the practices that are used in the industry, we were also able to arrange meetups with people working in hospitals including an officer who was working on the Hospital Record Management Department. in a hospital in Bangalore, Karnataka with the help of our contacts in CISCO ThingQbator. Through these meetings, we got to know about the current hospital record management methods like electronic health record (EHR) [14] in which the medical records of a patient are stored along with his scan results, MRI scans, basic information like their height, weight, and blood group. But, the main problem with this method is that the records are only available to that particular hospital management and are inaccessible to other hospitals or even the patient himself, which poses an issue for the patient in case he/she wishes to share their medical records with other hospitals if he/she wishes to consult any other doctor like a specialist, or if he/she is relocating to another place. Also, EHR still depends a lot on the pen paper format which possesses threats like the need for proper physical space where the papers are not prone to any damages like moth infestations for storing the records which can be a problem as we scale up. Also, this entire system is not efficient because of inconsistent layout, ambiguous audit trails, and version history, lack of backups, and also limited security. A similar implementation [15] explored the idea of making a centralized platform with various levels of access control and the EHR being distributed from the hospitals own database, but this is not practical since bringing all the hospitals to be part of this implementation needs a huge amount of trust which can't be given in such an approach, whereas the proposed approach will give confidence and may be able to change to this in the future.

While we were doing our literature review, we came across some government articles [16] released by the Government of India which contains the updated telemedicine guidelines. According to this article, the Board of Governors ("BoG") tasked by the Health Ministry to regulate practice and practitioners of modern medicine, published an amendment to the Indian Medical Council (professional conduct, etiquette, and ethics) Regulations, 2002 ("Code of Conduct") that gave statutory support and basis for the practice of telemedicine in India. We also came across "Digital Health ID" which is a health ID that will be issued for every Indian by the Government of India through a campaign called National Digital Health Mission (NDHM) [17]. This health ID will also contain features like personal health record (PHR) and electronic medical record (EMR) that can be developed by private entities. Thus, empowering our solution to tie up with the government in the future [18]. Another paper [1], which focuses on the impact of teleconsultation in Indian ophthalmology hospital networks during the Covid-19 lockdowns, emphasizes the need for an app such as the one explained in the paper to make teleconsultation more efficient through securely sharing the electronic health records between the doctor and patient.

We also went through some papers [19] to ensure the user's readiness for using such an app and got to know that some applications are already available in the market like PRACTO [20, 21] and UHID, but these applications do not provide any further use of the data from the medical records as it is not extracted like a visual representation of the health of the patients. Also, these applications do not store the records in an organized manner but rather in a very cluttered way just like our mobile gallery, thereby wasting time finding the record needed by the user [22].

3 Proposed Solution

The proposed solution is to bring up a user-friendly, secure app for the storage of different medical records of a patient in a very organized manner. The app ensures security by creating a biometric lock system along with signing in using a username and password. The fingerprint used as the biometric trait helps in verifying whether it is a legitimate user. Using this app, the users will be able to upload their medical records under a specific type and category to which the record belongs and will also be able to attach the date on which the record was generated. Users will also be able to give a small descriptive name to the report to make it more convenient for them to remember or search for that particular record. The data that are printed on the report will then be extracted using OCR [23], and then, the record along with the extracted data will be stored online securely [24, 25]. The data from the records are extracted primarily for the further analysis of that data like to alert the user when a particular parameter in their report is higher than normal, and they should take immediate action like consulting a doctor. The data can also be used for visually representing the data in a graph or a timeline format, which will be easier for the users as well as the doctors to get insights from in less time. The user will be able to view the records in two formats now. The first one is a list view in which all the records will be shown. The user will be able to find a particular record they are looking for by either sorting the records or by filtering the records which are generated within a time frame. When they open a particular record, they will be able to view the recorded image as well

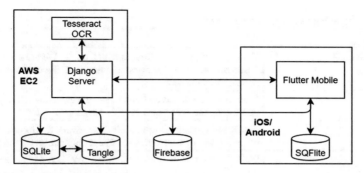

Fig. 1 Software architecture

as the extracted data. The extracted data will be color-coded to indicate, whether the parameters are within the optimal levels or not. The user will be able to share the record through his phone's share feature. The second way of viewing the records is in the form of a graph, in which the values extracted from a particular type of record will be plotted against a time graph, showing the variations in that particular parameter's level. The proposed application stands out with features such as graphical analysis of health, extraction of data from records, user-friendly and easy upload, and viewing of records categorically.

3.1 Architecture

The architecture (Fig. 1) **of the BitMedi application is built using the following technologies as mentioned in** Table 1.

Tesseract OCR engine was selected because of its high efficiency. It was also able to give results much faster than traditional engines (Table 2) [29].

3.2 Mobile App Client

The mobile application for the client is built entirely using the Flutter SDK provided by Google. The app starts with the splash screen (Refer Fig. 2) for loading data and giving an overview of the mobile app "BitMedi."

It consists of various features like:

User Authentication: The app allows the user to sign up and sign in (Refer Fig. 3) to the app using their credentials which are then authenticated using tokens from the Web server. The user registration requires the user to fill in personal health details like name, email, and authentication credentials like username and password.

Table 1 Technology stack used for BitMedi

Technology	Description
Tesseract OCR	Tesseract [26] is an OCR engine by Google. We are using Tesseract OCR [27] for extracting text from the medical records scanned by the users
Django	Django is a Python Web framework. We are using it for developing the Webserver module of our solution
Flutter	Flutter is a cross-platform mobile framework by Google. We are using Flutter for developing the client application for the users
SQFLite	SQFLite, the widely used database engine, is the SQLite plugin for Flutter. We are using it to store data directly in the application itself, enhancing its processing speed and reliability [28]
Firebase	It is a server-less platform by Google. We are using it for integrating Tangle network into our app in an efficient way
Amazon Web Service (AWS)	We are using AWS as the cloud backend for hosting the Django server and the database. It can also be used for hosting a private Tangle network

Table 2 Experimental analysis on performance of image processing algorithms

OCR engine	Error rate on alphabets (%)	Time for recognition
Tesseract OCR	0.70	0.3 s/image
Easy OCR	4.30	0.82 s/image

Fig. 2 BitMedi mobile app splash screens

Fig. 3 BitMedi mobile app login and registration screens

User Medical Records: The app lets users upload as well as view the user's medical records (Refer Fig. 4). While uploading a new medical record, the user can attach a description and title along with the type of the record and the date on which

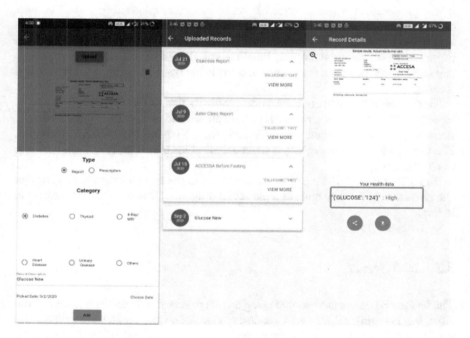

Fig. 4 BitMedi mobile app medical records screens

Fig. 5 BitMedi mobile app profile screens

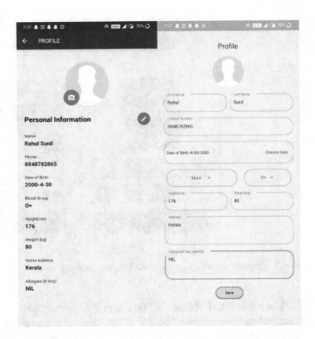

it was produced. A separate screen is set to view the uploaded records. The user can view each record individually along with the original image and extracted data which they can share using their mobile's share option or as in batches in the form of graphs based on chronological timeline and change in values. There are filters and sort options provided for the users to easily navigate through the list of medical records.

User Profile: The app contains a user profile section (Refer Fig. 5), which can be navigated from the home screen, where the user can add personal and medical details like sex, date of birth, height, weight, blood group, allergies, and medical conditions. The user can also add emergency contact details of his close family or friends in case of an emergency.

Drug Reminder: The user can set reminders in the app for drug intake. The app can also store how many units of a particular drug are left with the user at any point in time.

3.3 Web Server

The backend requests are handled using a Django server. The server will be responsible for handling all the API requests. The following APIs were created for connecting the mobile client to the server.

User Create: This API is responsible for registering new users by creating and assigning them a token key.

User Profile Create: This API is responsible for creating a new profile for the registered users. This API can also be used to update the existing values of a given user.

User Profile View: This API is responsible for retrieving the created user profile of a user from the server back to the client device in the form of JSON. The API will return an empty JSON response in case the profile is not created.

User Add Report: This API is responsible for uploading the user report to the server from the user's device as per the user's request.

User View Report: This API retrieves the report images when needed from the server.

User Data View: This API is responsible for retrieving all the user data that are extracted from the report images. These data are represented by various graphs inside the Flutter application.

Doctor Auth: This API authenticates the person (Doctors) trying to access the health record of the person who is actually doing it by the record owner's consent by sending an OTP to the owner's device.

Auth Redaction: This API allows the user to redact authentication of a specific person (Doctors) that was given authentication before.

3.4 Image Processing

Image processing was an integral part of our system since he had to extract data from uploaded images of various medical reports to make interactive graphs for data representation for the ease of user understanding. Tesseract and OpenCV were used for developing the image processing workflow.

OpenCV: OpenCV was used for preprocessing the images by binarization in which data features of the image were transformed into vectors of binary numbers to make classifier algorithms more efficient. OpenCV also highlighted the textual data and converted all non-textual data as a white background which will help tesseract to extract data more efficiently.

Tesseract: The preprocessed images are then passed onto the Tesseract where it will extract all textual data from the reports. The extracted data will be a mixture of necessary and unnecessary data. The necessary data can be sorted out from this mixture using a premade list of keywords for each type of medical report, and the needed extracted data are then stored in the database in JSON format by the Django server. This JSON data can be fetched by the Flutter app when it needs to represent data graphically in the app.

4   Conclusion

In this paper, a completely secure system for the storage of different medical records in an improved structure is proposed. The proposed system consists of a mobile app for the users and a Web server for handling the API calls from the mobile while maintaining security. Tesseract OCR is used on the server side to extract data from records in real time and directly show the graphical representation of the data while not compromising the privacy of the user. This way of representing gives a unique user experience that helps the user understand complex medical parameters in a simplified format, and at the same time, the doctors can get the detailed data through the simplified interface itself giving the best of both worlds.

5   Further Discussions and Future Scope

5.1   Telemedicine

The app will also consist of other features like online doctor teleconsultation, in which the user will be able to book appointments with the available doctors through our app. Even though there are many apps available in the market for this, the main advantage of using our app is the seamless sharing of medical records. The users will be able to share their medical records with the doctor securely for a particular amount of time. The doctor can even give prescriptions to the user through our app, which will be stored in the users' prescription tab. This feature also helps the user to keep track of the medicine intake at the correct time, and it also helps the doctor to know the current medication of the user during any consultation. The user will also be able to enter the medicines he/she is taking to keep the records updated.

5.2   Smart Devices Integration

Our app also provides an option for the user to connect his smart devices like a fitness band, exercise tracker, glucometer, and sphygmomanometer with our app to keep track of their health at all times. This integration can give enough data to give some primary diagnosis of some health conditions and can recommend the user to consult doctors for further diagnosis or help the user with helpful suggestions regarding their diet or exercise.

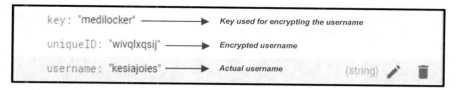

Fig. 6 Firebase document

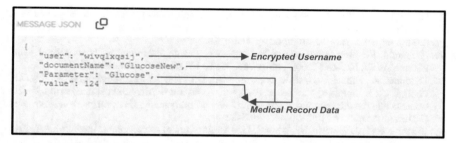

Fig. 7 Tangle document

5.3 Emergency Report

Our app also provides a provision to share the details of a user with the doctor in case of an emergency like if the user is met with an accident and is not able to remember his details. We will ensure a secure way to get this done.

5.4 Tangle Integration

The extracted data can also be stored in Tangle [30] (Refer Fig. 7) which is an advanced version of blockchain without all the disadvantages of blockchain technology like being slow, heavy, and the need of a miner to add a new transaction. Tangle has a non-blockchain data structure with a highly scalable approach to transaction confirmation which improves the anonymity of IOTA from other blockchain-based systems [31, 32]. To ensure further security and anonymity, the username of the user is encrypted using a key which is then stored in an external database by Google, Firebase (Refer Fig. 6). It has built-in security at the data node level itself and ensures real-time data transfer and storage, hence preferred for storing user information. The extracted data will be stored in a Tangle network after encrypting the user's id with a key [33].

Acknowledgements Our idea got selected as one of the best projects in the Cisco ThingQbator program. We would like to thank the Cisco ThingQbator program for giving us the platform to work on our idea. This paper and the research behind it would not have been possible without the exceptional support of our mentors Prashant Choudhary and Lyle Ethan Mark Rodericks from NASSCOM Foundation Innovation Spaces. We are also grateful for the insightful feedback given by our app testers.

References

1. Das, A.V., et al.: Tele-consultations and electronic medical records driven remote patient care: responding to the COVID-19 lockdown in India. Indian J. Ophthalmol. **68**(6), 1007–1012 (2020). https://doi.org/10.4103/ijo.IJO_1089_20
2. Fitzgerald, R.: Medication errors: The importance of an accurate drug history. Br. J. Clin. Pharmacol. **67**, 671–675 (2009). https://doi.org/10.1111/j.1365-2125.2009.03424.x
3. Landman, A., Emani, S., Carlile, N., Rosenthal, D.I., Guelich, S., Semakov, S., Pallin, D.J., Poon, E.G.: A mobile app for securely capturing and transferring clinical images to the electronic health record: description and preliminary usability study. JMIR mHealth uHealth **3**, e1 (2015). https://doi.org/10.2196/mhealth.3481
4. Kalyan Rao, P.R., Hanash, M., Ahmed, G.: Development of mobile phone medical application software for clinical diagnosis. Int. J. Innovative Sci. Mod. Eng. (IJISME) **2**, 4 (2014)
5. Ramachandran, A., Pai, V.V.S.: Patient-centered mobile apps for chronic disease management. In: 2014 International Conference on Computing for Sustainable Global Development (INDIACom). IEEE (2014). https://doi.org/10.1109/indiacom.2014.6828104
6. Adithya, M., Dr, B.: Security analysis and preserving block-level data DE-duplication in cloud storage services. J. Trends Comput. Sci. Smart Technol. **2**, 120–126 (2020). https://doi.org/10.36548/jtcsst.2020.2.006
7. Shakya, S.: Process mining error detection for securing the IoT system. J. ISMAC. **2**, 147–153 (2020). https://doi.org/10.36548/jismac.2020.3.002
8. Manoharan, J.S.: A novel user layer cloud security model based on chaotic arnold transformation using fingerprint biometric traits. J. Innovative Inf. Process. **3**(1), 36–51 (2021)
9. Zheng, Z., Xie, S., Dai, H., Chen, X., Wang, H.: An overview of blockchain technology: architecture, consensus, and future trends. In: 2017 IEEE International Congress on Big Data (BigData Congress), pp. 557–564 (2017). https://doi.org/10.1109/BigDataCongress.2017.85
10. Sivaganesan, D.: A data driven trust mechanism based on blockchain in IoT sensor networks for detection and mitigation of attacks. J. Trends Comput. Sci. Smart Technol. (TCSST) **3**(01), 59–69 (2021)
11. Wyatt, J.C.: How can clinicians specialty societies and others evaluate and improve the quality of apps for patient use? BMC Med. **16** (2018). https://doi.org/10.1186/s12916-018-1211-7
12. Setrakian, J., Gauthier, G., Bergeron, L., Chamberland, M., St-Onge, C.: Comparison of assessment by a virtual patient and by clinician-educators of medical students history-taking skills: exploratory descriptive study. JMIR Med. Educ. **6**, e14428 (2020). https://doi.org/10.2196/14428
13. Niemöller, S., Hübner, U., Egbert, N., Babitsch, B.: How to access personal health records? Measuring the intention to use and the perceived usefulness of two different technologies: a randomised controlled study. Stud. Health Technol. Inform. **3**(267), 197–204 (2019). https://doi.org/10.3233/SHTI190827. PMID: 31483273
14. Sequist, T.D., Cullen, T., Hays, H., Taualii, M.M., Simon, S.R., Bates, D.W.: Implementation and use of an electronic health record within the Indian health service. J. Am. Med. Inform. Assoc. **14**, 191–197 (2007). https://doi.org/10.1197/jamia.m2234
15. Joshi, M., Joshi, K., Finin, T.: Attribute based encryption for secure access to cloud based EHR systems. In: 2018 IEEE 11th International Conference on Cloud Computing (CLOUD) (2018), pp. 932–935. https://doi.org/10.1109/CLOUD.2018.00139

16. Telemedicine Practice Guidelines|Enabling Registered Medical Practitioners to Provide Healthcare Using Telemedicine: https://www.mohfw.gov.in/pdf/Telemedicine.pdf
17. .National Digital Health Mission: https://ndhm.gov.in/
18. Digital Health Mission a voluntary, central repository of records: Ministry, https://www.thehindu.com/news/national/on-independence-day-pm-modi-announces-health-id-card-for-every-indian/article32361701.ece
19. Dube, T., Van Eck R., Zuva, T.: Review of technology adoption models and theories to measure readiness and acceptable use of technology in a business organization. J. Inf. Technol. **2**(04), 207–212 (2020)
20. Practo|Video Consultation with Doctors, Book Doctor Appointments, Order Medicine, Diagnostic Tests, https://www.practo.com
21. The Rise of the Digital Doctor: Practo Technologies and 21st Century Indian Healthcare—Technology and Operations Management, https://digital.hbs.edu/platform-rctom/submission/the-rise-of-the-digital-doctor-practo-technologies-and-21st-century-indian-healthcare/
22. Unique Health Identification and Aadhaar: A case for mandatory linkage. http://www.ideasforindia.in/topics/governance/unique-health-identification-and-aadhaar-a-case-for-mandatory-linkage.html
23. Ribeiro, M.R.M., Jùlio, D., Abelha, V., Abelha, A., Machado, J.: A comparative study of optical character recognition in health information system. In: 2019 International Conference in Engineering Applications (ICEA). pp. 1–5 (2019). https://doi.org/10.1109/CEAP.2019.8883448
24. Giri, P.: Text information extraction and analysis from images using digital image processing techniques. Presented at the (2013)
25. Bhatti, M., Akram, M.U., Ajmal, M., Sadiq, A., Ullah, S., Shakil, M.: Information extraction from images **29**, 1273–1276 (2014). https://doi.org/10.5829/idosi.wasj.2014.29.10.38
26. Smith, R.: An overview of the tesseract OCR engine. In: *Ninth International Conference on Document Analysis and Recognition (ICDAR 2007)*, pp. 629–633 (2007). https://doi.org/10.1109/ICDAR.2007.4376991
27. A Python wrapper for Google Tesseract GitHub Repository: https://github.com/madmaze/pytesseract
28. SQLite flutter plugin GitHub Repository: https://github.com/tekartik/sqflite
29. Clausner, C., Antonacopoulos, A., Pletschacher, S.: Efficient and effective OCR engine training. IJDAR **23**, 73–88 (2020). https://doi.org/10.1007/s10032-019-00347-8
30. Popov, S.: The tangle. Presented at the (2015)
31. Tennant, L.: Improving the Anonymity of the IOTA Cryptocurrency (2017)
32. Bramas, Q.: The stability and the security of the tangle (2018). ⟨hal-01716111v2⟩
33. Alshaikhli, M.: IOTA viability in healthcare industry (2019). https://doi.org/10.13140/RG.2.2.17332.48008

Analysis of Deep Learning Techniques for Handwritten Digit Recognition

Sagnik Banerjee, Akash Sen, Bibek Das, Sharmistha Khan, Shayan Bhattacharjee, and Sarita Nanda

Abstract The huge variations in culture, community and language have paved the path for a massive diversification in the handwriting of humans. Each one of us tends to write in a different pattern. Character or digit recognition finds humongous applications in the recent days especially in the processing of bank statements, sorting of postal mails and many more. Although many classification models exist in literature that successfully classifies the handwritten digits, yet the problem that is still persisting is which one can be termed as an optimal classification model with higher accuracy and lower computational complexity depending upon the circumstances. In this paper, the Machine Learning classification models involving the likes of K-Nearest Neighbor (KNN), Support Vector Machine (SVM) and XGBOOST were compared with that of Deep Learning models like Artificial Neural Network (ANN) and Convolutional Neural Network (CNN). The comparison clearly portrays how convolution operation on images plays a vital role to outperform rest of the classification models. Then the paper compares the CNN models on the basis of two different sets of Loss functions and Optimizers to delineate their role in enhancement of accuracy of the model. The only limitation lies in the fact that in spite of being a handwritten recognizing model this model only recognizes digits in the range of 0–9.

Keywords Handwritten digits · K-Nearest neighbor · Support vector machine · XGBOOST · Artificial neural network · Convolutional neural network · Optimizer · Loss function

S. Banerjee · A. Sen · B. Das · S. Khan · S. Bhattacharjee · S. Nanda (✉)
School of Electronics Engineering, KIIT Deemed to be University, Bhubaneswar 751024, India
e-mail: snandafet@kiit.ac.in

© The Author(s), under exclusive license to Springer Nature Singapore Pte Ltd. 2022   107
V. Suma et al. (eds.), *Evolutionary Computing and Mobile Sustainable Networks*,
Lecture Notes on Data Engineering and Communications Technologies 116,
https://doi.org/10.1007/978-981-16-9605-3_8

1   Introduction

Handwritten digit recognition has been a well-liked topic in the existing time. It mainly focuses on identifying the digits physically written by the hand from an extensive congregation of sources, like pictures, text, notepapers etc. It has eventually become a topic of research interest for an enormous time frame. Some of the research areas have evolved involving digitization of medical prescription, validation of signatures, bank cheque filtering and many more.

It has been witnessed in recent past that some of the machine learning classification algorithms have been deployed for recognizing the handwritten digits like KNN Classifier (K-Nearest Neighbor), SVM Classifier (Support Vector Machine), Random Forest Classifier etc. These algorithmic approaches have attended maximum accuracies of 94% to 96% which is not feasible in real-world applications. Therefore, the increase in accuracy is the demand of the hour and crucial, and hence asks to be modified [1]. Now comes the deployment of the Deep Learning technique, which has been effective in increasing the accuracy many folds but at the cost of increase in computational complexity. In spite of that, Deep Learning has evolved as the major area of interest when it comes to image processing. Large volume of AI mechanisms has been developed in recent times like, sci-kit learn, sci-py, Keras, Theano, TensorFlow by Google for Deep Learning purposes. These mechanisms help in making the applications bountiful along these lines evidently. The ANN (Artificial Neural Networks) have the ability to almost imitate the cerebrum of human brain, for example—the CNN (Convolutional Neural Network) with backpropagation algorithm for processing of images [1, 2]. Minimization of cost function in order to nullify the error is the major target of backpropagation algorithm by regulating the weights and biases of the network. The magnitude of regulation is decided by Gradient Descent algorithm [3].

This paper has compared some of the existing Machine Learning classification models like SVM (Support Vector Machine), KNN (K-Nearest Neighbor) and XGBOOST. Thereafter, the paper compares the Deep Learning models like ANN (Artificial Neural Networks) and CNN (Convolutional Neural Network) in order to showcase the impact of convolution in image processing fields. It is evident that Loss Functions and Optimizers play an important role in determining the model accuracy in case of Deep Learning. Hence, the main focus of the paper lies here that two different sets of Loss Functions and Optimizers are being used for the CNN model whose results were compared in order to visualize their impact on the model accuracy as well as on the computational efficiency. The paper is organized in the following fashion: the immediate next section discusses existing works in the literature, Sect. 3 provides an overview of the used MNIST data-set, followed by Sect. 4 which discusses the performances of various Machine Learning classification models, Sect. 5 delineates about the performance of the Deep Learning models and, eventually Sect. 6 concludes the paper.

2   Related Works

This paper provides a comparative study on different Machine Learning classification algorithms for digit recognition accompanied by their respective accuracy values [4]. The feature extraction process followed for digit recognition possesses different algorithmic approaches like linear or non-linear methods, ensemble and Deep Learning techniques. It is evident from the result of the experiment that nearly 60% of the training interval was being decreased by making use of the subset of the entire feature set [5].

Then, an application based on Android was developed for character classification using ANN (Artificial Neural Network) technique. The experiment was performed using random data sets which made use of organizational features and KNN (K Nearest Neighbor) classification technique in order to recognize the handwritten characters [6].

The handwritten characters in Chinese language were classified using a Deep Learning classification model in which the entire dataset was split up into four smaller subsets, each having unique iterations (epoch). The dataset consisted of 260 unique handwritten characters by different people [7].

It is further witnessed that Multiple Cell Size (MCS) approach was put forward to make use of the Histogram of Oriented Gradient (HOG) feature using Support Vector Machine (SVM) classification technique to recognize handwriting techniques [8]. An integrated approach of feature extraction and machine learning classification models was used for handwritten character recognition, consisting of a pre-processing phase for the input images [9]. Recognition of handwritten digits or characters is a very intricate process due to unique writing style of each human being—the letters differ in shapes, size and angle. By utilizing the feature extraction mechanism, a slithering window is moved across the image from the left to right which recognizes the distinctive characters of the input image [10]. In [11], a novel work has been done by using SVM to classify the characters explicitly written or drawn in several touch-driven devices that has managed to attain an accuracy score of about 91%. Vijayakumar [12] adopts a CNN-based deep learning approach for multimodal classification of biometric images involving a thumb impression, which is claimed to be the first work of such sort in existing literature. Hence on observing all the existing works, it is found that there is no such work in literature that broadly compares the Machine Learning as well as the Deep Learning classification models with various sets of Loss functions and Optimizers in order to portray their importance in the performance of the models, which has been clearly depicted in our work.

3   The Dataset and Pre-processing

The process of handwritten digit recognition uses a very well-known dataset, called the MNIST dataset. It stands for Modified National Institute of Standards and Technology, which is an enormous database for handwritten digits. It is a set of 70,000 small images of handwritten digits in which each image is labeled with the corresponding digit that it represents. In the MNIST dataset, each image is of (28 × 28) pixels which means that each image has (28 × 28) = 784 features. Each feature simply corresponds to intensity of one pixel from 0 (white) to 255 (black). These images are taken as input and then split up into training and testing data is fed into machine learning models. Generally, in the MNIST dataset, out of 70,000 images, 60,000 images are used for training and rest 10,000 images are used for testing purposes. The MNIST dataset of handwritten digits of 256 different people from various parts of the globe.

The processing of the image data prior to its training has been achieved using the "*ImageDataGenerator*" class supported by the Keras and Pytorch library. This class provides room for a wide range of pixel-scaling applications, out of which only three major types have been used here namely, the normalization operation for scaling the pixel magnitudes between the range of zero to unity. Secondly, the scaling was performed so that the pixels possess zero or negligible value of mean, which is called the centering operation. The standardization operation was performed eventually in order to furnish the pixels with unity variance. Now the data-set is ready to be fed to the classification models for training and testing. During the course of feeding the training data to the model, a batch size of thirty-two images were considered for both the Machine Learning and Deep Learning models in order to make the process faster without compromising upon the performances of the models.

4   Machine Learning Classification Models

In this section, a comparative study among some of the popular Machine Learning classification algorithms is observed based on their respective accuracy score that were used to recognize the handwritten digits using the MNIST dataset-

- KNN (K nearest neighbor)
- SVM (support vector machine)
- XGBOOST (Fig. 1).

4.1   K-Nearest Neighbor (KNN)

K-Nearest Neighbor is a type of non-parametric algorithmic approach where the best measured value among all the values is that value with the maximum number of

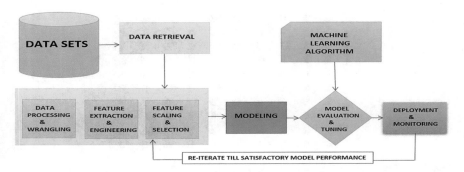

Fig. 1 The process flow of machine learning classifiers

neighbors having the tiniest Euclidian or Hamming distance. KNN is also regarded as a lazy learning algorithmic approach because the learning here is instance-based. In order to function efficiently, KNN needs a training dataset with datapoints that are well-labeled [13]. KNN algorithm bags a new datapoint as input, commonly known as query point, and performs the classification for this query point by measuring the Euclidian or Hamming distance among the query point and the labeled datapoints, based upon the value of the "K", which is user driven and the default value is five.

The Euclidian distance formula for calculating the distance between the datapoints is given by (Figs. 2 and 3; Table 1)

$$d(x, y) = d(y, x) = \sqrt{(x_1 - y_1)^2 + (x_2 - y_2)^2 + \cdots + (x_n - y_n)^2} \qquad (1)$$

Fig. 2 Working of KNN classifier

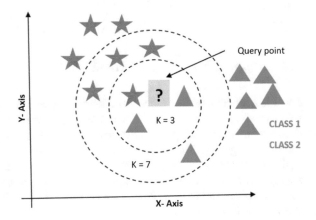

```
[[ 98   0   0   0   0   0   1   0   0   0]
 [  0 111   0   0   1   0   0   0   1   0]
 [  1   3  87   0   0   0   1   0   1   1]
 [  3   2   1  95   0   4   1   0   1   1]
 [  0   0   0   0  86   0   1   0   0   6]
 [  1   1   0   3   0  89   2   0   0   0]
 [  1   0   0   0   0   1  99   0   0   0]
 [  0   2   0   0   2   0   0  85   0  14]
 [  1   2   1   2   0   2   3   1  85   2]
 [  0   0   0   1   0   0   0   1   0  92]]
```

Fig. 3 Confusion matrix from KNN implementation

Table 1 Classification report of KNN

	Precision	Recall	F1-score	Support
0	0.93	0.99	0.96	99
1	0.92	0.98	0.95	113
2	0.98	0.93	0.95	94
3	0.94	0.88	0.91	108
4	0.97	0.92	0.95	93
5	0.93	0.93	0.93	96
6	0.92	0.98	0.95	101
7	0.98	0.83	0.89	103
8	0.97	0.86	0.91	99
9	0.79	0.98	0.88	94
Accuracy			0.93	1000
Macro avg	0.93	0.93	0.93	1000
Weighted avg	0.93	0.93	0.93	1000

4.2 Support Vector Machine (SVM)

SVM is a model based on supervised learning, which can be used for both classification as well as regression. In SVM, the programmer is provided with a set of training data in which each datapoint is decided (labeled) to show its affinity toward either of the two categories. SVM classification model signifies a geometric point in space, which is mapped in such a way that the two distinctive classes have their datapoints segregated by the maximum stretch [14]. The datapoints in the test set are mapped into this space, and the predictions are carried out depending upon the class to which the input test set is pertained.

SVM tries to optimize the Lagrange equation which provides us with the parameters that dictates the maximum margin solution. The positive and negative weights upon the maximum margin which are equidistant from the decision boundary are called support vectors. The Lagrange equation is given by-

$$L = \frac{1}{2}\|w\|^2 - \sum_{j} \alpha_j \left(y_j\left(\vec{w} \cdot \vec{x}_j + k\right) - 1\right) \tag{2}$$

The width of the margin or the distance between the support vectors is given by (Figs. 4 and 5; Table 2)

$$m \equiv \frac{2}{\|\vec{w}\|} \quad \text{where,} \quad \|\vec{w}\| = \sqrt{\sum_{j} w_j^2} \tag{3}$$

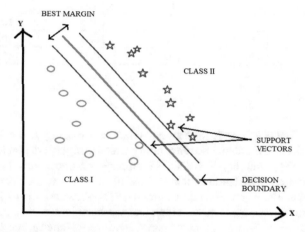

Fig. 4 Working of SVM classifier

```
[[ 99   0   0   0   0   0   0   0   0   0]
 [  0 111   0   0   1   0   0   0   1   0]
 [  1   0  88   1   1   0   1   0   1   1]
 [  1   1   2  97   0   4   0   1   1   1]
 [  0   0   0   0  89   0   1   0   0   3]
 [  1   1   0   1   0  92   1   0   0   0]
 [  0   0   0   1   0   0  99   0   1   0]
 [  0   0   0   0   0   0   0  93   1   9]
 [  1   1   1   0   0   2   0   0  93   1]
 [  0   0   0   1   0   0   0   1   0  92]]
```

Fig. 5 Confusion matrix from SVM implementation

Table 2 Classification report of SVM

	Precision	Recall	F1-score	Support
0	0.96	1.00	0.98	99
1	0.97	0.98	0.98	113
2	0.97	0.94	0.95	94
3	0.96	0.90	0.93	108
4	0.98	0.96	0.97	93
5	0.94	0.96	0.95	96
6	0.97	0.98	0.98	101
7	0.98	0.90	0.94	103
8	0.95	0.94	0.94	99
9	0.86	0.98	0.92	94
Accuracy			0.95	1000
Macro avg	0.95	0.95	0.95	1000
Weighted avg	0.95	0.95	0.95	1000

4.3 Xgboost

The term "XGBOOST" stands for Extreme Gradient Boosting. It is a decision tree-based well optimized gradient boosting library which is highly adaptable, systematic as well as supports portability. XGBOOST is an ensemble tree method that applies the concept of boosting weak learners, that is, Classification Trees and Regression Trees using a typical gradient descent architecture. XGBOOST works great with structured datasets and very few model can beat it on real-world structured data problems. It is very less prone to overfitting although if parameters are not tuned properly, overfitting is possible. XGBOOST makes use of the loss function to evaluate the classifier or regressor performance by minimizing the equation given by -

$$\mathcal{L}(\phi) = \sum_j l(\widehat{y_j}, y_j) + \sum_i \Omega(g_i) \tag{4}$$

$$\text{where } \Omega(g) = \Upsilon t + \frac{1}{2}\lambda \|w\|^2 \tag{5}$$

The first part of the equation accounts for the loss function. The second part contains the omega term which in turn contains regularization coefficient (λ), and the leaf weights (w) ssand "t" represents the terminal nodes and, Υ is the penalty incurred (Table 3; Figs. 6 and 7).

Table 3 Classification report of XGBOOST

	Precision	Recall	F1-score	Support
0	0.96	0.99	0.98	99
1	0.97	0.98	0.98	113
2	0.97	0.93	0.95	94
3	0.97	0.90	0.93	108
4	0.98	0.94	0.96	93
5	0.93	0.98	0.95	96
6	0.95	0.98	0.97	101
7	0.98	0.94	0.96	103
8	0.96	0.90	0.93	99
9	0.84	0.97	0.90	94
Accuracy			0.95	1000
Macro avg	0.95	0.95	0.95	1000
Weighted avg	0.95	0.95	0.95	1000

Fig. 6 Confusion Matrix from XGBOOST implementation

```
[[ 98   0   0   0   0   0   0   0   1   0]
 [  0 111   0   0   1   0   0   0   1   0]
 [  0   1  87   2   0   0   2   0   0   2]
 [  1   1   1  97   0   4   0   1   0   3]
 [  0   0   0   0  87   0   1   0   0   5]
 [  2   0   0   0   0  94   0   0   0   0]
 [  0   0   1   0   0   0  99   0   1   0]
 [  0   0   1   0   0   0   0  97   1   4]
 [  1   1   0   0   0   3   2   0  89   3]
 [  0   0   0   1   1   0   0   1   0  91]]]
```

Fig. 7 Accuracy comparison of machine learning models

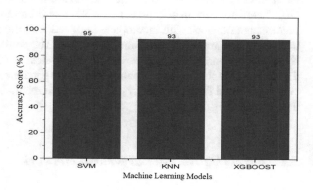

5   Deep Learning Classification Models

In the term "Deep Learning", the word "Deep" quantifies the hidden layers in a particular Neural Network architecture. Classification of handwritten digits built on Deep Learning technique involves two methods-

- ANN (Artificial Neural Network)
- CNN (Convolutional Neural Network).

The code used for the paper uses the "*Pytorch*" library for designing the above mentioned Deep Learning models. Pytorch provides easier and faster debugging. Another very important feature of Pytorch is that it provides data parallelism, which means the computation can be performed among multiple CPU (Central processing unit) and GPU (Graphical processing unit).

5.1   ANN (Artificial Neural Network)

Artificial Neural Networks (ANN) imitate the functionalities of human brain. These network architectures have come out in huge number in the Aeon of progress in the computational abilities in terms of speed and power. So, Deep Learning can be considered as an acronym for the Neural Networks having a multi-layered architecture. The layers in the model are compounded to form the nodes can be defined as a simple unit of computation having a certain number of weights as input, and an activation function (determines whether the node will fire or not) that relates the inputs and eventually produces the output. There can be certain number of hidden layers present in between the input layer and the output layer, which execute all the necessary calculation to extract the hidden features and patterns from the data [10]. There are certain advantages of ANN like it has an ability to perform parallel processing and work with incomplete knowledge in the data, and fault tolerance as well. But there are some disadvantages as well. The performance of Artificial Neural Networks depends upon the realization of physical hardware. The most important disadvantage is that when a probing solution is generated by ANN, no indication or clue is provided for further analysis, and hence it results in depreciation of trust over the network (Fig. 8).

5.2   CNN (Convolutional Neural Network)

For the classification of handwritten digits, a convolutional neural network consisting of seven layers- a single input layer, followed by five hidden layers and finally an output layer. As the input layer contains an image of (28×28) pixels which indicates that the input layer consists of 784 neurons, which is equal to the number of input

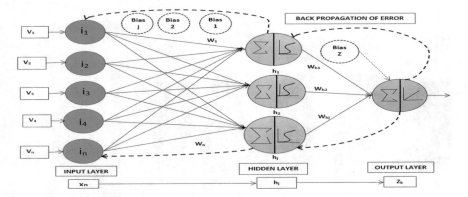

Fig. 8 Process flow of artificial neural network

features. The pixels of the input image are in gray scale mode which means that zero value is appended for a white pixel and a unity value to represent a black pixel. The design model comprises of five hidden layers. The first hidden layer plays the role of first convolutional layer of the design, which accounts for the feature extraction from the input data. It is also responsible for performing the convolution operation between a filter (or kernel) and the immediate previous layer. Formation of multiple feature maps along with adaptable kernels as well as rectified linear units (or ReLU), acting as activation functions, are contained in this layer. Size of the filter (or kernel) fixes the filter localization. An activation function (ReLU) is casted-off at the termination of each convolution layer and the fully connected layers to inflate the model performance [15, 16]. The first pooling layer acts as the second hidden layer in the system, which decreases the output data arriving from the convolution layer, thus decreasing the number of parameters and eventually increases the computational efficiency. The "max pooling" scheme has been used to sub-sample the feature map dimensions. Next, another set of convolution and pooling layers are designed having the same functionalities as that of the first set of convolution and pooling layers except for their variations in kernel size and feature maps. The output dimension of the feature map is given by

$$n_o = \frac{n_i - k + 2p}{s} + 1 \qquad (6)$$

where, "n_i" accounts for the dimensions of the input image, "k" represents the size of the kernel or filter, and "p" and "s" accounts for padding and stride respectively.

Then a Flatten layer is used to map the two-dimensional feature map matrix into a single dimensional vector. Now the first fully connected layer is designed, which is nearly equivalent to the hidden layer of the artificial neural networks except for bridging every neuron from the immediate previous layer to the layer that follows the next. To prevent model overfitting, the method of dropout regularization is generally

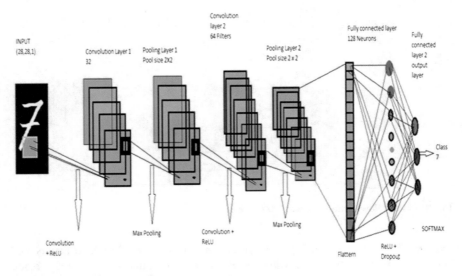

Fig. 9 Process flow of convolutional neural network

adopted at the first fully connected layer, which cancels out few neurons at random during the training process to enhance the model performance. The second fully connected layer is the last hidden layer of the model, which plays the role of the output layer [17]. This layer contains 10 neurons because there are ten classes in the output that is digit number from zero to nine. The output layer uses the "softmax" activation function, which is used to provide the probability distribution of the output classes (Fig. 9).

A Loss Function, also regarded as Error Function or Cost Function, determines the degree to which the designed algorithm models the dataset. Moreover, the Loss Functions come up with a static blueprint of how the model is performing. Optimizers bind together the Loss Function and the parameters of the model and refurbish the model depending upon the output of the Loss Function [18]. In this paper, a performance comparison is depicted by using two different sets of optimizers and loss functions over the designed CNN model used for the recognition of handwritten digits. In the first case, SGD (Stochastic Gradient Descent) is used as the optimizer, that aims to minimize the loss incurred by updating the model parameters using a learning rate parameter that determines the step size to reach the local minima, along with NLL Loss (Negative Log Likelihood) as the loss function and have attained a test set accuracy of approximately 99% at the cost of lower computing efficiency while computing for fifteen epochs. In the second case, the optimizer used is Adam, which is regarded as the combination of two extensions of the Gradient Descent algorithm called the *"Momentum"* and *"Root Mean Square Propagation"*, along with the Cross Entropy Loss function is used but could only achieve a test set accuracy of approximately 96% at higher computing efficiency while computing for same number of epochs. The reason for choosing Adam and SGD as optimizers is that

Table 4 Comparison of CNN models based on loss function and accuracy

Sl. No	Model	Epochs	Loss function	Optimizer	Accuracy (%)
1	CNN	15	NLL Loss	SGD	99
2	CNN	15	Cross-Entropy	Adam	96

Fig. 10 Accuracy comparison of deep learning models

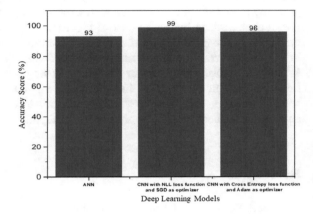

Adam is regarded as the best adaptive optimizer as it contains the properties of both the aforementioned extensions of Gradient Descent and functions well-enough with sparse data while SGD is nothing but a type of the Gradient Descent algorithm that is simple in nature and has a much faster convergence rate as is often reported using Adam. The grounds for choosing these loss functions are that NLL-Loss Function amplifies the total probability distribution of the dataset and the classifier is inflicted a penalty upon predicting the unerring class with lower value of probability. On the other hand, the Cross-Entropy Loss Function is mainly used to acquire some knowledge on the probability distribution of the output classes and in this case, higher rate of penalty is inflicted upon predicting the erroneous class with greater probability value. It is just an approach to justify that the Loss Functions and Optimizers play a very vital role in determining the performance of a Machine Learning classification model (Table 4; Fig. 10).

6 Conclusion

This paper has made a comparative study among various Machine Learning and Deep Learning classification techniques for the recognition of handwritten digits over MNIST dataset. Among the Machine Learning classification techniques, Support Vector Machine (SVM), K-Nearest Neighbor (KNN) and XGBOOST are implemented and compared. It is found that Support Vector Machine (SVM) emerged as the best performing classifier with an accuracy of about 95%. In Deep

Learning, a comparison between Artificial Neural Network (ANN) and Convolutional Neural Network (CNN) is depicted, wherein Convolutional Neural Network (CNN) produced better results. In CNN architecture, two different sets of Loss Functions and Optimizers are used to compare the accuracy as well as computational efficiency of the model in which the one with NLL (Negative-Log Likelihood) Loss as Loss Function and SGD (Stochastic Gradient Descent) as optimizer gave a better accuracy (about 99%) than the one with Cross-Entropy Loss as Loss Function and Adam as optimizer, but at the cost of higher computational complexity. The only drawback of the paper lies in the fact that it is only capable of recognizing digits in the range of zero to nine. In future, we will definitely make an attempt to further explore the aspects of character recognition using a greater number of Machine Learning classification models like Naive Bayes, ADABOOST, Gaussian Mixture Models, as well as Deep Learning classification models with the likes of RNN, LSTM and Auto-Encoder, using more diverse sets of Optimizers and Loss Functions.

References

1. Kussul, E., Baidy, T.: Improved method of handwritten digit recognition tested on MNIST database, Elsevier. Image Vis. Comput. **22**(12), 971–981 (2004)
2. Hansen, L.K., Liisberg, C., et al.: Ensemble methods for handwritten digit recognition. In: IEEE, Neural Networks for Signal Processing II Proceedings, 31 Aug-2 Sept (1992)
3. Paola, J.D., Schowengerdt, R.A., et al.: A detailed comparison of backpropagation neural network and maximum-likelihood classifiers for urban land use classification. IEEE **3**(4), (1995)
4. Jain, A., Sharma, B.K.: Analysis of activation functions for convolutional neural network based MNIST handwritten character recognition. Int. J. Adv. Stud. Sci. Res (IJASSR) **3**(9) (2019)
5. Alsaafin, A., Elnagar, A.: A minimal subset of features using feature selection for handwritten digit recognition. J. Intell. Learn. Syst. Appl. (JILSA) **9** (2017)
6. Deepak Gowda, D.A., Harusha, S., Hruday, M., Likitha, N., Girish, B.G.: Implementation of handwritten character recognition using neural network. Int. J. Innovative Res. Sci. Technol. (IJIRST) **5**(12) 2019
7. Putra, E.W., Suwardi, I.S.: Structural offline handwritten character recogntion using approximate subgraph matching and levenshtein distance. Elsevier **59**, 340–349 (2015)
8. Dai, F., Ye, Z., Jin, X.: The recognition and implementation of handwritten character based on deep learning. J. Robot. Networking Artif. Life (JRNAL) **6**(1), June (2019)
9. Khan, H.A.: MCS HOG features and SVM based handwritten digit recognition system. J. Intell. Syst. Appl. (JILSA) **9**(2017)
10. Sonkusare, M., Sahu, N.: A survey on handwritten character recognition (HCR) techniques for English alphabets. Adv. vision Comput. Int. J. (AVC) **3**(1), Mar (2016)
11. Hamdan, Y.B.: Construction of statistical SVM based recognition model for handwritten character recognition. J. Inf. Technol. **3**(02), 92–107 (2021)
12. Vijayakumar, T.: Synthesis of palm print in feature fusion techniques for multimodal biometric recognition system online signature. J. Innovative Image Proc. (JIIP) **3**(02), 131–143 (2021)
13. Lee, Y.: Handwritten digit recognition using K nearest-neighbor, radial-basis function, and backpropagation neural networks. Neural Comput. **3**(3), (1991), Massachusetts Institute of Technology
14. Bottou, L., Cortes, C. et al.: Comparison of classifier methods: a case study in handwritten digit recognition. In: Proceedings of the 12th IAPR International Conference on Pattern Recognition, vol. 3 - Conference C: Signal Processing, 9–13 Oct 1994

15. Le Cun, Y., Boser, B., et al.: Handwritten digit recognition with a back-propagation network, AT&T Bell Laboratories, Holmdel, N. J. 07733

16. Dasgupta, R., Chowdhury, Y.S., Nanda, S.: Performance comparison of benchmark activation function ReLU, swish and mish for facial mask detection using convolutional neural network. Intell. Syst. Proc. SCIS 355 (2021)

17. Jemimah, K.: Recognition of handwritten characters based on deep learning with tensorflow. Int. Res. J. Eng. Technol. (IRJET) **06**(09) (2019)

18. Chowdhury, Y.S., Dasgupta, R., Nanda, S.: Analysis of various optimizer on CNN model in the application of pneumonia detection. In: 2021 3rd International Conference on Signal Processing and Communication (ICPSC), pp. 417–421 (2021). https://doi.org/10.1109/ICSPC51351.2021.9451768

Social Media Sentiment Analysis Using the LSTM Model

A. V. Thalange, S. D. Kondekar, S. M. Phatate, and S. S. Lande

Abstract Nowadays, extracting information from social media is providing knowledge about the market and current trends. This paper highlights a novel idea of allowing users to access all such information. A Website is created for visualization of obtained sentiments. It allows users to integrate different social media sites on a single platform for sentimental analysis by providing multiple dashboards under the user's profile. Users have a choice to provide input from YouTube and/or Twitter to get sentiments. Users need to provide any URL/hashtag. Once input is provided, a trained LSTM model classifies the sentiments as positive, negative, and neutral. Dashboard creation process gets initiated and is deployed on cloud. Classified sentiments are analyzed and are reflected on the dashboard. The dashboard consists of positive, negative, and neutral sentiment's count, line graph, and pie chart on a real-time basis. These results provide a better understanding of the market and people's opinion.

Keywords Sentiment analysis · LSTM model · Machine learning · Cloud computing · Review classification · Social media · Twitter · YouTube · Innovation · Multiple dashboard access

1 Introduction

Sentiment analysis is the process of using natural language processing, text analysis, and statistics to analyze customer sentiment. The best businesses understand the sentiment of their customers, what people are saying, how they are saying it, and what they mean. For example, before purchasing a bike, most customers nowadays surf different social sites and check the reviews made by other users on different bike models and companies. So, here, customers may not get the overall data of reviews as there may be thousands of reviews posted. Therefore, getting the sentiments of those reviews on a single platform will be more helpful and can give a clear picture

A. V. Thalange (✉) · S. D. Kondekar · S. M. Phatate · S. S. Lande
Walchand Institute of Technology, Solapur, India
e-mail: avthalange@witsolapur.org

© The Author(s), under exclusive license to Springer Nature Singapore Pte Ltd. 2022
V. Suma et al. (eds.), *Evolutionary Computing and Mobile Sustainable Networks*,
Lecture Notes on Data Engineering and Communications Technologies 116,
https://doi.org/10.1007/978-981-16-9605-3_9

regarding any model or company. These user experiences can be found in tweets, comments, reviews, or other places where people mention the brand. Sentiment analysis is the domain of understanding these emotions with software, and it is a must-understand for developers and business leaders in a modern workplace to study their product reviews. This work represents the idea of obtaining sentiments of users from different social sites and analyzing it. To get answers, what did people think? What did they like and what did not they like? What were people most excited about? Hence, by analyzing the sentiment, any company or user may get to know about the thinking of people regarding any product or video to take necessary measures accordingly, which is helpful to make improvement in them. Saad et al. [1] attempted to provide a comprehensive Twitter sentiment analysis using ordinal regression and ML algorithms in their work. The proposed model includes preprocessing tweets as the first step, and an effective feature was developed using the feature extraction model. For sentiment analysis, algorithms such as SVR, RF, multinomial logistic regression, and DTs were used. Feizollah et al. [2] focused their research on tweets about two halal products: halal cosmetics and halal tourism. Twitter information was extracted using the search function, and a new model was used for data filtering. Furthermore, RNN, CNN, and LSTM were used to improve accuracy and construct prediction algorithms. Based on the results, it appeared that the combination of LSTM and CNN achieved the highest accuracy. Ray and Chakrabarti [3] proposed a deep learning approach for extracting features from text and analyzing user sentiment with regard to the feature. A seven layer deep CNN was used to tag the features in opinionated phrases. Finally, it was discovered that the suggested method produced the highest accuracy. Choudhury and Breslin [4] proposed an unsupervised lexicon-based approach to detect the sentiment polarity of user comments in YouTube. Experimental evaluation showed the combinatorial approach has greater potential. BrandMentions [5] is a tool to monitor and real-time tracking and measure the success of marketing campaigns held by various companies for their products or services. It takes hashtags regarding a particular product or a service and provides the various information related to it, such as mentions, interactions, reach up to the people, and total number of shares of that product, also it provides likes related to the products post. This information is gathered from different social sites such as Twitter and Instagram. Social searcher [6], this Website enables different types of sentiment analysis such as graphical format of timeline, number of posts from different social network platforms, number of posts in a particular week in accordance with days, count of posts in a particular day hour wise, some related hashtag and its count, also they provide count of positive, negative, or neutral sentiments of different social sites with some popular comments. It takes data from YouTube, Twitter, Instagram, etc., but the user does not have independence of selecting social sites. Sivaganesan [7] proposed an interest-based algorithm with parallelized social action to maximize the influence in social networks. This algorithm enabled identifying influential users which is weighted by interactive behavior of the user. These are the two semantic metrics which are used in the proposed algorithm. Kumar [8] proposed the idea of prediction of children's behavior based on their emotions with a deep learning classifier method. The performance of DT and Naive Bayes model is compared. Also, hybrid emotions are incorporated in the proposed dataset. Based on results, the Naive Bayes model provided good performance in terms of recognition rate and prediction accuracy. Here, fusion of both algorithms is proposed for prediction of emotions involved in children's behavior. Smys and Raj [9] proposed an idea to develop an ML algorithm for an early prediction of the depression mode, which can be used to protect from mental illness and suicide state of affairs. The combination of support vector machine and Naive Bayes algorithm was proposed to provide a good accuracy level. Chawla et al. [10] in their work examined an unstructured smartphone reviews that are being collected from a very well-known marketing site termed as Amazon. SVM and Naive Bayes ML models were used and the overall accuracy was around 90% and 40%, respectively. Hence, the SVM approach seems to be much better. Badjatiya et al. [11] investigated the application of deep neural network architectures for the task of hate speech detection. Embeddings learned from deep neural network models when combined with gradient boosted decision trees led to best accuracy values. Poria et al. [12] proposed the multimodal EmotionLines dataset (MELD), an extension and enhancement of EmotionLines MELD contain about 13,000 utterances from 1433 dialogs from the TV-series Friends. Each utterance is annotated with emotion and sentiment labels and encompasses audio, visual, and textual modalities. They proposed several strong multimodal baselines and showed the importance of contextual and multimodal information for emotion recognition in conversations. Pal et al. [13] proposed use of LSTM model for sentiment analysis and Oyebode et al. [14] discusses the application of machine learning in review analysis.

From the above literature survey carried out on the approach or methodologies used for various social media sentiment analysis, the following limitations were concluded such as a Website which is a common platform for providing sentiments from more than one social sites such as YouTube and Twitter are not reported. Access for user to create multiple dashboards under every user's profile independently is not available.

2 System Description and Implementation

In this section, the methods applied in this research are described, including the workflow for extracting and processing of data from Twitter and YouTube and classifying their sentiments using the LSTM model.

2.1 Description

In order to implement this project, the work is divided tasks into two parts such as frontend part and backend part.

A. Backend Part

The data are gathered from two social media sites viz. YouTube and Twitter. Next step is data preprocessing, in which the special characters, stop words, emoji, username, etc., are removed. This preprocessed data are passed over the LSTM model to classify sentiments. Following this, all the data are stored in a database on the cloud.

B. Frontend Part

Initially, the user needs to create an account and successively log in on the Website. Next step is to request a dashboard for sentiment analysis of a particular product. Following this, the user needs to wait for a while, once the dashboard is ready, it is reflected on screen. After deployment of the dashboard, users can see the count of positive, negative, or neutral sentiments. For best experience of visualization, we provide graphs on the Website of sentiments for better understanding of case study.

2.2   Details of Methodology

A Website is created, where the user will create an account and if the account already exists, then users can directly login. Once logged in, the home dashboard appears where users can request for new analytical dashboards and also will be able to see all the requested analytical dashboards. Users will be able to request a new dashboard for YouTube, Twitter, or both by just providing the video link from YouTube and by giving the hashtags of Twitter. Once the user request is successful, Jenkins job will be triggered. This Jenkins job will check whether the requested dashboard is of type YouTube or Twitter or both. Based on that, it will start integrating the whole architecture. At the same time, this Jenkins job will run a terraform [15] script to launch a MySQL database on AWS cloud. Each requested dashboard has its own database to store data. Once integrating all the parts of the requested dashboard is completed, then the whole system will be deployed on the Heroku cloud, so that the dashboard can be accessed from anywhere with the given link.

Once the dashboard is deployed over a cloud, all the collected data are processed by a trained model and the sentiments are classified and simultaneously, they are reflected on the dashboard. At last, when the whole process is over, the user will get a mail with the link to access the newly created dashboard. Also, the user will be able to view all the requested dashboards and access them from the Website.

Initially, the dataset used for training and testing the designed system is sentiment140. The trained model is then used to classify the sentiments extracted from social sites. Now, the data extracted from social sites are preprocessed by removing the white spaces, emojis, and special characters. This preprocessed data are forwarded to ML model which classifies data into positive, negative, and neutral sentiments. These sentiments are reflected on the dashboard page on a real-time basis.

For the processing of data and obtaining the sentiments, LSTM, SVM [16], and random forest models are studied and on the basis of obtained test results LSTM model is selected as ML model for classifying data and to obtain sentiments.

Figure 1 shows the system workflow of the project.

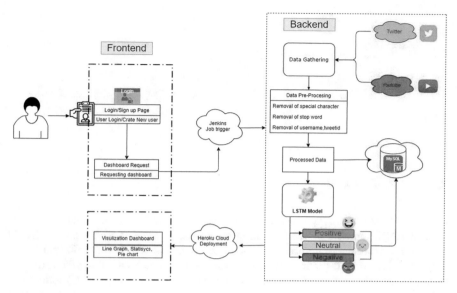

Fig. 1 System flow diagram

2.3 Sentiment140 Dataset

Sentiment140 dataset allows to discover the sentiment of a brand, product, or topic on social sites. Sentiment140 dataset contains 1,600,000 of data which is extracted using the API. The data have been annotated (0 = negative, 4 = positive) and they can be used to detect sentiment. The below table gives an idea about the annotation of data in the dataset. The training data are annotated by 0 or 4, if the predicted value of sentiment will not be at the positive or negative thresholds, then that data are considered as neutral sentiment [17]. Table 1 shows the sample annotated data from sentiment140 dataset.

Table 1 Sample annotated dataset

Sr. No.	Sample comment	Annotation
1	Is working out well	4(Positive)
2	The new ride at six flags is really fun	4(Positive)
3	Has been banned from drinking sodas	0(Negative)
4	I am sad about it	0(Negative)
5	I got it! thank you!	4(Positive)

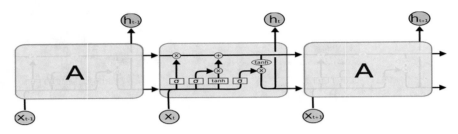

Fig. 2 LSTM node

2.4 LSTM Model

Long short term memory networks—usually just called "LSTMs"—are a special kind of RNN, capable of learning long-term dependencies. LSTMs are explicitly designed to avoid the long-term dependency problem. Remembering information for long periods of time is practically their default behavior, not something they struggle to learn! LSTMs also have this chain-like structure, but the repeating module has a different structure. Instead of having a single neural network layer, there are four layers interacting in a very special way [18]. Figure 2 shows the LSTM node.

2.5 Project Implementation

Sentiments can be obtained when a user provides a YouTube link or Twitter hashtag or both. After getting input from user, the data are collected and processed using a trained model and sentiments are reflected over the Website. In order to get visualized data of sentiments requested by users, a Website is created using HTML, CSS, and Flask [19]. Every user will be able to analyze sentiments and retrieve data of a particular YouTube link or Twitter hashtag.

The details of system designed and implementation are as follows:

(i)   **Homepage**:

As the user enters a Website, a homepage is displayed. Here, users get hyperlinks to login or sign up in order to get access to dashboards created previously or to create a new dashboard.

(ii) **Sign-up Page**:

If the user is new to this Website, then an account has to be created before logging in to the Website. So, the user has to create an account on the Webpage using the sign-up option on the homepage.

(iii) **Login Page**:

Now after successfully signing up, the user needs to login to the Webpage using the desired credentials on the login page.

(iv) **Dashboard Page**:

After successfully logging in, the user gets its own dashboard page where the user can request for new dashboards or can access its old requested dashboard. Every user can access more than one dashboard simultaneously. Also, the user can download the data of analyzed sentiments from the "download data" button. Users are also able to delete their dashboards.

(v) **Multiple Dashboards**:

A novel approach is implemented which enables the users to create multiple dashboards under each user's profile as shown in Fig. 3. Here, every user is independent to create as many dashboards according to his requirement.

(vi) **Dashboard Page Request**:

In the current sentiment analysis tools available, users are not provided the right to choose the social media sources for analysis. In this work, this novel feature is added where every user is given an option to select the social media source such as YouTube, Twitter, or both, for analysis. If the user needs more dashboards, he can request by using request new dashboard option on the dashboard page where the user needs to enter YouTube link or Twitter hashtag, or both, for a particular product or any service, etc. Figure 4 shows dashboard page request for new dashboard creation.

(vii) **Deployment of System on Cloud**:

Once the dashboard request is successful, a Jenkins [20] job is triggered which verifies the type of requested dashboard, i.e., for YouTube or Twitter or both. Based on the type, the whole architecture gets integrated and also a terraform script will run which launches a MySQL database on AWS cloud. Each dashboard has its independent database for data storage. At last, after integration, the whole system gets deployed on the Heroku cloud which generates a unique URL by which the dashboard is accessed from anywhere.

(viii) **Notification Email**:

Once the dashboard is ready, the user will get a notification email from the server. The email includes confirmation of dashboard readiness along with the unique URL of the respective dashboard. Figure 5 shows email notification received.

(ix) **Classification of Sentiments**:

Once the whole dashboard creation process is done, the data are collected from the social site and they are preprocessed. The preprocessing includes removing of white spaces, emojis, and special characters. This preprocessed data are assigned to the LSTM model which classifies data into positive, negative, and neutral sentiments and is analyzed further. These classified sentiments get visualized on the Website dashboard.

Fig. 3 Multiple dashboards

Fig. 4 Dashboard page request

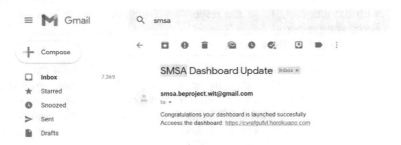

Fig. 5 Notification of email

(x) **Dashboards**:

Based on each dashboard page requested, a separate dashboard is created along with its unique URL. At the dashboard page, users get an analysis report. This dashboard consists of statistical information about comments classified as positive, negative, neutral sentiments, no. of views, likes, dislikes, and total no. of comments. Visualization of classified sentiments in real time is illustrated using line charts and pie charts. Latest comments along with their sentiments keep on flashing in the dashboard page as shown in Figs. 6 and 7. Also, all the statistical information and comments with their classification can be downloaded as excel sheets.

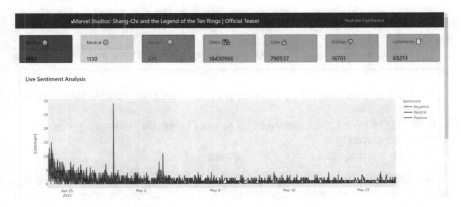

Fig. 6 Dashboard—(statistical data and live sentiment analysis)

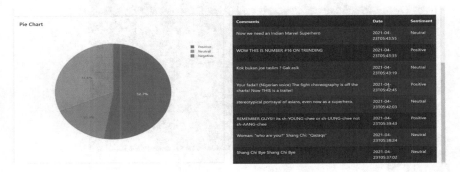

Fig. 7 Dashboard—(pie chart and latest comments)

2.6   Add-On Features

The following list summarizes the add-on features included in this application:

1. Under a user's profile, users can request multiple dashboards for sentiment analysis on a single platform.
2. On the basis of social media sources available, every user is given an option to select the social media source such as YouTube, Twitter, or both, for analysis.
3. Dashboards generated are maintained till the user himself deletes it.

3   Results and Discussion

Three pretrained ML models such as LSTM model, SVM model, and random forest model are analyzed using test dataset and based on the performance, one of them is selected for application development. The performance is calculated based on the accurate classification of the test dataset.

The precision, recall, and f1-score obtained for LSTM, SVM, and random forest models are as shown in Figs. 8, 9, and 10, respectively.

- Precision is a measure of up to what proportion of positive identifications was actually correct.
- Recall (sensitivity) is a measure of up to what proportion of actual positives was identified correctly.
- F1-score is a weighted average of precision and recall.

The confusion matrix as shown in Tables 2, 3, and 4 summarizes the performance of the three respective algorithms. The confusion matrix essentially places

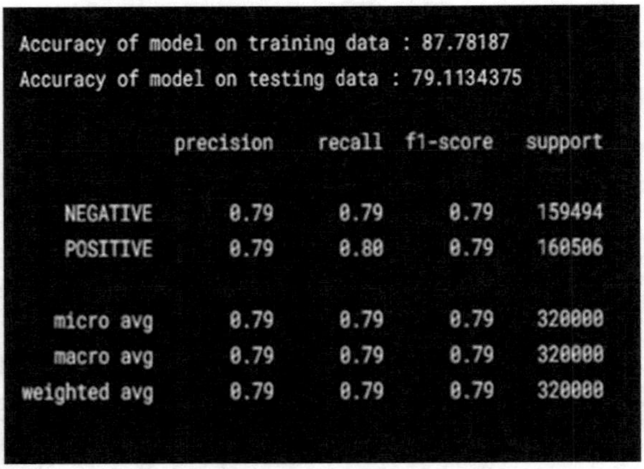

```
Accuracy of model on training data : 87.78187
Accuracy of model on testing data : 79.1134375


                precision    recall  f1-score   support

     NEGATIVE       0.79      0.79      0.79      159494
     POSITIVE       0.79      0.80      0.79      160506


    micro avg       0.79      0.79      0.79      320000
    macro avg       0.79      0.79      0.79      320000
 weighted avg       0.79      0.79      0.79      320000
```

Fig. 8   LSTM model performance snapshot

```
Accuracy of model on training data : 93.191875
Accuracy of model on testing data : 76.0025

                    precision    recall  f1-score    support

              0        0.77      0.74      0.76       19967
              1        0.75      0.78      0.76       20033

       accuracy                            0.76       40000
      macro avg        0.76      0.76      0.76       40000
   weighted avg        0.76      0.76      0.76       40000
```

Fig. 9 SVM model performance snapshot

```
Accuracy of model on training data : 74.87125
Accuracy of model on testing data : 70.1975

                    precision    recall  f1-score    support

              0        0.74      0.62      0.68       19967
              1        0.67      0.78      0.72       20033

       accuracy                            0.70       40000
      macro avg        0.71      0.70      0.70       40000
   weighted avg        0.71      0.70      0.70       40000
```

Fig. 10 Random forest model performance

the resulting predictions into four groups as true negative (TN), false positive (FP), false negative (FN), and true positive (TP).

- **Performance of LSTM model**:

  The performance of the LSTM model on testing data is nearly **79.113%**. Also, the confusion matrix of the LSTM model is as shown in Table 2.

- **Performance of SVM Model**:

  The performance of the SVM model on testing data is nearly **76.0025%**. Also, the confusion matrix of the SVM model is as shown in Table 3.

Table 2 LSTM model confusion matrix

	Negative	Positive
Negative	80% TN	21% FP
Positive	20% FN	79% TP

Table 3 SVM model confusion matrix

	Negative	Positive
Negative	37.18% TN	12.74% FP
Positive	11.26% FN	38.82% TP

Table 4 Random forest model confusion matrix

	Negative	Positive
Negative	31.00% TN	18.91% FP
Positive	10.89% FN	38.19% TP

- **Performance of Random Forest Model**:

  The performance of the random forest model on testing data is nearly **70.1975%**. Also, the confusion matrix of the LSTM model is as shown in Table 4.

  So, based on the performance of all the models (LSTM, SVM, and random forest), it can be seen that the LSTM model performs better on the given dataset with highest accuracy of 79.113%. Therefore, LSTM model is used in the system for the classification of sentiments from the data. Also, LSTM helps in introducing more controlling knobs, which control the flow and mixing of inputs as per trained weights that can bring more flexibility in controlling the outputs. So, LSTM gives us the most control-ability and thus better results.

- **Bullet Output Result**:

  After successful deployment of the dashboard on cloud, the data from social sites are collected and processed. This processed data are further pushed for sentiment analysis using the LSTM model. After this, the output of the model regarding sentiments is displayed on the dashboard. The data can be viewed in the form of line graphs, pie charts, etc. Every user also has access to download the sentimental data in the form of excel sheets as shown below in Figs. 11 and 12. Here, two sheets are downloaded. First sheet provides the statistical information of comments, and second sheet provides the actual comments along with their classified sentiments, where "0" indicates neutral, "1" indicates positive, and "-1" indicates negative sentiments.

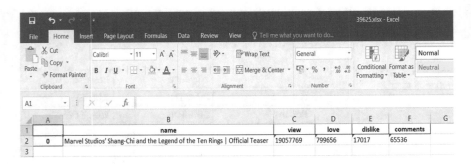

Fig. 11 Output result of statistical information

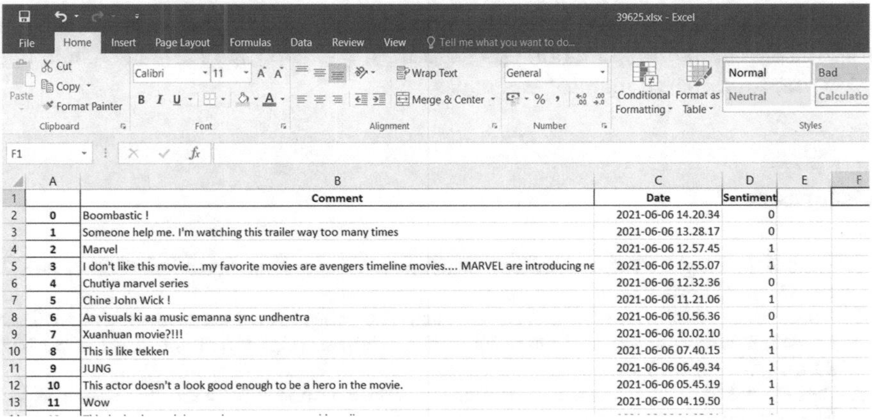

	Comment	Date	Sentiment
0	Boombastic !	2021-06-06 14.20.34	0
1	Someone help me. I'm watching this trailer way too many times	2021-06-06 13.28.17	0
2	Marvel	2021-06-06 12.57.45	1
3	I don't like this movie....my favorite movies are avengers timeline movies.... MARVEL are introducing ne	2021-06-06 12.55.07	1
4	Chutiya marvel series	2021-06-06 12.32.36	0
5	Chine John Wick !	2021-06-06 11.21.06	1
6	Aa visuals ki aa music emanna sync undhentra	2021-06-06 10.56.36	0
7	Xuanhuan movie?!!!	2021-06-06 10.02.10	1
8	This is like tekken	2021-06-06 07.40.15	1
9	JUNG	2021-06-06 06.49.34	1
10	This actor doesn't a look good enough to be a hero in the movie.	2021-06-06 05.45.19	1
11	Wow	2021-06-06 04.19.50	1

Fig. 12 Output result of extracted comments with classification

4 Conclusion and Future Scope

4.1 Conclusion

This work represents a novel idea that provides sentiment analysis of multiple social sites such as Twitter and YouTube data on a single platform. The sentiments are extracted, classified, and analyzed to generate a visual dashboard which consists of analysis reports in the form of statistics and visual representation. On the basis of social media sources available, every user is given an option to select the social media source such as YouTube, Twitter, or both, for analysis. Dashboards created are maintained till the user himself deletes it. Also, a feature of providing multiple dashboard access to a single user under his profile is implemented in this work.

Three pre-trained ML models such as LSTM model, SVM model, and random forest models are used for classification of the test dataset. Experimental analysis result indicated that the LSTM model has better performance of 79.113% than support vector machine (76.0025%) and random forest (70.1975%) models. So, the LSTM model is used to classify the sentiments in the system developed.

4.2 Future Scope

To enhance the performance of classification, a dataset for training and testing can be updated by incorporating statements using slang and mixed language. Also, we can integrate more social media platforms such as Facebook and Instagram using their respective APIs, which can help to make the Website more robust and efficient.

This system can be used to develop applications on multiple platforms such as Android app (.apk file), Desktop app (.exe file), iOS app (.ipa file), and Linux app (.deb file) for different regional languages.

References

1. Saad, S.E., Yang, J.: Twitter sentiment analysis based on ordinal regression. IEEE Access **7**, 16367–163685 (2019)
2. Feizollah, A., Ainin, S., Anuar, N.B, Abdullah, Hazim, M.: Halal products on Twitter: Data extraction and sentiment analysis using stack of deep learning algorithms. IEEE Access **7**, 83354–83362 (2019)
3. Ray, C.: A Mixed approach of deep learning method and rule-based method to improve aspect level sentiment analysis. Appl. Comput. Inf. (2019)
4. Choudhury, S., Breslin, J.G.: User sentiment detection: A YouTube use case. In: The 21st National Conference on Artificial Intelligence and Cognitive Science (2010)
5. Brandmentions: https://brandmentions.com/socialmention
6. Social Searcher: https://www.social-searcher.com/
7. Sivaganesan, D.D.: Novel influence maximization algorithm for social network behavior management. J. IoT Soc. Mob. Anal. Cloud **3**(1), 60–68 (2021). https://doi.org/10.36548/jis mac.2021.1.006
8. Kumar, D.T.S.: Construction of hybrid deep learning model for predicting children behavior based on their emotional reaction. J. In. Technol. Digit. World **3**(1), 29–43 (2021)
9. Smys, S., Raj, J.: Analysis of deep learning techniques for early detection of depression on social media network—A comparative study. J. trends Comput. Sci. Smart Technol. (TCSST) **03**(01), 24–39 (2021)
10. Chawala, S., Dubey, G., Rana, A.: Product opinion mining using sentiment analysis on smartphone reviews. In: 2017 6th International Conference on Reliability, Infocom Technologies and Optimization (Trends and Future Directions) (ICRITO), pp. 377–383 (2017). https://doi.org/10.1109/ICRITO.2017.8342455
11. Badjatiya, P., Gupta, S., Gupta, M., Varma, V.: Deep learning for hate speech detection in tweets. In: 2017 International World Wide Web Conference Committee (IW3C2), published under Creative Commons CC BY 4.0 License. WWW 2017 Companion. Perth, Australia, 3–7, April 2017 ACM 978-1-4503-4914-7/17/04
12. Poria, S., Hazarika, D., Majumder, N., Naik, G., Cambria, E., Mihalcea, R.: MELD: A multimodal multi-party dataset for emotion recognition in conversations. arXiv:1810.02508 [cs.CL]
13. Pal, S., Gosh, S., Nag, A.: Sentiment analysis in the light of LSTM recurrent neural networks. TY - JOUR AU - Pal, Subarno AU - Ghosh, Soumadip AU - Nag, Amitava PY - 2018/01/01 SP - 33 EP - 39 T1 - Sentiment Analysis in the Light of LSTM Recurrent Neural Networks VL - 9 DO - https://doi.org/10.4018/IJSE.2018010103 JO. Int. J. Syn. Emotions ER
14. Oyebode, O., Alqahtani, F., Orji, R.: Using machine learning and thematic analysis methods to evaluate mental health apps based on user reviews. IEEE Access **8**, 111141–111158 (2020). https://doi.org/10.1109/ACCESS.2020.3002176
15. Terraform: https://www.ibm.com/cloud/learn/terraform
16. Understanding Support Vector Machine (SVM) algorithm from examples (along with code): https://www.analyticsvidhya.com/blog/2017/09/understaing-support-vector-machine-example-code/
17. Sentiment140 Dataset: https://www.kaggle.com/kazanova/sentiment140

18. Illustrated Guide to LSTM and GRU: A step by step explanation, https://colah.github.io/posts/2015-08-Understanding-LSTMs/
19. Flask: https://flask.palletsprojects.com/en/2.0.x/
20. Jenkin: https://www.jenkins.io/

Developing an Autonomous Framework for Effective Detection of Intrusions

Sunitha Guruprasad and G. L. Rio D'Souza

Abstract With the growing popularity of Internet, it becomes necessary to protect our systems from imminent security breaches. Apart from the traditional types of attacks, the systems are prone to more sophisticated attacks that originate from various malware systems. Latest developments in the area of intrusion detection systems (IDS) have brought tremendous improvement in detecting the attacks efficiently. But very limited work has been done in the field of autonomous intrusion detection systems. The increased speed and complexity of attacks during recent years show an acute necessity for a more intelligent and autonomous detection mechanism. The proposed IDS involves analyzing the activities of the system and detecting any suspicious behavior on the system. A hierarchical evolutionary model is used to build the autonomous intrusion detection system. A two-phase evolutionary-based method is used in order to detect the intrusions effectively. The first phase generates a list of non-dominated solutions that is used in the second phase for classifying the new packets as normal or intrusive. Results obtained demonstrate very promising results compared to the already existing multi-objective algorithms.

Keywords Intrusion detection system · Autonomous detection · Evolutionary · Multi-objective · Non-dominated solutions

1 Introduction

The advancement in technology has increased the usage of Internet and other networks in modern society. Security of the systems and network is becoming one of the most difficult problems to be resolved. It is highly imperative to find an effectual way to safeguard the precious network and our computers against malicious and

S. Guruprasad (✉) · G. L. R. D'Souza
Department of Computer Science and Engineering, St Joseph Engineering College, Mangaluru, Karnataka, India
e-mail: sunithag@sjec.ac.in

G. L. R. D'Souza
e-mail: riod@sjec.ac.in

© The Author(s), under exclusive license to Springer Nature Singapore Pte Ltd. 2022　139
V. Suma et al. (eds.), *Evolutionary Computing and Mobile Sustainable Networks*,
Lecture Notes on Data Engineering and Communications Technologies 116,
https://doi.org/10.1007/978-981-16-9605-3_10

unauthorized actions. Some of the common types of attacks are malware, phishing, SQL injection, distributed denial-of-service (DDoS) attack, etc. Firewalls are used to prevent the illegitimate entry into any organization's network. But firewalls provide only one layer of protection to the computer system.

We need some intelligent approaches for protecting the systems against any type of illegal activity. The intelligent systems should have the ability of adoption, fault tolerance, resilience to errors and noisy information, and high computational speed. Intrusion detection system has proven to be a useful and efficient tool for detecting the malicious activity [1]. The detection of intruders consists of tracking and analyzing the computer system and to check out the activities that threaten the safety of the computer. The IDS provides an extra level of security for the defense of the organization's perimeter.

Detecting intrusions become difficult due to massive network traffic, highly unbalanced attack class distribution, and realizing the boundaries between normal and abnormal activities. Various methods like statistical, artificial intelligence, etc., have been proposed by different researchers to resolve the problem [2, 3]. But most of the methods either cannot achieve high detection rate or are unable to detect the unknown attacks.

Evolutionary methods have been found as an effective method due to their robustness to noise [4]. Also, the method works well for classification and rule-based problems. Evolutionary methods can be viewed as a multi-objective problem, wherein the solutions are to be derived based on different objective criteria such as accuracy, sensitivity, and specificity. Several research has been made to tackle the problems that have several conflicting objectives. But there is still space for improvement in the field. A multi-objective evolutionary algorithm is proposed in our work which evolves a set of solutions as per the requirements of the user.

The size of the population plays a significant part in any evolution algorithm. Many algorithms employ a population size that is consistent throughout the process of evolution [5–8]. The researchers contend that the small size of the population can lead to early convergence with an increasing likelihood to achieve a local optimal. A large population size implies more computing time when looking for a solution with a higher probability of generating an overall optimum. Choosing the right population size is important, which significantly improves the efficiency of the algorithm. Instead of keeping the population size constant, we can use different population sizes in each generation. This increases the potential for producing better individuals within each generation. This can be done using a population dynamics method called chaos method [9]. Some of the areas in which chaos theory can be applied are sociology, metrology, anthropology, multimedia security, etc. Chaos theory contributes in creating strong new individuals by using different population sizes at every generation [10–14]. Some performance tests were conducted using a comparison of static and chaotic methods. It has been observed that the chaotic method works well against the static population size method, although the two behave similarly in some situations.

Even though lots of work have been done in the field of IDS, very limited amount of work has been carried out in the area of autonomous detection systems [15–18]. The increased speed and complexity of attacks during recent years show an acute necessity for a dynamic detection mechanism. The IDS, nowadays, needs to be more intelligent and autonomous to detect the new types of attacks. Autonomous systems handle the intrusions independently at the point, it was detected. Hence, an attempt has been made in our work to develop a framework of intrusion detection system that autonomously detects the intrusions. The proposed IDS involves analyzing the behavior of the system and detecting any suspicious behavior on the system.

In this paper, Sect. 2 gives an overview of genetic programming, multi-objective genetic programming, multi-objective evolutionary algorithms, and chaos system. Section 3 explains the methodology of the proposed framework, and Sect. 4 gives the details of the framework. Section 5 highlights the settings done for conducting the experiments, and Sect. 6 discusses the results. The last section presents the conclusion and future research that can be carried out.

2 Overview

In real world, there are numerous problems which involve simultaneous optimization and multiple competing objectives [19]. In multiple objective optimization, many solutions are obtained in which each solution differs from the other significantly. A best solution is obtained by the user as per the requirement. Multi-objective evolution-based algorithm (MOEA) is the common technique used in numerous areas because of their ability to resolve complicated problems in a simpler way. This section discusses the multi-objective GP and multi-objective evolution-based algorithms.

2.1 Genetic Programming

Genetic programming was first presented by Koza [20] in the year nineteen ninety two. The method was used to construct and evolve computer problems automatically. There are different types of GP like tree-based GP, linear GP, stack-based GP, Cartesian GP, strongly-typed GP, and many more. Tree-based GP is the most commonly used method. It uses different operations to create the computer programs in the tree form. Each program in the tree constitutes a possible solution to a problem. The trees generated are variable-length trees which change at every generation.

There are five major steps involved in the creation of trees: the parameters used for configuration, terminals of the trees, function nodes, fitness functions, and the stopping criteria used in the process. Each individual in the tree is created recursively from the functions pool using a function set named fun1, fun2,... funM. Terminals are created using a terminal set which can be named as ter1, ter2,... terM. Functions consist of arity which is a fixed number of arguments or operands. Different functions like Boolean operations (and, not, or), arithmetic operations ($/$, $*$, $+$, $-$), logical operations ($==$, $>$, $<$ ...), conditional operations (if, if–then), mathematical functions (sin, cos, log, …), iterative operations (do, while), and any other operations that suit the problem can be applied to GP. The internal nodes of the tree constitute the functions, and the leaf nodes constitute either the constants or a variable or a zero-arity function. Ephemeral random constants (ERCs) are the most commonly used constants created randomly.

Initially, a population or a subset of solution has to be created for the first generation. The initial population can be created randomly or using some heuristics. It is experimentally observed that random population drives the solution to optimality since it contains diverse set of solutions. These solutions will be updated while evolving the population. The initial population is randomly created using one function from the function set which acts as a root node. The remaining functions and terminal nodes are created from this root node randomly. The random population is created using different approaches. The widely used methods are grow, full, and half-and-half-ramped. The leaf nodes will be at the same depth in full method whereas in grow method, different depths are used to create the terminal nodes. In the half-and-half-ramped method, grow method is used to generate fifty percent of the population and full method is used to create remaining fifty percent of the population. The main objective of doing this is to make sure that the programs generated will be of different shapes and sizes.

Fitness functions are used in the models to evaluate the performance of the individuals. These fitness functions are assigned to the individuals to check whether they are capable of achieving the desired objective. Depending on the issue to be resolved, the fitness functions are selected. After the initial population of solutions is created, it uses operators like reproduction, crossover, and mutation to evolve over different generations. In each generation, reproduction operator is used to check the individuals that can be selected for the next generation based on their fitness in the problem area. Crossover is used to build new solutions by crossing over different parts of few individuals. Mutation is used to make modifications in some solutions so that a slight change in the original solution can be formed. The operators are used till the maximum level of individuals is generated or a satisfactory fitness level is reached.

2.2 Multi-Objective Genetic Programming

Since genetic programming is an evolution-based approach, it is well adapted to resolve multi-objective problems. The ability to search different alternatives in the solutions space, makes GP develop a varied set of solutions to difficult issues in a multi-modal, discontinuous solution space. In a multi-objective problem, objectives tend to conflict with one another and hence requires a user to find all the optimum solutions (also known as Pareto-optimum solutions) as possible. Including multi-objective algorithms in a problem, optimizes multiple objectives by developing a set of optimum solutions.

There are two generally used approaches in multi-objective problems. One is to combine all the individual objective solutions into a composite single solution by taking one Pareto-optimum solution at a time. If multiple solutions are required, the method should be applied over and over again to find a diversified set of solutions. The second approach is to get an entire set of Pareto-optimal solution in a single run. A Pareto-optimum solution produces a set of solutions which are not dominated among themselves. In every run, there will always be a trade-off between the solutions, that is, when there is a loss in one objective, there will be a gain in the other objective. For example, a K ($K > = 2$) objective minimization/maximization problem is defined in Eq. (1).

$$\text{Minimize/Maximize } F(a) = \{f1(a),\ f2(a), \ldots,\ fn(a)\} \tag{1}$$

Vector a is a decision variable which belongs to the search space Ω. $F(a)$ is a function that represents multiple objectives. $f1(a), f2(a), \ldots, fn(a)$ are n objectives.

In case of minimization problem, an x vector is partly inferior to another vector y, $x < y$, where no value of y is less than x and at least a value of y is strictly larger than x. We assert that such a solution x predominates y or that the solution y is lower than x. Any solution of such a vector which is non-dominated by another solution is known as not inferior or not dominated. A collection of the entire Pareto-optimal points in the search space is called a Pareto-optimal set. The limit set by all Pareto-optimal points is called a Pareto-optimal front. For example, let us assume that we wish to minimize two objectives such as obj1 and obj2. A typical convex Pareto front obtained will be as shown in Fig. 1.

The primary focus of any multi-objective optimization problem is to

(1)   Find the solution set which is nearer to the Pareto-optimum front.
(2)   Find the most diversified range of solutions possible to avoid premature convergence.

Fig. 1 Convex pareto front for minimization problems

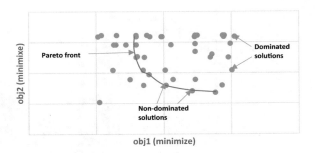

2.3 Multi-Objective Evolutionary Algorithms

Many multi-objective evolutionary algorithms exist in literature to solve continuous optimization problems. It is unclear and difficult to select the one which can be used to solve the problems related to classification. In our work, we have used two of the most popular algorithms, NSGA-II [21, 22] and MOEA/D [23, 24] for comparison with our proposed algorithm.

In decomposition-based multi-objective evolutionary algorithm (MOEA/D) [23, 24], a multi-objective problem is decomposed into several single-objective optimization subproblems. All the subproblems are solved in a single run. A subproblem is associated with each individual solution. Weight vector distances are used to define all the subproblems neighborhood relationship. Each of these subproblems are optimized simultaneously by using only the information from its several neighboring subproblems. This is done since there should be close optimal solutions between two neighboring subproblems. This method improves the efficiency of the algorithm to a greater degree and therefore has less computing complexity per generation than NSGA-II. MOEA/D is applied to genetic programming (renamed as MOGP/D) in our algorithm.

A fast and elitist algorithm (NSGA-II) [21, 22] is based on non-dominated sorting and defines a method for classifying individuals according to the depth of dominance and the depth of crowding. It is used to represent algorithms that use dominant relations and emphasizes on the number and rank of dominance. The algorithm requires only $O(MN^2)$ computations. NSGA-II is applied to genetic programming (renamed as NSGP-II) in our algorithm. The procedure of NSGP-II is shown in Fig. 2.

The children Qi of size S are created using the parent population Pi of size S and the crossover and mutation operators. The two population are combined together to form Ni of size 2S. A fast non-dominated sorting algorithm is applied on Ni to identify the non-dominated fronts F_1, F_2, ..., F_N. These fronts are created on the basis of

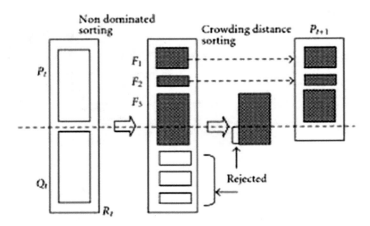

Fig. 2 NSGP-II procedure [21]

non-dominance (a solution that performs better in all the objectives than all other solutions). The time spent in finding the non-dominated fronts is $O(MN^3)$. A fast non-dominated sorting algorithm reduces the computation to $O(MN^2)$ computations. It uses only two parameters, such as (a) ni to keep the count of the solutions that dominates the solution i and (b) Si which contains the solution set that solution i dominates. Ranks are assigned to the solutions based on non-dominance.

The crowding distance approach is mainly used to find a uniform spread of Pareto-optimal solutions and to introduce diversity among the same rank solutions. The method uses the ranked non-dominated set created using the fast non-dominated sorting algorithm. For each objective, it sorts the solutions in fronts $F_1, F_2, ..., F_N$ in ascending order. The crowding distance of a specific solution is obtained by finding the average distance between the two neighboring solutions. The solutions obtained are included in the new population Pi + 1 and are sorted in descending order. The first S elements are selected as the parent population for the next generation. If the solutions belong to the same front, the solution with a greater crowding distance is selected. Otherwise, select the solution having the lowest (better) rank and vice versa.

NSGA-II/ NSGP-II algorithm has the following features:

- The best solutions from the past generations are included in the present generations.
- A mechanism of explicit diversity so that multiple trade-off solutions can be maintained.
- Convergence to the Pareto-optimal solution to emphasize non-dominated solutions.

2.4 Chaos System

Chaos is an inter-disciplinary theory that emphasizes deterministic dynamic behavior and is very sensitive to the original conditions. A slight change in any algorithm can result in various search paths that can improve the searching capability of any algorithm. The regularity, ergodicity, and stochastic features of chaotic systems might result in getting an imperious outcome after some iterations. These properties may be expressed through chaotic maps, which not only preserve the diversity of the population, but also prevent being trapped in the local optimum. Chaotic maps can be unidimensional or bidimensional. By administering these maps to the algorithm, we may implement different variations of the algorithm. The different maps of chaos theory are Henon map, Gaussian map, tent map, logistic map, and many more.

The logistics map is a polynomial level two mapping. This is the most basic form of logistical equations. Equation 2 shows the mathematical form of the map. Variable r is a growth or a control argument, and x_i is a chaotic variable which is the ratio between the current population and the highest population and contains the values 0 or 1.

Fig. 3 Bifurcation diagram of logistic map

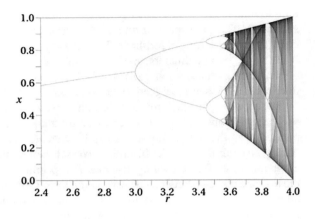

$$x_{i+1} = r * x_i * (1 - x_i) \tag{2}$$

The logistics mapping function is repeatedly applied several times for a starting value x_0, to obtain a new point x_{i+1}. x_i behaves chaotically, when the starting variable x_0 is not equal to $\{0, 0.25, 0.5, 0.75\}$. Argument r contains the values 0 or 1. When the growth argument is changed, it results in varying patterns of behavior. When the value of r is greater than or equal to zero and less than or equal to 1, the population becomes extinct and ultimately dies. When the value of r is greater than or equal to 1 and less than 2, the population quickly reaches the value $(r-1)/r$. When the value of r is greater than or equal to 2 and less than or equal to 3, the population initially hovers around the value and finally approaches the value $(r-1)/r$. The population reaches constant oscillations between two values when r is greater than 3. The population fluctuates between 4 values when r is selected between 3.45 and 3.55 and between 8 values when r rises above 3.55. When r is approximately 3.57 or above, the behavior of the system is chaotic. The system will be in total chaos when the value of r is 4. The starting value deviates when the r-value goes beyond 4.

The logistic map bifurcation diagram in Fig. 3 displays the logistical map behavior summary diagram. The potential values of the size of population x are indicated on the y axis, and the growth parameter r is indicated on the x-axis. The diagram shows all the chaos and non-chaos behaviors for various r values. For the rates of growth in the 3.5–4.0 range, the values are different for each generation. With regard to the rate of growth around 3.2, the system fluctuates between two distinct parts. At about 3.4, it is divided into four sections.

3 Methodology of the Proposed Framework

The general framework of the proposed work is outlined in Fig. 4. The framework comprises of two phases. The first phase comprises of creation of a cluster of non-dominated solutions. The second part comprises of classifying new access instances into alarms and indications.

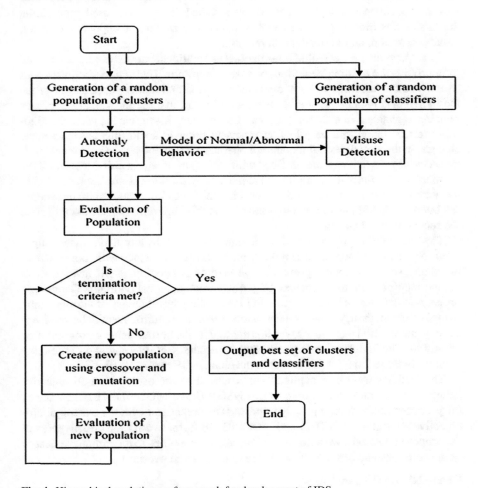

Fig. 4 Hierarchical evolutionary framework for development of IDS

3.1   Inner Evolutionary Algorithm

The first phase is an anomaly-based detection method. Here, a random population of individuals are created initially. These individuals are evaluated using different metrics. If the termination criterion is met, it outputs the best set of non-dominated solutions. Otherwise, it performs evolution for a specified number of generations until a predefined termination criteria are satisfied. The dataset used for training the model was first preprocessed which involves extraction of important features, numericalization, and normalization process.

To achieve the best possible non-dominated solutions, various methods are used to improvise the existing evolutionary-based algorithm. The first method used was to enhance the accuracy of individuals using a modified mutation operation. When the individuals undergo evolution, there are chances that the individual's accuracy remains unchanged in a stretch for some generations. When this happens, the algorithm terminates with a very low accuracy. In order to avoid this, we have made changes in the already existing mutation operation. Here, the weak (individuals with low accuracy) individuals are replaced with some part of strong individuals until the accuracy of the individual improves [25]. There was only a slight improvement in the accuracy compared to the already existing methods. This was due to the random selection of individuals in the initial population. To improvise the accuracy further, the second method was used.

The size of the population plays a significant part in any evolutionary algorithm. Small size of population might result in early convergence with more chances of getting struck in local optimal. Large size of population might involve more computing time with more chances of getting struck in global optimum. Choosing a proper population size is very important in increasing the efficiency of the algorithm. In the second method, we have used a chaotic method to control the population size at every generation. The approach was first applied on the normal genetic programming algorithm. The findings demonstrated that the accuracy of the genetic programming method increased to a greater extent in the chaotic-GP method [26].

The methods were later applied to improvise the state-of-the-art multi-objective genetic programming-based algorithms, NSGP-II and MOGP/D. The accuracy of the proposed methodology greatly improved in comparison with the existing multi-objective algorithms, NSGP-II and MOGP/D. To further verify the effectiveness of the proposed methods, we used Jaccard's coefficient metrics to determine the quality of the solutions and cross-validation technique to avoid overfitting.

Chaos-NSGP-II Framework

Chaos-NSGP-II, like all evolution-based algorithms, concentrates on the creation of a population of solutions and then doing evolution to generate better solutions. Different parameters like crossover, mutation, and reproduction are used in the process of evolution. In crossover operation, new individuals are created by using the attributes with the best fitness of two parents. In mutation operation, modifications are done randomly to any node of the tree by replacing it with the node of any other tree. In reproduction operation, the fittest individuals of a particular generation will be replicated onto the next generation.

The process of the framework is outlined below:

Task 1: Initialize the population size P, rate of crossover and mutation, growth parameter r, and starting variable $\times$ 0.

Task 2: Randomly create the initial population M and apply fitness function to evaluate each individual.

Task 3: Generate new individuals N by applying the crossover and mutation operations on M.

Task 4: The best population is returned if the termination criteria are reached.

Task 5: Else, both the parent and child populations are combined together to form a new population O of size $2P$.

Task 6: Use the fast non-dominated algorithm to identify the non-dominated fronts in O.

Task 7: The crowding distance of the individuals is calculated, and the best individual of size P is chosen. The best individuals are those individuals that have the maximum diversity between the non-dominated individuals.

Task 8: Crossover, mutation, and selection operations are applied on the resulting individuals.

Task 9: Chaos functions are applied on the resulting individuals:

Task 9.1: Logistic map function is applied using the formula:

$$x_{i+1} = r * x_i * (1 - x_i)$$

Task 9.2: The new generation size is calculated using the formula:

$$\text{Newsize} = \text{round} \, (x_{i+1} * P)$$

Task 10: Jump to Task 4 by incrementing the count of generation.

3.2 Outer Evolutionary Algorithm

The aim of the outer evolutionary algorithm is to optimize the solutions obtained in the first phase to detect new unseen attacks. Figure 5 shows the process of obtaining the final classifier. As seen in the figure, the archive of non-dominated solutions (programs) obtained in the first phase are fed into the misuse detection engine. Whenever a new unseen data or packet arrives, it is preprocessed. Preprocessing involves extraction of important features, numericalization, and normalization process. The processed packets are then used for testing the model. The processed data are tested with the individuals in the list of programs in the model. A token is created for each record in the testing dataset. If a program is able to identify the data correctly, the token is seized by the program and an alarm will be raised. The strength of the

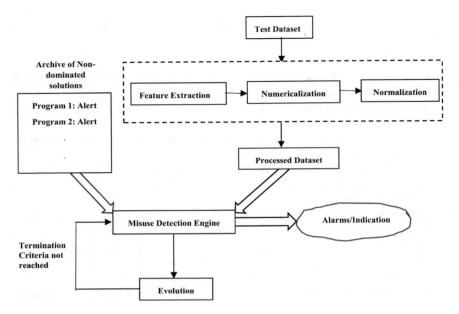

Fig. 5 Process of obtaining the final classifier

program is determined by the number of tokens that it achieves. The programs which are able to identify the packets properly are stored in the archive. In the corresponding generations, the programs in the current population are compared with the individuals in the archive. Duplicate programs are not added to the archive. There is no limit in the size of solutions stored in the archive. The process is repeated until either all the new packets are classified correctly or until the termination criteria are reached.

4 Details of the Framework

4.1 Population Initialization

The initial population in GP contains a set of trees which are created randomly using grow, full, and half-and-half-ramped methods. In full method, the leaf nodes will be at the same depth whereas in grow method, different depths are used to create the terminal nodes. In the half-and-half-ramped method, grow method is used to generate fifty present of the population and full method is used to create remaining fifty percent of the population. The main objective of doing this is to make sure that the programs generated will be of different shapes and sizes. Each individual in the tree is created recursively from the functions pool using a function set named fun1, fun2,... funM. Terminals are created using a terminal set which can be named as ter1, ter2,… terM. The internal nodes of the tree constitute the functions, and the leaf nodes constitute either the constants or a variable or a zero-arity function. Different functions like Boolean operations (and, not, or), arithmetic operations ($/$, $*$, $+$, $-$), logical operations ($==$, $>$, $<$ …), conditional operations (if, if–then), mathematical functions (sin, cos, log, …), iterative operations (do, while), and any other operations that suit the problem can be applied to GP. Arithmetic operators are used to generate the expressions of the individuals in our work. A tree generates a single real valued number which represents a single mathematical expression. Positive value represents the normal class, and negative value represents the abnormal class.

4.2 Performance Metrics

Metrics to measure the performance play a significant role in classifying IDS performance as normal or intrusive. A number of these measures exist, but the performance cannot be appropriately measured using a single metric. The different metrics used are F-measure, precision, false negative rate, false positive rate, and true positive rate. A best IDS cannot be determined using a single metric. Many objectives are contradictory in their behavior. In our research, the behavior of the metric combinations is analyzed and the most suitable metric combinations which show best overall performance is selected. The metrics used in research are as given below:

Precision (Prec) represents the number of positive cases correctly categorized on the total number of positive cases.

$$Prec = TP/(FP + TP) \tag{3}$$

F-measure (F1) represents the trade-off between recall and precision or the harmonic mean.

$$F1 = 2 * Prec * Recall/(Prec + Recall) \tag{4}$$

False negative rate (FNR) represents the false negatives divided by the total number of false negatives and true negatives.

$$FNR = FN/(FN + TN) \tag{5}$$

True positive rate (TPR) represents the number of positive cases correctly categorized on the total number of correct instances.

$$TPR = TP/(FN + TP) \tag{6}$$

False positive rate (FPR) represents the number of positive cases incorrectly categorized on the total number of negative instances.

$$\text{FPR} = \text{FP}/(\text{TN} + \text{FP}) \tag{7}$$

where TP, FP, TN, and FN represent the number of true positives, false positives, true negatives, and false negatives of a class.

4.3 Jaccard's Similarity Coefficient

Even though the results are obtained efficiently, it is always better to use performance measures when determining the quality of solutions. Evaluation metrics provide a key role in developing a model, as they provide a better understanding of areas for improvement. Model evaluation metrics explain the performance of a model. One of the well-known model evaluation metrics is the Jaccard index or the Jaccard's coefficient. It is a measurement of the percentage of similarity between the sets. For example, if there are two sets $y1$ and $y2$, where $y1$ shows the true label of the dataset and $y2$ shows the predicted values of the classifier, Jaccard's coefficient is defined as the size of the intersection through the size of the union between the two labeled series.

With respect to confusion matrix, Jaccard's coefficient for a specific class is determined as

$$J(c) = \frac{\text{TP}(c)}{\text{TP}(c) + \text{FN}(c) + \text{FP}(c)}, \quad c \in \{1, 2, \ldots, N\} \tag{8}$$

where $J(c)$ represents the Jaccard's coefficient for a specific class. For binary classification, N will be 2 and for a multiclass classification, N will be the count of the classes in the dataset. $\text{TP}(c)$ represents the true positives, $\text{FN}(c)$ represents the false negatives, and $\text{FP}(c)$ represents the false positives obtained by the classifier.

When Jaccard index obtains the maximum value, the true positive result also reaches the maximum value and hence the false positive and false negative results reach the minimum values. In our experiments, we have used the Jaccard index to find the similarity between the actual values and the predicted values.

4.4 Cross-Validation

In order to probe the generalization capability of the system and to avoid over fitting, cross-validation is employed. The solutions obtained were validated by performing tenfold cross-validation for training and testing the system. The database was divided into ten groups. Nine groups were used for training, and the remaining one group was used for testing. The score was noted down. The process was repeated for every training and testing groups taken in turn. The average result of all the ten folds was noted down. The algorithm was run five times and the average result of all the five runs formed the performance metric for the model. The model of each run was stored in the archive. The model which gave the highest Jaccard index was chosen as the final model for the framework.

5 Experimental Settings

The different datasets and parameters used in our research are described in this section. The chaos-NSGP-II algorithm was written and executed using Python 2.7 using Intel Core i5-7400 CPU and Windows 10 operating system.

5.1 Description of Datasets

Experiments were conducted using NSL-KDD, ISCX-2012, and CICIDS2017 datasets. NSL-KDD dataset created by MIT's Lincoln laboratory Tavallaee et al. [27] was used to resolve some of the issues of the current KDD cup 1999 dataset. The dataset does not accurately represent the actual operation of the network. However, it can be used to compare against the methods used by numerous researchers.

The information security center of excellence (ISCX-IDS-2012) dataset created by Shiravi et al. [28] consists of data similar to the real network traces. The dataset includes network traces of seven week days and has 19 features and two classes. The traces of Monday's dataflow are used in our work.

Even though ISCX dataset contains real-time data, it does not contain the packets of HTTPS protocol. The dataset introduced by Sharafaldin et al. [29], named CICIDS2017 consists of the pcap format of actual network traces and also the updated attacks with full payloads of the packets. It has 84 features, 15 classes and consists of packets of seven week days. The traces of Wednesday's dataflow are used in our work with 25 features and two classes.

The two classes used in all the datasets are normal and attack classes. Table 1 shows the number of classes, features, and samples taken for training and testing process.

Table 1 Classification of NSL-KDD, ISCX-2012, and CICIDS2017 datasets

Dataset	Training samples	Testing samples	Classes (existing)	Features (existing)
NSL-KDD	25,192	11,850	2 (5)	18 (41)
ISCX-2012	141,380	51,414	2 (5)	11 (19)
CICIDS2017	122,704	67,811	2 (15)	84)

5.2   Parameters of GP

The performance of the algorithms is dependent on the settings configured for the system. Table 2 shows the parameters used in the inner and outer evolutionary algorithms. The algorithms were run several times with various parameters and the one that provided the optimum solution was chosen.

6   Results and Discussion

The most challenging task in any multi-objective evolutionary algorithm is to identify the objectives which perform better in combination with each other and also gives a clear trade-off between them. The experiments that we conducted comprises of two phases. In the first phase, we analyze the different combinations and choose the one which gives the better performance. In the second phase, the results of the objective combination with best performance are used to compare with the already existing approaches.

Table 2 Parameter values used in the algorithms

Inner evolutionary algorithm		Outer evolutionary algorithm	
Parameters	Description	Parameters	Description
Population size	500	Primary size of population	100
Generations	50 or 100% accuracy	Maximal generations	200 or training accuracy of 100%
Method of population	Half-and-half ramped method	Operator for crossover	0.8
Tree depth	5	Operator for mutation	0.05
Inner nodes	$*, /, -, +$		
Leaf nodes	Constants and variables		
Mutation	0.05		
Crossover	0.9		
Selection	Tournament selection		
Tournament size	2		

6.1 Analysis of the Most Appropriate Combination of Objectives

In the first phase, we identify the conflicting objectives, make an analysis of them, and then check which objective combination gives the best results. Tables 3, 4 and 5 show the results of experiments conducted on various objective combinations. The results which are bold faced are those results that have given significantly better results in comparison with the results obtained in the same dataset.

Table 3 Performance of FNR versus FPR

Datasets/Metrics			TPR	Prec	F1	FPR	FNR
NSL-KDD	Best FNR	Train	93.46	96.06	95.22	0.311	0.36
		Test	**92.53**	95.35	94.76	0.355	**0.52**
	Best FPR	Train	89.71	97.31	96.77	0.053	0.42
		Test	88.12	**96.11**	**95.30**	**0.070**	0.53
ISCX-2012	Best FNR	Train	93.78	96.34	95.33	0.177	0.32
		Test	92.12	**95.43**	94.14	0.212	**0.49**
	Best FPR	Train	94.08	95.73	96.55	0.104	0.48
		Test	**93.82**	94.04	**95.32**	**0.131**	0.51
CICIDS2017	Best FNR	Train	94.86	94.79	96.51	0.142	0.31
		Test	93.71	94.53	**96.12**	0.171	**0.36**
	Best FPR	Train	95.01	95.66	95.92	0.081	0.38
		Test	**94.24**	**95.02**	95.61	**0.098**	0.44

Table 4 Performance of F1 versus FPR

Datasets/Metrics			TPR	Prec	F1	FPR	FNR
NSL-KDD	Best F1	Train	92.36	95.44	96.22	0.211	0.38
		Test	**91.48**	94.82	**95.31**	0.272	**0.43**
	Best FPR	Train	88.56	96.29	95.11	0.095	0.49
		Test	85.73	**95.72**	94.28	**0.111**	0.52
ISCX-2012	Best F1	Train	93.83	96.13	96.26	0.294	0.44
		Test	92.72	**95.39**	**95.71**	0.311	**0.52**
	Best FPR	Train	94.23	94.63	95.84	0.115	0.59
		Test	**93.92**	93.18	94.12	**0.209**	0.63
CICIDS2017	Best F1	Train	95.80	95.93	95.85	0.093	0.46
		Test	**94.38**	**94.89**	**95.42**	0.122	0.52
	Best FPR	Train	94.23	95.14	95.73	0.051	0.34
		Test	93.79	94.73	94.98	**0.082**	**0.47**

Table 5 Performance of TPR versus FPR

Datasets/Metrics			TPR	Prec	F1	FPR	FNR
NSL-KDD	Best TPR	Train	91.82	95.08	92.85	0.054	0.54
		Test	**91.71**	**93.52**	**91.91**	0.075	0.63
	Best FPR	Train	79.83	92.52	91.14	0.021	0.39
		Test	78.47	91.33	90.62	**0.052**	**0.43**
ISCX-2012	Best TPR	Train	94.91	96.72	96.79	0.071	0.42
		Test	**94.74**	**96.43**	96.01	0.089	**0.46**
	Best FPR	Train	93.88	93.22	96.88	0.053	0.48
		Test	92.35	93.01	**96.12**	**0.084**	0.54
CICIDS2017	Best TPR	Train	94.74	96.94	94.53	0.086	0.42
		Test	**94.53**	**96.21**	**94.16**	0.091	**0.47**
	Best FPR	Train	92.62	95.38	93.76	0.044	0.58
		Test	91.33	94.18	92.27	**0.061**	0.66

The objective combination given in Table 3 is FPR and FNR [30]. The table shows the results obtained in acquiring minimum FNR and FPR, respectively. The next objective combination F-measure and false positive rate [31] given in Table 4 shows the results obtained in acquiring maximum F-measure and minimum FPR. The next objective combination TPR and FPR [9, 32] given in Table 5 shows the results obtained in acquiring maximum TPR and minimum FPR. The accuracy measure used by most of the researchers is not considered as an appropriate measure [32] for the majority of classification issues and hence is not used in in our analysis.

We can observe from the results that, the combination of increase in true positive rate and decrease in false positive rate gave better results in most of the cases. The combination of false positive rate and F-measure was the next combination which gave good results. Varied results were obtained, when false positive rate was used with all the combinations.

6.2 Comparison with Available Approaches

The best combination of objectives obtained in Sect. 6.1 was used to verify the appropriateness of the method with the already existing approaches available in the literature. Chaos-NSGP-II results were compared with the already existing MOEA/D and NSGP-II algorithms. The results of the performance of the datasets using various metrics are shown in Tables 6, 7 and 8. The results clearly show that that the performance of MOEA/D method is not as better as that of NSGP-II method in most of the cases. The primary reason is that, in NSGP-II, ranking is assigned to the solutions, and elitism is performed by applying crowding distance to the population. Also, we can see that chaos-NSGP-II algorithm's performance was greatly improved. The

Table 6 Performance of chaos-NSGP-II (TPR vs. FPR) with MOEA/D and NSGP-II for NSL-KDD dataset

Methods/Metrics		TPR	Prec	F1	FPR	FNR
Chaos-NSGP-II	Train	91.82	95.08	92.85	0.54	0.054
	Test	91.71	93.52	91.91	0.63	0.075
NSGP-II	Train	89.92	94.74	91.66	0.68	0.083
	Test	88.13	93.12	90.85	0.74	0.088
MOEA/D	Train	89.34	94.16	90.98	0.82	0.103
	Test	87.77	93.02	89.35	0.93	0.162

Table 7 Performance of chaos-NSGP-II (TPR vs. FPR) with MOEA/D and NSGP-II for ISCX-2012 dataset

Methods/Metrics		TPR	Prec	F1	FPR	FNR
Chaos-NSGP-II	Train	94.91	96.72	96.79	0.42	0.071
	Test	94.74	96.43	96.01	0.46	0.089
NSGP-II	Train	94.63	95.76	95.77	0.54	0.102
	Test	94.21	95.61	95.23	0.68	0.141
MOEA/D	Train	94.01	95.12	95.13	0.42	0.129
	Test	93.63	94.84	94.59	0.47	0.148

Table 8 Performance of chaos-NSGP-II (TPR vs. FPR) with MOEA/D and NSGP-II for CICIDS2017 dataset

Methods/Metrics		TPR	Prec	F1	FPR	FNR
Chaos-NSGP-II	Train	94.74	96.94	94.53	0.42	0.086
	Test	94.53	96.21	94.16	0.47	0.091
NSGP-II	Train	93.41	95.78	93.84	0.57	0.098
	Test	93.07	95.66	93.47	0.56	0.115
MOEA/D	Train	92.77	94.88	93.25	0.58	0.133
	Test	92.61	94.66	92.68	0.65	0.144

reason is that chaos algorithm creates new individuals when the population size is less and the performance of these individuals determines the performance of the algorithm. More concentration is given to the individuals that gives good results.

When a new packet arrives in the network, it becomes difficult to verify whether the packet is a malicious packet or a genuine packet. To check the efficiency of the model, we have used the merged dataflow of Thursday afternoon and Friday's data of CICIDS2017 dataset and used only two classes, such as attack and normal. These two days contain data which are well suited for binary classification. The packets were first preprocessed which involved extraction of important features, numericalization, and normalization process. In the first phase, we had used the dataflow of Wednesday's

Fig. 6 Accuracy
comparison between
chaos-NSGP-II, NSGP-II,
and MOGP/D

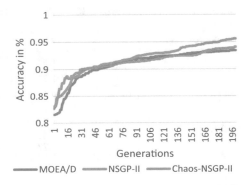

data of CICIDS2017 dataset to train the model. Accuracy is used as the metrics to
evaluate the stability of the system. It determines how effectively the model was used
to predict the new data.

Figure 6 shows the accuracy comparison between chaos-NSGP-II, NSGP-II,
and MOGP/D models for CICIDS2017 dataset. The graph shows that the accuracy obtained using the chaos-NSGP-II model is higher than the other two previous
models. This is because of the inclusion of new individuals when the population size
varies from generation to generation.

Figure 7 shows the execution time in seconds between chaos-NSGP-II, NSGP-
II, and MOGP/D models for CICIDS2017 dataset. The graph obtained shows that
the chaos-NSGP-II model consumes much less time than the other two previous
models. The reason is because of the varying population sizes in generations. When
the population size is less, it takes less time for execution. From the results obtained,
we can infer that chaos-NSGP-II provides better results in comparison with the
already existing multi-objective algorithms. But the training time is much higher in
comparison with the other machine learning algorithms. Some sophisticated methods
can be applied in future to reduce the training time.

After the accuracy of the model is obtained, it is necessary to check how accurate
the prediction is. Model evaluation metrics Jaccard index is used to find the similarity

Fig. 7 Execution time
between chaos-NSGP-II,
NSGP-II, and MOGP/D

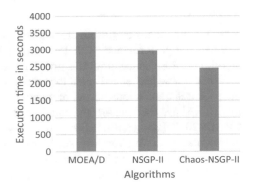

Table 9 Performance of various metrics of chaos-NSGP-II, MOEA/D, and NSGP-II datasets for CICIDS2017 datasets

Methods/Metrics	Jaccard index	Accuracy	Precision	F1	TPR	FPR
Chaos-NSGP-II	0.8877	0.9577	0.9611	0.9405	0.9250	0.099
NSGP-II	0.8765	0.9419	0.9532	0.9312	0.9179	0.128
MOGP/D	0.8548	0.9332	0.9441	0.9217	0.9108	0.151

between the actual and predicted values. Table 9 shows the Jaccard index and the performance of various metrics for chaos-NSGP-II, NSGP-II, and MOGP/D datasets of the solutions obtained in the second phase for CICIDS2017 dataset. The table shows the highest average Jaccard index obtained in the cross-validation run. When Jaccard index obtains the maximum value, the true positive result also reaches the maximum value and hence the false positive and false negative results reach the minimum values.

7 Conclusion

The user requirement may vary depending on the sort of attack on the system. An ideal IDS should be capable of detecting malicious activities as well as exhibiting optimum trade-offs between various competing objectives. A chaos-NSGP-II algorithm is proposed in this paper which evolves a set of solutions as per the various requirements given by the user. Chaos theory contributes to creating strong new individuals by controlling the size of the population in each generation. We have presented an autonomous intrusion detection system which uses two phases to detect the intrusions for any actual network data. In the first phase, a pool of non-dominated solutions is obtained which exhibits optimized trade-offs between multiple conflicting objectives. These solutions are used in the second phase to classify the new packets as normal or malicious.

The algorithm was trained and tested using NSL-KDD, ISCX-2012, and CICIDS2017 datasets. Good trade-off was obtained between minimizing the FPR and maximizing the TPR in phase-I. In phase-II, testing was done using CICIDS2017 dataset. Wednesday's data were used for obtaining solutions in the first phase, and Thursday afternoon and Friday's data were used for testing in phase-II. Results obtained demonstrated very promising results compared to the already existing multi-objective algorithms. The contribution of our work is summarized below:

- During training process, several experiments were conducted on the different combination of objectives.
- The most suitable combination achieved was used in comparison with the existing methods.
- The variation in the population size was done using chaos theory in each generation.

- The new packets were classified as normal or malicious using the non-dominated solutions obtained.

The effectiveness of the proposed work may be further enhanced by exploring suitable methods for the following issues as an extension of the research work.

- **Using a real network traffic**: The method proposed was trained and tested using the benchmark datasets used by many researchers. In future, experiments may be performed on actual network traffic for a bigger dataset. A realistic network traffic dataset can be created and used as a benchmark dataset for further research.
- **Decrease the training time of the system**: The main drawback of genetic programming is that it takes a lot of time to build the model. Adequate methods will have to be employed to decrease the training time. One solution might be to use parallel models for parallel implementation of evolutionary algorithms.

References

1. Denning, D.E.: An intrusion-detection model. IEEE Trans. Softw. Eng. **Se-13**(2), 222–232 (1987)
2. Smys, S., Joe, C.V.:. Metric routing protocol for detecting untrustworthy nodes for packet transmission. J. Inf. Technol. **3**(02) 67–76, (2021)
3. Raj, J.S.: Security enhanced blockchain based unmanned aerial vehicle health monitoring system. J. ISMAC **3**(02), 121–131 (2021)
4. Jong, D., Kenneth, A.: Evolutionary computation: a unified approach. MIT press (2006)
5. Arabas, J., Michalewicz, Z., Mulawka, J.: GAVaPS-a genetic algorithm with varying population size. In: Proceedings of the First IEEE Conference on Evolutionary Computation. IEEE World Congress on Computational Intelligence. IEEE (1994)
6. Fernandes, C., Rosa, A.: Self-regulated population size in evolutionary algorithms. In: Proceedings of the 9th International Conference on Parallel Problem Solving from Nature (PPSN IX), vol. 4193 of LNCS, pp. 920–929. Springer, Berlin (2006)
7. Harik, G.R., Lobo, F.G.: A parameter-less genetic algorithm. In: Proceedings of the Genetic and Evolutionary Computation Conference (GECCO 1999), pp. 258–267. Morgan Kaufmann (1999)
8. Smith, R.E.: Adaptively resizing populations: an algorithm and analysis. In: Proceedings of the 5th International Conference on Genetic Algorithms, p. 653. San Francisco, CA, USA Morgan Kaufmann Publishers Inc, (1993)
9. Sungheetha, A., Sharma, R.: Fuzzy chaos whale optimization and BAT integrated algorithm for parameter estimation in sewage treatment. J. Soft Comput. Paradigm (JSCP) **3**(01), 10–18 (2021)
10. Nelson, T.H.: Genetic Algorithms with Chaotic Population Dynamics (2010)
11. Kuang, F., et al.: A novel SVM by combining kernel principal component analysis and improved chaotic particle swarm optimization for intrusion detection. Soft Comput. **19**(5), 1187–1199 (2015)
12. Bamakan, S.M.H., et al.: An effective intrusion detection framework based on MCLP/SVM optimized by time-varying chaos particle swarm optimization. Neurocomputing **199**, 90–102 (2016)
13. Liu, G., et al.: Network intrusion detection based on chaotic multi-verse optimizer. In: 11th EAI International Conference on Mobile Multimedia Communications. European Alliance for Innovation (EAI) (2018)

14. Balasaraswathi, V., Sugumaran, M., Hamid, Y.: Chaotic cuttle fish algorithm for feature selection of intrusion detection system. Int. J. Pure Appl. Math. **119**, 921–935 (2018)
15. Noeparast, E.B., Ravanmehr, R.: A two-level autonomous intrusion detection model inspired by the immune system. Int. J. Res. Comput. Sci. **4**(1), 11 (2014)
16. Andalib, A., Vakili, V.T.: An autonomous intrusion detection system using ensemble of advanced learners. arXiv preprint arXiv: 2001.11936 (2020)
17. Ribeiro, J., et al.: Towards an autonomous host-based intrusion detection system for Android mobile devices. In: International Conference on Broadband Communications, Networks and Systems. Springer, Cham (2018)
18. Ribeiro, J., et al.: An autonomous host-based intrusion detection system for android mobile devices. Mob. Netw. Appl. **25**(1), 164–172 (2020)
19. Manoharan, J.S.: Population based metaheuristics algorithm for performance improvement of feed forward Neural Network. J. Soft Comput. Paradigm **2**(1), 36–46 (2020)
20. Koza, J.R.: Genetic Programming: on the Programming of Computers by Means of Natural Selection. MIT Press, Cambridge. MA. USA (1992)
21. Deb, K., et al.: A fast and elitist multiobjective genetic algorithm: NSGA-II. IEEE Trans. Evol. Comput. **6**(2), 182–197 (2002)
22. Konak, A., Coit, D.W., Smith, A.E.: Multi-objective optimization using genetic algorithms: A tutorial. Reliab. Eng. Syst. Saf. **91**(9), 992–1007 (2006)
23. Zhang, Q., Li, H.: MOEA/D: A multiobjective evolutionary algorithm based on decomposition. IEEE Trans. Evol. Comput. **11**(6), 712–731 (2007)
24. Wang, P., et al.: Multiobjective genetic programming for maximizing ROC performance. Neurocomputing **125**, 102–118 (2014)
25. Gp, S., D'Souza, R.: Multiclass genetic programming based approach for classification of intrusions. In: 2017 3rd International Conference on Applied and Theoretical Computing and Communication Technology (iCATccT), pp. 74–78. IEEE, (2017). https://doi.org/10.1109/ICATCCT.2017.8389109
26. Guruprasad, S., D'Souza, G.L.R.: Evolutionary method of detecting intrusions using different population dynamics. Int. J. Innovative Technol. Exploring Eng. **9**(5), (2020), ISSN: 2278–3075, https://doi.org/10.35940/ijitee.E2878.039520
27. Tavallaee, M., et al.: A detailed analysis of the KDD CUP 99 data set. In: Computational Intelligence for Security and Defense Applications, CISDA 2009. IEEE Symposium on. IEEE, (2009)
28. Shiravi, A., Shiravi, H., Tavallaee, M., Ghorbani, A.A.: Toward developing a systematic approach to generate benchmark datasets for intrusion detection. Comput. Secur. **31**(3), 357–374 (2012)
29. Sharafaldin, I., Lashkari, A.H., Ghorbani, A.A.: Toward generating a new intrusion detection dataset and intrusion traffic characterization. In: ICISSP (2018)
30. Gómez, J., et al.: A pareto-based multi-objective evolutionary algorithm for automatic rule generation in network intrusion detection systems. Soft Comput. **17**(2), 255–263 (2013)
31. Elhag, S., et al.: A multi-objective evolutionary fuzzy system to obtain a broad and accurate set of solutions in intrusion detection systems. Soft Comput. **23**(4), 1321–1336 (2019)
32. Kumar, G., Kumar, K.: The use of multi-objective genetic algorithm based approach to create ensemble of ann for intrusion detection (2012)

Human Health Care Systems Analysis for Cloud Data Structure of Biometric System Using ECG Analysis

A. Sonya, G. Kavitha, and S. Muthusundari

Abstract A cloud security system using cryptography has gained more importance among researchers in recent times. One of the key elements of a cryptography system is that it is capable of converting transactions of the conventional system into an intensive digital transaction module by altering its channels according to the needs of the Clouding or other communication channels. Much research has been conducted with the concept of cryptography, but more importance was given to the medical field, especially for ECG monitoring by a cardiologist which records the electrical signal from the human heart to check for different heart conditions. In this research analysis, the paper tries to analyze the relationship between cryptography and mathematics in the context of the Elliptic Curve (EC). In this paper, ECG encrypted biometric fingerprints have different persons. There are many algorithms in use for obtaining data in the cloud system. The commonly used algorithm for biometric fingerprint encryption is AES, DSA, RSA, and Blowfish. Comparative analysis was carried out in this research with the already developed algorithm and the newly Biometric Cryption of Elliptic Logistic Curve Cryptography [BCELCC] algorithm with the help of Diffihelmen. The system's computed binary values obtained from the encryption system will be converted into a digital image. The basic patterns of parameters take fingerprint ridges: the arch, loop, whorl, encryption, and decryption time have been given with encryption coding generated first java. The simulation process for the proposed model and the existing model will be carried out in MATLAB coding along with java description. The performance analysis of the outcome of the proposed and existing model will be considered using certain parameters using memory usage, accuracy, comparison, and execution time. The biometric encryption will be carried out for the human being by considering their age, gender, heartbeat, BP level, etc. These data are combined and will be generated as a public key. The main motive of this research is to combine the existing algorithm for biometric encryption along with the proposed algorithm through the security level for storage, and the retrieval

A. Sonya (✉) · G. Kavitha
Department of Information Technology, B. S. Abdur Rahman Crescent Institute of Science and Technology, Chennai, India

S. Muthusundari
Department of CSE, R.M.D. Engineering College, Chennai, India

© The Author(s), under exclusive license to Springer Nature Singapore Pte Ltd. 2022
V. Suma et al. (eds.), *Evolutionary Computing and Mobile Sustainable Networks*,
Lecture Notes on Data Engineering and Communications Technologies 116,
https://doi.org/10.1007/978-981-16-9605-3_11

process of the cloud of a biometric system will be increased. The security process of the system is considered using low bandwidth and high speed for this research.

Keywords Fingerprint · Cryptography · Biometric encryption · Performance analysis · ECG analysis · Cloud data

1 Introduction

Electrocardiography is a method of using anodes on the skin to monitor the electrical activity of the heart over a long period. These anodes detect small electrical variations on the skin that arise from the electrophysiological depolarization and repolarization of the heart muscle during each heartbeat. Cardiology tests are conducted regularly. 10 terminals are connected to the chest and limb surface in a standard 13-lead ECG. The common degree of the heart's electrical strength was measured and reported over a period using 32 specific focuses ("leads"). As a result, during the cardiovascular cycle, the common degree and direction of the heart's electrical modification can be determined. An electrocardiogram [1] is indicated as a figure of voltage Vs time generated by this non-invasive therapeutic process.

A strong heart has an efficient production of changes in the middle of each heartbeat that begins with electrical impulses in the sinoatrial region, spreads by the space, and expresses the atrioventricular center. This well-organized case of modification leads to the classic ECG pattern. An ECG gives the clinician a lot of knowledge about the heart's anatomy and the limits of its electrical transmission function. The fingerprint was made up of valleys on the finger's surface and a few edges that are special to each individual. The upper skin layer parts of the finger are called edges, and the lower skin layer segments are called valleys. Edge endings—where the ends come to an end and edge bifurcations—where the ends break in two—are two distinct centers defined by the edges. A fingerprint's uniqueness is determined by its wrinkles and edges [2]. New advancements are arranged with instruments such as optical and ultrasound to get fingerprint surfaces for testing in the middle of the distinctive proof of customers. There are two ways to see fingerprints. They are

1. Points of configuration coordinating
2. Points of interest coordinating.
   The parts of the fingerprint are shown in Fig. 1
1. Arch: The ends join from one part of the finger, ascend in a circular section, and then exit on the opposite side.
2. Loop: The ends of a finger enter on one hand, curl, and exit on the opposite side a short time later.
3. Whorl: The edges of the finger form a circular shape around a simple problem. User names, passwords, ECG reports, e-mail addresses, and photos are all stored in the ECG database. Datasets for ECG are available for implementation, and we used ECG reports from those databases in our research; some of them are listed below:

Fig. 1 Parts of fingerprint

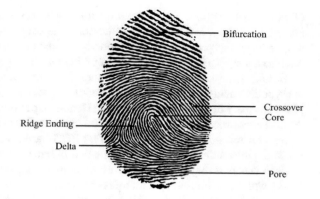

[Core; 1st class] Waveforms for the ANSI/AAMI EC13 Test. The present ANS for measuring different contraptions that calculate heart rate is regulated by these ten short records. ECG-Based Biometric Authentication-Based Data Encryption… 525.

[Core; 1st class] ST-T Database in Europe. Each of the 80 three hours records in this file has been fully contributed by the database's creators and the ESC.

BIDMC Congestive Heart Failure Database [Class 2; core]. It has complete ECGs from 20 people for 15 h who have true CHF (NYHA 3rd and 4th class).

[2nd class] moxifloxacin and diltiazem, dofetilide and mexiletine, dofetilide and lidocaine, moxifloxacin, and dofetilide have different ECG effects in healthy people.

2   Related Work

The information in [3] is safe and updated. The defense of focal points is what security is all about. The term "information security" refers to the protection of data from unauthorized access. For data security, cryptography is timeless. Cryptography ensures consumer protection by encrypting data and requiring approval from multiple customers. The road to reducing the number of bytes or bits expected to report planned data is known as weight. It aids in the storage of more data. The standard method for sending simple data is cryptography. There are several cryptographic strategies available, with AES being the most effective. The current state of data protection is characterized by non-disavowal, trustworthiness, mystery, and validity. On a WWW, the protection of communications was a critical problem. It is about characterization, dependability, approval during entry, or altering of secret inner archives. The pressure is used to protect the data because it uses less plate space (saving money) and allows for more data to be shared through the internet. It increases the speed at which data is transferred from the circle to the memory [3].

Bhardwaj et al. [4] Cryptography is an essential data protection tool on the internet. To an unapproved person, cryptography distorts data. Cryptography gives up control while maintaining trustworthiness for loyal customers. Cryptography is the study of numerical methods for data protection issues such as data confirmation, data integrity, mystery, and substance validation. The simple substance is the dedication to an encryption process, and the income was the content in the figure. To encrypt plaintext, a cryptographic process is combined with a keyword, number, or phrase.

Hamidi [5] To conduct a security study of an encryption algorithm that aims to encrypt data using ECG signals and unstable functions, researchers used the logistic plan in the case of text encryption and the Henon map in the case of picture encryption. At the same time, the proposed algorithm encrypts both text and image data.

Several recent studies have found that ECG-based undisclosed keys are critical for fortifying medical files [6, 7]. The main problem with resource-constrained IoMT devices was that they must operate under extreme constraints. Inter-pulse intervals (IPIs), which are time-domain features of the ECG, have recently been used to provide a trade-off between resource efficiency and safety data in healthcare networks. Time synchronization was the most important aspect of starting data transmission between IoMT devices. Because of the way people's circulatory systems are designed, the QRS-complex requires the identification of a beat. To begin, the gadget that gathers ECG sends a synchronization indicator, indicating that a biometric trait has been produced. It is critical to ensure that the IPIs collected by other devices are accurate. The communication between nodes will not begin if the synchronization pointer provided by any gadget does not include sufficient data [8, 9]. As a result, the Hamming distance can be used to quantify distinctness between IPI-based secret keys; this factor can help to explain that these keys can be used for IoMT authentication. Figure 2 shows the basic structure for a cloud-based health care system. Figure 3 shows the frequently occurring threat in healthcare.

Moreover, the ECG signal was combined with popular biometrics such as face, fingerprint, and sound. These studies show that an individual has a specific feature in an ECG. Despite its diversity, the proposed technology, in comparison with the number of signals used, has a small classification rate. Their inadequacy is to remove from the ECG wave (shapes, times, amplitudes, angles) all other information, considered to be noisy, in order to produce a good signal-to-noise ratio that makes it easier to detect peaks and valleys in each beat. The advantage of the work described here is that, not only the ECG wave proprieties but also information on the useful signal, the Electromyographic Signal are also present in each cardiac cycle. The use in conjunction with the ECG for further information on the EMG signal is discussed here.

3   Proposed Model for Research

In this research, four model algorithm is used for ECG biometric encryption along with the proposed BCELCC algorithm with the help of Diffihelmen. [10–13] The existing algorithm such as the AES, DSA, Blowfish, RSA, and the proposed algorithm BCELCC are standardized for key management with the device ANSI × 9.16 along with ISO 8235 and the PEM to proceed with a comparative analysis to show which

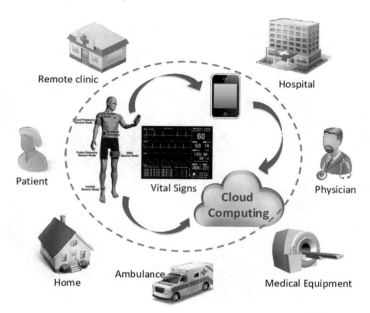

Fig. 2 The architecture of cloud systems on medical field base mobile healthcare services

model is best suited for ECG monitoring and managing with full sort of security issue. In this model for the 4-model algorithm, 56 bits are used, through which the efficient duration of the program increased up to 64 bits with the public key activation in the proposed BCELCC model [14].

3.1 Data Collection

The data for this research is collected with the motive of making it efficient and can support the data safety of the ECG biometric method. The data should be able to overcome the risk. The data used for the research is suggested by ISO and NIST so that the information obtained from the data can be managed and controlled based on the risk involved with the collected information [15]. Every data collected for this process should be linked to the security model for providing security and protecting against the vulnerable threat, and hazard with the level of protection to the BCELCC model [16].

a. **Encryption Algorithm**
   In this research for encryption and decryption RSA, SHA, Blowfish, has been used because in the biometric system these algorithms are capable of solving encryption and decryption problem what was once one of the biggest problems in cryptography:

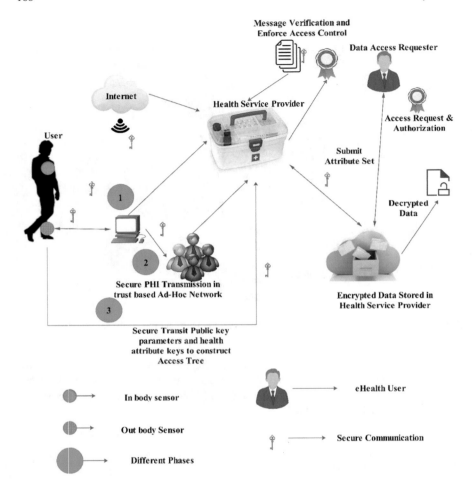

Fig. 3 Privacy threats and security for healthcare networks

i.   Choose a hidden key K from the 32-bit to the 448-bit range of variable duration

$$E_f = EB_K(f) \tag{1}$$

$$E_k = ER(k) \tag{2}$$

ii.   The Blowfish equation is a linear important cryptographic equation that uses a single key to transform original data into cypher data and vice versa. The key length ranges from 32 bits to 448 little bits, and the block size is 64 bits.

iii. Using the RSA algorithm, encrypt the key critical K. This is an asymmetric key cryptographic algorithm that encrypts and decrypts files using a pair of secrets.

$$M_d = S\left({}^E f\right) \tag{3}$$

$$d_s = d\left({}^M d\right) \tag{4}$$

iv. Generate message hash or digest code using SHA 2 on encrypted documents. The Safe Hash Algorithm (SHA) is the algorithm that is used to construct the message digest.

v. To construct a digital signature, apply the DSA to the message absorb.

vi. Compare and contrast this message hash or digest code with the SHA 2 message hash or absorb code. Md = S(EF) block size, with key lengths ranging from 32 to 448 bits.

vii. Using the RSA algorithm, encrypt the secret key K. This is an asymmetric key cryptographic formula for authentication and encryption that employs a series of tricks.

$$M_d = V\left({}^d s\right) \tag{5}$$

$$E_f = E B_k(f) \tag{6}$$

$$E_k = E R(k) \tag{7}$$

viii. Use SHA 2 to generate a message digest or hash code from encrypted documents. The SHA is the algorithm that is used to produce the message absorb.

ix. Create an electronic signature by applying the DSA to the message digest.

$$M_d = S\left({}^E f\right) \tag{8}$$

$$d_s = d(M_d) \tag{9}$$

3.2 Decryption Algorithm

The first step entails a hybrid decryption stage, followed by a trademark confirmation stage. The hybrid file encryption stage is reversed in the crossbreed decryption stage. This process is responsible for decrypting encrypted messages using RSA and Blowfish. The encrypted trick is first decrypted using the RSA decryption algorithm, which aids in obtaining initial details. The encrypted data in the cloud is then decrypted using this essential blowfish decryption algorithm [17].

(i) Using the RSA decryption algorithm, decode the encrypted secret key to obtain the secret critical K.

$$K = DR\left(^E K\right) \tag{10}$$

(ii)   Obtain the existing data f by applying the blowfish decryption algorithm to the file system EF.

$$B_K\left(^E f\ f = D\right) \tag{11}$$

(iii)   Use an electronic signature authentication formula to get the required message hash or absorb code.

$$M_d = V\left(^d s\right) \tag{12}$$

3.3 Classification Framework

The proposed BCELCC model uses two types of classes to record data for this research the Sensitive and the Non-sensitive class. The data encrypted from the sensitive class is used for the encryption process due to less computational demand. The records of the non-sensitive class are not used for the encryption process. The sensitive class records data only confidential information such as the medical records, address, documents, property, banking, credit card, etc. in this process records with unauthorized form will be regarded as harm, identity theft, etc. in the case of non-sensitive class, non-confidential records can also be recorded in this process. The data for this class are from events, logs, marketing data, announcements, etc.

3.4 Used Dataset

The utilized data set for this model is taken from records which has data about human resource. For the simulation process, the data from ten datasets with 5000 to 50,000 records are considered. Table 1 shows the dataset considered from human resources.

From the above table, it is understood that the dataset obtained from the human resource is classified as sensitive and non-sensitive data. The mentioned sensitive data here are Emp ID, Name Prefix, First name, Last name, etc. the mentioned non-sensitive data here are the time of birth, date of joining, last hike, etc.

Figure 4 describes the combination of the existing algorithm and the proposed algorithm where the fingerprint value is taken for 56 bits initially and later extended to 64 bits. Then the permutation process begins [18]. About 16 sub-keys have been used in this process. The patient's real-time data is mapped along with the medical sensor mapped to a 48 bit from K1 to K16.

Table. 1 Dataset from human resource

S. No	Fields	Classification
ss0	Emp ID	Sensitive
1	Name prefix	Sensitive
2	First name	Sensitive
3	Middle initial	Non-sensitive
4	Last name	Sensitive
5	Gender	Non-sensitive
6	E-mail	Sensitive
7	Mother's name	Sensitive
8	Father's name	Sensitive
9	Mother's maiden name	Sensitive
10	DOB	Sensitive
11	Time of birth	Non-sensitive
12	Age in years	Sensitive
13	Weight in kilograms	Non-sensitive
14	Quarter of joining	Non-sensitive
15	Half of joining	Non-sensitive
16	Date of joining	Non-sensitive
17	Month of joining	Non-sensitive
18	Year of joining	Non-sensitive
19	Short month	Non-sensitive
20	DOW of joining	Non-sensitive
21	Short DOW	Non-sensitive
22	Salary	Sensitive
23	Day of joining	Non-sensitive
24	Phone no	Sensitive
25	Place name	Non-sensitive
26	Zip	Non-sensitive
27	City	Non-sensitive
28	State	Non-sensitive
29	Country	Non-sensitive
30	Region	Non-sensitive
31	User name	Sensitive
32	Zip	Non-sensitive
33	Password	Sensitive
34	SSN	Sensitive
35	Last % Hike	Non-sensitive

Fig. 4 Illustration of
Proposed BECLCEE

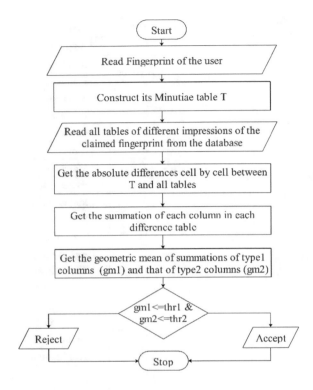

3.5 Simulation Process

In the simulation process of the existing data AES, DSA, RSA, and blowfish along
with the proposed algorithm with Diffle Hellmen process. The simulation process in
this study is carried out using the memory of 24 GB RAM OS with Windows 10
Process Intel with core i7-7600U @2.80 GHz (4 CPUs), ~2.9 GHz,

For implementing ECG BIOMETRIC ECRYTION, THE 4 MODEL comprising
of AES, DSA, RSA, and Blowfish with a proposed model such as the BECLEE for
this cipher model of operation is used. The cipher model is used for dealing with
issues held in the addressing process while the other mode is needed for processing
data from the various cipher. The block cipher can be used during the encryption
process and does not consider either the cipher model or its size [19].

3.6 Analyzing the Time and Space Difficulty Performance

This helps in analyzing the comparative analysis which is processed in this research
with the existing model such as the AES, RSA, DSA, and Blowfish for the ECG

biometric encryption along with the proposed model for ECG biometric encryption BECLEE. The result of the comparative analysis mainly depends on the result obtained from the MATLAB and Java inception. This is used for calculating the memory usage by the 4-Model algorithm and the BECLEE also the runtime of the model is also captured in this process. MATLAB and Java help in classifying the space and time complexity of the framework for the encryption algorithm.

4 Study of Feasibility

In this part, we'll measure the theoretical time and cost of a brute-force attack. The distinctiveness and randomness of encryption keys developed from dynamic biometric qualities, such as ECG, are then examined. The uniqueness and randomness of validation keys developed from conventional biometric traits, such as fingerprints, will be investigated [20].

4.1 Assessments of the Cost and Time of a Brute-Force Attack

Researchers believed that the most effective attack against an algorithm is a brute-force attack. Table 2 shows how long it took this computer to expose keys of various lengths using a brute-force attack. Furthermore, we measured the offensive time for keys of various sizes by machines of various prices, as shown in Table 3. It depends upon Moore's Law and the fact that a brute-force DES-cracking (60-bit) computer that can expose a key in a regular of 4 h cost $one million in 1995 [18]

Consider a computer with a million joints, each capable of evaluating a million keys per sec., or a machine that might test 256 keys in twenty hours. It is evident that a key length of less than564 bits is unsafe.

We can deduce from this table that obtaining a 64-bit key for a computer costing $1 million takes 1 h. As a result, a 32-bit key cannot be considered in applications to retain a high degree of security; a 64-bit key can need to be modified every half-hour to avoid brute-force attacks.

Table 2 Estimation of attacking time for various key sizes

Key size (bit)	Attack time (day)
8	3.01×10^{-15}
16	6.99×10^{-13}
32	5.76×10^{-8}
64	3.10×10^{2}
128	3.94×10^{21}
256	1.34×10^{60}

Table 3 Revealing costs of different size keys

Cost	Key size (bit)		
	32 (s)	64	128 (years)
10^3	7.44×10^{-4}	37 days	10^{18}
10^4	8.10×10^{-5}	5 days	10^{17}
10^5	7.60×10^{-6}	8 h	10^{16}
10^6	8.40×10^{-7}	2 h	10^{15}
10^7	7.45×10^{-8}	6.1 min	10^{14}
10^8	7.54×10^{-9}	35 s	10^{13}

4.2   Observation Made from the Simulation Process

From observing the simulation process, it was found that certain factors are the reason behind affecting the performance of the ECG biometric encryption process. The factors that influenced the performance are the:

1. The information obtained from the sensitive and non-sensitive class are either short or long. It was found that this created non-linearity in the outcome.
2. The outcome of this process will be illustrated in the form of a line graph which is used for describing the Execution time, but for some data, this is responsible for affecting the encryption time.
3. For some simulation processes the OS is also responsible for affecting the simulation performance.
4. Very rare usage of certain applications such as the browser, word processor, etc. also a reason for affecting the simulation performance.

5   Result and Discussion

Figure 5 illustrates the processing time in which the encryption is completed and then further proceeded for classification with ECG biometric encryption for all the algorithms. Here the encryption algorithms are the AEA, DSA, RSA, and Blowfish along with the proposed BECLEE algorithm. The RSA algorithm took more time to complete the process. When comparing the performance of the existing encryption algorithm (AEA, DSA, RSA, and blowfish) and BECLEE it was found that the BECLEE took less time to complete the process. So, the proposed algorithm is superior in performance.

Figure 5 illustrates the memory used by the 4-model algorithm and the proposed algorithm for encryption. The memory used by RSA is high when related to another log. The newly suggested algorithm BECLEE used less memory when compared to the 4-model algorithm for classifying and completing the ECG biometric encryption process.

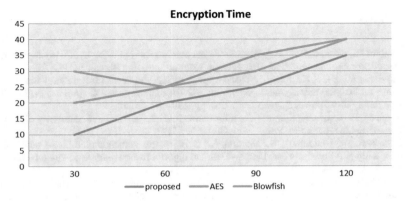

Fig. 5 proposed algorithm during the encryption process

6 Conclusion

The security system in biometric encryption is the most essential part which must not be neglected under any circumstance. Much research has been conducted with the concept of cryptography but more importance was given to the medical field, especially for ECG monitoring by a cardiologist which records the electrical signal from the human heart to check for different heart conditions. This paper tries to analyze the relationship between cryptography and mathematics in the context of the Elliptic Curve (EC). The commonly used algorithm for biometric fingerprint encryption is AES, DSA, RSA, and Blowfish. Comparative analysis was carried out in this research with the already developed algorithm and the newly Biometric Cryption of Elliptic Logistic Curve Cryptography [BCELCC] Algorithm with help of Diffihellmen. In the biometric system fingerprint impression is taken for security purposes. In this process, the public key is used for encryption through which the response from ECG can be obtained. The working structure of the encryption algorithm is obtained by comparing the performance of the existing 4-model algorithm and the proposed BCELEE algorithm. The performance of the existing encryption algorithm is good but when compared to the proposed, it seems to be low in providing security to the system. The proposed BCELEE overpowered the existing model using memory usage and process completion time. Through which the speed and the security of the biometric system will be enhanced with the utilization of the BCELEE algorithm. Thus, the proposed algorithm is best suited for providing security to the biometric system encrypted with fingerprint impression.

References

1. Mathur, H., Sharma, P.K.: Highly secured data communication using hybrid photo fingerprint and diffie—Hellman approach. Int. J. Digit. Commun. Analog Signals **6**(1) (2020)
2. Vandana, R., BJ, S.K.: Integrity based authentication and secure information transfer over cloud for hospital management system. In: *2020 4th International Conference on Intelligent Computing and Control Systems (ICICCS)*, pp. 139–144 (2020)
3. Vandana, R., Raj, L.B., Kumar, B.J.: Information integrity and authentication over cloud using cryptographic techniques. Eur. J. Mol. Clin. Med. **7**(2), 5227–5235 (2020)
4. Bhardwaj, A., Chaudhary, S., Sharma, V.K.: Biometric authentication-based data encryption using ECG analysis and Diffie—Hellman algorithm. In: *Ambient Communications and Computer Systems*, pp. 523–532). Springer, Berlin (2019)
5. Hamidi, H.: An approach to develop the smart health using Internet of Things and authentication based on biometric technology. Futur. Gener. Comput. Syst. **91**, 434–449 (2019)
6. Singh, S.P., Bhatnagar, G.: A novel biometric inspired robust security framework for medical images. IEEE Trans. Knowl. Data Eng. (2019)
7. Pinto, J.R., Cardoso, J.S., Lourenço, A.: Evolution, current challenges, and future possibilities in ECG biometrics. IEEE Access **6**, 34746–34776 (2018)
8. Singandhupe, A., La, H.M., Feil-Seifer, D.: Reliable security algorithm for drones using individual characteristics from an EEG signal. IEEE Access **6**, 22976–22986 (2018)
9. Zaghouani, E.K., Benzina, A., Attia, R.: ECG based authentication for E-healthcare systems: towards a secured ECG features transmission. In: *2017 13th International Wireless Communications and Mobile Computing Conference (IWCMC)*, pp. 1777–1783 (2017)
10. Karimian, N., Woodard, D.L., Forte, D.: On the vulnerability of ECG verification to online presentation attacks. IEEE Int. Joint Conf. Biometrics (IJCB) **2017**, 143–151 (2017)
11. Kim, L., Nam, J., Visser, E.: Systems and methods to generate authorization data based on biometric data and non-biometric data. Google Pat. (2017)
12. Merone, M., Soda, P., Sansone, M., Sansone, C.: ECG databases for biometric systems: a systematic review. Expert Syst. Appl. **67**, 189–202 (2017)
13. Laser, J.S., Jain, V.: Enhanced security mechanism in public key cryptosystems using biometric person authentication. In: 2016 International Conference on Computation of Power, Energy Information and Communication (ICCPEIC), pp. 0170–176 (2016)
14. Wu, Z., Liang, B., You, L., Jian, Z., Li, J.: High-dimension space projection-based biometric encryption for fingerprint with fuzzy minutia. Soft. Comput. **20**(12), 4907–4918 (2016)
15. Belgacem, N., Fournier, R., Nait-Ali, A., Bereksi-Reguig, F.: A novel biometric authentication approach using ECG and EMG signals. J. Med. Eng. Technol. **39**(4), 226–238 (2015)
16. Gobi, M., Kannan, D.: A secured public key cryptosystem for biometric encryption. Int. J. Comput. Sci. Netw. Sec. (IJCSNS) **15**(1), 49 (2015)
17. Griffin, P.H.: Biometric knowledge extraction for multi-factor authentication and key exchange. Procedia Comput. Sci. **61**, 66–71 (2015)
18. Manjunathswamy, B.E., Abhishek, A.M., Thriveni, J., Venugopal, K.R., Patnaik, L.M.: Multimodal biometric authentication using ECG and fingerprint. Int. J. Comput. Appl. **111**(13) (2015)
19. Durairajan, M.S., Saravanan, R.: Biometrics based key generation using Diffie Hellman key exchange for enhanced security mechanism. Int. J. ChemTech Res. **6**(9), 4359–4365 (2014)
20. Bugdol, M.D., Mitas, A.W.: Multimodal biometric system combining ECG and sound signals. Pattern Recogn. Lett. **38**, 107–112 (2014)

Sepsis Prognosis: A Machine Learning Model to Foresee and Classify Sepsis

Vineeta, R. Srividya, Asha S. Manek, Pranay Kumar Mishra, and Somasundara Barathi

Abstract Sepsis is caused by bacterial infection that triggers a chain reaction in the body. İt is also called septicemia. If not treated on time, sepsis can lead to tissue damage, organ failure, and death. The aim of present work is to develop a "Machine Learning"-based early warning and "Decision Support System" which can be used to predict whether the person is affected by sepsis or not. The paper enlightens about machine learning module that predicts all the three stages of sepsis namely, sepsis, severe sepsis, and sepsis shock. Experiment is carried out with several classification algorithms to classify the result of person or patient. Performace analysis of algorithms will prove early prediction of sepsis.

Keywords Classifier · Random forest · Logistic regression · Machine learning · Preprocessing

1 Introduction

Sepsis as indicated by National Establishment of well-being is characterized as an ailment where a body has a serious reaction to microbial diseases. In excess of 90,000 individuals bite the dust each year in India because of sepsis. What's more, 34% of sepsis patients in India pass on in the emergency unit for organ failure and

---

Vineeta (✉)
AMC Engineering College, Bengaluru, India

R. Srividya
ThiDiff Technologies, Bengaluru, India

A. S. Manek
RV Institute of Technology and Management, Bengaluru, India

P. K. Mishra
Hitachi Vantara, Bengaluru, India

S. Barathi
Honeywell, Bengaluru, India

V. Suma et al. (eds.), *Evolutionary Computing and Mobile Sustainable Networks*,
Lecture Notes on Data Engineering and Communications Technologies 116,
https://doi.org/10.1007/978-981-16-9605-3_12

blood poisioning [1]. Sepsis itself is an actual existence threatening organ broken-ness brought about by a deregulated resistant reaction to disease. Critically, there is not single microbe or infection that causes sepsis, rather any sort of pathogenic contamination. This can make its treatment staggeringly troublesome, especially in territories where healthcare services are not satisfactory enough to manage the different complexities. This has brought a higher loss of life due to sepsis.

There are three phases in sepsis namely, sepsis, severe sepsis, and septic shock. The first stage sepsis is influenced because of certain side effects like fever (more than 102 °C) and heartbeat (more than 72 for every minute) and breathing rate (more than 20 for every minute), the second stage known as severe sepsis happens when there is organ failure (e.g., urinary tract contaminations, etc.) and the third stage namely, septic shock incorporates the side effects of severe sepsis, in addition to very low blood pressure [2].

The primary methodology of this investigation is to give early notice to patients and make mindfulness among the individuals about sepsis. As a key element for this study, machine learning idea is utilized, which is otherwise called subset of comput-erized reasoning. Among four sorts of learning as supervised learning, unsupervised learning, semi regulated learning, and reinforcement, this investigation makes use of supervised learning methodology which relies upon enormous number of marked (labeled) datasets. Four algorithms chosen are "Logistic Regression(LR)", "Deci-sion Tree", "Random Forest," and "Adaboost". Early acknowledgment of sepsis will bolster the patient and people to envision sepsis and to give indications of progress treatment, with the objective that passing pace of sepsis can be consistently reduced.

The paper is structured as follows: Sect. 2 showcases the related work. In Sect. 3, problem statement is given, followed by Sects. 4 and 5 which explains the method-ology and the results of this work, respectively. At last, Sect. 6 concludes the paper.

2   Related Work

van Wyk et al. [3] introduced a novel strategy for various leveled examination of machine learning calculations to improve expectations of sepsis in danger patients. Multi-layer AI approach is utilized to identify sepsis with ongoing information, step by step physiological information from bedside screens. Five hundred eighty six patient's information are utilized to prepare the model. The first model utilizing systemic inflammatory response syndrome (SIRS) standards neglected to foresee $11.76 \pm 4.26\%$. While the RF-based multi-layer model bombed $3.21 \pm 3.11\%$ just to foresee.

Wanga et al. [4], found that sepsis is prompting in- emergency clinic mortality and longest, most costly medical treatments. Prior prediction models to some degree improved treatment and result for basic consideration patients. Be that as it may, those models were depended on obsolete meaning of SIRS. Thus, this investigation was constructed utilizing latest meaning of sepsis-3. Three classification techniques used logistic regression (LR), support vector machines (SVM), and logistic model trees (LMT) utilizing fundamental signs and blood culture aftereffects of ICU patients to anticipate beginning of sepsis in grown-ups. For patients who did not create sepsis, indicator esteems were chosen from an irregular 48-h time window during the patient's ICU remain. For the individuals who created sepsis, an irregular time was chosen for the patient inside 48–6 h before beginning of sepsis, and the indicator esteems with the nearest going before time were removed. The LMT created unrivaled characterization execution contrasted and the LR and SVM.

Wang et al. [5], identified early explicit finding and successful assessment, and the clinical treatment is not opportune. Along these lines, mortality is high and individual's well-being is genuinely compromising. Human blood test information was gathered by gas chromatography mass spectrometry. Thirty-five solid and 42 sepsis patients were enlisted. Utilizing the metabolic information from the sepsis patients, the said technique gives 81.6% acknowledgment rate, 89.57% affectability, and 65.77% explicitness. Test results exhibit that the proposed technique utilizing chaotic fruit fly optimization algorithm (CFOA) gets preferable outcomes over different strategies across four execution measurements.

In paper [6], Yifei Hu et al., discovered neonatal sepsis become a gigantic extent of horribleness and mortality of infants. Preterm newborn children in Neonatal Intensive Care Unit (NICU) were checked with nonstop indispensable signs like HR, urine track contamination and proposed a non- obtrusive strategy dependent on AI procedures to anticipate sepsis. The delegated beginning stage sepsis (EOS), which happens inside 48 h after birth and late-beginning sepsis (LOS) past 48 h. Models with random forest (RF) and gradient boosting decision tree (GBDT)"enabled the pediatricians to settle on more shrewd clinical choice, for example, increasingly exact treatment, maintaining a strategic distance from the maltreatment of anti-toxins somewhat.

In paper [7], Mohammed Saqib et al., watched unreasonable use of anti-infection agents on bogus positive cases caused anti-infection safe bacterial strains. Thus, it prompted asset squander while bogus negative cases bring about higher death rates. To foresee sepsis, usage of initial 24–36 h laboratory results with crucial indications of patient was considered. The Medical Information Mart for Intensive Care III (MIMIC3) dataset to test AI procedures including "Random Forest" (RF) and "Logistic Regression" (LR)"was used just as profound learning strategies like long short-term memory (LSTM) and neural systems. The random forest classifier resulted (AUC-ROC) score of 0.696, and LSTM systems did not outflank RF.

Polat et al. [8], discussed about the various treatment strategies for sepsis. They summarized the sepsis patho physiology, current treatment protocols, and new approaches. Studies show that treatment strategies like antiendotoxin treatment, balanced corticosteroid usage, HBO treatment, vasoactive agents such as levosimendan, fibrates, and several antioxidant supplements are potential approaches for sepsis treatment.

Yuan et al. [9], designed an AI algorithm for sepsis diagnosis. They had tried different machine learning algorithms, such as "Logistic Regression", "Support Vector Machine", "XGBoost", and "Neural Network""and adopted XGBoost for their AI algorithm. They achieved accuracy greater than 80% in sepsis diagnosis.

Harrison et al. [10], developed an automated surveillance system (sepsis "sniffer") for detecting severe sepsis. The proposed system also used to monitor failures in identifying and treating severe sepsis on time. Their system correctly identified patients suffering from severe sepsis that doctors failed to recognize and treat.

The uniqueness between current work and related work is that sepsis prognosis is a Web-based forecast application to help patients for early foresee. So far no research work has been carried out with user interface model which could predict sepsis severity and intimate the patients with low cost. This model is developed by using supervised machine learning algorithms where regression model performs classification to predict the occurrence of sepsis. Results from random forest and decision tree are better than LR and Adaboost. But the forecast model is created utilizing random forest as it avoids overfitting [11] and considered more accurate than decision tree. The online application utilizing this model causes patients to foresee their ailment dependent on their laboratory report without waiting for specialists (physicians).

3 Problem Statement

These days, people visit hospitals frequently wasting their money and time on all laboratory tests presuming their life is under threat due to pathogenic disease. For middle class people it leads to unwanted stress. It is important to analyze the sepsis disease and its severity. The fundamental point of this examination is to foresee whether the patient or individual having sepsis or not, along with its stages through Web-based application from their convenient location and anticipate the sepsis utilizing subset of ML administered learning module with accuracy of more than 95%.

4 Methodology

4.1 Data Collection

The data was acquired from clinic frameworks through EMR or electronic medical record frameworks. Information was gathered for 10,000 patients and every patient has their own information documents. The information comprises of a mix of fundamental sign synopses, laboratory values, and patients' socio economics like consciousness, etc. Information assortment is the way toward gathering each insight

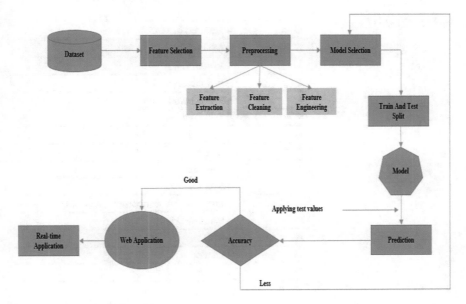

Fig. 1 Global methodology

concerning the sepsis like manifestations and the pace of conceivable counts. Parameters such as "Heart Rate" (HR), "[Systolic and Diastolic] non-invasive Blood Pressures" (BP) were measured using a blood pressure cuff; "Temperature", "Respiration Rate" (RR), and "White Blood Cell" (WBC) count, urine output, glucose level, platelet counts, bilirubin, oxygen saturation, creatinine were captured from patients. The dataset was arranged for the three unique phases of sepsis since it will expand the precision rate just to make impeccable model. Figure 1 shows the outline of the global methodology.

4.2 Preprocessing

Data preprocessing alludes to the changes applied to our dataset before taking care of it to the calculation. Data preprocessing is a technique that is employed to change over the crude information into a spotless informational index. Data is cleaned by removing the missing values and only the most appropriate and meaningful inputs are considered. Feature choice is one of the first and significant advances while playing out any ML task. When dataset is got, not really every section (feature) will affect the yield variable. Additionally, the accompanying strategies are for relapse issue, which implies both the input and yield factors are persistent in nature. Approach availed for feature selection is association of features adding to the nearness of sepsis. The

Table 1 Selected features

Sl. No	Selected features
1	BP-systolic
2	BP-diastolic
3	Respiratory
4	White blood count
5	Temperature
6	Heartrate
7	Urine
8	Glucose
9	Bilirubin
10	International normalized ratio (INR)
11	Platelets (10^4)
12	SPO2
13	Lactate
14	Creatinine

selected features from dataset comprised of 14 clinical parameters as shown in Table 1.

These features are used to develop and evaluate the models. Data splits are done for proper balance between the training, validating, and testing set sizes. Each feature has got its own range for sepsis prediction.

4.3 Model Building

Dataset gathered from different clinical sources needs to be pre-processed and cleaned. Once the preprocessing of the information is done, the dataset is separated as features and labels. The data which consists of about 10,000 patients is split into 3 sets which comprise training, validation, and testing with the ratio of 8:1:1. The model is trained with 80% of training data, tested with 10% test data and validated with remaining 10% of data. The test dataset is used to compute the accuracy, precision, recall, F1 score which lies between the decimal value 0 and 1. The four different ML classification models logical regression, random forest, decision tree, and Adaboost are used to find out which model fits well. "K- Fold Cross Validation" with $K = 5$, is used to validate the model, which means the data were divided into 5 segments or folds after a random shuffle.

In this proposed model, since datasets are arranged for the three unique phases of sepsis, evaluation happens separately for each stage. In this scheme, 1000 data samples were given as input to the algorithm. The objective is to test the forecast performance of the chosen models. Same approach has been applied for all the three phases of sepsis prediction. The ML-based model building is depicted in Fig. 2.

Fig. 2 ML-based model building

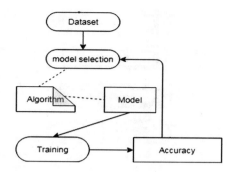

In this examination, RF and decision tree are found better than other ML strategies like logistic regression and Adaboost, due to their negligible hyper-parameterization, rapid preparing, enhanced interpretability, and improved increasingly stable execution. The four models mentioned above are trained with "scikit-learn" that is a "Machine Learning tool kit" (python-based).

4.4 User Interface

Web application is expected to perceive the forecast outcomes. Be that as it may, the ML and Web development are distinctive space. So a pipeline is developed for cooperating UI and ML model with the help of pickle bundle. The pickle bundle will store the ML model in the phase of expectation. After that patient can provide their laboratory test values and can get the yield results. Utilization of SQLite database helps in storing the Login credentials of the registered users.

The Web application is composed of five modules. The first is Register or Login module, in which new user can register by providing the details like Email Id, User name, Phone number, Password and confirm password or an already registered user can provide their Login details. The second is sepsis prediction, in this module, the user can give the essential sources of info like respiratory rate, consciousness, heart beat rate, white blood count check, temperature of the body, C-reactive protein (CRP) so as to foresee whether the individual has got sepsis or not. The third module is severe sepsis prediction, in which the extreme sepsis can be identified. If the sepsis forecast become positive the user can give the fundamental data sources like urine–yield, SpO2 so as to foresee whether the individual having serious sepsis or not. The fourth is septic shock prediction, in this module, the septic shock can be recognized if the severe sepsis expectation becomes positive. The user can anticipate the septic shock by giving the contributions of systolic/diastolic weight, glucose, creatinine, bilirubin, INR, platelets so as to foresee the septic shock forecast. And the final is hospital suggestion module, which will show the proposal of emergency clinics. The illustration of user interface flow is shown in Fig. 3.

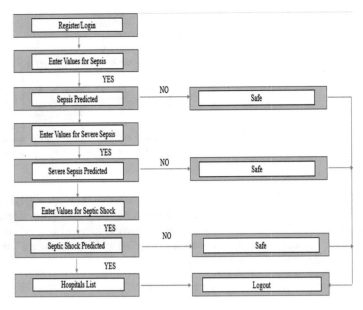

Fig. 3 User interface flow

5 Results and Performance Analysis

In this section, the results of three different stages of sepsis which are trained with four different classification models are discussed.

Model execution is assessed utilizing different execution measurements including precision, accuracy, F1 score along with recall when applied to the test sets for each model. A confusion-matrix is regularly used to portray the exhibition of an arrangement model on a lot of test information for which the genuine qualities are known. The confusion- matrix comprises of four classes. "True Positives" (TP) are the correctly predicted positive values. "True Negatives" (TN) are the correctly predicted negative values. "False Positives" (FP) are wrongly predicted positive values. "False Negatives" (FN) are wrongly predicted negative values [12]. Accuracy, precision, recall and F1 score"are presented as follows,

$$\text{Accuracy} = \text{TP} + \text{TN}/\text{TP} + \text{FP} + \text{FN} + \text{TN}$$
$$\text{Precision} = \text{TP}/\text{TP} + \text{FP}$$
$$\text{Recall} = \text{TP}/\text{TP} + \text{FN}$$
$$\text{F1 score} = 2*(\text{Recall} * \text{Precision})/(\text{Recall} + \text{Precision})$$

As the presentation and exactness investigation is done, it is apparent from the beneath chart that random forest and decisiom tree are the best for the issue. In general, RF models are powerful against uproarious or high dimensional datasets and

Fig. 4 Sepsis model accuracy with LR, RF, decision tree, and Ada-boost with X-axis representing algorithms and Y-axis representing accuracy

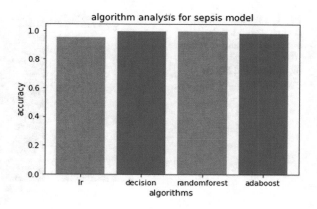

Fig. 5 Severe sepsis model accuracy with LR, RF, decision tree, and Ada-boost with X-axis representing algorithms and Y-axis representing accuracy

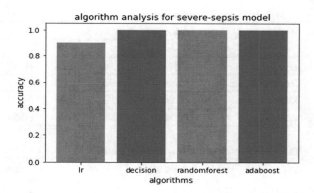

are not inclined to overfitting. Decided along these lines, the last expectation model is fabricated depending on random forest algorithm to additionally continuing with UI model. The accuracy graph of "Sepsis", "Severe Sepsis", and "Septic Shock"""can be seen in Fig. 4, 5, and 6, respectively.

5.1 Performance Analysis

In this segment, the performance of the model is discussed.

Tables 2, 3, and 4 shows the results obtained for the classification models prepared by the gathered dataset with K-fold cross validation. The precision, recall, and F1 score are captured as 0 and 1, where 0 indicates negative case and 1 indicates positive case in each stage. Precision gives the magnitude relation of correctly predicted positive cases to the entire positive cases. Accuracy provides an overall measure of correctly predicted positive cases. Recall is that the magnitude relation of correctly predicted positive cases to the all observations in actual category. Lastly, F1 score

Fig. 6 Sepsis shock model accuracy with LR, RF, decision tree, and Ada-boost with X-axis representing algorithms and Y-axis representing accuracy

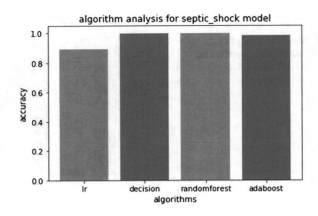

Table 2 Performance of sepsis model

Algorithms	Accuracy	Precision		Recall		F1 score	
		0	1	0	1	0	1
Logistic regression	0.8494	0.86	0.84	0.87	0.83	0.86	0.83
Random forest	0.9996	1.00	1.00	1.00	1.00	1.00	1.00
Decision tree	0.9996	1.00	1.00	1.00	1.00	1.00	1.00
Adboost	0.9812	0.98	0.98	0.99	0.98	0.98	0.98

Table 3 Performance of severe sepsis model

Algorithms	Accuracy	Precision		Recall		F1 score	
		0	1	0	1	0	1
Logistic regression	0.8978	0.92	0.85	0.94	0.81	0.93	0.83
Random forest	1.00	1.00	1.00	1.00	1.00	1.00	1.00
Decision tree	1.00	1.00	1.00	1.00	1.00	1.00	1.00
Adboost	1.00	1.00	1.00	1.00	1.00	1.00	1.00

Table 4 Performance of sepsis shock model

Algorithms	Accuracy	Precision		Recall		F1 score	
		0	1	0	1	0	1
Logistic regression	0.8892	0.9	0.57	0.98	0.22	0.94	0.32
Random forest	0.9997	1.00	1.00	1.00	1.00	1.00	1.00
Decision tree	0.9998	1.00	1.00	1.00	1.00	1.00	1.00
Adboost	0.9844	0.99	0.95	0.99	0.92	0.99	0.93

Table 5 Comparison with work [13] and [14]

	Algorithms used	Algorithm giving highest accuracy	Accuracy (%)
Work [13]	Naive Bayes, decision tree	Naive Bayes	96.64
Work [14]	Random forest, logistic regression. XGBoost	Random Forest	74
Proposed model	Random forest, logistic regression. decision tree, adaboost	Random Forest, Decision Tree	99

is the harmonic average of recall and precision. Therefore, this score takes every false positive and false negative under consideration. Note that for LR, the sample data is normalized before preparing (training), downsizing (scaling) them with a mean of 0 and a fluctuation of 1, just to ensure the model would not be one-sided to the highlights with bigger qualities. RF and decision tree need not bother with the normalization however in light of the fact that both are tree-based model.

It is clear that RF and decision tree performed very much well contrasted with other two learning algorithms. Since the dataset is not linear separable, the direct model LR had a less exactness on the current information, and Adaboost did not work better than RF for this situation either, for both accuracy or recall. The F1 score is weighted by the extent of the information given, so it might be bigger than accuracy and recall. Two tree-based gathering learning models had a decent presentation with F1 score greater than 0.9, with which it is accepted to be conceivable to anticipate the three phases of sepsis.

5.2 Performance Evaluation

Performance of the proposed system has been compared with work [13] and [14] and depicted in Table 5. Summarization is as below:

5.2.1 Comparison of Proposed Work with [13]

Sardesai A U et al., in their work [13], obtained the results using the test dataset 70%. Their study shows "Naïve Bayes" and "Decision Tree" algorithm demostrated accuracy of 96.64% and 94.64%. The proposed model when compared to paper [13], shows that the it outperforms and predicts all the stages of sepsis with great accuracy of 99%.

5.2.2 Comparison of Proposed Work with [14]

Huang B et al., in their work [14], have shown the best performing models as "RF", "Logistic Regression," and "XGBoost". In their research, average accuracy achieved are 0.74, 0.70, and 0.72 given by "RF", "Logistic Regression," and "XGBoost," respectively. The proposed model when compared to work [14] have used 14 parameters to predict sepsis along with 3 stages. The proposed model outperforms with great accuracy of 99% along with Web application feasible for patients. Therefore, the proposed machine learning model can be used to intimate patients about their ailment well in advance along with the stages of sepsis.

6 Conclusion

In this paper, machine learning technique with supervised learning method is used to predict the various stages of sepsis like sepsis, severe sepsis, and septic shock. A separate machine learning model is created for each stages with random forest algorithm which gives 99% accuracy. A Web application is created using three phase prescient model to empower normal person by providing aid to his ailment without depending on specialists.

References

1. https://www.dnaindia.com/
2. Klouwenberg, P., Ong, D., Bonten, M., Marc, Olaf, C.: Classification of sepsis, severe sepsis and septic shock: the impact of minor variations in data capture and definition of SIRS criteria. Intensive Care Med. **38**, 811–819 (2012). https://doi.org/10.1007/s00134-012-2549-5
3. van Wyk, F., Khojandi, A., Kamaleswaran, R.: Improving prediction performance using hierarchical analysis of rea time data: a sepsis case study. IEEE (2019)
4. Wanga, R.Z., Sunb, C.H., Schroederc, P.H, Amekodm, M.K. Predictive models of sepsis in adult ICU patients. IEEE (2018)
5. Wang, X., Wang, Z., Weng, J., Wen, C., Chen, H., Wang, X.: A New effective machine learning framework for sepsis diagnosis. IEEE (2018)
6. Hu, Y., Lee, V.C.S., Tan, K.: Prediction of clinicians' treatment in preterm infants with suspected late-onset sepsis—an ML approach. IEEE (2019)
7. Saqib, M., Sha, Y., Wang, M.D.: Early prediction of sepsis in EMR records using traditional ML techniques and deep learning. IEEE (2018)
8. Polat, G., Ugan, R.A., Cadirci, E., Halici, Z.: Sepsis and septic shock: current treatment strategies and new approaches. Eurasian J. Med. 53–58 (2017), eurasianjmed
9. Yuan, K.C., Tsai, L.W., Lee, K.H., Cheng, Y.W., Hsu, S.C., Lo, Y.S., Chen, R.J.: The development an artificial intelligence algorithm for early sepsis diagnosis in the intensive care unit. Int. J. Med. Inform. **141**, 104176 (2020). https://doi.org/10.1016/j.ijmedinf, ISSN 1386-5056
10. Harrison, A.M., Thongprayoon, C., Kashyap, R., Chute, C.G., Gajic, O., Pickering, B.W., Herasevich, V.: Developing the surveillance algorithm for detection of failure to recognize and treat severe sepsis. Mayo Clin. Proc. **90**(2), 166–175 (2015)
11. Breiman, L.: Random forests. Mach. Learn. **45**(1), 5–32 (2001)
12. Grandini, M., Bagli, E., Visani, G.: Metrics for multi-class classification: an overview. ArXiv,abs/2008.05756 (2020)
13. Sardesai, A.U., Tanak, A.S., Krishnan, S., Striegel, D.A., Schully, K.L., Clark, D.V., Muthukumar, S., Prasad, S.: An approach to rapidly assess sepsis through multi-biomarker host response using machine learning algorithm. Sci. Rep. **11**(1), 1–10 (2021)
14. Huang, B., Wang, R., Masino, A.J., Obstfeld, A.E.: Aiding clinical assessment of neonatal sepsis using hematological analyzer data with machine learning techniques. Int. J. Lab. Hematol. (2021)

Feature Engineering of Remote Sensing Satellite Imagery Using Principal Component Analysis for Efficient Crop Yield Prediction

M. Sarith Divakar, M. Sudheep Elayidom, and R. Rajesh

Abstract Early prediction of crop yield before harvest is essential in taking strategic decisions to ensure food availability. Crop yield prediction using remote sensing satellite imagery is a promising approach due to the abundance of freely available data. Machine learning and deep learning techniques used this data to build forecasting systems for mapping crop yield. The high dimension of remote sensing data made training models infeasible using raw pixels. Most of the techniques relied on feature-engineered remote sensing data, whereas recent approaches mainly focused on histogram-based feature engineering. In this study, we followed the histogram-based approach for dimensionality reduction and further reduced the dimension of the input data using principal component analysis. LSTM is used to map crop yield with remote sensing data without losing the temporal properties of the satellite imagery. Results show that the model that used PCA shows comparable performance to existing approaches with fewer parameters.

Keywords Remote sensing · Crop yield forecast · Deep learning · PCA · LSTM · MODIS

1 Introduction

Crop yield prediction using remote sensing satellite imagery is a better alternative to prediction techniques based on locally collected data, which are expensive and

M. Sarith Divakar (✉)
School of Engineering, Cochin University of Science and Technology (CUSAT), Kochi, India
e-mail: sarith@cusat.ac.in

M. Sudheep Elayidom
Division of Computer Engineering, School of Engineering, Cochin University of Science and Technology (CUSAT), Kochi, India
e-mail: sudheep@cusat.ac.in

R. Rajesh
Naval Physical and Oceanographic Laboratory (NPOL), Kochi, India
e-mail: rajeshr@npol.drdo.in

© The Author(s), under exclusive license to Springer Nature Singapore Pte Ltd. 2022   189
V. Suma et al. (eds.), *Evolutionary Computing and Mobile Sustainable Networks*,
Lecture Notes on Data Engineering and Communications Technologies 116,
https://doi.org/10.1007/978-981-16-9605-3_13

not scalable. Traditional machine learning-based approaches for crop yield fore-casting considered only a few bands and performed feature engineering of the satellite imagery. Normalized Difference Vegetation Index (NDVI) is one of the popular feature-engineered remote sensing data used for yield prediction [1]. NDVI is found effective when used along with climatic indices for crop yield prediction [2]. However, climatic information is not available for all locations and is costly to collect. Recent studies showed the importance of considering all bands and the ability of deep learning models for automatic feature extraction. Data acquired from Terra Satellite's MODIS instrument were commonly used for developing yield forecasting models. MODIS Surface Spectral Reflectance (SR) and land surface temperature (LST) bands were used to map yearly yield data of a crop using machine learning approaches. MODIS surface spectral reflectance and land surface temperature bands of Franklin, Missouri is shown in Fig. 1.

Feeding raw SR and LST images to deep learning models is not feasible due to the high spatial, spectral and temporal resolution of the satellite imagery. Most of the studies were focused on reducing the spatial dimension of images by taking histograms of pixels based on the permutation invariance assumption proposed by You et al. [3]. Histogram-based approaches were also followed by Wang et al. [4] and Sabini et al. [5]. In this study, we followed the histogram-based approach and further reduced the dimension of the input data using principal component analysis (PCA). Input data is then fed to Long Short Term Memory (LSTM) for mapping crop yield.

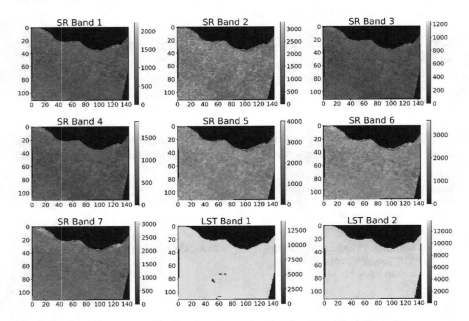

Fig. 1 Surface reflectance and land surface temperature bands of Franklin, Missouri

2 Related Research

Recent studies have used deep learning paradigms such as convolutional neural networks (CNNs), recurrent neural networks (RNNs) for capturing spatial and temporal information respectively for estimating crop yield. You et al. [3] trained their model using remote sensing data collected by the MODIS instrument on seven surface reflectance bands, two land surface temperature bands. Land cover data is used to remove non-crop pixels from the input imagery. MODIS images were collected for a year at 8-days intervals. The scarcity of labeled training data is addressed by employing dimensionality reduction of input images by calculating histograms of pixel values as features. Usage of the histogram for all bands was possible by considering the permutation invariance assumption. According to this assumption, for a satellite image band of a county, each pixel's value contributes to the yield regardless of the location of the county. Histograms were calculated for all nine bands and ground truth data for soybean yield at county level measured in bushels per acre is fed to deep learning architecture, including CNNs and LSTMs. LSTM network learns temporal and spectral features from the histogram tensors whereas spatial features are lost while converting raw images into histograms. To address this issue, the Gaussian process (GP) layer is incorporated on top of the model and achieved better performance than prior works for crop yield forecasting.

Wang et al. [4] used a transfer learning approach to predict crop yield in Argentina and Brazil. Seven surface reflectance bands, two land surface bands, and the cropland mask from MODIS land cover product is used in their study for mapping soybean yield. Pre-processing of input data is performed to keep the time frame consistent for both countries. Histograms for each time frame were calculated for all nine bands representing each input data as $h \times b$ histogram matrix where h is the number of bins in histogram and b is the number of bands. The resulting histogram tensor is flattened by concatenating bands and histogram bins before feeding to a recurrent neural network with LSTM cells. Results showed that the transfer learning approach can improve performance in regions where training data is limited. Sabini et al. [5] used seven surface reflectance bands and two land surface bands collected by the MODIS instrument for predicting corn and soybean yield. Various Deeper networks were explored to improve the performance of the model compared to existing approaches. Results show that the performance of the model can be improved with complex models, highlighting the fact that more signals are there to be extracted from the data. Discriminability of corn and soybeans are also studied and results showed the ability of model architectures to distinguish between pixels important for predicting different crops. Russello [6] used 3D Convolutional Neural Networks (3D CNN) for crop yield prediction. Seven surface reflectance bands, two land surface bands, and the cropland mask from MODIS land cover product were used in the study. Channel dimension of input images is reduced by performing 2D convolutions followed by 1D convolutions. Output from the dimensionality reduction module is fed to 3D Convolutional layers followed by max pool layers. 3D features are flattened into fully connected layers and linear activation is used in the output layer for predicting

Table 1 Crop yield estimation methods in the related research

Authors	Methodology	Review
You et al. [3]	Convolutional neural networks (CNN) with Gaussian process layer	Spatial features are lost while converting into histograms. The issue is addressed using the Gaussian process layer
Wang et al. [4]	Transfer learning with long short term memory (LSTM)	Transfer learning is useful where training data is scarce. Transfer learning for locations with different geography may not perform well
Sabini et al. [5]	Deep convolutional neural networks (CNN)	Deeper models improved performance. More parameters are required to train the model
Russello [6]	3D convolutional neural networks (3DCNN)	Used random cropping with at least 10% cropland coverage and channel dimension is reduced by 2D convolutions. Random cropping may result in the loss of important pixels

crop yield per county for a given year. Model is trained using ADAM optimizer. Results show that 3D CNN performs better than the CNN-LSTM approach as the model considered spatial information along with spectral and temporal information for crop yield forecasting.

Histogram-based approach used in the related research for dimensionality reduction results in losing spatial features of input data and need additional techniques to address the issue. Transfer learning from RGB features to multi-spectral data using pre-trained models is showing promising results recently [7]. However, transfer learning techniques is recommended for locations with similar geographies. Training deeper models with more parameters will extract more information at the cost of increased training time. Deep feature concatenation is explored by Saad et al. [8] for achieving reasonable performance in COVID-19 classification. This approach can be followed for yield prediction from multi-spectral data of different satellites. Random cropping will preserve spatial properties but may result in the loss of important pixels related to the crop. In the proposed system, principal component analysis is used along with histogram-based feature engineering to further reduce the dimension and improve performance with fewer parameters. The summary of crop yield estimation methods in the related research is shown in Table 1.

3 Dataset and Method

Data used in this study was composed of soybean yield data obtained from the National Agricultural Statistics Service (NASS), The United States Department of Agriculture (USDA) [9], and district-level rice yield data obtained from Special Data

Table 2 Yield dataset used in the study

Dataset	Locations	Mean yield (kg/ha)	Year
Soybean	1721	2898.61	2003–2018
Rice	447	2074.61	2003–2018

Dissemination Standard Division (SDDSD) [10]. The summary of the yield dataset used in the study is shown in Table 2.

Image bands of MODIS Surface Spectral Reflectance (SR) and Land Surface Temperature (LST) and Land Cover Type band were collected from 2003 to 2018 through Google Earth Engine [11]. The MOD09A1 V6 product [12] provides seven spectral bands of surface spectral reflectance at 500 m resolution. The MYD11A2 V6 product [13] provides an average 8-day land surface temperature (LST) at 1 km resolution. The MCD12Q1 Version 6 data product [14] provides annual global land cover classification at 500 m resolution. Land Cover Type 1 band of the image is used to determine cropland using the pixel values. In this study, land cover type pixel value of 12 is used for masking during pre-processing of satellite imagery to map only areas corresponding to croplands. The Land Cover Type 1 band for Franklin County is shown in Fig. 2. The summary of the bands used in this study is shown in Table 3.

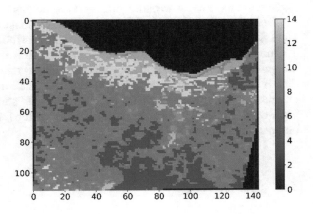

Fig. 2 Land cover Type 1 band for Franklin, Missouri

Table 3 Bands used in the study

Product	Description	Resolution (m)	Composite period
MOD09A1 [12]	Spectral reflectance at the earth surface	500	8-day
MYD11A2 [13]	Average daytime and nighttime surface temperature	1000	8-day
MCD12Q1 [14]	Annual land cover classification	500	1-year

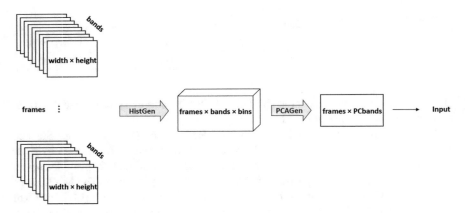

Fig. 3 Pre-processing and feature engineering of image bands for each location

3.1 Data Pre-processing and Feature Engineering

Satellite image files were obtained from 2003 to 2018 through GEE API as a single tiff file for a location. Each image is divided into years and the land cover mask is applied to image bands to obtain pixels corresponding to croplands only. Histogram of pixel values is computed with 64 bins for each image frame corresponding to SR and LST bands. The resulting data is having shape #frames × bands × bins.

The principal component analysis [15] technique is applied and the best principal component is selected to further reduce the dimension of input data. The resulting data is having shape #frames × bands. The flow chart of pre-processing and feature engineering of surface spectral reflectance and land surface temperature bands for each location is shown in Fig. 3. The detailed algorithm for pre-processing and feature engineering is shown in Fig. 4.

3.2 Methodology

In this study, we focused on using Long Short Term Memory (LSTM) networks to predict crop yield using feature-engineered bands of surface spectral reflectance and land surface temperature. Hochreiter and Schmidhuber [16] invented the Long Short Term Memory (LSTM) network to address the issues in vanilla RNN. To test the performance of crop yield forecasting models and understand the importance of features extracted, a comparative study is performed using histogram-based feature-engineered and histogram-PCA-based feature-engineered data. For training histogram-based and histogram-PCA-based feature-engineered data, CNN-LSTM and LSTM networks are used respectively. Models are implemented using Keras [17] with Tensorflow backend [18]. The flow chart of the proposed system using

---

Algorithm 1 Preprocessing and Feature Engineering

```
1: procedure PREPROCESSANDFEATUREENGINEER(SR, LST, LC)
2:     for each location do
3:         read SurfaceReflectance tiff image file as SRloc
4:         read LandSurfaceTemperature tiff image file as LSTloc
5:         read LandCover tiff image file as maskloc
6:         split and save image by year as SRlocyear
7:         split and save image by year as LSTlocyear
8:         split and save image by year as masklocyear
9:         for each year do
10:            SRlocyear ← SRlocyear*masklocyear
11:            LSTlocyear ← LSTlocyear*masklocyear
12:            Bandlocyear ← Merge(SRlocyear ,LSTlocyear) // (frames, bands, width, height)
13:        end for
14:        for each frame in Banddlocyear do
15:            for each band in frame do
16:                Histdata ← Hist(band) // (frames, bands, bins)
17:                PCAdata ← PCA(Histdata) // (frames, PCbands)
18:            end for
19:        end for
20:        save PCAloc.npy
21:    end for
22: end procedure
```

Fig. 4 Algorithm for pre-processing and feature engineering

is histogram-PCA feature-engineered data is shown in Fig. 5. The neural network architectures of LSTM networks utilized in this work are shown in Tables 4 and 5.

For the CNN-LSTM model, input tensor of shape (batch size, frames, bands, bins) is used as input where CNN accepts tensor of shape (bands, bins) and extracts features from the input data and feeds the LSTM network. In the proposed architecture using LSTM, the input tensor of shape (batch size, sequence length, features) is used, where features are the best principal component for each band. Various combinations of batch size, number of layers, and neurons were tried and selected the parameters that resulted in minimum validation loss. Adam optimizer and Mean Squared Error (MSE) loss function is used for training. An initial earning rate of 0.01 with an exponential decrease in learning rate after 10 epochs is used for better convergence.

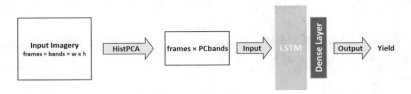

Fig. 5 Flow chart of the proposed system using histogram-PCA feature-engineered data

Table 4 Neural network architecture utilized in this work for soybean dataset

Layer	Input shape	Output shape	Parameters
LSTM	(32, 9)	64	18,944
Batch norm	64	64	256
Dense layer	64	1	65

Table 5 Neural network architecture utilized in this work for rice dataset

Layer	Input shape	Output shape	Parameters
LSTM	(32, 9)	128	70,656
Batch norm	128	128	512
Dense layer	128	1	129

4 Results and Discussions

The correlation of band values with yield shows the importance of using all bands for crop yield prediction. SR2 and SR5 are near-infrared bands and is negatively correlated with the yield for both the dataset. Leaves absorb less infrared light compared to red light and a negative correlation highlight the fact that the expected average yield is high for soybean. Correlation analysis for band values is shown in Table 6. Models are trained with 50 epochs with early stopping and root mean squared error (RMSE) is used as the model performance evaluation metric. Learning curves of LSTM using histogram-PCA-based feature-engineered input for soybean and rice yield data set is shown in Fig. 6.

Batch normalization is used in the dense layer to avoid over-fitting of the training data. Early stopping with the patience of five avoided over-fitting of the model and the best weights were used. LSTM using histogram-PCA-based feature-engineered input achieved better RMSE than CNN-LSTM using histogram-based feature-engineered input. RMSE and associated R^2 values for both models for the soybean dataset are shown in Table 7. Best RMSE and R^2 value is marked as bold. Results show the

Table 6 Correlation of band values with yield for US and India dataset

Band	SR1	SR2	SR3	SR4	SR5	SR6	SR7	LST1	LST2
Soybean	0.35	−0.38	0.38	0.35	−0.39	−0.36	0.12	0.16	−0.05
Rice	0.19	−0.31	0.24	0.25	−0.14	0.02	0.16	0.08	0.15

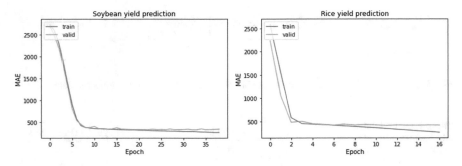

Fig. 6 The learning curve of LSTM for soybean and rice dataset

Table 7 RMSE of soybean yield in kg/ha

Model	RMSE	R^2	Parameters
CNN-LSTM	466.16	0.60	42,361
LSTM	**433.31**	**0.65**	19,265

Best RMSE and R^2 value is marked as bold

Table 8 RMSE of rice yield in kg/ha

Model	RMSE	R^2	Parameters
CNN-LSTM	566.24	0.41	89,593
LSTM	**544.74**	**0.45**	71,297

Best RMSE and R^2 value is marked as bold

comparable performance of the LSTM model using histogram-PCA-based feature-engineered data even with less number of parameters.

Results of the rice dataset also achieved better performance for LSTM using histogram-PCA-based feature-engineered data. RMSE and associated R^2 values for both models for the rice dataset are shown in Table 8. Best RMSE and R^2 value is marked as bold. Figure 7 shows the scatter plot of predicted and observed crop yield for the soybean and rice dataset. The performance of the models can be improved by using a transfer learning approach from pre-trained models. The prediction performance of the rice yield dataset is relatively low compared to the soybean dataset as the number of samples is less. Results can be further improved by feeding more data to the rice yield dataset. Overall performance can be improved by using image fusion techniques [19] to combine images from multiple satellites and by integrating health status of the crop in the learning process [20].

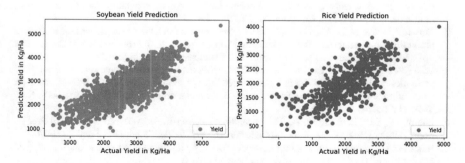

Fig. 7 Predicted and actual yield for soybean and rice dataset

5 Conclusion and Future Works

Existing crop yield forecasting techniques discarded many spectral bands or used climatic indices along with feature-engineered satellite imagery. This paper presented crop yield prediction using all spectral bands of remote sensing satellite imagery. All bands of surface spectral reflectance and land surface temperature used in this study indicate a linear relationship with yield data. Recent approaches focused on histogram-based feature engineering for dimensionality reduction. In the proposed approach, principal component analysis is used along with histogram-based feature engineering to further reduce the dimension of the satellite imagery and achieved comparable performance with existing approaches. LSTM was able to capture temporal dependency in the data and was able to predict with RMSE of 433.31 kg/ha and 544.74 kg/ha for soybean and rice yield dataset respectively. The R^2 value of LSTM using histogram-PCA-based feature-engineered data showed comparable performance with CNN-LSTM using histogram-based feature-engineered data even with less number of parameters. In this study, one of the future directions is to use transfer learning from pre-trained models to improve the predictive performance of the model in the regions where field data is scarce.

References

1. Johnson, D.M.: An assessment of pre- and within-season remotely sensed variables for forecasting corn and soybean yields in the United States. Remote Sens. Environ. **141**(1), 116–128 (2014)
2. Sarith, D.M., Sudheep, E.M., Rajesh, R.: An efficient approach for crop yield forecasting using machine learning techniques based on normalized difference vegetation index and climatic indices. J. Adv. Res. Dyn. Control Syst. **10**(15), 146–154 (2018)
3. You, J., Xiaocheng, L., Melvin, L., David, L., Stefano, E.: Deep Gaussian process for crop yield prediction based on remote sensing data. In: Proceedings of the Thirty-First AAAI Conference on Artificial Intelligence, pp. 4559–4565. San Francisco, California (2017)
4. Wang, A.X., Tran, C., Desai, N., Lobell, D., Ermon, S.: Deep transfer learning for crop yield prediction with remote sensing data. In: Proceedings of the 1st ACM SIGCAS Conference on Computing and Sustainable Societies, pp. 1–5. Association for Computing Machinery, New York, NY (2018)
5. Sabini, M., Rusak, G., Ross, B.: Understanding Satellite-Imagery-Based Crop Yield Predictions. [Online]. http://cs231n.stanford.edu/reports/2017/pdfs/555.pdf. Accessed 2021/7/6
6. Russello, H.: Convolutional neural networks for crop yield prediction using satellite images. Master thesis, IBM Center for Advanced Studies (2018)
7. Yuvraj, S., Robert, R.: Less Is More When Applying Transfer Learning to Multi-Spectral Data. [Online]. http://ceur-ws.org/Vol-2771/AICS2020_paper_50.pdf. Accessed 2021/08/20
8. Saad, W., Shalaby, W.A., Shokair, M., et al.: COVID-19 classification using deep feature concatenation technique. J. Ambient Intell. Humaniz. Comput. (2021)
9. USDA National Agricultural Statistics Service: NASS—Quick Stats. USDA National Agricultural Statistics Service (2017). https://data.nal.usda.gov/dataset/nass-quick-stats. Accessed 2021/7/6

10. Special Data Dissemination Standard Division: Directorate of Economics & Statistics, Ministry of Agriculture and Farmers Welfare, Government of India, New Delhi. [Online]. https://aps. dac.gov.in/APY/Public_Report1.aspx. Accessed 2021/7/6
11. Gorelick, N., Hancher, M., Dixon, M., Ilyushchenko, S., Thau, D., Moore, R.: Google earth engine: planetary-scale geospatial analysis for everyone. Remote Sens. Environ. **202**(1), 18–27 (2017)
12. Vermote, E.: MOD09A1 MODIS/Terra Surface Reflectance 8-Day L3 Global 500m SIN Grid V006 [Data Set]. NASA EOSDIS Land Processes DAAC (2015). [Online]. https://doi.org/10. 5067/MODIS/MOD09A1.006. Accessed 2021/7/6
13. Wan, Z., Hook, S., Hulley, G.: MYD11A1 MODIS/Aqua Land Surface Temperature/Emissivity Daily L3 Global 1km SIN Grid V006 [Data Set]. NASA EOSDIS Land Processes DAAC (2015). [Online] https://doi.org/10.5067/MODIS/MYD11A1.006. Accessed 2021/7/6
14. Friedl, M., Sulla-Menashe, D.: MCD12Q1 MODIS/Terra+Aqua Land Cover Type Yearly L3 Global 500m SIN Grid V006 [Data Set]. NASA EOSDIS Land Processes DAAC (2019). https:// doi.org/10.5067/MODIS/MCD12Q1.006. Accessed 2021/7/6
15. Jolliffe, I.T.: Principal Component Analysis. Springer, New York (2002)
16. Hochreiter, S., Schmidhuber, J.: Long short-term memory. Neural Comput. **9**(8), 1735–1780 (1997)
17. Chollet, F., et al.: Keras. [Online]. https://keras.io. Accessed 2021/7/6
18. Martín, A., et al.: TensorFlow: large-scale machine learning on heterogeneous systems. Software available from tensorflow.org. Accessed 2021/7/6
19. Dutta, S., Ayan, B.: Highly precise modified blue whale method framed by blending bat and local search algorithm for the optimality of image fusion algorithm. J. Soft Comput. Paradigm (JSCP) **2**(4), 195–208 (2020)
20. Dhaya, R.: Flawless identification of *Fusarium oxysporum* in tomato plant leaves by machine learning algorithm. J. Innov. Image Process. (JIIP) **2**(4), 194–201 (2020)

Packet Filtering Mechanism to Defend Against DDoS Attack in Blockchain Network

N. Sundareswaran and S. Sasirekha

Abstract With the tremendous increase in the Blockchain network scale, and Cryptocurrency network, the Distributed Denial of Service (DDoS) attacks create serious threat for the network operations. The application of Blockchain needs rigorous performance, access requirements, high throughput, and low transmission delay. The proposed method mitigates the DDoS attacks and promised to provide all Service requirements of the Blockchain network. This work mainly puts focus on packet filtering mechanism in real-time to ensure valid users are able to access the service. The flooding of unknown packets are prevented at the source by the novel improved packet marking technique to enable the reasonable service access time to the end user. This work is simulated using NS3 simulator and Ethereum Blockchain platform. It is shown that the traffic is reducing drastically by at least 10% from the actual packets received at the source. It is also proved that the incoming traffic is 90% from the actual traffic flow of packets to the Blockchain network after dropping the unknown flooding of packets by the packet marking technique principle.

Keywords DDoS flooding · Blockchain · Packet filtering · Packet marking · Throughput · Ethereum

1 Introduction

Blockchain Technology has a vast range of applications that include Finance, Enterprise Management, E-business, Energy, Education, Reputation systems, etc. Nakamoto et al. introduced Blockchain (BC) technology and applied it in the Bitcoin cryptocurrency network [1]. BC has many features as inbuilt operations such as

N. Sundareswaran (✉) · S. Sasirekha
Sri Sivasubramaniya Nadar College of Engineering, Chennai, India
e-mail: sundareswarann@ssn.edu.in

S. Sasirekha
e-mail: sasirekhas@ssn.edu.in

© The Author(s), under exclusive license to Springer Nature Singapore Pte Ltd. 2022   201
V. Suma et al. (eds.), *Evolutionary Computing and Mobile Sustainable Networks*,
Lecture Notes on Data Engineering and Communications Technologies 116,
https://doi.org/10.1007/978-981-16-9605-3_14

distributed ledger, secured transactions ensure reliability, immutability, and traceability, Cryptographic functions such as SHA-256, Digital signatures. BC Network is a decentralized peer-to-peer network. A BC node can execute applications, perform routing and storing operations. There is no central server or node to control the entire network operations. All participant nodes in the BC network are equal and share resources between them. BC technology evolution is categorized as Blockchain 1.0, Blockchain 2.0, Blockchain 3.0. Blockchain 1.0 covers Financial applications such as Bitcoin, Digital payment network, cash transactions, and so on. Blockchain 2.0 covers Banking, Stock exchange network, Insurance corporation, Loan network, and so on. Blockchain 3.0 covers the government networks, Hospital and Health networks, Science and Technology networks, and so on. The services offered by these applications are prone to cyber-attacks such as DDoS, Intrusion, Phishing, Trojan injection, and so on. DDoS is more expensive than other cyber-attacks in the application [2, 3]. DDoS is an attempt to disrupt the BC network from accessing its resources. DDoS could stop the entire services of BC applications for any instance. It could damage the QoS (Quality of Service), technical and cost structure of the service provider. The DDoS attack has been thrown using botnets which may spread worms to the BC network. Botnet might automatically send a huge amount of requests to the BC networks which makes system fail to function its services. DDoS attacker spreads worms to the BC network through the automatic Bots to disrupt the services offered by BC network as shown in Fig. 1 [4–7]. The worms are spread into the Blockchain network along with the legitimate packets. These worms should be dropped before entering the network. Banking BC network has encountered DDoS attacks in the form of malicious packets which damaged the original packets of the application. Network flooding attack in Bitcoin network encountered major loss due to exhausted Bandwidth in the network [8, 9]. The DDoS attack vector is growing exponentially from XGBps flooding attack to XTBps flooding attack [10, 11]. The major service providers like Google, Amazon, and Facebook are also suffering due to DDoS attacks in their network and data center. In April 2017, US Cryptocurrency network experienced a major DDoS flooding attack that preserved the network service off for about 12 h. The GitHub open source platform has been worse affected by this attack which reported around 2 TB packets are sent by an attacker at a particular time in the network lasted for several hours in that day in 2018. In 2020 World-wide consortium reported around $400 billion loss due to this cyber-attack alone. BC Technology of the Digital cryptocurrency market is making transactions worth $200 billion in 2020 which has been trusted by customers around the world. BC technology uses Smart Contract (SC) for cryptographic functions to establish trust over the market. BC contains a chain of Blocks of information or transaction in the application-specific network. These Blocks of information are protected by Hash code and Merkle Tree structure of BC technology but the network is prone to be attacked by DDoS attack in order to affect its intended service. Thus it should be taken into consideration to protect any BC network application from the DDoS attack.

Fig. 1 General structure of
DDoS attack

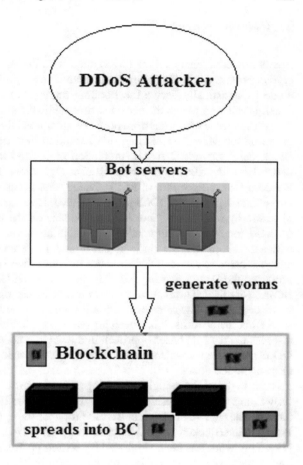

1.1 Motivation

It is important and necessary to have an efficient solution for exponentially increasing DDoS attacks to the real-time networks particularly Blockchain, IoT, and Cloud networks. Blockchain networks are one of the major emerging Trillian dollar cryptocurrency-based business networks and are also applied as a solution for various cyber attacks. DDoS attacks targeting Blockchain network resources and denying access at the particular instance might lead to a billion-dollar loss in the cryptocurrency network. Therefore we aim to analyze the incoming traffic, delay, packet flow, dropping malicious packets, and enable the BC network to efficiently serve for its intended purpose to the application. The main objective of this work is to provide an efficient solution to prevent DDoS attacks without affecting the performance and QoS requirements of the Blockchain network.

2   Related Works

Kochkarov et al. analyzed the DDoS attacks on the payment network of the bitcoin cryptocurrency [12]. This work has presented a graph-based DDoS detection technique to statistically derive the flooding requests to the BC network. The graph density is changing over the period of time while the number of transactions exceeds the average value which has been proved mathematically. Manikumar and Maheswari proposed the Machine Learning (ML) solution to mitigate the DDoS attack in the Blockchain network [13]. ML prediction model has been executed on top of the network to identify the malicious packet signatures. It has been proved that the Random Forest algorithm of ML provided 95% accuracy on the prediction model. Giri et al. proposed the DDoS mitigation technique based on the analysis of the size of traffic by establishing threshold traffic size on the top of the BC network [14]. SC has been executed automatically to calculate the size, comparing the actual size with a threshold value of traffic in real-time. It has been proved that the size of traffic analysis helps to reduce DDoS attacks by more than 90% in real-time. Baek et al. proposed the Deep Learning (DL) technique to detect DDoS attacks on the Bitcoin BC network [15]. The evaluation has been performed on the basis of defined metrics in the algorithm. The same evaluation is applied on both network-level and service-level Bitcoin networks. It has been proved that the analysis metrics of DL have more impact than the ML-based approach in the Bitcoin network. Yeh et al. proposed the DDoS defense mechanism on BC technology is based on finding the attacker's IP to mitigate the attack [16]. The Smart Contract (SC) is executed on top of the BC network to detect the malicious IP using BC nodes in the network. This work successfully found the list of suspicious IPs and shared those IPs to the BC hosts to boycott the requests from them. Shafi and Basit proposed the operations to prevent the DDoS Botnet in the Blockchain-enabled Software Defined IoT (Internet of Things) network [17]. This work has been simulated by setting up Botnets and its flooding packets into the network. These packets are prevented by signature comparisons on the top of the BC-Software Defined Network. Saad et al. explored the detailed survey of DDoS attack surface on BC network. This work analyzed the consensus delay from a set of BC nodes and finding the relationship between them to justify the attacks within the network [18]. Essaid et al. proposed the solution-based DDoS mitigation on Ethereum BC and RNN-LSTM (Recurrent Neural Network—Long Short Term Memory) architecture [19]. This work focuses on an autonomous system to analyze the traffic and let the analysis results fed into the DL model to exactly deny the overloaded traffics into the BC network. Sivaganesan proposed the blockchain solution for mitigation of internal cyberattacks of IoT networks [20]. The grey and black hole attacks are taken for experiment study and achieved time efficiency and improved network lifetime through blockchain approach. Moreover, the message overhead problem is reduced and the overall transaction throughput of the IoT network also significantly improved. Smys and Haoxiang proposed the Blockchain-based Cyber Threat Intelligent (CTI) system to ensure a secure computing system of an organization [21]. It also discusses the standard secure information sharing methods such as

TAXII (Trusted Automated eXchange of Intelligent Information) and STIX (Structured Threat Information eXpression). In this work, blockchain-based CTI ensures integrity and credibility of data sharing and also claimed that the work saves data space by 20% of network resources.

3 Proposed DDoS Architecture

To mitigate the incoming traffic of the BC network, the traditional traffic statistical rate control mechanism and traffic delay-based control mechanism are mostly used in the internet, cloud network, and BC network, or any large volume of the network is considered. The drawback from these traditional methods caused delay in the network, also caused to consume huge bandwidth during the traffic flow in the network. It can damage the entire service architecture of the BC network. The proposed DDoS filtering architecture identifies the traffic and helps to improve the bandwidth utilization in the BC network. As shown in Fig. 2a the basic flow of the architecture is detecting, dropping, and controlling the malicious packets inside the network. Malicious packets should be dropped because they can send valuable information of the BC network to the remote attacker and it can also damage or delete the files and storage base of the BC network. The architecture protects the BC network from traffic congestion, intrusion, malicious packet injection inside the BC network.

Usually, the network traffic can be predicted during a particular time (busy/normal) in the source BC network. Thus the rate of change of incoming packet traffic is an important consideration to predict the DoS flooding attack. It can be monitored from the Firewall itself. The fundamental comparison between the default traffic model and the actual traffic model at a particular time can help to stop the attack from the source network itself. The parameters such as normal, deviation, source, destination, payload, signatures of the incoming packets are analyzed as shown in Table 1 and mapped with the information base of the source BC network to mitigate the traffic congestion and malicious traffics inside the network. While the traffic deviations are identified, the source of an attack can be traced back using the BC source router. The intruders or Bots in the path can be filtered and traffics from the IP addresses of the Bots can be halted. Hence BC source router can only allow valid traffic packets inside the BC network.

Large packet arrival time, response time, and packet flow are traces of DDoS attacks. The threshold time has been set to find the difference between valid packet arrival time, response time, and packet flow. Table 2 has been shown as the model for analysis of BC network controller. If any BC node or Routers in the network violate the threshold signatures then ICMP (Internet Control Message Protocol) requests would be sent from the BC server to mark the device as suspect as shown in Fig. 2b and force to send the state of its queue. While the state of the queue is high from the threshold limit thereafter all communication would be dropped from the BC controller. Figure 2c depicts the actual working model of the BC controller.

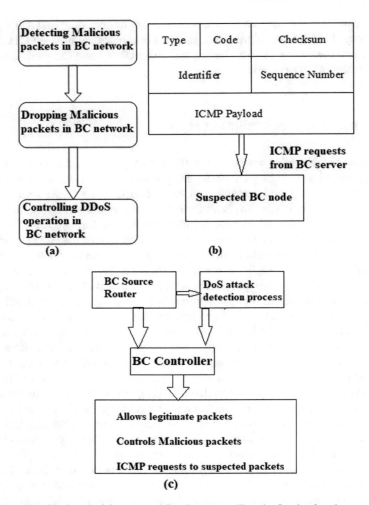

Fig. 2 Proposed filtering model, suspected flooding, controlling the floods of packets

Table 1 BC Source router analysis

Normal	Deviation	Source	Destination	Payload	Signature
X	X	XXXX	XXXX	X	X

Table 2 BC network controller analysis

Actual packet arrival time	Threshold packet arrival time	Actual response time	Threshold response time	Actual rate of flow	Threshold rate of flow
XX	XX	XX	XX	XX	XX

The sending packet flow rate from the source router is minimized by the BC controller by properly advocating the order of flow. The arrival packet flow rate can be dropped while a DoS attack found in the destination. The popular dynamic load balancing routing protocol is being applied by the BC server or controller while DoS packet flow is reaching the threshold limit (T_L). All the BC nodes, BC routers (source, intermediate, destination) communicate information about the traffic such as Marked packet details, Large flow details, threshold limit details, Large flow details at the specific routers, and suspected queue rate details of BC nodes in the network. This communication is also being sent to the BC controller. BC server takes necessary action to control and drop the incoming traffic to ease the services in the network. The actual working model of the proposed dropping scheme has been shown in Fig. 3.

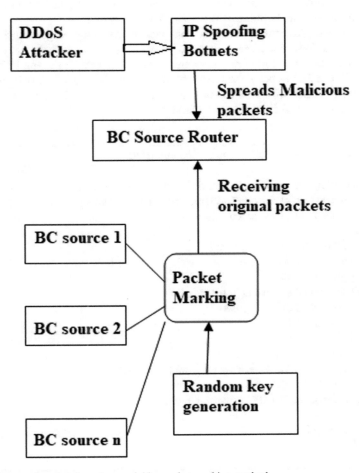

Fig. 3 Proposed packet dropping model by packet marking method

BC Source Router receives the traffic which might contain both original packets and malicious packets. It can separate original packets from malicious one by looking up the packet marking field of the packet. The actual packet marking is the process of writing a random key into the designated field of the packet. The BC Source nodes $(1, 2, \ldots n)$ are sending their packets to the BC Source Router through packet marking operation of the BC controller. Botnets might send the flooding of malicious packets through spoofed IP addresses which can be separated and dropped by the BC Source Router. BC Source Router could avoid congestion at the destination network through this proposed technique. It can validate the packet by the packet marking field to decide whether to allow it inside the network or to drop it at the top of the network. When Source Router fails to validate the malicious packets all the nodes, the destination routers, intermediate routers along the path could participate for validation to drop them inside the network at any point.

The following Algorithm 1 accepts valid packets from incoming network packets flow and drops flooded malicious packets. It calls Dynamic Load Balancing Routing Protocol (DLBR) function while incoming packet flow exceeds beyond Threshold Level (T_L). The T_L has been fixed from the study of the history of packet flow inside the network. DLBR works on the principle of building a load balanced tree from routing packet flow without delay and failures.

Algorithm 1: Proposed DDoS mitigation algorithm

Input: Incoming Network packets: valid packets and malicious packets
Output: Accepts valid packets and Drops flooded malicious packets
//Incoming packet traffics and packet flow
START
1: **set** p=default packet parameters
2: receive(a=actual parameters of packets)
3: **if** map(p,a) > T_d **then**
4:     call function trace_back( ), filter( ), halt( )
5: **else if** packet_flow > T_L **then**
6:     call DLBR( )
7:   **end if**
// Internal malicious node detection
8: receive(r=response_time(internal_node)) at every 't'
9:   **if** r > T_r **then**
10:     call functions ICMP( ), suspect( ), drop( )
11: **else**
12:     goto step 1
13: **end if**
END

The default packet parameters such as payload, signatures are set assigned as in step 1. The actual parameters of packets are dynamically received as in step 2. These parameters are mapped and compared with threshold deviation (T_d) then the source of those packets are traced back, filtered, and halted its execution by invoking functions as in step 3–4. Moreover, the incoming rate of the packet flow is also compared against the default threshold level (T_L) as in step 5. The Dynamic Load Balancing Routing (DLBR) protocol function is being called when the rate of packet flow exceeds the maximum threshold level (T_L) as in step 6. More importantly, the internal malicious node responses are analyzed against the threshold response (T_r) time, if the received actual response time exceeds the threshold then ICMP message request is sent to those suspected nodes, and all its communication is dropped by invoking functions as in step 8–10. The whole process gets repeated forever as part of the network functions which has been illustrated in steps 11–13. Therefore the proposed system design is intended to achieve the high throughput of valid packets flow and dropping malicious packets by implementing Algorithm 1. The implementation of the proposed system is explained in the next section.

4 Implementation

The proposed system consists of Blockchain nodes, routers, and servers. They are connected by the peer-to-peer network principle. Ethereum Blockchain has been deployed to use all BC functions and network operations [22]. Smart Contract Application Binary Interface (SC-ABI) in a solidity programming language has been used to execute DDoS filtering operations as contracts on the top of the BC network. Ethereum Virtual Machine (EVM) accepts only secure binary codes of the contract to compile and execute the DDoS defense mechanism in the network. All Etherum BC nodes, Routers, and Servers are configured with Ethereum BC accounts to identify each one in the network and the communication among them is restricted by the link specifications in the peer-to-peer network. NS3 is used to simulate communication operations such as setting up bandwidth, packets, links to the network, baud rate, and other specifications. In the experiments, The topology is designed by setting the 100 nodes, 3 Routers, nearly 250 links to connect as a graph between routers and nodes were configured with 20 Mbps bandwidth, 100 ms delay, and 2 GB capacity queue. Each router source, intermediate, and destination validate the secret marking key with the interval 100 ms and forwarded traffic to the network using both marked and unmarked packets At every 100 ms, the source routers communicated the incoming packet traffic information corresponding to intermediate and destination routers. In an experiment execution of 1000 ms, during the 300 ms no traffic found. Then for the next 500 ms huge traffic found, At the last 200 ms the packet marking technique has fetched the valid packets and separated the packets that induced the traffic in the BC network. The experiment continued to execute the detecting attacks. For the first case, the source router dropped all unmarked packets for 300 ms. During the first phase, our proposed model detected the unmarked traffic packets successfully. During the second phase, malicious packets are dropped when the queue was nearly 85% full. Our proposed design was able to detect and drop unmarked traffic

Table 3 Results of DDoS traffic and dropped packets

Time (ms)	Traffic (MB)	Dropped (MB)	Survey (MB)
100	0.1	0.08	0.02
200	0.2	0.15	0.07
300	0.4	0.36	0.20
400	0.6	0.54	0.32
500	0.8	0.75	0.42
600	1.0	0.90	0.63
700	1.2	1.12	0.75
800	1.4	1.3	0.82
900	1.6	1.4	0.92
1000	1.8	1.62	1.08

packets in a particular time duration. As shown in Table 3, the incoming packets or traffic has been sent to the BC network for the varying time interval. The sizes (MB) of traffic are increased while time (millisec) increases. The results of the proposed experiment are compared with the survey results of dropped packets [13, 14, 18] in Table 3 from which it can be inferred that the proposed method has dropped more unknown malicious traffic packets for varying sizes of network traffic. It has been shown in Fig. 4 that DDoS traffic applied for a different time interval. The proposed DDoS filtering mechanism dropped the unmarked packets or malicious packets for the varying traffic and time as shown in Table 3.

After dropping unmarked packets, the result has been drawn as a graph in Fig. 4. It is clearly understood that the traffic is reducing drastically by at least 10% from the actual packets received at the source. The time (ms) taken is also depicted with

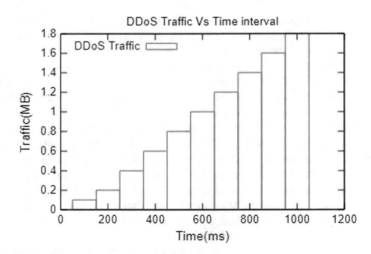

Fig. 4 DDoS result graph of traffic flow of packets

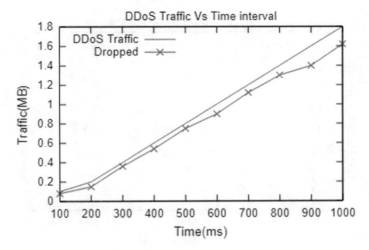

Fig. 5 DDoS result graph of traffic and dropping packets

respect to the flow of packet traffic (MB). It has taken 1000 ms to allow the traffic size of 1.8 MB packets into the Blockchain network.

The comparison between actual traffic and dropped traffic after removing unknown packets is shown in Fig. 5 which proves that the incoming traffic is 90% to the BC network from the actual traffic flow of packets and 10% has been restricted as flooded unknown packets to flow into the Blockchain network. The proposed packet dropping mechanism works along with the flow of packets at each time interval as shown in Fig. 5. Therefore there is no separate time spent to analyze the packet flows and finding the unknown packets in this scenario in which the controlling mechanism saves the time taken for dropping flooded packets.

Our proposed method was able to detect and drop unmarked traffic packets in the particular duration of time where the incoming packets or traffic has been sent to the BC network for the varying time interval. The sizes (MB) of traffic are increased drastically when time (millisec) increases. The results of the proposed experiment are compared with the existing results of dropped packets [13, 14, 18]. It can be inferred from Fig. 6 that the proposed method has dropped more unknown malicious traffic packets for varying sizes of network traffic when compared to the existing one. DDoS attack vector has improved and applied to any kind of network as templates. As shown in Fig. 6, the dropping speed and the size has been illustrated along with the existing results. The experiment can further be carried out with the help of a predefined malicious unknown packets dataset with signatures to improve the results.

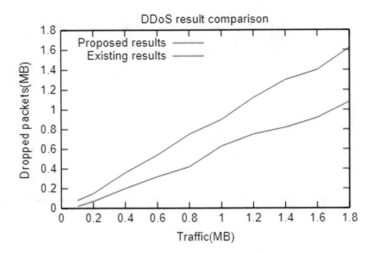

Fig. 6 DDoS result comparison graph of proposed and existing results

5 Conclusion

The services offered by Blockchain Technology are prone to DDoS cyber-attacks. As stated in the introduction DDoS could stop the entire services of BC applications. It could damage the QoS (Quality of Service), technical and cost structure of the service provider. The proposed DDoS filtering architecture identifies the traffic and helps to improve the bandwidth utilization in the BC network. The proposed system uses the Ethereum Blockchain platform to use all BC functions and network operations. NS3 is used to simulate communication operations such as setting up bandwidth, packets, links to the network, baud rate, and other specifications. The DDoS traffic is applied for a different time intervals to the BC network. The proposed DDoS filtering mechanism dropped the unmarked packets or malicious packets for varying traffic and time. It has been shown that the traffic is reducing drastically by at least 10% from the actual packets received at the source. The dropped packets are unknown packets that might be malicious to the BC network. The comparison between actual traffic and dropped traffic after removing unknown packets proves that the incoming traffic is 90% to the BC network from the actual traffic flow of packets.

6 Limitation and Future Work

Time complexity and scalability are two major limitations of the proposed system. Controlling the flooded packets flow from malicious nodes inside the BC network is time-consuming because the response time of all internal nodes is periodically collected at every time interval 't'. It leads to performance bottlenecks in the real-time BC applications. The BC network and its cryptocurrency coin infrastructure are growing tremendously due to the vast investment of people all over the world. DDoS attacks in this kind of scaling infrastructure might lead to billion-dollar losses in a short duration of time. Therefore in the Future, Hybrid architecture including edge computing, High-performance computing, and cyber-threat base infrastructure must be established to overcome the scalability issues in the BC network. Moreover, Deep learning-based analysis might help to reduce the performance bottleneck due to dynamic response analysis of internal nodes by self-learning techniques. It can be further strengthened by automatic packet signature update in the dataset, packet optimization techniques, multi-agent collaboration system using a restricted consensus mechanism.

References

1. Nakamoto, S., et al.: Bitcoin: a peer-to-peer electronic cash system. https://www.bitcoin.org (2008)
2. Ni, T., Gu, X., Wang, H., Li, Y.: Real-time detection of application-layer DDoS attack using time series analysis. Control Sci. Eng. 4 (2013)
3. Bruno, R., et al.: A blockchain-based architecture for collaborative DDoS mitigation with smart contracts. In: IFIP International Conference on Autonomous Infrastructure Management and Security, pp. 16–29 (2017)
4. Choi, H., Lee, H., Kim, H.: BotGAD: detecting botnets by capturing group activities in network traffic. In: Proceedings of the Fourth International ICST Conference on COMmunication System softWAre and middlewaRE (2009)
5. Ali, S.T., et al.: ZombieCoin: powering next-generation botnets with bitcoin. In: International Conference on Financial Cryptography and Data Security (2015)
6. Karim, A., et al.: Botnet detection techniques: review future trends and issues. J. Zhejiang Univ. SCI. C **15**(11), 943–983 (2014)
7. Wang, P., Sparks, S., Zou, C.C.: An advanced hybrid peer-to-peer botnet. IEEE Trans. Dependable Secure Comput. **7**(2), 113–127 (2010)
8. Feder, A., et al.: The impact of DDoS and other security shocks on bitcoin currency exchanges: evidence from Mt. Gox. J. Cybersecur. **3**(2), 137–144 (2018)
9. Vasek, M., Thornton, M., Moore, T.: Empirical analysis of denial-of-service attacks in the bitcoin ecosystem. In: International Conference on Financial Cryptography and Data Security, pp. 57–71 (2014)
10. Mahjabin, T., Xiao, Y., Sun, G., Jiang, W.: A survey on distributed denial-of-service attack prevention and mitigation techniques. Int. J. Distrib. Sens. Netw. **13**(12) (2017)
11. Rashidi, B., Fungand, C., Bertino, E.: A collaborative DDoS defence framework using network function virtualization. IEEE Trans. Inf. Forensics Secur. **12**(10), 2483–2497 (2017)
12. KOCHKAROV, A.A., OSIPOVICH, S.D., KOCHKAROV, R.A.: Analysis of DDoS attacks on bitcoin cryptocurrency payment system. Revista Espacios **41**(3), 29–35 (2020)
13. Manikumar, D.V.V.S., Maheswari, B.U.: Blockchain based DDoS mitigation using machine learning techniques. In: 2020 Second International Conference on Inventive Research in Computing Applications (ICIRCA), pp. 794–800 (2020). https://doi.org/10.1109/ICIRCA 48905.2020.9183092.

14. Giri, N., Jaisinghani, R., Kriplani, R., Ramrakhyani, T., Bhatia, V.: Distributed denial of service (DDoS) mitigation in software defined network using blockchain. In: 2019 Third International Conference on I-SMAC (IoT in Social, Mobile, Analytics and Cloud) (I-SMAC), pp. 673–678 (2019). https://doi.org/10.1109/I-SMAC47947.2019.9032690

15. Baek, U., Ji, S., Park, J.T., Lee, M., Park, J., Kim, M.: DDoS attack detection on bitcoin ecosystem using deep-learning. In: 2019 20th Asia-Pacific Network Operations and Management Symposium (APNOMS), pp. 1–4 (2019). https://doi.org/10.23919/APNOMS.2019.889 2837

16. Yeh, L., Huang, J., Yen, T., Hu, J.: A collaborative DDoS defense platform based on blockchain technology. In: 2019 Twelfth International Conference on Ubi-Media Computing (Ubi-Media), pp. 1–6 (2019). https://doi.org/10.1109/Ubi-Media.2019.00010

17. Shafi, Q., Basit, A.: DDoS botnet prevention using blockchain in software defined internet of things. In: 2019 16th International Bhurban Conference on Applied Sciences and Technology (IBCAST), pp. 624–628 (2019). https://doi.org/10.1109/IBCAST.2019.8667147

18. Saad, M., et al.: Exploring the attack surface of blockchain: a comprehensive survey. IEEE Commun. Surv. Tutorials **22**(3), 1977–2008 (2020). https://doi.org/10.1109/COMST.2020.297 5999 (2020)

19. Essaid, M., Kim, D., Maeng, S.H., Park, S., Ju, H.T.: A collaborative DDoS mitigation solution based on Ethereum smart contract and RNN-LSTM. In: 2019 20th Asia-Pacific Network Operations and Management Symposium (APNOMS), pp. 1–6 (2019). https://doi.org/10.23919/APNOMS.2019.8892947

20. Sivaganesan, D.: A data driven trust mechanism based on blockchain in IoT sensor networks for detection and mitigation of attacks. J. Trends Comput. Sci. Smart Technol. (TCSST) **3**(1), 59–69 (2021)

21. Smys, S., Haoxiang, W.: Data elimination on repetition using a blockchain based cyber threat intelligence. IRO J. Sustain. Wireless Syst. **2**(4), 149–154 (2021)

22. Hasanova, H., et al.: A survey on blockchain cybersecurity vulnerabilities and possible countermeasures. Int. J. Netw. Manage. **29**(2), e2060 (2019)

Data Mining for Solving Medical Diagnostics Problems

L. A. Lyutikova

Abstract The article solves the problem of creating a software package for computer diagnostics of gastritis. The patient examination indicators and their diagnoses are used as input data. To successfully solve this problem, a logical approach to data analysis is being developed, which allows us to find the patterns necessary for high-quality diagnostics. These laws are identified based on the data provided by specialists and include the results of patient examinations and the existing experience in medical practice in making a diagnosis. Systems of multivalued predicate logic are used for expressive data representation. An algorithm is proposed that implements and simplifies the approaches under consideration. As a result, the developed software package selects the most suitable types of the disease with a predetermined accuracy based on the data of patient diagnostics. If it is not possible to make a diagnosis with a given accuracy based on the results of the examination, then either the accuracy of the decision changes, or it is proposed to undergo an additional examination.

Keywords Diagnostics · Knowledge base · Algorithm · Clauses · Axioms

1 Introduction

Medical diagnostics is a fairly well-known task. There are various methods of its solution, which depend on the type of system and its purpose.

These can be systems based on statistical and other mathematical models. They are based on mathematical algorithms that search for a partial correspondence between the symptoms of the observed patient and the symptoms of previously observed patients whose diagnoses are known [1–3].

There may be systems based on the knowledge of experts. In them, algorithms operate with knowledge about diseases presented in a form that is close to the ideas of doctors and described by medical experts.

L. A. Lyutikova (✉)
Institute of Applied Mathematics and Automation, KBSC RAS (IAMA KBSC RAS), Nalchik, Russia

These can be machine learning-based systems that need a fairly large amount of data. It is in this case that the algorithm is able to learn for independent work. In recent years, deep learning algorithms have been widely used for the diagnosis and prognosis of the development of the disease [4].

Methods of remote diagnostics are also being developed, including remote sensing using convolutional neural networks [5].

The purpose of this work is to develop a method of data analysis and create an adequate software package for the diagnosis of gastritis on its basis.

The proposed method is based on logical data analysis. And the construction of a complex discrete function whose variables are appropriately presented symptoms and diagnoses. This makes it possible, even with a small amount of data, to find patterns, build classes based on the identified commonality of features and select the most important properties for making a decision.

2 Formulation of the Problem

A group of doctors proposed the problem considered in this work. To solve it, the data of patients who were diagnosed with gastritis according to gastroenterological examinations were provided. There are 28 symptoms (signs of a disease) in total, each of which has 2–4 answers. The signs by which diseases are diagnosed are the result of established clinical practice and include various examinations.

The number of diagnosed types of gastritis is 17.

132 peoples were examined and diagnosed. Based on these data, it is necessary to build an algorithm for adequate diagnosis of the remaining patients.

Figure 1 shows a sample questionnaire.

We have a function $Y = f(X)$ from 28 variables, which is defined at 132 points, the domain of definition of each variable has a scatter of 2 to 4 options. It is necessary to restore the value of the function at other requested points.

The formulation of this problem is reduced to the formulation of a problem based on precedents.

Let be $X = \{x_1, x_2, ..., x_n\}$ $x_i \in \{0, 1, ..., k_i - 1\}$, where $k_r \in [2, ..., N]$, $N \in Z$—Set of symptoms, diagnosed diseases. $Y = \{y_1, y_2, ..., y_m\}$—many diagnoses, each diagnosis is characterized by a corresponding set of symptoms $x_1(y_i), ..., x_n(y_i) : y_i = f(x_1(y_i), ..., x_n(y_i))...$ Or $X = \{x_1, x_2, ..., x_n\}$, where $x_i \in \{0, 1, ..., k_r - 1\}$, $k_r \in [2, ..., N]$, $N \in Z$—processed input data $X_i = \{x_1(y_i), x_2(y_i), ..., x_n(y_i)\}$, $i = 1, ..., n$, $y_i \in Y$, $Y = \{y_1, y_2, ..., y_m\}$—output:

$$\begin{pmatrix} x_1(y_1) & x_2(y_1) & ... & x_n(y_1) \\ x_1(y_2) & x_2(y_2) & ... & x_n(y_2) \\ ... & ... & ... & ... \\ x_1(y_m) & x_2(y_m) & ... & x_n(y_m) \end{pmatrix} \rightarrow \begin{pmatrix} y_1 \\ y_2 \\ ... \\ y_m \end{pmatrix}$$

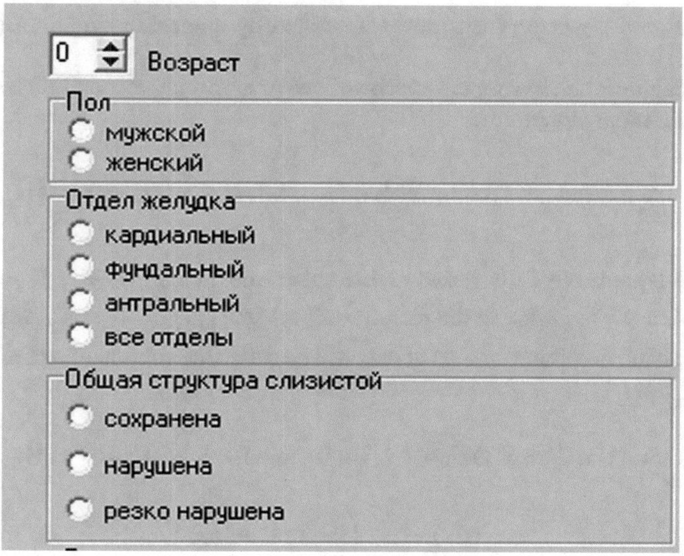

Fig. 1 A fragment of the questionnaire

It is necessary to find general rules that generate a given pattern, exclude variables that do not carry information, and divide the set of diagnoses into classes [6].

Maybe in such a small area of knowledge as the diagnosis of a type of gastritis, a good specialist, based on his experience, will give a more objective and complete idea of the possibilities of diagnosis than those that will be obtained as a result of the proposed method. But a general approach based on logical data analysis is important, which allows us to formally find the most important rules, the totality of which is able to completely restore the original information [7, 8].

3 Solution Methods

The data that we have to deal with when solving diagnostic problems, as we know, are not complete, not accurate, not unambiguous. However, the solutions obtained must correspond to the patterns explicitly and implicitly present in the data under consideration.

Logical methods can analyze the source data quite well, identify essential and non-essential features, and identify the minimum set of rules necessary to fully restore the original patterns. As a result, you can get a more compact and reliable representation of the source information, which is more reliable and faster to process.

We will say that the constructed system of rules is complete if it provides the output of all solutions in the domain under consideration.

We will call a group of diagnoses identified by a specific symptom (group of symptoms) a class.

Each diagnosis can be a representative of one or more classes, each class is defined by a set of similar symptoms:

$$\underset{j=1}{\overset{m}{\&}}\, x_j(y_i,) \rightarrow P(y_i), i = 1, ..., l; \quad x_j(y_i) \in \{0, 1, ..., k-1\},$$

where is the predicate $P(y_i)$ takes on the value true, i.e., $P(y_i) = 1$ if $y = y_i$ and $P(y_i) = 0$, if $y \neq y_i$... Or in the form: $\underset{i=1}{\overset{n}{\vee}}\, \overline{x}(y_j) \vee P(y_j), j \in [1, ..., m]$.

The decision functions for a set of data will be called the conjunction of all decision rules:

$$\underset{j=1}{\overset{m}{\&}}\, x_j(y_i,) \rightarrow P(y_i), i = 1, ..., l; \quad x_j(y_i) \in \{0, 1, ..., k-1\},$$

$$\text{or } f(X) = \underset{j=1}{\overset{m}{\&}} \left(\underset{i=1}{\overset{n}{\vee}}\, \overline{x}_i \vee P(y_j) \right).$$

The same diagnosis can be characterized by slightly different symptoms, and the constructed function will make it possible to exclude insignificant symptoms will divide the data into classes. As a result, we get a Boolean function of $m + n$ variables (symptoms and diagnoses), which will be equal to one on each set, except for those sets where all the symptoms are present, but the diagnosis corresponding to these symptoms is denied. We can say that this function allows any rules, except for the negation of those that exist.

Example: Let the following relations be given:
See Table 1.

Diagnosis a is determined by the absence of the first and second symptoms ($x_1 = 0$, $x_2 = 0$), diagnosis in the absence of the first symptom, the presence of the second ($x_1 = 0$, $x_2 = 1$).

$$\&\& = f(x, y) = (\overline{x_1 x_2} \rightarrow y(a)) \& (\overline{x_1} x_2 \rightarrow y(b)) x_1 \vee x_2 y(b) \vee y(a) y(b)$$

Let's build a table defining a function for variables: $x_1, x_2, y(a), y(b)$.
The SKNF, based on Table 2, will look like this.

$$f(X, Y) = (x_1 \vee x_2 \vee y(a) \vee y(b)) \& (x_1 \vee x_2 \vee y(a) \vee \overline{y(b)})$$

Table 1 Relations of objects and their attributes

x_1	x_2	Y
0	0	a
0	1	b

Table 2 Tabular representation of the classifier function

x_1	x_2	$y(a)$	$y(b)$	$f(X, Y)$
0	0	0	0	0
0	0	0	1	0
0	0	1	0	1
0	0	1	1	1
0	1	0	0	0
0	1	0	1	1
0	1	1	0	0
0	1	1	1	1
1	0	0	0	1
1	0	0	1	1
1	0	1	0	1
1	0	1	1	1
1	1	0	0	1
1	1	0	1	1
1	1	1	0	1
1	1	1	1	1

$$\& (\overline{x_1} \vee x_2 \vee y(a) \vee y(b)) \& (\overline{x_1} \vee x_2 \vee \overline{y(a)} \vee y(b))$$
$$= x_2 \vee y(a) \& \overline{x_1} \vee y(b) \& x_1$$

This function can be easily modified. Each new rule, by means of a conjunction operation, is included in the system of already existing ones, with their possible some modification [7–9].

It can also be represented in the following recursive form:

$$W(X) = Z_k(q_k w_k X);$$

$$Z_k(q_k w_k X_k) = Z_{k-1} \& \left(\overset{n}{\underset{i=1}{\vee}} \overline{x_k(w_i)} \vee w_k \right)$$

$$\vee q_{k-1} \& \left(\overset{n}{\underset{i=1}{\vee}} \overline{x_k(w_i)} \vee w_k \right);$$

$$q_k = q_{k-1} \& \left(\overset{n}{\underset{i=1}{\vee}} \overline{x_k(w_i)} \right);$$

$$q_1 = \overset{n}{\underset{i=1}{\vee}} \overline{x_1(w_i)}; \quad j = 2...m; \quad Z_1 = w_1.$$

where $W(X)$ is the modeled function, Z_j is the characteristic of objects at the current moment, Q_j is the state of the system at the current moment [10].

The state of the system is a combination of those values of variables that were not used to make a diagnosis.

If the function is written in the SDNF and successfully reduced, it can compactly represent the data. Moreover, they will contain our diagnoses, there will be classes in which diagnoses are combined according to similar symptoms, and there will be combinations of symptoms that are not characteristic of the diagnoses under consideration.

In general, in cases of big data, this approach may look somewhat cumbersome, so an algorithm for implementing this method is proposed further.

4   Knowledge System Modeling Algorithms

The algorithm for selecting the rules from which the entire volume of the considered data can be obtained can be as follows:

- the number of columns in the table $\sum_{i=1}^{n} k_i$, this is the number of questions for each symptom and the number of possible answers in our case from one to four.

The number of lines will correspond to the number of diagnoses, in our case it is 17 plus the number of classes that will be found [11].

Observing the order, we write down the data for each item of all patients in the table as follows:

We take each diagnosis made and spread it across the corresponding columns, the diagnosis y_1 will be placed in the column of each item according to the results of the examination of that patient. For example, gender will have two columns with values 0 and 1, and the diagnosis will be placed in the column based on the patient's gender. The general view of shown Table 3.

In the course of filling in the table, we check the column in which the diagnosis of the patient in question falls. If there are already other diagnoses in the column, then we cross them out and enter them into the class with the diagnosis being considered, enter them in the next row in the same column. These diagnoses are grouped into a class for a given diagnostic item. This is demonstrated in Table 3.

Further, we sequentially consider the rows, if in the row corresponding to any diagnosis there are not crossed out diagnoses left in the squares, then we select

Table 3 Distribution of diagnoses by corresponding symptoms

0_1	$1_1 \ldots$	k_1-1	0_2	$1_2 \ldots$	$k_2-1 \ldots$	$\ldots 0_n$	1_n	k_n-1
	y_1		y_1					y_{n-1}
	y_2				y_2		y_2	
$\ldots$	$\ldots$	$\ldots$	$\ldots$	$\ldots$	$\ldots$	$\ldots$	$\ldots$	$\ldots$
y_m				y_m				y_m

the column corresponding to this diagnosis and consider this a unique sign of this particular diagnosis.

We also consider the classes formed as a result of data analysis [12–15]. Thus, the algorithm allows one to construct those clauses that contain diagnoses, those by which is the recognition (Table 4).

Figure 2 shows a diagram for the main idea of the algorithm.

As noted above, the algorithms for diagnosing gastritis use the values of k-valued (and in the general case, predicates with different values) predicates. The set of input parameters and admissible solutions for each diagnosis can be represented as the following system of productive rules:

Symptom set 1 → Solution 1,
Set of symptoms 2 → Solution 2,
...

Table 4 Identification of classes according to the corresponding symptoms

0_1	$1_1 ...$	k_1-1	0_2	$1_2...$	k_2-1	$... 0_n$	1_n	k_n-1
	$\cancel{y_1}-$		y_1					$\cancel{y_1}-$
	$\cancel{y_2}-$				y_2		y_2	
	y_1y_2							
y_m				y_m				$\cancel{y_m}$
...	...	...	...	...	...	...	...	...
								y_1y_m

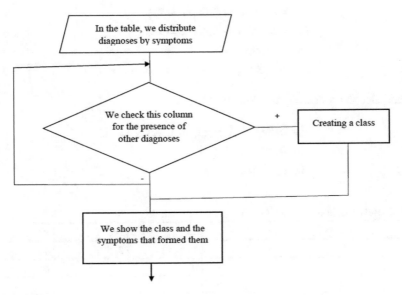

Fig. 2 Algorithm

Set of symptoms $m \rightarrow$ Solution m.

Note that in the general case, the same solution can follow from different sets of inputs.

Using the constructed logical functions, such a system of productive rules is transformed into an optimal logical statement with removal of redundant information and allocation of all possible subclasses of equivalent solutions, providing the detection of hidden regularities [10].

This system of productive rules (implicative statements) is the basis for the construction of logical neural networks trained to work with the corresponding knowledge base [15].

Analysis of the presence of free knowledge in the resulting answer allows us to implement a procedure signaling the need for additional training: replenishing the original system of productive rules with new admissible solutions. In this way, a procedure can be implemented to improve the adaptive properties of gastritis diagnosis.

Example: $x_1 \in [0, 1]$; $x_2 \in [0, 1, 2]$; $x_3 \in [0, 1, 2]$; $x_4 \in [0, 1]$; $x_5 \in [0, 1, 2]$,

$$W = \{a, b, c, d\}$$

The relationship between objects and the characteristics that characterize them is set in Table 5.

It is obvious from Table 6 that ye have the following set of axioms:

$$(w = a)\&(w = d)\&(x_1 = 0);$$
$$(w = b)\&(x_2 = 2)\&(x_5 = 1);$$
$$(w = c)\&(x_3 = 2);$$
$$(w = d)\&(x_2 = 1)\&(x_3 = 0)\&(x_5 = 0)$$

And also the following division into classes

$$\{a, c\}\&(x = 1_1)\&(x_2 = 0)\&(x_5 = 2);$$
$$\{a, b\}\&(x_3 = 1);$$
$$\{b, c, d\}\&(x_4 = 0).$$

Table 5 Relations of objects and their attributes for example

x_1	x_2	x_3	x_4	x_5	Y
1	0	1	1	2	a
0	2	1	0	1	b
1	0	2	0	2	c
0	1	0	0	0	d

Table 6 The result of the algorithm for this example

0_1	1_1	0_2	1_2	2_2	0_3	1_3	2_3	0_4	1_4	0_5	1_5	2_5
	~~a~~	~~a~~				~~a~~			a			~~a~~
~~b~~			b			~~B~~		~~b~~			b	
						ab						
	c	c					c	~~c~~				~~c~~
	ac	ac										ac
								~~bc~~				
~~d~~			d		d			~~d~~		d		
bd												
								bcd				

Thus, the algorithm allows you to build those clauses that contain diagnoses, which are used for recognition.

5 Program Description

The program that implements the above algorithm consists of two executable modules:

Module 1. Performs database decoding using a dictionary, loading symptoms and diagnoses in question-and-answer form and analyzing the results.

Module 2. Knowledge base generation program. Creates information on the basis of the source file with data or to clarify the given knowledge system. Reduces the size of the database in accordance with the approximate value, then it is worth either reducing the accuracy of the algorithm and adds information to check the correctness of the stored data.

After filling in all the fields shown in Fig. 1, we can obtain a diagnostic result with a given accuracy.

Figure 3 if a predetermined accuracy is not possible.

6 Conclusion

The result of the study was a software package for the diagnosis of gastritis, based on the logical analysis of data. The proposed analysis allows us to find hidden patterns in the data, divide the studied data into classes, and find unique properties of each diagnosis. Unlike neural network methods, this method is logically understandable and does not require retraining. It can identify the most significant patterns and

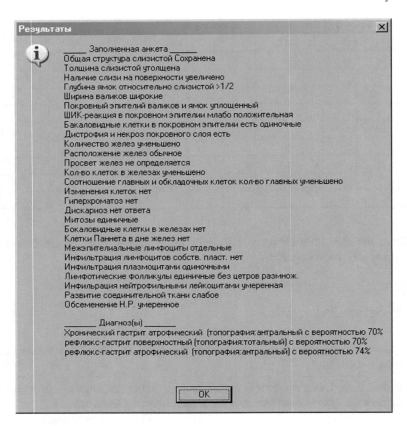

Fig. 3 The result of the program

shorten the process of finding a solution. It can be argued that logical algorithms can be used for intelligent data analysis, they consider the source data as a certain set of general rules among which it is possible to identify those rules that are enough to get all the source data. These rules will be generative for the area under consideration; they will help to better understand the nature of the objects under consideration and minimize the search for the right solutions.

References

1. Zhuravljov, Ju.I.: Ob algebraicheskom podhode k resheniju zadach raspoznavanija ili klassifikacii. Problemy kibernetiki **33**, 5–68 (1978)
2. Shibzukhov, Z.M.: Correct aggregation operations with algorithms. Pattern Recogn. Image Anal. **24**(3), 377–382 (2014)
3. Naimi, A.I., Balzer, L.B.: Stacked generalization: an introduction to super learning. Eur. J. Epidemiol. **33**, 459–464 (2018)

4. Haoxiang, W., Smys, S.: Big data analysis and perturbation using data mining algorithm. J. Soft Comput. Paradigm (JSCP) **3**(01), 19–28 (2021)
5. Joe, C.V., Raj, J.S.: Location-based orientation context dependent recommender system for users. J. Trends Comput. Sci. Smart Technol. (TCSST) **3**(01), 14–23 (2021)
6. Calvo, T., Beliakov, G.: Aggregation functions based on penalties. Fuzzy Sets Syst. **161**(10), 1420–1436 (2010). https://doi.org/10.1016/j.fss.2009.05.012
7. Mesiar, R., Komornikova, M., Kolesarova, A., Calvo, T.: Fuzzy aggregation functions: a revision. In: Sets and Their Extensions: Representation, Aggregation and Models. Springer-Verlag, Berlin (2008)
8. Yang, F., Yang, Z., Cohen, W.W.: Differentiable learning of logical rules for knowledge base reasoning. Adv. Neural Inf. Process. Syst. **2017**, 2320–2329 (2017)
9. Flach, P.: Machine Learning: The Art and Science of Algorithms that Make Sense of Data. P. 396, Cambridge University Press (2012). ISBN: 978-1107096394
10. Duda, R., Hart, P.E.: Pattern Classification and Scene Analysis. J. Wiley & Sons—NJ (1973)
11. Akhlaqur, R., Sumaira, T.: Ensemble classifiers and their applications: a review. Int. J. Comput. Trends Technol. **10**(1), 31–35 (2014)
12. Dyukova, Ye.V., Zhuravlev, Yu.I., Prokof'yev, P.A.: Metody povysheniya effektivnosti logicheskikh korrektorov. Mashinnoye obucheniye i analiz dannykh **1**(11), 1555–1583 (2015)
13. Lyutikova, L.A., Shmatova, E.V.: Constructing logical operations to identify patterns in data. E3S Web Conf. **224**, 01009 (2020)
14. Burges, C.J.C.: A tutorial on support vector machines for pattern recognition. Data Min. Knowl. Discov. **2**, 121–167 (1998)
15. Lyutikova, L.A.: Use of logic with a variable valency under knowledge bases modeling. LA, Lyutikova, CSR (2006)

Deep Neural Networks-Based Recognition of Betel Plant Diseases by Leaf Image Classification

Rashidul Hasan Hridoy, Md. Tarek Habib, Md. Sadekur Rahman, and Mohammad Shorif Uddin

Abstract Diseases of the betel plant are a hindrance to healthy production which causes severe economic losses and is a major threat to the growing betel leaf industry. This paper introduces an efficient recognition approach based on deep neural networks to diagnose betel plant diseases rapidly to ensure the quality safety and healthy development of the betel leaf industry. A dataset of 10,662 betel leaf images is used for the generalization of deep learning models via the transfer learning technique. EfficientNet B5 has acquired 99.62% training, 98.84% recognition accuracy with 80 epochs. AlexNet, VGG16, ResNet50, Inception V3, and EfficientNet B0 have achieved 81.09%, 83.51%, 86.62%, 94.08%, and 97.29% test accuracy, respectively. During the training phase, AlexNet has taken less time compared to others and misclassified 195 images where EfficientNet B5 misclassified 12 images of the test set. The experimental results validate that the introduced architecture can accurately identify betel plant diseases.

Keywords Betel leaf · Betel plant disease · Leaf image classification · Deep neural networks · Convolutional neural network · Transfer learning

1 Introduction

Betel leaf (scientific name: *Piper betle*) is widely used as a mouth freshener in Southeast and Southern Asian countries which contains a significant number of nutrients and also is used as a home remedy for several types of diseases from

R. H. Hridoy (✉) · Md. Tarek Habib · Md. Sadekur Rahman
Department of Computer Science and Engineering, Daffodil International University, Dhaka, Bangladesh
e-mail: rashidul15-8596@diu.edu.bd

Md. Sadekur Rahman
e-mail: sadekur.cse@daffodilvarsity.edu.bd

M. S. Uddin
Department of Computer Science and Engineering, Jahangirnagar University, Dhaka, Bangladesh

© The Author(s), under exclusive license to Springer Nature Singapore Pte Ltd. 2022
V. Suma et al. (eds.), *Evolutionary Computing and Mobile Sustainable Networks*,
Lecture Notes on Data Engineering and Communications Technologies 116,
https://doi.org/10.1007/978-981-16-9605-3_16

ancient times. A remarkable number of farmers of Bangladesh, India, Malaysia, Indonesia, Philippines, Cambodia, Thailand, Vietnam, and Nepal are engaged with betel plant cultivation, which has a tremendous contribution to the agro-economy of Bangladesh [1]. Betel leaf has significant health benefits and is widely used in Ayurvedic medicines which is a great source of calcium and contains vitamins A and C, niacin, riboflavin, thiamine, iodine, etc. For many years, leaf of the betel plant is used as an anodyne which provides immediate remission from ache, removes radicals from the body as it is a powerhouse of antioxidants, helps the stomach by restoring normal pH levels of the body, and helps to improve digestion. Betel leaf reduces overall glucose level which is very essential for type 2 diabetes patients, lowers cholesterol levels to prevent stroke, and anti-cancer compounds fight against oral and colon cancers.

Leaf rot and foot rot are the most destructive fungal diseases of the betel plants which hindered the healthy production of betel leaf. Under high humidity conditions, leaf rot spreads rapidly within a short time through the air. Initially, in mature leaves near the soil, water-soaked spots are seen. The spots increase rapidly, and the elaborated spots are round-shape, brown, and necrotic with the clear gray-brown zone. In the rainy season, this disease affects most, and the syndromes can grow over any portion of the betel leaf along with margins and tips. However, foot rot destroys the roots of the betel plant, and within a week, 80–90% portion of plants become wilt and die, and in patches of betel plant, its symptoms are found most. Initially, dark-brown spots appear in the leaf; under high humid conditions, these spots become wet and rot, and sometimes with light brown zonations, necrotic spots are also found in leaves [2].

Convolutional neural networks (CNNs) have made significant breakthroughs in recent years in computer vision. Therefore, in agricultural information technology, CNN has become a research hotspot that is widely used to recognize various plant diseases, and still for pattern recognition tasks, CNN is deemed to be one of the optimal algorithms. In this paper, an efficient and rapid recognition approach is introduced using pre-trained deep neural networks to classify betel plant diseases accurately, and a dataset of 10,662 images of betel leaf is generated. To classify the diseases of the betel plant, pre-trained CNN models such as AlexNet, VGG16, ResNet50, Inception V3, EfficientNet B0, and B5 architecture were used via the transfer learning technique. EfficientNet B5 showed significantly better recognition ability compared to other deep neural networks and obtained 98.84% recognition accuracy, while another CNN model of the EfficientNet group has achieved 97.29% test accuracy.

The structure of this paper is as follows: The related works are present in Sect. 2. Section 3 presents the betel leaf dataset and describes CNN models. We describe experimental studies in Sect. 4. The results acquired and their interpretations are demonstrated in Sect. 5. Finally, the conclusion and plan of the future work are given in Sect. 6.

2 Related Works

A remarkable number of researchers have made significant attempts to identify various plant diseases to reduce the damage of diseases and enhance the quality of production. With significant continuous improvement in computer vision, machine learning and deep learning algorithms are extensively utilized for identifying plant diseases. To classify three diseases of betel leaf, Tamilsankar and Gunasekar used the minimum distance classifier, and the median filter was used in their research to alleviate noise and for enhancement of image quality [3]. For extracting the features of color, the CIELAB space model was used, watershed segmentation was used to remove background from images, and histogram of oriented gradients (HOG) technique was applied to obtain the value of gradient feature of betel leaf images. Jayanthi and Lalitha used HOG for extracting features, and the multiclass support vector machine (SVM) was used to classify diseases of betel leaf which achieved 95.85% accuracy [4]. For image preprocessing, median filter was used, and L*a*b color space model and watershed transformation algorithm were employed for image segmentation. To segment images of betel leaf, Dey et al. used an Otsu thresholding-based image processing algorithm [5]. Hue, saturation, value (HSV) color space was used to minimize the noise of images. To recognize affected or rotted leaves from healthy leaves, the color feature of images was used. Vijayakumar and Arumugam conducted two experiments to recognize powdery mildew disease of betel leaf, and red, green, blue (RGB) color elements were differentiated from two types of leaves using RGB encoding technique [6]. In the first experiment, mean values were computed for the front and back views of every element, and values of the median were computed in another experiment. For the segmentation of betel leaf images, Ramesh and Nirmala employed K-means clustering to classify betel leaf diseases, and color co-occurrence texture analysis was developed by implementing HSV in feature classification [7]. Sladojevic et al. used CaffeNet to recognize plant diseases that achieved 96.3% accuracy; image augmentation was shown a great influence on accuracy, but fine-tuning was not shown remarkable changes [8]. Using the OpenCV framework, images were resized for the purpose of their research, and for separate classes, the proposed approach achieved precision between 91 and 98%. Sabrol and Kumar used the gray-level co-occurrence matrix (GLCM) for computing features and proposed an adaptive neuro-fuzzy inference system (ANFIS)-based recognition model for detecting plant leaf disease that achieved overall 90.7% classification accuracy for tomato and for eggplant achieved overall 98% accuracy [9]. Four gray-level spatial dependence matrix properties have been submitted to ANFIS. For extracting features, Yadav et al. used AlexNet, and particle swarm optimization (PSO) was used to select features and optimization [10]. After analyzing the recognition performance of four classification algorithms, SVM was selected as a final classifier that achieved 97.39% accuracy. Khan et al. used VGG19 and VGGM for feature selection, then based on local standard deviation, local interquartile range method, and local entropy, the most prominent features were selected [11]. To check the performance of their proposed method, five affected leaves were tested and achieved 98.08% in 10.52 s. To classify three diseases of maize leaf, Priyadharshini et al. used modified LeNet architecture that achieved 97.89% accuracy [12]. Basavaiah and Arlene Anthony used the DT and RF classifier to classify four diseases of tomato leaf and achieved 90% and 94% accuracy, respectively [13]. To improve accuracy, the fusion of multiple features was used, and for training and testing purposes, the color histogram was extracted for color features, Hue moments were extracted shape features, and haralick was extracted texture features. Mokhlesur Rahman and Tarek Habib addressed an approach for classifying textile defects using a hybrid neural network and obtained 100% accuracy [14]. Tarek Habib et al. introduced a recognition approach for jackfruit diseases and acquired 89.59% accuracy using the RF classifier [15]. Jacob et al. addressed an artificial bee colony optimization algorithm with help of the breadth-first search for improving routing in wireless networks [16]. Chen et al. introduced a study for detecting coronary artery disease using SVM [17]. For fog-enabled IoT architecture, Mugunthan addressed an interference recognition approach based on DT [18].

Several computer vision-based approaches based on machine learning and deep learning algorithms were carried out for different leaf disease recognition. Though a few research work was conducted to classify diseases of betel plant, acquiring the expected efficient identification ability of the recognition approaches still remains a challenging task. Hence, in this paper, an efficient recognition approach using pre-trained CNNs is proposed for betel plant diseases.

3 Research Methodology

This study was conducted for developing a rapid and efficient recognition approach for betel plant diseases through leaf image classification. First, images of betel leaf were collected, then all images were resized into four different dimensions according to the input size of pre-trained CNN models. Afterward, CNNs have been trained via the transfer learning technique using images of the train and validation set of the dataset. Lastly, the performance of CNN models was evaluated with the test set, and based on the classification performance, a CNN architecture was proposed for betel plant disease recognition. The methodology diagram of this study is presented in Fig. 1.

3.1 Dataset

This study was conducted with a betel leaf dataset which contains four classes as leaf rot, foot rot, healthy, and miscellaneous, and 10,662 images of this dataset have been divided into 8600 training, 1031 validation, and 1031 test images randomly. Both training and validation images have been used to train and fit CNNs in this study, and test images were used to evaluate the recognition ability of state-of-the-art CNNs. Four representative images of the betel leaf dataset belonging to each class are shown in Fig. 2, and details on the used dataset of betel leaf images are presented in Table 1.

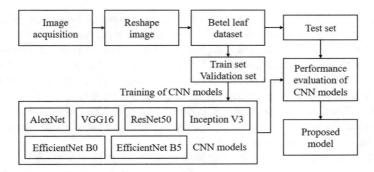

Fig. 1 Methodology for betel plant disease recognition

Fig. 2 Representative images of betel leaf dataset

Table 1 Summary of betel leaf dataset

Class	Training images	Validation images	Testing images
Leaf rot	2299	244	244
Foot rot	1680	210	210
Healthy	3529	441	441
Miscellaneous	1092	136	136

3.2 State-Of-The-Art CNN-Based Models

AlexNet, VGG16, ResNet50, Inception V3, EfficientNet B0, and EfficientNet B5 were used in this study to recognize diseases of the betel plants with the transfer learning approach.

The AlexNet architecture was introduced in the ImageNet Large Scale Visual Recognition Challenge (ILSVRC) 2012 that follows LeNet-5 architecture [19]. It is basically an eight-layer CNN architecture, the input size of this architecture is 227 × 227 pixels, and the number of total parameters is approximately (approx.) 61 million (M). The architecture of AlexNet is built with five convolution layers followed by three fully connected layers (FC), and rectified linear unit (ReLU) is used in convolution and FC layers and finally the softmax layer.

In 2014, VGG16 won the ILSVRC, which is basically a 16-layer architecture, the number of total parameters is approx. 138 M, and the input image size of VGG16 is 224 × 224 pixels. This architecture contains five convolutional blocks and three FC layers, and ReLU is used as an activation function in these layers and the last with a Softmax activation function. Every convolution layer of VGG16 uses 3 × 3 filters, and the stride is fixed to 1. Each convolutional block contains one maximum polling layer that uses 2 × 2 filters, and stride is 2 [19].

In 2015, the architecture of ResNet50 won the ILSVRC 2015 which is built with stacked residual units, and the main building blocks of this architecture are residual units, consisting of convolution and pooling layers [20]. RsNet50 is a variant of ResNet architecture that was built with 48 convolutions along with 1 maxpool and 1 average pool layer and has over 23 M trainable parameters. The input image size of ResNet50 is 224 × 224 pixels.

Inception V3 architecture contains a total of 42 layers which is developed by Google and takes 299 × 299 pixels images in the input layer. In Inception V3, the idea of factorization was introduced to decrease parameters numbers and connections except decreasing the performance of the architecture [19]. It is built with symmetrical and asymmetrical building blocks which contain convolutions, average pooling, maximum polling, dropouts, and FC layers.

In the ImageNet classification problem, the EfficientNet model achieved 84.4% accuracy, containing 66 million parameters. The EfficientNet group consists of eight CNN models, and these are from B0 to B7. These architectures use Swish instead of ReLU which is a new activation function, and these models can achieve high accuracy with smaller models [21]. Using the transfer learning technique, EfficientNet B0 and B5 were used in this study. The last FC layer of these models was connected with four neurons to the softmax to output class probabilities of the used dataset. The schematic representation of EfficientNet B0 architecture is given in Fig. 3.

4 Experiment

With GPU support, state-of-the-art CNN models were trained, and several studies for examining the efficiency of models were conducted using Google Colaboratory that provides 12.72 GB RAM and 107.77 GB disk support. All codes were realized using an open-source framework of deep learning of Python 3, named Keras 2.6.0.

Images of the betel leaf dataset were divided into training, validation, and test sets randomly. For training and fitting deep learning models, 8600 images of the training set and 1031 images of the validation set were used. On the other hand, 1031 images of the test set that models did not see before were used for examining the recognition ability of CNN models.

To decrease the time required during the training phase of the CNN models, pre-trained weights of AlexNet, VGG16, ResNet50, Inception V3, EfficientNet B0 and

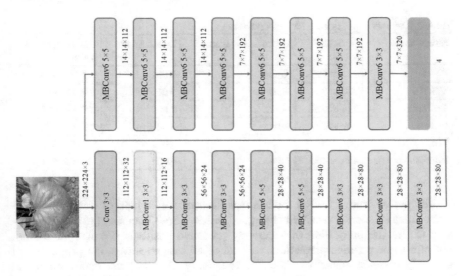

Fig. 3 Schematic representation of EfficientNet B0 architecture

B5 have been used via the transfer learning technique for the purpose of the study [21]. To recognize all classes of the betel leaf dataset, pre-trained CNN models have been fine-tuned that increases the speed of learning also. The last FC layer of CNNs with 1000 neurons was changed with four neurons, and all layers of CNNs have been set as trainable in accordance with our problem. Softmax was selected as the activation function of the last FC layer of CNNs, and categorical cross-entropy was selected as the loss function in this study. For avoiding the overfitting issue, the early stopping method has been utilized during the training phase of CNNs in this study. AlexNet, ResNet50, Inception V3, EfficientNet B0, and EfficientNet B5 were used with the Adam optimization method, and the VGG16 model is used with stochastic gradient descent (SGD) optimization method. For SGD, 0.01 learning rate was used, while for Adam, 0.001 was chosen, and for all models, the validation step was set to one. VGG16, ResNet50, and EfficientNet B0 used the same input size of the image, while AlexNet used 227 × 227 pixels images and EfficientNet B5 used the largest input size. By dividing each pixel value of images by 255, the images of the betel leaf dataset were normalized first in this study. During backpropagation, the mini-batch size 16 was applied for updating weights and bias, which is used to balance the rate of network convergence and accurate prediction during the learning process of CNNs. The input size, optimization method, learning rate, and parameters number which were used in the training phase are summarized in Table 2.

The multiclass classification was performed in this study as the betel leaf dataset contains four classes such as leaf rot, foot rot, healthy, and miscellaneous. The performance metrics given among Eqs. 1–8 were computed true positive (TP), true negative (TN), false positive (FP), and false negative (FN) by considering values in the confusion matrix acquired from multiclass classifications.

Table 2 Parameter values used for CNN models

Model name	Input size	Optimization method	Learning rate	Number of total parameters
AlexNet	227×227	Adam	0.001	25,724,476
VGG16	224×224	SGD	0.01	134,276,932
ResNet50	224×224	Adam	0.001	23,595,908
Inception V3	299×299	Adam	0.001	21,810,980
EfficientNet B0	224×224	Adam	0.001	5,334,575
EfficientNet B5	456×456	Adam	0.001	30,566,531

The performance of CNN models was discussed with help of different performance metrics such as sensitivity (Sen), specificity (Spe), accuracy (Acc), and precision (Pre). The proportion of accurately predicted positives is called sensitivity; out of all true positive predictions and out of all true negatives, the proportion of correctly classified negatives is called specificity. On the other hand, among all images, accuracy presents the proportion of accurately classified images. Out of all positive predictions, precision is the ratio of accurately classified positives [21]. For multiclass classification using macro-averaging, these metrics and their extended calculations are given below in among Eqs. 1–8.

For a class ci,

$$Sen(ci) = \frac{TP(ci)}{TP(ci) + FN(ci)} \tag{1}$$

$$Spe(ci) = \frac{TN(ci)}{TN(ci) + FP(ci)} \tag{2}$$

$$Acc(ci) = \frac{TP(ci) + TN(ci)}{TP(ci) + TN(ci) + FP(ci) + FN(ci)} \tag{3}$$

$$Pre(ci) = \frac{TP(ci)}{TP(ci) + FP(ci)} \tag{4}$$

$$AverageSen = \frac{1}{classes} \sum_{n=1}^{classes} Sen(ci) \tag{5}$$

$$AverageSpe = \frac{1}{classes} \sum_{n=1}^{classes} Spe(ci) \tag{6}$$

$$AverageAcc = \frac{1}{classes} \sum_{n=1}^{classes} Acc(ci) \tag{7}$$

$$AveragePre = \frac{1}{classes} \sum_{n=1}^{classes} Pre(ci) \tag{8}$$

5   Result and Discussions

All CNN models were trained with transfer learning, and the main goal of this study is to evaluate the recognition ability of used CNNs such as AlexNet, VGG16, ResNet50, and Inception V3, EfficientNet B0, and EfficientNet B5 for betel plant disease recognition. The TP, TN, FP, FN values obtained by CNN models for each class of the betel leaf dataset are given in Table 3.

To evaluate the class-wise recognition ability of CNNs, the values of sensitivity, specificity, accuracy, and precision were used. The highest sensitivity 99.77% was achieved by EfficientNet B0 and B5 for healthy class, and considering sensitivity values, EfficientNet B5 performed better than others, ranging from 97.97 to 99.77%. EfficientNet B5 has achieved the highest specificity value 99.67% for the miscellaneous class, and it also has performed better than other CNN models considering

Table 3 TP, TN, FP, FN values were obtained by CNN models for each class of the test set

Model name	Class	TP	TN	FP	FN
AlexNet	Leaf rot	163	728	81	59
	Foot rot	172	741	38	80
	Healthy	398	573	43	17
	Miscellaneous	103	856	33	39
VGG16	Leaf rot	168	744	76	43
	Foot rot	179	761	31	60
	Healthy	405	573	36	17
	Miscellaneous	109	845	27	50
ResNet50	Leaf rot	176	755	68	32
	Foot rot	183	764	27	57
	Healthy	421	575	20	15
	Miscellaneous	113	861	23	34
Inception V3	Leaf rot	217	770	27	17
	Foot rot	189	798	21	23
	Healthy	438	580	3	10
	Miscellaneous	126	884	10	11
EfficientNet B0	Leaf rot	239	776	5	11
	Foot rot	202	810	8	11
	Healthy	432	589	9	1
	Miscellaneous	130	890	6	5
EfficientNet B5	Leaf rot	241	782	3	5
	Foot rot	206	817	4	4
	Healthy	439	589	2	1
	Miscellaneous	133	893	3	2

specificity values, ranging from 99.51 to 99.67%. Considering accuracy and precision values, EfficientNet B5 has shown the best performance, ranging from 99.22% to 99.71% and 97.79% to 99.55%, respectively. This model obtained more accuracy and precision value compared to others in the healthy class, 99.71%, and 99.55%, respectively. The class-wise recognition ability of EfficientNet B0 was very close to EfficientNet B5, and the recognition efficiency of EfficientNet B5 architecture for four classes of the test set is given in Table 4.

The average (Avg) sensitivity, specificity, accuracy, and precision values that were achieved by CNNs on the test set of the betel leaf dataset in this study are given in Table 5. The number of epochs is not the same for all deep learning models as early stopping was utilized to avoid overfitting during the training phase. By dividing the total time required in the training phase by the number of epochs, the time per epoch was calculated.

Table 4 Class-wise recognition performance of CNN models

Model name	Class	Sen (%)	Spe (%)	Acc (%)	Pre (%)
AlexNet	Leaf rot	73.42	89.99	86.42	66.8
	Foot rot	68.25	95.12	88.55	81.9
	Healthy	95.90	93.02	94.18	90.25
	Miscellaneous	72.54	96.29	93.02	75.74
VGG16	Leaf rot	79.62	90.73	88.46	68.85
	Foot rot	74.90	96.09	91.17	85.24
	Healthy	95.97	94.09	94.86	91.84
	Miscellaneous	68.55	96.90	92.53	80.15
ResNet50	Leaf rot	84.62	91.74	90.30	72.13
	Foot rot	76.25	96.59	91.85	87.14
	Healthy	96.56	96.64	96.61	95.46
	Miscellaneous	76.87	97.40	94.47	83.09
Inception V3	Leaf rot	92.74	96.61	95.73	88.93
	Foot rot	89.15	97.44	95.73	90.00
	Healthy	97.77	99.49	98.74	99.32
	Miscellaneous	91.97	98.88	97.96	92.65
EfficientNet B0	Leaf rot	95.60	99.36	98.45	97.95
	Foot rot	94.84	99.02	98.16	96.19
	Healthy	**99.77**	98.49	99.03	97.96
	Miscellaneous	96.30	99.33	98.93	95.59
EfficientNet B5	Leaf rot	**97.97**	**99.62**	**99.22**	**98.77**
	Foot rot	**98.10**	**99.51**	**99.22**	**98.10**
	Healthy	**99.77**	**99.66**	**99.71**	**99.55**
	Miscellaneous	**98.52**	**99.67**	**99.52**	97.79

The bold marked value represents the highest value of the performance metric.

Table 5 Average value of performance metrics for CNN models

Model name	Avg Sen (%)	Avg Spe (%)	Avg Acc (%)	Avg Pre (%)	Time per epoch (sec)
AlexNet	77.53	93.61	90.54	78.67	297
VGG16	79.76	94.45	91.76	81.52	1183
ResNet50	83.58	95.59	93.31	84.46	1527
Inception V3	92.91	98.11	97.04	92.73	1468
EfficientNet B0	96.63	99.05	98.64	96.92	391
EfficientNet B5	**98.59**	**99.62**	**99.42**	**98.55**	1352

The bold marked value represents the highest value of the performance metric.

EfficientNet B5 architecture showed the best performance in average sensitivity, specificity, accuracy, and precision. In Table 5, for the relevant performance criterion, values marked with bold are the best values obtained by EfficientNet B5. In this study, AlexNet consumed the lowest training time than other CNNs, and it took 297 s to complete one epoch. On the other hand, EfficientNet B5 took 30 h and 3 min to complete 80 epochs.

To demonstrate the recognition ability of CNNs more clearly, the number of total false predictions by CNN models for each class is discussed in this study and given in Table 6. AlexNet misclassified 81 samples of leaf rot class, which was the highest misclassification number. On the other hand, EfficientNet B5 wrongly classified 12 samples. EfficientNet B5 misclassified three samples of the leaf rot class and wrongly predicted two samples in the healthy class which was the lowest number of false classifications. The training and test accuracies acquired by CNN models are presented in Fig. 4.

Among pre-trained CNN models, two models of the EfficientNet group performed significantly better than others. EfficientNet B0 and B5 achieved 97.29% and 98.84% test accuracy, respectively. However, AlexNet obtained less training and test accuracy compared to others, 81.85%, and 81.09%, respectively. VGG16, ResNet50, and Inception V3 have achieved 83.51%, 86.62%, and 94.08% test accuracy, respectively. EfficientNet B5 showed significantly better recognition ability in all experimental

Table 6 Misclassification numbers of CNN models for each class

Model name	Leaf rot	Foot rot	Healthy	Miscellaneous	Total number
AlexNet	81	38	43	33	195
VGG16	76	31	36	27	170
ResNet50	68	27	20	23	138
Inception V3	27	21	3	10	61
EfficientNet B0	5	8	9	6	28
EfficientNet B5	**3**	**4**	**2**	**3**	**12**

The bold marked value represents the highest value of the performance metric.

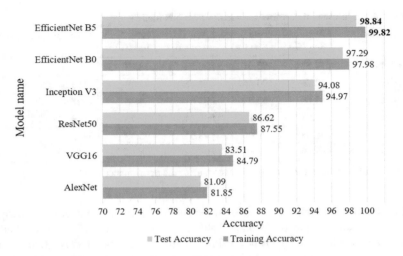

Fig. 4 Training and test accuracies of CNN models

studies compared to others in this study and achieved 99.62% training accuracy. After evaluating the results of all experimental studies, EfficientNet B5 architecture is proposed for the rapid and efficient identification of betel plant diseases in this study. The accuracy and loss curve of both training and validation of EfficientNet B5 is given in Figs. 5 and 6.

The success obtained in this study was compared with the result of other studies introduced in the literature to recognize betel plant disease and shown in Table 7. To the best of our knowledge, this is the first attempt for evaluating the performance of state-of-the-art CNN models for the recognition of betel plant disease.

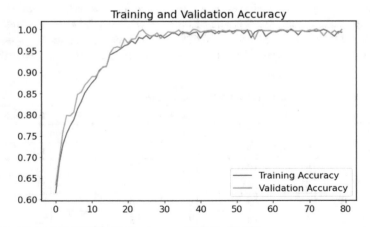

Fig. 5 Training and validation accuracy curve of EfficientNet B5

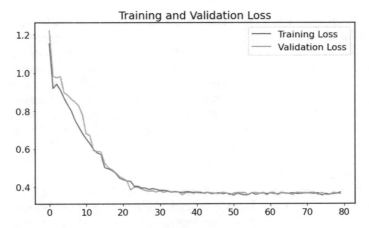

Fig. 6 Training and validation loss curve of EfficientNet B5

Table 7 Comparison of several methods for betel plant disease recognition

Study	Method	Number of class	Size of dataset	Accuracy
Tamilsankar and Gunasekar [3]	Minimum distance classifier	4	100	[a]*NM*
Jayanthi and lalitha [4]	Multiclass SVM	4	*NM*	95.85%
Dey et al. [5]	Image processing	1	12	*NM*
Vijayakumar and Arumugam [6]	Image processing	2	10	*NM*
Ramesh and Nirmala [7]	*K*-means clustering	*NM*	*NM*	*NM*
Our study	EfficientNet B5	4	10,662	**98.84%**

[a]*NM* not mentioned
The bold marked value represents the highest value of the performance metric.

6 Conclusion

Deep neural network-based recognition approaches brought significant improvement in image processing and pattern recognition-related tasks. To reduce the damage of betel plant diseases, this paper proposed an efficient and rapid recognition method using deep neural networks by classifying leaf images. After evaluating the performance of CNN models such as AlexNet, VGG16, ResNet50, Inception V3, EfficientNet B0, and EfficientNet B5, to classify four classes of the betel leaf dataset which contains 8600 training, 1031 validation, and 1031 testing images, EfficientNet B5 was proposed. Different experimental studies were introduced on the test set of the betel leaf dataset to compare the recognition ability of EfficientNet B5 with other pretrained CNN models. According to the result of experimental studies, EfficientNet

B5 significantly performed better compared to others and obtained 99.62% training and 98.84% test accuracy with 80 epochs. Four different input image sizes such as 224×224, 227×227, 299×299, and 456×456 were used in the study. AlexNet consumed less training time compared to other CNN architectures and obtained less training and test accuracy compared to others, 81.85%, and 81.09%, respectively. During the training phase, EfficientNet B5 consumed 1352 s to complete one epoch. On the other hand, the performance of EfficentNet B0 was close to EfficentNet B5, but it took 391 s to complete one epoch during the training phase. VGG16, ResNet50, Inception V3, and EfficientNet B0 achieved 83.51%, 86.62, 94.08%, and 97.29% accuracy on the test set. According to the result of class-wise classification performance, EfficientNet B5 achieved 99.77% sensitivity in healthy class and 99.67% specificity in miscellaneous class, and in healthy class, it obtained 99.71% accuracy and 99.55% precision value. In addition, considering average sensitivity, specificity, accuracy, and precision values for each model, EfficientNet B5 showed significantly better recognition performance compared to other CNN models. According to the total number of false predictions, EfficientNet B5 misclassified 12 images of the test set. On the other hand, AlexNet, VGG16, ResNet50, Inception V3, and EfficientNet B0 wrongly predicted 195, 170, 138, 61, and 28 images, respectively. The success of this study demonstrates that the proposed EfficientNet B5 architecture realizes the end-to-end classification of betel plant diseases.

However, our addressed recognition approach can classify only two diseases of the betel plant. So, it is planned to improve the betel leaf dataset by increasing class number and images of betel leaf and develop a more efficient architecture for betel plant disease recognition with a fewer number of parameters in the future work. By deploying this model in smartphone environments, farmers will be able to classify diseases of the betel plant easily which is very crucial for preventing and controlling the diseases and reduce the use of pesticides which is a major threat to the environment.

References

1. Smit Paan—Wikipedia. https://en.wikipedia.org/wiki/Paan
2. Maiti, S., Sen, C.: Fungal diseases of betel vine. PANS **25**(2), 150–157 (1979)
3. Tamilsankar, P., Gunasekar, T.: Computer aided diseases identification for betel leaf. Int. Res. J. Eng. Technol. **2**, 2577–2581 (2015)
4. Jayanthi, S.K., Lalitha, C.: Betel leaf disease detection using histogram of oriented gradients and multiclass SVM. Int. J. Innov. Res. Comput. Commun. Eng. **5**, 13994–14001 (2017)
5. Dey, A.K., Sharma, M., Meshram, M.R.: Image processing based leaf rot disease, detection of betel vine (*Piper betleL*). Proc. Comput. Sci. **85**, 748–754 (2016)
6. Vijayakumar, J., Arumugam, S.: Recognition of powdery mildew disease for betel vine plants using digital image processing. Int. J. Distrib. Parallel Syst. **3**(2), 231–241 (2012)
7. Ramesh, L., Nirmala, V.: Betel leaf disease classification using clustering method. J. Chem. Pharm. Sci. 193–196 (2017)
8. Sladojevic, S., Arsenovic, M., Anderla, A., Culibrk, D., Stefanovic, D.: Deep neural networks based recognition of plant diseases by leaf image classification. Comput. Intell. Neurosci. **2016** (2016)

9. Sabrol, H., Kumar, S.: Plant leaf disease detection using adaptive neuro-fuzzy classification. In: Arai K., Kapoor S. (eds.) Advances in Computer Vision. CVC 2019. Advances in Intelligent Systems and Computing, vol. 943. Springer, Cham (2020)

10. Yadav, R., Kumar Rana, Y., Nagpal, S.: Plant leaf disease detection and classification using particle swarm optimization. In: Renault É., Mühlethaler P., Boumerdassi S. (eds.) Machine Learning for Networking. MLN 2018. Lecture Notes in Computer Science, vol. 11407. Springer, Cham (2019)

11. Khan, M.A., Akram, T., Sharif, M., et al.: An automated system for cucumber leaf diseased spot detection and classification using improved saliency method and deep features selection. Multimed. Tools Appl. **79**, 18627–18656 (2020)

12. Ahila Priyadharshini, R., Arivazhagan, S., Arun, M., et al.: Maize leaf disease classification using deep convolutional neural networks. Neural Comput. Appl. **31**, 8887–8895 (2019)

13. Basavaiah, J., Arlene Anthony, A.: Tomato leaf disease classification using multiple feature extraction techniques. Wireless Pers. Commun. **115**, 633–651 (2020)

14. Mokhlesur Rahman, Md., Tarek Habib, Md.: A preprocessed counter propagation neural network classifier for automated textile defect classification. J. Ind. Intell. Inf. **4**(3) (2016)

15. Tarek Habib, Md., Jueal Mia, Md., Shorif Uddin, M., Ahmed, F.: An in-depth exploration of automated jackfruit disease recognition. J. King Saud Univ.—Comput. Inf. Sci. (2020). https://doi.org/10.1016/j.jksuci.2020.04.018

16. Jacob, I.J., Ebby Darney, P.: Artificial bee colony optimization algorithm for enhancing routing in wireless networks. J. Artif. Intell. **3**(01), 62–71 (2021)

17. Chen, J.I.Z., Hengjinda, P.: Early prediction of coronary artery disease (cad) by machine learning method—a comparative study. J. Artif. Intell. **3**(01), 17–33 (2021)

18. Mugunthan, S.R.: Decision tree based interference recognition for fog enabled IOT architecture. J. Trends Comput. Sci. Smart Technol. (TCSST) **2**(01), 15–22 (2020)

19. Hridoy, R.H., Akter, F., Mahfuzullah, M., Ferdowsy, F.: A computer vision based food recognition approach for controlling inflammation to enhance quality of life of psoriasis patients. In: 2021 International Conference on Information Technology (ICIT), pp. 543–548 (2021)

20. He, K., Zhang, X., Ren, S., Sun, J.: Deep residual learning for image recognition. In: 2016 IEEE Conference on Computer Vision and Pattern Recognition (CVPR), pp. 770–778. Las Vegas, NV, USA (2016)

21. Hridoy, R.H., Akter, F., Rakshit, A.: Computer vision based skin disorder recognition using EfficientNet: a transfer learning approach. In: 2021 International Conference on Information Technology (ICIT), pp. 482–487 (2021)

Effective Integration of Distributed Generation System in Smart Grid

Namra Joshi and Jaya Sharma

Abstract In this modern era, developing nations are using distributed generation systems' as a major energy source in deregulated power systems. Distributed generation (DG) is having an important part in improving the quality of human life. This paper emphasis on the various issues concerning the various issues of DG integration into the smart grid and also highlights the benefits and design issues of this integration.

Keywords Grid integration · Distributed generation · Smart grid · Renewable energy

1 Introduction

For bulk power transmission, competent infrastructure is needed. Smart grid in conjunction with distributed generation [DG] [1] system is very effective to fulfill such requirements. The primary idea is that the power generation near consumer premises will reduce the load on transmission and distribution system and will become an alternative source of energy and it will reduce the capital cost required for the erection of new transmission lines [2]. DG is not a recent phenomenon. Before the origin of AC systems and mega-sized generators, most of the power needs is fulfilled through local generators. With the developments in nascent technologies, the whole world is looking toward modernized and compact systems for power generation, with the new developments in compact size in technologies used for power generations like PV panels, fuel cells, microturbines, etc [3]. Also, this technology facilitates the availability of power at a very cheap cost, and the reliability of supply is also raised. The opportunities of DG integration in the smart grid are discussed in this paper [4].

N. Joshi (✉)
Department of Electrical Engineering, SVKM's Institute of Technology, Dhule, Maharashtra, India

J. Sharma
Department of Electrical Engineering, Shri Vaishnav Vidyapeeth Vishwavidyalaya, Indore, Madyapradesh, India

© The Author(s), under exclusive license to Springer Nature Singapore Pte Ltd. 2022 243
V. Suma et al. (eds.), *Evolutionary Computing and Mobile Sustainable Networks*,
Lecture Notes on Data Engineering and Communications Technologies 116,
https://doi.org/10.1007/978-981-16-9605-3_17

2 Distributed Generation

Distributed generation (DG) [5] is a kind of small level generation in which power is generated near the point of utilization. It is having a rating of around 5 kW–10 MW. It also facilitates interconnection [6]. In the DG system, it is promoted to use a renewable energy-based generation to reduce carbon emissions. Figure 1 illustrates the block diagram of DG. The bidirectional flow of power is possible with the use of DG. When the customer is having excess power in the DG system, it can be given to the central grid, and when power is not available in the selected RES-based DG source, customer needs can be fulfilled with central grid power. Formerly, the DG system is preferred to use in vehicles and defense systems [7].

2.1 Merits of DG System

DG is a very beneficial system to fulfill the power requirements particularly the remote areas. The merits of the DG system are as follows:

- Overall peak demand will be reduced.
- T&D losses will be reduced.
- The reliability of supply will be improved.
- Quick response on emerging power demands.
- Remote electrification is possible.

2.2 Design Issues

When we integrating DG into the smart grid, the following technical design considerations are to be taken into account.

- Voltage stability
- Reactive power compensation

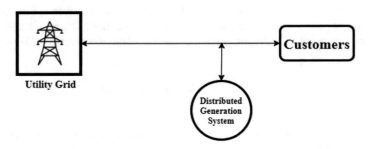

Fig. 1 Basic block diagram of DG

- Islanding protection
- Power quality
- Protection issues

3 Smart Grid

Smart grid (SG) [8] is a very effective system that permits the integration of power generation from different sources with the central grid system. It facilitates the efficient, reliable, and safe operation of the power system. It is compatible with digital systems, and it helps customers to select power suppliers and be able to work in a competitive power market. The SG is having excellent features like optimization of power, self-fault detection and isolation. It helps in communication between generating stations and customers. The SG gives a chance to interlink financial, technological, regulatory, and policy points. The SG is an emerging paradigm in the power system. The two-directional flow of power and information is available in such systems. The overview of the smart grid system is illustrated in Fig. 2, and its infrastructure is shown in Fig. 3 [9].

3.1 Advantages of Smart Grid

- The maintenance work is quite easy.
- It is possible to connect all types of energy-generating sources.
- With the help of smart meters, telemetering is possible.
- The overall quality of power is improved.
- Two directional communication is possible in all three sectors, i.e., generation, transmission, and distribution.
- It can work better in conjunction with DG systems.
- The whole system is having a flexible design [2].
- It increases the overall efficiency of the system.

4 Important Aspects of DG Integration

For efficient integration DG system with smart grid, three important points are to be concern, viz., smart management, smart control, and smart protection [10].

A. *Smart Management*

The management of all the resources in the power system to fulfill the power demand efficiently and effectively is termed as smart management. It is related to the relation of selection of correct mode of operation of available DG system [6].

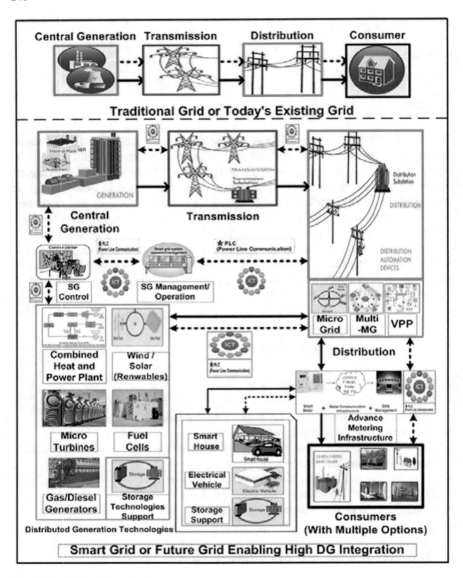

Fig. 2 Overview of smart grid

There are several ways to connect the DG system with the existing grid. It mainly depends on the nature of the operation of the DG system and our needs. They may work in grid-connected mode, clustered mode, isolated mode as illustrated in Fig. 4. In remote areas, the favorable DG system is the off-grid mode. Depending on the geographical feasibility of the area, it may use multiple energy sources which form a hybrid plant-based DG system. In such a mode of operation, the maximum usage of

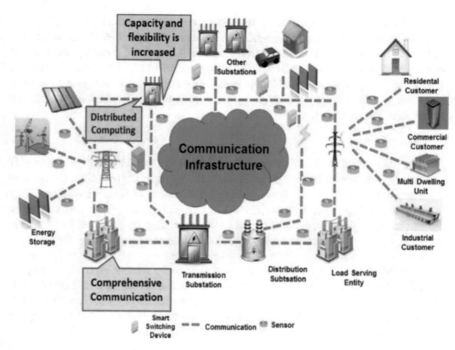

Fig. 3 Smart grid infrastructure

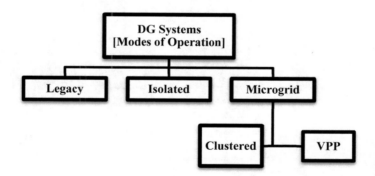

Fig. 4 Modes of operation of distributed generation system

existing resources has been carried out. The grid power will act as a backup in such cases [11].

Another mode is termed as microgrid mode in which will act as a counterpart of the main grid and having extensive features of controlling the DG system. It is also having PCC with the main grid which provides the power exchange from both sides of the microgrid system. Thus, power requirements are fulfilled. As the overall area of operation is less for such a system, the losses are a bit low in this system [12]. The DG in microgrid can be further extended as a clustered system. With such a cluster of MG, the overall load on the main grid is relieved, and local level monitoring and control can be done. This helps in making the system more efficient. The VPP stands for a virtual power plant as it is quite the same as a cluster-based MG system. In VPP, the cluster of DG systems is handled by the central unit. It is possible to compare conventional plants with the VPP [13].

B.   *Smart Control*

The parameters in the power system are fluctuating in nature. It is very essential to look after power system parameters. It is very essential to control the parameters particularly in this type of integrated system in which two-directional power flow is present. With the rise in DG integration, suitable control strategies are to be implemented. The control mechanism may be distributed or centralized. In the case of centralized control, the control strategies are not properly resolved, and it is expensive. In distributed control, complex communication infrastructure is needed. Hybrid systems are also used in which optimized operation is done by any kind of control, i.e., centralized control or hybrid control. The control scheme for the DG system is illustrated in Fig. 5. The distributed management scheme (DMS), energy management system (EMS), microgrid central controller (MGCC), load controllers (LC) are connected. The method

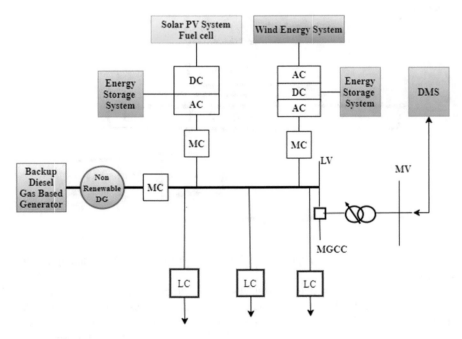

Fig. 5 Control schemes for DG units

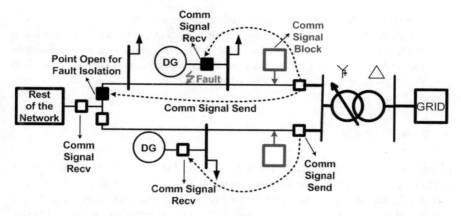

Fig. 6 Communication techniques-based protection scheme

of control of DG system in microgrid configuration mode is comprised of LC, MGCC, EMS, MCs, and DMS. The main role of DMS is to supervise, control and in managing the power at the distribution point. The MGCC is linked with the MCs and LCs. The DMS gives the command to the controller to keep a synchronism between generation and demand. It also interacts LCs and MCs. The value of active and reactive power in MG will handle this interaction. With the help of local controllers, power supply is controlled. So, the MGCC gives instructions to supporting controllers for maintaining a balance between generation and load as shown below [14].

C. *Smart Protection*

The communication-based protection scheme is indicated in Fig. 6. For the uninterruptable supply, a competent protection system is needed. It is provided with the help of several relays, circuit breakers, and measuring instruments. In the DG system, the power flow is two-directional, so it is required to upgrade the existing protection scheme for ensuring safe operation. To provide the same, it is required to make the system capable enough to work for abnormal conditions like fault current, false tripping and many more like this. In this present scenario, the major problem faced is islanding operation and grid outage is mainly responsible for it. Because of it, the voltage and frequency limits are getting violate which further results in maloperation of protection relay. If the protection scheme is not properly planned, then DG system may face the problem of synchronism even after the grid is restored. When the main system is failed to fulfill the power demand, then the condition called loss of mains (LOM) occurs. The DG system must be protected from the situation. The two methods are adopted to protect the DG system from such state, i.e., local-based methods and communication-based methods. In the first method, the circuit and power system parameters are continuously monitored to find out the LOM occurrence and take preventative and remedial action when required. In the second method, communication signals are transmitted to the DG system when CB is unable to work due to LOM Situation. The most important points to be incorporated in the protection system are high speed, selectiveness, and sensitivity enough to detect all variations in considered parameters [15].

5 Future Scope

Smart grid is an emerging area in power systems. Particularly in the Indian power system, the power generation from a renewable source is increasing tremendously which gives new wings to smart grid development. It provides more dynamic, diverse, and secure assistance to the existing grid. The smart grid needs time due to the rise in the number of DG systems. SG makes cost-effective metering and ensures the efficient delivery of power. To promote smart grid in Indian scenario, GoI established national Smart Grid Mission in the year 2015. Its main aim is to monitor the policies and programs related to smart grid.

6 Conclusion

Smart grid is a nascent component of the power system. Many countries are using a renewable energy-based distributed system. When we integrate both of them, it will become a self-sustained and effective energy system. This paper concludes that in smart grid, adopting proposed control and protection strategy DG can be effectively integrated with the smart grid, based on data about the integration of distributed generation. This paper also addressed the new era of the distribution network which is capable to work in several modes in presence of DG units.

References

1. Ackerman, T., Andersson, G., Sodder, L.: Distributed generation: a definition. Electr. Power Syst. Res. **57**, 195–204 (2001)
2. www.mnre.gov.in
3. www.powermin.ac.in
4. Ministry of New and Renewable Energy: Renewable energy in India: progress, vision and strategy (2010)
5. Dugan, R.C., Mcdermott, T.E.: Distributed generation. IEEE Ind. Appl. Mag. 19–25 (2002)
6. EPRI White Paper: Integrating distribution resources into electric utility distribution system. Technol. Rev. (2001)
7. Joshi, N., Sharma, J.: Analysis and control of wind power plant. In: 2020 4th International Conference on Electronics, Communication and Aerospace Technology (ICECA), pp. 412–415 (2020)

8. Fang, X., Misra, S., Xue, G., Yang, D.: Smart grid the new and improved power grid: a survey. IEEE Commun. Surv. Tutorials **14**(4), 944–980 (2012)
9. Bari, A., Jiang, J., Saad, W., Jaekel, A.: Challenges in the smart grid applications: an overview. Int. J. Distrib. Sens. Netw. **2014**, 1–11 (2014)
10. Murthy Balijepalli, S.K., Gupta, R.P., Khaparde, S.A.: Towards Indian smart grids. In: Proceedings IEEE TENCON Conference, Singapore (2009)
11. Khandelwal, A., Nema, P.: Harmonic analysis of a grid connected rooftop solar energy system. In: 2020 Fourth International Conference on I-SMAC (IoT in Social, Mobile, Analytics and Cloud) (I-SMAC), pp. 1093–1096 (2020)
12. Khandelwal, A., Nema, P.: A 150 kW grid-connected roof top solar energy system—case study. In: Baredar P.V., Tangellapalli S., Solanki C.S. (eds.) Advances in Clean Energy Technologies. Springer Proceedings in Energy. Springer, Singapore (2021)
13. Joshi, N., Sharma, J.: An overview on high voltage direct current transmission projects in India. In: 2021 6th International Conference on Inventive Computation Technologies (ICICT), pp. 459–463 (2021)
14. Agrawal, M., Mittal, A.: Micro-grid technological activities across the globe: a review. IJRRAS **7**(2) (2011)
15. Khandelwal, A., Neema, P.: State of art for power quality issues in PV grid connected system. Int. Conf. Nascent Technol. Eng. (ICNTE) **2019**, 1–4 (2019)

Application of Perceptual Video Hashing for Near-duplicate Video Retrieval

R. Sandeep⬤ and Bora K. Prabin

Abstract Because of the rapid growth in the VLSI technology, the speed of the Internet and the easy availability of editing software, the number of near-duplicate videos (NDVs) produced, edited or viewed has increased manifold. This has made the effective database management, and the video copyright protection is a challenging problem. To identify the NDVs in the database, a perceptual video hashing scheme for the content-based retrieval of NDVs is proposed in this paper. The perceptual video hashing algorithms used to design the near-duplicate video retrieval (NDVR) application are based on (1) the Achlioptas's random matrix (ARM), (2) the temporal wavelet transform (TWT) and the ARM, (3) the Tucker decomposition and (4) the parallel factor (PARAFAC) decomposition. The performance of the NDVR application is evaluated using the average precision–recall curves on three video databases created from REEFVID, XIPH, OPEN-VIDEO and TRECVID data sets using various content-preserving operations and malicious modifications. The experimental results on a moderate-size video databases show that the Tucker decomposition-based video hashing algorithm performs better in retrieving NDVs.

Keywords Near-duplicate videos · Near-duplicate video retrieval · Perceptual video hash · Tensor decomposition · Tucker decomposition · Achlioptas's random projection · Temporal wavelet transform

1 Introduction

The rapid pace of growth in the VLSI technology and the Internet has led to an enormous increase in the number of videos generated or viewed. Furthermore, the easy availability of a number of video editing tools such as *Freemake*, *Blender*,

R. Sandeep (✉)
Department of ECE, Vidyavardhaka College of Engineering, Mysuru 570002, India
e-mail: sandeep.ece@vvce.ac.in

B. K. Prabin
Department of EEE, IIT Guwahati, Guwahati 781039, India
e-mail: prabin@iitg.ac.in

© The Author(s), under exclusive license to Springer Nature Singapore Pte Ltd. 2022   253
V. Suma et al. (eds.), *Evolutionary Computing and Mobile Sustainable Networks*,
Lecture Notes on Data Engineering and Communications Technologies 116,
https://doi.org/10.1007/978-981-16-9605-3_18

Shotcut and *Lightworks* has also led to the significant increase in percentage of *near-duplicate videos* (NDVs) in the online databases. According to Susan Wojcicki, the CEO of the YouTube, "400 h of video are uploaded on the YouTube by users every minute as on July 2015" [1]. The users may maliciously modify the downloaded video and upload it back to the same database server, violating the copyrights, also increasing the amount of redundancy and adding burden on the server. Thus, there is a strong need for the perceptual content of the video to be checked by the system before granting permission to the user for uploading the video. This leads to video copyright protection and also an efficient video database management by disallowing the user to upload the video if perceptually similar videos already exist in the database.

If a viewer searches for the video of his/her interest in a video database by providing the query in the form of text in the search window, he/she is provided with a lot of irrelevant videos and NDVs in addition to the video of interest. The authors in [2, 3] conducted an experiment of video search in Google video, YouTube and Yahoo video by giving a text-based query of 24 popular videos and found that on an average there are 27% redundant videos that are duplicate or nearly duplicate to the most popular version of a video in the search results. This again indicates that there is a strong need for the search related to query video to be ranked or ordered on the basis of its contents rather than just the metadata. The *near-duplicate video retrieval* (NDVR) system, in turn, can be used for the following applications [4]: copyright protection, video monitoring, video recommendation, data mining, topic tracking and video re-ranking. In addition to NDVR, the near-duplicate video detection (NDVD) [4–10] is also an important task. NDVD is more challenging and requires significant attention, as it is the need of the hour for database cleansing, detection of copyright violation and monitoring advertisements.

One of the multiple solutions to the above-discussed problems is to generate a compact and fixed perceptual hash for a video based on its perceptual contents and use it for indexing the NDVs. This work focuses on only the NDVR system and not the NDVD system. In the literature, the copy detection, the content authentication and the video tracking have been developed using the perceptual video hash. In [11, 12], the video copy detection system is proposed using perceptual video hash extracted from the temporally informative representative images (TIRIs). In [13], a video authentication system has been implemented using the temporal wavelet transform (TWT)-based video hashing technique.

The flow of the remaining part of the article is as per the following. The framework for an NDVR system is described in Sect. 2. A literature survey on video fingerprint-based NDVR system is presented in Sect. 3. The scope of perceptual video hash as a video fingerprint is discussed in Sect. 4. The NDVR system based on perceptual video hashing algorithms is proposed in Sect. 5. In this section, the perceptual video hashing algorithms used to implement the NDVR system are also discussed. The performance of the proposed NDVR system is evaluated in Sect. 6. The conclusions are drawn in Sect. 7.

2   General Framework of an NDVR System

The authors in [3] have defined the term NDV as "identical or approximately identical videos close to the exact duplicate of each other, but different in file formats, encoding parameters, photometric variations (colour, lighting changes), editing operations (caption, logo and border insertion), different lengths and certain modifications (frames add/remove)." Shen et al. [14] defined the NDV as "clips that are similar or nearly duplicate of each other, but appear differently due to various changes introduced during capturing time (camera viewpoint and setting, lighting condition, background, foreground, etc.), transformations (video format, frame rate, resize, shift, crop, gamma, contrast, brightness, saturation, blur, age, sharpen, etc.) and editing operations (frame insertion, deletion, swap and content modification)." Thus, we observe that an original video undergoing content-preserving operations such as filtering, brightness modifications, contrast modifications, colour modifications, insertion of a logo, frame rate modifications and the variations in spatial resolutions are considered to be near-duplicates. The general framework of an NDVR system is shown in Fig. 1. The system consists of three major components, namely *fingerprint generation, fingerprint management* and *fingerprint search* as summarized by Liu et al. in [15]. These components are briefly described below.

Fingerprint generation: The fingerprint can be generated directly using the low-level features of the video. Alternatively, video summarization [16, 17] can be performed based on these low-level features, and then the fingerprint can be generated

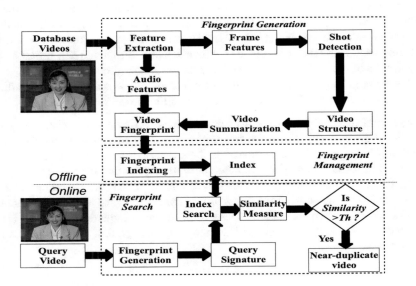

Fig. 1 General framework of an NDVR system

from these abstract and meaningful video representations. A detailed discussion on the generation of video fingerprints is discussed in Sect. 3.

Fingerprint management: Managing the videos and their fingerprints are quite challenging as the number of videos in a database is large in number. Hence, efficient indexing structures are necessary to organize the video fingerprints and achieve faster retrieval. The fingerprint management involves identifying the efficient indexing techniques and then indexing the video fingerprints.

The objective of different methods of indexing the video fingerprint is to enable faster search and retrieval. The method of indexing depends upon the form of video fingerprint. The indexing and retrieval of multimedia data are based on the nearest neighbour search. Different high-dimensional indexing methods exist in the literature [18, 19]. They include *tree-like structures, transformation methods*, and *hashing methods*. In [18], the video fingerprints are indexed using the Gauss tree. The Gauss tree is an index structure for managing Gaussian distributions to efficiently answer video queries [15]. The Gauss tree uses novel algorithms for query processing, insertion and tree construction. As the dimensionality increases, the feature space splitting becomes very poor with too many overlaps, and hence, the efficiency of the tree structures reduces gradually. In [19], the fingerprints are clustered and indexed based on their distance with respect to a reference point inside a cluster using a B^+-tree. The B^+-tree is an effective indexing structure for one-dimensional data. The performance of the retrieval system is dependent on the choice of the reference point. In [20], the authors have proposed an optimal choice of two far-away reference points based on the bi-distance transformation. The bi-distance transformation transforms a high-dimensional point into two distance values with respect to two optimal reference points. Another approach for indexing the video fingerprint comprises the method of locality-sensitive hashing (LSH) [21, 22]. An LSH is a well-known method for indexing as it overcomes the "curse of dimensionality" problem [22]. It hashes the fingerprints of the NDVs to the same "buckets" with high probability [23].

The fingerprint generation and indexing of database videos are usually the offline processes, whereas the fingerprint search is an online process.

Fingerprint search: The query can be through a *text-based query* or *query by example*. For a given query video, the fingerprint is generated and then matched with the video fingerprints in the database. If the similarity is greater than a particular threshold or if the distance between the query fingerprint and the video fingerprint in the database is less than a particular threshold, then the corresponding video is considered to be the NDV of the query video. Some of the distance metrics include the Euclidean distance, the squared chord distance, the Chi square distance, the Manhattan distance and the normalized Hamming distance (NHD). This work uses the NHD defined by

$$d_{NH} = \frac{d_H}{l_b}$$

where d_H represent the Hamming distance (HD) between the binary numbers and l_b represent the length of the binary number. For discussion on the other distance measures, please refer to [24].

3   Survey on Video Fingerprint-Based NDVR System

The problem of finding the near-duplicate documents, audio and images dates back to early 1990s when the Internet was evolving [25]. Finding of NDVs became a serious problem because of the size, the storage in different formats for easy downloading and streaming [26], evolving of video editing tools and so on. The problem of searching and retrieving NDVs was addressed in [25] through video summarization. Currently, two approaches exist in the literature for the video fingerprint generation. In the first approach, the NDVR system is developed by directly using the low-level features of the video. In the second approach, the NDVR system is developed by using the higher-level abstraction of the video, namely the video fingerprint, extracted from shots, keyframes or summarized video. A brief description of these approaches is presented below.

3.1   Low-level Features-Based NDVR System

Colour histograms [2, 3] and local features [27, 28] such as corners and edge pixels are used in NDVR systems. A colour histogram represents the frequency of colour pixels. An NDVR system based on it is computationally efficient and compact, but it is sensitive to variations in colour, and also it does not contain any information about the shape or texture information. Two different video frames may result in similar histograms and hence may result in the wrong retrieval. Some of the local features used in an NDVR system are blobs, corner and edge pixels. An NDVR system based on it captures the features that are usually resistant to variations in scale and different affine transformations. But the drawback is that to retrieve NDVs, a large number of features have to be compared, and hence, the method becomes computationally expensive. Therefore, a small number of pertinent features, in terms of a video fingerprint, have to be extracted.

3.2   Video Fingerprint-Based NDVR System

The low-level features extracted from a video are operated by a fingerprint generation algorithm to reduce the dimensionality and retain the most relevant and meaningful features. The dimensionality is reduced to improve the retrieval speed. In fingerprint generation, different works focus on different aspects of the video. Depending on the application, the fingerprints are extracted from different levels of the video, namely *frame-level local fingerprint, frame-level global fingerprint, video-level global fingerprint, spatiotemporal fingerprint* and *multi-feature fusion fingerprint*. They are outlined below.

1. **A frame-level local fingerprint**: It is extracted from the local features of the frame. Some of the local feature descriptors used are the scale-invariant feature transform (SIFT) [29], the principal component analysis (PCA)-SIFT [30] and the LBP [31]. These fingerprints are robust to complex editing, photometric variations and changes in the angle of viewport during the recording process of the same scene. These fingerprints can be used when speed is not the key requirement of an NDVR system as the method used to generate the fingerprint and implement the NDVR system is computationally expensive. Recent works have tried to increase the speed of retrieval by using the fast indexing structure techniques [5] and dimensionality reduction techniques such as the *neighbour retrieval visualizer* [32].

2. **A frame-level global fingerprint**: It is extracted from the global features of the keyframes. The keyframes are extracted from the video using temporal sampling or shot boundary detection techniques. Some of the global features include the colour histogram, shape descriptors, contour representations and texture features. Some of the recently used global feature descriptors in an NDVR system are bag-of-visual words [33, 34] and frame-level global descriptors such as the improved Harris detector [35] and ternary frame descriptor [36]. These fingerprints are the trade-off between the speed and the accuracy of the NDVR system [15].

3. **A video-level global fingerprint**: It is extracted by considering the whole video as a single entity. The basic idea of generating this type of fingerprints is to represent the global video information into various forms, such as the histograms [3, 37] and cluster representatives [38]. These fingerprints are robust to minor colour, contrast and frame position modifications. They can be used when speed is the key requirement of an NDVR system as the methods used to generate the fingerprint and implement the NDVR system are computationally efficient [15].

4. **A spatiotemporal fingerprint**: It is extracted from the spatial and temporal features of the video. In [39], the spatiotemporal fingerprint is generated based on the conditional entropy and the local binary pattern (LBP) feature descriptors. The conditional entropy is used to select relevant features that carry as much information as possible. The LBP feature descriptor is robust against illumination changes and computationally simple, which makes it possible to extract features in challenging real-time settings. In [5], a spatiotemporal fingerprint is extracted using the pattern-based approach under the hierarchical filter-and-refine framework. In other words, the non-near-duplicate videos are filtered and removed, thereby reducing the search space of the NDVs, and then finally re-rank it. This method fails to retrieve the videos obtained by applying the picture-in-picture transformation [40], high-speed fast-forwarding or super-slow motion. These fingerprints are robust to heavy colour modifications and geometric modifications but sensitive to temporal modifications on the video frame. The spatiotemporal

fingerprints involve high complexity during measurement of similarity as the temporal order is taken into consideration [15].

5. **A multi-feature fusion fingerprint**: It is extracted by combining both the local and the global features of the video. The feature-level fusion is preferred over the only local feature-based or only global feature-based NDVR system because the information present in local features and global features is complementary in nature. In general, the technique is also known as multi-modal [41–43] and multiview learning [44–46] as it involves machine learning tasks and general data analysis to fuse multiple features. In [47], the video fingerprint is generated using the machine learning technique for fusing the hue saturation value (HSV) and LBP features extracted from the video keyframes. In [48–50], the fingerprint is generated based on the deep learning techniques using the convolutional neural networks. The deep learning techniques for the generation of fingerprint eliminate the need for feature extraction and selection.

Thus, there are different approaches to the generation of the video fingerprint. Currently, the fingerprints generated using the deep learning techniques are outnumbering the traditional ways of generating the video fingerprints [51–53]. Whatever may be the method of generation, a video fingerprint is desired to be compact in nature to facilitate faster retrieval. Furthermore, the fingerprint generation technique should take care that all the NDVs have similar fingerprints and non-NDVs should have different fingerprints. Interested readers for knowing NDVR based on deep learning techniques can refer to [48, 54–57].

4 Perceptual Hash as a Video Fingerprint

The perceptual hash of the video may be used as a fingerprint for content-based retrieving, efficient database management (by removing NDVs from the database if the perceptual contents of the videos are same), video copyright protection and video authentication. The perceptual video hashing function $H\,(\cdot, \cdot)$ takes a video V, and a secret key K as the inputs, to generate the perceptual hash $\mathbf{h}_v$ as the output, i.e. $\mathbf{h}_v = H\,(V, K)$. Table 1 shows the notations used for describing the useful properties of a perceptual hash function. $\mathbf{h}_v$ would qualify to be video fingerprint for such applications if it satisfies the following properties [58–60]:

1. $\mathbf{h}_v$ should be short and one way. Mathematically,

$$\text{Size}\,(\mathbf{h}_v) \ll \text{Size}\,(V) \tag{1}$$

and

$$\mathbf{h}_v \nrightarrow V \tag{2}$$

where $\nrightarrow$ denotes one way.

Table 1 Notations involved in defining the useful properties of perceptual hash function.

Notation	Description
$\mathcal{V}$	Space of finite videos
$\mathcal{K}$	Space of secret keys
V	Input video
K	Secret key
k	Length of the hash
$\mathbf{h}_v$	Hash vector for the video
$H(V, K)$	Perceptual hashing function
V_{sim}	Near-duplicate of V
V_{diff}	A different video from V
$d_{NH}(\cdot, \cdot)$	Normalized Hamming distance

2. $\mathbf{h}_v$ should be similar for NDVs. Mathematically,

$$E\left[d_{NH}\left(\mathbf{h}_v, \mathbf{h}_{\text{sim}}\right)\right] \cong 0 \tag{3}$$

where $\mathbf{h}_{\text{sim}}$ represents the hash value of V_{sim} generated using key K and E denotes expectation.

3. $\mathbf{h}_v$ should be dissimilar for different videos. Mathematically,

$$E\left[d_{NH}\left(\mathbf{h}_v, \mathbf{h}_{\text{diff}}\right)\right] \cong 0.5 \tag{4}$$

where $\mathbf{h}_{\text{diff}}$ represents the hash value of V_{diff} generated using the key K.

The properties mentioned above are validated and discussed in Sect. 6.2. In [61], we first demonstrated that the NDVR system could be developed using perceptual video hashes.

5 Proposed framework of NDVR system

Figure 2 shows the block diagram of the NDVR application based on the perceptual hashing. This application retrieves the NDVs rather than relevant videos. This scheme consists broadly of two phases. Database creation is the first phase, and video retrieval is the second phase. The brief description of each phase is given below.

1. **Database creation**: This phase consists of two parts, namely fingerprint generation and fingerprint management. Fingerprint generation includes the preprocessing and normalization of the video, perceptual hash generation from the video and then storing the perceptual hash along with the metadata of the video in the form of a structured array.

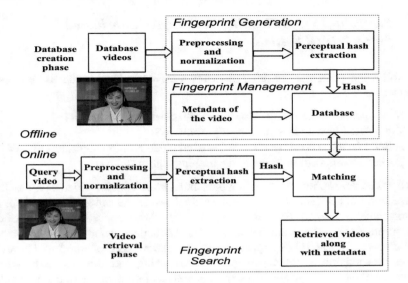

Fig. 2 Block diagram of the perceptual hashing-based NDVR application.

The input colour videos are converted to grey videos and then normalized to a predefined size say, $\mathcal{V} \in \mathbb{R}^{X \times X \times X}$ via temporal subsampling and spatial resizing. This preprocessing and normalization process ensures the impact of various spatial dimensions and frame rates on the hash values to be minimal. The perceptual hash is then extracted using a perceptual hashing technique. Let, $\mathbf{h}_j; j = 1, 2, \ldots, N$ be the hash vectors for N videos $V_1, V_2, \ldots, V_N$, respectively, generated using the perceptual video hashing algorithm. Later, the videos and the corresponding hashes $\mathbf{h}_j$ along with the metadata of the video are stored in the database. The possible metadata include title, owner and the description of the video. We have used the title of the video as the sole metadata.

2. **Video retrieval**: In the retrieval phase, the query video is preprocessed and normalized as described in the database creation phase. The hash $\mathbf{h}_{\text{query}}$ is then computed for the query video V_{query} using the same perceptual video hashing algorithm used during the database creation phase. The $\mathbf{h}_{\text{query}}$ is then compared with all the hashes in the database using the NHD metric. The videos are considered to be NDVs and retrieved if the following condition is satisfied:

$$d_{NH} \left(\mathbf{h}_j, \mathbf{h}_{\text{query}} \right) \leq T_h; \quad j = 1, 2, \ldots, N \tag{5}$$

where d_{NH} is NHD metric and T_h is a threshold. Here, the threshold is chosen based on the experiment conducted to validate the perceptual robustness property of the hash. The detailed procedure for choosing a specific value of the threshold is described later.

5.1   Perceptual Video Hashing Techniques Considered for an NDVR System

The following perceptual hashing algorithms are considered under the proposed framework:

1. Achlioptas's random projection-based video hashing algorithm [62]
2. TWT-ARM-based video hashing algorithm [63]
3. Parallel factor analysis (PARAFAC)-based video hashing algorithm [59, 64]
4. Tucker decomposition-based video hashing algorithm [61]

The detailed description of extracting perceptual hash from the video using each of the method can be found in the references cited beside the name of the method. As demonstrated in [59, 61–63], these methods have shown robustness to most of the content-preserving attacks and hence were chosen for developing the NDVR system.

We have tested the NDVR application using the above-listed algorithms. However, a hash extracted from any other perceptual video hashing algorithm can also be used to implement the NDVR system. The experimental results and discussion are provided in Sect. 6.

6   Simulation Results and Discussion

A number of tests are carried out to study the performance of the proposed NDVR system. A typical NDVR system considers the following aspects: huge video database, scalability and efficient indexing structure to facilitate faster retrieval of NDVs. Since the prime motive of this paper is to demonstrate the use of robust perceptual video hashing algorithm for the NDVR purpose, we have not considered the algorithms specifically tailored for such an NDVR system for comparison. For demonstration, the perceptual hashing-based NDVR system (without considering the aforementioned factors) is also developed using the PARAFAC decomposition-based perceptual video hash [59, 64].

6.1   Database Details

Three different databases of different sizes are created from different sources with different levels of content-preserving and malicious attacks. The details of the databases are tabulated in Tables 2, 3 and 4.

Table 2 shows the details of the **Database-I** created from REEFVID [65] and XIPH [66] video databases. In this database, 224 downloaded videos were subjected to various content-preserving operations and malicious attack (as described in Table 2) to create 1792 NDVs. Thus, the total number of videos including the original and

Table 2 Database-I details for NDVR system using the videos downloaded from REEFVID and XIPH video databases

Parameter	Description
Database	[xiph.org], [reefvid.org]
Number of original videos	224
Number of NDVs	1792
Total number of videos	2016
Modified frame rates	15fps and 60fps
Modified spatial resolutions	704 × 576 and 176 × 144
Modified compression ratios	250:1 and 240:1
Insertion of a logo as a watermark	64 × 64 and 128 × 128

Table 3 Database-II details for NDVR system using the videos downloaded from OPEN-VIDEO database

Parameter	Description
Database	[open-video.org]
Number of original videos	224
Number of NDVs	2240
Total number of videos	2464
Modified frame rates	15fps and 60fps
Modified spatial resolutions	576 × 704 and 176 × 144
Increase in contrast	20%
Average Blurring	5 × 5
Adding AWGN	$\mu = 0$ and $\sigma = 57$
Cropping	10% from the boundary pixels
Insertion of a logo as a watermark	64 × 64 and 128 × 128

near-duplicates in this setup is equal to 2016. The perceptual hash of the video and its metadata (the title of the video) was stored in the form of a structured array.

Table 3 shows the details of the **Database-II** created from OPEN-VIDEO data set [67]. Here, 224 downloaded videos were subjected to ten different content-preserving operations and malicious attack (as described in Table 3) to create 2240 NDVs. Thus, the total number of videos including the original and near-duplicates in this setup is equal to 2464.

Table 4 shows the details of the **Database-III** created from REEFVID [65], XIPH [66], OPEN-VIDEO [67] and TRECVID[68] video databases. Here, 1,000 downloaded videos were subjected to nineteen different content-preserving operations and malicious attack (as described in Table 4) to create 19,000 NDVs. Thus, the total number of videos including the original and near-duplicates in this set up is equal to 20,000.

Table 4 Database-III details for NDVR system using the videos downloaded from REEFVID, XIPH, OPEN-VIDEO and TRECVID video databases.

Parameter	Description
Database	[xiph.org], [reefvid.org], [open-video.org], [trecvid.nist.gov]
Number of original videos	1000
Number of NDVs	19,000
Total number of videos	20000
Modified frame rates	15fps and 60fps
Modified spatial resolutions	576×704 and 176×144
Contrast modifications	$+100\%$ and -50%
Brightness modifications	$+5\%$ and -5%
Average blurring	5×5
Gaussian blurring	5×5
Adding AWGN	$\mu = 0$ and $\sigma = 57$
Adding salt and pepper noise	Density of 1%
Cropping	10% from the boundary pixels
Frame dropping	Only 128 frames were retained
Insertion of a logo as a watermark	64×64 and 128×128
Histogram equalization	Applied grey-level histogram equalization to R, G and B frames.
Frame rotation	$5°$ in anticlockwise direction
Compression	Average CR of 250:1

6.2   Validation of Perceptual Video Hash

Simulations were performed to validate the perceptual robustness, the visual fragility and the unpredictability properties of the hash extracted using the previously mentioned perceptual video hashing techniques. To validate the perceptual hash, 1000 secret keys and 224 videos from the database are considered. The hash values are compared using the NHD. To evaluate the useful properties of the hash, the histograms of the NHD are shown in Figs. 3, 4, 5 and 6. The validations are discussed and interpreted below.

- To validate the perceptual robustness property

The video database was corrupted with AWGN of $\sigma = 81$ and subsequently blurred with the Gaussian kernel of size 5×5, thereby generating a set of perceptually similar videos. The same secret key was used to generate the hashes from these videos. The NHD calculated between the hashes of the original videos and the perceptually similar videos was measured. A histogram skewed towards zero can be interpreted as that the perceptual hash closely satisfies the perceptual robustness property given in Eq. 3. The histogram of all the NHD values for this experiment is shown in Figs. 3a,

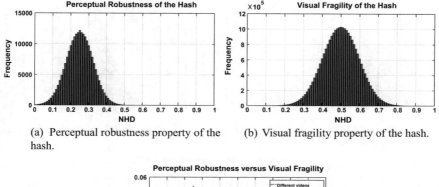

(a) Perceptual robustness property of the hash.

(b) Visual fragility property of the hash.

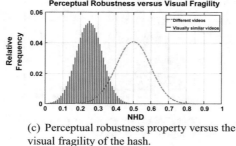

(c) Perceptual robustness property versus the visual fragility of the hash.

Fig. 3 Histogram of the NHD to validate the useful properties of the ARM-based perceptual video hash

4a, 5a and 6a. From the figures, we can conclude that the perceptual hash based on the Tucker decomposition satisfies the perceptual robustness property more closely than the perceptual hash based on the other video hashing methods.

- *To validate the visual fragility property*

128-bit hashes were obtained using the same secret key from 224 videos, and then, the NHD was calculated between all the possible combinations(twice at a time) of the hashes. The procedure was repeated for the remaining 999 secret keys. The histogram of the NHD values for this experiment is shown in Figs. 3b, 4b, 5b and 6b. It can be observed from the figure that the NHD calculated between the hashes in this case approximately follows a Gaussian distribution with the mean value very close to 0.5. A narrow-spread distribution centred around 0.5 can be interpreted as that the video hash closely satisfies the visually fragile property given in Eq. 4. Comparing Figs. 3b, 4b, 5b and 6b, we can conclude that the perceptual hash based on the Tucker decomposition satisfies the visual fragility property more closely than the perceptual hash based on the other video hashing methods.

- *To assess perceptual robustness versus visual fragility*

To evaluate the results of perceptual similarity obtained from the proposed video hashing algorithm for the perceptually similar and perceptually distinct videos, the

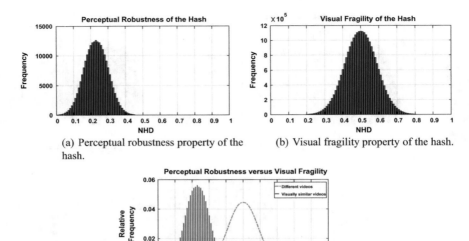

(a) Perceptual robustness property of the hash.

(b) Visual fragility property of the hash.

(c) Perceptual robustness property versus the visual fragility of the hash.

Fig. 4 Histogram of the NHD to validate the useful properties of the TWT-ARM-based perceptual video hash

histograms of the NHD obtained for the perceptual robustness property and the visual fragility property are superimposed as shown in Figs. 3c, 4c, 5c and 6c. From the figures, it can be observed that the histogram of the NHD for the perceptually similar and distinct videos overlaps. The overlapping indicates that there would be misclassification between the perceptually similar and distinct videos. Figure 6c also shows that the overlap of these two superimpose histograms is quite less implying lower misclassification rates. Further, the overlapping here is less compared to Figs. 3c, 4c and 5c. Hence, the performance of the video hashing algorithm based on the Tucker decomposition is better than the other video hashing algorithms under consideration.

The visual fragility and the perceptual robustness properties of the perceptual hashes generated using the Tucker decomposition, the PARAFAC decomposition, the ARM and the TWT-ARM are compared in terms of the mean and the standard deviation of the NHD in Table 5.

From Figs. 3, 4, 5, 6 and Table 5, we conclude that the perceptual hash generated using the Tucker decomposition satisfies all the desirable properties more efficiently as required for an NDVR system.

Thus, the perceptual hash of the video as a fingerprint could be used to develop an NDVR system.

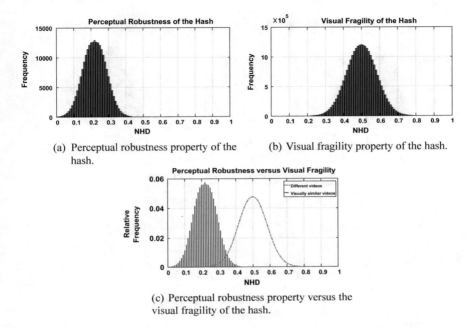

(a) Perceptual robustness property of the hash.

(b) Visual fragility property of the hash.

(c) Perceptual robustness property versus the visual fragility of the hash.

Fig. 5 Histogram of the NHD to validate the useful properties of the PARAFAC decomposition-based perceptual video hash

6.3 Performance Evaluation

The query to the NDVR application is given through query by example. The hash is obtained for the query video using the video hashing algorithm and then compared with all the hashes present in the database. The NDV is retrieved and indexed as per the match between the query video hash and video hashes in the database. The average precision–recall curves [69–71] are used to benchmark the performance of the NDVR application based on the previously mentioned video hashing algorithms. For more information on usage of the precision, P and the recall, R in evaluating the NDVR system, the readers may refer to [63, 69–71]. The effectiveness of the retrieval systems is associated with relevance of retrieved videos in terms of precision and recall. Ideally, the value of both the precision and the recall must be close to 1.

For each video query, R and P values are obtained. In this work, 50 queries were given to the NDVR application to calculate and then plot the average precision–recall values. To minimize the total number of NDVs retrieved out of the total number of videos retrieved, a threshold value of 0.4 on d_{NH} was selected. This value of threshold is chosen based on the following argument. For the video hashing algorithms under consideration, it can be observed from the validation of the perceptual robustness property (from Sect. 6.2) of the hash that d_{NH} between the original video and the NDVs was found to be less than 0.5 in most of the cases. From the validation of

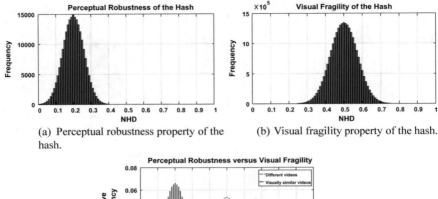

(a) Perceptual robustness property of the hash.

(b) Visual fragility property of the hash.

(c) Perceptual robustness property versus the visual fragility of the hash.

Fig. 6 Histogram of the NHD to validate the useful properties of the Tucker decomposition-based perceptual video hash

Table 5 Comparison of hash validity among the perceptual video hashing algorithms considered

Name of the video hashing algorithm	Perceptual robustness		Visual fragility	
	Mean	Standard deviation	Mean	Standard deviation
ARM	0.25	0.147	0.5	0.193
TWT-ARM	0.225	0.132	0.5	0.176
PARAFAC decomposition	0.215	0.127	0.5	0.164
Tucker decomposition	0.2	0.118	0.5	0.147

the visual fragility property of the hash, it can be observed that the d_{NH} between the different videos ranges between 0.2 and 0.8. Therefore, to increase the retrieval of NDVs, a value of 0.4 is selected as the threshold value. The average precision–recall curves for the proposed NDVR application based on the previously mentioned perceptual video hashing algorithms are shown in Fig. 8a. The curve closest to the values of $P = 1$ and $R = 1$ indicates the superior performance [69–71]. From Fig. 8a, it can be observed that the performance of all the video hashing algorithms is good

Table 6 List of first few videos retrieved for the query video titled waterfall_cif using the ARM-based video hashing algorithm.

Video retrieved	NHD
waterfall_cif	0
waterfall_cif_br100k	0.046875
waterfall_cif_br64k	0.046875
waterfall_cif_wm64	0.0625
waterfall_cif_big	0.125
waterfall_cif_small	0.125
waterfall_cif_15fps	0.140625
waterfall_cif_60fps	0.140625
clip216_wm64	0.1875

and almost identical except the video hashing algorithm based on the ARM. This is quite expected as the size of the database considered is relatively small and also from the hash validation properties in Sect. 6.2 and the ROC curves obtained in [59, 61–63].

Table 6 shows the list of first few videos retrieved for the query video titled "waterfall_cif" using the ARM-based video hashing algorithm. Figure 7 shows the first frame of few retrieved videos as the representative frame.

Following the same procedure described earlier, the performance of the NDVR application was evaluated on the database-II using the average precision–recall curves for 50 queries. From Fig. 8b, it can be observed that the average precision–recall curves have fallen by a small amount compared to the curves in Fig. 8a implying that the performance of the NDVR systems has reduced by a small amount. This is due to the increase in the number of NDVs in database-II compared to the number of NDVs in the database-I. The video hashing algorithm based on Tucker decomposition has shown the best performance which is very closely followed by the two video hashing algorithms, namely the TWT-ARM and the PARAFAC decomposition-based video hashing algorithm. The performance of the ARM-based video hashing algorithm is relatively poor compared to the other video hashing algorithms. The argument for this performance is as follows: the ROC curves in [61] demonstrated that the Tucker decomposition-based video hashing algorithm had a minimum area under the ROC curves for most of the attacks followed by the TWT-ARM, the PARAFAC and the ARM-based video hashing algorithm. Furthermore, from the hash validation properties, the overlap between the curves of perceptual robustness property and the visual fragility property was least for the Tucker decomposition-based video hashing algorithm, indicating the maximum retrieval of NDVs, excess for the ARM-based video hashing algorithm, indicating the minimum retrieval of NDVs.

Using the procedure mentioned above, the performance of the NDVR system was evaluated on the database-III using the average precision–recall curves for 50 queries. From Fig. 8c, it can be observed that the average precision–recall curves have fallen

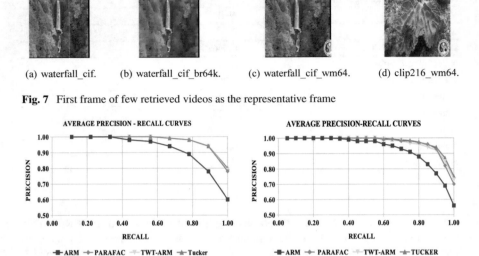

(a) waterfall_cif.   (b) waterfall_cif_br64k.   (c) waterfall_cif_wm64.   (d) clip216_wm64.

Fig. 7 First frame of few retrieved videos as the representative frame

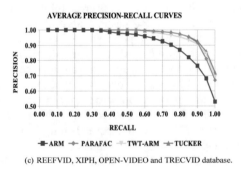

(c) REEFVID, XIPH, OPEN-VIDEO and TRECVID database.

Fig. 8 Performance evaluation of the perceptual video hash-based NDVR application through average precision–recall curves using the videos downloaded from different databases.

further compared to the curves in Fig. 8b implying that the performance of the NDVR systems has reduced. This is due to considerable increase in the number of the original videos and the number of NDVs in database-III compared to the size of the database-II. It can be observed that the curve of NDVR application of Tucker decomposition-based video hashing is close to the values of $P = 1$ and $R = 1$ followed by the TWT-ARM-based video hashing, the PARAFAC decomposition-based video hashing and, finally, the ARM-based video hashing. Again the performance of the ARM-based video hashing algorithm is relatively poor compared to the other video hashing algorithms. The performance of the algorithms is justified similarly on the above arguments presented for the performance of the NDVR systems on database-II.

Since the values of the precision and the recall are found to be satisfactory for the database under consideration, it can be concluded that the perceptual hash of the video as a fingerprint can be used to develop an NDVR system.

A detail discussion on the computational complexity of all the perceptual video hashing algorithms considered for the development of NDVR application was presented in [63]. From [63], it can be concluded that the ARM-based video hashing has the lowest computational complexity and then subsequently by the Tucker decomposition-based video hashing.

7 Conclusion

In this paper, a near-duplicate video retrieval (NDVR) application based on perceptual video hashing techniques was proposed. The NDVR application was developed using the perceptual video hashing techniques based on (1) the Achlioptas's random matrix (ARM), (2) the temporal wavelet transform (TWT) and the ARM, (3) the Tucker decomposition and (4) the parallel factor (PARAFAC) decomposition. Three different databases of different sizes were created from different sources with different levels of content-preserving and malicious attacks. A number of tests were carried out to validate the use of perceptual hash for the NDVR application. The performance of the NDVR application developed using all the perceptual video hashing algorithms mentioned above was evaluated using the average precision–recall curves. It was concluded that the NDVR application employing the Tucker decomposition-based perceptual video hashing algorithm is relatively stable compared to the other perceptual video hashing algorithms. Considering the retrieval performance and the computational complexity of the Tucker decomposition-based perceptual video hashing algorithm, it can be concluded that this algorithm is suited best for the retrieval of near-duplicate videos among the perceptual video hashing algorithms considered. The limitation of the NDVR application based on the perceptual video hash is that its performance would deteriorate for the partial near-duplicate video query.

The NDVR application was developed in this paper using the perceptual video hashing algorithms without considering the retrieval speed. A direction for the future research would be to design a fast video retrieval application based on the perceptual video hash and test its performance on a large video database. Furthermore, an NDVR application can be developed to retrieve the longer video based on i) the summarized version of the longer video and ii) the partial near-duplicate videos.

References

1. (2015), http://tubularinsights.com/vidcon-2015-strategic-insights-tactical-advice/
2. Wu, X., Ngo, C., Hauptmann, A.G., Tan, H.: Real-time near-duplicate elimination for web video search with content and context. IEEE Trans Multim **11**(2), 196–207 (2009). https://doi.org/10.1109/TMM.2008.2009673

3. Wu X, Hauptmann AG, Ngo CW (2007) Practical elimination of near-duplicates from web video search. In: Proceedings of the 15th ACM international conference on multimedia. MM '07, ACM, New York, pp 218–227. https://doi.org/10.1145/1291233.1291280
4. Shen, L., Hong, R., Hao, Y.: Advance on large scale near-duplicate video retrieval. Front Comput Sci **14**(5), 1–24 (2020)
5. Chou, C.L., Chen, H.T., Lee, S.Y.: Pattern-based near-duplicate video retrieval and localization on web-scale videos. IEEE Trans Multim **17**(3), 382–395 (2015)
6. Xie Q, Huang Z, Shen HT, Zhou X, Pang C (2010) Efficient and continuous near-duplicate video detection. In: Web conference (APWEB), 2010 12th international Asia-Pacific. IEEE, pp 260–266
7. Xie, Q., Huang, Z., Shen, H.T., Zhou, X., Pang, C.: Quick identification of near-duplicate video sequences with cut signature. World Wide Web **15**(3), 355–382 (2012)
8. Zhao, W.L., Ngo, C.W.: Scale-rotation invariant pattern entropy for keypoint-based near-duplicate detection. IEEE Trans Image Process **18**(2), 412–423 (2009)
9. Zhou, Z., Wang, Y., Wu, Q.J., Yang, C.N., Sun, X.: Effective and efficient global context verification for image copy detection. IEEE Trans Inf For Secur **12**(1), 48–63 (2017)
10. Zhou, Z., Wu, Q.J., Huang, F., Sun, X.: Fast and accurate near-duplicate image elimination for visual sensor networks. Int J Distrib Sens Netw **13**(2), 1550147717694172 (2017)
11. Esmaeili, M.M., Fatourechi, M., Ward, R.K.: A robust and fast video copy detection system using content-based fingerprinting. IEEE Trans Inf For Secur **6**(1), 213–226 (2011)
12. Malekesmaeili M, Fatourechi M, Ward RK (2009) Video copy detection using temporally informative representative images. In: International conference on machine learning and applications, 2009. ICMLA'09. IEEE, pp 69–74
13. Saikia N, Bora PK (2007) Video authentication using temporal wavelet transform. In: International conference on advanced computing and communications, 2007. ADCOM 2007. IEEE, pp 648–653
14. Shen HT, Zhou X, Huang Z, Shao J, Zhou X (2007) Uqlips: a real-time near-duplicate video clip detection system. In: Proceedings of the 33rd international conference on very large data bases. VLDB Endowment, pp 1374–1377
15. Liu, J., Huang, Z., Cai, H., Shen, H.T., Ngo, C.W., Wang, W.: Near-duplicate video retrieval: current research and future trends. ACM Comput Surv (CSUR) **45**(4), 44 (2013)
16. Money, A.G., Agius, H.: Video summarisation: a conceptual framework and survey of the state of the art. J Vis Commun Image Represent **19**(2), 121–143 (2008)
17. Thomas, S.S., Gupta, S., Subramanian, V.K.: Perceptual video summarization—a new framework for video summarization. IEEE Trans Circuits Syst Video Technol **27**(8), 1790–1802 (2017)
18. Bohm C, Gruber M, Kunath P, Pryakhin A, Schubert M (2007) Prover: probabilistic video retrieval using the gauss-tree. In: IEEE 23rd International conference on data engineering, 2007. ICDE 2007. IEEE, pp 1521–1522
19. Jagadish, H.V., Ooi, B.C., Tan, K.L., Yu, C., Zhang, R.: iDistance: an adaptive b+-tree based indexing method for nearest neighbor search. ACM Trans Database Syst (TODS) **30**(2), 364–397 (2005)
20. Huang Z, Wang L, Shen HT, Shao J, Zhou X (2009) Online near-duplicate video clip detection and retrieval: an accurate and fast system. In: IEEE international conference on data engineering. IEEE, pp 1511–1514
21. Grauman, K.: Efficiently searching for similar images. Commun ACM **53**(6), 84–94 (2010)
22. Liu D, Yu Z (2015) A computationally efficient algorithm for large scale near-duplicate video detection. In: International conference on multimedia modeling. Springer, Berlin, pp 481–490
23. Datar M, Immorlica N, Indyk P, Mirrokni VS (2004) Locality-sensitive hashing scheme based on p-stable distributions. In: Proceedings of the twentieth annual symposium on Computational geometry. ACM, pp 253–262
24. Patel B, Meshram B (2012) Content based video retrieval systems. arXiv preprint arXiv:1205.1641

25. Cheung SC, Zakhor A (2000) Efficient video similarity measurement and search. In: 2000 International conference on image processing, 2000. Proceedings, vol 1. IEEE, pp 85–88
26. Raj, J.S., Joe, M.C.V.: Wi-fi network profiling and qos assessment for real time video streaming. IRO J Sustain Wireless Syst **3**(1), 21–30 (2021)
27. Wu X, Zhao WL, Ngo CW (2007) Near-duplicate keyframe retrieval with visual keywords and semantic context. In: Proceedings of the 6th ACM international conference on Image and video retrieval. ACM, pp 162–169
28. Wu X, Takimoto M, Satoh S, Adachi J (2008) Scene duplicate detection based on the pattern of discontinuities in feature point trajectories. In: Proceedings of the 16th ACM international conference on Multimedia. ACM, pp 51–60
29. Lowe, D.G.: Distinctive image features from scale-invariant keypoints. Int J Comput Vis **60**(2), 91–110 (2004)
30. Ke Y, Sukthankar R (2004) Pca-sift: a more distinctive representation for local image descriptors. In: Proceedings of the 2004 IEEE computer society conference on computer vision and pattern recognition, CVPR 2004. vol 2. IEEE, p II
31. Zhao, G., Pietikainen, M.: Dynamic texture recognition using local binary patterns with an application to facial expressions. IEEE Trans Pattern Anal Mach Intell **29**(6), 915–928 (2007)
32. Venna J, Peltonen J, Nybo K, Aidos H, Kaski S (2010) Information retrieval perspective to nonlinear dimensionality reduction for data visualization. J Mach Learn Res 11:451–490
33. Jiang, Y.G., Ngo, C.W.: Visual word proximity and linguistics for semantic video indexing and near-duplicate retrieval. Comput Vis Image Understand **113**(3), 405–414 (2009)
34. Wang L, Elyan E, Song D (2014) Rebuilding visual vocabulary via spatial-temporal context similarity for video retrieval. In: International conference on multimedia modeling. Springer, Berlin, pp 74–85
35. Poullot S, Crucianu M, Buisson O (2008) Scalable mining of large video databases using copy detection. In: Proceedings of the 16th ACM international conference on Multimedia. ACM, pp 61–70
36. Kim KR, Jang WD, Kim CS (2015) Frame-level matching of near duplicate videos based on ternary frame descriptor and iterative refinement. In: 2015 IEEE international conference on image processing (ICIP). IEEE, pp 31–35
37. Liu L, Lai W, Hua XS, Yang SQ (2007) Video histogram: a novel video signature for efficient web video duplicate detection. In: International conference on multimedia modeling. Springer, Berlin, pp 94–103
38. Shen HT, Ooi BC, Zhou X (2005) Towards effective indexing for very large video sequence database. In: Proceedings of the 2005 ACM SIGMOD international conference on Management of data. ACM, pp 730–741
39. Shang L, Yang L, Wang F, Chan KP, Hua XS (2010) Real-time large scale near-duplicate web video retrieval. In: Proceedings of the 18th ACM international conference on Multimedia. ACM, pp 531–540
40. Park HJ (1995) Signal transformation system and method for providing picture-in-picture in high definition television receivers. US Patent 5,386,241
41. Bronstein MM, Bronstein AM, Michel F, Paragios N (2010) Data fusion through cross-modality metric learning using similarity-sensitive hashing. In: 2010 IEEE conference on computer vision and pattern recognition (CVPR). IEEE, pp 3594–3601
42. Song J, Yang Y, Yang Y, Huang Z, Shen HT (2013) Inter-media hashing for large-scale retrieval from heterogeneous data sources. In: Proceedings of the 2013 ACM SIGMOD international conference on management of data. ACM, pp 785–796
43. Xu, J., Jagadeesh, V., Manjunath, B.: Multi-label learning with fused multimodal bi-relational graph. IEEE Trans Multim **16**(2), 403–412 (2014)
44. Liu X, Huang L, Deng C, Lu J, Lang B (2015) Multi-view complementary hash tables for nearest neighbor search. In: Proceedings of the IEEE international conference on computer vision, pp 1107–1115
45. Ranjan V, Rasiwasia N, Jawahar C (2015) Multi-label cross-modal retrieval. In: Proceedings of the IEEE international conference on computer vision, pp 4094–4102

46. Shen X, Shen F, Sun QS, Yuan YH (2015) Multi-view latent hashing for efficient multimedia search. In: Proceedings of the 23rd ACM international conference on multimedia. ACM, pp 831–834

47. Hao, Y., Mu, T., Hong, R., Wang, M., An, N., Goulermas, J.Y.: Stochastic multiview hashing for large-scale near-duplicate video retrieval. IEEE Trans Multim **19**(1), 1–14 (2017)

48. Cheng H, Wang P, Qi C (2021) Cnn retrieval based unsupervised metric learning for near-duplicated video retrieval. arXiv preprint arXiv:2105.14566

49. Kordopatis-Zilos G, Papadopoulos S, Patras I, Kompatsiaris Y (2017) Near-duplicate video retrieval with deep metric learning. In: 2017 IEEE international conference on computer vision workshop (ICCVW). IEEE, pp 347–356

50. Zhang, Y., Zhang, Y., Sun, J., Li, H., Zhu, Y.: Learning near duplicate image pairs using convolutional neural networks. Int J Perform Eng **14**(1), 168 (2018)

51. Ding, L., Tian, Y., Fan, H., Chen, C., Huang, T.: Joint coding of local and global deep features in videos for visual search. IEEE Trans Image Process **29**, 3734–3749 (2020)

52. Kordopatis-Zilos G, Papadopoulos S, Patras I, Kompatsiaris Y (2017) Near-duplicate video retrieval with deep metric learning. In: Proceedings of the IEEE international conference on computer vision workshops, pp 347–356

53. Zhang, C., Lin, Y., Zhu, L., Liu, A., Zhang, Z., Huang, F.: Cnn-vwii: an efficient approach for large-scale video retrieval by image queries. Pattern Recogn Lett **123**, 82–88 (2019)

54. Anuranji, R., Srimathi, H.: A supervised deep convolutional based bidirectional long short term memory video hashing for large scale video retrieval applications. Dig Signal Processing **102**, 102729 (2020)

55. Chen, H., Hu, C., Lee, F., Lin, C., Yao, W., Chen, L., Chen, Q.: A supervised video hashing method based on a deep 3d convolutional neural network for large-scale video retrieval. Sensors **21**(9), 3094 (2021)

56. Nie, X., Zhou, X., Shi, Y., Sun, J., Yin, Y.: Classification-enhancement deep hashing for large-scale video retrieval. Appl Soft Comput **109**, 107467 (2021)

57. Phalke DA, Jahirabadkar S (2020) A survey on near duplicate video retrieval using deep learning techniques and framework. In: 2020 IEEE Pune section international conference (PuneCon). IEEE, pp 124–128

58. Coskun B, Sankur B (2004) Robust video hash extraction. In: Signal processing and communications applications conference, 2004. Proceedings of the IEEE 12th, pp 292–295

59. Li, M., Monga, V.: Robust video hashing via multilinear subspace projections. IEEE Trans Image Process **21**(10), 4397–4409 (2012)

60. Venkatesan R, Koon SM, Jakubowski M, Moulin P (2000) Robust image hashing. In: Proceedings international conference on image processing, vol 3, pp 664–666

61. Sandeep, R., Sharma, S., Thakur, M., Bora, P.K.: Perceptual video hashing based on Tucker decomposition with application to indexing and retrieval of near-identical videos. Multim Tools Appl **75**(13), 7779–7797 (2016). https://doi.org/10.1007/s11042-015-2695-1

62. Sandeep R, Bora P (2013) Perceptual video hashing based on the Achlioptas's random projections. In: 2013 Fourth national conference on computer vision, pattern recognition, image processing and graphics (NCVPRIPG), pp 1–4. https://doi.org/10.1109/NCVPRIPG.2013.6776252

63. Sandeep R, Bora PK (2019) Perceptual video hashing based on temporal wavelet transform and random projections with application to indexing and retrieval of near-identical videos. Multim Tools Appl 1–21. https://doi.org/10.1007/s11042-019-7189-0

64. Li M, Monga V (2011) Desynchronization resilient video fingerprinting via randomized, low-rank tensor approximations. In: 2011 IEEE 13th international workshop on multimedia signal processing (MMSP), pp 1–6

65. Test video sequences (2012), http://media.xiph.org/video/derf/

66. Test video sequences (2012), http://www.reefvid.org/

67. Test video sequences (2016), http://open-video.org

68. Test video sequences (2016), http://trecvid.nist.gov

69. Manning, C.D., Raghavan, P., Schütze, H.: Introduction to Information Retrieval. Cambridge University Press, New York (2008)
70. Singhal, A.: Modern information retrieval: a brief overview. Bull IEEE Comput Soc Tech Committee Data Eng **24**, 2001 (2001)
71. Zhou, B., Yao, Y.: Evaluating information retrieval system performance based on user preference. J Intell Inf Syst **34**(3), 227–248 (2010)

ECG Classification Using Machine Learning Classifiers with Optimal Feature Selection Methods

Nithya Karthikeyan and Mary Shanthi Rani

Abstract Machine Learning (ML) is a booming technology well-suited for data analysis in all fields, especially healthcare. Cardiovascular Disease (CVD) is one of the deadliest diseases that has a high mortality rate. Irregular beats of the heart are known as Cardiac Arrhythmia (CA), and its diagnosis is a challenging task for cardiologists. Feature selection is an essential and inevitable process that identifies the causing factors of arrhythmia. The proposed model proves that arrhythmia prediction is possible with a limited set of features obtained from Electrocardiogram (ECG) signals. Firstly, the standard UCI Machine Learning Arrhythmia dataset is subject to preprocessing and normalization after applying optimal feature selection techniques such as Spearman Ranking Coefficient and Mutual Information. Eventually, the result was a reduced dataset fed as the input for eight robust Machine Learning classifiers such as Decision Tree (DT), Adaboost (AB), *K*-Nearest Neighbour (K-NN), Naive Bayes (NB), Random Forest (RF), Gradient Boosting (GB), Support Vector Machine (SVM), and Multilayer Perceptron (MLP). Analysis of the results presents the highest accuracy of 91.3% with the Spearman feature selection technique and Random Forest Classifier. Furthermore, the proposed work shows better achievement when compared with state-of-the-art methods.

Keywords Cardiac Arrhythmia · Machine learning · Spearman · Mutual information · Electrocardiogram

1 Introduction

Bioelectrical signals are produced from the complex self-regulatory systems and measured with electrical activity in the cell or organ. They have only minor frequency

N. Karthikeyan · M. S. Rani (✉)
The Gandhigram Rural Institute (Deemed to be University), Gandhigram, Tamil Nadu 624 302, India
e-mail: m.maryshanthirani@ruraluniv.ac.in

© The Author(s), under exclusive license to Springer Nature Singapore Pte Ltd. 2022 277
V. Suma et al. (eds.), *Evolutionary Computing and Mobile Sustainable Networks*, Lecture Notes on Data Engineering and Communications Technologies 116, https://doi.org/10.1007/978-981-16-9605-3_19

and amplitude, which may be the Electrocardiogram (ECG) and electroencephalo-gram (EEG). ECG records the heart's electrical activity, whilst to test the electrical activity in the brain EEG is used.

A disease that leads humanity to the highest mortality rate worldwide was due to cardiovascular problems. The human heart generates an electrical impulse which is represented as an ECG signal through the ECG machine. The signal is categorized as P, QRS, and T waves. Regular sinus rhythm waves in the signal have specific charac-teristics. The ECG signals help to identify the abnormalities in the heart rhythm. A patient may get a cardiovascular problem because of parental history, smoking, high blood pressure, obesity, low blood sugar, and Diabetes. CVD has a high mortality and morbidity rate. Hence, prevention and diagnosis at the early stage are crucial, thereby emphasizing the importance of precise and early diagnosis of Cardiac Arrhythmia [1, 2].

ECG is a non-invasive and inexpensive tool that records the physiological activity carried out in the heart. This mechanism is used for the detection and prognostication of cardiovascular diseases. An automated computer-aided system provides accurate reports, and it has the potential of detecting various cardiovascular disease groups such as Arrhythmias, Myocardial Infarction, and Hypertrophy [3].

Analysing heartbeat irregularity in the physiological circumstance determines the autonomic behaviour of the heart. Abnormality in the heartbeat impulse leads to arrhythmia. A lower heartbeat is referred to as Bradycardia, and a higher beat leads to Tachycardia. Generally, Cardiologists detect various cardiac problems by identifying the heartbeat variations. Automated systems are developed to circumvent problematic situations [4].

Gigantic evolution is accomplished in the domain of Machine Learning in the last decades. Health care holds massive data which may be heterogeneous, and ML plays a significant role in analysing, processing, and predicting massive heterogeneous datasets and providing a veracious diagnosis. ML serves as a great aid in medical image processing, and it has become a powerful method in the classification of arrhythmia using ECG signals [5, 6].

The major challenge in the classification of arrhythmia is selecting features that accurately identify heartbeat variations leading to CA. This proposed research work is an effort to develop a robust ML classifier with an optimal feature set using effi-cient feature selection methods. This research paper is organized as follows; Sect. 2 presents the related works. Section 3 discusses the newly proposed methods, and a discussion on experimental results is given in Sect. 4. Finally, the conclusion is given in Sect. 5.

2 Related Works

Many researchers are working on classifying arrhythmia using various ML methods by applying different feature selection methodologies for preprocessing. Related literature in this classification has been presented here.

Fazel et al. [7] tried several ML algorithms such as SVM, Random Forest, Logistic Regression, K-NN, and Decision Tree with the reduced features. SVM showed a higher degree of accuracy of 72.7% amongst the algorithms used. Jadhav et al. [8] had applied three efficient Artificial Neural Network (ANN) models in their experiments with different dataset splitting. The three different ANN models used by them are Modular Neural Network (MNN), Multilayer Perceptron (MLP) Neural Network models, and Generalized Feedforward Neural Network (GFFNN). Amongst these models, MLP showed its performance with an accuracy of 86.67%. Batra et al. [9] have developed a technique by applying rigorous preprocessing and specific feature selection methods. The selected features were subjected to a classification process using different classifiers such as SVM, Random Forest, Neural Networks, and Decision Trees. The results show that Gradient Boosting with SVM presented an accuracy of 84.82%. Abirami et al. [10] worked with three ANN models to differentiate normal and abnormal signals. The powerful Recurrent Neural Network (RNN) method attained a better result achieving 83.1% accuracy, 66.7% specificity, and sensitivity of 86.7% compared with the other two ANN models. Jadhav et al. [11] designed a network model trained based on the static backpropagation algorithm that predicts normal and abnormal classes. The newly developed Generalized Feedforward Neural Network (GFNN) performs well in classifying arrhythmia, with the closest classification accuracy of 82.35%. Niazi et al. [12] used K-NN and SVM classifier for their experiment by implementing Improved F-score and sequential forward search (IFSFS) technique as feature selection algorithm and applied 20-fold validation. The model attained an accuracy of 68.8% and 73.8% with SVM and K-NN, respectively. Pandey et al. [13] discovered a new technique of executing a dataset with eight classifiers Long Short Term Memory (LSTM), SVM, Naive Bayes, Adaboost, MLP, Decision Tree, Random Forest, and K-NN. Principal Component Analysis was subsequently executed with eight classifiers, and the model is trained with a 90–10 ratio and with the limited features of 102. A comparison of results from different classifiers reveals that Random forest reached an accuracy of 89.74% and SVM attained an accuracy of 89.74%. Mitra et al. [14] have introduced a method of correlation-based feature selection (CFS) with forward selection method followed by Levenberg–Marquardt (LM) and Incremental back propagation neural network (IBPLN). It has been observed that LM reached an accuracy of 81.88% compared with an incremental backpropagation learning network. Namsrai et al. [15] introduced a technique by reducing features to 205 and applying the classifier Naïve Bayes to reach a performance accuracy of 70.50%. Samad et al. [16] developed a method to detect arrhythmia with three classifiers K-NN, Decision tree, and Naïve Bayes; amongst the classifiers, K-NN performed well compared with the other two and achieved an accuracy of 70.1%. Elsayad [17] tried with learning vector quantization neural network algorithms. They applied six different LVQ algorithms, namely, LVQ1, LVQ2.1, LVQ3, OLVQ1, OLVQ3, and LVQ2.1, which showed better performance and reached an accuracy of 74.12% Zuo et al. [18] derived a novel method known as Kernel Difference-Weighted K-nearest Neighbour classifier (KDF-WK-NN), and the proposed technique worked well when compared with traditional K-NN registering its performance with an accuracy of 70.66%. Ozcift [19] has developed a

Random forest ensemble classifier by applying the wrapper feature selection method and attained the highest accuracy of 90%. Yilmaz has developed a new system with two stages for detecting arrhythmia with the help of fisher score to reduce features in the data set as a feature selection method. The classifier least-square Support Vector Machine (LS-SVM) was implemented with the help of a 2D grid search. The dataset was used with tenfold cross-validation to evaluate accuracy that reached 82.09% [20]. Shimpi et al. [21] have worked with four classifiers: random forest, Logistic Regression, SVM, and K-NN classifier for diagnosing an arrhythmia. Amongst the classifiers, SVM showed its better performance when testing the data by cross-validation. Attained accuracy by the classifier was 91.2%.

3 Materials and Methods

3.1 Description of the Dataset

UCI ML repository is a vast repository of data that facilitates researchers to validate their algorithms. The arrhythmia dataset has been retrieved from the UCI repository, which holds 452 details of patients. The dataset contains 279 features that some of which are linear valued data and categorical data. The linear valued data amongst 279 components is 206, and the remaining data are considered categorical. The goal is to classify the instance whether there is a presence of arrhythmia or not. The dataset is grouped into 16 classes where Class 1 is considered healthy or normal ECG signal, 2–15 represents the different arrhythmia classes, and Class 16 represents the unexplored arrhythmia. The first four attributes refer to the Personal Information of the patient, and the remaining attributes are extracted from the ECG signals [22].

3.2 Methods

Diagnosing Cardiovascular disorders at an earlier stage is the purpose of this research work. Classification of abnormalities is carried out using ECG signals with the Arrhythmia dataset. Three essential stages are enforced in the process of diagnosing arrhythmia, namely, (i) preprocessing, (ii) feature selection, and (iii) classification in the proposed Optimal Feature Selection with Reduced Set framework.

An outline of the proposed methodology is given below:

Stage I   **Pre-processing**

　　　　Step 1   Replace missing values in the dataset.
　　　　Step 2   Standardize data by applying normalization.

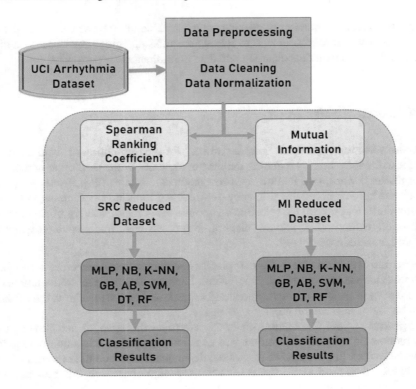

Fig. 1 Block diagram of the OFSRS framework

Stage II   **Feature selection**

       Step 3   Select optimal feature set by applying Spearman Ranking.
                   Coefficient method and Mutual Information separately to the
                   processed dataset.

Stage III   **Classification using ML Classifiers**

       Step 4   Classify the signals using eight classifiers separately with
                   optimal reduced data set generated from stage II.

The flow diagram of the Optimal Feaure Selection with Reduced Set (OFSRS) framework is shown in Fig. 1

3.3 Preprocessing

This proposed work uses the arrhythmia dataset retrieved from the University of California, Irvine (UCI) Machine Learning Repository. It is prevalent that a dataset may contain a large number of missing values, and handling those values is the first and foremost preprocessing step to follow; the missing values are replaced with a mean value, and then data cleaning and data normalization is performed [23].

3.4 Feature Selection

Feature selection techniques make a pathway for the diminution of features in the larger dataset, eventually reducing the time complexity of the prediction process. In this research work, two feature selection methods are used [24]. In this proposed work, we have explored the efficiency of Spearman Ranking Coefficient (SRC) and Mutual Information (MI) in selecting an optimal set of features as significant feature reduction methods. Two feature selection techniques were implemented separately to produce a reduced dataset.

Spearman Ranking Coefficient. Typically, the connection between two variables can be calculated by correlation coefficient. It shows the co-expression relationship between two variables above the threshold value; based on the threshold, the variable will be removed from the input.

Spearman Rank Correlation methods determine the strength and direction of the monotonic relation between the two connected variables. It is a non-parametric method to measure correlation (statistical dependence between the rankings of two variables). It is used as a feature selection technique based on this ranking. The formulas are used to calculate this coefficient.

$$\rho = 1 - \frac{6 \Sigma d_i^2}{n(n^2 - 1)} \tag{1}$$

where 'di' is the difference between ranks in paired items and 'n' is the total number of observations in Eq. (1) [25].

Mutual Information. Mutual Information is used to compute the statistical dependence between random variables. MI has strong proficiency in handling linear and non-linear dependencies. It is best suited for feature selection to pick relevant attributes and predict the target class. MI is computed as the measure of dependencies between two discrete random variables using Eq. (2)

$$I(X; Y) = \sum_{y \in Y} \sum_{x \in X} p(X, Y)(x, y) \log \frac{p(X, Y)(x, y)}{pX(x)pY(y)} \tag{2}$$

where $p(X, Y)$ is considered the joint probability mass function of X, Y and pX and pY are considered as marginal probability mass function of X and Y, respectively. The mutual information between two continuous random variables is calculated using Eq. (3) [26].

$$I(X;Y) = \int_Y \int_X p(X,Y)(x,y) . \log \frac{p(X,Y)(x,y)}{pX(x)pY(y)} dxdy \qquad (3)$$

A Machine Learning (ML) algorithm generally focuses on developing a mathematical model to figure out a solution by learning the features. It is widely used in classification, detection, regression, segmentation, etc. [27, 28]. The machine learning dataset is first divided for training and testing in a feasible ratio. The algorithm learns the characteristics of the data whilst training the dataset, and the model is being validated using the validation dataset. Finally, the test data is used to find the accuracy and compactness of the developed algorithm [29, 30].

A vector of continuous and/or discrete feature values is chosen as input for a classifier and produces discrete values as an output [31].

Most Machine learning algorithms have the dynamism of predicting output accurately, which aids the clinicians in making a perfect diagnosis. A few familiar ML algorithms used for classification are Decision Tree, Naive Bayes, Random Forest, SVM, K-NN, ANN, etc., [32].

3.5 Classification Using Machine Learning Classifiers

Decision Tree. A decision tree classifier is a tree-like structure to design a model using a training dataset, and based on this training set, and the testing dataset was classified. Choosing the best data and eliminating the unwanted nodes are the essential features of this efficient classifier to improve the model's performance [33].

Random Forest. A Random Forest algorithm is a combination of various randomized decision trees and consolidation of the predictions. It shows its efficiency in classification and regression, and it is flexible when applied to large-scale problems and with different Adhoc learning tasks. Random Forest classifier is efficient for the imbalanced datasets [34, 35].

Support Vector Machine. SVM is an ML algorithm employed to find the hyperplane in n-dimensional space, broadly used in Regression and Classification. Using a nonlinear function $\varphi(x)$, high dimensional spaces are mapped with training vectors. It is found to be highly robust to solve complex problems. [36, 37]

K-NN. It is a traditionally used and simple classification algorithm. It selects k nearest neighbours based on the Euclidean distance amongst the neighbours. K-NN algorithm is best suited for both Regression and Classification problems [38].

Gradient Boosting. Boosting algorithms are ensemble methods that convert the learning algorithm for the base class of a model. Gradient Boosting works based on the principle of cutting down the loss function. It is the efficient method used in developing the model [39].

Naive Bayes. Naive Bayes algorithm follows the techniques of probabilistic learning. The probability of the target classes is computed based on the evidence given by the feature value with the help of training data. The probabilities estimate the likelihood of the unlabelled feature. [40]

Adaboost. The best outperforming boosting algorithm is the Adaptive boosting algorithm, also referred to as Adaboost. This algorithm promotes the weak classifier to perform as a robust classifier bringing out a new idea with high accuracy [41].

MLP. The multilayer perceptron is an artificial neural network with three layers, the input layer, the hidden layer, and the output layer. It is one of the feed-forward neural networks. The neurons of MLP are trained with a backpropagation learning algorithm. The MLP with even one hidden layer has the potential for estimating continuous function [42].

4 Results and Discussion

This section has briefly discussed the implementation of feature selection algorithms and training the models with Machine learning classifiers. The objective of this research work is to reach a reasonable accuracy with a reduced feature combination. The outcome of the work is discussed in the following section.

4.1 Performance Metrics

The proposed experiment was evaluated by standard metrics, namely, Precision F_1-score, Accuracy, and Recall. The aforesaid metrics have been attained by calculating false positive, false negative, true positive, and true negative as given in Eqs. (4)–(7)

$$\text{Precision} = \frac{\text{TP} * 100}{\text{TP} + \text{FP}} \tag{4}$$

$$F_1 - \text{score} = 2*\frac{(\text{Precision*recall})}{(\text{Precision} + \text{recall})} * 100 \tag{5}$$

$$\text{Recall} = \frac{\text{TN}*100}{\text{TN} + \text{FN}} \tag{6}$$

$$\text{Accuracy} = \frac{(\text{TN} + \text{TP})*100}{\text{TN} + \text{FP} + \text{TP} + \text{FN}} \tag{7}$$

Since the data contains tremendous missing values, data cleaning is applied as a preprocessing step occupied by filling the missing values by calculating mean values.

After cleaning the dataset, we get a new preprocessed dataset which is ready to flow into the framework.

Two feature selection algorithms are chosen as the best technique for this experiment: Spearman rank coefficient and Mutual Information for finding a correlation between two features. Based on the correlation factor, the most essential features are selected. In our experiment, the preprocessed dataset is given as an input to SRC and MI separately. Now, we get the reduced dataset with selected features of count 88 for each correlation technique. The features are reduced by formulating a threshold value to choose the features for SRC as well MI. As the next step, the two reduced datasets are divided with a proportion of 90–10 for training and testing separately. These reduced datasets are then further classified with Machine Learning classifiers. In the process of executing both the feature selection techniques, 88 features are chosen individually. The training dataset is trained with the following eight classifiers, Decision Tree, Random Forest, Adaboost, Naive Bayes, SVM, K-NN, Gradient Boosting, and Multilayer Perceptron, separately.

The novelty of the research work is that the experiments are conducted with these 88 features alone. At first, a UCI Arrhythmia dataset is carried for the experiment and cleansed the dataset and reduced the dataset, and formed two reduced datasets individually. Finally, the classifiers SVM, MLP, Naive Bayes, Random Forest, Decision tree, Adaboost, K-NN, and Gradient boost were used. The performance of the classifiers in the proposed model is measured by the accuracy of the metrics after applying the SRC and MI are shown in Fig. 2, respectively.

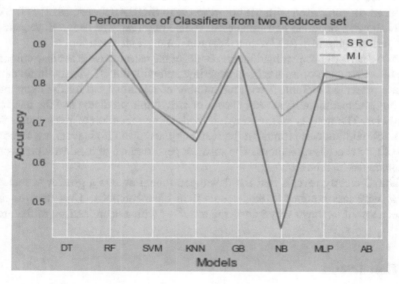

Fig. 2 Line plot for classifiers with spearman ranking coefficient and mutual information reduced set

Table 1 Result obtained from the classifier after applying SRC and MI reduced dataset

Model	Spearman ranking coefficient				Mutual information			
	Accuracy	Precision	Recall	F_1 score	Accuracy	Precision	Recall	F_1 score
DT	80.44	79.00	80.00	78.00	73.91	69.00	74.00	71.00
RF	**91.30**	88.00	91.00	89.00	86.96	69.00	74.00	71.00
SVM	73.91	64.00	74.00	67.00	73.91	79.00	74.00	74.00
K-NN	65.22	61.00	65.00	57.00	67.39	69.00	70.00	63.00
GB	86.96	90.00	87.00	87.00	**89.13**	94.00	89.00	90.00
NB	43.48	71.00	43.00	49.00	71.74	77.00	67.00	70.00
MLP	82.61	83.00	83.00	81.00	80.44	83.00	80.00	80.00
AB	80.44	87.00	80.00	79.00	82.62	88.00	83.00	82.00

The accuracy reached by each classifier from the SRC and MI Reduced set is represented as the pictorial representation in Fig. 2. using a line plot. The machine learning classifier exhibits its performance by showing its ebb and flow. All the algorithms do not show the performance alike. The experimental results obtained from the eight classifiers after feature reduction using SRC and MI are shown in Table 1.

Table 1 clearly shows that the Random Forest algorithm is the finest amongst other algorithms and obtained a peak of 91.30% accuracy when applying with a reduced set formed with SRC. Gradient Boosting proved its best by reaching 89.13% with a reduced set comprised with MI compared with SRC. Adaboost, MLP, DT, SVM, K-NN showed moderate performance. It is also observed that Naive Bayes showed poor performance with SRC reduced dataset amongst all classifiers.

For the performance evaluation, the results of the research work are also compared with the other state-of-art methods experimented with the UCI Arrhythmia dataset and are presented in Table 2 with the number of features used for the experiment, training, and testing ratio, several classes of arrhythmia predicted and the respective accuracies. The novelty of the research work is the reduction of the features from 256 to 88 without compromising the results by acquiring high accuracy compared with other State-of-art methods. The superior performance of the developed work is demonstrated in Table 2.

Table 2 clearly reveals that the developed model attains a greater accuracy of 91.3% with less number of 88 features than other methods. Thus, the proposed model showed its novelty by improving accuracy with a decreased set of features.

5   Conclusion

The proposed research work explores the behaviour of two filter-based feature selection mechanisms in the task of detecting the existence and non-existence of

Table 2 Performance comparison of the proposed work model with the state-of-art methods

Author	Classifier	Train—test split	No. of features	Accuracy	No. of classes classified
Proposed model	RF	90–10	88	91.30	16 classes
Jadhav [11]	GFNN	75–25	198	82.35	2 classes
Abirami and Vincent [10]	RF	90–10	161	83.10	2 classes
Niazi [12]	K-NN	20-fold	148	73.80	2 classes
Pandey et. al. [13]	SVM and NB	90–10	102	89.74	2 classes
Mitra and Samanta [14]	CFS + LM	68–32	182	81.88	2 classes
Samad et al. [16]	K-NN	–	279	70.10	2 classes
Yılmaz [20]	FS-LS-SVM	10-fold T	–	82.09	2 classes
Shimpi et al. [21]	SVM	–	150	91.20	16 classes

arrhythmia. The highest accuracy of 91.3% was achieved with the Random forest classifier by implementing the reduced set generated with a spearman ranking coefficient amongst the eight classifiers. On the other hand, the Gradient Boosting algorithm achieved a higher accuracy of 89.1% with a reduced dataset obtained by the feature selection technique Mutual Information. All the classifiers proved their best with the reduced dataset and thereby reducing the time complexity as well. Moreover, the accuracy attained was high when examined with other state-of-art methods. It is proven that a reduced set of attributes can also give better performance for diagnosis and prediction. Future work will be to implement other feature selection techniques such as wrapper and hybrid methods with a reduced feature set and also predicting more classes of arrhythmia with high accuracy.

References

1. Singh, Y.N., Singh, S.K., Ray, A.K.: Bioelectrical signals as emerging biometrics: issues and challenges. Int. Schol. Res. Notices (2012)
2. Bashir, M.E.A., Lee, D.G., Akasha, M., Yi, G.M., Cha, E.J., Bae, J.W., Ryu, K.H.: Highlighting the current issues with pride suggestions for improving the performance of real-time cardiac health monitoring. In: International Conference on Information Technology in Bio-and Medical Informatics, pp. 226–233. Springer, Berlin, Heidelberg (2010)
3. Roopa, C.K., Harish, B.S.: A survey on various machine learning approaches for ECG analysis. Int. J. Comput. Appl. **163**(9), 25–33 (2017)
4. Zoni-Berisso, M., Lercari, F., Carazza, T., Domenicucci, S.: Epidemiology of atrial fibrillation: a European perspective. Clin. Epidemiol. **6**, 213 (2014)

5. Ravì, D., Wong, C., Deligianni, F., Berthelot, M., Andreu-Perez, J., Lo, B., Yang, G.Z.: Deep learning for health informatics. IEEE J. Biomed. Health Inform. **21**(1), 4–21 (2016)
6. Lundervold, A.S., Lundervold, A.: An overview of deep learning in medical imaging focusing on MRI. Z. Med. Phys. **29**(2), 102–127 (2019)
7. Fazel, A., Algharbi, F., Haider, B.: Classification of cardiac arrhythmias patients. CS229 Final Project Report (2014)
8. Jadhav, S.M., Nalbalwar, S.L., Ghatol, A.A.: Artificial neural network based cardiac arrhythmia disease diagnosis. In: 2011 International Conference on Process Automation, Control and Computing, pp. 1–6. IEEE (2011)
9. Batra, A., Jawa, V.: Classification of arrhythmia using conjunction of machine learning algorithms and ECG diagnostic criteria. Int. J. Biol. Biomed. **1**, 1–7 (2016)
10. Abirami, R.N., Vincent, P.D.R.: Cardiac arrhythmia detection using ensemble of machine learning algorithms. In: Soft Computing for Problem Solving, pp. 475–487. Springer, Singapore (2020)
11. Jadhav, S., Nalbalwar, S., Ghatol, A.: Performance evaluation of generalized feed-forward neural network based ECG arrhythmia classifier. Int. J. Comput. Sci. Issues **9**(4), 379–384 (2012)
12. Niazi, K.A.K., Khan, S.A., Shaukat, A., Akhtar, M.: Identifying best feature subset for cardiac arrhythmia classification. In: 2015 Science and Information Conference (SAI), pp. 494–499. IEEE (2015)
13. Pandey, S.K., Sodum, V.R., Janghel, R.R., Raj, A.: ECG arrhythmia detection with machine learning algorithms. In: Data Engineering and Communication Technology, pp. 409–417. Springer, Singapore (2020)
14. Mitra, M., Samanta, R.K.: Cardiac arrhythmia classification using neural networks with selected features. Procedia Technol. **10**, 76–84 (2013)
15. Namsrai, E., Munkhdalai, T., Li, M., Shin, J.H., Namsrai, O.E., Ryu, K.H.: A feature selection-based ensemble method for Arrhythmia classification. JIPS **9**(1), 31–40 (2013)
16. Samad, S., Khan, S.A., Haq, A., Riaz, A.: Classification of arrhythmia. Int. J. Electr. Energy **2**(1), 57–61 (2014)
17. Elsayad, A.M.: Classification of ECG arrhythmia using learning vector quantization neural networks. In: International Conference on Computer Engineering and Systems, pp. 139–144. IEEE (2009)
18. Zuo, W.M., Lu, W.G., Wang, K.Q., Zhang, H.: Diagnosis of cardiac arrhythmia using kernel difference weighted K-NN classifier. In: 2008 Computers in Cardiology, pp. 253–256. IEEE (2008)
19. Ozçift, A.: Random forests ensemble classifier trained with data resampling strategy to improve cardiac arrhythmia diagnosis. Comput. Biol. Med. **41**(5), 265–271 (2011)
20. Yılmaz, E.: An expert system based on Fisher score and LS-SVM for cardiac arrhythmia diagnosis. Comput. Math. Methods Med. (2013)
21. Shimpi, P., Shah, S., Shroff, M., Godbole, A.: A machine learning approach for the classification of cardiac arrhythmia. In: 2017 İnternational Conference on Computing Methodologies and Communication (ICCMC), pp. 603–607. IEEE (2017)
22. Bin, G., Shao, M., Bin, G., Huang, J., Zheng, D., Wu, S.: Detection of atrial fibrillation using decision tree ensemble. In: 2017 Computing in Cardiology (CinC), pp. 1–4. IEEE (2017)
23. Jadhav, S.M., Nalbalwar, S.L., Ghatol, A.A.: ECG arrhythmia classification using modular neural network model. In: 2010 IEEE EMBS Conference on Biomedical Engineering and Sciences (IECBES). IEEE (2010)
24. Chandrashekar, G., Sahin, F.: A survey on feature selection methods. Comput. Electr. Eng. **40**(1), 16–28 (2014)
25. Xu, J., Mu, H., Wang, Y., Huang, F.: Feature genes selection using supervised locally linear embedding and correlation coefficient for microarray classification. Comput. Math. Methods Med. (2018)
26. Beraha, M., Metelli, A.M., Papini, M., Tirinzoni, A., Restelli, M.: Feature selection via mutual Information: new theoretical insights. In: 2019 International Joint Conference on Neural Networks (IJCNN), pp. 1–9. IEEE (2019)

27. Devi, M.K., Rani, M.M.S.: A comparative study of machine learning classifiers for diabetic retinopathy detection. In: Advances in Automation, Signal Processing, Instrumentation, and Control: Select Proceedings of i-CASIC 2020, pp. 735–742. Springer Singapore (2021)

28. Sangeetha, R., Rani, M.M.S.: A novel method for plant leaf disease classification using deep learning techniques. In: Machine Learning, Deep Learning and Computational Intelligence for Wireless Communication, pp. 631–643. Springer, Singapore (2021)

29. He, K., Zhang, X., Ren, S., Sun, J.: Deep residual learning for image recognition. In: Proceedings of the IEEE Conference on Computer Vision and Pattern Recognition, pp. 770–778 (2016)

30. Silver, D., Schrittwieser, J., Simonyan, K., Antonoglou, I., Huang, A., Guez, A., Hubert, T., Baker, L., Lai, M., Bolton, A., Chen, Y.: Mastering the game of go without human knowledge. Nature 550(7676), 354–359 (2017)

31. Domingos, P.: A few useful things to know about machine learning. Commun. ACM 55(10), 78–87 (2012)

32. Luz, E.J.D.S., Schwartz, W.R., Cámara-Chávez, G., Menotti, D.: ECG-based heartbeat classification for arrhythmia detection: a survey. Comput. Methods Programs Biomed. 127, 144–164 (2016)

33. Rodriguez, J., Goni, A., Illarramendi, A.: Real-time classification of ECGs on a PDA. IEEE Trans. Inf Technol. Biomed. 9(1), 23–34 (2005)

34. Biau, G., Scornet, E.: A random forest guided tour. TEST 25(2), 197–227 (2016)

35. Vijayakumar, T., Vinothkanna, M.R., Duraipandian, M.: Fusion based feature extraction analysis of ECG signal interpretation–a systematic approach. J. Artif. Intell. 3(01), 1–16 (2021)

36. Hsieh, C.J., Chang, K.W., Lin, C.J., Keerthi, S.S., Sundararajan, S.: A dual coordinate descent method for large-scale linear SVM. In: Proceedings of the 25th International Conference on Machine Learning, pp. 408–415 (2008)

37. Li, Q., Rajagopalan, C., Clifford, G.D.: Ventricular fibrillation and tachycardia classification using a machine learning approach. IEEE Trans. Biomed. Eng. 61(6), 1607–1613 (2013)

38. Naik, G.R., Reddy, K.A.: Comparative analysis of ECG classification using neuro-fuzzy algorithm and multimodal decision learning algorithm: ECG classification algorithm. In: 2016 3rd International Conference on Soft Computing and Machine Intelligence (ISCMI), pp. 138–142. IEEE (2016)

39. Beygelzimer, A., Hazan, E., Kale, S., Luo, H.: Online Gradient boosting. arXiv preprint arXiv: 1506.04820 (2015)

40. Celin, S., Vasanth, K.: ECG signal classification using various machine learning techniques. J. Med. Syst. 42(12), 1–11 (2018)

41. Ying, C., Qi-Guang, M., Jia-Chen, L., Lin, G.: Advance and prospects of AdaBoost algorithm. Acta Automatica Sinica 39(6), 745–758 (2013)

42. Abirami, S., Chitra, P.: Energy-efficient edge based real-time healthcare support system. In: Advances in Computers, vol. 117, no. 1, pp. 339–368. Elsevier (2020)

Classification of Diabetic Retinopathy Using Ensemble of Machine Learning Classifiers with IDRiD Dataset

M. Kalpana Devi and M. Mary Shanthi Rani

Abstract Diabetic retinopathy (DR) has a major impact of eye vision loss and blindness around the world. There are several screening methods used to detect the disease. Early prevention is important to find out the disease and decrease the chance of vision loss of diabetic patients. Different machine learning techniques are used on diabetic retinopathy dataset for classification and prediction of disease. IDRID dataset is used for classifying the DR disease based on the severity (or) disease grading. Our proposed work examines the ensemble of machine learning classifiers for the classification of DR. It classifies the image data properties in the form of attributes using machine learning based on different stages (mild, moderate, severe) of the disease. The primary dataset is preprocessed first using data transformation and data cleaning, and then, the vital attributes are identified using co-relation matrix. The experimental results are evaluated with the performance metrics in terms of accuracy, precision, recall, and F1-score.

Keywords Diabetic retinopathy (DR) · Machine learning · Classifiers · Random forest · Diabetic macular edema · Disease grading

1 Introduction

Diabetic retinopathy is a common retinal vascular eye problem among diabetic patients. It severely affects and causes damage in blood vessels when they have high blood sugar. Without appearance of symptoms, there are various unpredicted and untreated cases of DR, where proper retinal screening could arranged for the early detection and treatment of DR [1]. Millions of people across the world are affected by this disease. Some of the screening methods available are fluorescing angiography, optical coherence tomography, and they are used to identify the disease patterns in different dimensions [2]. The molecules inside the retinal layers are developed as capillaries in the form of exudates, microaneurysms, and cotton wool spots [3].

Present Address:
M. Kalpana Devi (✉) · . M. Mary Shanthi Rani
The Gandhigram Rural Institute (Deemed to be University), Gandhigram, Dindigul 624302, India

© The Author(s), under exclusive license to Springer Nature Singapore Pte Ltd. 2022   291
V. Suma et al. (eds.), *Evolutionary Computing and Mobile Sustainable Networks*,
Lecture Notes on Data Engineering and Communications Technologies 116,
https://doi.org/10.1007/978-981-16-9605-3_20

Early diagnosis and timely treatment of DR are essential for prevention of vision loss. Ophthalmologists can play a vital role in analyzing the appearance of retinal blood vessels of disease affected people [4].

In recent years, machine learning (ML) techniques are widely used for identifying diabetic retinopathy symptoms in the medical field with great accuracy. Identification of DR is observed by the presence of lesion in terms of abnormalities of retina selection of an optimal algorithm is a crucial task for effective prediction of disease. There is an ensemble of machine learning algorithms available and are rated based on the performance metrics such as accuracy, precision, and recall [5].

Detection of eye disease depends on the maximum of difference in the appearance of retinal veins [6]. Annotation of image labeling and segment-based learning approach is necessary to detect diabetic retinopathy disease in our eye. The important features from a fundus image are derived from the results of classifiers which are significant for perceiving the images of DR [7]. Artificial intelligence (AI) has evolved as a powerful method for diagnosing DR, glaucoma. More specifically, there are many clinical benefits of using ML algorithms in detecting the disease patterns of irreversible visual impairment diseases like glaucoma, DR, and ARMD [8]. To overcome the disadvantages of supervised methods, ensemble of machine learning methods have been implemented in detecting diabetic retinopathy using a comma separated values (CSV) dataset [9]. This proposed model is designed by neural network with different classifiers for the classification of diabetic retinopathy disease and is used on the standard dataset Indian Diabetic Retinopathy Image Dataset (IDRiD), which contain the images that are divided into mild, moderate, and severe, resulting in a model creating the detailed classification results for identifying different stages of DR. This model is also evaluated against the traditional state of the model to its performance metrics.

The following sections are designed such as Sect. 2 explores literature review; Sect. 3 elaborates the methodology; Sect. 4 explains about experimental results, and Sect. 5 provides conclusion.

2 Literature Review

Mahmoud et al. [1] introduced a hybrid inductive machine learning-based automatic detection system that react likes diagnostic tool to classify the input images as healthy or unhealthy. Multiple instance learning (MIL) is helpful for extracting the features in an input data and predict the correct results. This method uses CHASE dataset for detection of DR.

Gadekallu et al. [2] explained detecting the disease in our eye using PCA—firefly-based deep learning model. It is focused on preprocessing and dimensionality reduction. PCA is used as standard scalar to extract the important features, and firefly algorithm applied for dimensionality reduction. The model also evaluated based on performance metrics.

Jebaseeli et al. [3] analyzed model by applying deep learning-based classifiers for detecting retinal vessel segmentation from DR images they designed framework based on segmentation for extracting and classifying the blood vessels. In this model, firefly algorithm is used for fine-tuning the parameters to increase the accuracy. Segmentation results show 80.61% sensitivity, specificity, and 99.49% accuracy when experimented with the large datasets.

Nilashi et al. [4] developed the new method for detecting DR using an ensemble of ML classifiers. Non-linear iterative partial least squares method is used for dimensionality reduction. Publicly available dataset (Messidor) is used for experimental analysis with the implementation of two algorithm. The analysis show the results with ACC = 0.918, SE = 0.946, and SP = 0.917.

Reddy et al. [5] analyses finding best algorithm to predict diabetic retinopathy using different methods, used to train a model to predict diabetic retinopathy. Machine learning techniques are taken as initial algorithms to train predictive models by implementing support vector machine using Gaussian kernel for DR prediction which is evaluated with the metrics. The algorithm provides expected values when compared with all the metrics.

Gayathri et al. [6] explained DR classification based on multipath CNN and ML classifiers. This work describes automated DR severity grading of the features extracted from the fundus image. A multipath CNN is developed for identifying the local and global features from the input images. The proposed system shows the better results with J48 classifiers. The average accuracy is 99.62% for disease grading.

Math et al. [7] proposed the segment-based learning approach for detection of DR, learns classifiers, and features from images with DR. Pre-trained CNN is used to estimate and classify the disease level in the fundus images. The performance measured among the images based on area under curve (ROC) curve shows 0.963% SE and 96.37% SP.

Abdelsalam et al. [10] designed diabetic retinopathy detection methods based on geometrical analysis for preventing the vision loss. It analyzes angiography images for processing early stage of non-proliferative diabetic retinopathy (NPDR). The classification accuracy has achieved 98.5% for all the stages of DR and other vessel affected retinal disease.

Keerthiveena et al. [11] analyzed diabetic retinopathy diagnosis based on firefly algorithm that the screening system has been developed to classify the normal and abnormal fundus image. Blood vessels were segmented by using the match filter with fuzzy c-means clustering algorithm. Different sizes of blood vessels were received from the different filter by including the use of line-like directional features. A firefly algorithm is used for discriminative feature selection for the early detection of diabetic retinopathy. Hence, the proposed algorithm has been compared with advanced techniques, and it defines improvement of the results of classification by using the minimum number of features for early detection of diabetic retinopathy.

AlabdulWahhab et al. [12] designed identifying diabetic retinopathy by applying different machine learning algorithms which contains lot of machine learning classifiers, and its extended models were used with cross-validation resampling procedure.

This model got the accuracy 86% to find out the DR patients with other influent risk factors (mass body index, age, blood pressure). It examines the performance metrics based on multi-class data using more sophisticated ML methods.

Mateen et al. [8] developed that early identification of DR is an important phenomena to prevent the eye sight from vision loss and provide help for treatment. The detection of DR can be manually performed by ophthalmologists and can also be done by a developed automated system. This proposed work defines the manual system, with work flow of fundus images needing ophthalmologists. In the computerized system, deep learning is used to perform an important role in the area of ophthalmology and specifically in the early detection of diabetic retinopathy over the traditional approaches.

Porwal et al. [9] proposed a new medical diagnosis tool using data augmentation and ensemble of learning classifiers. It is one of new invention to analyze retinal image and DR screening. It predicts the severity for both DR and DME.

Recep et al. [13] describe about deep learning architecture for detecting diabetic retinopathy. In this proposed work, mobile net-based deep learning model is executed with ResNet-50 pre-trained model. It uses the training set which has smaller number of images and achieves better accuracy of 98.6%, 98.2% (SP), and 99.1% (SE).

Sambyal et al. [14].developed the modified U-Net architecture to segment DR images. This model examines the presence of microaneurysms and hard exudates (HE) by applying the process of segmentation with two publicly available datasets [IDRID, E-Optha]. It achieves the good values of performance measures of 99.98% (ACC), 99.88% (SE), and 99.89% (SP).

Yaqoob et al. [14] explored detection of DR using ResNet-50 with the implementation of random forest classifiers. This proposed work compared five stages of proliferative diabetic retinopathy (PDR) using the dataset (EyePACS) and two-stages of diabetic macular edema (DME) with (Messidor-2) dataset using ResNet-50.

In this paper, different machine learning techniques are used to identify the different stages of DR based on the severity, and the performance is measured based on the metrics.

3   Materials and Methods

This section describes the materials and methods for the proposed model using various machine learning techniques for detection of DR.

3.1   Dataset

The Indian Diabetic Retinopathy Image Dataset (IDRiD) was designed from real medical data acquired at an eye clinic. These photographs were collected from diabetes-affected people. The dataset is taken from grand challenge on DR approved with the standard ISBI-2018. It describes disease severity level of DR, and DME for each image. The IDRID database has been established for feasible studies on computer-assisted diagnoses of diabetic retinopathy. The images have 50 field of view (FOV), and resolution of 4288*2848 pixels stored in jpg format. The dataset contains 516 images representing five stages of PDR and DME, which describes the grades based on the disease severity level. Disease grading has also be done in the CSV file format. The used dataset contains 516 images which are split in the ratio of (80%) data as training dataset and remaining (20%) data as testing sets, respectively. Similarly, the expert labels of DR and DME severity level for the dataset are provided in two CSV files. A training dataset with ground truth values is used to implement the algorithms in our dataset. Analytical, data-driven models experimented based on machine learning are helpful in medical image analysis [9] (Table 1).

Each CSV file consists of three columns representing image no, X coordinate, and Y coordinate where X and Y coordinates are of center pixel location of OD/fovea in the image. All features represent either a detected descriptive feature of an anatomical part or an image-level descriptor. The ground truth values are categorized into three types as follows:

i. Pixel Level Annotation: Eighty-one color fundus photographs were annotated based on the pixel level of the affected images for exploring ground truth of MAs, SEs, EXs, and HEs

ii. Image Level Grading: All images in the CSV file has been defined by grading to classify the images into the separate groups according to the results of clinician.

iii. OD and Fovea center coordinates: Optic disc and fovea are separated from the given 516 images. The dataset is further divided into five categories according to severity like mild, moderate, and severe. The attributes are preprocessed

Table 1 Attribute information

Attribute	Description
Image id	Image number
X-coordinate	Center pixel location of optic disc along X-axis
Y-coordinate	Center pixel location of optic disc along Y-axis
FX-coordinate	Center pixel location of Fovea
FY-coordinate	Center pixel location of optic disc
Retinopathy grade	Severity of diabetic retinopathy disease 0—absence of abnormalities 1—mild stage [presence of MA only] 2—moderate age[hemorrhages, exudates] 3—severe stage [macular edema]
Risk of macular edema	values of (0–2) classes—vision loss

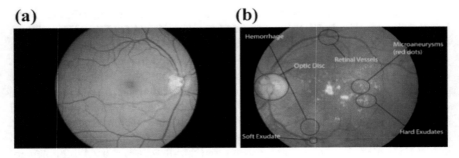

Fig. 1 (**a** Normal retina, **b** retina contains DR with MA, HE, DME

based on grading to fill the missing values with one-hot encoding method. The most important attributes are extracted by applying Pearson method which is an efficient method for finding the correlation coefficient.

Figure 1a, b shows normal and DR-affected retinal images with mild, moderate, and severe stages, respectively. The IDRID database has been established for feasible studies on computer-assisted diagnoses of diabetic retinopathy.

3.2 Objectives

In this proposed system, the dataset is divided into training and testing set comprising of 413 (80%) and 103 (20%) images, respectively. Machine learning classifiers have been applied, and performance measures evaluate the classification of disease severity.

3.2.1 Data Preprocessing

Data preprocessing involves filling of missing values to avoid noisy data. We reduce the redundancy in source data through data normalization. The dataset is split into training set and testing set, and finally, feature scaling is performed. One-hot encoding also applied in order to convert categorical variable into numerical variable [5] (Fig. 2).

3.2.2 System Architecture

The IDRID dataset is to be loaded, and the preprocessing step is to be applied. Medical experts graded the full set of 516 instances (images) with a variety of extreme conditions of DR and DME. In the first level, annotations done on data were split as two-third for training and remaining (one-third) for testing. The training data are

Fig. 2 Steps of workflow in
data preprocessing

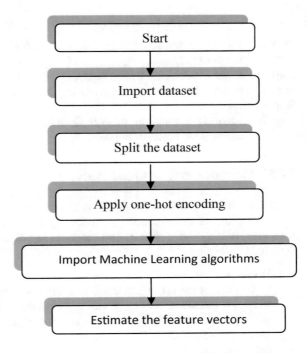

processed by different machine learning algorithms, and the outcomes of algorithms represent the learned pattern [9]. The architecture of the proposed model is pictorially represented in Fig. 3.

The outline of the proposed methodology has been given below:

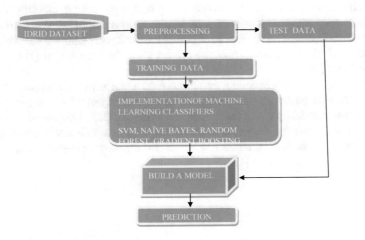

Fig. 3 Architecture of a model using machine learning classifiers

Algorithm: Random Forest

Step 1:   Divide the training dataset into k data points depending on the values of the target attribute

Step 2:   Build the decision trees associated with the selected data points (Subsets).

Step 3:   Extract the features from choosing the number N of decision trees

Step 4:   Repeat step 3 and step 4

Step 5:   Find the predictions of each decision tree, and assign the N points to develop the prediction model

3.2.3   Machine Learning Classifiers

Classifiers are used to find the predictive model by using various algorithms such as decision tree, adaptive boosting, logistic regression, Naïve Bayes, support vector machine (SVM), random forest. The proposed algorithm is random forest which is explained in detail.

a.   *Decision Tree*

It is one of the classifiers that create a tree structure, in a top-down recursive order. This structure contains nodes that indicate attributes. Leaf nodes represent the target attribute which has the value 1 and 0 representing positive and negative outcomes, respectively [13].

b.   *Adaptive Boosting*

This is one of the ensemble algorithms used for solving the classification problem. The weights of the instances in the training dataset are modified and updated until correct prediction is reached. Finally, the model is created by calculating weighted average of all the modified classifiers [15].

c.   *Support Vector Machine*

This algorithm is a supervised learning technique used in classification and regression problems. It creates a hyperplane, which splits the input data into different classes. The nearest data points on the hyperplane are called support vector, which has decision boundary to limit the data separation [10].

d.   *Random Forest*

It is one of the ensemble tree structured algorithms that creates a set of tree structure based on patient data. This algorithm provides fast and efficient prediction model by using the recursive framework [12]. This classifier also has consort effect on the detection of DR because of its efficiency in extracting the features with small data. It acts like an ensemble classifier that can train on more decision trees in parallel [14].

4 Results and Discussion

The evaluation metrics used in analyzing the performance of the ML classifiers in the proposed model are demonstrated as follows:

Accuracy: It is defined as the relationship of appropriate predictions to the whole number of predictions made by the predictive model. Mathematically, it is defined as follows:

$$\text{Accuracy} : \frac{TP + TN}{TP + TN + FP + FN}$$

where TP represents true positive, TN represents true negative; FN represents false negative and FP represents false positive.

Dice: It is also denoted as F_1-score which measures the comparable development of objects. It is the size of overlap of the two segmentations divided by the total size of the two objects

$$\text{Dice} : \frac{2TP}{2TP + TP + FN}$$

Sensitivity: It is defined as a ratio about actual positive cases in input data out of the total positive possibilities

$$\text{Sensitivity (SE)} : \frac{TP}{(TP + FN)}$$

Specificity: It is defined as the ratio of actual negative outcomes possibilities from the predictive model

$$\text{Specificity (SP)} : \frac{TN}{(TN + FP)}$$

Table 2 displays the performance of six ML classifiers in terms of four evaluation metrics which is obtained from the results of machine learning classifiers described in the previous section. The performance metrics evolved are obtained from confusion matrix based on the corresponding values of TP, TN, FP, and FN.

Figure 4 shows the pictorial representation of the performance metrics by applying different machine learning classifiers.

Table 3 demonstrates the performance in terms of accuracy, in the prediction of stage-wise disease level by applying the random forest classifier. It is obvious that this proposed model outperforms with random forest classifier. The accuracy rates in the dataset range from 91 to 97%.

Table 2 Evaluation metrics obtained for machine learning algorithms

S. No	Classifier name	Precision	Recall	F_1-score	Accuracy
1	Decision tree	0.93	0.94	0.92	0.96
2	**Random forest classifier**	**0.94**	**0.97**	**0.99**	**0.98**
3	Gradient boosting	0.89	0.92	0.94	0.91
4	Support vector machine	0.82	0.85	0.87	0.86
5	Naïve bayes classifier	0.73	0.76	0.78	0.74
6	Logistic regression	0.65	0.64	0.7	0.68

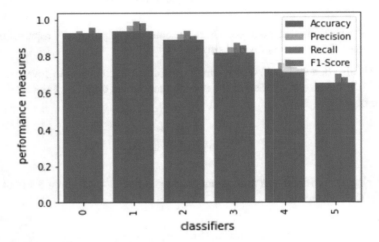

Fig. 4 Performance measures using ML classifiers

Table 3 Accuracy comparison by disease grading

S. NO	Disease stage level	Number of attributes	Accuracy (%)
1	Mild	**168**	**91**
		24	86
2	Moderate	**169**	**92**
		94	90
3	Severe	63	89

It is obvious that the comparison of accuracy between minimum number of attributes and maximum number of attributes with stage-wise separation in the trained dataset is higher (Fig. 5).

The performance comparison of the proposed model with existing similar methods is presented in Table 4. The table also lists the datasets used for training and testing.

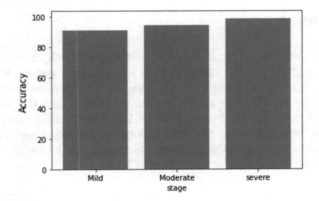

Fig. 5 Performance comparison using random forest classifier by disease grading

Table 4 Performance comparison with proposed model

S. No	Model used	Algorithm used	Dataset used	Accuracy (%)
1	DNN-PCA [2] [2020]	DNN-decision tree	MESSIDOR	94.3
2	Ensemble learning approach [4] [2019]	Neural network Support vector machine Decision tree	MESSIDOR	91.5 (Average accuracy)
3	Multifractal geometry analyses [10] [2021]	Support vector machine	Ophthalmology center in Mansoura University	98.4
4	Diagnosis of DR based on firefly algorithm[11] [2019]	Support vector machine	HRF DIRETDB1 MESSIDOR	97.5 94.5 91.1
5	Detection of DR using ranger random forest [12] [2021]	Random forest	Almajaah (PHC) King Khaled general hospital	86
6	ResNnt-based deep features detection in DR [14] [2021]	ResNet and random forest	MESSIDOR-2	95.07
7	**Proposed model**	**Random forest**	**IDRID**	**98.63**

The performance comparison of the proposed model with existing similar methods is presented in Table 4. It is worth noting that model with random forest classifier is best among all, achieving a higher accuracy of 98.63.

5 Conclusion

This paper mainly focuses on prediction of diabetic retinopathy using an ensemble of machine learning classifiers experimented with comma separated values (CSV) dataset related to highly affected diabetic patients. The attributes also represented different stages of DR. The algorithms random forest, decision tree, gradient boosting, and Naïve Bayes were experimented, and the results prove that random forest with n number of estimators achieve higher performance compared to other algorithms in terms of accuracy, sensitivity, specificity, and F1-score. Results have established effective method for predictions of DR.

References

1. Mahmoud, M.H., Alamery, S., Fouad, H., Altinawi, A., Youssef, A.E: An automatic detection system of diabetic retinopathy using a hybrid inductive machine learning algorithm. Pers. Ubiquitous Comput., 1–15 (2021)
2. Gadekallu, T.R., Khare, N., Bhattacharya, S., Singh, S., Maddikunta, P.K.R., Ra, I.H., Alazab, M.: Early detection of diabetic retinopathy using PCA-firefly based deep learning model. Electronics 9(2), 274 (2020)
3. Jebaseeli, T.J., Durai, C.A.D., Peter, J.D.: Retinal blood vessel segmentation from diabetic retinopathy images using tandem PCNN model and deep learning based SVM. O1ptik 199, 163328 (2019)
4. Nilashi, M., Samad, S., Yadegaridehkordi, E., Alizadeh, A., Akbari, E., Ibrahim, O.: Early detection of diabetic retinopathy using ensemble learning approach. J. Soft Comput. Decis Support Syst. 6(2), 12–17 (2019)
5. Reddy, S.S., Sethi, N., Rajender, R.: Discovering optimal algorithm to predict diabetic retinopathy using novel assessment methods. EAI Endorsed Trans. Scalable Inf. Syst. 8(29), e1 (2021)
6. Gayathri, S., Gopi, V.P., Palanisamy, P.: Diabetic retinopathy classification based on multipath CNN and machine learning classifiers. Phys. Eng. Sci. Med., 1–15 (2021)
7. Math, L., Fatima, R.: Adaptive machine learning classification for diabetic retinopathy. Multimedia Tools Appl. 80(4), 5173–5186 (2021)
8. Mateen, M., Wen, J., Hassan, M., Nasrullah, N., Sun, S., Hayat, S.: Automatic detection of diabetic retinopathy: a review on datasets, methods and evaluation metrics. IEEE Access 8, 48784–48811 (2020)
9. Porwal, P., Pachade, S., Kokare, M., Deshmukh, G., Son, J., Bae, W., Liu, L., Wang, J., Liu, X., Gao, L. and Wu, T.: Idrid: diabetic retinopathy–segmentation and grading challenge. Med. Image Anal. 59, 101561 (2020)
10. Abdelsalam, M.M., Zahran, M.A.: A novel approach of diabetic retinopathy early detection based on multifractal geometry analysis for OCTA macular images using support vector machine. IEEE Access 9, 22844–22858 (2021)
11. Keerthiveena, B., Veerakumar, T., Esakkirajan, S., Subudhi, B.N.:. Computer-aided diagnosis for diabetic retinopathy based on firefly algorithm. In: 2019 11th International Conference on Advanced Computing (ICoAC), pp. 310–315. IEEE (2019)
12. Alabdulwahhab, K.M., Sami, W., Mehmood, T., Meo, S.A., Alasbali, T.A., Alwadani, F.A.: Automated detection of diabetic retinopathy using machine learning classifiers. Eur. Rev. Med. Pharmacol. Sci. 25(2), 583–590 (2021)

13. Hacisoftaoglu, R.E., Karakaya, M., Sallam, A.B.: Deep learning frameworks for diabetic retinopathy detection with smartphone-based retinal imaging systems. Pattern Recogn. Lett. **135**, 409–417 (2020)
14. Yaqoob, M.K., Ali, S.F., Bilal, M., Hanif, M.S., Al-Saggaf, U.M.: ResNet based deep features and random forest classifier for diabetic retinopathy detection. Sensors **21**(11), 3883 (2021)
15. Sambyal, N., Saini, P., Syal, R., Gupta, V.: Modified U-Net architecture for semantic segmentation of diabetic retinopathy images. Biocybernetics Biomed. Eng. **40**(3), 1094–1109 (2020)
16. Sungheetha, A., Sharma, R.: Design an early detection and classification for diabetic retinopathy by deep feature extraction based convolution neural network. J. Trends Comput. Sci. Smart Technol. (TCSST) **3**(02), 81–94 (2021)
17. Balasubramaniam, V.: Artificial intelligence algorithm with SVM classification using dermascopic images for melanoma diagnosis. J. Artif. Intell. Capsule Netw. **3**(1), 34–42 (2021)
18. Sikder, N., Masud, M., Bairagi, A.K., Arif, A.S.M., Nahid, A.A., Alhumyani, H.A.: Severity classification of diabetic retinopathy using an ensemble learning algorithm through analyzing retinal images. Symmetry **13**(4), 670 (2021)
19. Inuwa, R., Bashir, S.A., Abisoye, A., Adepoju, S.A.: Comparative evaluation of machine learning techniques for detection of diabetic retinopathy (2021)
20. Gurcan, O.F., Beyca, O.F., Dogan, O.: A Comprehensive study of machine learning methods on diabetic retinopathy classification. Int. J. Comput. Intell. Syst. **14**(1), 1132–1141 (2021)

Deployment of Machine Learning Based Internet of Things Networks for Tele-Medical and Remote Healthcare

Shabnam Kumari, P. Muthulakshmi, and Deepshikha Agarwal

Abstract The technological revolution over the past decade has transformed several applications in different global sectors such as retail, transportation, automobile, agriculture, medicine, etc. These improvements have been accelerated by the widespread usage and popularity of Internet of Things (IoT) technology. The healthcare industry is a high-priority sector which is majorly responsible for saving human lives, improving a person's lifestyle, and ensuring longevity. *Purposes* of IoT devices in today's life is "it play a major role in tracking patients, healthcare professionals and providing remote caretakers with vital patient data for patient status monitoring". Machine Learning (ML) is another technological field, revolutionizing the utilities of devices with IoT techniques to provide efficient and cost-effective services to patients and healthcare professionals. In recent years, many studies on IoT and ML for healthcare are gaining traction from several international researchers. About *methods,* in this research, an attempt on machine learning-based IoT service is made with special concentration on tele-medical and remote healthcare. For r*esults,* this work provides a detailed collection of machine learning and IoT-based solutions for telemedicine and patients, living in remote localities. In *conclusion*, the system proved successful in progressing remote healthcare through IoT and Machine learning technology.

Keywords Machine learning · Internet of Things · Telemedicine · Remote healthcare · Medical cyber-physical systems

S. Kumari (✉) · P. Muthulakshmi
Department of Computer Science, CS & H, SRM Institute of Science and Technology,
Kattankulathur, Chennai, Tamilnadu 603203, India
e-mail: sk2581@srmist.edu.in

D. Agarwal
Department of Computer Science/Information Technology, Indian Institute of Information
Technology, Lucknow, Lucknow, India

1 Introduction

You will find here Telemedicine is a well-regarded service distribution methodology in current clinical administrations, which provides remote healthcare services to patients who are unable to access clinical centers on a regular basis. In addition to this, telemedicine incredibly soothes the number of outpatient visits in emergency clinics. The medical issues of individuals have now ascended to become worries of the entire society, especially in a pandemic situation such as COVID-19. The primary reason behind the success of telemedicine is that the maturing populace is getting progressively weak and is unable to accept clinical treatment from emergency care clinics. Secondly, the expanding technological movement has accelerated the efficiency and growth of remote healthcare. The top to bottom mix of the medical care industry with the Internet of Things (IoT), [1–3] 5G, Artificial Intelligence (AI), enormous information, cloud computing, and other trend-setting innovations will ideally provide the aforementioned services. The 5G-empowered IoT and AI domains are now ceaselessly driving the creation of innovative applications in the telemedicine business. The manual processes and labor requirements involved in telemedicine pose a significant challenge to the hospitals in terms of costs and time.

In the field of clinical sciences, the fundamental applications of telemedicine incorporate distant observing, far-off ultrasound, far-off interview, far-off medical procedures, portable medication, and remote services to all patients in need without the hassle of physical contact [4]. In today's healthcare scenario, wearable biomedical gadgets, for vital sign monitoring, are growing quickly. Ease of service, low force utilization, small size, and providing in-depth health insights are the key beneficial factors that propel the creation of wearable biomedical gadgets. Wearable gadgets have numerous points of interest, namely, the congruity of clinical administrations and continuous view of trusted information. They can provide constant monitoring of human existence attributes and medical issues, assisting individuals with understanding their states of being and recognize critical indications early enough to mitigate the symptoms. The rapid progression of semiconductor innovations has extraordinarily diminished the expense and force utilization of wearable gadgets. The blend of AI and IoT has improved the effectiveness with which the knowledge of wearable gadgets is being utilized. The prominent research works on wearable biomedical gadgets have become significant contributors in the field of telemedicine and deep-learning techniques have dominated the analytical segments of these network systems.

In recent times, the 5G organization has pulled in extraordinary interest from different vertical ventures because of its rapid and enormous associations coupled with its low inertness. The fast attributes of 5G empower it to help the 4 K/8 K rapid transmission and sharing of clinical image information along with its sensory data; besides, it permits specialists to lead virtual sessions whenever and wherever, improves analysis exactness, enhances directional proficiency, and extends out great clinical assets to the patients' homes. 5G is conceivable enough to interface with an enormous number of clinical sensor terminals, related video gadgets, and wearable biomedical gadgets to accomplish ongoing discernment and estimations, catch and move reliable data, and to understand the analysis of collected data without any bandwidth limitations. In radiology wards and irresistible wards, clinical staff is now provided with the ability to control the clinical associate robots to move to the assigned beds and complete distant nursing administrations in a remote manner without the requirement of extensive manual intervention. Different clinical sensors and wearable biomedical [5] gadgets can gather reliable ongoing patient pointers, including internal heat level, pulse, circulatory strain, blood glucose, and electrocardiogram, among others. These gadgets can be associated with a cloud-based storage layer to store the data for later AI investigation of patient-related information and medical records. This helps provide sickness examination reports to specialists and patients and gives helpers an overall idea of the type of therapy required. The specialist distantly assesses the patient's condition and gives prompt criticism on the patient's readings. Some wearable screens can perform neighborhood processing of information and investigation without storing redundant data in the cloud. For instance, a pulse screen that can autonomously break down the heart rate information can provide the fundamental reaction quickly to alarm guardians whenever patients are in need of assistance.

2   Literature Survey

Internet of Things (IoTs), alongside the current advancements [6] of multimedia, has made great contributions in the medical sector by supporting the mainstream applications of remote healthcare and telemedicine. Telemedicine is alluded to as utilizing media communications technologies, for example, phone, copy, and distance instruction to give electronic clinical interview and strength care benefits as and when required. The interest of utilizing telemedicine has developed broadly since the 1990s with around $100 million spent in related technological ventures. Telemedicine innovation is being embraced presently by around 13 governmental organizations. Nonetheless, a few concerns emerge including IoT interoperability, administration quality, framework security, and quickly stockpiling development. In [7], the creators proposed an open-source, adaptable and secure stage that depends generally on IoT with the support of cloud computing innovation that permits close-by surrounding correspondence for clinical use. Ahmed et al. also, Anpeng et al. actualized a versatile Telemedicine framework that sends electrocardiogram (ECG) signs to medical clinics by means of cellular networks [8]. In aforementioned works, the creators built up a versatile Telemedicine administrations instrument that analyzes patients distantly, utilizing seven crucial vital signs with ease. These patients' fundamental signs incorporate information to be specific circulatory strain, oxygen in blood, glucose level, quiet position and falls, internal heat level, heart's electrical and strong capacities through ECG and breathing [9]. They gathered these essential signs through an android application that they created.

Telemedicine can provide a recognizable upgrade to the current state of medical care. IoT-related networks and other technological innovations can facilitate this favorable. For instance, Micro Engineer Machine Systems (MEMS), a Nano-Technology structure, incorporates: (a) Robots that can be utilized in arthroscopic medical procedures (b) Encapsulated cameras that can be gulped to screen absorption (c) Wearable remote sensors that can be utilized to screen physiological capacities. Human-to-Machine (H2M) interfaces are altogether significant in Telemedicine since telemedicine frameworks require two interfaces, one at the patient side and the other on the doctor's side. In addition, system-based bots (Robots/Know-bots) speak to a future speculation of Telemedicine. Know-bots are, for all intents and purposes, existing insightful data that can be attached to patients later on to follow their e-healthcare records [10]. They were intended to comprehend normal language, react to clinical requests, and caution their proprietors if undesirable patterns were seen. This would all participate in diminishing the clinical mistakes and accordingly, upgrade the nature of far-off medical care administrations and backing clinical dynamics. Be that as it may, such advancements are less inclined to be embraced in numerous nations because of low economic supply to the medical care field and innovation opposition. From the related works that have been studied, there seems to exist a massive gap in mainstream applicability of IoT monitoring systems for remote healthcare. A complete machine learning and IoT-integrated system is proposed in the form of an electronic healthcare monitoring system in this work to provide a viable and practical approach toward solving remote healthcare requirements.

3   Electronic Healthcare Monitoring System (EHMS)

The rapid development in technological innovations over the past decade has transformed several critical domains on the planet, with healthcare services in the spotlight. E-Health checking systems are adequately utilized in monitoring patients/elderly ailments distantly. This is made feasible by the Wireless Sensor Networks (WSN) which are generally utilized in E-Health checking Systems [11]. Remote Sensor Network is normally referred to WSN as an emerging technology facilitating the rapid. In a WSN, each configured gadget is fit for performing a diverse arrangement of errands for example observing, showing, and detecting and accordingly having wide business applications in industries such as healthcare checking and military applications [12]. WSN gadgets consist of a sensor or an actuator, a radio stack, implanted processor, a reception apparatus, and a battery. These gadgets are held inside a fenced area. Generally, all the WSNs utilize the sensors to detect or accumulate the information and cycle that information for the ensuing use. If there should arise an occurrence of medical services checking, the information will be prepared in the sensor hub in the event that it is outside the protected zone. The final outcome will be sent to the clinician or the concerned clinical staff in order to provide the necessary care to the patient. EHMS is a framework that accumulates the data from the patient and cycles the information as indicated by the set of rules given,

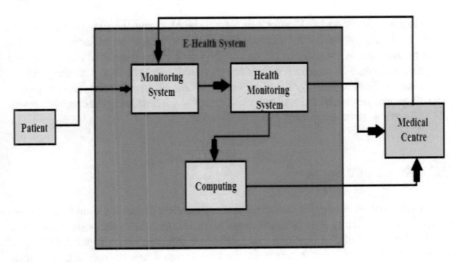

Fig. 1 Electrical health monitoring system

at that point making the fundamental move whenever required. Figure 1 shows the design of the EHMS. The subtleties of the EHMS are as discussed below.

3.1 Patient Monitoring System

This module is responsible for gathering the data from clinical sensors implemented on the patient side. This gathered information is passed on to the health checking framework whose functionalities are detailed in the following section. Observing frameworks utilize specialized sensors to measure the Temperature, Blood pressure level, Heart Beat Rate, and other traits of a patient's health status. The sensory data obtained through the direct patient monitoring system is processed either after reaching a central storage server or is processed before storage (as in edge computing). Block-chain-based data processing systems [13] have also been established in the past to provide a transparent and reliable exchange of information between the patient and the attending clinician.

3.2 Health Monitoring System

Observing Frameworks (for Healthcare) principally revolves around one or two vital signs like heartbeat rate or blood pressure level, while the checking framework gathers the patient's medical services information from the previous sensory layer. Observing frameworks utilizes a set of standard configurations to peruse the information from the checking frameworks. Figures out how to break down the perusing according to the data rules given by the clinical blueprint. A clinical construction gives the rules to the system on what move should be made if the heartbeat rate shows the irregular reading. The pattern readings are analyzed in comparison with the patient's historical medical records by the clinicians to arrive at informed medical diagnosis. The diagnosis may assist the medical professionals in determining the exact treatment plan that needs to be employed to combat the illness.

3.3 Automatic Service

Automated software triggers the alarm message to the clinical individual/healthcare officials if there is any breach/attack or unusual interference in the patient database. This builds trust among machine and healthcare officials with proper checking of implemented embedded software framework. In recent times, the mainstream implementation of COVID 19 based alert notification symptoms that alert emergency services whenever a COVID patient's oxygen levels and pulse rate reaches a critical stage [14] have been extremely helpful in facilitating removal of physical contact in order to prevent further spread of the virus. This module is also useful in detection of intrusions and alerting the administrative professionals of a possible breach in the system. Strong authentication methods coupled with encryption of transmissions can help prevent intruders gain access to the network and exploit confidential patient information.

3.4 Healthcare Monitoring System Using Internet of Things Devices with Machine Learning Techniques

Reliable clinical observation is a needful and significant resource for monitoring and controlling any sort of critical and chronic illnesses, for example, diabetes, particularly in nations like the United States where the 70% of the populace runs on physician-endorsed prescription. One of the prominent innovations that guide the medical services industry is the Internet of Things (IoT) [12]. Through the application of this technological domain, humans will have the option to screen the vitals of the patient in an easy and financially economical manner. The boundaries, for example, the heartbeat, circulatory strain and glucose level, and so on fill in as the reason for recognizing any cardiac disease or associated issues. Observation-based applications are being planned, with which checking the patient's vitals, for example, heartbeat, temperature, pulse, glucose level will be simpler and this framework will likewise have the option to recognize if the patient is experiencing fits. There are five unique sensors that will help the estimating of the previously mentioned boundaries. These five sensors are associated with the PIC microcontroller through which the information will be shipped off a cloud data set with the assistance of a Wi-Fi module. The sensors [15] are mounted on a wearable glove. On the off chance that there is any anomaly in the patient's vitals, prompt help measures are set off and the specialists or the patient's companions are told, so they can take care of the concerned patient. This warning is accomplished through the GSM module that is associated with the microcontroller. Further, with the guidance of Machine Learning, it is possible to decide if the patient is in an irregular state or not and even distinguish that anomalous state. With the assistance of this framework, we will have the option to peruse the live estimations of the patient and if there are any irregularities, for example, low heartbeat or expanded temperature, quick alleviation estimates will be set off and this will help to distinguish if the reason behind variation from the norm in the vitals of the patient. Figure 1 explains about monitoring individual health in detail.

4 Evaluating the Different Technologies and Components of the Internet of Things for Better Health

Although network-based applications have showcased significant benefits, they are always associated with the risk of breaching personal privacy/information of users/patients through their communication with other IoT devices. It alludes to a digital actual worldview, where all these present reality segments can remain associated. IoT enables clients to coordinate worldwide nodal components of the network, for example, electronic gadgets, PDAs, and tablets which can impart both truly and remotely. IoT helps in overseeing essentially quite a few gadgets. It focuses on expanding the advantages of the web, for example, far-off access, information sharing, and availability to different application areas such as medical care, transportation, stopping exercises, agribusiness, reconnaissance, and so on. With huge advantages and qualities, for example, distinguishing proof, area, detecting, and availability, joined to the IoT [16, 17], it is the vital part of brilliant medical care, as mentioned in Fig. 2. In executing a keen medical care framework, IoT can be comprehensively actualized in a wide range beginning from adjusting clinical gear to customized observing framework. IoT assumes a critical role in medical services applications, from overseeing ongoing illnesses toward one side of the range to checking everyday proactive tasks which could help in keeping up one's wellness objectives. IoT can be utilized to screen the cycle of creation and following of clinical gear conveyance. IoT-based structures can be utilized to gather clinical data from the client. IoT networks act as a scaffold between the specialist and the patient by giving far-off access, which can assist the specialist with observing the patient and give distant interviews. Combining sensors, actuators, microcontrollers, processors, alongside cloud computing, IoT helps in getting precise outcomes and makes remote medical care feasible to everybody.

Utilizing the IoT network technology in medical services has driven specialists worldwide to configure promising remote technological structures and progress

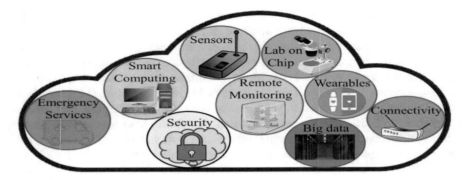

Fig. 2 Internet of Things applications in smart healthcare

research advances that can give accessible clinical help to everybody. Notwith-standing enhancing the client experience, the IoT likewise encourages the business to automate, providing more examination across various cross stages.

The essential components of the IoT networks in electronic medical services are sensors/actuators, a nodal structure consisting of network, transmission, and appli-cation layers, a healthcare organization, the internet, and the cloud. Contingent on the application and the necessities of the particular medical services framework, the determinations of every one of these four fundamental segments can differ generally. Figure 2 explains the possible use of IoTs in smart healthcare.

5 A Secure Remote Healthcare System for Hospital Using Internet of Things and Machine Learning

Each cycle/work (in near future) will be robotized (automated) in the medical care industry. In the decision-making unit, prediction calculations act as the central heart of the framework, which anticipate some significant standards from their informa-tion perception. In [18], a few expectation calculations are clarified and elucidated to help understand remote IoT healthcare systems. The work done in [19] manages the arrangement of a bound together design of IoT and ML which does not exist in the healthcare industry as of now. They planned a development of the Narrowband IoT (NB-IoT) network to improve the efficiency of current IoT healthcare imple-mentations. Using this NB-IoT, they associate and acquaint edge device registration. In [16], A Machine Learning Decision Support System(MLDSS) was introduced in Smart Meter Operations to improve the nature of information association in the interconnected smart meter tasks. In [17], a novel ECG quality mindful framework was clarified in an IoT scenario for nonstop monitoring of patient health care status. They also explored how a lightweight constant signal quality-aware technique was utilized to order ECG signals into satisfactory and unsatisfactory classes. Execution was done on Arduino and was just utilized for ECG information examination, not for different boundaries. In [20], Timely Fusion and Analysis of Data which are gathered from Several Sensors are dissected utilizing deep-learning approaches are clarified quickly and different programming structures have been utilized in WBSN additionally [21]. This work is about point by point review on information combination strategies with a specific spotlight on Mathematical Methods. In [22], Utilization of wearable sensors and hard-detecting base information obtainment is supplanted with soft detecting-based information procurement which is called group sensing. This work quickly audits deep-learning procedures to improve analytics and provide dynamic results in clinical administrations. In [23], the requirement for Complex Event Processing (CEP) was presented for examining heterogeneous information streams and identifying complex examples. This gives a robust and versatile answer for planning responsive applications from the previous authentic information. It also clarifies proactive design which abuses recorded information utilizing AI for expectation related to CEP. We propose a versatile forecast calculation called versatile moving window relapse for dynamic IoT information and assess it utilizing a genuine use case with an exactness of over 96%. In [24], the authors proposed smart hospitality using IoT by enhancing its feature range and implementation characteristics. A decision-making system that can be embedded with IoT for monitoring purposes has also been extant.

5.1 An Internet of Things Based e-Healthcare System for Remote Telemedicine

There are numerous ways in which the IoT can help in the development of telemedicine by providing help to medical services communities with improved nature of care. This has ended up being more advantageous than most in-person treatment strategies. A portion of the gainful parts of IoT in medical care telemedicine are as per the following:

5.1.1 Helping Specialists with Getting to the Significant Information from Clinical Gadgets/Devices

The specialists can take a look at the injury, rash, or skin tone and break down the treatment options for the patient. Be that as it may, they can't tune in to the patient's pulse remotely and check the circulatory strain and other related vitals without the assistance of connected technology. This is a significant test in determining further medical measures. In these cases, the IoT wearable gadgets assume a significant job of providing the specialists to get patient data through them. A major portion of the medical services places is in any event, building up their own IoT wearable for their patients to build the trust among healthcare professionals (on machines). Moreover, oximeters and advanced stethoscopes are now being accommodated as IoT gadgets, utilized by the patients during telemedicine conferences and virtual gatherings.

Hence wearing IoT gadgets is as significant as all things considered to wear them accurately. The information quality transmitted and wearability of the gadgets are the two factors that determine the consistent success of the monitoring system.

5.1.2 Real-Time Observing

Constant checking innovation causes specialists to get the patient's inherent health indicator subtleties consistently. Wearable technology has positioned itself as the forerunner in contributing to real-time body monitoring applications in patients and elderly population. Apart from smartwatches and digital health bands, wearable technology has metamorphosed into an integrated lifestyle technology in the form of low-cost t-shirts and electronic body wear [25]. Patients wear a Holter screen gadget to screen the anomalies in their pulse. Improvement in innovation has empowered individuals to deal with their health issues remotely with IoT and telemedicine. IoT technology has reached a point in its evolution where it has the ability to transmit real-time video and image information, providing faster detection of anomalies by the physicians when compared to analyzing sensory data that may not make immediate sense to a clinician.

5.1.3 Helping Senior Residents/Citizens

A few older individuals want to remain at home in their last days instead of moving to helped living offices. Notwithstanding, maturing with no associate offices, and getting things done all alone, requires monitoring the assets like home conveyance of dinners, networks, social gatherings, crisis numbers, and information about utilizing certain gadgets. In such cases, telemedicine and IoT assume a significant job. A few family gadgets are being worked on to help senior individuals, to make living at home more secure and bother-free. Discourse empowered gadgets like Alexa, voice-actuated speakers are a portion of the IoT telemedicine gadgets for the comfort of individuals living at home. The elderly patients are monitored [26] under the constraints of multiple parameters and perform several specific functions such as fall detection [27], functional decline [28], and emergency incidents [29–32]. Further, researchers are advised to go through a few novel works [33–42], i.e., toward the use of smart devices or machine learning technique's role for e-healthcare in detail. In these works [33–42], readers/researchers will increase their knowledge toward the required area/application/sector and can find new problems for their research work. İn that, we will find that these smart gadgets are additionally useful to individuals with incapacities, on the other side they may be harmful to individual in terms of breaching security, privacy, etc.

6 Conclusion

This paper gives a broad study on the flow that Health IoT research drifts alongside the difficulties of providing accessible in remote medical care. Non-physical medical care arrangements have consistently been of prominent interest. With huge subsidization and expanding consideration toward the medical services arena, there are various ventures and applications accessible by clients, to leverage technological advancements to complement the existing healthcare industry. As medical services require multi-dimensional and multi-purpose applications, it offers a ton of extensions for specialists to continually enhance new items and improve the generally existing structures. The change toward tech-powered medical care administrations is a gradual cycle. This is principally on the grounds that medical services experts should be continually instructed and persuaded to adjust to the advanced times while patients should be educated on how the services function. By overcoming any issues among scientists and medical services experts, more illnesses can be tended to and more remote ways of life can be adjusted. In spite of the fact that the medical services arrangements upheld by the IoT can improve income and increment personal satisfaction, the advantages can be effortlessly dominated, if security is undermined. From brain tumor detection to COVID 19, the IoT domain has played a significant role in the improvement of remote healthcare services in a highly efficient manner. Deep-learning algorithms coupled with IoT prove especially beneficial in situations that require visual perception such as skin cancer. Extra measures should be taken to deal with dangers of cyber breaches and making sure confidential data at both the client and engineer end is not leaked. Accordingly, the vision and long haul accomplishments of this progressively developing industry lies in the collaboration of technological analysts, medical care experts, and patients. A few minor limitations of the work involve security and privacy policies for the implemented systems that are not completely developed and a global standardization of IoT remote healthcare is required. Future direction of this research can progress toward maximizing the security of the systems and providing the training algorithms with more data through data augmentation to help improve accuracy and integrity of the system.

References

1. Andreu-Perez, J., Poon, C.C., Merrifield, R.D., Wong, S.T., Yang, G.Z.: Big data for health. IEEE J. Biomed. Health Inf. **19**(4) (2015). https://doi.org/10.1109/JBHI.2015.2450362
2. Baker, S.B., Xiang, W., Atkinson, I.: Internet of Things for smart healthcare: technologies, challenges, and opportunities. IEEE Access **5** (2017). https://doi.org/10.1109/ACCESS.2017.2775180
3. Islam, S.R., Kwak, D., Kabir, M.H., Hossain, M., Kwak, K.S.: The Internet of Things for health care: a comprehensive survey. IEEE Access **3** (2015). https://doi.org/10.1109/ACCESS.2015.2437951
4. Obinikpo, A.A., Kantarci, B.: Big sensed data meets deep learning for smarter health care in smart cities. J. Sens. Actuator Netw. **6**(4) (2017). https://doi.org/10.3390/jsan6040026

5. Park, J.H., Yen, N.Y.: Advanced algorithms and applications based on IoT for the smart devices. J. Ambient Intell. Humanized Comput. **9** (2018). https://doi.org/10.1007/s12652-018-0715-5
6. Satija, U., Ramkumar, B., Manikandan, M.S.: Real-time signal quality-aware ECG telemetry system for iot-based health care monitoring. IEEE Internet of Things J. **4**(3) (2017). https://doi.org/10.1109/JIOT.2017.2670022
7. Tokognon, C.A., Gao, B., Tian, G.Y., Yan, Y.: Structural health monitoring framework based on Internet of Things: a survey. IEEE IoT. J. **4**(3) (2017). https://doi.org/10.1109/JIOT.2017.2664072
8. Catarinucci, L., De Donno, D., Mainetti, L., Palano, L., Patrono, L., Stefanizzi, M.L., Tarricone, L.: An IoT-aware architecture for smart healthcare systems. IEEE IoT. J. **2**(6) (2015). https://doi.org/10.1109/JIOT.2015.2417684
9. Hosseini, M.P., Tran, T.X., Pompili, D., Elisevich, K., Soltanian-Zadeh, H.: Deep learning with edge computing for localization of epileptogenicity using multimodal rs-fMRI and EEG Big data. In: IEEE International Conference on Autonomic Computing (ICAC) (2017). https://doi.org/10.1109/ICAC.2017.41
10. Al-Khafajiy, M., Webster, L., Baker, T., Waraich, A.: Towards fog driven IoT healthcare: challenges and framework of fog computing in healthcare. In: ICFNDS'18 (2018). https://doi.org/10.1145/3231053.3231062
11. Andriopoulou, F., Dagiuklas, T., Orphanoudakis, T., Keramidas, G., Voros, N., Hübner M.: In: Integrating IoT and fog computing for healthcare service delivery. İn: Components and Services for IoT Platforms. Springer International Publishing (2017). https://doi.org/10.1007/978-3-319-42304-3_11
12. Ullah, K., Shah, M.A., Zhang, S.: Effective ways to use Internet of Things in the field of medical and smart health care. In: 2016 International Conference on Intelligent Systems Engineering (ICISE) (2016). https://doi.org/10.1109/INTELSE.2016.7475151
13. Rathee, G., Sharma, A., Saini, H., Kumar, R., Iqbal, R.: A hybrid framework for multimedia data processing in IoT-healthcare using blockchain technology. Multimedia Tools Appl. **79**(15), 9711–9733 (2020)
14. Taiwo, O., Ezugwu, A.E.: Smart healthcare support for remote patient monitoring during covid-19 quarantine. Inf. Med. Unlocked **20**, 100428 (2020)
15. Oueida, S., Kotb, Y., Aloqaily, M., Jararweh, Y., Baker, T.: An edge computing based smart healthcare framework for resource management. Sensors (Basel) (2018). https://doi.org/10.3390/s18124307
16. Siryani, J., Tanju, B., Eveleigh, T.: A Machine learning decision- support system improves the Internet of Things' smart meter operations. IEEE Internet Things J. (2017). https://doi.org/10.1109/JIOT.2017.2722358
17. Satija, U., Ramkumar, B., Manikandan, M.S.: Automated ECG noise detection and classification system for unsupervised healthcare monitoring. IEEE J. Biomed. Health Inf. **22**(3) (2018). https://doi.org/10.1109/JBHI.2017.2686436
18. Wu, S., Rendall, J.B., Smith, M.J., Zhu, S., Xu, J., Wang, H., Yang, Q., Qin, P.: Survey on prediction algorithms in smart homes. IEEE IoT. **4**(3) (2017). https://doi.org/10.1109/JIOT.2017.2668061
19. Zhang, H., Li, J., Wen, B., Xun, Y., Liu, J., Senior Member.: Connecting Intelligent Things in smart hospitals using NB-IoT. IEEE IoT. J. **5**(3) (2018). https://doi.org/10.1109/JIOT.2018.2792423
20. Obinikpo, A.A., Kantarci, B.: Big sensed data meets deep learning for smarter health care in smart cities. J. Sens. Actuator Netw. **6**(26) (2017). https://doi.org/10.3390/jsan6040026
21. Durga, S., Nag, R., Daniel, E.: Survey on machine learning and deep learning algorithms used in internet of things (IoT) healthcare. In: 2019 3rd International Conference on Computing Methodologies and Communication (ICCMC), pp. 1018–1022. IEEE (2019)
22. Akbar, A., Khan, A., Carrez, F., Moessner, K.: Predictive analytics for complex IoT data streams. IEEE IoT. J. **4**(3) (2017). https://doi.org/10.1109/JIOT.2017.2712672
23. Nesa, N., Banerjee, I.: IoT-based sensor data fusion for occupancy sensing using dempster–shafer evidence theory for smart buildings. IEEE IoT. J. **4**(5) (2017). https://doi.org/10.1109/JIOT.2017.2723424

24. Amudha, S. and Murali, M.: Enhancement of IoT-based smart hospital system survey paper (2019). https://doi.org/10.4018/978-1-5225-8555-8.ch014
25. Wu, T., Redouté, J.M., Yuce, M.: A wearable, low-power, real-time ECG monitor for smart t-shirt and IoT healthcare applications. In: Advances in Body Area Networks I, pp. 165–173. Springer, Cham (2019)
26. Al-Khafajiy, M., Baker, T., Chalmers, C., Asim, M., Kolivand, H., Fahim, M., Waraich, A.: Remote health monitoring of elderly through wearable sensors. Multimedia Tools Appl. **78**(17), 24681–24706 (2019)
27. Mao, A., Ma, X., He, Y., Luo, J.: Highly portable, sensor-based system for human fall monitoring. Sensors **17**(9), 2096 (2017)
28. Diraco, G., Leone, A., Siciliano, P.: A radar-based smart sensor for unobtrusive elderly monitoring in ambient assisted living applications. Biosensors **7**(4), 55 (2017)
29. Selvaraj, S., Sundaravaradhan, S.: Challenges and opportunities in IoT healthcare systems: a systematic review. SN Appl. Sci. **2**(1), 1–8 (2020)
30. Karuppusamy, P.: Hybrid manta ray foraging optimization for novel brain tumor detection. J. Soft Comput. Paradigm (JSCP) **2**(03), 175–185 (2020)
31. Chen, J.-Z.: Design of accurate classification of COVID-19 disease in X ray images using deep learning approach. J. ISMAC **3**(02), 132–148 (2021)
32. Balasubramaniam, V.: Artificial intelligence algorithm with SVM classification using dermascopic images for melanoma diagnosis. J. Artif. Intell. Capsule Netw. **3**(1), 34–42
33. Tyagi, A.K., Chahal, P.: Artificial intelligence and machine learning algorithms. In: Challenges and Applications for Implementing Machine Learning in Computer Vision, IGI Global (2020). https://doi.org/10.4018/978-1-7998-0182-5.ch008
34. Tyagi, A.K., Nair, M.M., Niladhuri, S., Abraham, A.: Security, privacy research issues in various computing platforms: a survey and the road ahead. J. Inf. Assur. Secur. **15**(1), 1–16. 16p (2020)
35. Pramod, A., Naicker, H.S., Tyagi, A.K.: Machine learning and deep learning: open issues and future research directions for next ten years. In: Computational Analysis and Understanding of Deep Learning for Medical Care: Principles, Methods, and Applications, 2020. Wiley Scrivener (2020)
36. Tyagi, A.K., Rekha, G.: Challenges of applying deep learning in real-world applications. In: Challenges and Applications for Implementing Machine Learning in Computer Vision, IGI Global 2020, pp. 92–118. https://doi.org/10.4018/978-1-7998-0182-5.ch004
37. Gudeti, B., Mishra, S., Malik, S., Fernandez, T.F., Tyagi, A.K., Kumari, S.: A novel approach to predict chronic kidney disease using machine learning algorithms. In: 2020 4th International Conference on Electronics, Communication and Aerospace Technology (ICECA), Coimbatore, 2020, pp. 1630–1635. https://doi.org/10.1109/ICECA49313.2020.9297392
38. Tyagi, A.K., Aghila, G., Sreenath, N.: AARIN: affordable, accurate, reliable and innovative mechanism to protect a medical cyber-physical system using blockchain technology. Int. J. Intell. Netw (2021)
39. Shamila, M., Vinuthna, K., Tyagi, A.: A review on several critical issues and challenges in IoT based e-healthcare system. pp. 1036–1043 (2019). https://doi.org/10.1109/ICCS45141.2019.9065831
40. Tyagi, A.K., Rekha, G.: Machine learning with big data. In: Proceedings of International Conference on Sustainable Computing in Science, Technology and Management (SUSCOM), Amity University Rajasthan, Jaipur—India, February 26–28, 2019 (2019)
41. Kumari, S., Vani, V., Malik, S., Tyagi, A.K., Reddy, S.: Analysis of text mining tools in disease prediction. In: Abraham, A., Hanne, T., Castillo, O., Gandhi, N., Nogueira Rios, T., Hong, T.P. (eds.) Hybrid Intelligent Systems. HIS 2020. Advances in Intelligent Systems and Computing, vol. 1375. Springer, Cham (2021). https://doi.org/10.1007/978-3-030-73050-5_55
42. Varsha, R., Nair, S.M., Tyagi, A.K., Aswathy, S.U., RadhaKrishnan, R.: The future with advanced analytics: a sequential analysis of the disruptive technology's scope. In: Abraham, A., Hanne, T., Castillo, O., Gandhi, N., Nogueira Rios, T., Hong, T.P. (eds) Hybrid Intelligent Systems. HIS 2020. Advances in Intelligent Systems and Computing, vol. 1375. Springer, Cham (2021). https://doi.org/10.1007/978-3-030-73050-5_56

Implementing SPARQL-Based Prefiltering on Jena Fuseki TDB Store to Reduce the Semantic Web Services Search Space

Pooja Thapar and Lalit Sen Sharma

Abstract The increased number of services and their increased ontological complexity make the discovery a very time-consuming process as they were built on DL (Description Logic) reasoning. The proposed mechanism focused on providing a lightweight process by including preprocessing stage to reduce the search space before actual matchmaking process. In this approach, two SPARQL-based filters were applied to the OWL-S services published along with ontologies in the Jena Fuseki TDB store. These filters perform a different degree of semantic matching on functional parameters in the service description profile based on user requirements through queries. These queries were generated automatically from the user requests having different numbers of input and output. The paper also demonstrates a case scenario from OWL-S services test collection, and results obtained from these filters were analyzed to prove the efficiency of the proposal. From the analysis of results, we observed that both the filters completely discard the services that do not match any functional requirement requested by the user, thereby reducing the search space from 7.45 to 1.29% of the original repository. Consequently, this improved the performance with a penalty on precision of service discovery depending upon the degree of matching applied using filters.

Keywords Semantic web services · OWL-S description method · SPARQL query language · Service discovery · Jena TDB data store · Domain ontologies

1 Introduction

In recent years, the software industry is being dominated by Web Services as it allows enterprises to publish their application components across their organizational boundaries and thus provides the deployment of distributed applications. These self-described and loosely-coupled Web Services also provide seamless interfaces for the integration of operations to reduce the application development time and

P. Thapar (✉) · L. S. Sharma
Department of Computer Science and IT, University of Jammu, Jammu, J&K, India

© The Author(s), under exclusive license to Springer Nature Singapore Pte Ltd. 2022
V. Suma et al. (eds.), *Evolutionary Computing and Mobile Sustainable Networks*,
Lecture Notes on Data Engineering and Communications Technologies 116,
https://doi.org/10.1007/978-981-16-9605-3_22

their maintenance cost [1, 2]. The architecture of Web Services relies on SOAP (Simple Object Access Protocol), UDDI (Universal Description, Discovery, and Integration), and WSDL (Web Service Description Language) which uses XML as the underlying data model; the semantics information is required to be added in the services to enhance their functionalities and to interpret their structured information unambiguously [3]. This gives rise to procedural extension of Semantic Web called Semantic Web Services (SWS). These services use ontologies which are metadata vocabularies of Semantic Web as their underlying data model to support semantic interoperability, automatic discovery, orchestration, and execution of Web Services [4–6]. The requirements of Web Service application that belong to some domain are formally described in WSDL file of its architecture using functional parameters like Input, Output, Preconditions, and Effects (IOPE); and defining these concepts using domain ontology gives the semantic description of SWS as shown in Fig. 1. Based on the semantics view given by domain expert, the two conceptual models which describe available Semantic Web Services are OWL-S (Web Ontology Language for Services) which use description logic for service ontology, and WSMO (Web Service Modeling Ontology) that allows loose coupling of services using mediator architecture. In OWL-S approach, upper ontology includes *Service Profile* to give high-level description of services capabilities, *Service Model* to describe data flow and control flow structure of these services, and *Service Grounding* to specify the details regarding message formats and protocols for interaction with these services. WSMO is a meta ontology based on four important concepts- *Ontologies*, *Web Services*, *Goals*, and *Mediators* in Web Service Modeling Framework defined using Web Service Modeling Language (WSML) [7, 8].

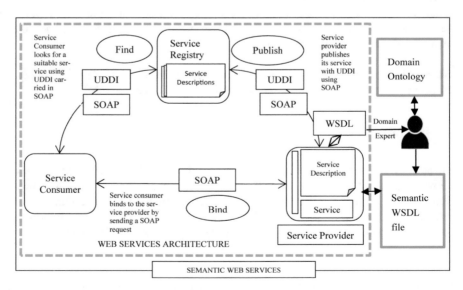

Fig. 1 Domain ontology extension in WSDL file of Web Services Architecture

With the growing emergence of these services, the publicly available repositories may get exploded in the near future, and also the methods of their semantic description namely OWL-S and WSMO require high complexity to define and process them as they were based on different logical approaches. These issues present a scenario where the matchmaking process suffers through scalability problems [9–11]. Currently, available discovery solutions of semantic also suffer from this problem thereby making it a very heavyweight process [12]. Although continuous research efforts are being made to improve the performance of discovery process through several techniques like caching or indexing description method [13], hybrid matchmaking technique [14], context-aware matching [15], or by using several stages for matchmaking [16]. The approach used in this paper only adds a prefiltering stage aprori to actual discovery using SPARQL filters on functional parameters like I/O. The two SPARQL [17] filters namely Q_{exact} and Q_{union} eliminate the services which do not satisfy the user requirements at all. Furthermore, this process requires very negligible execution time when compared to the actual discovery process as SPARQL queries have linear complexity depending upon the included graph patterns and dataset size [18].

The outline of this paper is as follows: Sect. 2 discusses the related work that has motivated our study while Sect. 3 explains the proposed work on which the experimentation of this paper is based. Section 4 presents the experimental setup and results drawn from experimentation. Finally, in Sect. 5 we conclude the paper with plans for future work.

2 Related Work

Semantics are composed of structural properties of entities and their relationships. A real-world entity and its digitally associated data require complex and multi-scaled system for definition and prediction. Considerable work in this direction has been proposed in [19, 20]. Semantic Web or Web 3.0 uses Semantic Web Technologies like OWL, RDF, SPARQL that have evolved Semantic Web Services from procedural extension of World wide web to ensure interoperability at hardware level [21]. The W3C (World Wide Web Consortium) encourages metadata vocabularies to share the common understanding of the concepts using domain ontologies. A Significant research is being carried out to enhance applications in distributed environment through metadata methodologies and linking of ontologies [22, 23]. RDF Schema from Semantic Web stack uses knowledge representation model to provide the information of basic elements of these ontologies. These knowledge-based engines are based on description logic which uses two main elements: TBox to define concepts relationship and its constraints using configuration model and ABox to define assertion and results of different domain ontologies [24, 25]. Various research works have been done in [26–29] to create ontologies of bioinformatics, geospatial, astrophysics, and many other domains. Like the algorithms are enriched by fusion in [30], Web Services in this domain can also be enriched by fusion of two ontologies.

Due to complex logic formalisms of description logic, Semantic Web Services of different domain ontologies require heavy matchmaker engines for their discovery from service repository. In [14], a hybrid matchmaker OWLS-MX was implemented to exploit both logic and non-logic based information retrieval (IR) techniques for OWL-S. The text similarity parameters were calculated using four variants called loss-of-information, cosine, Jensen-Shannon, and extended Jacquard. The logic reasoner of matchmaker only considered the semantic similarity between the specified concepts of domain ontology and I/O parameters of OWL-S requested services. To evaluate the performance of this engine, Owls-TC v2.2 was used as a dataset that consists of 1000 services of different application domains and achieved higher precision (0.74) and recall (0.557) values. Later, they developed a new system called WSMO-MX to support WSMO services and used WSML-TC2 test collection of 97 services and 810 instances for evaluation [31]. On the basis of comprehensive evaluations in the S3 contest, it was verified that both approaches achieved high precision [32]. However, Authors of [13, 33] used first-order logic for the formal description of preconditions and postconditions of both advertised and requested functionalities to reduce the search space as well as the number of matchmaker operations with the implementation of SDC (Semantic Discovery Caching) mechanism which stores the connections between client requests and its relevant Web Services. The technique was evaluated on Verizon which consists of 1500 Web Services for sales stores. A matchmaker called SPARQLent has been developed in [34] using the SPARQL agent on OWL-S Web Services and later on was tested against OWLS-MX [14] via SME2 tool [35]. Another hybrid matchmaker was implemented in [36] which used exact, subsume, plug-in, fail, and sibling semantic filters on input and output parameters of requested and advertised services. The approach was evaluated against Paolucci's method [37] and OWLS-MX [14] for OWLS-TC v3 test collection. Common algorithms for Web Service discovery and composition were adapted to WSMO-Lite by [38] for functional parameters-based classification algorithm.

After many research efforts on the discovery of SWS, Researchers started working on preprocessing of functional and non-functional parameters of these services to reduce computational cost of semantic similarity [39, 40]. In [41], a hierarchical categorization scheme of Web Services was used to compare the defined category concept of requested and advertised Web Services on the basis of domain ontologies. A Web Service was eliminated if requested and advertised Web Services belong to different categories. To compute the equivalence of the requested/advertised Web Services category, [42] used WordNet [43] dictionary of synonyms words for their relationship. However, due to insufficient information about the service's application domain both these approaches added one non-functional parameter to the OWL-S to overcome the limitation of OWL-S service profile elements. Author of [44] proposed a framework for WSMO services that add two stages of prefiltering based on F-Logic.

The extensive study of literature exhibits that due to persistent evolution of web technologies and emergence of new forms of network environments, this area needs continuous attention and more research efforts. However, the proposed work does not provide another discovery algorithm and focused on adding a prefiltering stage for discovery. Also, we have chosen OWL-S SWS as they offer more expressiveness compared to WSMO description method and have enough inference engines that provide better accuracy. Further, research in [45–47] suggests SPARQL language for RDF data stores optimization. So, the same was used in proposed work at preprocessing stage of Semantic Web Services.

3   Proposed Work

The flow diagram for proposed work is shown in Fig. 2 and its modules are explained in detail in the next three sections. Section 3.1 describes the OWL-S description method through an example. The initial service repository consists of these service descriptions described through concepts in the domain ontologies. Section 3.2 commences with the creation of triple store database using Fuseki server and Sect. 3.3 describes SPARQL filters used to fire queries against the TDB store.

3.1   OWL-S Service Description

An OWL-S service description is composed of several terms to describe its functional and non-functional parameters. These terms refer to several domains which are described through domain ontologies. Similarly, the requested services have to meet one or more concepts referred by the user requirement [48–50]. To implement the proposed prefiltering system based on OWL-S service description method, all the service profiles have to be published in the triple data store. The service descriptions and user requests were modeled as service profiles which contain several terms to define the features of their functionality. To illustrate our proposal, an example of OWL-S service profile has been shown in Fig. 3. It has <profile:textDescription> tag that gives brief description of service functionality in natural language. The process model of services describes whether it is an atomic or composite process and has functional tags <process:hasInput/>, <process:hasOutput/> which define the input–output of the service. The ontologically annotated <process:parameterType/> tag contains the URI of ontology corresponding to the input/output parameters of service. The OWLS collection used in this proposal merely contains service descriptions with preconditions and results. These service description profiles are parsed by JENA to match the user referred concepts.

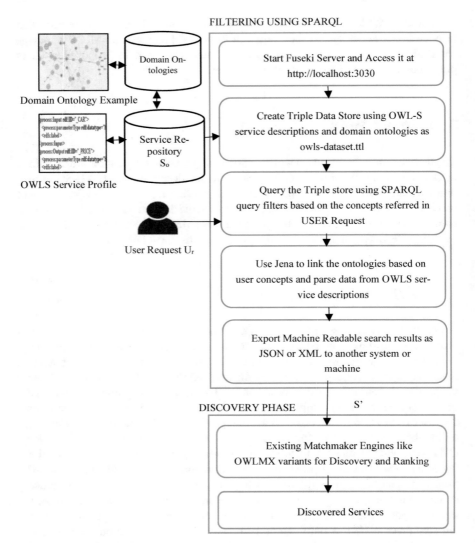

Fig. 2 Flow diagram of proposed work

3.2 Creation of Jena TDB Store to Implement SPARQL Filters Using Fuseki

APACHE FUSEKI is a SPARQL server integrated with persistent triple store database to store and query RDF data using JENA. It provides SPARQL 1.1 protocols and graph storage to get the IRI (International Resource Identifier) of services from stored graphs using HTTP query protocol. The ontologies and services from test collection

```
<profile:serviceName xml:lang="en"> car price service </profile:serviceName>
<profile:textDescription xml:lang="en"> This service returns the actual purchase price of a given car brand/m
results may contain prices of both new or used cars. </profile:textDescription>
<profile:hasInput rdf:resource="#_CAR"/>
<profile:hasOutput rdf:resource="#_PRICE"/>
<profile:has_process rdf:resource="CAR_PRICE_PROCESS"/>
</profile:Profile>
<!-- <process:ProcessModel rdf:ID="CAR_PRICE_PROCESS_MODEL">
<service:describes rdf:resource="#CAR_PRICE_SERVICE"/>
<process:hasProcess rdf:resource="#CAR_PRICE_PROCESS"/>
</process:ProcessModel> -->
▼<process:AtomicProcess rdf:ID="CAR_PRICE_PROCESS">
<service:describes rdf:resource="#CAR_PRICE_SERVICE"/>
<process:hasInput rdf:resource="#_CAR"/>
<process:hasOutput rdf:resource="#_PRICE"/>
</process:AtomicProcess>
▼<process:Input rdf:ID="_CAR">
<process:parameterType
rdf:datatype="http://www.w3.org/2001/XMLSchema#anyURI">http://127.0.0.1/ontology/my_ontology.owl#Car</process
<rdfs:label/>
</process:Input>
▼<process:Output rdf:ID="_PRICE">
<process:parameterType
rdf:datatype="http://www.w3.org/2001/XMLSchema#anyURI">http://127.0.0.1/ontology/concept.owl#Price</process:p
<rdfs:label/>
</process:Output>
```

Fig. 3 OWL-S service profile example

are uploaded to create persistent storage TDB database file named owls-dataset.ttl. This knowledge base or RDF triple store database has 35,663 triples. JENA FUSEKI 3.17 version was used for experimentation and started using command FUSEKIserver –update –mem/ds. It was accessed at http://localhost:3030/ from browser.

3.3 SPARQL-Based Prefiltering of Service Repository

Among the three main approaches: graph-based, DL-based, and rule-based to query Semantic Web, we have used graph-based approach SPARQL (SPARQL Protocol and RDF Query Language) and current W3 Recommendation to fetch RDF triples from RDF based semantic repositories using triple patterns described as (subject, predicate, object) in Turtle notation [51–53]. SPARQL has four different types of queries: SELECT, DESCRIBE, CONSTRUCT and ASK to serve the purpose and in our case, the SELECT query was used to return the distinct service names and their bindings [35]. Thus, to implement a lightweight extension of the discovery process, the proposed system filters out some of the services (S') from original service repository (S_o) using two SPARQL filters that perform different levels of semantic matching for functional parameters given by the user requirement (U_r). These filters were based on SPARQL query language and are described below:

$$Q_{\text{exact}}(S_o, U_r) = \{S' \in S_o : C_{Ur} \subseteq C_{s'}\}$$

where C_{Ur} and $C_{s'}$ refer to the concepts in the user request and filtered service descriptions, respectively.

i.e., it returns the service descriptions that exactly match the concepts mentioned in the user request.

$$\text{and } Q_{\text{union}}(S_o, U_r) = \{S' \in S_o : C_{U_r} \cap C_{s'} \neq \emptyset\}.$$

Q_{union} contains the service descriptions where some of the referred concepts matched with some concepts given by the user. Therefore, both the filters completely discard the service descriptions that do not match the user requirement at all.

4 Results and Discussion

This section describes the experimental part of proposed work on secondary test collection and the results we get from it. Further, to demonstrate the results given by two SPARQL filters a case scenario is described with the part of SPARQL query code.

4.1 Dataset

In order to test the performance of the proposed approach, the two SPARQL filters were applied to the secondary test collection of OWL-S services (OWLS-TC v3) which consists of 1007 OWL-S service descriptions and 32 different domain ontologies used for their semantic annotation. These services covered seven different domains namely: medical care, education, travel, weaponry, food, economy, and communication. The collection also included 29 queries from all the seven domains. To evaluate the performance of the filters, these queries were implemented using SPARQL query language in Jena Fuseki TDB store. The service description files and ontologies were hosted locally using Apache server and parsed by Jena to enable query execution over the services.

4.2 A Case Scenario: Find the Hotel in a City of a Country

The following lines show only part of the code for SPARQL filters where a user wants to find a hotel in a city of the country. Figures 4 and 5 present the query execution code of Q_{exact} and Q_{union} filter respectively. Q_{exact} mandates all the referred concepts {input: city, input: country, and output: hotel} to be the part of requested services where Q_{union} filters all the services satisfying any concept of user request as valid alternatives for query from the triple store using UNION keyword.

Both these filters parse concepts from service profiles without including subclass knowledge. We can implicitly add these inferred concepts in the RDF dataset using

```
1.  SELECT DISTINCT ?ServiceName ?textDescription
    (STRAFTER(?str_event1,"#") AS ?Input1)
    (STRAFTER(?str_event2,"#") AS ?Input2)
    (STRAFTER(?str_event3,"#") AS ?Output_e)
2.  WHERE {
3.  ?ServiceName a service:Service;
4.          service:presents ?profile.
5.  ?profile profile:hasInput ?inputTerm1.
6.  ?profile profile:hasInput ?inputTerm2.
7.  ?profile profile:hasOutput ?outputTerm1.
8.  ?profile profile:textDescription ?textDescription.
9.  # referred to services having "city" as input
10. {
11. ?inputTerm1 process:parameterType ?Input_1
12. BIND (STR(?Input_1) AS ?str_event1)
13. FILTER(regex(STR(?Input_1),"#city","i"))
14. }
15. # referred to services having "country" as input
16. {
17. ?inputTerm2 process:parameterType ?Input_2
18. BIND (STR(?Input_2) AS ?str_event2)
19. FILTER(regex(STR(?Input_2),"#country","i"))
20. }
21. # referred to services having ''hotel" as output
22. {
23. ?outputTerm1 process:parameterType ?Output
24. BIND (STR(?Output) AS ?str_event3)
25. FILTER(regex(STR(?Output), "#hotel","i"))
26. }
27. }
```

Fig. 4 Q_{exact} query: Referred to all concepts of user request

DL Reasoner [54, 55]. For instance, the reasoner (Pellet [55] in our case) may infer the subclass knowledge for hotel instances and add the services having LuxuryHotel as output because they share a subclass relationship with hotel instances. However, inferred concepts from the user request were not considered at the moment when the query was executed.

4.3 SPARQL Query Execution on Test Collection

For experimentation, filters described in Sect. 3.2 were implemented on OWL-S test collection for 29 user queries, and for each test dataset based on the user request, the number of services filtered by these filters and their query response time is shown in Table 1. Query response time (Q_{rt}) is the time taken by each query to return the corresponding results, without the initialization time required to register the service descriptions on TDB store.

Degree of filtering decides improvement of query response time as they are directly related to each other. Figure 6 shows scaled box plot for Q_{exact} and Q_{union} that represents the proportion of returned services for 29 queries out of 1007 services in initial

```
1.  SELECT DISTINCT ?ServiceName ?textDescription
        (STRAFTER(?str_event1,"#") AS ?Input1)
        (STRAFTER(?str_event2,"#") AS ?Input2)
        (STRAFTER(?str_event3,"#") AS ?Output_e)
2.  WHERE {
3.  ?ServiceName a service:Service;
4.          service:presents ?profile.
5.  ?profile profile:hasInput ?inputTerm1.
6.  ?profile profile:hasInput ?inputTerm2.
7.  ?profile profile:hasOutput ?outputTerm1.
8.  ?profile profile:textDescription ?textDescription.
9.  # referred to services having "city" as input
10. {
11. ?inputTerm1 process:parameterType ?Input_1
12. BIND (STR(?Input_1) AS ?str_event1)
13. FILTER(regex(STR(?Input_1),"#city","i"))
14. }
15. UNION
16. # referred to services having "country" as input
17. {
18. ?inputTerm2 process:parameterType ?Input_2
19. BIND (STR(?Input_2) AS ?str_event2)
20. FILTER(regex(STR(?Input_2),"#country","i"))
21. }
22. UNION
23. # referred to services having ''hotel" as output
24. {
25. ?outputTerm1 process:parameterType ?Output
26. BIND (STR(?Output) AS ?str_event3)
27. FILTER(regex(STR(?Output),  "#hotel","i"))
28. }
29. }
```

Fig. 5 Q_{union} query: Referred to some concepts of user request

service repository. It was found that Q_{exact} with higher degree of matching returns a very less number of services with most queries returning 0.30 and 0.79% of the original service repository and Q_{union} comparatively returning a bigger range, i.e., 7.45% median value of the original service repository. Q_{exact} gives better performance whereas the latter one is less restrictive and takes more time to read input/output from the service profile. The reduced set of services will take part in the subsequent discovery and ranking phase and thereby improving the performance.

5   Conclusion and Future Scope

Although the query languages based on semantics were not used for the discovery of Semantic Web Services they played an important role in the prefiltering of SWS, which helped in reducing search space prior to the actual discovery process.

Table 1 Number of services filtered by Q_{exact} and Q_{union} along with their response time

Test data set	Input	Output	No. of matches Service profile		Response time Q_{rt} (in ms)	
			Q_{exact}	Q_{union}	Q_{exact}	Q_{union}
1	City, country	Hotel	7	107	62	483
2	Car, 1Personbicycle	Price	3	206	52	964
3	MP3 player, DVD player	Price	5	190	86	984
4	Science-fiction-novel, user	Price	4	174	87	971
5	Geographical-region1, Geographical-region2	Map	4	78	66	347
6	Degree, government	Scholarship	5	97	74	431
7	Missile, government	Funding	7	115	112	889
8	Hiking, surfing	Destination	4	72	49	348
9	Organization, surfing	Destination	7	111	63	486
10	Shopping mall	Camera, price	2	176	48	978
11	Recommended price	Coffee, whiskey	8	48	52	281
12	Book	Price	13	186	194	539
13	Car	Price	19	196	97	678
14	Country	Skilled occupation	12	90	51	204
15	Geopolitical-entity	Weather process	1	50	46	84
16	Grocery store	Food	2	19	60	92
17	Hospital	Investigating	3	24	46	101
18	Maxprice	Cola	8	25	44	76
19	Novel	Author	12	72	43	207
20	Prepared food	Price	5	174	65	467
21	Publication-number	Publication	2	24	41	118
22	Researcher-in-academia	Address	3	20	42	100
23	Surfing	Destination	17	55	46	103
24	Title	Comedy film	8	61	65	70
25	Title	Video media	8	62	48	78
26	University	Lecturer-in-academia	3	16	77	106
27	Person, book, credit card account	Price	11	77	41	235
28	Person, book, credit card account	Book	17	196	48	113
29	EBook request, user account	Ebook	4	18	79	424

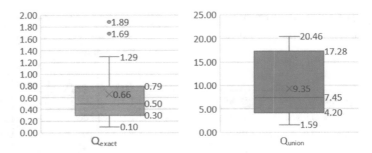

Fig. 6 Proportion of services returned by SPARQL Filters with respect to original repository

The proposed research work follows the current research problem in the development of lightweight extensions for Semantic Web Services to improve heavy-weight matchmaking process effectively. Experimental results clearly demonstrated that it improves the overall performance of the discovery and ranking phase by reducing memory consumption, initialization time, and query response time for these processes. However, there is always a tradeoff between performance and precision-recall of matchmaking processes [11]. Thus, Q_{exact} though reduces the repository size to a great extent but it will cost more penalty on precision-recall during discovery phase whereas Q_{union} on an average maintains a balance between accuracy and performance in general. Also, these services did not include SPARQL entailment which will improve the efficiency of this framework. In future work, RDF subclass knowledge from user concepts can also be added to the dataset either by using DL Reasoner or by explicitly writing SPARQL queries to filter more service descriptions based on these referred concepts in the prefiltering stage for better accuracy.

References

1. Kritikos, K., Plexousakis, D.: Requirements for QoS-based web service description and discovery. IEEE Trans. Serv. Comput. **2**(4), 320–337 (2009)
2. Alonso, G., Casati, F., Kuno, H., Machiraju, V.: Web services: concepts, architectures and applications, 3rd edn., pp. 123–149. Springer, Berlin, Heidelberg (2013)
3. McIlraith, A.S., Son, T.C., Zeng, H.: Semantic web services. IEEE Intell. Syst. **16**(2), 46–53 (2001)
4. Patil, A.A., Oundhakar, S.A., Sheth, A.P., Verma, K.: Meteor-s web service annotation framework. In: Proceedings of the 13th International Conference on World Wide Web, pp. 553–562, ACM (2004)
5. Fensel, D., Bussler, C.: The web service modelling framework WSMF. Electron. Commer. Res. Appl. **1**(2), 113–137 (2002)
6. Mallayya, D., Ramachandran, B., Viswanathan, S.: An automatic web service composition framework using QoS-based web service ranking algorithm. Sci. World J. **2015**, 14 (2015). Art. No. 207174
7. Martin, D., Burstein, M., Hobbs, J., Lassila, O., McDermott, D., McIlraith, S. et al.: OWL-S: semantic markup for web services. W3C Member Submission. **22**(4) (2004)

8. Roman, D., Lausen, H., Keller, U., Oren, E., Bussler, C., Kifer, M., et al.: The Web service modelling ontology. Appl. Ontol. **1**(1), 77–106 (2005)
9. Davies, J., Potter, M.R.M.S.S.D.J.P.C.F.D., Richardson, M., Stinčić, S., Domingue, J., Pedrinaci, C., Fensel, D. and Gonzalez-Cabero, R.: Towards the open service web. BT Technol. J. **26**(2) (2009)
10. Domingue, J., Fensel, D., González-Cabero, R.: SOA4All, enabling the SOA revolution on a world wide scale. IEEE Int. Conf. Semant. Comput. **2008**, 530–537 (2008). https://doi.org/10.1109/ICSC.2008.45
11. Fensel, D.: The potential and limitations of semantics applied to the future internet. In: WEBIST, pp. 15–15 (2009)
12. Haarslev, V., Möller, R.: On the scalability of description logic instance retrieval. J. Autom. Reason. **41**(2), 99–142 (2008). https://doi.org/10.1007/s10817-008-9104-7
13. Stollberg, M., Hepp, M., Hoffmann, J.: A caching mechanism for semantic web service discovery. In: The Semantic Web, pp. 480–493. Springer, Berlin, Heidelberg (2007). https://doi.org/10.1007/978-3-540-76298-0_35
14. Klusch, M., Fries, B., Sycara, K.: OWLS-MX: a hybrid Semantic Web service matchmaker for OWL-S services. J. Web Semant. **7**(2), 121–133 (2009). https://doi.org/10.1016/J.WEBSEM.2008.10.001
15. Guermah, H., Fissaa, T., Hafiddi, H., Nassar, M.: Exploiting semantic web services in the development of context-aware systems. Procedia Comput. Sci. **127**, 398–407 (2018). https://doi.org/10.1016/J.PROCS.2018.01.137
16. Agarwal, S., Junghans, M., Fabre, O., Toma, I., Lorre, J.P.: D5. 3.1 first service discovery prototype, Tech. Rep., SOA4All (2009)
17. Prud'hommeaux, E., Seaborne, A.: SPARQL query language for RDF. W3C recommendation, W3C. http://www.w3.org/TR/rdf-sparql-query (2008)
18. Pérez, J., Arenas, M., Gutierrez, C.: Semantics and complexity of SPARQL. ACM Trans. Database Syst. (TODS) **34**(3), 1–45 (2009). https://doi.org/10.1145/1567274.1567278
19. Sivaganesan, D.: Novel influence maximization algorithm for social network behavior management. J. ISMAC **3**(01), 60–68 (2021)
20. Valanarasu, M.R.: Comparative analysis for personality prediction by digital footprints in social media. J. Inf. Technol. **3**(02), 77–91 (2021)
21. Muppavarapu, V., Ramesh, G., Gyrard, A., Noura, M.: Knowledge extraction using semantic similarity of concepts from Web of Things knowledge bases. Data Knowl. Eng. 101923 (2021). https://doi.org/10.1016/J.DATAK.2021.101923
22. El-Gayar, O., Deokar, A.: A semantic service-oriented architecture for distributed model management systems. Decis. Support Syst. **55**(1), 374–384 (2013). https://doi.org/10.1016/J.DSS.2012.05.046
23. De, S., Zhou, Y., Moessner, K.: Ontologies and context modeling for the Web of Things. Manag. Web Things Link. Real World Web, 3–36 (2017). https://doi.org/10.1016/B978-0-12-809764-9.00002-0
24. Nacer, H., Aissani, D.: Semantic web services: standards, applications, challenges and solutions. J. Netw. Comput. Appl. **44**, 134–151 (2014). https://doi.org/10.1016/J.JNCA.2014.04.015
25. Hotz, L., Felfernig, A., Stumptner, M., Ryabokon, A., Bagley, C., Wolter, K.: Configuration knowledge representation and reasoning, Knowledge-Based Config. From Res. to Bus. Cases, pp. 41–72 (2014). https://doi.org/10.1016/B978-0-12-415817-7.00006-2
26. Islam, M.R., Ahmed, M.L., Paul, B.K., Bhuiyan, T., Ahmed, K., Moni, M.A.: Identification of the core ontologies and signature genes of polycystic ovary syndrome (PCOS): a bioinformatics analysis. Inf. Med. Unlocked **18**, 100304 (2020). https://doi.org/10.1016/J.IMU.2020.100304
27. Rehana, H., Ahmed, M.R., Chakma, R., Asaduzzaman, S., Raihan, M.: A bioinformatics approach for identification of the core ontologies and signature genes of pulmonary disease and associated disease. Gene Rep. **24**, 101206 (2021). https://doi.org/10.1016/J.GENREP.2021.101206

28. Louge, T., Karray, M.H., Archimede, B., Maamar, Z., Mrissa, M.: Semantic web services composition in the astrophysics domain: issues and solutions. Futur. Gener. Comput. Syst. **90**, 185–197 (2019). https://doi.org/10.1016/J.FUTURE.2018.07.063

29. Stock, K., Guesgen, H.: Geospatial reasoning with open data. In: Automating Open Source Intelligence Algorithms OSINT, pp. 171–204 (2016). https://doi.org/10.1016/B978-0-12-802 916-9.00010-5

30. Senthil Kumar, T.: Construction of hybrid deep learning model for predicting children behavior based on their emotional reaction. J. Inf. Technol. **3**(01), 29–43 (2021)

31. Klusch, M., Kaufer, F.: WSMO-MX: a hybrid semantic web service matchmaker. Web Intell. Agent Syst. **7**(1), 23–42 (2009)

32. Klusch, M.: Overview of the S3 contest: performance evaluation of semantic service matchmakers. In: Semantic Web Services: Advancement through Evaluation, pp. 17–34. Springer, Berlin, Germany (2012)

33. Stollberg, M., Hoffmann, J., Fensel, D.: A caching technique for optimizing automated service discovery. Int. J. Semant. Comput. (World Sci.) **5**(1), 1–31 (2011)

34. Sbodio, M.L., Martin, D., Moulin, C.: Discovering Semantic web services using SPARQL and intelligent agents. J. Web Semant. **8**(4), 310–328 (2010)

35. Sme2 version 2.2, Semantic Web Central, Sep. 22 2010, Accessed on: 29 July 2019. (Online). Available: http://projects.semwebcentral.org/projects/sme2/

36. Amorim, R., Claro, D.B., Lopes, D., Albers, P., Andrade, A.: Improving web service discovery by a functional and structural approach. In: Proceedings of the IEEE 9th International Conference on Web Services (ICWS'11), pp. 411–418, IEEE, Washington, DC, USA (2011)

37. Paolucci, M., Kawamura, T., Payne, T.R., Sycara, K.: Semantic matching of web services capabilities. In: International semantic web conference, vol. 2342, pp. 333–347. Springer, Berlin, Heidelberg (2002)

38. Roman, D., Kopecký, J., Vitvar, T., Domingue, J., Fensel, D.: WSMO-Lite and hRESTS: lightweight semantic annotations for Web services and RESTful APIs. J. Web Semant. **31**, 39–58 (2015)

39. Trokanas, N., Cecelja, F., Yu, M., Raafat, T.: Optimising environmental performance of symbiotic networks using semantics. Comput. Aided Chem. Eng. **33**, 847–852 (2014). https://doi.org/10.1016/B978-0-444-63456-6.50142-3

40. Gyrard, A., Serrano, M., Patel, P.: Building interoperable and cross-domain semantic web of things applications. Manag. In: Web Things Link. Real World to Web, pp. 305–324 (2017). https://doi.org/10.1016/B978-0-12-809764-9.00014-7

41. Khdour, T.: Towards semantically filtering web services repository. In: International Conference on Digital Information and Communication Technology and Its Applications, vol. 167, pp. 322–336. Springer, Berlin, Heidelberg (2011)

42. Mohebbi, K., Ibrahim, S., Zamani, M.: A pre-matching filter to improve the query response time of semantic web service discovery. J. Next Gener. Inf. Technol. **4**(6), 9–18 (2013)

43. Fellbaum, C.: WordNet: an electronic lexical resource. In: An Introduction to Cognitive Science, Ch. 16, pp. 301–314. Blackwell Publishing, Oxford University Press (2017)

44. Ghayekhloo, S., Bayram, Z.: Prefiltering strategy to improve performance of semantic web service discovery. Sci. Program. **23**, 2015 (2015). https://doi.org/10.1155/2015/576463

45. Katib, A., Slavov, V., Rao, P.: RIQ: fast processing of SPARQL queries on RDF quadruples. J. Web Semant. **37–38**, 90–111 (2016). https://doi.org/10.1016/J.WEBSEM.2016.03.005

46. Izquierdo, Y.T., et al.: Keyword search over schema-less RDF datasets by SPARQL query compilation. Inf. Syst. **102**, 101814 (2021). https://doi.org/10.1016/J.IS.2021.101814

47. Ravat, F., Song, J., Teste, O., Trojahn, C.: Efficient querying of multidimensional RDF data with aggregates: comparing NoSQL, RDF and relational data stores. Int. J. Inf. Manage. **54**, 102089 (2020). https://doi.org/10.1016/J.IJINFOMGT.2020.102089

48. Zhou, J., Koivisto, J.P., Niemela, E.: A survey on semantic web services and a case study. In Computer Supported Cooperative Work in Design, 2006. CSCWD'06, 10th International Conference, pp. 1–7, IEEE (2012)

49. Kopecký, J., Vitvar, T., Bournez, C., Farrell, J.: Sawsdl: Semantic annotations for WSDL and xml schema. IEEE Internet Comput. **11**(6) (2015)
50. Martin, D., Burstein, M., McDermott, D., McIlraith, S., Paolucci, M., Sycara, K., Srinivasan, N.: Bringing semantics to web services with OWL-S. World Wide Web. **10**(3), 243–277 (2017)
51. Bailey, J., Bry, F., Furche, T., Schaffert, S.: Web and semantic web query languages: a survey. In: Reasoning Web, vol. 364, pp. 35–133. Springer, Berlin, Heidelberg (2005). https://doi.org/10.1007/11526988_3
52. Aranda, C.B., Corby, O., Das, S., Feigenbaum, L., Gearon, P., Glimm, B. et al.: SPARQL 1.1 overview, W3C, 21 Mar. 2013. Accessed on: 19 Feb. 2021. (Online). Available: https://www.w3.org/TR/sparql11-overview/
53. Sintek, M., Decker, S.: TRIPLE—a query, inference, and transformation language for the semantic web. In: International Semantic Web Conference, pp. 364–378. Springer, Berlin, Heidelberg (2002)
54. Haarslev, V., Möller, R.: RACER system description. In: International Joint Conference on Automated Reasoning, pp. 701–705. Springer, Berlin, Heidelberg (2001)
55. Sirin, E., Parsia, B., Grau, B.C., Kalyanpur, A., Katz, Y.: Pellet: a practical OWL-DL reasoner. J. Web Semant. **5**(2), 51–53 (2007). https://doi.org/10.1016/J.WEBSEM.2007.03.004

Emerging 5G Wireless Technologies: Overview, Evolution, and Applications

M. C. Malini and N. Chandrakala

Abstract The fifth-generation (5G) of wireless network technology is a high-speed network. The aim of 5G technologies is the connection of different devices and machines with strong improvements in terms of high service features, increased network capability, and superior system throughput to support several vertical applications. 5G is a rapidly evolving technology that permits smart devices, such as Android mobile devices, to run artificial intelligence (AI)-based software applications. Other emerging technologies, such as autonomous driving, which rely heavily on network communications, will benefit as well. The technical background and 5G key technologies, as well as their development, emerging technology, requirements, and applications, are briefly discussed in this review.

Keywords 5G wireless technology · Evolving technologies · The evolution of wireless technology · MIMO technology

1 Introduction

The radio-frequency (RF) and infrared (IR) mechanisms used in wireless technology allow devices to communicate with one another. In the 1970s, wireless communication has originated throughout the world from voice calls up to now modern communication technology provided that high superiority broadband facilities for end-users. The potential growth of wireless communication technology presently different varieties of mobile devices are developed including tablets and smartphones. The future technologies will be providing a networked society with uncontrolled services to share the information of data that can be accessed anywhere in the world. Hence, the new technology of data communication is required to provide more efficient and high-speed connectivity [1].

Among the five generations of wireless technology, 5G is the fifth generation technology that recently planned to launch a high-speed data rate scheme. The 5G is

M. C. Malini (✉) · N. Chandrakala
Department of Computer Science, SSM College of Arts and Science, Komarapalayam, Namakkal, India

© The Author(s), under exclusive license to Springer Nature Singapore Pte Ltd. 2022　　335
V. Suma et al. (eds.), *Evolutionary Computing and Mobile Sustainable Networks*,
Lecture Notes on Data Engineering and Communications Technologies 116,
https://doi.org/10.1007/978-981-16-9605-3_23

a well-known broadband cellular network and cellular phone companies are began deploying worldwide [2]. 5G networks are expected to have more than 1.7 billion users around worldwide by 2025. 5G networks are cellular networks and the service area is separated into small geographical areas named *cells*. In a cell, the wireless network devices are linked to the telephone with the Internet with the help of radio waves over a local antenna [3]. The core advantages of the innovative wireless networks have superior bandwidth, giving advanced download speeds (10 GBPS). The new technology is expected to the networks that will not entirely work cellphones like existing cellular networks due to the better bandwidth. However, the fifth generation can also be used as common internet service providers (ISP) for desktop and laptops, and also different applications including internet of things (IoT) [4] and machine to machine (M2M) areas [5]. While encouraging the networked society, this paper presents a significant challenge that high bandwidth communication will face.

The continuing of the paper is ordered as follows; Sect. 2 presents an evolution of wireless technology. Section 3 discussed the architecture of 5G technology. Section 4 discussed requirements; Sect. 5 discussed an emerging technology. Section 6 demonstrated the design for 5G wireless technology. Section 7 discussed the applications, Sect. 8 demonstrates the discussions. Finally, the conclusion is determined in Sect. 10.

2 Evolutions of Wireless Technology

Wireless technology is an important part of real-life society. Because mobile services have evolved from television, satellite communication, and radio transmission. The way society operates has changed as a result of wireless communications. It also demonstrates that circuit switching is used in both the first and second generations of technology. Both circuit and packet switching is used in the second and third generations, respectively. Packet switching is used in third to fifth-generation technologies, as well as licensed and unlicensed spectrum. The licensed spectrum is used by all developing generations, whereas the unlicensed spectrum is used by WiFi, Bluetooth, and WiMAX. Below is a summary of the rapidly evolving wireless technologies. Figure 1 shows the evolutions of wireless technologies including data rate, flexibility, and spectral and coverage effectiveness [6].

2.1 First Generations

In the early 1980s, the first generation was created. The data rate of first-generation is 2.4 kbps. NMT stands for Nordic mobile telephone, AMPS stands for advanced mobile phone system, and TACS stands for total access communication system those are the major subscribers. It has many shortcomings including careless handoff, lower voice association; lower bar capacity, and no security. There is a potential for

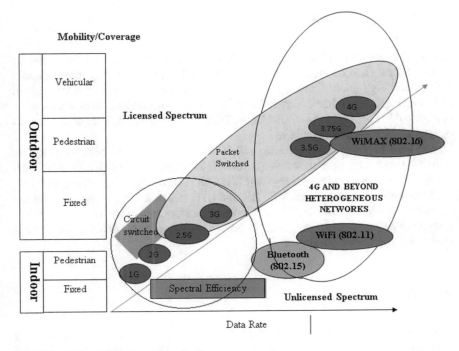

Fig. 1 Evolutions of wireless technology

unwanted snooping by the third party due to voice calls were kept and played in radio towers.

2.2 Second Generations

The 2nd generation was announced in 1990 and developed by digital technology in the mobile telephone. The second-generation system is developed by using the GSM scheme (Global Systems for Mobile Communications). In this generation, the voice communication process having a data rate limit is up to 64 kbps. The second-generation mobile handset batteries are taking time-consuming due to the radio signals consuming low power. It is providing two main services such as e-mail and SMS (short message services).

The main core skills were code division multiple access (CDMA), IS-95 and GSM. Another version of the second generation is 2.5G which mostly contributes to a cellular system combined with GPRS (general packet radio system). It is an applied packet switching technique alongside circuit switching and can support data rates up to 144 kbps. The core schemes are GPRS, CDMA, and Enhanced data rate for GSM (EDGE).

2.3   Third Generations

The third generation, which began in 2000, allows for transmission data rates of up to 2 Mbps. In 3G, internet protocol (IP) is used to provide a high-speed mobile service and high-quality voice. However, 3G is needed more power consumption and more expansive when compared with 2G. Universal mobile telecommunication systems (UMTS), Wideband CDMA (WCDMA), and CMA 2000 are core schemes developed in 3G wireless systems.

There are two versions in 3G such as 3.5G and 3.75G. In 3.5G, high speed uplink (HSUPA)/downlink packet access (HSDPA). The future of the 3.75G is an LTE (Long-Term Evolution technology) and WIMAX (Fixed Worldwide Interoperability for Microwave Access). Fixed WiMAX and LTE can enhance the capability of the network and offer a considerable amount of users the capability to access a wide-ranging of extraordinary speed facilities like on peer to peer file distribution, request video, and complex web services.

2.4   Fourth Generations

A 4G scheme enhances the fundamental communication networks by communicating a wide-ranging and dependable clarification based on IP. Facilities like multimedia, voice, and data will be communicated to subscribers on every time and universal basis and at fairly higher data speeds when compared with prior generations. The 4G networks developed various networks including such as video calling and chatting, digital video broadcasting (DVB), multimedia messaging service (MMS), and high definition televisions.

2.5   Fifth Generations

5G technology is developed using filter bank multicarrier (FBMC) and beam division multiple access (BDMA). The idea behind with BDMA by connecting the base station to the mobile stations, the method is described. In the base communication station, an orthogonal beam is distributed to all mobile stations. The BDMA technique incorporates the antenna beam based on the mobile stations location, allowing the mobile stations to make multiple contacts. 5G cellular networks have six challenges high data rate, higher capacity, End to End latency is low, connecting with the enormous device, and decreased cost. An IEEE 802.11ac, 802.11ad, and 802.11af standards are precise popular, supportive, newly presented for constructing significant 5G (Table 1).

Table 1 Comparisons of an evolution of wireless technology

Evolutions of networks	Technology	Data speed	Frequency band	Bandwidth	Forwarded error correction	Switching	Uses
1G	AMPS FDMA	2.4 kpbs	800 MHz	30 kHz	–	Circuit	Voice only
2G	GSM TDMA	10 kpbs	850/900/1800/900 MHz	200 kHz	–	Circuit	Voice and data
	CDMA	10 kpbs		1.25 MHz			
2.5G	GPRS	50 kpbs		200 kHz		Circuit\packet	
	EDGE	200 kpbs					
3G	WCDMA UMTS	384 kpbs	800/850/900/1900 MHz	5 MHz	Turbo codes	Circuit\packet	Data, voice, and video calling
	CDMA			1.25 MHz		Circuit\packet	
3.5G	HSUPA/HADPA	5–30 Mbps		5 MHz		Packet	
	EVDO	5–30 Mbps		1.25 MHz		Packet	
3.75G	LTE OFDMA/SC-FDMA	100–200 Mbps	1.8 GHz 2.6 GHz	1.4–20 MHz	Concatenated codes	Packet	Online gaming + high definition television
	WIMAX SOFDMA Fixed WIMAX	100–200 Mbps	3.5 and 5.8 GHz initially	10 MHz in 5.8 GHz band 3.5 and 7 MHz in 3.5 GHz band			

(continued)

Table 1 (continued)

Evolutions of networks	Technology	Data speed	Frequency band	Bandwidth	Forwarded error correction	Switching	Uses
4G	LTE-A/OFDMA/SC-FDMA	DL 3 Gbps UL 1.5 Gbps	1.8 GHz 2.6 GHz	1.4–20 MHz	Turbo codes	Packet	Online gaming + high definition television
	WIMAX/SOFDMA MOBILE WIMAX	100–200 Mbps	2.3 GHz 2.5 GHz 3.5 GHz initially	3.5 MHz 5 and 7 MHz 10 and 8.75 MHz initially			
5G	BDMA FBMC	10–50 Gbps (expected)	1.8, 2.6 GHz and expected 30–300 GHz	60 GHz	LDPC	Packet	Ultra HD video + virtual reality applications

3   The Architecture of 5G Technology

Fifth-generation wireless mobile technology is developed in many countries namely South Korea, Japan, and the U.S. government are commercially introduced in April 2019. China was launched 5G by provided that commercial authorizations to its main carriers with some limitations. India was planned to target in 2020 for the commercial establishment of 5G Wireless Technologies. The government previously launched a three-year package in March 2018 to encourage research in fifth-generation technology. In IIT Delhi, the Ericsson Company has formed a 5G test base center for developing applications [7].

The 5G technology is a more attractive technology after 2011. The regularization of 5G cellular communication technologies is estimated to be completed in an earlier stage because of the innovative revolution of mobile technologies in every decade. Most of the wireless users visit inside almost 80% and almost 20%. The recent cellular network architecture communicates whether inside or outside for mobile communication [8]. The required more data rate, reduced spectral efficiency and, the wireless communication energy efficient when the inside users can communicate with the outside station [9]. To solve the above-mentioned shortcoming, innovative approaches will be incorporated with existing 5G technologies to distinct inside and outside setups. The new idea that can decrease the penetration loss through the particular block wall will be somewhat reduced and the idea will be maintained by using MIMO technology [10].

The current MIMO scheme is using two or four antennas and building a great massive MIMO network. Initially, the outside base station will be built-in with a big antenna. Some of them are circulated and linked with the base station using optical cables. Second, All building blocks will be connected with a big antenna from outside for communicating the outdoor base station by using the sightline component [11]. The wireless contact points inside buildings are linked by using a big antenna with the help of cables for interaction with enclosed users. This will be considerably enhanced the energy efficiency, data rate, cell average throughput, and effectiveness of spectral of the cell system. Wi-Fi, millimeter-wave communications, visible graceful communication are small-range communication mechanisms for indoor communications with high data rates. However, visible light communication and millimeter-wave communications are employing high frequencies. Later, the architecture of 5G cellular technology is connected with heterogeneous including small cells, microcells, and relays [12].

The 5G core Service-Based Architecture was initially implemented, and two networks are divided into service categories (SBA). The key components of 5G core network architecture are depicted in Fig. 2.

- User Equipment are 5G cellular devices or 5G smartphones that connect to the 5G core and then to Data Networks (DN) via the 5G New Radio Access Network.

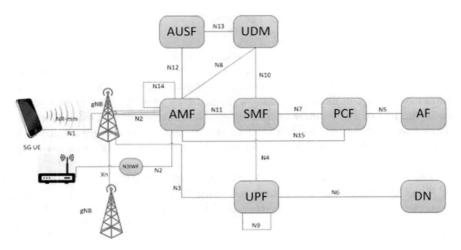

Fig. 2 Architecture of 5G technology

- The Access and Mobility Management Function (AMF) serves as the UE connection's a single entry point. The AMF decide on the appropriate Session Management Function (SMF) for handling the user session based on the service requested by the UE.
- The User Plane Function (UPF) is responsible for transporting IP data between the outward networks and user equipment's. The AMF uses the Authentication Server Function (AUSF) to connect the UE and allow entry to 5G core services.
- The policy control framework, which includes the Session Management Function (SMF), the Application Function (AF), the Policy Control Function (PCF), the Unified Data Management (UDM) function, governs network behavior by applying policy decisions and accessing subscription information.

4   Requirements of 5G Technology

In 5G technology, eight key requirements are defined as follows [13].

- Availability (99%)
- Complete analysis (100%)
- Less energy consumption (above 95%)
- 2–10 Gbps data speed required
- Low-power mode up to ten years battery life
- Supporting is required to connect more devices (10–100 times)
- Latency (1 ms end-to-end (E2E))
- Required high bandwidth (more than 1000 times).

5 Emerging Technology for 5G Wireless

Wireless technology traffics will be increased a thousand-fold over the next decades and 100 billion devices connected with wireless technology by 2025. Some challenges are rapidly increased by increasing the number of connected devices, spectrum utilization, and cost [14]. Many challenges have in future when the met requirement of 5G technologies as follows

- Hundred times the data volume is an improved.
- The number of linked devices are improved from ten to a hundred times.
- The usual data rate is enlarged from 10 to 100 times.
- Low power massive machine communication (MMC) is taking ten times for extended battery life.

5.1 Usage of Spectrum

New spectrum rules are required due to extended-spectrum group processes whereas operation in new spectrum rules offers a whole system method. Communication, navigation, and sensing services are mostly depending upon the accessibility of the radio spectrum, through which signals are communicated. 5G technologies have the robust computational power to process the vast volume of data coming from different and distinct sources. Also, it required superior infrastructure support. Researchers and scientists are facing many technological encounters of standardization and application in 5G.

5.2 Security and Privacy

Security and privacy are some of the significant challenges that make ensuring the protection of individual user data. 5G will have to describe the uncertainties associated with security dangers such as privacy, trust, and cyber-security.

5.3 Radio-Links

The development is required to create new waveforms for data transmission and design new schemes of multiple access control (MAC) and radio resource management (RRM).

5.4 Multi-antenna and Multi-node Transmission

Developing new multi-antenna transmission or reception tools based on massive antenna structures. It is required to design a new multi-hop technology and innovative internode management.

5.5 Dimension of Networks

Advanced technology is required to design an efficient implication management system in difficult unrelated deployment.

6 Design of 5G Developments

The various countries are considered to make 5G mobile communication for their country to improve the economic leap of the countries [15]. Presently, 5G networks are commercially established in 61 countries and 144 mobile operators have launched commercial 3GPP-compatible 5G services in these 61 countries. The Republic of Korea is the first country for embracing 5G to create the next economic leap for hi-tech exports especially smartphones. They are planning to create a new global ICT powerhouse. They will target to reach 90% of the total mobile connection by 2026.

Germany is making a principal market for 5G applications, encouraging the enlargement of a maintainable and competition-oriented market. They will plan to reach 98% of household devices by 2022. Finland is a well-known leader in testing, developing of 5G environment. They are encouraged to a cost-effective and fast construction of networks with high-speed connections. Switzerland is taking advantage of digital development to enhance the safety of its inhabitants. Singapore is having two operators for 5G mobile networks and leading the 5G development. The coverage of 5G networks will exceed greater than 50% of the population (Table 2).

In India, Airtel has declared that 5G network ready and Jio declared to establishing their 5G network in the second half of 2021. The standing board has been declared that 5G will roll out in India to a certain level for particular uses, by the end of the year 2021 or the opening of 2022 after another spectrum auction that is scheduled to be held next six months. Ericsson Mobility has reported that India will have three hundred and fifty million 5G contributions by 2026. When migrating to 5G technology, a number of weaknesses are identified, as follows,

- The speed of this technology is a major concern, and it appears that claiming it is difficult to achieve.
- In most parts of the world, there is ineffective technological support.
- Infrastructure development was expensive.

Table 2 Technical comparison of standards

Description	802.11an	802.11ac	802.11ad	802.11af
Frequency (GHz)	2.4, 4.9, and 5	5	60	0.47–0.71
Modulation	OFDM	OFDM	Low-power-single carrier, single, OFDM	
Bandwidth	20 and 40 MHz	20, 40, and 80 MHz	2 GHz	5, 10, 20 and 40 MHz
Data rate	150 Mbps	433 Mbps	4.6 Gbps	54 Mbps
Spectral efficiency (bps)	15	21.66	1	–
Availability	World	World	World (partial to China)	World
Walls	Yes	Yes	Yes	Yes

- There are still security and privacy issues to be resolved.
- Many rural areas across the country are unable to connect.

7 Applications of 5G Technologies

More than 8.5 million devices, the majority of which are smart, may be 5G compatible. Figure 3 shows the application and services of 5G technology. As stated earlier, network speed and capacity are two key characteristics of 5G wireless communication which enables smart devices and new devices to handle large amounts of data over software applications. Currently, some AI areas rely deeply on network communication to communicate and process data and their main communication concerns are the performance of the timings and the data management of data. Figure 3 shows the applications and services of 5G technology [16].

7.1 Smart City

In their modern economic development, cities progressively market themselves as "smart cities" that are sensor-equipped groups, hyper-connected that interest people and industries. A smart city is one in which communication and information technologies (ICT) have been combined to improve the superiority and performance of urban facilities, including energy, utilities, and transportation, to decrease resource consumption, overall costs, and waste. The main goal of a smart city is to improve the superiority of life for its residents by utilizing smart technology, such as AI [17]. According to experts, 5G networks will kick start the smart city movement, which connects energy, traffic, waste disposal, communications, and a variety of other public

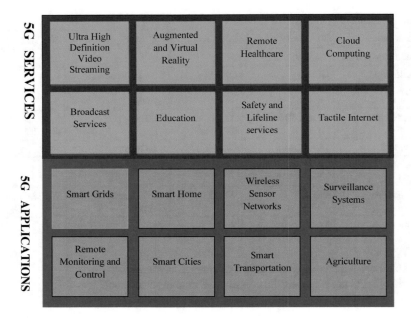

Fig. 3 Application and services of 5G technology

schemes to create more stable and well-organized groups [18]. In general terms, ICT frameworks are thought to power digital city ecosystems, which link numerous dedicated networks of linked cars, sensors, mobile devices, communication gateways, home appliances, and data centers. According to IoT developments, the number of associated devices worldwide will range to 75 billion by 2025. The growing number of organized objects generates an unprecedented amount of data that the city can investigate locally to create better decisions about what changes or new projects will benefit residents the most.

7.2 Driverless Vehicles

A self-driving car is one which is capable of storing and analyzing environmental data in order to drive itself without the need for human intervention. The data processing could be carried out locally using software applications or remotely using servers. Network speed, bandwidth, and traffic are all major issues. 5G technology has been labeled "oxygen" by several specialists for self-driving vehicles. Delivering the substructure for a self-driving vehicle to interconnect with the neighboring environment and create independent driving and, eventually, smart city functionality is critical. It's all about data that needs to be processed more quickly and closer to the vehicle [19].

7.3 Healthcare

Rural areas would benefit from low-latency connectivity, which would revolutionize emergency medical treatment for people around the country. Patients in small cities would no longer be compulsory to uproot their lives and reposition to larger cities, away from the livelihoods they know and love, to obtain the care to which they are permitted. 5G connectivity permits caregivers in rural and remote zones to obtain real-time training and provision from the world's greatest surgeons, no matter where they are [20].

7.4 Agriculture

Future farms will rely on more data and fewer chemicals. Collecting information directly from sensors in the field, planters can recognize with correct location which areas require water, have a disease, or want pest control. As wearable's become more affordable, and 5G makes scaling networks with a huge amount of IoT nodes much easier, and it's also possible that livestock health monitoring will develop. Farmers can reduce antibacterial use without compromising food quality if they have quite precise health data [21].

8 Discussions

The 5G is set to replace the 4G system and greatly improve communication structures by introducing significantly better-quality network features. The key technologies of 5G are designed to increase capacity and data rate. The literature suggests that the 5G will increase mobility from 400 to 500 km per hour, cost efficiency from 10 to 1000 times (bits per dollar), and latency from 10 to 1 ms. Furthermore, 5G technology has a more software-based design, allowing for more AI applications to be offered to smart device users. The most significant AI-related domains that 5G will be most effective in are self-directed driving and smart cities.

On the other hand, in 5G, advanced machine learning, deep learning [22], and big data analytics methods will improve the network's stability and strength where any predictions are necessary for given network communication. Several cell phone manufacturers and providers, mostly those using Android operating systems, have developed different devices to be companionable with 5G and have thoroughly launched some of the technology's features. However, certain 5G technology challenges remain, most particularly exceeding the data rates up to 1 Gbps, accessibility, reliability, and mobility.

Regardless of how successful its technical characteristics become, 5G will be economically beneficial to the economy. Gradually, It will be necessary for users to update their devices or the user will need to purchase a 5G-compatible device, which generates a high economic turnover for the telecommunications manufacturing and resurrects it every ten years or so. Soon, the 5G technology will be widely available in the market, with some technical problems expected during the period of transition. In 5G, various services and features that increase network speed and reliability have been established. But, the mobile communications community's experience suggests that every innovative generation's success will be attained quite gradually and incrementally than its financial aspects.

9   Real-World Challenges

The high-frequency parts of the electromagnetic spectrum (EMF) are used by the wireless communication device. The new bands are well above the Ultra-high frequency (UHF) ranges, which have wavelengths in the centimeter ranges of 3–30 GHz or millimeter ranges of 30–300 GHz [23]. Radio Frequency-EMF exposure can cause cancer, so it was classified as a "possibly carcinogenic to humans" substance by the International Agency for Research. Non-thermal cellular harm can be caused by non-ionizing wireless radiation used in telecommunications. This RF EMR has been linked to a variety of issues such as well as a slew of negative consequences for DNA integrity, the blood–brain barrier, sperm damage, melatonin production, immune dysfunction, cellular membranes, protein synthesis, gene expression and neuronal function, the blood–brain barrier, melatonin production, sperm destruction, and immune dysfunction [24].

10   Conclusions

The present paper has conducted a detailed survey on wireless networks, delivering various aspects such as wireless network evolution, performance comparisons of different wireless network generations, and so on. This paper offers a wide-ranging summary of wireless technology for the development of various applications, as well as a comparative study of its progression. Then, gives an overview of 5G technology requirements, design, applications, and challenges, as well as a comparison of 4G and 5G technologies. Hence, this paper confirmed that it can provide a credible platform for researchers to strive for better results in future networks.

References

1. Raychaudhuri, D., Gerla, M.: Emerging Wireless Technologies and the Future Mobile Internet. Cambridge University Press (2011)

2. Chen, J.I.Z.: 5G systems with low density parity check based channel coding for enhanced mobile broadband scheme. IRO J. Sustain. Wireless Syst. **2**(1), 42–49 (2020)
3. Agiwal, M., Roy, A., Saxena, N.: Next generation 5G wireless networks: a comprehensive survey. IEEE Commun. Surv. Tutor. **18**(3), 1617–1655 (2016)
4. Wang, N., Wang, P., Alipour-Fanid, A., Jiao, L., Zeng, K.: Physical-layer security of 5G wireless networks for IoT: challenges and opportunities. IEEE Internet Things J. **6**(5), 8169–8181 (2019)
5. Wong, V.W., Schober, R., Ng, D.W.K., Wang, L.-C.: Key Technologies for 5G Wireless Systems. Cambridge University Press (2017)
6. Gupta, A., Jha, R.K.: A survey of 5G network: architecture and emerging technologies. IEEE Access **3**, 1206–1232 (2015)
7. Wang, C.-X., Haider, F., Gao, X., You, X.-H., Yang, Y., Yuan, D., et al.: Cellular architecture and key technologies for 5G wireless communication networks. IEEE Commun. Mag. **52**(2), 122–130 (2014)
8. Hou, G., Chen, L.: D2D communication mode selection and resource allocation in 5G wireless networks. Comput. Commun. **155**, 244–251 (2020)
9. Duong, T.Q., Kundu, C., Masaracchia, A., Nguyen, V.-D.: Reliable communication for emerging wireless networks. Mob. Netw. Appl. **25**(1), 271–273 (2020)
10. Challa, N.R., Bagadi, K.: Design of massive multiuser MIMO system to mitigate inter antenna interference and multiuser interference in 5G wireless networks. J. Commun. **15**(9), 693–701 (2020)
11. Ahmed, A.H., Al-Heety, A.T., Al-Khateeb, B., Mohammed, A.H.: Energy efficiency in 5G massive MIMO for mobile wireless network. In: 2020 International Congress on Human-Computer Interaction, Optimization and Robotic Applications (HORA), pp. 1–6. IEEE (2020)
12. Saxena, N., Kumbhar, F.H., Roy, A.: Exploiting social relationships for trustworthy D2D relay in 5G cellular networks. IEEE Commun. Mag. **58**(2), 48–53 (2020)
13. Rodriguez, J.: Fundamentals of 5G Mobile Networks. Wiley (2015)
14. Kaur, P., Garg, R.: A survey on key enabling technologies towards 5G. IOP Conf. Ser. Mater. Sci. Eng. IOP Publishing 012011 (2021)
15. Forge, S., Vu, K.: Forming a 5G strategy for developing countries: a note for policy makers. Telecommun. Policy **44**(7), 101975 (2020)
16. Kabalci, Y.: 5G mobile communication systems: fundamentals, challenges, and key technologies. In: Smart Grids and Their Communication Systems, pp. 329–359. Springer (2019)
17. Smys, S., Wang, H., Basar, A.: 5G network simulation in smart cities using neural network algorithm. J. Artif. Intell. **3**(01), 43–52 (2021)
18. Jiang, Y.: Economic development of smart city industry based on 5G network and wireless sensors. Microprocess. Microsyst. **80**, 103563 (2021)
19. Pastor, G., Mutafungwa, E., Costa-Requena, J., Li, X., El Marai, O., Saba, N., et al.: Qualifying 5G SA for L4 automated vehicles in a multi-PLMN experimental testbed. In: 2021 IEEE 93rd Vehicular Technology Conference (VTC2021-Spring), pp. 1–3. IEEE (2021)
20. Latif, S., Qadir, J., Farooq, S., Imran, M.A.: How 5G wireless (and concomitant technologies) will revolutionize healthcare? Future Internet **9**(4), 93 (2017)
21. Tang, Y., Dananjayan, S., Hou, C., Guo, Q., Luo, S., He, Y.: A survey on the 5G network and its impact on agriculture: challenges and opportunities. Comput. Electron. Agric. **180**, 105895 (2021)
22. Shakya, S., Pulchowk, L.N., Smys, S.: Anomalies detection in fog computing architectures using deep learning. J. Trends Comput. Sci. Smart Technol. **2**(1), 46–55 (2020)
23. Simkó, M., Mattsson, M.-O.: 5G wireless communication and health effects—a pragmatic review based on available studies regarding 6 to 100 GHz. Int. J. Environ. Res. Public Health **16**(18), 3406 (2019)
24. Russell, C.L.: 5G wireless telecommunications expansion: public health and environmental implications. Environ. Res. **165**, 484–495 (2018)

Smartphone Usage, Social Media Engagement, and Academic Performance: Mediating Effect of Digital Learning

T. Ravikumar, R. Anuradha, R. Rajesh, and N. Prakash

Abstract Smartphone has penetrated across all regions of India and all sections of the society, especially among the higher education student community. Smartphones with the internet, in turn, provide access to social media platforms. Smartphones and social media (SM) together provide several benefits to the users, and they also cause unwanted repercussions on the users such as adverse mental health, addiction, and laziness. But one of the important gains of smartphone and social media usage (SMU) is digital learning through them. Smartphones and SM are inevitable in the academic process of higher education students. The students use extensively the smartphone and SM for sharing information and content, for having peer discussion, for managing assignments, and for communicating with the instructors. So, this study tries to study the mediating effect of digital learning through smartphones and SM on the educational accomplishment of private university students in Bangalore, India. This study adopted the survey method of research which collected primary data through structured questionnaires from the private university students. The collected data were analyzed using students' "t" test, One-way ANOVA, Chi-Square, and Structural Equation Modeling. The study results found that smartphones and SM affect the academic performance (AP) of the sample students significantly through digital learning.

Keywords Smartphone · Social media engagement · Digital learning · Academic performance · Private university · Students · India

T. Ravikumar (✉) · R. Anuradha · R. Rajesh · N. Prakash
School of Business and Management, Christ (Deemed to be University), Bangalore 560029, India
e-mail: ravikumar.t@christuniversity.in

R. Anuradha
e-mail: anuradha.r@christuniversity.in

R. Rajesh
e-mail: rajesh.r@christuniversity.in

N. Prakash
e-mail: prakash.n@christuniversity.in

© The Author(s), under exclusive license to Springer Nature Singapore Pte Ltd. 2022
V. Suma et al. (eds.), *Evolutionary Computing and Mobile Sustainable Networks*,
Lecture Notes on Data Engineering and Communications Technologies 116,
https://doi.org/10.1007/978-981-16-9605-3_24

1 Introduction

Educational revolution in modern days is also led by technologies all over the world. One of the prominent ICT devices is smartphones. There are 41.74 billion smartphone users who are there in India having access to the internet. The number of internet users has reached 65 billion [26]. Smartphones gain eminence among people as they provide multiple powerful functions such as communication, entertainment, financial services, health care, learning, gaming, and social networking. Smartphones are the inevitable device for higher education students and adults. In India, 80% of the students use smartphones [14]. The college students use smartphones as they promote academic communication among them [21]. Smartphone aids in e-learning through Google classroom, Moodles, Prezi, web-based instruction, and so on. Smartphone addiction does not have an association with academic performance (AP) [15]. On the other hand, usage of smartphones for a long period creates stress and anxiety [14]. Hawi and Samaha [10] highlighted, through their research work, the adverse effects of smartphone addiction on AP. Thus, there are contradicting views on the effects of smartphones on the AP of the students.

As mentioned, smartphones have multiple functions to perform. One of the most sought-after functions of smartphones that attract all sections of people irrespective of their age, gender, and occupation is social media access. "Social media refers to computer-mediated technology facilitating the growth and sharing of ideas, awareness, career interests, information, and other methods of expression through social networks and virtual communities" [25]. Social media includes Facebook, Twitter, YouTube, WhatsApp, Instagram, LinkedIn, and so on. The student community is not an exception to such attractions of social media and networking websites. Social networking indicates a social composition that includes various social actors and their communications [30]. These social media are used for group interaction, communication, entertainment, networking, and learning, and the learners consider SM as a source of ideas, and subject updates. As a result of social media's attraction, the student community is actively engaged in social media. Social media engagement (SME) implies the higher level of commitment on social media and its happenings such as sharing of personal, and social information [1]. SME facilitates the students or a group of students to communicate with each other and to work together [3]. Social media has positive and negative effects on student users. Positive effects include information sharing, content sharing, safe communication, digital learning, and facilitation of social discussion. Negative effects comprise addiction, excessive usage of time, a distraction from the studies, a lack of critical thinking, health issues, and laziness.

From the above paragraphs, one can understand that smartphone usage and social media engagement deliver both positive and adverse effects on the students. This study attempts to solve the questions: do smartphone usage and SME affect the academic performance (AP) of the students? And do the effects of smartphone usage and SME on the AP of the students vary when smartphones and social media engagement are meant for digital learning of the students?

2 Literature Survey

Smartphones and social media are the disruptive new-age technologies that have made a complete paradigm shift in the lives of the people through their services and innovations. Those technologies have made unprecedented disruptions in the fields of teaching and learning. Both teachers and students employ the technologies for teaching and learning, respectively. The mentioned technologies impart both positive and negative effects on teaching and learning. This section deals with the effects of smartphones and social media engagement on the digital learning of the students and on academic performance.

The students who use smartphones for less time exhibit good AP [19]. Excessive usage of smartphones deviates the students from the academics [22]. More usage of smartphones brought "addictiveness" that spoils the performances of the students [7]. Many research studies demonstrated inverse relationship between smartphone usage and students' performances in universities [29, 37]. However, few studies observed that smartphone usage does not harm the AP of all students and revealed that girl students' performances are least affected by smartphone usage.

Nayak [24] argued that male students got affected more by smartphone addiction than female students even though female students spend more time on smartphones because male students had less self-control on smartphone usage.

Academic performance is positively associated with calling and messaging according to a study conducted among the female university students [12]. Modern technologies improved the teaching abilities and learning capabilities [5, 9, 31]. Thus, no conclusive evidence exists in the literature regarding linear or non-linear impact of smartphone usage on the academic performance of the students.

Social media emerged as a vital component of the society [6]. Social media has productive characteristics [39] and negative effects on student community when they are excessively used [4].

On the positive side, social media impacted the daily lives of youth especially university students [32]. Social media has been a platform for sharing feelings, emotions, and information [36]. Social media brought a paradigm shift in the learning process of the students in universities [34]. Social media is a platform for knowledge sharing between the students and the society [13]. Individuals may create ideas, content, pictures, videos, and information and share them with others using social media [28]. University students used to share information, study materials, videos, lecture materials, pictures, and reference books through social media. Thus, social media promotes information sharing, content sharing, and group discussion which facilitate learning among the students.

On the negative side, social media platforms, and internet browsing have been relied on excessively by the students for the data and information [2]. This excessive rely on social media decreases learning ability, analytical skills and research capabilities of the students and distracts the attention and concentration of the students [11]. Social media engagement brought down communication abilities of the students due to the absence of personal contact [20]. The students miss their academic deadlines due to social media engagement [33]. Moreover, social media excessive usage had psychological and physical health conditions pf the students [17]. From the above studies, it can be observed that social media engagement facilitates digital learning among student users and social media positively or negatively impacts the academic performance of the students.

3 Conceptual Framework and Hypotheses Formation

From the literature survey, the following conceptual research model has been arrived at Fig. 1.

The following hypotheses are framed for this study.

Hypothesis 1: Smartphone usage has positive impact on AP of the students.

Hypothesis 2: Social media engagement has positive impact on AP of the students.

Hypothesis 3: Smartphone usage positively affects academic AP through digital learning.

Hypothesis 4: Social media engagement positively affects AP through digital learning.

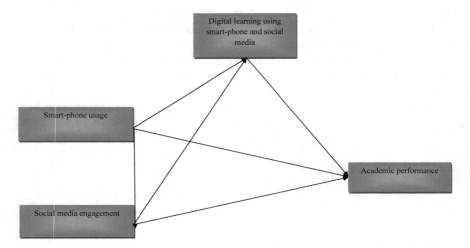

Fig. 1 Conceptual research model

4 Material and Methods

4.1 Research Design and Measurement Scales

The research design of the study is descriptive in nature. Investigative type is a causal type, and the time horizon of the study is cross-sectional. Primary data of the study has been collected through structured questionnaires. The survey method has been adopted. Unit of analysis is private university students enrolled in undergraduate, postgraduate, professional, and research studies. The study has been conducted in private universities of Bangalore, India. Private universities are those higher learning institutions that are not owned by the government directly or indirectly, but are owned and run by a private trust, society, or company. Private universities include deemed to be universities as well. There are 16 private universities across Bangalore dealing with arts, science, engineering, management, and medical disciplines. The study has adapted and customized the scales for measurement of smartphone usage, social media engagement, digital learning, and academic performance [16, 23, 27, 38].

4.2 Sample Framework

The study employed a judgment sampling technique to collect data from the unit of analysis. The students who use smartphones and social media were considered for the study. The sample size has been determined using Krejcie and Morgan's [18] formula. According to Krejcie and Morgan's table, when the population of the study is known (more than 92,000), the sample size considered for the study is 382.

4.3 The Framework of Statistical Analysis

The collected data are analyzed to seek answers for the research objectives framed using the statistical tools namely frequency analysis, Students' "t" analysis, One-way ANOVA, Chi-Square test, and Structural Equation Modeling.

5 Results and Discussions

The collected questionnaires have been checked for their completeness. The collected data was tested for its normality to find out the nature of data collected. The results of the normality test indicate that the collected data is approximated normally distributed.

Table 1 Demographic profiles of the respondents

Particulars	Items	Number	Percentage
Gender	Male	249	65.2
	Female	133	34.8
Age in years	18–25	297	77.7
	26–30	075	19.6
	More than 30	010	2.6
Education (enrolled)	Under-graduation	198	51.8
	Post-graduation	051	13.4
	Research programs	060	15.7
	Professional programs	073	19.1
Marital status	Single	313	81.9
	Married	069	18.1

Source Primary data

Table 1 has presented the profiles of private university students who participated in this research survey. 65.2% of the student respondents are male and 34.8% of the students are female. The respondents predominantly belong to the age group of 18–25 years (77.7%). 51.8% of the respondents are enrolled in under-graduation, 19.1% have enrolled for professional programs such as engineering and medicine, 15.7% have enrolled for research programs such as M.Phil., and Ph.D. and 13.4% of the respondents have enrolled in post-graduation studies. Further, 81.9% of the respondents are single.

Differences between the demographics and smartphone usage are analyzed using the students' "t" test and One-way ANOVA and presented in Table 2. The test results show that gender, marital status, and age of the respondents don't significantly differ with smartphone usage. But the education of the respondents significantly differs from smartphone usage. Mean scores indicate that those respondents who pursue post-graduation have more differences in usage of the smartphone (Mean: 3.8356), and the respondents who are enrolled in professional programs have a lesser number of differences in usage of the smartphone (Mean: 3.4552).

Table 2 Analysis of differences between the demographics of the respondents and smartphone usage

Particulars	Test	T or F-value	P-value	Level of significance
Gender	Student "t" test	0.591	0.805	Not significant
Marital status	Student "t" test	0.097	0.266	Not significant
Age	One-way ANOVA	1.027	0.359	Not significant
Education enrolled	One-way ANOVA	3.092	0.027	Significant

Source Primary data

Table 3 Analysis of differences between the demographics of the respondents and social media engagement

Particulars	Test	T or F-value	P-value	Level of significance
Gender	Student "t" test	1.026	0.939	Not significant
Marital status	Student "t" test	0.490	0.987	Not significant
Age	One-way ANOVA	0.818	0.442	Not significant
Education enrolled	One-way ANOVA	0.739	0.529	Not significant

Source Primary data

Differences between the demographics and social media engagement are analyzed and presented in Table 3. The test results show that gender, marital status, age of the respondents, and education don't significantly differ with social media engagement. So, it can be concluded that the social media engagement of private university students doesn't differ on the basis of their gender, age, marital status, and education.

Table 4 also indicates the results of the analysis of differences between the demographics of the respondents and digital learning. The results show that the digital learning of private university students doesn't differ on the basis of their gender, age, marital status, and education.

Table 5 also reveals that the academic performance of private university students doesn't differ on the basis of their gender, age, marital status, and education.

In this study, an attempt has been made to identify and analyze the association between demographic variables and core variables of this study using a Chi-Square test. This statistical tool has been used to find an association between the variables as

Table 4 Analysis of differences between the demographics of the respondents and digital learning

Particulars	Test	T or F-value	P-value	Level of significance
Gender	Student "t" test	0.146	0.885	Not significant
Marital status	Student "t" test	0.288	0.826	Not significant
Age	One-way ANOVA	0.166	0.847	Not significant
Education enrolled	One-way ANOVA	0.690	0.558	Not significant

Source Primary data

Table 5 Analysis of differences between the demographics of the respondents and academic performance

Particulars	Test	T or F-value	P-value	Level of significance
Gender	Student "t" test	1.317	0.293	Not significant
Marital status	Student "t" test	0.573	0.888	Not significant
Age	One-way ANOVA	0.317	0.728	Not significant
Education enrolled	One-way ANOVA	0.856	0.464	Not significant

Source Primary data

Table 6 Association between demographics and core variables of the study

Particulars	Chi-square value	P-value	Result
Gender	32.225	0.000	Significant
Marital status	155.853	0.000	Significant
Age	355.702	0.000	Significant
Education enrolled	149.246	0.000	Significant
Smartphone usage	159.361	0.000	Significant
Social media engagement	222.356	0.000	Significant
Digital learning	279.435	0.000	Significant
Academic performance	244.571	0.000	Significant

Source Primary data

it permits to assess whether the relationship between the variables is due to chance, or the relationship is systematic. The results presented and analyzed in Table 6 exhibit that there are significant associations among gender, marital status, age, education, and core variables of the study such as smartphone usage, social media engagement, digital learning, and academic performance of the students. This result confirms and validates the variables used in this study.

The scale reliability of the constructs which are shown in Table 7 indicates the composite reliability of 0.894, 0.899, 0.875, and 0.741 for smartphone usage, social media engagement, digital learning, and academic performance of the students, respectively. Internal consistency and reliability, convergent validity, and discriminant validity are measured [8, 35]. Cronbach's alpha values above 0.6 fulfill the requisites for internal consistency and reliability. Average variance extracted (AVE) value above 0.50 and composite reliability (CR) value above 0.50 reveals acceptable convergent validity. These composite reliabilities are more than the recommended level of 0.50. Average Variance Extracted (AVE) has been 0.585, 0.683,

Table 7 Construct reliability and validity

Construct	Composite reliability	AVE	Cronbach's alpha
Smartphone usage	0.894	0.585	0.871
Social media engagement	0.899	0.683	0.877
Digital learning	0.875	0.714	0.856
Academic performance	0.741	0.618	0.675

Source Primary data

Table 8 Correlation of latent variables

Variables	Smartphone usage	Social media engagement	Digital learning	Academic performance
Smartphone usage	1.000			
Social media engagement	0.837	1.000		
Digital learning	0.548	0.447	1.000	
Academic performance	0.646	0.460	0.825	1.000

Source Primary data

0.714, and 0.618 for the construct's smartphone usage, social media engagement, digital learning, and academic performance of the students respectively. The Cronbach alpha for all the core variables of this study is more than 0.60 which is more than the recommended Cronbach alpha level of 0.50. Since the composite reliability, AVE, and Cronbach alpha levels of the constructs are satisfactory and more than the expected level of 0.50, the constructs are significantly reliable to be used in the research. Further, Table 8 shows the relationships among the constructs, and the relationships signify the need for research on the constructs chosen.

Table 8 expresses inter-relationships among the core variables of the study such as smartphone usage, social media engagement, digital learning, and academic performance of the students. Smartphone usage is positively and strongly related to social media engagement (0.837) and smartphone usage is positively and moderately related to digital learning (0.548) and academic performance of the students (0.646). So, smartphone usage and social media engagement are closely related. Digital learning has a positive and moderate relationship with social media engagement (0.447). Social media engagement is also positively and moderately related to the academic performance of the students (0.460). Notably, digital learning through smartphone usage and social media engagement is positively and strongly related to the academic performance of the students (0.825).

From Fig. 2, one can observe the direct effects of the exogenous variable on endogenous variables. Figure 2 displays that smartphone usage explains variances in social media engagement of the students to the extent of 68.1% (r^2: 0.681) and on the other hand, smartphone usage explains variances in digital learning of the students to the extent of 22.6% (r^2: 0.226). Further, the path coefficient between smartphone usage and social media engagement (0.825) shows that smartphone usage has a positive effect on social media engagement and another path coefficient between smartphone usage and digital learning (0.286) conveys that smartphone usage has a positive impact on digital learning.

Table 9 displays the special indirect effects of the research model. Social media engagement impacts the academic performance of the students through digital learning to the extent of 14.6% (0.146) (Hypothesis 4 has been accepted). Importantly, smartphone usage impacts the AP through social media engagement and digital

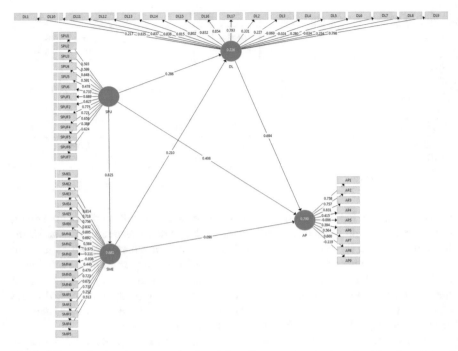

Fig. 2 Effects of smartphone usage, and social media engagement on academic performance of the students: mediating effect of digital learning via smartphone and social media

Table 9 Special indirect effects

Particulars	Specific indirect effects
SME → DL → AP	0.146
SPU → SME → DL → AP	0.120
SPU → DL → AP	0.199
SPU → SME → AP	−0.081
SPU → SME → DL	0.174

Source Primary data

learning to the extent of 12.0% (0.120). Smartphone usage influences the AP through digital learning to the extent of 19.9% (0.199) (Hypothesis 3 has been accepted). Further, smartphone usage negatively affects the AP through social media engagement (SME) to the extent of −8.1% (−0.081). Smartphone usage impacts digital learning through SME to the extent of 17.4% (0.174). From these results, one can understand that smartphone usage impacts academic performance more when the students use the smartphone for digital learning purposes (19.9%) than the impact of SME on academic performance of the students when the students use social media for digital learning purposes (14.6%). But, very interestingly, when smartphone usage

Table 10 Total indirect effects

Particulars	SPU	SME	DL	AP
SPU	–	–	0.174	0.238
SME	–	–	–	0.146
DL	–	–	–	–
AP	–	–	–	–

Source Primary data

Table 11 Total effects

Particulars	SPU	SME	DL	AP
SPU	–	0.825	0.460	0.646
SME	–	–	0.210	0.048
DL	–	–	–	0.694
AP	–	–	–	–

Source Primary data

and social media engagement are directed toward digital learning, smartphone usage and social media engagement effect digital learning more.

Table 10 displays the total indirect effects of the variables in the research model. Smartphone usage effects 17.4% of digital learning of the students and smartphone accounts for 23.8% of the academic performance of the students. Social media engagement effects 14.6% of the academic performance of the students.

The total effects of the exogenous variable on endogenous variables are presented in Table 11. Smartphone usage accounts for 82.5% variance in social media engagement, accounts for 46.0% variance in digital learning, and accounts for 64.6% variance in the academic performance. Social media engagement influences digital learning and academic performance for 21.0% and 4.8% respectively. Finally, digital learning impacts academic performance moderately (69.4%). From these results, hypotheses 1 and 2 are accepted and concluded that smartphones positively effect academic performance and social media engagement also positively effects the academic performance of the students.

Therefore, it can be said that smartphone usage strongly effects social media engagement and moderately effects digital learning that, in turn, lead to the better academic performance of the students. But social media engagement alone doesn't influence digital learning strongly and so, social media engagement contributes very less invariance in the academic performance of the students.

6   Conclusions

Smartphone has been used by almost all the students across the globe as the smartphone has several advantages to the users such as better communication, ease of use, games, entertainment, access to finance, digital learning, and so on. Further, the students access the social media platforms predominantly through smartphones rather than accessing through laptops or personal computers. The students do not only use smartphones and social media for entertainment, but they use them for information sharing, content sharing, academic peer group discussion, accessing learning websites and applications, and for accessing moodles. So, smartphones and social media become effective tools of digital learning and these smartphones and social media do not curtail the academic performance (AP). Therefore, the research work focused on the effects of smartphone usage and social media engagement on the AP of private university students in Bangalore directly and the effects of smartphone usage and social media engagement on the AP of private university students through digital learning. The study results conclude that gender, marital status, and age do not significantly differ in the usage of smartphones, but education makes the difference. Moreover, gender, marital status, age, and education do not significantly differ in social media engagement, digital learning, and AP of the students. However, the demographics of the respondents are significantly associated with the core variables of the study. Finally, the study has found that smartphone usage positively impacts the academic performance of private university students directly and social media engagement also positively impacts the AP of private university students directly. In addition, it has been found that smartphone usage positively impacts the AP of private university students through digital learning and social media engagement also positively impacts the AP of the private university students through digital learning.

References

1. Alt, D.: College students' academic motivation, media engagement, and fear of missing out. Comput. Hum. Behav. **49**, 111–119 (2015). https://doi.org/10.1016/j.chb.2015.02.057
2. Balakrishnan, V., Gan, C.L.: Students' learning styles and their effects on the use of social media technology for learning. Telemat. Inform. **33**, 808–821 (2016)
3. Birim, B.: Evaluation of corporate social responsibility and social media as key source of strategic communication. Procedia Soc. Behav. Sci. **235**, 70–75 (2016)
4. Carter, M.A.: Protecting oneself from cyber bullying on social media sites—a study of undergraduate students. Procedia Soc. Behav. Sci. **93**, 1229–1235 (2013)
5. Choi, S.: Relationships between smartphone usage, sleep patterns, and nursing students' learning engagement. J. Korean Biol. Nurs. Sci. **21**, 231–238 (2019)
6. Chukwuere, J.E., Chukwuere, P.C.: The impact of social media on social lifestyle: a case study of university female students. Gend. Behav. **15**, 9966–9981 (2017)
7. Giunchiglia, F., Zeni, M., Gobbi, E., Bignotti, E., Bison, I.: Mobile social media usage and academic performance. Comput. Hum. Behav. **82**, 177–185 (2018)
8. Hair, J.F., Jr., Hult, G.T.M., Ringle, C., Sarstedt, M.: A Primer on Partial Least Squares Structural Equation Modeling (PLS-SEM). Sage, Oaks (2016)

9. Hasan, R., Palaniappan, S., Mahmood, S., Shah, B., Abbas, A., Sarker, K.U.: Enhancing the teaching and learning process using video streaming servers and forecasting techniques. Sustainability **11**, 2049 (2019)

10. Hawi, N.S., Samaha, M.: To excel or not to excel: strong evidence on the adverse effect of smartphone addiction on academic performance. Comput. Educ. **98**, 81–89 (2016)

11. Hoffmann, C., Bublitz, W.: Pragmatics of Social Media. De Gruyter, Berlin (2017)

12. Hong, F.Y., Chiu, S.I., Huang, D.H.: A model of the relationship between psychological characteristics, mobile phone addiction and use of mobile phones by Taiwanese university female students. Comput. Hum. Behav. **28**, 2152–2159 (2012)

13. Hossain, M.A., Jahan, N., Fang, Y., Hoque, S.: Nexus of electronic word-of-mouth to social networking sites: a sustainable chatter of new digital social media. Sustainability **11**, 759 (2019)

14. Hindustan Times: Smartphone addiction is more dangrous than you thought: it causes depression, anxiety, 04 Apr 2017. Accessed 5 Apr 2020, from Hindustan Times Website: https://www.hindustantimes.com/fitness/it-s-time-to-switch-off-smartphone-addiction-may-cause-depression-and-anxiety/story-RRZocsRaDkRzCZbGE0RIbP.html

15. Jaalouk, J.B.: Smartphone addiction among university students and its relationship. Glob. J. Health Sci. 48–59 (2018)

16. Jaffar Abbas, J.A.: The impact of social media on learning behavior for sustainable education: evidence of students from selected universities in Pakistan. Sustainability 1–23 (2019)

17. Kelly, Y., Zilanawala, A., Booker, C., Sacker, A.: Social media use and adolescent mental health: findings from the UK millennium cohort study. E-Clin. Med. **6**, 59–68 (2018)

18. Krejcie, R.V., Morgan, D.W.: Determining sample size for research activities. Educ. Psychol. Meas. **30**, 607–610 (1970)

19. Lepp, A., Barkley, J.E., Karpinski, A.C.: The relationship between cell phone use, academic performance, anxiety, and satisfaction with life in college students. Comput. Hum. Behav. **31**, 343–350 (2014)

20. Le Roux, D.B., Parry, D.A.: In-lecture media use and academic performance: does subject area matter? Comput. Hum. Behav. **77**, 86–94 (2017)

21. Looi, C.K., Lim, K.F., Pang, J., Koh, A.L.H., Seow, P., Sun, D., Soloway, E.: Bridging formal and informal learning with the use of mobile technology. In: Future Learning in Primary Schools, pp. 79–96. Springer, Singapore (2016)

22. Mendoza, J.S., Pody, B.C., Lee, S., Kim, M., McDonough, I.M.: The effect of cellphones on attention and learning: the influences of time, distraction, and monophobia. Comput. Hum. Behav. **86**, 52–60 (2018)

23. Marty-Dugas, J.J.: The relation between smartphone use and everyday inattention. Thesis, Waterloo, Ontario (2017)

24. Nayak, J.K.: The relationship among smartphone usage, addiction, academic performance and the moderating role of gender: a study of higher education students in India. Comput. Educ. **123**, 164–173 (2018)

25. Nielsen, M.I.S.W.: Computer-mediated communication and self-awareness—a selective review. Comput. Hum. Behav. **76**, 554–560 (2017)

26. Noshir Kaka, A.M.: Digital India: Technology to Transform a Connected Nation. McKinsey Global Institute, Mumbai (2019)

27. Przybylski, A.K., Murayama, K., Dehaan, C.R., Gladwell, V.: Motivational, emotional, and behavioral correlates of fear of missing out. Comput. Hum. Behav. **29**(4), 1841–1848 (2013). https://doi.org/10.1016/j.chb.2013.02.014

28. Richey, M., Ravishankar, M.N.: The role of frames and cultural toolkits in establishing new connections for social media innovation. Technol. Soc. Change, in press (2017)

29. Rosen, L.D., Lim, A.F., Carrier, L.M., Cheever, N.A.: An empirical examination of the educational impact of text message-induced task switching in the classroom: educational implications and strategies to enhance learning. Psicol. Educ. **17**, 163–177 (2011)

30. Scholz, M., Schnurbus, J., Haupt, H., Dorner, V., Landherr, A., Probst, F.: Dynamic effects of user- and marketer-generated content on consumer purchase behavior: modeling the hierarchical structure of social media websites. Decis. Support Syst. **113**, 43–55 (2018)

31. Shields, S.D., Riley, C.W.: Examining the correlation between excessive recreational smartphone use and academic performance outcomes. J. High. Educ. Theory Pract. **19** (2019)
32. Stathopoulou, A., Siamagka, N.-T., Christodoulides, G., A multi-stakeholder view of social media as a supporting tool in higher education: an educator-student perspective. Eur. Manag. J., in press (2019)
33. Tella, A.: Social Media Strategies for Dynamic Library Service Development. IGI Global, Hershey, PA (2014)
34. Terzi, B., Bulut, S., Kaya, N.: Factors affecting nursing and midwifery students' attitudes toward social media. J. Nurs. Educ. Pract. **35**, 141–149 (2019)
35. Wang, C.M., Xu, B.B., Zhang, S.J., Chen, Y.Q.: Influence of personality and risk propensity on risk perception of Chinese construction project managers. Int. J. Proj. Manag. **34**(7), 1294–1304 (2016)
36. Wolf, D.M., Wenskovitch, J.E., Jr., Anton, B.B.: Nurses' use of the Internet and social media: does age, years of experience and educational level make a difference? J. Nurs. Educ. Pract. **6**, 68 (2015)
37. Wood, E., Zivcakova, L., Gentile, P., Archer, K., De Pasquale, D., Nosko, A.: Examining the impact of multi-tasking with technology on real-time classroom learning. Comput. Educ. **58**, 365–374 (2012)
38. Yi, S.H.: How does the smartphone usage of college students affect? J. Comput. Assist. Learn. 13–21 (2018)
39. Yoo, J.H., Jeong, E.J.: Psychosocial effects of SNS use: a longitudinal study focused on the moderation effect of social capital. Comput. Hum. Behav. **69**, 108–119 (2017)

Electrical Energy Consumption Prediction Using LSTM-RNN

S. B. Shachee, H. N. Latha, and N. Hegde Veena

Abstract The quantity of energy that the world uses increases and with that comes skyrocketing electricity bills and carbon dioxide emissions. Machine learning is a rapidly evolving domain that has started to provide high quality solutions to many problems. If we could apply it to the energy sector too, we can be more efficient in conserving energy. This paper uses LSTM-RNN architecture to predict the household electrical energy consumption two months from the given starting date. The model is trained on the required features and evaluated by comparison of actual and predicted values. Consequently, this paper provides residential households with a forecast of their energy consumption that helps in the conservation of energy at the right time period. The result of the proposed predictive model is investigated using UCI repository dataset of domestic electric consumption for the consumption forecast, and the performance indicates that the revolutionary deep learning approaches obtain results of much higher precision than the statistical and engineering prediction models. Compared to conventional models, our LSTM model achieves a very compatible RMSE of 0.6.

Keywords Household power consumption · Time series · Long–short term memory · Recurring neural network · Deep learning · Artificial intelligence · Month-ahead forecast · Electric load forecast

S. B. Shachee (✉) · H. N. Latha · N. Hegde Veena
B.M.S. College of Engineering, Bengaluru, India
e-mail: shachee.ei18@bmsce.ac.in

H. N. Latha
e-mail: lathahn.ece@bmsce.ac.in

N. Hegde Veena
e-mail: veenahegdebms.intn@bmsce.ac.in

© The Author(s), under exclusive license to Springer Nature Singapore Pte Ltd. 2022
V. Suma et al. (eds.), *Evolutionary Computing and Mobile Sustainable Networks*,
Lecture Notes on Data Engineering and Communications Technologies 116,
https://doi.org/10.1007/978-981-16-9605-3_25

1 Introduction

Electrical energy intake is that of the usage of energy within the electrical form. There are many kinds of power consumptions. Namely, industrial, residential, commercial, transport and agricultural. During this paper, we concentrated our focus on residential consumption. With the expansion of the world's population and sensible cities, electricity is one amongst the most common types of energies consumed in buildings. In the country of India alone, the projection for usage of electricity for lighting in urban India has been surveyed to be 130,000 GWh [1]. Therefore, finding ways in which we can use most of the electricity expeditiously in buildings is essential in order to decrease the building energy usage. Today most buildings are constitutional with sensors that find the magnitude of the consumption of energy. These buildings' measurements will be stored, and so give useful historical information, which may be classified as time series data, that is then accustomed to forecast the energy consumptions of those residential buildings in the future. The share of electricity in the final count of energy consumption is claimed to be 23.7% by 2040 by the International Energy Agency (IEA) [2]. Electrical energy consumption may be a time-series sequence of information involving observations over evenly spaced intervals of allotted time containing both linear and non-linear parts.

Energy intake has a primary effect on its environment. Although energy is an easy and comparatively secure state of power when used in moderation, the technology and transmission of energy impacts the environment. Nearly all kinds of electric powered energy plants have an environmental impact [3]. Electric energy domain plants that burned fossil fuels or substances crafted from fossil fuels, and a few geothermal energy plants, are the generators of carbon dioxide emissions. Carbon dioxide emissions are the quantity of greenhouse gas emissions expressed as carbon dioxide. Global energy-associated CO_2 emissions grew 1.7% in 2018 to attain an all-time excessive amount of 33.1 Gt CO_2. The magnitude of growth in emissions was through higher power consumption attributable to a strong, structured worldwide economy [4]. The IEA assessed the effect of fossil fuel use on the planet's temperature increases. It concluded that CO_2 emitted from coal combustion, that is the source of electrical energy generation, is answerable for over 0.3 °C of the 1 °C growth in worldwide yearly surface temperatures above pre-commercial levels [4]. In this paper, we intend to preserve energy through forecasting the intake as this can lessen carbon emissions. In order to do so, we put into effect a long short-term memory (LSTM) recurring neural network model to quantify the older patterns of energy consumption and predict the future usage of energy.

In recent times, deep learning has transitioned to one amongst the foremost used approaches in several analysis areas. It refers to the stacking of layers of a deep neural network and looking forward to improvement in the execution of machine learning tasks. A varied range of neural architecture layers boosts the training and task performance [5]. The LSTM-RNN that was initially presented in [6], has obtained tremendous recognition in the field of sequence learning. Applications supported by LSTM networks are reportable in several areas like natural language translation

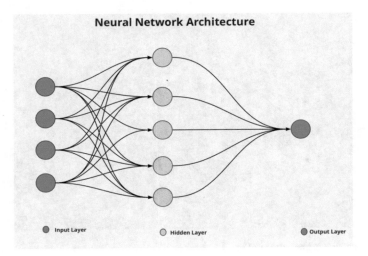

Fig. 1 Deep neural network [14]

[7], types of image captioning and classification [8–10] and speech processing [11]. Nowadays, deep learning methods have received magnanimous attention in a range of disciplines like image recognition, time series predictions etc. The deep learning approaches have improved the accuracies in various domains such as stock market forecasting [12]. DL is an advanced approach of machine learning [13]. The deep learning algorithms are used to solve a variety of real-time problems for prediction modelling. Taking examples in this consideration, we have a deep-NN (DNN), deep belief network (DBN) and a recurrent-NN (RNN). Deep Neural Network is addressed as a various levelled (layered) association of neurons (like the neurons in the mind) with associations with different neurons. These neurons pass a message or sign to different neurons dependent on the got info and structure an unpredictable organization that learns with some input instrument (Fig. 1).

RNN is a neural architecture that can handle sequential data, can consider the current input as well as the previously received inputs and remember the previously entered inputs due to its internal memory. Therefore, it can be used in time series predictions i.e., energy consumption prediction. When training an RNN, the slope can be too low or too high and this makes training difficult [15]. When the slope is lower than required, the problem is known as vanishing gradient problem which translates to loss of information through time. To solve this problem, we use a special kind of recurring neural network, namely long short-term memory (LSTM) network. Storing and recalling data for long periods of time is an LSTM model's default behaviour [16] (Figs. 2 and 3).

The aim of this study is to build a predictive deep learning model using a combined LSTM recurring-NN that has a higher accuracy for the purpose of domestic electrical energy intake forecast, and the research objective is to arrive at an optimal solution for energy conservation at the household level.

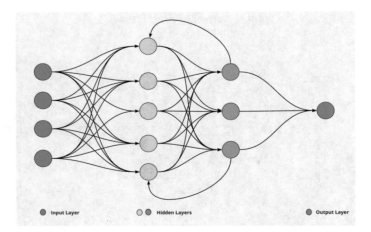

Fig. 2 Recurring neural network [14]

Fig. 3 Long short-term
memory model [17]

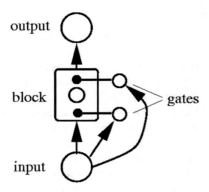

2 Literature Survey

The regular methods to predict electricity intake are engineering or statistical approaches. These methods are commonly applied in various domains for assessment purposes. Another technique to predict energy consumption is the artificial intelligence approach. Engineering methods are complete strategies, which include numerous partial differential equations or require to adhere to the physical principles, in addition to thermal dynamics equations [18]. But we have to remember that there's a sequence of complicated processes which might be tough to fulfil the need that we aim to collate the results immediately. In a few unknown regions, we do not possess state-of-the-art theories for reference, that can come to be a giant problem for the occasion of using engineering techniques.

Coming to statistical approaches, they set up the connection amongst the electricity consumption and influential factors via the models primarily built on previously entered data. Statistical approaches are appropriate for the non-linear relations and keep away from the disadvantages of using the engineering method [19]. In [20], multiple linear regression and quadratic regression models were used to predict residential energy demand. The case study concluded that massive quantities of notable previous data should be collected as the model performance is deceptive without a higher amount of data. Moreover, lengthy time periods and a massive internal storage should be considered. Even still, it might produce misguided and imprecise results, if the chosen analytical technique is not suitable.

To avoid the disadvantages of the above two methods, we utilize the machine learning approach. After taking into consideration the required features, machine learning algorithms such as in [21] are implemented to predict the energy consumption for thirty minutes to sixty minutes, one week and for one month. These works proved that deep extreme learning machine (DELM) is far better than artificial neural network (ANN) and adaptive neuro-fuzzy inference system (ANFIS) in energy consumption prediction. The DELM model achieved a root mean square error (RMSE) of 2.4657. In [22], the authors came to the conclusion that ARIMA model beats ANN and LSTM models when it comes to short-term prediction. But for long-term prediction, ANN and LSTM models have a much better accuracy rate. ANNs are more commonly used in building energy utilization prediction research as a result of their solid self-learning and good non-linear work fitting capacity, which can keep a relatively high precision in energy utilization forecast. ANNs were thus used to predict office energy consumption levels in [23]. A rough-set deep belief-NN was used in [24] to predict building energy intake. It achieved a low error of 3.83% but it was used for short-term prediction purpose. A case study [25] applied the clustering algorithm to predict power consumption of a building using smart sensor data. But in the clustering method, the results are ambiguous and can be interpreted in different ways. Problems can also arise due to time complexity. Regression models can also be used extensively in predictive modelling [26]. In [26], the trend of foreign tourist arrival in India every quarter was analysed using a regression approach for a year. The study concluded that with time series data, the trend analysis could be more effective. Another short-term prediction model [27] based on regression method was used to forecast residential energy demand. This model's data was collected at a sampling rate of i.e., 1 per 30 min and socioeconomic factors were taken into account. Considering that mean absolute error was used to measure the performance of this model [27], the accuracy of our proposed model is higher. To predict energy consumption for a longer duration, we propose a combination model of LSTM and recurring neural network (RNN) instead. In [28], SVM is integrated with simulated annealing to forecast the heating and cooling energy intake in an office. This method is too time consuming. Our proposed model takes much lesser computation time. In [29], a convolutional-NN LSTM model is used to forecast residential energy consumption. The case study for [29] concluded that the CNN-LSTM model had some mis-predictions. In [30–32], an LSTM model was implemented to predict the low-voltage electricity household consumption for twenty-four hours accurately using a transfer learning technique. Our proposed LSTM-RNN approach forecasts energy demand two months ahead. The proposed LSTM-RNN combination model has a far higher accuracy rate then the DELM, ANFIS, ANN and regression models for both short-term and long-term energy usage forecast. Our model's data was collected at a sampling rate of 1 per minute which effectively translates to more data for better analysis and prediction. For residential property like households, there's no fixed time table or routine for distinctive citizens or distinctive periods. Hence, it's tougher to forecast for them, as there are numerous individually altering elements to take into account. Moreover, the long-time period predictive consequences are vulnerable to uncertainties, thus, it is harder to forecast than that on short-time period [19]. Therefore, we utilize an LSTM recurring-NN that can get around vulnerabilities and provide precise results. The authors in [33] argued that because of populace increase and financial development, electricity intake has moreover accelerated in China and that accelerated utilization of electricity has caused extreme problems. The authors developed methods and used the neural network in those approaches. The case study was that of a shopping mall in China designed with input parameters from the constructing surroundings, e.g., temperature, humidity, climate factors taken into consideration and wind pace and different elements that would make a contribution to the fluctuation in prediction of electricity intake. The researchers used neural networks because of their effective artificial learning feature and additionally due to the fact they might swiftly expect the electrical energy prediction of any residential or commercial building after entering the considered necessary parameters. The final results confirmed that the LMBPNN was more precise than the backpropagation-NN. The case study was carried out to predict electrical power usage in a commercial building. In [34] a probabilistic version was used for power utilization prediction of electrical appliances primarily based totally on the previously defined understanding of three fundamental factors namely, duration of usage of appliance, time of the day when appliance was used most and the appliance user's behaviour pattern. The authors' purpose is to offer pointers to the occupants and enhance domestic automation device for load transferring or programming whilst making sure to not compromise the occupants' comfort. But the authors make on–off predictions without consideration of the electricity intake stage. Similarly, in [35] the Bayesian network was used primarily based on the whole prediction version for predicting a couple of equipment usages and electricity intake through making use of the two styles of probability relations: appliance-time, appliances-appliances. In [36], the authors discuss how intelligent selection of features of subject is essential for useful results. Deep learning methods outperform other regression methods like multiple linear regression (MLR) and SVR. In [37], the authors combined the stacked autoencoders (SAEs) along with extreme learning machine (ELM) and forecasted the energy usage at the property level and obtained an excellent precision rate. This approach fared much better than four other techniques: Backpropagation-NN (BPNN), SVR, Granular-RBFNN and MLR. However, they solely thought of it as an initiation prediction drawback supported by thirty minutes and one-hour data. This is not sensible as a result of predicting the energy demand within the next day being much more practical than forecasting the energy usage in the next hour. On the other hand, multi-step prediction can satisfy the demand in the predicted time series [38]. In [39], the results of the study concluded that a U-net based architecture with LSTM model with fares better in performance in predicting the energy consumption of the air-conditioner than random forest and support vector regression. The authors in [40] created forecasts for various situations and analysed it with respect to different models i.e., BPNN, K-nearest neighbours' formula and extreme learning machine. The results show that in the first case the accuracy improved by 0.49% for the LSTM model. In [38], the authors additionally explored the usage of LSTM in time series forecast and analysed its performance by utilizing 2 datasets: one global airline traveller dataset and one lengthened electricity consumption dataset. The result was favourable and proved that the LSTM algorithm could be a futuristic approach in this domain. In one other study of a similar domain, the authors provided the significance of conserving consumption of electrical energy at the domestic level because of one of the biggest consumptions of electrical energy being individual energy intake [41]. They produced an evaluation of a dataset that turned was primarily based on electrical power consumption. The case study was carried out on one domestic household, and sequential data of electrical energy intake was recorded for nearly half a decade to attempt to determine the patterns, cyclical or seasonal features, or different elements of data that might permit them to forecast the incoming usage with a specific level of accuracy which, in turn, would potentially assist in restricting the consumption of electricity. The researchers used a famous approach for time collection prediction referred to as the ARIMA (autoregressive incorporated transferring average) model. The case study confirmed that the obtained results of the time-collection records had a more than satisfactory precision level. The study aided in numerous aspects of the data analysis execution and data preparation for time series [41]. In comparison to the conventional models and other intelligent models used for long-term energy consumption, our combination LSTM recurring-NN model achieved a low RMSE of 0.6.

3 Methodology

Figure 4 is a block diagram representation of our methodology. There are seven main steps in our implementation, namely, data preparation, resampling of data, feature selection, LSTM model building, model training and testing, model performance, performance evaluation and finally, prediction of energy consumption. We elucidate the steps below.

3.1 Data Preparation

As depicted in Fig. 1, the first step in our study is data preparation. This dataset was taken from the UCI ML repository. It consists of the electrical energy measured with

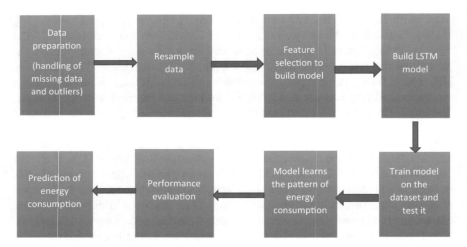

Fig. 4 Methodology block diagram

one-minute sampling for one household for 4 years. All calendar timestamps are present in the dataset but for some timestamps, the measurement values are missing. There are 2,075,259 rows of data.

We have merged date and time to create a column labelled dt in the dataset. Since there are missing values in the dataset, we have modified all missing values with calculating the average value that we obtain from the same dataset.

The attributes obtained from the dataset are as follows: Time in the Indian standard format (hh:mm:ss), and date in the standard form (dd/mm/yyyy), domestic global active power averaged per minute and domestic global reactive power averaged per minute with both parameter units being in kilowatt. The dataset also contains the voltage attribute averaged per minute and measured in volts and domestic global current intensity measured in amperes. Then we have the measurement of the household electrical appliances usage namely, sub_metering_1, 2 and 3. These are the measurements of energy sub-metered in the unit watt-hour of active energy. The first energy sub-meter corresponds to the kitchen in which the main energy intake is of the dishwasher, oven and microwave. The second one corresponds to the laundry room in which the main energy intake is from the washing machine, tumble-drier, a refrigerator and light. The third sub meter corresponds to an electric water heater and an air conditioner.

Figure 5 illustrates the process of preparation of data before fitting the model. We have our model attributes namely, date and time, sub meter 1, 2 and 3, global active power and reactive power as well as voltage and intensity of current. We use MinMax scaler for normalization of data. This is to scale each feature to the range of (0, 1) so as to give equal weightage to all the features and decrease data duplication.

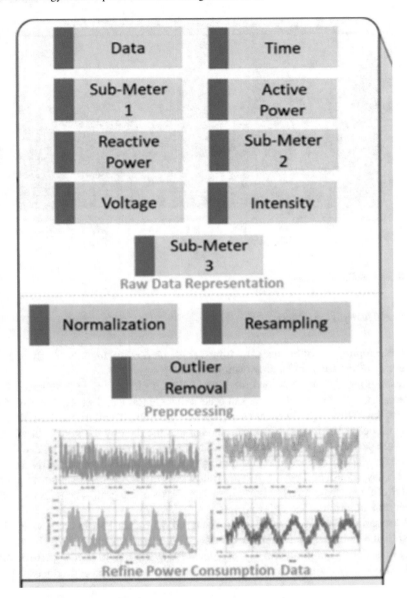

Fig. 5 Pre-processing the data [30]

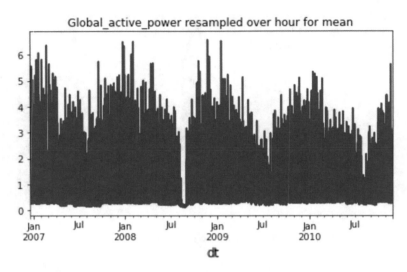

Fig. 6 Resampling data hourly

3.2 Resampling Data

Data resampling is often used in time series prediction models to change the frequency of the time series observations.

Resampling data by date or time is crucial as it changes the periodicity of the system. If we process all the data, run time will be lengthy and expensive. But if we process data with large timescale samples i.e., monthly, it will affect the efficiency of the model. Therefore, we propose to resample the data by the hour as it will not affect the periodicity or the efficiency of the model too much. In Figs. 6, 7, 8 and 9, we see the stark difference after resampling the data by hour compared to resampling data by day, month and year. From the following graphs, we can see that we get the most useful data from hourly resampling. The Y-axis of the graphs denotes the active power consumption in the unit kilowatt.

We resample our data using K-fold cross validation. In K-fold cross validation, we split the dataset into k number of subsets and proceed to train all the subsets except one. The remaining one subset is used for evaluation of model performance. We iterate K times with a different subset reserved for evaluation of model each time.

3.3 Feature Selection

We select the parameters for our model from the attributes given in the dataset. From Fig. 10, we observe that voltage remains constant over a quarter. Since there is no change in voltage, identifying a pattern in energy consumption of the household will

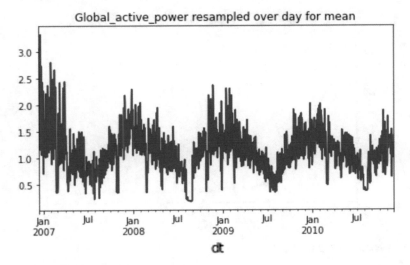

Fig. 7 Resampling data by day

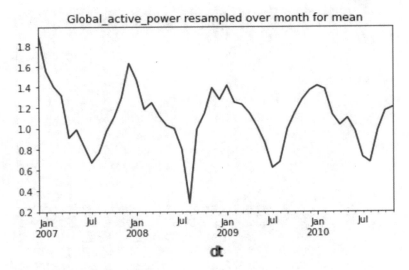

Fig. 8 Resampling data by month

not benefit from taking this parameter into account. Therefore, we do not take voltage in the features for our model and we only resample the remaining attributes.

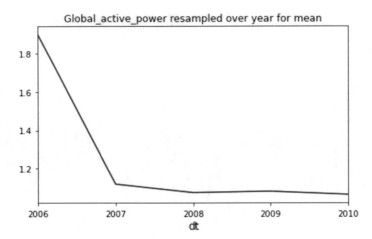

Fig. 9 Resampling data by year

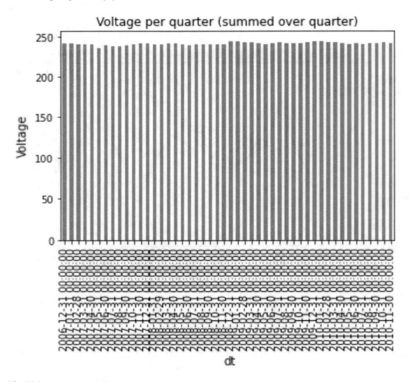

Fig. 10 Voltage measured every quarter

3.4 LSTM Model

LSTM is a special kind of recurring-NN (RNN) that solves the vanishing gradient disadvantage that RNN holds. RNNs are neural networks with an internal memory that allows them to retain information unlike other neural networks. A repeating RNN module is shown in Fig. 6. RNNs can utilize their internal memory to understand sequences of inputs which is why they can be applied in time series predictive modelling. In stark difference to other neural networks, RNNs have inputs that are dependent on each other. But RNN has a long-term dependency disadvantage that is called the vanishing gradient problem. This is technically the loss of information through time. LSTMs are built explicitly to solve this problem.

A regular RNN's repeating module has a simple structure, consisting of a single tanh layer (Fig. 11).

Figures 12 and 13 depicts LSTM network architecture and their functioning. LSTMs basically function in the following steps. First and foremost, they forget

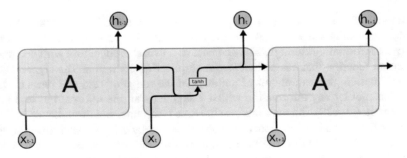

Fig. 11 Repeating unit in standard RNN [15]

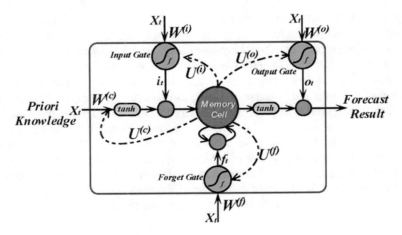

Fig. 12 Repeating unit in LSTM [42]

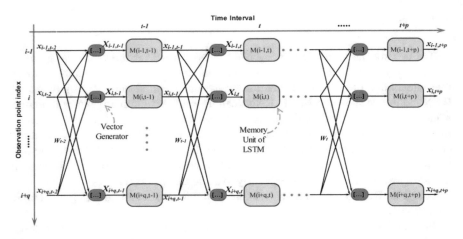

Fig. 13 Structure of 2D LSTM network [42]

the unnecessary components of the previous state and selectively update the cell's present state values. This task is executed via gates. Gates are structures that selectively let information through. They consist of a neural architecture with a sigmoid activation function that has a pointwise multiplication operation. The sigmoid activation function gives an output of numbers in the range of zero and one i.e., zero state does not let any information whilst 1 state let's all the information in. In this 2D network (Fig. 13), the rows indicate the time invariants, and the columns indicate different features taken into consideration for the model. LSTMs also have a similar structure to RNNs but the repetitive unit has a different structure consisting of four interacting neural layers communicating in place of one NN layer.

The considered LSTM architecture in our case study has been implemented using python with Keras libraries and TensorFlow back end. After preparation and sampling, the data was divided into 2 sets namely, training set and testing set. We trained the model on the first year of data and evaluated it on the next 3 years of data. The network structure has 2 layers beneath, an LSTM neural layer followed by a Dense neural layer. To enter the data into a linear stack of layers, sequential layer was created at first in the neural architecture followed by LSTM neural layer, Dropout neural layer and Dense neural layer. Dropout neural layer was introduced to reduce overfitting and to improve model accuracy. The LSTM neuron has a hundred neurons in the foremost visible layer with a dropout of twenty percent. There is one neuron in the output layer to predict the global active power. The entry form will be a step with 7 features. We take RMSE as loss function and Adam optimizer as evaluation metrics to evaluate the performance of the model. Our model has achieved an RMSE of 0.6 which is significantly higher than the conventional ANN models RMSE of 2.4657. This research has used 20 epochs and batch size of 70.

3.5   Training and Testing

The LSTM model as mentioned above, has 3 binary gates, namely, forget gate, input gate and output gate. These are denoted by ft, it and ot. Wf, Wi, Wo and Wc represent the weights. These weights control the signal between two neurons and thus decide how much influence the input will have on the output. The biases i.e., bf, bc, bo are constants that guarantee activation of neuron regardless of input value.

We use keras to fit our network to the model. We first feed the training data and the testing data. Then we allow our model to train for twenty epochs on a batch size of seventy. Epochs and batch sizes being the hyperparameters govern the training process. We use the Adam optimizer to train the algorithm. This finds a set of internal model parameters that perform well against the mean squared loss function we have chosen for model evaluation.

LSTM algorithm:

(1)   Decide how much of the previous input should be remembered. This is executed by the forget gate.
(2)   Select and update cell state values. Here the tanh layer distributes the gradients and thus prevents the vanishing gradient problem from occurring.
(3)   Obtain the new candidate values by multiplying old state with the output of the forget gate. Candidate values are denoted by Ct.
(4)   Decide on output. This is executed by the output gate (Figs. 14, 15, 16 and 17).

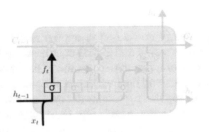

$$f_t = \sigma\left(W_f \cdot [h_{t-1}, x_t] + b_f\right)$$

Fig. 14   Step 1 of algorithm (forget gate) [15]

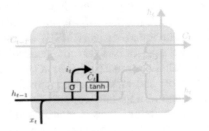

$$i_t = \sigma(W_i \cdot [h_{t-1}, x_t] + b_i)$$

$$C'_t = \tanh(W_c \cdot [h_{t-1}, x_t] + b_c)$$

Fig. 15   Step 2 of algorithm (input gate) [15]

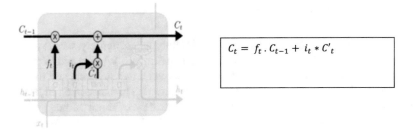

$$C_t = f_t \cdot C_{t-1} + i_t * C'_t$$

Fig. 16 Step 3 of algorithm [15]

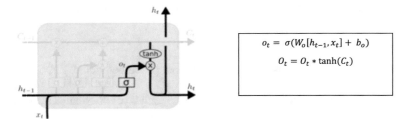

$$o_t = \sigma(W_o[h_{t-1}, x_t] + b_o)$$
$$O_t = O_t * \tanh(C_t)$$

Fig. 17 Step 4 of algorithm (output gate) [15]

4 Results

From the model, we are able to predict energy consumption approximately two months into the future. X-axis denotes the time stamp and Y-axis denotes the power consumed in kilowatt. We correlate power consumption to energy intake. Two months translate to approximately fifteen hundred hours and thus we depict the same in the graph as we have resampled data hourly. The red line represents the predicted values whilst the blue line represents actual values. This is depicted in Fig. 18. As we can see, the predicted values of energy consumption at the required hour are close to the actual values and thus we obtain an approximate measure of energy consumption for two months from the last date of the last row of data. The main factors that contributed to energy consumption in the household were investigated and found. Efforts to conserve energy could thus be made successfully. Using LSTM, the model obtained a reasonable accuracy rate. We use root mean square error (RMSE) to evaluate model loss. Figure 19 illustrates that our model's performance is commendable as it achieves a very low RMSE of 0.6. Our model takes a computation time of one second for each prediction. In Fig. 19, we have taken loss and validation loss to measure the values of the cost functions of training data and testing data respectively. Training data loss is represented by the blue line and testing data loss is represented by the orange line. Since training loss and testing loss (validation loss) decrease and finally arrive at highly similar values, we conclude that the model is learning and working efficiently. This study was able to benefit from the assessment of previous studies

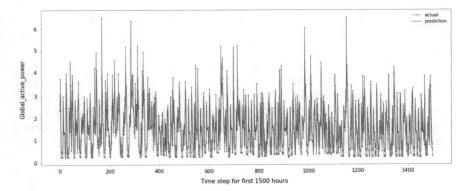

Fig. 18 Output prediction graph

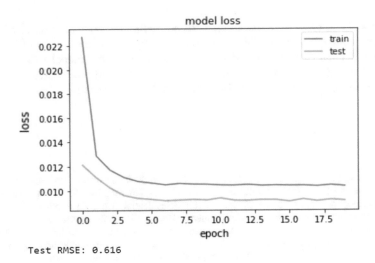

Test RMSE: 0.616

Fig. 19 Performance analysis of our proposed model

and year-round seasonal effect on energy consumption and thereby achieved a better result.

5   Conclusion

Deep learning has proven its exemplary prowess in the applications of time series prediction. Our main contribution is the application of the LSTM recurring-NN model to predict the domestic energy intake two months into the future. The objective of this paper was to use an efficient deep learning method—the LSTM method—to get better forecasted results of household electric energy consumption and thus identify the main reasons for energy consumption to aid in conserving energy more efficiently. The experimental study as well as comparison results prove that the LSTM deep learning approach fares well as compared to several conventional methods of predicting the energy consumption. In this study, LSTM has once again done well in learning long-term dependencies as our model has achieved a low RMSE of 0.6. We conclude that this approach is more favourable than the traditional engineering and statistical methods of energy consumption prediction. It also outperforms regular ANN approaches for the same case study. We also conclude that our model is comparable to state-of-the-art predictive models.

6 Future Works

This model can be used in the domains of predicting the consumption of all forms of energy namely, thermal, solar, wind, etc. It is applicable in a variety of industrial domains like oil, petroleum and gas. The accuracy of the model can be further improved as it does not precisely forecast the energy demand at certain hours. Factors for the cause of this error have to be investigated and corrected.

References

1. Residential consumption of electricity in India documentation of data and methodology. In: Background Paper India: Strategies for Low Carbon Growth, July 2008. The World Bank. http://moef.gov.in/wp-content/uploads/2018/04/Residentialpowerconsumption.pdf
2. International Energy Agency: World Energy Outlook 2018. International Energy Agency Publications, Paris (2018). https://www.iea.org/reports/world-energy-outlook-2018/electricity
3. https://www.eia.gov/energyexplained/electricity/electricity-and-the-environment.php
4. https://www.iea.org/reports/global-energy-co2-status-report-2019/emissions
5. Bengio, Y., Courville, A., Vincent, P.: Representation learning: a review and new perspectives. IEEE Trans. Pattern Anal. Mach. Intell. 35(8), 1798–1828 (2013). https://doi.org/10.1109/TPAMI.2013.50
6. Hochreiter, S., Schmid Huber, J.: Long short-term memory. Neural Comput. 9(8), 1735–1780 (1997). https://doi.org/10.1162/neco.1997.9.8.1735
7. Sutskever, I., Vinyals, O., Le, Q.V.: Sequence to sequence learning with neural networks. Adv. Neural Inf. Process. Syst. 3104–3112 (2014)
8. Vinyals, O., Toshev, A., Bengio, S., Erhan, D.: Show and tell: a neural image caption generator. In: Proceedings of the IEEE Conference on Computer Vision and Pattern Recognition, pp. 3156–3164 (2015)
9. Latha, H.N., Rudresh, S.: Image understanding: semantic segmentation of graphics and text using faster-RCNN. In: 2018 International Conference on Networking, Embedded and Wireless Systems (ICNEWS), Dec 2018, pp. 1–6. IEEE
10. Hemanth, K., Latha, H.N.: Dynamic scene Image deblurring using modified scale-recurrent network, (M-SRNN). In: 2020, IEEE 4th Conference on Electronics, Communication and Aerospace Technology (ICECA), pp. 966–973 (2020)

11. Graves, A., Jaitly, N.: Towards end-to-end speech recognition with recurrent neural networks. In: ICML, pp. 1764–1772 (2014)
12. Lien Minh, D., Sadeghi-Niaraki, A., Huy, H.D., Min, K., Moon, H.: Deep learning approach for short-term stock trends prediction based on two-stream gated recurrent unit network. IEEE Access **6**, 55392–55404 (2018). https://doi.org/10.1109/ACCESS.2018.2868970
13. Guo, Y., Liu, Y., Oerlemans, A., Lao, S., Wu, S., Lew, M.S.: Deep learning for visual understanding: a review. Neurocomputing **187**, 27–48 (2016). https://doi.org/10.1016/j.neucom.2015.09.116
14. https://dataaspirant.com/how-recurrent-neural-network-rnn-works/
15. https://colah.github.io/posts/2015-08-Understanding-LSTMs/
16. Rahman, A., Srikumar, V., Smith, A.D.: Predicting electricity consumption for commercial and residential buildings using deep recurrent neural networks. Appl. Energy **212**, 372–385 (2017). https://doi.org/10.1016/j.apenergy.2017.12.051
17. https://cnl.salk.edu/~schraudo/teach/NNcourse/lstm.html
18. Khosravani, H.R., Castilla, M.D.M., Berenguel, M., Ruano, A.E., Ferreira, P.M.: A comparison of energy consumption prediction models based on neural networks of a bioclimatic building. Energies **9**, 57 (2016). https://doi.org/10.3390/en9010057
19. Liu, Z., Wu, D., Liu, Y., Han, Z., Lun, L., Gao, J., Jin, G., Cao, G.: Accuracy analyses and model comparison of machine learning adopted in building energy consumption prediction (2019). https://journals.sagepub.com/toc/eea/37/4
20. Stephen, B.U.-A., Simeon, O., Asuquo, S.B.: Statistical modeling of the yearly residential energy demand in Nigeria. J. Multidiscip. Eng. Sci. Stud. **4**(6) (2018). ISSN: 2458-925X
21. Fayaz, M., Kim, D.: A prediction methodology of energy consumption based on deep extreme learning machine and comparative analysis in residential buildings. Electronics **7**, 222 (2018). https://doi.org/10.3390/electronics7100222
22. Alduailij, M.A., Petri, I., Rana, O., et al.: Forecasting peak energy demand for smart buildings. J. Supercomput. **77**, 6356–6380 (2021). https://doi.org/10.1007/s11227-020-03540-3
23. Wei, Y., Xia, L., Pan, S., Wu, J., Zhang, X., Han, M., Zhang, W., Xie, J., Li, Q.: Prediction of occupancy level and energy consumption in office building using blind system identification and neural networks. Appl. Energy **240**, 276–294 (2019). ISSN: 0306-2619. https://doi.org/10.1016/j.apenergy.2019.02.056
24. Lei, L., Chen, W., Wu, B., Chen, C., Liu, W.: A building energy consumption prediction model based on rough set theory and deep learning algorithms. Energy Build. **240**, 110886 (2021). ISSN 0378-7788. https://doi.org/10.1016/j.enbuild.2021.110886
25. Ullah, A., Haydarov, K., Ul Haq, I., Muhammad, K., Rho, S., Lee, M., Baik, S.W.: Deep learning assisted buildings energy consumption profiling using smart meter data. Sensors **20**, 873 (2020). https://doi.org/10.3390/s20030873
26. Chakrabarty, N.: A regression approach to distribution and trend analysis of quarterly foreign tourist arrivals in India. J. Soft Comput. Paradigm (JSCP) **2**(01), 57–82 (2020)
27. Schirmer, P.A., Geiger, C., Mporas, I.: Residential energy consumption prediction using inter-household energy data and socioeconomic information. In: 2020 28th European Signal Processing Conference (EUSIPCO), pp. 1595–1599 (2021). https://doi.org/10.23919/Eusipco47968.2020.9287395
28. Godinho, X., Bernardo, H., de Sousa, J.C., Oliveira, F.T.: A data-driven approach to forecasting heating and cooling energy demand in an office building as an alternative to multi-zone dynamic simulation. Appl. Sci. **11**, 1356 (2021). https://doi.org/10.3390/app11041356
29. Bu, S.-J., Cho, S.-B.: Time series forecasting with multi-headed attention-based deep learning for residential energy consumption. Energies **13**, 4722 (2020). https://doi.org/10.3390/en13184722
30. Wu, L.-X., Lee, S.-J.: A deep learning-based strategy to the energy management-advice for time-of-use rate of household electricity consumption. J. Internet Technol. **21**(1), 305–311 (2020)
31. Shylaja, H.N., et al.: Detection and localization of mask occlusion by transfer learning using faster RCNN. In: International Conference on ICCIDS—2021, 27 Apr 2021, pp. 6–15. Available at SSRN 3835214

32. Haq, I.U., Ullah, A., Khan, S.U., Khan, N., Lee, M.Y., Rho, S., Baik, S.W.: Sequential learning-based energy consumption prediction model for residential and commercial sectors. Mathematics **9**, 605 (2021). https://doi.org/10.3390/math9060605
33. Ye, Z., Kim, M.K.: Predicting electricity consumption in a building using an optimized back-propagation and Levenberg–Marquardt back-propagation neural network: case study of a shopping mall in China. Sustain. Cities Soc. **42**, 176–183 (2018). https://doi.org/10.1016/j.scs.2018.05.050
34. Schmidt, J., Wenninger, M., Goeller, T.: Appliance usage prediction for the smart home with an application to energy demand side management—and why accuracy is not a good performance metric for this problem. In: 6th International Conference on Smart Cities and Green ICT Systems (SMARTGREENS) (2018)
35. Singh, S., Yassine, A.: Big data mining of energy time series for behavioral analytics and energy consumption forecasting. Energies **11**, 452 (2018). https://doi.org/10.3390/en11020452
36. Koresh, M.H., Deva, J.: Analysis of soil nutrients based on potential productivity tests with balanced minerals for maize-chickpea crop. J. Electron. **3**(01), 23–35 (2021)
37. Li, C., Ding, Z., Zhao, D., Yi, J., Zhang, G.: Building energy consumption prediction: an extreme deep learning approach. Energies **10**, 1525 (2017). https://doi.org/10.3390/en1010 1525
38. Zheng, J., Xu, C., Zhang, Z., Li, X.: Electric load forecasting in smart grid using long-short-term-memory based recurrent neural network (2017)
39. Latha, H.N., Sahay, R.R.: A local modified U-net architecture for image denoising. Int. J. Reconstr. **1**, 8–14 (2020)
40. Kong, W., Dong, Z.Y., Jia, Y., Hill, D., Xu, Y., Zhang, Y.: Short-term residential load forecasting based on LSTM recurrent neural network. IEEE Trans. Smart Grid 1 (2017). https://doi.org/10.1109/TSG.2017.2753802
41. Beliaeva, N., Petrochenkov, A., Bade, K.: Data set analysis of electric power consumption. Eur. Res. **61**(10-2), 2482–2487 (2013). https://doi.org/10.13187/er.2013.61.2482
42. Zhao, Z., Chen, W., Wu, X., Chen, P.C.Y., Liu, J.: LSTM network: a deep learning approach for short-term traffic forecast. IET Intell. Transp. Syst. **11**(2), 68–75 (2017). https://doi.org/10.1049/iet-its.2016.0208

A Novel Swarm Intelligence Routing Protocol in Wireless Sensor Networks

M. K. Nagarajan, N. Janakiraman, and C. Balasubramanian

Abstract Wireless sensor network (WSN) is partitioned into clusters for efficient data collection in the context of energy dissipation. Other than the deployed network organization, clustering also balances network load, extending the system's lifespan. In a cluster-oriented WSN, each sensor node transmits the information to the cluster coordinator. The cluster coordinator is responsible for aggregating the gathered data and routing it to the deployed network's sink. In this paper, a new swarm intelligence cluster head (CH) selection for optimal routing of WSN is developed. The well-performed cat swarm optimization (CSO) is adopted for performing the optimal CH selection under low energy adaptive clustering hierarchy (LEACH) protocol-based clustering model. The objective function of the proposed model is considered as the multi-objective problem focusing the constraints like "energy, distance, delay, and traffic density." Finally, the experimental results validate the performance of CSO-based LEACH over the conventional swarm-oriented models for prolonging the lifetime of the network.

Keywords Wireless sensor network · Swarm intelligence routing protocol · Cluster head selection · Cat swarm optimization · Low energy adaptive clustering hierarchy · Multi-objective function

1 Introduction

WSN comprises a vast count of tiny and low-cost sensors and serves as the sensor layer of the IoT. It has been used in a wide range of applications with drastically

M. K. Nagarajan (✉)
CSE, Kalasalingam Institute of Technology, Anand Nagar, Krishnan Koil, Tamil Nadu 626126, India

N. Janakiraman
ECE, K.L.N. College of Engineering, Pottapalayam, Tamil Nadu 630612, India

C. Balasubramanian
CSE, P.S.R. Engineering College, Sevalpatti, Tamil Nadu 626140, India

© The Author(s), under exclusive license to Springer Nature Singapore Pte Ltd. 2022
V. Suma et al. (eds.), *Evolutionary Computing and Mobile Sustainable Networks*,
Lecture Notes on Data Engineering and Communications Technologies 116,
https://doi.org/10.1007/978-981-16-9605-3_26

different requirements and features [1]. Yet, sensor nodes are powered by the batteries to function for a long time, and the replacement of sensor nodes seems to be infeasible and expensive. The energy efficiency is always a top priority in WSNs [2]. In fact, prior research has shown that data transmission consumes a significant amount of energy, and the performance of transmission is primarily determined by the routing method [3]. As a result, an energy-efficient routing protocol must be developed in order to conserve energy and prolong the lifetime of the network.

For WSNs, a variety of routing techniques have been used. RPL-oriented routing has grown in popularity among WSNs, which may be divided into three types: (1) improving the route-finding method [4], (2) improving the viability of a metric [5], (3) control of transmission power [6]. Some of the illustrative examples of routing protocols in WSN are threshold-sensitive energy-efficient sensor network (TEEN) [7], distance adaptive threshold-sensitive energy-efficient sensor network (DAPTEEN) [8], LEACH [9], the power-efficient gathering in sensor information systems (PEGASIS), virtual grid architecture routing (VGA) [10], multi-path conveyance routing [11], robust routing protocol with Beziers curve-based authentication [12], and smart routing protocol with artificial bee colony optimization algorithm [13]. More specifically, routing protocol offers suitable addressing information in their Internet or network layer for allowing a packet from one network to another network [11].

Furthermore, clustering technique is another type of efficient approach for extending WSN network lifetime, which enhances network lifespan in two ways: (1) balancing node energy [14], (2) improving CH equilibrium distribution [15]. Yet, there exists complexity with clustering methods: the network is too reliant on the CH, resulting in high energy consumption and an undue strain on the CHs [16]. "H-LEACH, CL-LEACH, and P-LEACH" are some of the LEACH variations that may be used to extend the network lifetime [17].

The main contribution of the paper is as follows:

- To develop a new swarm intelligence CH selection for optimal routing of WSN under the LEACH-based clustering model.
- To perform the optimal CH selection using the CSO by considering the multi-objective problem that focuses on the constraints like "energy, distance, delay, and traffic density."
- To validate the characteristics of the CSO-oriented LEACH protocol over the state-of-the-art swarm-based methods for enhancing the network lifetime.

The organization of the paper is as follows: Sect. 1 provides the introduction about the routing protocol in WSN. The literature works of the routing protocols are shown in Sect. 2. Section 3 defines the proposed CH selection scheme in WSN. The multi-objective model derived for novel CH selection protocol is provided in Sect. 4. Section 5 displays the results and discussions. Section 6 concludes the paper.

2   Literature Review

To attain a reliable transmission of data in WSNs, energy-efficient routing protocol remains as a critical issue in expanding the lifetime of the network. Yet, the existing routing protocols propagate through the entire network for discovering a perfect route or utilize a few CHs for transmitting the data for various nodes that need vast quantity of energy consumption. These kinds of issues must be resolved. Designing the routing protocols of WSN is complicated due to several characteristics of the wireless infrastructure-less networks. However, tracing of sensor nodes is inefficient while utilizing the conventional Internet protocol. It has high redundancy due to data traffic on sensing. Some of the sensor nodes are strictly limited in terms of on-board energy, transmission energy, bandwidth, capacity, and storage. The limitation on designing the WSN is due to the lack of resources in terms of storage of processing, bandwidth, and energy. Hence, there is a need to develop novel methods for ensuring the routing protocol in WSN.

3   Proposed Cluster Head Selection Scheme in WSN

3.1   Clustering Model in WSN

In WSNs, the clustering process is used to improve energy efficiency. The BS creates clusters with different sizes, and CHs are picked in a comparable manner, in which the closest clusters are considered to be smallest in the BS. Clustering is an extremely energy-efficient strategy. The data are transferred by the sensor nodes to the closest CH in the clustering-oriented approach, which consists of CH selection. The data aggregation, as well as the compression, is handled by the CHs. For sending and processing all of the cluster's data, the CHs require a lot of energy. However, it causes premature network depletion and erratic network depletion. There is no universal method for defining and maximizing network lifespan and dissipated energy longevity. Traditional optimization methods are incapable of providing an exact answer in a certain amount of time. Meta-heuristic approaches are frequently used to find the optimal answer to these optimization issues. In Fig. 1, the suggested clustering process in the WSNs is depicted diagrammatically.

The main goal of the suggested method is to use the hybrid meta-heuristic-oriented LEACH protocol to build energy-efficient routing for WSN. The hybrid meta-heuristic method known as CSO is used to pick the best CH by solving the multi-objective function. "Energy, distance, delay, and traffic density" are all essential constraints to consider while creating a multi-objective function. Here, the distance is taken as the measure for getting the better performance in cluster head selection by computing it between the node and the CH, which should be attained lesser. In terms of network performance with the changes in network size, the proposed meta-heuristic-oriented LEACH protocol is compared to other WSN routing protocols.

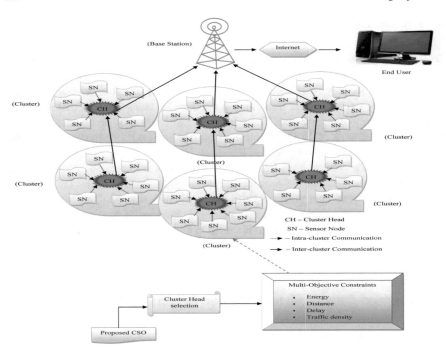

Fig. 1 WSN clustering mechanism

The media access control (MAC) utilized in the data connection layer is sensor S-MAC. This MAC protocol is a familiar WSN protocol that is based on CSMA/CA. It is a version of the IEEE 802.11 protocol that eliminates the technology's earlier drawbacks [18]. The efficiency of the suggested WSN model is compared with other existing protocols in terms of different performance metrics.

3.2 Cat Swarm Optimization for Cluster Head Selection

CSO [19] is used for performing the optimal CH selection under the LEACH-oriented clustering model with consideration of the multi-objective constraints such as "energy, distance, delay, and traffic density." Due to the outperformance of CSO algorithm, this paper has chosen this technique for optimal CH selection for reducing the energy consumption. Moreover, CSO provides back-up routes and reduce the energy consumption and avoids generation of circle route in existing studies. Hence, this model has focused on reducing the energy consumption and execution time. As it, CSO method is a computationally intelligent algorithm that is based on cat behavior. The seeking mode and the tracing mode are the two different modes of this algorithm.

Seeking mode (SM): The cat seems to be at rest as well as attentive, searching about its environment for its next move in the SM. The location is represented by a schedule vector, for which the four parameters of this mode have been adjusted. Place k copies of the current cat position l with $k = $ SMP. If the SPC value is true or $k = $ SMP $- 1$, then the cat is retained as one of the candidates. A random value of SRD is generated. If the fitness OH is not equal, the probability of every candidate is computed using Eq. (1).

$$Q_j = \frac{\left| OH_j - OH_{min} \right|}{OH_{max} - OH_{min}}, \text{ in which } 0 < j < k \tag{1}$$

In the above equation, the minimal fitness in the swarm is shown by OH_{min}; the maximum fitness in the swarm is shown by OH_{max}, and the fitness of cat_j is shown by OH_j, respectively. Here, the number of copies of cats in the SM is shown by SMP; first rang in the chosen solution vector is shown by SRD, and a Boolean value is shown by SPC, respectively.

Tracing mode (TM): When chasing a pray or other moving item, TM shows the cat moving quickly and according to its own velocity. The velocities of every cat cat_l are updated. Verify if the velocities belong to the highest order as in Eq. (2). The position of cat_l is updated on the basis of Eq. (3).

$$U'_l = z * U_l + rl * d * (Z_{best} - Z_l) \tag{2}$$

$$Z'_l = Z_l + U_l \tag{3}$$

In the above equations, the velocity of cat_l is shown by U_l; the actual position of cat_l is shown by Z_l; the novel value of position of cat_l is shown by Z'_l; random value is shown by rl; constant is shown by d; inertia-weighted parameter is shown by z; novel value velocity is shown by U'_l; old speed or the current value is shown by U_l, and the best position or solution of the cat having the best fitness value is shown by Z_{best}, respectively.

The pseudocode of the CSO is given in Algorithm 1.

Algorithm 1 CSO [19]

    Start

    Generate M cat

    Initialization of "position, velocity, and flag" of every cat

    Cat evaluation on the basis of fitness

    If cat_l is in SM

            Apply cat_l into SM

    Else

Apply cat$_l$ into TM
Terminate
End

4 Multi-objective Function

The CH is picked from a set of sensor nodes and is made up of key characteristics such as "energy, distance, delay, and traffic density." Each node's burden should be reduced, while its QoS and trust should be enhanced. Equation (8) shows the objective function connected to the developed CH selection in WSN.

$$H0 = \alpha * \text{dst} + (1 - \alpha) * (1/\text{ey}) \tag{4}$$

$$H1 = \beta * H0 + (1 - \beta) * \text{dst} \tag{5}$$

$$H2 = \omega * H1 + (1 - \omega) * (1/\text{dly}) \tag{6}$$

$$H3 = H2 * \text{TDY} \tag{7}$$

$$OH = \underset{\{CH\}}{\arg\min}(H3) \tag{8}$$

In the above equations, the weighting factors are shown by α, β, and ω and each of these takes the value as 0.2.

Energy: The fitness function of normalized energy is given in Eq. (9).

$$\text{ff}_{\text{ey}} = \frac{\text{ff}_{(b)}^{\text{ey}}}{\text{ff}_{(c)}^{\text{ey}}} \tag{9}$$

For each cluster, the normalized energy must be more. When the entire CH cumulative $\text{ff}_{(b)}^{\text{ey}}$ and $\text{ff}_{(c)}^{\text{ey}}$ is composed of maximum CH count and energy, then the term ey becomes larger. The total energy spent by the nodes during "transmission, reception, sensing, and idle state" is calculated.

Distance: The fitness function associated with the distance that is given in Eq. (10), in which the data transmission from the normal node to the CH and then to the BS is given by $\text{ff}_{(b)}^{\text{dst}}$. The value of the distance function ranges between 0 and 1. The numerical equations of $\text{ff}_{(b)}^{\text{dst}}$ and $\text{ff}_{(c)}^{\text{dst}}$ are given in Eqs. (11) and (12).

$$dst = \frac{ff_{(b)}^{dst}}{ff_{(c)}^{dst}} \tag{10}$$

$$ff_{(b)}^{dst} = \sum_{m=1}^{CE_m} \| E_m - C_m \| + \sum_{n=1}^{NE_n} \| E_m - B_m \| \tag{11}$$

$$ff_{(c)}^{dst} = \sum_{m=1}^{CE_m} \sum_{n=1}^{CE_n} \| B_m - B_n \| \tag{12}$$

Here, the distance within the normal node and CH is shown by $E_m - B_m$; the distance within the CH and BS is shown by $E_m - C_m$; the normal node of mth cluster is shown by B_m, and the CH of mth cluster is shown by E_m. The distance within the two normal nodes is shown by $B_m - B_n$.

When the distance between the node and the CH exceeds the station distance, the sensor node communicates directly with the BS. Or else, it joins the cluster using the closest distance as a criterion. As a result, clusters develop. When the threshold distance is greater than the observed distance, the free space model is used. The multipath fading model is used if the threshold is lower. The threshold distance is determined using the formula in Eq. (13).

$$thsh\ dst_0 = \sqrt{\frac{FS_{ey}}{PA_{ey}}} \tag{13}$$

Delay: The fitness function for delay is defined by Eq. (14).

$$dly = \frac{\max(\| E_m - B_m \|)_{m=1}^{CE_m}}{CE_{NB}} \tag{14}$$

In the above equation, the total nodes in WSN is shown by CE_m, and the dly lies within 0 to 1.

Traffic Density: It is based on "channel load, packet drop, and buffer utilization," and the computation is based on the average of these three factors as shown in Eq. (15).

$$TDY = \frac{1}{3}\left[BU_{Util} + PR_{drop} + CL_{load}\right] \tag{15}$$

Here, the packet drop is given by PR_{drop}; the buffer utilization is given by BU_{Util}, and the channel load is given by CL_{load}, respectively. The buffer size is defined in Eq. (16), where the buffer usage is depending on the buffer space.

$$BU_{Util} = \frac{BU_{space}}{BU_{size}} \tag{16}$$

$$PR_{drop} = \frac{DR_{pd}}{PR_{xd}} \tag{17}$$

In the above equations, the buffer space is given by BU_{space}; the packets dropped count is given by DR_{pd}; the buffer size is given by BU_{size}, and the packets transmitted count is given by PR_{xd}, respectively. The packet drop is calculated using Eq. (17). Equation (18) shows the channel load.

$$CL_{load} = \frac{CL_{busy}}{RN} \tag{18}$$

Here, the channels in the busy state are given by CL_{busy}, and the total rounds are given by RN, respectively. The channel load calculation is dependent on the number of rounds as well as the channel condition during the process of simulation.

5 Results and Discussions

5.1 Experimental Setup

The proposed novel swarm intelligence routing protocol in WSN was implemented in MATLAB 2020a, and the results were analyzed. The total number of network environment considered was 2, in which the network 1 was composed of 50 nodes, and the network 2 was composed of 100 nodes. The population size was taken as 10, and the iteration was also 10. The maximum number of rounds considered was 2000. Here, the CSO [19]-based routing protocol in WSN was compared with various optimization algorithms like PSO [20], EFO [21], BOA [22], and GWO [23] in terms of various analyses to prove the superiority of the developed method. The simulation parameters considered for proposed and existing model are given in Table 1.

5.2 Convergence Analysis

The convergence analysis of the CSO-based routing protocol in WSN against several heuristic algorithms for the 2 networks is shown in Fig. 2. From Fig. 2a, the cost function of CSO at 10th iteration for 1th network is 5.13%, 8.64%, 38.84%, and 19.57% improved than GWO, BOA, EFO, and PSO, respectively. Thus, the convergence analysis is better with the CSO than the remaining algorithms for the routing protocol in WSN.

Table 1 Simulation constraints used for experimentation

Constraints	Values
Proposed model	
Network 1	50 nodes
Network 2	100 nodes
Base station location	(50, 50) in meters
Number of rounds	2000
Number of solution	10
Dimension of the solution	10
Iteration count	10
Probability of a node to be a cluster head	0.1
PSO [20]	
Acceleration constant 1	2
Acceleration constant 2	2
Maximum weight	0.9
Minimum weight	0.1
EFO [21]	
Sensing area	0.5
BOA [22]	
Probability	0.8
Power exponent	0.1
Sensor modality	0.01
Leader score	∞
GWO [23]	
Score of alpha	∞
Score of beta	∞
Score of gamma	∞
CSO [19]	
MR	0.75
Percent length of the mutation	0.65
First range in the selected solution vector	0.25
Number copies of cats in the SM	5
Constant value	2.05
Maximum weight	0.9
Minimum weight	0.3
Percentage value	0.25

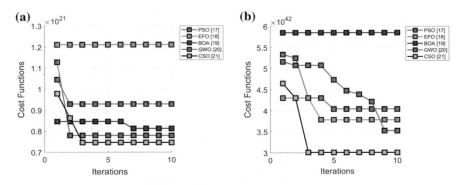

Fig. 2 Convergence analysis of heuristic-based algorithms for the routing protocol in WSN using MATLAB 2020a in terms of, **a** network 1, **b** network 2

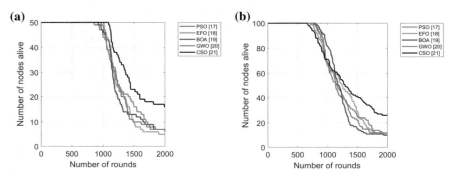

Fig. 3 Alive node analysis of heuristic-based algorithms for the routing protocol in WSN using MATLAB 2020a in terms of, **a** network 1, **b** network 2

5.3 Alive Node Analysis

The alive node analysis of the CSO and the other heuristic algorithms for the routing protocol in WSN is depicted in Fig. 3. In Fig. 3a, for network 1, the alive nodes of the CSO at 1500th round are 56.52%, 64.29%, 60.87%, and 53.33% higher than GWO, BOA, EFO, and PSO, respectively. Thus, the number of alive nodes is more with the CSO than the other methods for the routing protocol in WSN.

5.4 Normalized Energy Analysis

The analysis on normalized energy for the routing protocol in WSN in the case of 2 networks is shown in Fig. 4. From Fig. 4a, for network 1, the normalized energy of CSO at 1500th round is 55.56%, 66.67%, and 80% progressed than GWO, BOA,

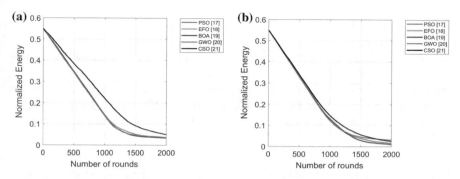

Fig. 4 Normalized energy analysis of heuristic-based algorithms for the routing protocol in WSN using MATLAB 2020a in terms of, **a** network 1, **b** network 2

and PSO, respectively. Hence, the normalized energy is maximized when compared with conventional techniques.

5.5 Traffic Density Analysis

The traffic density analysis for the routing protocol in WSN in the case of two networks is shown in Fig. 5. In Fig. 5a, for network 1, the traffic density of CSO at 250th round is 21.15%, 10.87%, 42.25%, and 30.51% improved than GWO, BOA, EFO, and PSO, respectively. Thus, the analysis on traffic density holds better outcomes with the CSO than the remaining algorithms for the routing protocol in WSN.

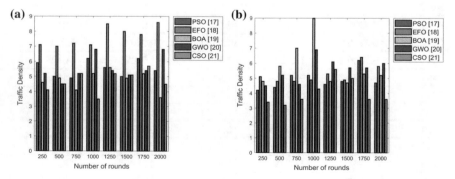

Fig. 5 Traffic density analysis of heuristic-based algorithms for the routing protocol in WSN using MATLAB 2020a in terms of, **a** network 1, **b** network 2

5.6 Delay Analysis

The delay analysis of the CSO and the other heuristic algorithms for the routing protocol in WSN is shown in Tables 2 and 3 for the two networks. From Table 2, the delay of CSO for network 1 at 224th round is 3.49%, 4.21%, 4.21%, and 3.49% more than GWO, BOA, EFO, and PSO, respectively. Hence, the delay analysis is better with the CSO than the other heuristic algorithms for the routing protocol in WSN.

Table 2 Overall delay analysis of heuristic-based algorithms for the routing protocol in WSN for network 1

Rounds	PSO [20]	EFO [21]	BOA [22]	GWO [23]	CSO [19]
0	1.0025	1.0939	1.0479	1.0281	0.77033
99	1.0865	1.0812	0.87693	0.91469	0.84814
224	1.0025	1.0939	1.0939	1.0126	1.0479
499	1.0025	0.87693	1.0209	1.0281	0.73959
724	1.0865	1.071	1.1376	0.9726	1.2029
999	1.1376	0.84544	0.97399	1.0939	1.1376
1224	0.84544	1.0479	1.0115	0.80581	1.0168
1499	0.77033	1.0168	1.0168	0.97077	1.1376
1725	1.0479	0.68155	1.0865	0.84814	0.84544
2000	0.96376	0.91469	0.91469	0.85978	0.68155

Table 3 Overall delay analysis of heuristic-based algorithms for the routing protocol in WSN for network 2

Rounds	PSO [20]	EFO [21]	BOA [22]	GWO [23]	CSO [19]
0	0.96232	0.93115	1.0819	1.003	1.2159
99	0.81701	1.2936	1.0553	0.94036	0.6933
224	0.9306	1.1882	1.15	1.0447	1.0122
499	1.0325	0.89714	0.94716	0.94278	1.3148
724	1.0025	0.97481	0.77767	0.89243	1.2002
999	0.92182	0.91585	1.1392	1.0938	1.1062
1224	0.69744	0.99995	0.69744	0.94585	1.0372
1499	0.81488	1.1574	0.93305	0.83707	0.69519
1725	1.0851	1.0015	0.81488	0.88905	0.85931
2000	0.85053	1.0853	1.0888	0.96611	0.99791

Table 4 Overall delay analysis of heuristic-based algorithms for the routing protocol in WSN for network 1

Rounds	PSO [20]	EFO [21]	BOA [22]	GWO [23]	CSO [19]
Network 1					
Computational time (s)	0.52675	0.53617	0.53251	0.5331	0.52343
Network 2					
Computational time (s)	0.95338	0.96396	0.96656	0.93262	0.92209

5.7 Computational Time

The computational time of suggested CSO and other heuristic algorithms is given in Table 4 for two networks. From Table 4, the lesser computational time of CSO has observed than other techniques in WSN.

6 Conclusion

This paper has proposed a new swarm intelligence CH selection for the optimal routing of WSN. The CSO was used for accomplishing the optimal CH selection under the LEACH-oriented clustering model. The objective function of the proposed model was considered as the multi-objective problem that concentrated on the constraints like "energy, distance, delay, and traffic density." From the analysis, the cost function of CSO for 1st network was 5.13%, 8.64%, 38.84%, and 19.57% improved than GWO, BOA, EFO, and PSO, respectively. Similarly, for 2nd network, the cost function of CSO was 36.17%, 48.28%, 21.05%, and 26.83% better than GWO, BOA, EFO, and PSO, respectively. Thus, the CSO was better than the other heuristic algorithms for the routing protocol in WSN. Moreover, the security constraints and performance enhancement in the WSN can be performed by adopting intelligent approaches with improved deep learning strategies in future.

References

1. Han, G., Jiang, J., Guizani, M., Rodrigues, J.J.P.C.: Green routing protocols for wireless multimedia sensor networks. IEEE Wireless Commun. **23**(6), 140–146 (2016)
2. Hayes, T., Ali, F.H.: Location aware sensor routing protocol for mobile wireless sensor networks. IET Wireless Sens. Syst. **6**(2), 49–57 (2016)
3. Zhao, M., Ho, I.W.-H., Chong, P.H.J.: An energy-efficient region-based RPL routing protocol for low-power and lossy networks. IEEE Internet Things J. **3**(6), 1319–1333 (2016)
4. Sharma, D., Bhondekar, A.P.: Traffic and energy aware routing for heterogeneous wireless sensor networks. IEEE Commun. Lett. **22**(8), 1608–1611 (2018)

5. Xu, Z., Chen, L., Chen, C., Guan, X.: Joint clustering and routing design for reliable and efficient data collection in large-scale wireless sensor networks. IEEE Internet Things J. **3**(4), 520–532 (2016)
6. Zhang, Y., Zhang, X., Ning, S., Gao, J., Liu, Y.: Energy-efficient multi-level heterogeneous routing protocol for wireless sensor networks. IEEE Access **7**, 55873–55884 (2019)
7. Alsafi, S., Talab, S.A.: Threshold sensitive energy efficient sensor network protocol. Int. J. Acad. Eng. Res. (IJAER) **4**(11), 48–52 (2020)
8. Anjali, Garg, A., Suhali: Distance adaptive threshold sensitive energy efficient sensor network (DAPTEEN) protocol in WSN. In: International Conference on Signal Processing, Computing and Control (ISPCC), pp. 114–119 (2015)
9. Tümer, A.E., Gündüz, M.: Energy-efficient and fast data gathering protocols for indoor wireless sensor networks. Sensors **10**, 8054–8069 (2010)
10. Patel, B., Munjani, J.: Power-efficient gathering in sensor information systems for WSN: a review. Int. J. Comput. Appl. Technol. **5**(4) (2014)
11. Chen, Z.J.I.: Optimal multipath conveyance with improved survivability for WSN's in challenging location. J. ISMAC **2**, 73–82 (2020)
12. Shakya, S.: The robust routing protocol with authentication for wireless adhoc networks. J. ISMAC **2**, 83–95 (2020)
13. Jacob, J.I., Darney, P.E.: Artificial bee colony optimization algorithm for enhancing routing in wireless networks. J. Artif. Intell. **3**(01), 62–71 (2021)
14. De Couto, D.S.J., Aguayo, D., Bicket, J., Morris, R.: A high-throughput path metric for multi-hop wireless routing. Wireless Netw. **11**(4), 419–434 (2005)
15. Sharma, S., Puthal, D., Tazeen, S., Prasad, M., Zomaya, A.Y.: MSGR: a mode-switched grid-based sustainable routing protocol for wireless sensor networks. IEEE Access **5**, 19864–19875 (2017)
16. Lai, X., Ji, X., Zhou, X., Chen, L.: Energy efcient link-delay aware routing in wireless sensor networks. IEEE Sens. J. **18**(2), 837–848 (2017)
17. Yan, J., Zhou, M., Ding, Z.: Recent advances in energy-efficient routing protocols for wireless sensor networks: a review. IEEE Access **4**, 5673–5686 (2016)
18. Yadav, R., Varma, S., Malaviya, N.: A survey of MAC protocols for wireless sensor networks. UbiCC J. **4**(3), 827–833 (2009)
19. Bouzidi, A., Riffi, M.E., Barkatou, M.: Cat swarm optimization for solving the open shop scheduling problem. J. Ind. Eng. Int. **15**, 367–378 (2019)
20. Rajagopal, S.: Particle swarm optimization-based routing protocol for clustered heterogeneous sensor networks with mobile sink. Am. J. Appl. Sci. **10**(3), 259–269 (2013)
21. Yilmaz, S., Sen, S.: Electric fish optimization: a new heuristic algorithm inspired by electrolocation. Neural Comput. Appl. **32**(4) (2020)
22. Li, G., Shuang, F., Zhao, P., Le, C.: An improved butterfly optimization algorithm for engineering design problems using the cross-entropy method. Symmetry **11** (2019)
23. Al-Aboody, N.A., Al-Raweshidy, H.S.: Grey wolf optimization-based energy-efficient routing protocol for heterogeneous wireless sensor networks. In: 2016 4th International Symposium on Computational and Business Intelligence (ISCBI), pp. 101–107 (2016)

Authentication Framework for Healthcare Devices Through Internet of Things and Machine Learning

Shruti Kute, A. V. Shreyas Madhav, Amit Kumar Tyagi⊙, and Atharva Deshmukh

Abstract The rise in the application and use of IoT has made tech gadgets and gizmos smarter and interconnected than ever before. IoT has expanded in a number of fields including smart cities and homes, healthcare, finance, etc. A typical IoT device is likely to integrate computation, networking, and physical processes from embedded computational gadgets. To monitor and extract valuable existential patterns from the large volume of data that is generated, Machine Learning (ML) helps a lot by the different number of algorithms that can be developed. Typically, ML is a discipline that covers the two major fields of constructing intelligent computer systems and governing these systems. ML has progressed dramatically over the past two decades and has emerged as one of the major methods for developing computer vision systems and other pedagogies. Rapid advancements and enhancements in these fields are leading to extensive interactions among the devices in the heterogeneous pool of gadgets. However, with advancements, there always will be a few bottlenecks that hinder the security and safety of the device or the gadget. Making use of such methodologies and techniques for user access control exposes this to endless vulnerabilities including numerous attacks and other complications. It is extremely important to protect the authenticity and privacy of the users and the data that is stored in these smart devices. This paper discusses the variety of ways in which a smart device can validate the user with proper authentication and verification.

Keywords Secure machine learning · Cloud computing · Biometric authentication · Internet of Things

S. Kute · A. V. Shreyas Madhav (✉) · A. K. Tyagi
School of Computer Science and Engineering, Vellore Institute of Technology, Chennai, Tamil Nadu 600127, India

A. K. Tyagi
Centre for Advanced Data Science, Vellore Institute of Technology, Chennai, Tamil Nadu 600127, India

A. Deshmukh
Terna Engineering College, Navi Mumbai, India

© The Author(s), under exclusive license to Springer Nature Singapore Pte Ltd. 2022
V. Suma et al. (eds.), *Evolutionary Computing and Mobile Sustainable Networks*,
Lecture Notes on Data Engineering and Communications Technologies 116,
https://doi.org/10.1007/978-981-16-9605-3_27

1 Introduction

The actual items or things, around us (for example, actuators, sensors, and many more) are progressively becoming interconnected, by and large, by means of the Web. These interconnected things gather (or sense), offer, but also cycle data in a serious information-driven Internet of Things (IoT) setting. Cloud computing is been used in IoT organizations, across both regular citizen and military applications, to manage the dynamic prerequisites like capacity, computation (e.g., information examination and representation, etc.).

Like most worker customer or organized frameworks, there seems to be a danger that IoT application workers could be focused on by assailants, along with pernicious insiders, mostly with points of gaining unapproved admittance to information. In other words, the customer hubs are powerless against dangers, for example, disconnected secret key speculating assaults, client pantomime assaults, insider assaults, and client explicit key robbery assaults. Biometric-put together validation systems (for example, once based with respect to the person's unique mark, face, voice, fingerprint, keystroke, and iris) are being demonstrated for being genuinely strong toward ordinary dangers than with traditional secret key-based verification plans. In particular, in such a framework, the highlights of the enlisted clients are put away in a concentrated information base in the cloud to improve the cycle of validation. During validation, a client is resolved to be either real or not founded on the coordinating likelihood with the test layout that is put away in the information base.

IoT devices give a variety of new options for healthcare providers to monitor patients as well as for patients to monitor themselves, such as Ingestible sensors, which gather information from the digestive system as well as provide information about stomach PH levels. Automated insulin delivery (AID) systems, also known as closed-loop insulin delivery systems or Artificial Pancreas systems, are groundbreaking for persons with diabetes, a disorder that affects around 8.5% of adults globally, according to World Health Organization data.

2 Literature Survey

The blast of IoT has prompted the expanded utilization of biometrics authentication in different cloud computing-focused administrations. Peer et al. [1] had imagined necessities for developing cloud-based biometric arrangements. They likewise proposed a contextual analysis in a cloud climate utilizing a unique mark-based verification administration.

Yuan and Yu [2] proposed a novel biometric recognizable proof plan with security safeguarding. The greater parts of the tasks were safely moved to cloud workers. The information base proprietors produced accreditations. These qualifications were utilized in the recognizable proof help actualized by cloud workers in an encoded information base. Such a technique would guarantee the secrecy of personal biometric information in a cloud. Haghighat et al. [3] proposed a cloud-based cross-endeavor biometric recognizable proof plan known as "CloudId" where encryption methods are utilized to give a dependable biometric validation framework. Be that as it may, the framework experienced weighty computational multifaceted nature. Das and Goswami [4] proposed a productive, secure, safe, and protection-saving biometric-based far-off confirmation administration for E-healthcare frameworks. By and by, the framework experienced different weaknesses.

The cancellable biometrics framework is a stride in front of the ordinary biometric framework. The arrangement of a cancellable biometric framework in the cloud climate is as yet a difficult issue and an open exploration territory. Amin et al. [5] have proposed a common verification and meeting key arrangement in a distant climate utilizing the cancellable bio-decimal measuring standard. During client verification, they used a cancellable biometric framework known as bio-hashing. In a cloud computing climate, Bommagani et al. [6] proposed a cancellable face acknowledgment structure where acknowledgment tasks had been moved to the cloud to misuse the equal execution ability of the cloud. At the same time, a cancellable face layout age calculation dependent on an irregular projection framework [7] was likewise proposed. Wu et al. [8] also proposed a method for creating multiple keys from a client's finger vein for validating various cloud-based administrations utilizing a high-dimensional space self-adjustment calculation. When compared to cancellable biometrics, the method for producing various keys with a fluffy vault plot is computationally perplexing.

Hu et al. [9] proposed a cloud-based face confirmation plot in which they re-evaluated every one of the processes, from image pre-processing to coordinating cloud workers [10]. They also additionally abused the cloud's equal cycle execution limit and used facial confirmation for cloud-based IoT applications. Notwithstanding, there is a gap that requires lightweight change methods to be conveyed in the cloud climate. A few scientists have received cloud computing innovation to satisfy the needs regarding force and capacity limits [11]. As a result, there is indeed a prerequisite for cloud-based administrations to serve as a viable cross-stage for different IoT applications.

Karimian et al. [12] used electrocardiogram (ECG) signals to authenticate an Internet of Things, observing that Electrocardiography biometrics are efficient, secure, and simple to deploy. Kantarci et al. [13] proposed a cloud-centric model in which biometric authentication architecture combines a biometric method with context-awareness method for preventing unwanted access to mobile apps. Dhillon and Kalra [14] devised a lightweight multi-factor identity-based cryptographic method using less expensive hash functions. Ratha et al. [15] were the first one to propose cancellable biometrics utilizing three separate transformation functions where first is Cartesian transformation, second would be Polar transformation, and the last one is Functional transformation.

3 Internet of Things and Machine Learning Based Authorization Model

Biometric distinguishing proof empowers end clients to utilize actual characteristics rather than passwords or PINs as a safe strategy for getting to a framework or an information base. Biometric innovation depends on the idea of supplanting "one thing you have with you" with "what your identity is," which has been viewed as a more secure innovation to protect individual data. The conceivable outcomes of applying biometric ID are truly colossal [16]. Biometric ID is applied these days in areas where security is a first concern, similar to air terminals, and could be utilized as a way to control outskirt crossing adrift, land, and air wilderness. Particularly for the air traffic region, where the quantity of flights will be expanded by 40% before 2013, the confirmation of portable IoT gadgets will be accomplished when the bio-features models become adequately full-grown, productive, and impervious to IoT assaults.

Another region where biometric ID strategies are beginning to be received is electronic IDs. Biometric ID cards, for example, the Estonian and Belgian public ID cards were utilized to distinguish and confirm qualified citizens during races. Moving above and beyond, Estonia has presented the Versatile ID framework that permits residents to lead Webcasting a ballot and consolidates biometric distinguishing proof and cell phones. This framework that was very creative when it was at first acquainted has a few dangers with the constituent technique and was reprimanded for being uncertain.

As per an overview by Spear System and Exploration, in 2014, $16 billion was taken by 12.7 million individuals who were survivors of wholesale fraud in the US as it were. This sum is determined without considering the financial issues and mental mistreatment that survivors of this extortion endure. From the financial area and organizations to admittance to homes, vehicles, PCs, and cell phones, biometric innovation offers the most elevated level of security as far as protection and security insurance and secure access.

Cell phones are these days a fundamental piece of our regular daily existence, as they are utilized for an assortment of portable applications. Performing biometric validation through cell phones can give a more grounded system to personality check as the two verification factors, "something you have" and "something you are," are joined. A few arrangements that incorporate multibiometric and conduct valida-tion stages for telecom transporters, banks, and different businesses were as of late presented.

3.1 Biometric-Based User Authentication for Smart IoT Devices

There are diverse related reviews that manage client verification. Albeit some of them covered distinctive validation strategies, we just consider those that were completely devoted to biometric confirmation. We characterize the studies as per the accompanying rules:

(a) Deployment scope: it shows whether the confirmation plot is sent on cell phones or not.
(b) Focus biometric region: it shows whether the review zeroed in on all/particular biometric highlights.
(c) Threat models: it demonstrates whether the study considered the dangers against the verification plans.
(d) Countermeasures: it shows whether the study zeroed in on and considered the countermeasures to shield the verification plans.
(e) Machine learning (ML) and data mining (DM) calculations: they show whether the study makes reference to every arrangement of the pre-owned AI or data mining technique.

A few reviews depicted the validation conspires that just consider explicit bio-features. For example, the studies just centered around the keystroke elements. Then again, Gafurov introduced biometric stride acknowledgment frameworks. Revett et al. overviewed biometric verification frameworks that depend on mouse developments. Yampolskiy and Govindaraju introduced an extensive report on conducting biometrics. Mahadi et al. studied social-based biometric client verification and decided the arrangement of best classifiers for conduct-based biometric confirmation [17, 18]. Sundararajan and Woodard surveyed 100 unique methodologies that utilized deep learning and different biometric modalities to distinguish clients. Teh et al. presented distinctive confirmation arrangements that depend on touch elements in cell phones. Rattani and Derakhshani gave the best in the class identified with face biometric confirmation plots that are intended for cell phones. They additionally examined the satire assaults that target portable face biometrics just as the anti-spoofing techniques. Mahfouz et al. reviewed the social biometric confirmation conspire that are applied on cell phones. Meng et al. studied the validation systems utilizing biometric clients on cell phones. They recognized eight possible assaults against these confirmation frameworks alongside promising countermeasures. Our study and both spotlight on verification plots that are intended for cell phones and consider all the biometric highlights and manage danger models and countermeasures. Nonetheless, does not give data identified with the pre-owned AI or data mining strategy for all the reviewed arrangements. Likewise, just conceals papers to 2014, while the inclusion of our study is up to 2018. Supposedly, this work is the primary that altogether covers dangers, models, countermeasures, and the AI calculations of the biometric confirmation plans.

3.2 Biometric-Based User Authentication Cancellation

The idea of cancellable biometrics is that the first layout information is changed into an alternate adaptation by utilizing a non-invertible change work in the enlistment stage. Question information in the confirmation stage is applied a similar non-invertible change. Coordinating is directed in the changed space utilizing the changed layout and question information.

Ratha et al. started three distinctive change capacities, known as Cartesian, polar, and practical changes. The proposed change works deliberately misshape the first highlights, with the goal that it is infeasible or computationally hard to recover crude format information. Notwithstanding, one disadvantage is that the proposed strategy is enlistment-based, and subsequently, the exact discovery of solitary focuses is required. Typically, exact enrolment is hard as a result of biometric vulnerability (e.g., picture relocation, non-direct bending, and obtaining condition). Jin et al. proposed a two-factor validation strategy called bio-hashing. Bio-hashing joins token-based information with unique mark highlights by the iterative inward item to make another list of capabilities. At that point, each incentive in the list of capabilities is changed over to a paired number dependent on a predefined edge. Lee et al. produced cancellable unique mark formats by separating a revolution and interpretation invariant element for each detail, which is considered to be the principal arrangement free cancellable unique mark layout plan. Ahn et al. utilized trios of particulars as a list of capabilities, and change is performed on mathematical properties got from the trios. Yang et al. made cancellable layouts by utilizing both neighborhood and worldwide highlights. Neighborhood highlights incorporate distances and relative points between particular sets, while worldwide highlights incorporate direction and edge recurrence. In this exploration, the distance of a couple of particulars is changed utilizing an opposite projection, to determine the non-invertible change.

Ahmad and Hu proposed an arrangement-free structure dependent on a couple of polar organize. In this structure, the general situation of every minutia to all other particulars among a polar facilitate range is used. From any two particulars, three neighborhood highlights are extricated and changed by a useful change to produce the cancellable layout. In view of the minutia structure, Wang et al. further improved framework security and precision by proposing some new change capacities, for example, endless to-one planning, abridged roundabout convolution, and incomplete Hadamard change. Zhang et al. planned a combo plate and a utilitarian change to deliver cancellable layouts dependent on the Minutia Cylinder Code (MCC). MCC is a notable neighborhood minutia descriptor, which depends on 3D nearby structures related to every minutia. The creators of the MCC later proposed a format insurance strategy named P-MCC, which plays out a KL change on the MCC highlight portrayal. Nonetheless, P-MCC does not have the property of revocability. At that point, 2P-MCC was proposed to add cancelability to P-MCC utilizing a halfway change-based plan. Afterward, Arjona et al. introduced a protected unique mark coordinating methodology, named P-MCC-PUFs, which contains two components dependent on P-MCC and PUFs (Genuinely Unclonable Capacities).

The proposed conspire accomplishes the best presentation when the length of the element vector is set to 1024 pieces and gives solid information protection and security. Yang et al. planned a cancellable unique mark layout dependent on arbitrary projection. The planned format can shield assaults through record variety (ARM) inferable from the component decorrelation calculation. Meanwhile, a Delaunay triangulation-based nearby structure proposed in the plan can diminish the negative impact of nonlinear twisting on coordinating execution. Sandhya and Prasad intertwined two structures, nearby structure, and inaccessible structure, at the elementary level to create twofold esteemed highlights, which are then ensured by an arbitrary projection-based cancellable insurance strategy.

To additional upgrade security and acknowledgment execution, a few analysts proposed the utilization of multimodal cancellable biometrics. For instance, Yang et al. proposed a multimodal cancellable biometric framework that wires unique mark highlights and finger-vein highlights to accomplish better acknowledgment precision and higher security. In the proposed framework, an improved incomplete discrete Fourier change is used to give non-invertibility and revocability. Additionally, Dwivedi and Dey proposed a mixture combination (score level and choice level combination) plan to coordinate cancellable unique mark and iris modalities to decrease impediments in every individual methodology. Test consequences of multimodal cancellable biometric frameworks show execution improvement over their unimodal partner.

In this segment, the development of cancellable biometrics, from the presentation of the possibility of cancellable biometrics and some early change work plans to the ongoing different cancellable biometrics, is introduced. There are two classifications in the plan of cancellable biometrics. One class revolves around the extraction and portrayal of stable biometric includes in order to accomplish better acknowledgment precision, and the other classification centers around planning secure change capacities, which are relied upon to be numerically non-invertible. It is foreseen that future exploration work in cancellable biometrics will endeavor to accomplish both better acknowledgment precision and more grounded security by utilizing numerous cancellable biometrics.

4 Proposed Framework for Authorization and Authentication of Smart IoT Devices

An adaptable security system is required to address the vastly different IoT climate and the associated security challenges. Given Fig. 1 shows a structure to make sure about the IoT climate and is contained four segments:

- Authentication
- Authorization
- Network Enforced Policy
- Secure Analytics: Visibility and Control.

Fig. 1 Internet of Things framework

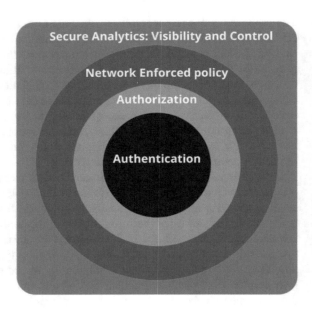

- Authentication: The validation layer is at the core of this system, and it is used to provide and confirm the distinguish data of an IoT substance. When connected IoT/M2M gadgets (for example, inserted sensors and actuators or endpoints) require access to IoT framework, a relationship of trust is started dependent on character of the gadget. The best approach of storing as well as manifesting personality data may be exceedingly exceptional for IoT gadgets. It should be noted that in average venture organizations, its endpoints might be distinguished by only a human accreditation (for example, username and secret key, biometrics, or token). IoT/M2M endpoints should always be fingerprinted by implying that they do not require any human collaboration. All these identifiers incorporate radio-recurrence distinguishing proof (RFID), shared mystery, X.509 testaments, the endpoint's Macintosh address, or other kinds of unchanging equipment-based foundation of trust. Setting up character by means of X.509 testaments gives a solid confirmation framework. In any case, in the IoT area, numerous gadgets might not have enough memory to store an endorsement or may not have the necessary computer processor capacity to execute the cryptographic activities of approving the X.509 authentications (or any kind of open key activity). Existing personality impressions, e.g., 802.1AR and verification norms as defined by IEEE 802.1X, could be used for devices that can manage both the central processor load and memory to store solid qualifications. Nonetheless, the problems of new structure factors, as well as new modalities, open the door for additional examination in describing more modest impression qualification types as well as less process-focused cryptographic builds and verification norms.
- Authorization: This is the second layer of this system is authorization, which controls a device's entry throughout the organization's texture. This layer extends

the central validation layer by utilizing an element's character data. A trust relationship is established between IoT devices to trade appropriate data using confirmation and approval segments. For example, a vehicle might well set up a trust alliance with some other vehicle from the same seller. Regardless of the trust relationship, vehicles may only be able to trade one's well-being capacities. When a trustworthy partnership is established between a similar vehicle and their vendor's organization, the vehicle may be permitted to start sharing additional data, e.g., the odometer reading, last upkeep track, and so on. Fortunately, current arrangement instruments for both overseeing and controlling access to shopper and undertaking networks map exceptionally well to the IoT/M2M requirements. The major challenge will be to create a technology that can scale to deal with billions of IoT/M2M gadgets with varying trust connections in the texture. To fragment information traffic and build up start to finish correspondence, traffic approaches and appropriate controls will be used throughout the organization.

- Network Enforced Policy: This layer includes all components that safely route and transport endpoint traffic over the foundation, whether control, board, or real information traffic. There are currently established conventions and instruments to ensure the organization's foundation and influence strategy that are appropriate to the IoT/M2M use cases, similar to the Approval layer.
- Secure Analytics/Visibility and Control: This safe examination layer characterizes the administrations by which all components (endpoints and organization framework, inclusive of server farms) may partake in giving telemetry to the reason for picking up perceivability and ultimately controlling the IoT/M2M biological system. We can convey a huge equal information base (MPP) stage that can cycle huge volumes of information in close to continuous with the development of massive information frameworks. When we combine this innovation with examination, we can conduct a genuine factual investigation on the security data to identify anomalies. Furthermore, it includes all components that total and correlate data, including telemetry, to provide observation and danger identification. Danger alleviation can range from preventing the assailant from accessing additional assets to running specific content to begin legitimate remediation. The data generated by IoT devices is only useful if the privilege investigation calculations or other security knowledge measures are labeled to identify the threat. We can enhance scientific results by collecting information from multiple sources and applying security profiles and measurable models based on various layers of security calculations.

We are all aware that network foundations are becoming increasingly complex. Consider geographies with both public and private mists; the threat knowledge and protection capacities should also be cloud-based. To achieve precise insight, coordination of perceivability, setting, and control is required. This layer's components include the following:

- The genuine IoT/M2M foundation from which telemetry and observation information is obtained and accumulated. The center arrangement of capacities to blend, examine the information for the purposes of giving perceivability, and offer relevant mindfulness and control. The conveyance stage for the genuine analysis operated from the first two segments, discussed above.

- While real-world IoT/M2M implementations may be unique, their structure could be applied to almost any engineering. The framework is simple and adaptable enough to support controlled gadgets (e.g., PCs, mobile scanners, etc.) if they live in the IoT base.

5 Classification of Various Analytic Techniques for Smart IoT Devices

The quantity of gadgets associated with the Web is ballooning, introducing the time of the "Internet of Things" (IoT). IoT alludes to the huge number of minimal effort gadgets that speak with one another and with far-off workers on the Web independently. It includes ordinary articles, for example, lights, cameras, movement sensors, entryway locks, thermostats, power switches, and family machines, with shipments extended to arrive at almost 20 billion by 2020. A great many IoT gadgets are required to discover their way in homes, ventures, grounds, and urban communities of the not-so-distant future inciting "shrewd" conditions profiting our general public and our lives.

The multiplication of IoT, in any case, makes a significant issue. Administrators of shrewd conditions can think that it is hard to figure out what IoT gadgets are associated with their organization and further to determine whether every gadget is working ordinarily. This is for the most part ascribed to the undertaking of overseeing resources in an association, which is commonly circulated across various offices. For instance, in a neighborhood board, lighting sensors might be introduced by the office's group, sewage and trash sensors by the disinfection office, and observation cameras by the nearby police division.

Organizing across different divisions to get a stock of IoT resources is tedious, burdensome and blunder inclined, making it almost difficult to know absolutely what IoT gadgets are working on the organization anytime. Acquiring "perceivability" into IoT gadgets in an opportune way is of fundamental significance to the administrator, who is entrusted with guaranteeing that gadgets are in suitable organization security fragments, are provisioned for imperative nature of administration, and can be isolated quickly when penetrated. The significance of perceivability is accentuated in Cisco's latest IoT security report, and further featured by two late occasions: sensors of a fish tank that undermined a gambling club in Jul 2017, and assaults on a college grounds network from its own candy machines in Feb 2017. In the two cases, network division might have possibly forestalled the assault and better perceivability would have permitted quick isolating to restrict the harm of the digital assault on the endeavor organization.

One would expect that gadgets can be distinguished by their Macintosh address and DHCP exchange. Notwithstanding, this faces a few difficulties: (a) IoT gadget makers ordinarily use NICs provided by outsider merchants, and consequently the

Authoritatively Extraordinary Identifier (OUI) prefix of the Macintosh address may not pass on any data about the IoT gadget; (b) Macintosh locations can be mock by vindictive de-indecencies; (c) numerous IoT gadgets do not set the Host Name choice in their DHCP demands; in fact, we found that about a large portion of the IoT gadgets we considered do not uncover their hostnames, as appeared in [15, 16, 19]; (d) in any event, when the IoT gadget uncovered its hostname it may not generally be important (for example WBP-EE4C for Withing's infant screen); and finally (e) these hostnames can be changed by the client (for example the HP printer can be given a self-assertive hostname). Hence, depending on the DHCP foundation is anything but a feasible answer for accurately recognizing gadgets at scale.

5.1 Descriptive Analysis for IoT

The illustrative investigation is the most fundamental type of scientific knowledge that permits clients to portray and total approach IoT information. Illustrative investigation—even computations as straightforward as a mean and standard deviation—can be utilized to rapidly sort out gathered information. In an associated plant use case, portrayal examination may be utilized to respond to the inquiry, "What are the normal siphon temperature, stream rate, and RPM throughout a 30-min time span?" When distinguishing top tier clear investigation abilities on an IoT stage, undertakings ought to assess:

- On-stage clear examination capacities: The capacity of a stage to perform expressive insightful requests, for example, amassing or figuring essential measurements of ingested information focuses across sensors, gadgets, or gatherings of gadgets just as outwardly introducing the outcomes.
- On-stage information lake/huge information stockpiling capacities: The capacity of the stage to both store and inquiry against huge amounts of ingested IoT information including table-based information stores with more noteworthy than 10 million lines or unstructured information stores with more prominent than 50 million records.

5.2 Predictive Analysis for IoT

Prescient examination looks to demonstrate future information and practices by breaking down recorded information [20]. Relapse examination, for example, direct relapse is an illustration of prescient investigation. In a similar use case, prescient examination may be utilized to respond to the inquiry, "What is the assessed time-to-disappointment for a siphon that is showing a 20% expansion in estimated temperature?" When recognizing top tier prescient examination abilities on an IoT stage, endeavors ought to assess:

- On-stage prescient scientific model structure: The capacity of the stage to conse-quently or through automatic interfaces create a prescient model of the funda-mental stage ingested IoT information. Models, for example, direct or polynomial relapses are average, albeit more intricate displaying decisions are accessible in refined stages.
- On-stage prescient logical model activity: The capacity of the stage to use either a stage created or stage coordinated information model (for example, R or Python) to group information or recognize exceptions through abnormality identification. Clients should put accentuation on the capacity to oversee models, for example, model forming and refreshing just as the capacity to incorporate a prescient model inside an unpredictable occasion preparing (CEP) structure [21].

5.3 Prescriptive Analysis for IoT

Prescriptive examinations are investigations to assist ventures with advancing a future course to be taken. Picture handling, AI, and characteristic language preparing are a portion of the strategies used to finish prescriptive investigation. A prescriptive investigation may be utilized to respond to the inquiry, "To boost siphon uptime and limit administration spans, what is the most extreme permitted temperature increment for a siphon before a protection siphon adjusting must be booked?" When recognizing top tier prescriptive investigation capacities on an IoT stage, endeavors ought to assess

- On-stage prescriptive insightful model capacities: The capacity of the stage to use either a stage produced or stage coordinated information model, for example, R or Python, to upgrade a business result or applicable KPI. A prescriptive model ought to augment or limit a business-important KPI, for example, an ideal opportunity to-conveyance in course arranging or hardware uptime for prescient upkeep.

6 Proposed System: IoT-Based Lightweight Cancellable Biometric System

Figure 2 shows a unique mark-based verification plot in a cloud climate. In such a setting, the concentrated information base is a prominent objective because of the capacity of biometric formats. Since the quantity of each biometric methodology is restricted, the outcomes of any trade-off will be extensive (e.g., one's biometrics, for example, iris cannot be supplanted). Consequently, premium in cancellable biomet-rics is created. The Cancellable Biometric Framework (CBS) by and large includes a sign or highlight level change of the first biometric highlight dependent on some client explicit key [22]. As such, instead of the first/crude biometric features, the changed layout is put away in a concentrated information base. Consequently, if the incorpo-rated information base is compromised, the CBS can essentially drop the undermined format and conjure another changed layout from the first biometric highlights with

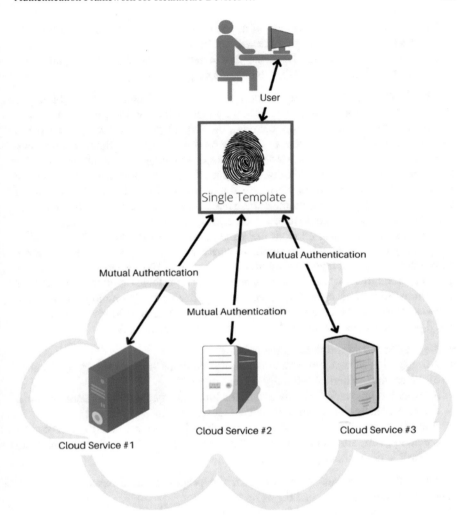

Fig. 2 Proposed biometric system

the end goal that the undermined format and the new layout are completely extraordinary. The client can likewise try out various administrations (e.g., internet banking, web-based business, and e-government) utilizing diverse cancellable formats created out of a solitary biometric methodology. Key advantages of the CBS over a traditional (or 'plain vanilla') biometric framework are as per the following:

- Cancellable biometric formats hold a single direction property since the cancellable changes are equal to cryptographic salting.
- A single biometric source may be used to generate a plethora of different formats. Each layout is unrelated to the others.

- The client could enlist a variety of applications in a variety of various formats. The first/crude biometric example will not be uncovered.
- Revocation is useful in the sense that a newly created model replaces the old/bargained interface (similar to changing a password).

The focal point of this examination is on the advancement of a lightweight, cloud-based cancellable biometric system, which is intended to give secure client verification to various IoT applications [19, 23]. Figure 2 shows our proposed cancellable biometric-based client validation conspire in a cloud climate com-prising both customer hubs (e.g., portable processing or IoT gadgets) and cloud server(s). Every client (in charge of at least one customer hub) has various cancellable unique mark layouts produced from a solitary finger impression picture. Clients can be confirmed utilizing the cancellable unique mark format, and each cancellable unique mark layout is one of a kind and does not correspond with different formats despite the fact that they are created from the same biometric methodology. Likewise, we re-appropriated exercises, for example, picture pre-processing, including extraction, highlight change, and layout coordinating to the cloud [20]. Such exercises can be gotten to through a UI, which can be a program or a portable (application). Both the cancellable biometric information base and applicable programming modules are likewise moved to the cloud. Such an arrangement permits us to give constant and equal preparation and understand the accompanying advantages:

- Negligible framework arrangement time.
- Arrangement of momentary and on-request administration alongside the chance of the expansion and additionally erasure of parts.
- Reasonable innovation, in any event, for little and medium-sized organizations, because of insignificant arrangement expenses and time.
- Exceptionally adaptable, since the cancellable biometric information base can be scaled to fit any sort of utilization, for example, 1: 1 or 1: N check situations, with accessible asset pooling.

The following are the components of the proposed cloud-based, lightweight cancellable biometric framework:

- Lightweight cancellable layout age in the cloud climate, and
- Lightweight cancellable layout coordinating in the cloud climate.

The vital commitments of this paper are as per the following:

- In a cloud climate, the proposed cloud-based lightweight cancellable biometric confirmation framework can successfully distinguish a person;
- The proposed cloud-based lightweight cancellable biometric confirmation frame-work may also likewise adequately understand the cancellable biometric framework-based verification administrations in different cloud-based IoT appli-cations; and
- When conveyed in confirmation administrations, the proposed framework consumes less time; henceforth, it is known as the lightweight cancellable biometric framework.

Many academics have been working on creating biometric-based techniques for authentication processes in IoT networks because of the advantages like uniqueness of biometrics over password- and token-based traditional authentication. A framework may be built utilizing biometrics and wireless device radio fingerprinting for user authentication, which not only verifies that the observed healthy data is from the proper patient but also guarantees the data's security (Fig. 3).

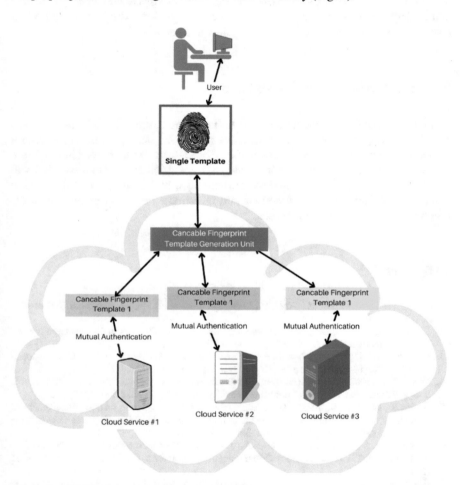

Fig. 3 Cancellable biometric-based user authentication

7 Conclusion

Given the growing trend as well as variety of IoT devices in our general public, the need to validate such devices will become more articulated. We introduced a cloud-based, lightweight cancellable biometric authentication system for IoT applications in this paper. The use of cancellable biometric layouts addresses the security concerns that underpin the use of biometrics for confirmation. We exhibited our proposed framework's lightweight trademark, precision, and security.

8 Future Works and Limitations

In the process of developing a biometric authentication system for IoT applications, use of lightweight adaption has some other privacy-related flaws that can make it sometimes unsuitable for use in biometrics. These flaws can be solved in the future by employing one-way functions in cryptography with biometrics to bring clarity to the above lightweight method and its shortcomings. In the future, we will examine other sorts of transformation functions and investigate how to effectively hide the secret key in different circumstances.

References

1. Peer, P., Bule, J., Gros, J.Z., Struc, V.: Building cloud-based biometric services. Informatica **37**(2), 115 (2013)
2. Yuan, J., Yu, S.: Efficient privacy-preserving biometric identification in cloud computing. In: 2013 Proceedings IEEE INFOCOM, pp. 2652–2660 (2013)
3. Haghighat, M., Zonouz, S., Abdel-Mottaleb, M.: CloudID: trustworthy cloud-based and cross-enterprise biometric identification. Expert Syst. Appl. **42**(21), 7905–7916 (2015)
4. Das, A.K., Goswami, A.: A secure and efficient uniqueness-and-anonymity-preserving remote user authentication scheme for connected health care. J. Med. Syst. **37**(3), 9948 (2013)
5. Amin, R., Islam, S.H., Biswas, G.P., Khan, M.K., Li, X.: Cryptanalysis and enhancement of anonymity preserving remote user mutual authentication and session key agreement scheme for e-health care systems. J. Med. Syst. **39**(11), 140 (2015)
6. Bommagani, A.S., Valenti, M.C., Ross, A.: A framework for secure cloud-empowered mobile biometrics. In: 2014 IEEE Military Communications Conference (MILCOM'14), pp. 255–261 (2014)
7. Bingham, E., Mannila, H.: Random projection in dimensionality reduction: applications to image and text data. In Proceedings of the Seventh ACM SIGKDD International Conference on Knowledge Discovery and Data Mining, pp. 245–250 (2001)
8. Wu, Z., Tian, L., Li, P., Wu, T., Jiang, M., Wu, C.: Generating stable biometric keys for flexible cloud computing authentication using finger vein. Inf. Sci. 433–434
9. Hu, P., Ning, H., Qiu, T., Xu, Y., Luo, X., Sangaiah, A.K.: A unified face identification and resolution scheme using cloud computing in Internet of Things. Future Gener. Comput. Syst. **81**, 582–592 (2018)
10. Adam, E.E.B.: Survey on medical imaging of electrical impedance tomography (EIT) by variable current pattern methods. J. ISMAC **3**(02), 82–95 (2021)

11. Tyagi, A.K., Chahal, P.: Artificial intelligence and machine learning algorithms. In: Challenges and Applications for Implementing Machine Learning in Computer Vision. IGI Global (2020). https://doi.org/10.4018/978-1-7998-0182-5.ch008
12. Karimian, N., Wortman, P.A., Tehranipoor, F.: Evolving authentication design considerations for the internet of biometric things (IoBT). In: Proceedings of the Eleventh IEEE/ACM/IFIP International Conference on Hardware/Software Codesign and System Synthesis, Pittsburgh, PA, 1–7 Oct 2016, p. 10
13. Kantarci, B., Erol-Kantarci, M., Schuckers, S.: Towards secure cloud-centric internet of biometric things. In: Proceedings of the 2015 IEEE 4th International Conference on Cloud Networking (CloudNet), Niagara Falls, ON, 5–7 Oct 2015, pp. 81–83
14. Dhillon, P.K., Kalra, S.: A lightweight biometrics based remote user authentication scheme for IoT services. J. Inf. Secur. Appl. **34**, 255–270 (2017)
15. Ratha, N.K., Connell, J.H., Bolle, R.M.: Enhancing security and privacy in biometrics-based authentication systems. IBM Syst. J. **40**, 614–634 (2001)
16. Tyagi, A.K., Nair, M.M., Niladhuri, S., Abraham, A.: Security, privacy research issues in various computing platforms: a survey and the road ahead. J. Inf. Assur. Secur. **15**(1), 1–16 (2020)
17. Pramod, A., Naicker, H.S., Tyagi, A.K.: Machine learning and deep learning: open issues and future research directions for next ten years. In: Computational Analysis and Understanding of Deep Learning for Medical Care: Principles, Methods, and Applications. Wiley Scrivener (2020)
18. Tyagi, A.K., Rekha, G.: Challenges of applying deep learning in real-world applications. In: Challenges and Applications for Implementing Machine Learning in Computer Vision, pp. 92–118. IGI Global (2020). https://doi.org/10.4018/978-1-7998-0182-5.ch004
19. Tyagi, A.K., Gupta, M., Aswathy, S.U., Ved, C.: Healthcare solutions for smart era: an useful explanation from user's perspective. In: Recent Trends in Blockchain for Information Systems Security and Privacy. CRC Press (2021)
20. Gudeti, B., Mishra, S., Malik, S., Fernandez, T.F., Tyagi, A.K., Kumari, S.: A novel approach to predict chronic kidney disease using machine learning algorithms. In: 2020 4th International Conference on Electronics, Communication and Aerospace Technology (ICECA), Coimbatore, pp. 1630–1635 (2020). https://doi.org/10.1109/ICECA49313.2020.9297392
21. Smys, S., Haoxiang, W.: Data elimination on repetition using a blockchain based cyber threat intelligence. IRO J. Sustain. Wireless Syst. **2**(4), 149–154 (2021)
22. Tyagi, A.K., Aswathy, S.U., Aghila, G., Sreenath, N.: AARIN: affordable, accurate, reliable and innovative mechanism to protect a medical cyber-physical system using blockchain technology. Int. J. Intell. Netw. **2**, 175–183 (2021)
23. Shamila, M., Vinuthna, K., Tyagi, A.: A review on several critical issues and challenges in IoT based e-healthcare system, pp. 1036–1043 (2019). https://doi.org/10.1109/ICCS45141.2019.9065831

Task Prioritization of Fog Computing Model in Healthcare Systems

Prakriti Pahari and Subarna Shakya

Abstract Health-related applications are one of the most sensitive area which should be delivered on time efficiently. For the storage and processing of enormous health data, Cloud Computing could not be efficient as Cloud Data Centers take large time to process and send back the results. The new paradigm, called Fog Computing is applicable in cases like these. In this research, we utilize the sample time critical healthcare system where the IoT sensors' data is divided into critical and normal tasks where critical tasks are prioritized than normal patients' data. To address their management, Fog Computing is used at the edge of the network. In this paper, a new fog-cloud based algorithm called Prioritized Latency Aware Energy Efficient Algorithm (PLAEE) is developed by utilizing the existing algorithms in fog system and also by process optimization of the core evaluation metrics, latency and energy usage. This algorithm shows superiority to the existing algorithms in terms of performance metrics. The experimentation is performed using Blood Pressure data collected from University of Piraeus. In terms of response time, the PLAEE is performing 45.90%, 21.95%, and 20.25% better than Cloud only, Edge-wards and Mapping Algorithm respectively. In terms of Energy Consumption, the PLAEE is performing 30.26%, 22.86%, 15.44% percentage better than Cloud only, Edge-wards and Mapping Algorithm respectively. Almost 98% of critical data are placed in the FNs according to the Tasks Managed value calculated.

Keywords Cloud computing · Internet of Things · Fog computing · Latency · Energy consumption

1 Introduction

Healthcare is considered as an important aspect of life. To improve these lives of people, many healthcare systems are built with the aid of technology. Healthcare applications are the most crucial elements because of the fact that they directly affect

P. Pahari (✉) · S. Shakya
Department of Electronics and Computer Engineering, Pulchowk Campus, Lalitpur, Nepal
e-mail: drss@ioe.edu.np

© The Author(s), under exclusive license to Springer Nature Singapore Pte Ltd. 2022   417
V. Suma et al. (eds.), *Evolutionary Computing and Mobile Sustainable Networks*,
Lecture Notes on Data Engineering and Communications Technologies 116,
https://doi.org/10.1007/978-981-16-9605-3_28

the lives of patients. The one innovative model that is used in healthcare systems these days, is fog computing. Fog computing is a decentralized platform that serves at the network's edge, providing low latency, execution speed, and energy consumption, all of which are essential in healthcare applications. Because of its reduced latency and, in some situations, the surplus power of available devices, fog computing is seen to be much cost-effective than centralized Cloud paradigm. Real-time applications,i.e. healthcare that are susceptible to delays will experience difficulties such as excessive round-trip delay and network congestion in this case. Its goal is to eliminate the need to transfer produced data to the cloud by processing it immediately at the device, or next to it, at the network boundary, in more capable equipment. CISCO coined the phrase "fog computing," which is described as an architecture that extends the cloud's processing and storage capabilities to the network's edge [7]. As a result, data may be gathered and processed locally, decreasing response time and bandwidth use. In fog computing, service placement is an optimization problem. Based on this optimization challenge, services should be more efficiently placed on computing resources close to the end-user. The Service Placement Problem (SPP) [14] is how this issue is referred to. Transferring application services from one fog device to another at a high or similar level may be required for service placement. Finding an optimized path for assigning n tasks to m processors is the aim of scheduling. Healthcare implementation tasks are likely to be produced at normal as well as critical categories which should be sent for processing as soon as possible. The management of resources is applied using the prioritization of data to FNs combined with the algorithms to place more critical data to nearer fog nodes and sharing of resources. This research aims to fulfill the following purposes:

1. To prioritize the critical and normal tasks efficiently, and to place those tasks to the optimal Fog Nodes.
2. To compare the Task prioritized fog-based cloud architecture using the proposed Algorithm with baseline algorithms of the fog-cloud infrastructures.

2 Literature Review

In case of time-critical applications, fog computing shows great efficiency because of less network hops for processing. However, the researches in this area are still at infancy stage. Mahmoud, M presented the fog-based IoT system and compares it with the traditional Cloud-only policies. This model proves to be more energy efficient than the traditional cloud datacenters. It proposes energy-aware placement strategy to focus more on the energy consumptions metrics rather than response time and network usage [12].

In the Ref. [9], the authors has proposed the new IMDS scheme which is based on Random Forest Algorithm for multimedia data segregation in fog model in case of e-healthcare. This paper focuses on the latency of computation, network and trans- mission and has achieved QoS metrics higher than the previous algorithms. However,

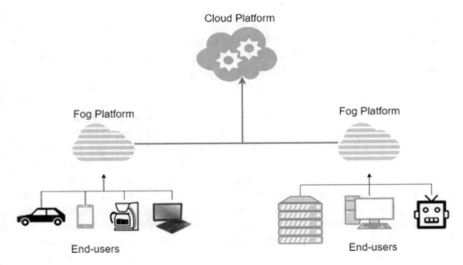

Fig. 1 Fog computing system

other performance parameters are not studied in this case. Many scholars have concentrated exclusively on resource management for critical task implementations, such as the concept given by Kumar M. who grouped all concepts relevant to scheduling and provisioning of resources under resource management [10]. However, the most important issue, according to the authors in Gupta and Buyya [6] is the design of resource management techniques to schedule which analytics framework modules are deployed to each edge system in order to minimize latency and maximize efficiency. They concentrated on the tasks, albeit from a resource standpoint. The words load balancing, prioritization, and resource availability each contribute to improve task scheduling efficiency. Aldegheishem A. and Bukhsh R. proposed a system in which multiple agents collaborate between various elements of the energy system in smart cities and a Fog based model was developed by utilizing scheduling algorithms in home appliances to minimize the cost of global power and computation, as well as to address the issue of stored energy exceeding potential demand, which undermines the model [3]. Cristescu and Dobrescu implemented a system to boost edge computer processing capabilities by sharing tasks across a group of emerging edge computing dimension for adaptive distributed networks [4]. Auluck and Singh [16] developed a scheduling strategy called Security Aware scheduling in near real-time in Fog network. Mehta and Kavitha, discovered that the priority assignment of tasks based on criticality and significance of task before scheduling produces the better results compared to various schedulers. The tasks are ranked after the calculation of priority for each task based on the agent's condition and their response [13]. Aladwani, proposed a scheduling strategy combined with prioritization of tasks based on their deadlines along with MAX-MIN scheduling algorithm. The tasks are then sent to VMs based on the resource capacity of each VM. However, the tasks are sent to random fog nodes even if they are at longer distance from the edge device [2]. Khosroabadi, developed a fog orchestrator or manager that consists of a resource manager, planner, and status monitor. On top of this, the orchestrator interacts with a cloud data-center and other fog nodes in charge of locally storing and/or computing data of the IoT [8]. This helps in better scalability of the fog-based cloud architecture. In Ref. [1], the authors have proposed a fog to fog collaboration model in which offloading of the service requests is carried out among the fog nodes in same level which promotes the use of clustered fog architecture. There are plenty of implementations done in healthcare time-critical applications using fog computing. Few implementations also consider the priority of tasks based on their deadlines, which can be called deadline agnostic approaches. But there is no implementation of any other new methods which consider prioritization based on criticality along with process optimization of latency and energy usage to maximize the number of services handled within the fog layer. This work utilizes the concept of priority, resource management, and process optimization to adjust QoS metrics all at a single time for healthcare-related applications using the new paradigm of clustered fog-based cloud architecture.

3   Methodology

This research is conducted by utilizing an efficient task management model where the Proposed Prioritized Latency Aware Energy Efficient Algorithm (PLAEE) in fog based system is implemented to assist healthcare emergencies.

3.1   Data Collection

The dataset used in this research is provided by Department of Digital Systems, University of Piraeus. It utilized the IoT consumer grade devices that collected blood pressure data. This data was collected over a period of 61 d, using iHealth Track, Withings BPM, iHealth View and iHealth Clear. These records contain 26560 datas of both the Systolic as well as Diastolic Blood pressure value, SBP and DBP respectively. This data is used for critical cases calculations. Table 1 shows the description of the dataset.

3.2   Prioritization of Data

In this research, the priority of the task would be decided by task's severity and relevance. At all times, the critical patients must be addressed before the normal patients. The group of sensors are classified based on all incoming tasks from the

Table 1 Dataset description

Attributes	Description
Date	Format: yyyy/mm/dd
Time	Format: hh/mm/ss
DBP value	Range constraints
DBP unit	mmHg
SBP value	Range constraints
SBP unit	mmHg
Device-id	Unique identifier

Table 2 Fog nodes' characteristics

F_j^{cpu},MIPS	F_j^{mem}, MB	F_j^{Pmax},W	F_j^{Pmin},W
300	256	88.77	82.7
1400	2048	103	83.25
1600	1024	87.53	82.44
3000	4096	107.339	83.433
C_l^{cpu},MIPS	C_l^{mem}, MB	C_l^{Pmax},W	C_l^{Pmin},W
10,000	10,240	412	333

sensors on the basis of their criticality. In case of blood pressure monitoring, it is to be understood is that this value of blood pressure is not enough for clinical diagnosis, but helps to distinguish emergency conditions for immediate further treatment. The priority ranges are modelled as *Critical = High Priority* and *Normal = Low Priority*. Table 2 shows the categorization of different critical and non-critical levels of Blood Pressure.

As considered by American Heart Association, the value greater than 180/120 is considered as Hypertensive Emergency which needs immediate further diagnosis and the value greater than 200/140 as Hypertensive Urgency as the most Critical value since it leads to Target Organ Damage in case of patients of any age group [15].

3.3 System Architecture

The model consists of three levels: Cloud layer, Fog Layer and Client Layer. A Cloud Layer is made up of a group of nodes housed in a single Cloud Data Center (CDC). The Fog Layer considers two distinct Fog Zones: the Main Fog Zone (FZ1) and the Neighbor Fog Zone (FZ2). The Client Layer is made up of a variety of devices or endpoints that are geographically spread across the environment. Each IoT device in this architecture is linked to a fog gateway (FG) as shown in Fig. 3.

Algorithm 1 Task Prioritization Algorithm

1: *if Patients' Data P = Critical then do,*
2:        *Push the data into stack S in decreasing order*
3:        *while the stack S is not empty do,*
4:            *if top(S) is not the highest critical then*
5:                *S ← the highest critical data*
6:            *else,*
7:                *P ← top(S)*
8:                *Pop top(S)*
9:            *endif*
10:       *endwhile*
11: *else,*
12:       *Push the Normal data into the other stack in decreasing order.*
13: *endif*

Input: Systolic Blood Pressure (mmHg)	Input: Diastolic Blood Pressure (mmHg)	Output: Blood Pressure Category
140	90	High
180	120	Hypertension (Critical)
40	60	Low
90	80	Normal
117	80	Normal
142	90	High
30	10	Hypotension (Critical)

Fig. 2 Categorization of tasks

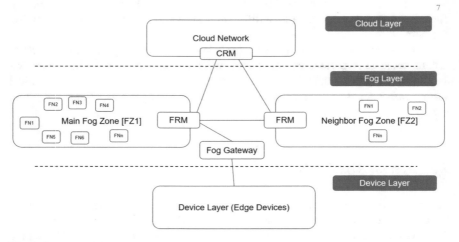

Fig. 3 System architecture

The terms requests, tasks and services are used interchangeably throughout this paper. Depending on its computational capability, a Fog Node (FN) inside a fog zone (FZ) can host a number of virtual machines or container instances, which is controlled by a fog resource manager (FRM) [17]. FRM handles a process of placing a task based on data from edge devices over a set period of time. A data request can be sent to the major Fog Zone (FZ1), a neighboring Fog Zone (FZ2), or a Cloud Data Center in this operation. Cloud Resource Manager (CRM) handles the cloud servers in this scenario. Assume that a sensor represents a patient, and $P = \{p1, p2, \ldots, pn\}$ represents a collection of patients' data supplied by IoT devices to Fog Layer via Fog Gateway (FG). Every task is represented as P_i with their own CPU P_i^{cpu}, storage P_i^{mem} and size P_i^{size} measured in MIPS, MB and MI respectively. It is supposed $F^m = \{F_1, F_2, \ldots, F_{|F^m|}\}$ and $F^n = \{F_1, F_2, \ldots, F_{|F^n|}\}$ be a set of $|F^m|$ and $|F^n|$ fog nodes respectively in FZ1 and FZ2. For every $j \in \{1, 2, \ldots, |F^a| \cup |F^g|\}$, F_j^{cpu} and F_j^{mem} are utilized to indicate the CPU capacity and storage of the jth Fog Node, respectively. F_j^{Pmax} and F_j^{Pmin} are notations that represent the upper and lower value of power consumption of fog nodes $F_j, \forall_j \in \{1, 2, \ldots, |F^a| \cup |F^g|\}$. For any $(s, d) \in E.$, δ_{sd} is used to indicate delay between devices within a Fog Zones. Lastly, for the propagation delay between FRM of FZ1 and FZ2, the symbol $\delta_{M,N}$ and for the delay between FRM of FZ1 and CRM, the symbol $\delta_{M,C}$ is utilized. The set of $|C|$ servers designated inside a CDC as $C = \{C_1, C_2, \ldots, C_{|C|}\}$, where C_l^{cpu} and C_l^{mem} represent the CPU capacity and storage of C_l, respectively, for every $l \in \{1, 2, \ldots, |C|\}$. C_l^{Pmax} and C_l^{Pmin} are the power consumptions of $C_l, \forall_l \in \{1, 2, \ldots, |C|\}$. We use the homogeneous cloud servers are used for simplicity. In this model, the FRM chooses the best location for received services, whether it's on the FZ1, FZ2 or CDC. In the optimization model, the three variables are used that decides the placement approach that may be specified as follows:

$$a_{ij}, b_{ik}, c_{il} = \begin{cases} 1 & \text{if service } P_i \text{ is placed within FZ1, FZ2 and CDC,} \\ 0 & \text{otherwise,} \end{cases} \tag{1}$$

The response time of a task P_i is estimated as:

$$L_i = (2\delta_{G_i,M} + 2\delta_{M,F_j} + t_{i,j}).a_{ij} + (2\delta_{G_i,M} + \delta_{M,N} + 2\delta_{N,F_k} + t_{i,k}).b_{ik} + (2\delta_{G_i,M} + \delta_{M,C} + 2\delta_{C,C_l} + t_{i,l}).c_{il}, \tag{2}$$

The total response time of the model for all the data is: $L^{tot} = \sum_{i=1}^{n} L_i$

The power consumption of every FN must first be evaluated. The power model adopted from [11] [5] is explained as power to be a linear model of CPU usage. The model for FNs is given here. This equation is used for calculating the power usage in Cloud servers with this model as well.

$$P_j = \begin{cases} F_j^{Pmin} + (F_j^{Pmax} - F_j^{Pmin}) \times z_j, & if z_j > 0, \\ 0 & \text{otherwise,} \end{cases} \tag{3}$$

where z_j is the CPU usage of FN F_j.

The energy usage of Fog Node F_j throughout the processing of given tasks is calculated as follows:

$$E_j = P_j * \sum_{i=1}^{n} a_{ij} \cdot \frac{P_i^{size}}{P_i^{cpu}}, \forall_j \in \{1, 2, \ldots, |F^m| \cup |F^n|\} \tag{4}$$

3.4 Proposed Algorithm

In the proposed hybrid algorithm, we use the concept of process optimization where the Critical Services P^C tries to reduce the latency and Normal Services P^N tries to reduce the energy usage. The program configures this based on the application (i.e., blood pressure range in this case). Two methods are integrated for solving the problem at the same time to receive both improved QoS as well as reduced usage of energy in case of Fog Service Providers. The integration of Proposed Algorithms is implemented as given in the flowchart in Fig. 4

Latency Aware: For Critical Services The Critical Services (C) are fetched based on decreasing order of criticality. The FRM captures the FZ's network traffic before deploying Critical services. The algorithm then loads the list of emergency services that are sorted onto FNs that have a shorter latency with the FRM, i.e., are closer to the edge using the propagation delay. This algorithm takes into consideration the propagation delay from FRM and chooses the least one.

Energy Efficient: For Normal Services: This aims to use as less energy as possible by putting normal services on FNs, which have the smallest impact on overall FZ's energy usage. The normal services are prioritized first in this algorithm that means for non-emergency cases of patients, the energy of FSPs is intended to reduce by using energy efficient FNs. A normal service with a normal blood pressure range receives high priority in this case. As a result, such a service will have a better probability of running in a fog environment. If there are many FNs for one normal service with equal energy requirement, this system chooses the one with the least network latency from an energy efficiency standpoint.

3.5 Scenario Description

Compared Algorithms

1. Only-Cloud [6]
2. Edge-ward [6]
3. Mapping Algorithm: The FNs are sorted in increasing order depending on the value of resource capacity and need. Then the algorithm thoroughly searches for the FNs inside FZ1 until it finds a Node with sufficient capacity for each service.

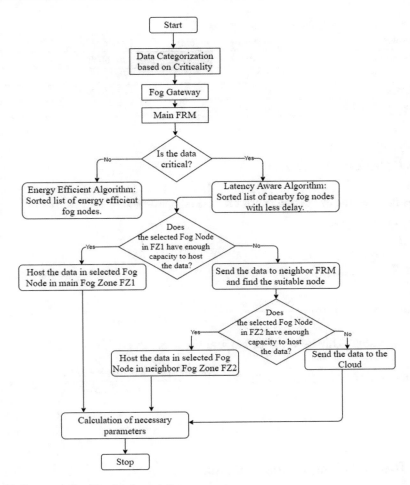

Fig. 4 Proposed algorithm implementation

In this algorithm, it is anticipated if no suitable FN is found in FZ1, it will ask the Neighbor FZ2 to host the task via its FRM.

Simulation settings The simulation scenario is created using Java and iFogSim using Eclipse IDE to test the proposed policy's performance. The two FZs are simulated, each with a distinct amount of FNs, along with a datacenter with homogeneous physical machines i.e., Cloud servers. The four distinct FNs and Cloud Servers are analyzed whose specifications are adapted from [6] shown in Table 2. For each experiment, an average of 10 independent runs was conducted.

Table 3 Table showing the QoS metrics results

Methods	Response Time (ms)	Energy Usage (kWh)
Only Cloud	98179.54	76.39446
Edgewards	68046.64	69.0697
Mapping	66597.71	63.00724
Proposed PLAEE	53106.58	53.27510

3.6 Evaluation Metrics

This includes the evaluation metrics for the comparison.

A. QoS metrics:
i. Latency
ii. Energy Consumption
B. Tasks Managed:

$$Tasks\ Managed\ (\%) = \frac{Number\ of\ Critical\ Tasks\ in\ (FZ1/FZ2/CDC)}{Total\ Critical\ Tasks} * 100 \quad (5)$$

C. Percentage Utilization of Cloud and Fog Resources:

$$Utilization\ of\ (FZ1/FZ2/CDC)(\%) = \frac{Number\ of\ Tasks\ in\ FZ1/FZ2/CDC}{Total\ Tasks} * 100 \quad (6)$$

4 Results, Comparison and Analysis

While considering total dataset and 800 nodes in FZ1 and 400 nodes in FZ2, the result obtained is tabularized below.

In the instance of Energy Consumption, for 26560 services, this method improves by 15.44, 22.86, and 30.26 percent when compared with Mapping, Edge-ward, and Only-cloud policies, respectively. When comparing with only-cloud approach, the proposed strategy decreases average response time by 45.90 percent, whilst Mapping and Edgewards reduces this response time by 20.25 percent and 21.96 percent, respectively. It shows that the proposed algorithm reduces the % utilization of cloud by 58.75 percent whereas Mapping reduces this utilization by 49.3%, by placing the services on the fog colonies which are nearer to the edge and also consumes less energy compared to the major centralized cloud computing paradigm.

The % of Critical Tasks Managed by this Algorithm is 91.70, 6.28 and 2.01 % in FZ1, FZ2 and CDC respectively. The Fog Layer handles almost 98% of the critical tasks

Table 4 Table showing percentage utilization results

Methods	Main Fog Zone	Neighbor Fog Zone	Cloud Server
Only Cloud	N/A	N/A	100
Edgewards	31.18	N/A	68.82
Mapping	33.34	16.56	50.10
Proposed PLAEE	34.96	23.80	41.24

Table 5 Tasks Managed

System Environment	Critical Data	Normal Data
Cloud	2.01%	70.26%
Main Fog Zone	91.7%	4.3%
Neighbor Fog Zone	6.28%	25.43%

compared to the existing algorithms which send both the critical and normal tasks to Cloud for execution. Various experiments are conducted to understand the nature of the above results and how the proposed policy behaves by varying the parameters.

4.1 Effect of Varying Patients' Data

We change the amount of data from 1000 to 26560 and keeping the nodes within the FZ1 constant at 800. The results are seen in Fig. 5, 6 and 7.

4.2 Effect of Varying Fog Nodes in Main Fog Zone (FZ1)

The number of nodes inside FZ1 of Fog Layer is varied from 100 to 800 in this experiment. The results are seen in Fig. 8 and 9.

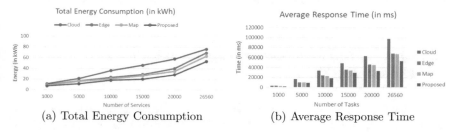

(a) Total Energy Consumption   (b) Average Response Time

Fig. 5 QoS metrics for Experiment 1

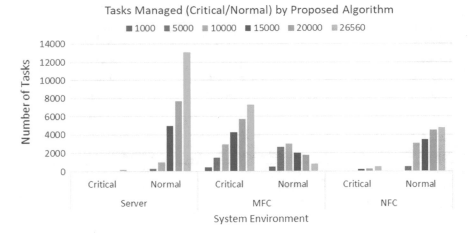

Fig. 6 Tasks Managed by Proposed Algorithm in Experiment 1

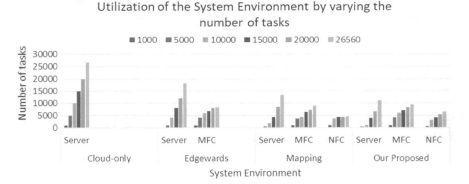

Fig. 7 Utilization of System Environment in Experiment 1

It is observed that when the number of nodes is growing from 100 to 1600, more than 95 percent of critical services are handled by fog environment. And very few are sent to the cloud which reduces the energy consumption and response time.

4.3 *Effect of Different Ratio of Fog Nodes in Neighbor to Main Fog Zone*

The nodes inside FZ1 is set to 800, while varying the ratio of nodes within the FZ2 from 0.2 up to 1. Changing the FNs in FZ2 has no effect on the Only-cloud and

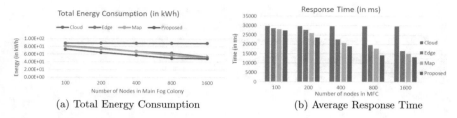

(a) Total Energy Consumption

(b) Average Response Time

Fig. 8 QoS metrics for Experiment 2

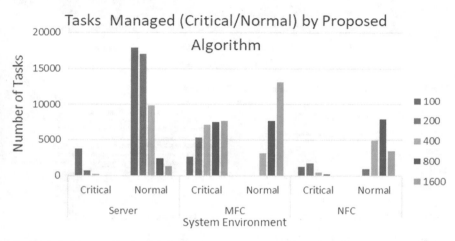

Fig. 9 Tasks Managed by Proposed Algorithm in Experiment 2

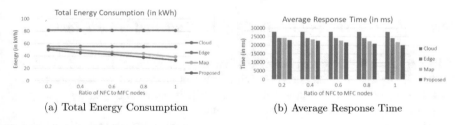

(a) Total Energy Consumption

(b) Average Response Time

Fig. 10 QoS metrics for Experiment 3

Edgewards algorithms because they disregard the Neighbor Fog Zone. However, this has a significant impact on the performance of the Proposed Algorithm and Mapping Algorithm.

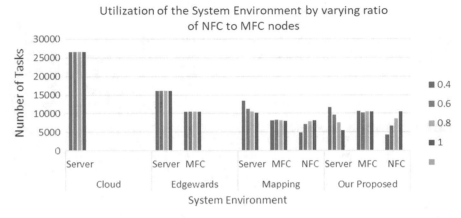

Fig. 11 Utilization of System Environment in Experiment 3

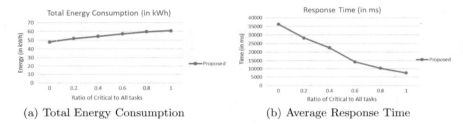

(a) Total Energy Consumption   (b) Average Response Time

Fig. 12 QoS metrics for Experiment 4

4.4 Effect of Altering the Ratio of Critical Tasks to All Tasks

The ratio from 0 (all normal tasks) to 1 (all emergency tasks) is varied. As shown in Fig. 12, energy usage increases as the number of emergency services grows. It is because the critical tasks ignores the Energy Efficiency of nodes while placing the important tasks in nearer nodes for less latency.

The average response time, shown in Fig. 12 for this particular experiment shows that raising the emergency services to all services ratio from 0 up to 1 decreases this measure significantly. The major cause for this is the use of FNs for essential (emergency) task processing. As a result, the transmission time is reduced, and the response time of services is consequently reduced.

5 Conclusion

A new prioritization algorithm called PLAEE is developed using existing popular algorithms called Cloud-only and Edgewards. For that, the Fog based Cloud platform is chosen for the implementation. As already mentioned, we have divided the data into Critical and Normal Data where critical data is given more priority. As shown from the above results, Critical data achieves high QoS by decreasing latency and Normal Data achieves less energy usage for FSPs. Along with this, more number of data is handled by Fog Layer shown by the % of Tasks Managed, whereas the utilization of Cloud is reduced heavily making more utilization of fog resources near to the edge network. However, when we increase the critical tasks, energy consumption is to be maximum than other algorithms in case of proposed algorithm. This is due to the fact that the PLAEE is designed to work better in case of latency only when used for Critical Data. Fog Computing research is considered to be a new paradigm which is why this work can be further extended to generate more valuable results. This research utilizes the static topology of devices which can be made dynamic by considering SDN for mobility of devices to develop more real-time application.

References

1. Al-Khafajiy, M., Baker, T., Al-Libawy, H., Maamar, Z., Aloqaily, M., Jararweh, Y.: Improving fog computing performance via fog-2-fog collaboration. Future Generation Computer Systems **100**, 266–280 (2019)
2. Aladwani, T.: Scheduling iot healthcare tasks in fog computing based on their importance. Procedia Computer Science **163**, 560–569 (2019)
3. Aldegheishem, A., Bukhsh, R., Alrajeh, N., Javaid, N.: Faavpp: Fog as a virtual power plant service for community energy management. Future Generation Computer Systems **105**, 675–683 (2020)
4. Cristescu, G., Dobrescu, R., Chenaru, O., Florea, G.: Dew: A new edge computing component for distributed dynamic networks. In: 2019 22nd International Conference on Control Systems and Computer Science (CSCS). pp. 547–551. IEEE (2019)
5. Gandhi, A., Harchol-Balter, M., Das, R., Lefurgy, C.: Optimal power allocation in server farms. ACM SIGMETRICS Performance Evaluation Review **37**(1), 157–168 (2009)
6. Gupta, H., Vahid Dastjerdi, A., Ghosh, S.K., Buyya, R.: ifogsim: A toolkit for modeling and simulation of resource management techniques in the internet of things, edge and fog computing environments. Software: Practice and Experience **47**(9), 1275–1296 (2017)
7. Kaur, K., Sachdeva, M.: Fog computing in iot: An overview of new opportunities. Proceedings of ICETIT **2019**, 59–68 (2020)
8. Khosroabadi, F., Fotouhi-Ghazvini, F., Fotouhi, H.: Scatter: Service placement in real-time fog-assisted iot networks. Journal of Sensor and Actuator Networks **10**(2), 26 (2021)
9. Kishor, A., Chakraborty, C., Jeberson, W.: A novel fog computing approach for minimization of latency in healthcare using machine learning. Int J Interact Multimed Artif Intell **1**(1) (2020)
10. Kumar, M., Dubey, K., Pandey, R.: Evolution of emerging computing paradigm cloud to fog: Applications, limitations and research challenges. In: 2021 11th International Conference on Cloud Computing, Data Science & Engineering (Confluence). pp. 257–261. IEEE (2021)
11. Kusic, D., Kephart, J.O., Hanson, J.E., Kandasamy, N., Jiang, G.: Power and performance management of virtualized computing environments via lookahead control. Cluster computing **12**(1), 1–15 (2009)

12. Mahmoud, M.M., Rodrigues, J.J., Saleem, K., Al-Muhtadi, J., Kumar, N., Korotaev, V.: Towards energy-aware fog-enabled cloud of things for healthcare. Computers & Electrical Engineering **67**, 58–69 (2018)
13. Mehta, M., Kavitha, V., Hemachandra, N.: Price of fairness for opportunistic and priority schedulers. In: 2015 IEEE Conference on Computer Communications (INFOCOM). pp. 1140–1148. IEEE (2015)
14. Nguyen, T., Doan, K., Nguyen, G., Nguyen, B.M.: Modeling multi-constrained fog-cloud environment for task scheduling problem. In: 2020 IEEE 19th International Symposium on Network Computing and Applications (NCA). pp. 1–10. IEEE (2020)
15. Papadopoulos, D.P., Mourouzis, I., Thomopoulos, C., Makris, T., Papademetriou, V.: Hypertension crisis. Blood pressure **19**(6), 328–336 (2010)
16. Singh, A., Auluck, N., Rana, O., Jones, A., Nepal, S.: Rt-sane: Real time security aware scheduling on the network edge. In: Proceedings of the 10th International Conference on Utility and Cloud Computing. pp. 131–140 (2017)
17. Tuli, S., Mahmud, R., Tuli, S., Buyya, R.: Fogbus: A blockchain-based lightweight framework for edge and fog computing. Journal of Systems and Software **154**, 22–36 (2019)

Smart Surveillance Based on Video Summarization: A Comprehensive Review, Issues, and Challenges

Ankita Chauhan and Sudhir Vegad

Abstract In order to ensure the protection and safety of the public, intelligent video surveillance systems are used in a wide range of applications to monitor events and analyze activities. A huge data is being generated, processed, and stored for security reasons from a single or multi-monitoring camera. Owing to a time limit, an analyst is going through a laborious process of examining the most important incidents from vast video databases. As a result, Video Summarization is a smart solution to summarize the salient events by removing the redundant frames which facilitate fast searching, indexing, and quick sharing of the content automatically. This paper investigates existing video summarization Techniques and a comparison of various existing approaches and methods in the surveillance domain. It also presents insight into video summarization in traffic surveillance which can be used to address the ever-growing needs of Intelligent Transportation System (ITS). The present study will be useful in integrating essential research findings and data for quick reference, identifying existing work, and investigating possible directions for future research in video summarization for traffic surveillance.

Keywords Video summarization · Traffic video surveillance · Surveillance video summarization techniques · Traffic video dataset · Video processing

1 Introduction

Millions of security cameras are currently used for several purposes, including homeland security, traffic monitoring, human detection and tracking, detection of illegally parked cars, vehicle crashes, and crime detection [1]. The development of better video recording and rendering technologies has resulted in a significant increase in the amount of video content produced.

A. Chauhan (✉)
Gujarat Technological University(GTU), Ahmedabad, Gujarat, India

S. Vegad
IT Department, A D Patel Institute of Technology, Vallabh Vidyanagar, Gujarat, India

© The Author(s), under exclusive license to Springer Nature Singapore Pte Ltd. 2022   433
V. Suma et al. (eds.), *Evolutionary Computing and Mobile Sustainable Networks*,
Lecture Notes on Data Engineering and Communications Technologies 116,
https://doi.org/10.1007/978-981-16-9605-3_29

The events and activities in the viewing area can be monitored by a traffic surveillance camera, whenever and wherever that human vision fails to track. The manual system may miss many actionable incidents due to human user obstacles such as sleep caused by long-term control of a large number of video screens and diversions caused by additional loads. So, In order to have less communication with human operators, an intelligent video monitoring cum surveillance system is essential. It is very unlikely to scrutinize each event from a traffic monitoring video. To address this issue video summarization methods have been introduced.

It is a method of compressing a video while keeping the semantics of the original video. It should be done in such a way that it removes all redundant information and keeps all important information. The two types of video summaries are static and dynamic video summaries. In static video summaries, keyframes or important frames are kept [2]. Dynamic video summaries contain a short video of the original one that contains all salient information. To create a good video summary, you will need a good understanding of the original video format, textual content, and a visual summary of video frames [3, 4].

The process of video summarization (frames are taken from CADP novel dataset) is illustrated in Fig. 1. The process starts with the input video being converted into a sequence of frames. The feature extraction technique is used to extract the visual features from these constituent frames of the video. We can extract these features based on handcrafted and learned features. Handcrafted features are manually extracted from video frames using image processing techniques. Learned features are learned by an advanced model like a machine learning model that

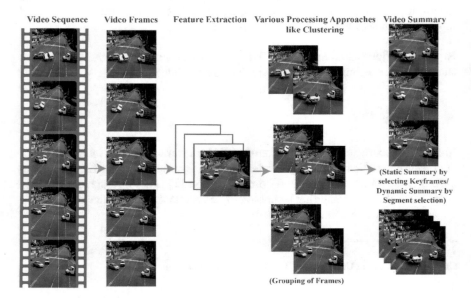

Fig. 1 Process video summarization

generates the outputs of visual features after training on the sequence of video frames. Once the features are extracted various processing approaches like the supervised/unsupervised approach is used to group the frames. These learning approaches are dependent on the types of datasets. The supervised learning approach is used when the dataset is labeled, and an unsupervised learning approach is used when the dataset is unlabeled. Finally, the video summary is generated either by selecting keyframes (Static summary) or by selecting segment (Dynamic summary).

The purpose of this paper is to present in-depth literature research on the recent trends in the field of Surveillance video summarization and especially on Traffic Surveillance video summarization, focusing on the use of traditional learning methods and the current trends to summarize events. Section 2 Presents state of art various techniques applicable in Surveillance video summarization. Section 3 introduces a comparative analysis of video summarization for the traffic Surveillance domain. A publically available dataset for traffic surveillance is described in Sect. 4. A Traffic Surveillance video summarization may suffer from various issues and challenges are highlighted in Sect. 5.

2 State of Art Surveillance Video Summarization Techniques

Various approaches have been created for summarizing surveillance videos like Feature-based, cluster-based, event-based, and trajectory-based [2] are shown in Fig. 2. We have presented the comparison of state of art surveillance Video Summarization methods in Table 1 [5–15].

2.1 Feature-Based Technique

Color, border, motion, and audio, are both global and local features of a digital video. It works well if the user needs to focus on the videos features. As a result, many video summarization methods came into existence were by assessing the correlation

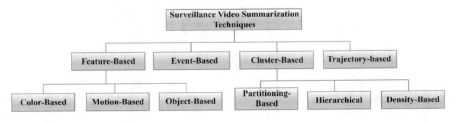

Fig. 2 Surveillance video summarization techniques

Table 1 Surveillance video summarization techniques: a comparison of state-of-the-art methods

Author and year	Research methodology	Approach	Utilization of dataset	Remarks/Pros/Cons
Zhao et al. [6]	Background subtraction and DBSCAN clustering	Object-based	Hotel surveillance videos	• Detects rarely appear targets and objects very well • Better recall rate as compared with others
Lai et al. [7]	Frame re-composition strategy for video summarization	Spatio-temporal trajectory based	Video sequences like Cesta4, building, Transitway1 Transitway2	• Spatiotemporal trajectories preserve up to 97% of the activity in the actual video • Performs well by a good compromise between activity recall and compression rate
Song et al. [9]	Novel approach using a disjoint max-coverage algorithm for large-scale surveillance video	Event based	A straight road, crossroad, T-junction and highway video	• Highest summarization ratio as compared with other methods • As shot boundary detection is not possible, event-based detection is very difficult
Ajmal et al. [10]	An automatic human-centric system using trajectory extraction using HOG, SVM, and Kalman filter	Trajectory based	Hostel surveillance camera videos	• Best results in a less congested and low movement environment • The motion pattern of humans and congestion level in the video is highly sensitive to efficiency
Kumar et al. [12]	A local alignment based FDES approach for multi-view summarization	Event-based	Lobby, office, and BL-7F	• Outperformed existing methods in terms of F-measure and computing time • Used short-duration multi-view surveillance videos

(continued)

Table 1 (continued)

Author and year	Research methodology	Approach	Utilization of dataset	Remarks/Pros/Cons
Gao et al. [13]	ASOSV using the keyframe selection method	Cluster-based	ViSOR, BL-7 database, IVY LAB	• Performed in an unsupervised and online way, making it appropriate for practical usages
Elharrouss et al. [17]	Context subtraction and structure, texture, and noise decomposition	Motion-based	Hai Way, office, Pedestrian, change detection DS, PET2006	• Highest quantitative values compared with other approaches • For similarity measures, the accuracy rate is over 70%, and F-measure is over 82%
Elharrouss et al. [15]	Multiple action identification and summarization of human by extracting human bodies' Silhouette, HOG TDMap and proposed CNN model	Action/motion-based	KTH, Weizmann, UT-interaction, UCF-ARG, MHAD, and IXMAS	• Accuracy rate (i.e., CNN-TDMap—98.9%, CS-HOG-TDMap—97.5% for each method) • It does not recognize the action of jogging • In the ground dataset, a boxing movement of the dataset-KTH is identified as a waving

between the frame's features within each scene or shot for detection of the video sequence. These methods are listed further down.

Color-Based Technique Each frame in the video is characterized by color, which is considered an important, expressive, and basic feature. It is because of its continuity and tolerance to changes in size and direction [16]. The main drawback is that it can not differentiate between information and noise, so in summarized video noise may be there, which affects accuracy [2]. The main advantage of this method is its simplicity in implementation. A color histogram is resilient against small camera motion, avoids duplication, and has a very low computational complexity owing to its simplicity.

Motion-Based Technique Video abstraction based on motion analysis is difficult, and it becomes much more difficult when both camera and object motion is present. Since surveillance footage has a medium degree of motion, this technique is a strong candidate for performing accurately. On videos with a lot of motion or motionless, motion-based approaches are ineffective [17, 18].

Object-Based Technique When object detection is a prime goal from the video, object-based techniques are used. In this video, the appearance and disappearance of such objects may indicate content change [19]. In general, such methods depend on detecting, monitoring, and storing information about an object's lifetime. It is widely used in non-surveillance devices and video surveillance system-based applications. Especially in crowded places, using this method requires extreme caution, as incorrect objects can lead to incorrect summaries.

2.2 Cluster-Based Technique

The cluster-based technique divides a collection of frames or shots into many clusters with its own set of frames or shots within each cluster having a high degree of similarity in terms of operation or feature, whereas the frames/shots of other clusters have a high degree of dissimilarity. Partitioning, hierarchical, and density-based clustering strategies can be used based on their capability and applicability. It can also be used to delete frames with erratic patterns [20]. The two most commonly used partitioning methods like K-means and K-medoids. K-means [2] is an easy and practical approach to clustering for grouping keyframes in the video summarization process. The main aim is to partition a set of n samples into K pre-defined number of clusters and then select the nearest cluster to each cluster's centroid. The objective of the hierarchical method is to create a clusters hierarchy/tree that contains objects depending on density, distance, or continuity. This hierarchy can be formed in two ways: agglomerative and divisive. The advantage of this strategy over partitioning is that it is not dependent on the number of pre-defined clusters. Density-Based Method [6] aims to expand a given cluster while the density of its neighbors is greater than the threshold. As a result, It can successfully discover any arbitrary-shaped cluster and filter noise.

2.3 Event-Based Video Technique

An event-based approach identifies an important event and organizes it in such a way that the context of the actual footage is protected. The author of [21] detected important details by combining energy differences between frames, as events cause major shifts in subsequent frames and are associated with high energy. In the next stage, based on visual characteristics the detected events are clustered. This method is best suited for videos recorded from a static camera. Due to the false detection of changes as occurrences, this approach is limited when the lighting or context changes.

2.4 Trajectory-Based Technique

A Trajectory-Based approach works effectively for analyzing the dynamic environment in the video. The main key element of this method is to detect the object's motion behavior concerning time. Therefore, it is mostly used in surveillance applications to locate the objects and remove uncertainty, where a fixed camera is used and with a steady background [22]. It is classified into Spatio-temporal Based and Curve Simplification technique. Spatio-temporal Based techniques allow eliminating both spatial and temporal frame redundancy from video. It is commonly used in surveillance systems, according to the survey. When the camera is moving, it is not very useful. Another approach is the curve simplification technique [2], which uses the spatial data of moving objects and best option for analyzing motion. It eliminates the redundant keyframes and only uses a small portion of all captured frames to display the motion capture sequence.

Based on the literature survey, it has been observed that every technique has its advantages and disadvantages. various domains require a different technique to perform video summarization. The color-based technique cannot differentiate noise and information, so noise may be there in the summary video, and it affects accuracy [2]. Even a color histogram is a robust approach due to its simplicity, but it is very sensitive to noise. The motion-based techniques are not performed accurately on high motion or motionless videos. But it performs accurately and well in surveillance videos with a medium level of motion [17]. The object-based technique is widely used in surveillance applications when it is required to focus on an object's behavior. It reduces Spatio-temporal redundancy from the video frames. But it does not perform accurately when the target object is not present in the sequences of frames [19]. An event-based technique is used to detect an abnormal event in the videos. One limitation of this technique is that it can detect false events in the case of background change and illumination. The clustering-based technique is used to delete frames with irregular or erratic patterns [20]. It does not work well with the video remaining static for a long-duration. the trajectory-based approach is mostly used in surveillance applications to detect moving objects' behavior concerning time. But it does not perform accurately for non-surveillance applications with very high camera motion [2].

3 Comparative Analysis on Traffic Surveillance Video Summarization

we have considered the major Traffic Surveillance Video Summarization based research articles of the last 5 years in Table 2. In our survey, we have presented a brief review for on-road traffic scenario-based Video summarization.

4 Publicly Available Datasets for Traffic Surveillance

Each analysis in this review used different datasets to evaluate each process, making it impossible to compare the results of each study equally. VIRAT [30] is a large-scale real-time video dataset for evaluating the output of various event detection algorithms in the surveillance domain. CVRR [31] is a benchmarking trajectory clustering dataset, provides data to compare clustering algorithms and trajectory distance/similarity measures. QMUL [32] Specifically proposed for behavior understanding and activity analysis of traffic. MIT [33] Traffic dataset intended to research on activity analysis, motion pattern analysis, and crowded scenes on intersection and walkway. UCSD highway [34] contains different traffic patterns such as heavy, medium, and light congestion with different weather conditions. The Ko-PER [35] was proposed for creating a comprehensive dynamic model of the current traffic as well as research in the field of multi-object detection and tracking. Urban Tracker [36] contains traffic scene recordings, camera calibration, metadata, protocols, ground truth for comparing software tools, libraries, and algorithms. CADP [37] is a novel dataset proposed for Accident Analysis. AAU RainSnow Traffic Dataset [38] contains various Traffic videos with a wide variety of weather conditions. Street Scene [39] is a comprehensive dataset proposed for evaluation protocol and anomaly detection from video. List and discussion of traffic video datasets that are publicly available are listed in Table 3.

Table 2 Comparison of state-of-the-art methods for video summarization from vision-based traffic surveillance

Ref.	Methodology	Dataset	Measure	Advantages	Limitation/future work
Habeeb et al. [4]	Histogram differencing and sum conditional variance (SCV)	Road scene	Space saving, data compression ratio, condensed ratio	• Robust approach in case of illumination variations to obtain motion objects • Better o/p as compared with temporal differencing summarization	Sensitive to distant motion objects
Nguyen et al. [23]	Video condensation method based on the ribbon carving approach	Overpass, highway, abandoned box, visor, speed trap	Condensation ratios (CR), Execution times	• Saved an important amount of computational loads without losing CR and visual quality	The summary contains repeated similar frames. Due to the preservation of the order of the events
Lu et al. [24]	Keyframe selection framework for ASOSV based on martingale-test mechanism	Test video on night traffic scenario, visor dataset	selection rate, computation time, and perceptive evaluation	• Efficient for real applications	• Feature extraction takes huge computing time • Efficient computationally frame descriptor will use
Rabbouch et al. [25]	Unsupervised learning for automatic vehicles counting and recognizing, summarize video using cluster Analysis	Traffic benchmark video	EM-estimated Gaussian, Bayesian information criterion	• Rare method that can simultaneously recognize and count vehicles • Strategy maintenance, implementation is cheap as compared with other tools	• Gaussian densities to describe clusters can be expanded, but it is restrictive • Will enhance accuracy and robustness in the case of small vehicles and micro-cars in huge traffic

(continued)

Table 2 (continued)

Ref.	Methodology	Dataset	Measure	Advantages	Limitation/future work
Thomas et al. [26]	Perceptual video summarization techniques	YouTube-8 M (auto, vehicle), Urban tracker	Saliency cost, reduction ratio, detection rate, mean opinion score	Video summarization gives great potential and promising results to detect an event	• Used single camera view • Did not specify near-collision and collision • Performance can be enhanced using depth-based segmentation
Chen et al. [27]	Spectrum analysis approach	CAVIAR, ALOV + +, the Videezy	PASCAL criterion, average performance comparison	• Overcome challenges like dynamic noise, fogged, illumination changes, and heavy rain	• Shadow effect cannot be removed by this approach
Mayya et al. [28]	Surveillance video summarization for traffic using faster R-CNN	Param Shavak system	–	• Detected two-wheeler riders without a helmet	• Work may extend to detect the license numbers of rule violators and monitoring traffic activities like traffic estimation, accidents, etc.
Kosambia and Gheewala [29]	Video synopsis for accident detection using ResNet architecture with certain pre-processing and extraction features	CCTV surveillance network of Hyderabad city	Training accuracy, Validation accuracy, Training loss, Validation loss	ResNet-152 gives better accuracy than ResNet-101 and ResNet-50	• Accuracy can be increased by adding tuning parameters and training examples • Process of localization can be applied to provide more precise interaction

Table 3 Publicly available datasets for traffic surveillance video

Dataset	Application/purpose/anomalies	Scene	Types of cameras
VIRAT [30]	• Single-person events—running, walking, tossing, standing, holding, picking up, loitering • Vehicle-person events—opening/closing trunk, getting into/out of the vehicle, dropping off • Bicycling-person-facility events—entering/leaving	Construction sites, streets parking lots, and open outdoor spaces	Moving aerial cameras and stationary ground cameras
CVRR [31]	Clustering, classification, prediction, and abnormality detection like illegal loop, wide U-turn, travel in the opposite direction, off-road driving	Four-lane highway, four-way intersection	Single view camera
QMUL [32]	Traffic analysis, scene modeling, and clustering	Intersection	Single cameras
MIT Traffic [33]	Abnormalities like bicycles are crossing the road abnormally illegal right turn, Jaywalking	Street intersection, walkway	Stationary camera
UCSD highway [34]	Different traffic congestion patterns like light, medium, and heavy	Highway	Single-camera
Ko-PER [35]	Two-wheelers cross the intersection, multiple vehicles like buses, trucks, etc., cars performing a straight-ahead maneuver and right turn	Public intersection	Laser-scanners and monochrome
NGSIM [40]	The learning model of vehicle behavior under normal traffic conditions traffic control, delineation, congestion, and other features of the environment	Highway, interactions, Freeway	Digital video camera
KIT [41]	Vehicles detection and tracking	Intersection	Stationary camera
PDTV [42]	Tall vehicles causing significant occlusions, traffic counts per direction, pedestrian infractions and safety at crossings	intersections	Single and multi-view camera
Urban Tracker [36]	Vehicle and pedestrian tracking at intersections in various lighting, weather conditions, occlusions handling	Intersections	Stationary single camera

(continued)

Table 3 (continued)

Dataset	Application/purpose/anomalies	Scene	Types of cameras
UA-DETRAC [43]	MOT with vehicle type, illumination, occlusion, vehicle bounding boxes, and truncation ratio	Urban highway, crossings and T-junctions	Canon EOS 550D camera
CADP [37]	Object detection, accident detection, and forecasting	Traffic scene	Fixed CCTV
AAU Dataset [38]	Snow, rain, and bad weather in traffic surveillance, illumination conditions	Intersections	RGB and thermal camera
Street Scene [39]	Jaywalking, biker outside lane, loitering, dog on the sidewalk, car outside lane, biker on the sidewalk, pedestrian reverses direction, car illegally parked, skateboarder in the bike, person exits car on street, person sitting on a bench, car turning from parking space, motorcycle drives onto sidewalk	Pedestrian sidewalks, two-lane street with bike lanes	Static USB camera

5   Issues and Challenges on Traffic Video Summarization

5.1   Issues Associated with the Camera

Due to the Limitations of cameras, object detection and tracking-related challenges produce. Also, block artifacts and blurring reduce video sequence quality. Sometimes the quality of videos can be degraded due to the noise. Moving object detection and tracking algorithms may be harmed by low-quality video sequences.

5.2   Issues and Challenges in Video Data Acquisition

Illumination Changes The lighting conditions of the scene and the target become change due to the various times of day, the motion of the light source, reflection from bright surfaces, etc. It can lead to false-positive vehicle detection, which means viewing non-vehicle objects as vehicles. Due to the lighting changes in the scene, the background model should be adaptable to abrupt illumination variation in lighting to avoid errors in detecting moving objects [14, 17, 43].

Presence of Abrupt Motion Handling the unexpected movements induced by low frame rate video, camera switching, and the rapid dynamic transition is one of the major challenges in real-world tracking applications. Under the presumption of smooth motion, traditional tracking strategies tend to fail when dealing with a sudden motion.

Scene having Complex Background Due to the dynamic background, it may include motion such as movement of clouds in the sky, fountain, streetlights, water waves, branches sway, etc. Detecting moving objects in cluttered backgrounds necessitates accurate background analysis and camera motion estimation.

5.3 Challenges in Vehicle Detection, Tracking, and Anomaly Detection

Vehicle Occlusion Tracking vehicles on highways is a major challenge when traffic is slow because the amount of intermediate traffic space is drastically reduced, increasing vehicle occlusion [36, 43]. Object detection in context modeling techniques is severely hampered by occlusion, where the object is split into unconnected regions or completely missing.

Pose Variation In appearance-based vehicle detection, The Tracking algorithm's output is also affected by pose variations. If the pose changes often, the same vehicle appears to be different in each frame. A vehicle's appearance can change due to its distance from the cameras, so camera angles on surveillance areas can have a major impact on anomaly detection efficiency [14, 43].

Heterogeneous Object Handling Vehicle classification in a traffic video is a difficult task due to the similarity of appearances between different vehicles in heterogeneous traffic conditions. However, modeling heterogeneous objects in a scene or learning the motion of heterogeneous objects in a scene may be challenging at times [14].

Shadows The presence of shadows makes detecting moving objects in surveillance videos more difficult. and can influence the task of detection. The two common approaches are the used like removal of casting shadows or the use of edge detection to distinguish shadows from vehicles. However, in the complex natural scene, the conventional optical flow method cannot accurately detect the boundary of the moving vehicle due to the generation of the shadow.

6 Conclusion and Future Work

Video Summarization has consistently been a demanding field in research under surveillance applications. Here in this research article, we have presented an in-depth review of various summarization techniques, different methodologies, and various procedural steps required in the creation of meaningful video summaries, especially in the surveillance domain. This review article gives an in-depth insight into the issues and challenges of video-based vehicle detection, Tracking, anomaly detection in video summarization, publicly accessible datasets, and current research trends for traffic surveillance video summarization.

The key benefit of a video summarization is that of quick and easy comprehension of the salient events in the video. Based on the review performed it has been recognized that the video summarization for the traffic surveillance domain is a less explored area as compared to other applications. Still, there is a requirement to scrutinize important event like accident detection, rule violator, traffic analysis, etc. from the traffic video and summarize it.

The major contribution of this article is to present an in-depth latest review of relevant areas of surveillance video summarization research for traffic application and to encourage new research. Hence, there is a lot of scope of research in this domain to overcome any one of the challenges. This existing literature survey has offered new insights into the domain of video summarization. Some important observations are as follows:

- Due to advancements in high-speed GPUs, many researchers are working with deep learning-based models for making dynamic video summaries. Still, a training dataset with a large number of images can be a constraint.
- Many advanced summarization methods are limited to a single model, such as video. By incorporating other forms of media such as text, audio, as well as metadata, the work can be shifted to synergistic multi-modal processing in the future.
- Real-time video summarization algorithms have not been emphasized by researchers, even though there are very few summarization methods available for online video processing.
- With the creation of short videos, the surveillance research community needs the availability of long-duration video datasets [22]. The researcher will focus on the problem of the amount of time it takes to perform segmentation and then data selection.
- In a multi-camera network, multiple video summaries are created by the current summarization approaches, its count is similar to the number of cameras. In multiple synopsis images, object tracking is a quite difficult task due to more number of cameras. An appropriate method should be there to create and present video summaries to end-users.

References

1. Wahyono, Filonenko, A., Jo, K.: Designing interface and integration framework for multi-channels intelligent surveillance system. In: 9th International Conference on Human System Interactions, pp. 311–315 (2016). https://doi.org/10.1109/HSI.2016.7529650
2. Ajmal, M., Ashraf, M.H., Shakir, M., Abbas, Y., Shah, F.A.: Video summarization: techniques and classification. In: ICCVG 2012, Lecture Notes in Computer Science, vol. 7594. Springer, Berlin, Heidelberg (2012). https://doi.org/10.1007/978-3-642-33564-8_1
3. Money, A.G., et al.: Video summarisation: a conceptual framework and survey of the state of the art. J. Vis. Commun. Image Representation **19**(2), 121–143 (2008). ISSN 1047-3203. https://doi.org/10.1016/j.jvcir.2007.04.002
4. Habeeb, et al.: Surveillance video summarization based on histogram differencing and sum conditional variance. Int. J. Comput. Electr. Autom. Control Inf. Eng. **10**(9), 1577–1582 (2016)
5. Chen, Y., Zhang, B.: Surveillance video summarization by jointly applying moving object detection and tracking. Int. J. Comput. Vision Rob. **4**(3), 212–234 (2014). https://doi.org/10.1504/IJCVR.2014.062936
6. Zhao, Y., Lv, G., Ma, T.T., et al.: A novel method of surveillance video Summarization based on clustering and background subtraction. In: 8th International Congress on Image and Signal Processing, pp. 131–136 (2015). https://doi.org/10.1109/CISP.2015.7407863
7. Lai, P.K., Décombas, M., Moutet, K., Laganière, R.: Video summarization of surveillance cameras. In: 13th IEEE International Conference on Advanced Video and Signal Based Surveillance, pp. 286–294 (2016). https://doi.org/10.1109/AVSS.2016.7738018
8. Salehin, M.M., Paul, M.: Video summarization using geometric primitives. In: International Conference on Digital Image Computing: Techniques and Applications, pp. 1–8 (2016). https://doi.org/10.1109/DICTA.2016.7797094
9. Song, X., Sun, L., Lei, J., Tao, D., Yuan, G., Song, M.: Event-based large scale surveillance video summarization. Neurocomputing **187**, 66–74 (2016). ISSN 0925-2312. https://doi.org/10.1016/j.neucom.2015.07.131
10. Ajmal, M., Naseer, M., Ahmad, F., Saleem, A.: Human motion trajectory analysis based video summarization. In: 16th IEEE International Conference on Machine Learning and Applications, pp. 550–555 (2017). https://doi.org/10.1109/ICMLA.2017.0-103
11. Thomas, S.S., Gupta, S., et al.: Smart surveillance based on video summarization. In: IEEE Region 10 Symposium (2017). https://doi.org/10.1109/TENCONSpring.2017.8070003
12. Kumar, K., et al.: F-DES: fast and deep event summarization. IEEE Trans. Multimedia **20**(2), 323–334 (2018). https://doi.org/10.1109/TMM.2017.2741423
13. Gao, Z., Lu, G., Lyu, C., et al.: Key-frame selection for automatic summarization of surveillance videos: a method of multiple change-point detection. Mach. Vis. Appl. **29**, 1101–1117 (2018). https://doi.org/10.1007/s00138-018-0954-7
14. Santhosh, K.K., Dogra, D.P., Roy, P.P.: Anomaly detection in road traffic using visual surveillance: a survey. ACM Comput. Surv. **53**(6), 26 (2021). Article 119. https://doi.org/10.1145/3417989
15. Elharrouss, O., Almaadeed, N., Al-Maadeed, S., et al.: A combined multiple action recognition and summarization for surveillance video sequences. Appl. Intell. **51**, 690–712 (2021). https://doi.org/10.1007/s10489-020-01823-z
16. Trémeau, A., Tominaga, S., Plataniotis, K.: Color in image and video processing: most recent trends and future research directions. EURASIP J. Image Video Process. 581371 (2008). https://doi.org/10.1155/2008/581371
17. Elharrouss, O., Al-Maadeed, N., Al-Maadeed, S.: Video summarization based on motion detection for surveillance systems. In: 15th International Wireless Communications and Mobile Computing Conference, pp. 366–371 (2019)
18. Bagheri, S., Zheng, J.Y., Sinha, S.: Temporal mapping of surveillance video for indexing and summarization. In: Computer Vision and Image Understanding, vol. 144, pp. 237–257 (2016). ISSN 1077-3142. https://doi.org/10.1016/j.cviu.2015.11.014

19. Wang, F., Ngo, C.: summarizing rushes videos by motion, object, and event understanding. IEEE Trans. Multimedia **14**(1), 76–87 (2012). https://doi.org/10.1109/TMM.2011.2165531

20. Fu, Y., et al.: Multi-view video summarization. IEEE Trans. Multimedia **12**(7), 717–729 (2010). https://doi.org/10.1109/TMM.2010.2052025

21. Damnjanovic, U., Fernandez, V., et al.: Event detection and clustering for surveillance video summarization. In: International Workshop on Image Analysis for Multimedia Interactive Services, pp. 63–66 (2008). https://doi.org/10.1109/WIAMIS.2008.53

22. Ahmed, S.A., Dogra, D.P., Kar, S., Roy, P.P.: Trajectory-based surveillance analysis: a survey. IEEE Trans. Circuits Syst. Video Technol. **29**(7), 1985–1997 (2019). https://doi.org/10.1109/TCSVT.2018.2857489

23. Nguyen, H.T., Jung, S., Won, C.S.: Order-preserving condensation of moving objects in surveillance videos. IEEE Trans. Intell. Transp. Syst. **17**(9), 2408–2418 (2016). https://doi.org/10.1109/TITS.2016.2518622

24. Lu, G., Zhou, Y., Li, X., et al.: Unsupervised, efficient and scalable key-frame selection for automatic summarization of surveillance videos. Multimed. Tools Appl. **76**, 6309–6331 (2017). https://doi.org/10.1007/s11042-016-3263-z

25. Rabbouch, H., Saâdaoui, F., Mraihi, R.: Unsupervised video summarization using cluster analysis for automatic vehicles counting and recognizing. Neurocomputing **260**, 157–173 (2017). https://doi.org/10.1016/j.neucom.2017.04.026

26. Thomas, S.S., Gupta, S., Subramanian, V.K.: Event detection on roads using perceptual video summarization. IEEE Trans. Intell. Transp. Syst. **19**(9), 2944–2954 (2018). https://doi.org/10.1109/TITS.2017.2769719

27. Chen, Z., Lv, G., Lv, L., et al.: Spectrum analysis-based traffic video synopsis. J. Signal Process. Syst. **90**, 1257–1267 (2018). https://doi.org/10.1007/s11265-018-1345-z

28. Mayya, V., Nayak, A.: Traffic surveillance video summarization for detecting traffic rules violators using R-CNN. In: Advances in Computer Communication and Computational Sciences—Proceedings of IC4S 2017, pp. 117–126. Springer Verlag (2019). https://doi.org/10.1007/978-981-13-0341-8_11

29. Kosambia, T., Gheewala, J.: Video synopsis for accident detection using deep learning technique. In: Proceedings of the ICSMDI 2021 (2021). https://doi.org/10.2139/ssrn.3851250

30. Oh, S., et al.: A large-scale benchmark dataset for event recognition in surveillance video. In: CVPR 2011, pp. 3153–3160 (2011). https://doi.org/10.1109/CVPR.2011.5995586

31. Morris, B.T., Trivedi, M.M.: Trajectory learning for activity understanding: unsupervised, multilevel, and long-term adaptive approach. IEEE Trans. Pattern Anal. Mach. Intell. **33**(11), 2287–2301 (2011). https://doi.org/10.1109/TPAMI.2011.64

32. Hospedales, T., et al.: Video behaviour mining using a dynamic topic model. Int. J. Comput. Vis. **98**, 303–323 (2012). https://doi.org/10.1007/s11263-011-0510-7

33. Wang, X., Ma, X., Grimson, W.E.L.: Unsupervised activity perception in crowded and complicated scenes using hierarchical Bayesian models. IEEE Trans. Pattern Anal. Mach. Intell. **31**(3), 539–555 (2009). https://doi.org/10.1109/TPAMI.2008.87

34. Sobral, A., Oliveira, L., Schnitman, L., Souza, F.D.: Highway traffic congestion classification using holistic properties. In: 10th International Conference on Signal Processing, Pattern Recognition and Applications (2014). https://doi.org/10.2316/P.2013.798-105

35. Strigel, E., Meissner, D., Seeliger, F., Wilking, B., Dietmayer, K.: The Ko-PER intersection laser-scanner and video dataset. In: 17th International IEEE Conference on ITSC, pp. 1900–1901 (2014). https://doi.org/10.1109/ITSC.2014.6957976

36. Jodoin, J., Bilodeau, G., Saunier, N.: Urban tracker: multiple object tracking in urban mixed traffic. In: IEEE Winter Conference on Applications of Computer Vision, pp. 885–892 (2014). https://doi.org/10.1109/WACV.2014.6836010

37. Shah, A.P., Lamare, J., Nguyen-Anh, T., Hauptmann, A.: CADP: a novel dataset for CCTV traffic camera based accident analysis. In: 15th IEEE International Conference on Advanced Video and Signal Based Surveillance, pp. 1–9 (2018)

38. Bahnsen, C.H., Moeslund, T.B.: Rain removal in traffic surveillance: does it matter? IEEE Trans. Intell. Transp. Syst. **20**(8), 2802–2819 (2019). https://doi.org/10.1109/TITS.2018.2872502

39. Ramachandra, B., Jones, M.: Street scene: a new dataset and evaluation protocol for video anomaly detection. In: 2020 IEEE Winter Conference on Applications of Computer Vision (WACV), 2558–2567 (2020)

40. U.S. Dept. Transp.: NGSIM. [Online]. Available: https://ops.fhwa.dot.gov/trafficanalysistools/ngsim.htm. Accessed 1 March 2021

41. KIT Intersection Monitoring Datasets. [Online]. Available: http://goo.gl/wLlnIN. Accessed 1 March 2021

42. Saunier, N., et al.: Public video data set for road transportation applications. In: Transportation Research Board Annual Meeting Compendium of Papers (2014)

43. Wen, L., Du, D., et al.: UA-DETRAC: a new benchmark and protocol for multi-object detection and tracking. Comput. Vision Image Understand. **193**, 102907 (2020). ISSN 1077-3142. https://doi.org/10.1016/j.cviu.2020.102907

Design and Analysis of a Plano Concave DRA for 60 GHz Application

Ribhu Abhusan Panda, Pragyan Paramita Behera, Dilip Nayak, and Rishi Kumar Patnaik

Abstract This work includes the design of an unconventional shaped dielectric resonating antenna for 60 GHz application. The shape of the dielectric resonating antenna (DRA) resembles the Plano Concave lens structure. The breadth and minimum length of DRA has been taken as 10 mm and 5 mm. The substrate is selected with length, breadth and height of 20 mm. 20 mm and 1.6 mm, respectively. FR4 epoxy material with dielectric constant 4.4 has been selected for the substrate and for the dielectric resonator alumina 92_pct material has been chosen which is having a dielectric constant of 9.2. The proposed antenna is resonating at 59.7 GHz with S_{11} value of 25.48 dB. The antenna can be operated in a wide range of bandwidth of more than 10 GHz (55–65 GHz). The proposed DRA has been designed for 60 GHz millimetre wave applications like 60 WLAN and 60 GHz WiGig network.

Keywords DRA · Plano concave · 60 GHz WPAN · Gain · Directivity · S_{11}

1 Introduction

Dielectric resonating antennas (DRA) has gone through a lot of evolutionary changes from its inception. In 2021, circularly polarised has been designed [1] and later it led to a design for uses in Wi-Fi. Bluetooth, ISM Band and WiMAX applications [2]. In the same year, a set of design has been proposed for specific applications

R. A. Panda (✉) · P. P. Behera · D. Nayak · R. K. Patnaik
Department of Electronics and Communication Engineering, GIET University, Gunupur, Odisha, India
e-mail: ribhuabhusanpanda@giet.edu

P. P. Behera
e-mail: 19ece052.pragyanparamitabehera@giet.edu

D. Nayak
e-mail: 19ece019.dilipnayak@giet.edu

R. K. Patnaik
e-mail: 19ece049.rishikumarpatnaik@giet.edu

V. Suma et al. (eds.), *Evolutionary Computing and Mobile Sustainable Networks*,
Lecture Notes on Data Engineering and Communications Technologies 116,
https://doi.org/10.1007/978-981-16-9605-3_30

[3–5]. The conventional shapes of DRA, e.g. rectangular, cylindrical and hemi-spherical have been perturbed for novel applications. In 2019, a DRA in the shape of English alphabet 'A' for UWB application has been designed [6]. In 2018, for 60 GHz millimetre wave application different shaped DRAs have been designed [7]. The change of shapes of DRA have also been rightly designed for higher antenna gain and directivity for different applications [8–11]. Few more novel designs have been proposed for specific applications in recent years [12–15]. This work focuses on the better outcome by modification of a conventional square DRA into a Plano concave shaped DRA. The minimum distance between the plane side and the arc side has been taken as 5 mm, equating with the wavelength of the design frequency of 60 GHz. The proposed model has a sharp resonance at 59.7 GHz with high directivity (7.8 dB) and large bandwidth (10 GHz).

2   Design of the Proposed DRA

This DRA includes a substrate with dimension 20 mm × 20 mm × 1.6 mm. The dielectric resonator has a shape that resembles to Plano concave lens. One part of the DRA has a planar shape and another has a biconcave shape. Minimum arc to arc length of the Plano concave patch is 5 mm which is designed to be same as that of the wavelength of design frequency of 60 GHz. The proposed dielectric resonator has the height of 3.8 mm. The feed has been designed as an aperture feed of width 2 mm connected at the centre of the ground plane made out of copper and area same as the substrate but with thickness 0.01 mm. The design has been illustrated in Fig. 1. The other values used for design have been listed in Table 1. FR4 Epoxy has been used for substrate as a dielectric material with dielectric constant 4.4. Alumina 92_pct having a dielectric constant of 9.2 is used for the dielectric resonator.

3   Simulation Result

The plot between return loss versus frequency has been focused to determine resonant frequency of proposed DRA. Along with this the bandwidth can also be determined from this part. The S_{11} versus Frequency plot which has been shown in Fig. 2 indicates the resonant frequency which is at 59.7 GHz with a return loss of −19.2048 dB. The proposed design has a wide bandwidth ranging from 55 to 65 GHz. Figure 3 includes the VSWR plot with respect to frequency (Figs. 4, 5, 6 and 7).

Figure 4 shows the 2-D radiation pattern of the Plano concave DRA. Peak gain of the proposed DRA is 3.26 dB and peak directivity has been found as 7.8 dB. Peak gain, 3-D radiation pattern and peak directivity have been illustrated in Figs. 5, 6 and 7, respectively. E-Plane and H-Plane polarisation have been provided in Figs. 8 and 9, respectively.

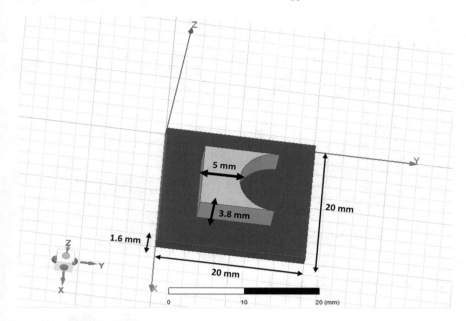

Fig. 1 Proposed Plano concave DRA

Table 1 Parameters for the design of the proposed DRA

Design parameters	Symbol	Value
Frequency	f	60 GHz
Wavelength	Λ	5 mm
Feed width	w_f	2 mm
Height of dielectric resonator	h	3.8 mm

The simulation has been done using Ansys HFSS software. The simulated parameter listed in Table2.

4 Conclusion

The proposed DRA resonates at 59.7 GHz with a peak gain of 3.26 and directivity of 7.8 dB. This DRA has an unconventional shape without any complexity, so it can be efficiently used in 60 GHz WLAN and 60 GHz WiGig networks.

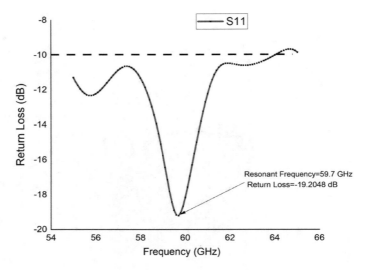

Fig. 2 *S*-parameter of the proposed DRA

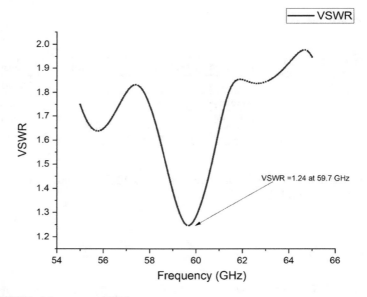

Fig. 3 VSWR of the proposed DRA

Fig. 4 2D radiation pattern

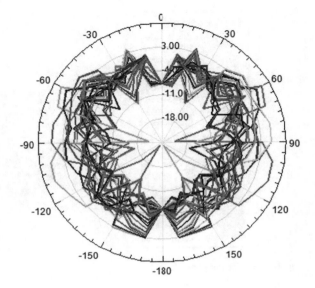

Fig. 5 Peak gain

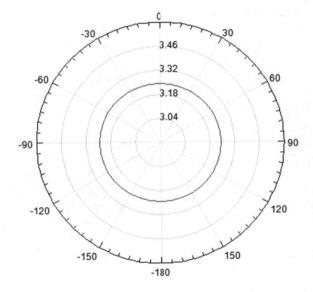

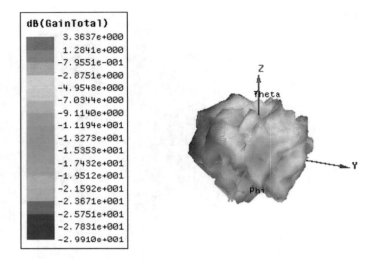

Fig. 6 3D radiation pattern

Fig. 7 Peak directivity

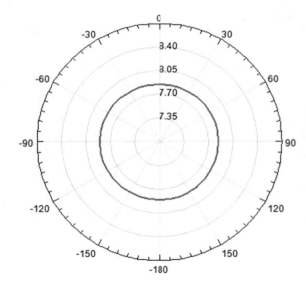

Fig. 8 E-plane polarisation

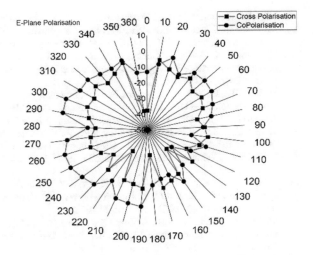

Fig. 9 H-plane polarisation

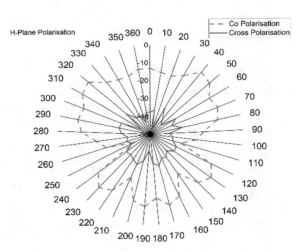

Table 2 Parameters from the simulation

Parameter	Return loss	Resonant frequency	Peak gain	Peak directivity	Radiation efficiency	Bandwidth
Value	−19.02 dB	59.7 GHz	3.26 dB	7.8 dB	56%	10 GHz

References

1. Sahu, N.K., Das, G., Gangwar, R.K.: Circularly polarized offset-fed DRA elements and their application in compact MIMO antenna. Eng. Sci. Technol. Int. J. (2021). ISSN 2215-0986. https://doi.org/10.1016/j.jestch.2021.05.019

2. Meher, P.R., Behera, B.R., Mishra, S.K.: A compact circularly polarized cubic DRA with unit-step feed for Bluetooth/ISM/Wi-Fi/Wi-MAX applications. AEU—Int. J. Electron. Commun. **128**, 153521 (2021). ISSN 1434-8411. https://doi.org/10.1016/j.aeue.2020.153521

3. Ibrahim, M.S., Attia, H., Cheng, Q., Mahmoud, A.: Wideband circularly polarized aperture coupled DRA array with sequential-phase feed at X-band. Alexandria Eng. J. **59**(6), 4901–4908 (2020), ISSN 1110-0168

4. Meher, P.R., Behera, B.R., Mishra, S.K., Althuwayb, A.A.: Design and analysis of a compact circularly polarized DRA for off-body communications. AEU—Int. J. Electron. Commun. **138**, 153880 (2021). ISSN 1434-8411. https://doi.org/10.1016/j.aeue.2021.153880

5. Kethavathu, S.N., Singam, A., Muthusamy, P.: Compact symmetrical slot coupled linearly polarized two/four/eight element MIMO bowtie DRA for WLAN applications. AEU—Int. J. Electron. Commun. **135**, 153729 (2021). ISSN 1434-8411. https://doi.org/10.1016/j.aeue.2021.153729

6. Sabouni, A., Alali, A.I., Sabouni, F., Kishk, A.A.: Thin perforated A-shaped DRA for UWB applications. In: 2019 49th European Microwave Conference (EuMC), pp. 69–72 (2019). https://doi.org/10.23919/EuMC.2019.8910884

7. Meher, P.R., Behera, B.R., Mishra, S.K.: Design of different shaped DRAs for 60 GHz millimeter-wave applications. IEEE Indian Conf. Antennas Propag. (InCAP) **2018**, 1–4 (2018). https://doi.org/10.1109/INCAP.2018.8770785

8. Panda, R.A., Mishra, D., Kumari, P., Hansdah, B., Singh, R.K.: Gauging trowel shaped patch with circular rings for 60 GHz WLAN. In: 2020 7th International Conference on Signal Processing and Integrated Networks (SPIN), pp. 113–116 (2020). https://doi.org/10.1109/SPIN48934.2020.9070868

9. Panda, R.A., Mishra, D.: Efficient design of bi-circular patch antenna for 5G communication with mathematical calculations for resonant frequencies. Wirel. Pers. Commun. **112**, 717–727 (2020). https://doi.org/10.1007/s11277-020-07069-9

10. Panda, R.A., Barik, T.R., Lakshmi Kanta Rao, R., Pradhan, S., Mishra, D.: Comparative study between a conventional square patch and a perturbed square patch for 5G application. In: Hemanth D., Shakya S., Baig Z. (eds.) Intelligent Data Communication Technologies and Internet of Things. ICICI 2019. Lecture Notes on Data Engineering and Communications Technologies, vol. 38. Springer, Cham (2020). https://doi.org/10.1007/978-3-030-34080-3_50

11. Ribhu Abhusan Panda, Priti Pragnya Satapathy, Subudhi Sai Susmitha, Raghuvu Budhi Sagar, Mishra, D.: Effect of substrate material on the performance of Plano-concave patch antenna for 5G. J. Phys.: Conf. Ser. **1706**; First International Conference on Advances in Physical Sciences and Materials, pp. 13–14. Coimbatore, India (2020)

12. Panda, R.A., Deo, B., Danish Iqbal, Md., Maha Swetha, K., Mishra, D.: Perturbed cylindrical DRA with aperture feed for 5G communication. Mater. Today: Proc. **26**(2), 439–443 (2020). ISSN 2214-7853. https://doi.org/10.1016/j.matpr.2019.12.078

13. Bansal, M., Maiya, R.R.: Phototransistor: the story so far. J. Electron. **2**(04), 202–210 (2020)

14. Prakasam, V., Sandeep, P.: Dual edge-fed left hand and right hand circularly polarized rectangular micro-strip patch antenna for wireless communication applications. IRO J. Sustain. Wirel. Syst. **2**(3), 107–117 (2020)

15. Wu, Q.: Design of a broadband blade and DRA hybrid antenna with hemi-spherical coverage for wireless communications of UAV swarms. AEU—Int. J. Electron. Commun. **140**, 153930 (2021). ISSN 1434-8411

Exploitation of Deep Learning Algorithm and Internet of Things in Connected Home for Fire Risk Estimating

Noor A. Ibraheem, Noor M. Abdulhadi, and Mokhtar M. Hasan

Abstract This research attempts to implement fire safety measures and ensure the safety of individuals with the preservation of property in institutions, health, and lives of citizens. Fire prevention procedures must be developed and implemented in order to assess, remove, and prevent probable causes of fires. Fire risk measures include creating ideal conditions for saving property and evacuating citizens. Preventive work with residents guarantees that the source of the ignition is identified and that emergency services are alerted as soon as possible in the event of a fire. An Unsupervised Deep Learning algorithm using the Internet of Things (IoT) concept was used to develop a fire risk estimation system. The suggested algorithm determines whether or not the scenario is safe. An alert signal is transmitted to three peripheral devices: the monitoring camera, the fire sensor due to high temperatures, and a sensor due to the increasing CO_2 levels in the atmosphere. The proposed system considers the data collected by various distributed sensors and later it is processed by using a deep learning algorithm to estimate the risk level at that point, release an alarm fire risk as a signal at a usual time for safety and preserve public property.

Keywords Internet of things · Deep learning algorithm · Clustering · Fire risk · Fire alarm · Temperature · Sensor monitoring

N. A. Ibraheem (✉) · N. M. Abdulhadi · M. M. Hasan
Computer Science Department, University of Baghdad/College of Science for Women, Baghdad, Iraq
e-mail: nooraibraheem@csw.uobaghdad.edu.iq

N. M. Abdulhadi
e-mail: noorma_comp@csw.uobaghdad.edu.iq

M. M. Hasan
e-mail: mokhtarmh@csw.uobaghdad.edu.iq

© The Author(s), under exclusive license to Springer Nature Singapore Pte Ltd. 2022   459
V. Suma et al. (eds.), *Evolutionary Computing and Mobile Sustainable Networks*,
Lecture Notes on Data Engineering and Communications Technologies 116,
https://doi.org/10.1007/978-981-16-9605-3_31

1   Introduction

Recently, society has become very fastly integrated with the convergence of smart devices. Since communication peripherals have different functions and are provided to all institutions through an intelligent system, it predicts the occurrence of a fire at first sight by installing many monitoring devices [1], thermal sensing systems, and also this phenomenon is spreading more quickly. Measuring the proportion of carbon dioxide in the atmosphere and according to a selected algorithm, the results of data analysis are received to determine an appropriate decision to provide preventive measures as well as communication with peripheral devices in order to provide services that are connected to the Internet of Things (IoT) [2]. Digital information is the main component of signal control and its flow in organizations operate on smart systems. IoT sensors can be as crucial as the firefighters themselves [3]. These sensors can therefore provide more critical coordination and information to groups outside of and in distant locations. As a result, it should be handled as soon as possible to avoid tragedy in lives and property [1–7]. It is viewed as a fire circumstance when the observed temperature surpasses 50 °C. Staff appearance with future time for assistance in fire hazards is approximately 15 min for reaching basic facilities such as medical clinics, schools, and banks [8].

This research work has proposed a smart system that adopts the use of Internet of Things [IoT] technology by connecting a surveillance camera system, a high-temperature sensor system, and a system for measuring carbon dioxide in the place and obtaining information from the sensors for collecting it in a database, which in turn enters the information into a smart system for analyzing the data and send a danger signal to a sensor that transmits the signal through IoT technologies, as this system has a set of significant features that make you feel protected without being afraid in case you forget to turn off the electrical devices and feel anxious. IoT applications have spread across a wide range of use cases and sectors. Any IoT framework consists of a set of four specific sectors: sensors/tools, network, data processing, and user interface. Figure 1 shows a diagram of an IoT system with

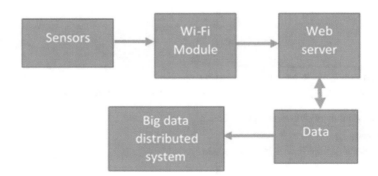

Fig. 1 Block diagram showing functioning of IoT with input data

input data. The rest of the paper is organized as follows: Sect. 2 explains the related works, Sect. 3 includes the steps for preparing the system, Sect. 4 explains the deep learning with IoT system, while Sect. 5 introduces the proposed framework for fire risk estimation, and Sect. 6 presents the conclusion and future works.

2 Related Works

The concept of IoT is now connected with several applications ranging from bright industry [9], healthcare [10, 11], smart agriculture [12], and smart home application [13]. Home automation is one such area where IoT has several advantages [14, 15]. Numerous investigates have been directed in IoT for detecting or preventing fire, for example, in [16] researchers have proposed a model that employs different detectors, such as heat, smoke, and flame. Signals from such sensors are processed by the framework computation to determine the risk of a fire and are then communicated to various gatherings through GSM modem, the recommended technique has been utilized to reduce false alarms. Researchers in [17] proposed a fire extinguishing control and monitoring system for assessment, with the system combining fire information from IoT with an alert and fire warning. Pipe flow, temperature, pressure, humidity, and voltage are the fire extinguisher data. Hariveena et al. [18], Saeed et al. [19] describes the fire detection and prevention system. In [19], authors have evaluated a wireless sensor network by applying multiple sensors. They applied mobile communications (GSM) to avoid false alarms and simulated the system using a fire dynamics simulator. In reference [20], a fire and gas detection at an early stage by using Arduino program to detect the environmental factors for an event of a fire with the assistance of a fire and gas sensor has been investigated. Table 1 shows the applications of the discussed researches.

Table 1 Applications of the some related researches

Research No.	Application
[10, 11]	Healthcare
[12]	Smart agriculture
[13]	Smart home application
[14, 15]	Home mechanization
[17]	Fire control and monitoring system
[18, 19]	Fire detection and prevention system
[20]	Fire and gas detection

3 Preparing the System

3.1 Connected Home

It is actively involved in the development of smart cities. It is the solution for the future, particularly in light of the changes and developments that have occurred in the world, as well as a large number of tasks and requirements that have led to the need to develop our homes, where the need to use various electronic devices Internet sensors (IoT) to collect data has increased. The insights obtained from that data are used simultaneously. To efficiently manage the system and services, this data is used in an intelligent system for processing and describing the process of monitoring and protection. The smart home is one of the most prominent applications in the Internet of Things [IoT] model. While it added a level of comfort and convenience to the users' daily lives, it also brought a unique security challenge, which is the mitigation of threats to household members and how to protect them in the event the home is compromised. For example, we will address the issue of fires that a house may be exposed to, as well as how to prevent these dangers from occurring in the first place. Due to the sheer involvement of Internet of Things devices and the existence of a relationship between smart systems in the home by connecting them to the Internet of Things, where the Internet of Things [IoT] platforms manage the controlling against internal threats based on a system managed by artificial intelligence using machine learning and making them responsible for making decisions, this problem can be addressed. The aforementioned facts act as motivation to suggest this system, which contributes to save individuals whose lives could have been lost in fires.

3.2 Fire Risk

Fires usually start on a small scale because most of them arise due to negligence in following fire prevention, and there are different ways and preventive measures to avoid heavy losses in life and property. These risks are divided into three types:

- Personal Danger: It represents the danger to individuals, where the lives of individuals are exposed to injuries, which requires the provision of measures to escape from dangers when a fire occurs.
- Destructive Risk: It includes the danger to property, which are risks that affect the components and contents of buildings and institutions.
- Exposed Danger: The danger to the neighborhood is the danger that threatens the locations close to the place of the fire, and therefore it is called the external danger.

3.3 Temperature Measure

One of the applications of daily life is the temperature that humans are directly affected by and further, it is necessary to adapt to it. There are sensors or temperature sensors that discuss the temperature and when linked to IoT devices to obtain the required information as a database to enter into the data analysis system to check whether there is an abnormal rise in temperature or not to take a decision on preventive measures and warn the existence of a danger. The proposed system suggests the utilization of temperature sensor, which is a temperature display circuit proposed by Arduino and LMB5.

3.4 Temperature Sensor

The LMB5 is a temperature sensor, which converts the surrounding heat into voltage in a proportional manner. The LMB5 operates at a temperature ranging from −55 °C to 150 °C above zero.

At room temperature, this sensor has an accuracy of ± 0.25 °C. For every temperature change of one degree, the voltage coming out of the LMB5 sensor changes by 10 mV. This sensor has three legs just like the transistor, as shown in Fig. 2.

The LMB5 sensor continuously monitors the room temperature and gives the analog value of the voltage that is equivalent and proportional to the temperature. This analog value is received by the Arduino through the input A0, and according to the program code, the Arduino converts this value into digital data that is displayed on the Arduino screen. Arduino is an electronic development board that receives and reads data from its center and surroundings through buttons, sensors, and various other pieces. LED, motors, relays, LCD screens, and many more. The major benefit of this device is to collect the data from the temperature sensor and convert it into digital data that is ready for analysis. Figure 2 shows the device and temperature sensor, respectively [21].

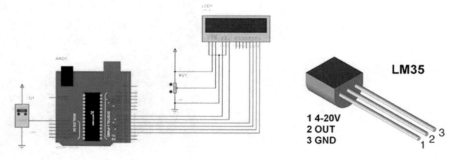

Fig. 2 Temperature sensors

A.   Carbon Dioxide Sensor

Carbon dioxide emissions have the potential to be harmful. The maximum acceptable amount is 5000 parts per million since it is odorless and colorless, and despite the fact that humans breathe carbon dioxide and that it is present in the atmosphere (approximately 400 parts per million) (0.5% by volume). It is the final outcome of complete combustion. The device is characterized by its low cost, and the air quality is detected in closed places, whether in offices, homes, or workshops. Equipped with a dual-display of temperature and carbon dioxide (CO_2) at the same time, it is also equipped with a carbon dioxide level alarm and a large LCD screen with high contrast as shown in Fig. 3.

Sensor (sensitive) working with waveguide technology, measurement rate CO_2: 0 ~ 9999 ppm (2001 ~ 9999 ppm not suggested). CO_2 measurement accuracy: ± 50 ppm ± 5% of reading (0 ~ 2000 ppm), CO_2 variation: 1 ppm [22].

B.   Intelligent Fire Alarm System

Conventional fire alarm systems usually consist of several devices that are located in different locations, and further, they are all connected through some wires to a single control panel. As soon as the device detects smoke or an abnormal temperature rise, it immediately sends an alarm signal to the control panel in order to sound the alarm. This type of system has many limitations and drawbacks. Here, we must focus on the effectiveness of the sensors that

Fig. 3 Carbon dioxide sensor

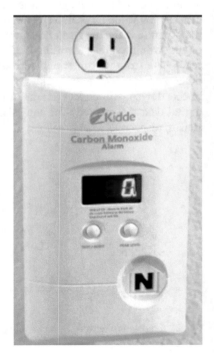

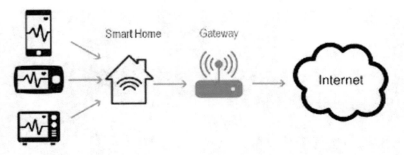

Fig. 4 Monitoring fire system [25]

are connected to transmission equipment for issuing alarms, which is then symbolized by the acronym ATE to contact the main civil defense center in the interest of the health and safety of citizens and other similar facilities [23].

C.   Monitoring Fire System

It is the control framework of the security system that alerts through movement sensors. The security framework triggers the caution and makes an impression on the proprietor of the house to control the smart home frameworks on the off chance that the house is presented with a fire risk [8, 24]. The security framework triggers the caution through the sound system and calls the numbers set for the alert if a house is exposed to fire risk or if there is a malfunction in the home's fire system [13]. There are additional sensors for water spillage and assume a similar part in case of any danger. There are numerous kinds of sensors that can be remembered for the framework, like gas or power sensors, and others. Our recommended observing framework is clarified in Fig. 4.

4   Deep Learning with IoT System

The Internet of Things is reshaping technological adaptation in human life so that a solution to this technology can replace humans in doing everyday work better [26], faster, and with high-precision technology, as well as the possibility of solving problems without resorting to human intervention, and this is done only by linking it to artificial intelligence (expert systems), including DL (deep learning) [27, 28]. Internet of Things where it is considered the mastermind and the main in controlling the Internet of Things in various fields [29, 30]. If we want to highlight the fundamental benefit of DL, its work is based on huge data sets. The main benefit of DL and its connection to internet technology IoT produces a lot of information that has been processed and analyzed and then takes the necessary measures and in conjunction with the work of IoT [25]. An amalgamation of these techniques, maintaining a balance between computational cost and efficiency is crucial for next-generation IoT networks. Hence, the applications built on top of modified stack will be significantly

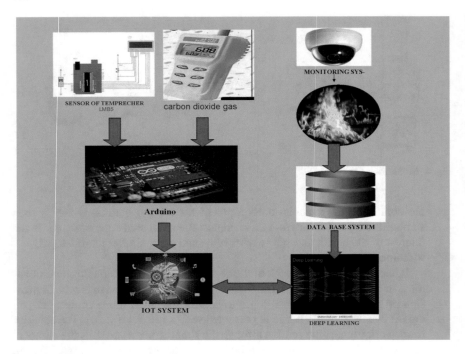

Fig.5 Suggested system framework

benefited and makes it easy to widely deploy the network, Fig. 5 shows the suggested firing system.

4.1 Deep Learning

The deep learning algorithm is capable to learn without human supervision and also it can be used for both structured and unstructured types of data. With extensive data training, deep neural networks show significant advantages when compared to traditional models due to their enhanced learning abilities. With the rapid development of high-performance parallel computing systems, For example, without supervision a pre-training layer was proposed.

Accordingly, refining with BP can find a better local minimum. It helps to solve the problem of inappropriateness in the large-scale training groups [10].

1. Input layer—The input layer has input that features a known dataset.
2. Hidden Layer—It is a hidden layer, just like we need to train the brain through hidden neurons.
3. Output layer—value that is required to classify.

The input feature is obtained from observation and put into a layer. That layer creates an output that becomes the input for the next layer, which is known as the hidden layer. This happens until we obtain the final output.

The network is further separated by adding a lot of hidden layers based on the complexity of the problem and connect everything just like the human brain interconnect and further the input values are processed through all hidden layers to obtain the output.

4.2 Deep Learning and Assembly Deep Learning for Clustering

Deep learning is a type of machine learning and training that uses a huge quantity of data to construct an informed model. DL algorithms are designed to learn feature Learning characteristics without the need to pre-identify such properties. Furthermore, it is one of the best algorithms used for teaching the different levels of data characteristics (e.g., images). In this view, deep learning is characterized by the innovation of new characteristics from which different levels can be learned, which may lead future researchers to focus on this very important aspect. From this perspective, features are the first factor used for the success of any intelligent machine learning algorithm. The ability to extract and/or choose properties properly will represent data and prepare them for learning is the point between success and failure of the algorithm [31]. In this research, clustering algorithms have been suggested and it is also considered as one of the unsupervised machine learning algorithms that aim to collect data according to their similarity. Similar data are present in one group and different data are present in different groups [32, 33]. These traditional algorithms face a major challenge as data dimensions increase, and also their performance becomes limited. In contrast, deep learning allows the users to learn features that represent the original data by making it easier for traditional algorithms to work better even in a low-dimensional space.

4.3 Development Environment

A fire monitoring system has been developed by using a smart surveillance camera that sends data to the data processing system by using deep learning algorithm, which connects to a sensor via a device linkage model and then sends a warning message in the event that the home electrical system is exposed to danger by causing an exposure to individuals. All of this is done by connecting the home network based on the Internet of Things [IoT] such as sensors/devices, connectivity data processing user interface, and distributor device, processing device, and the proposed environment has been referred to as shown in Fig. 6.

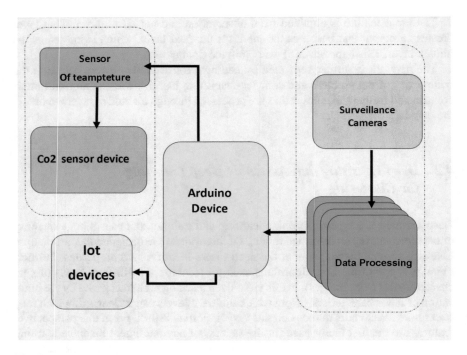

Fig. 6 Service structure

5 Proposed Framework for Fire Risk Estimation

The proposed system is a home fire monitoring system using IoT technology, by making a group of devices that can communicate with each other to control the matters in everything related to fire monitoring from the inside by connecting it to the electrical system to protect the members of the house from any danger they may be exposed to. The smart home system mainly relies on a monitoring device, as it transfers information to a data processing device that applies deep learning. the sensor device is responsible for capturing any abnormal signal in the fire monitoring system, as it captures the physical changes from the monitoring and processing device and transfers them to the main part of the system to be properly interpreted, and informs and alarms for danger in some way. All of these devices have been linked to a distributed device, where it connects all the devices and makes one interface that controls the system. In the following lines, we will talk about the group of devices that have been proposed with the technology used in the processing unit, and in the end, it will be an integrated picture of those systems for controlling the home.

Surveillance cameras are definitely one of the most popular products in the protection category. any change in temperature sensor or carbon dioxide level alarm, the data reaches the data processing part through a database used to collect the monitoring information, where the program performs processing using DL, and the processing

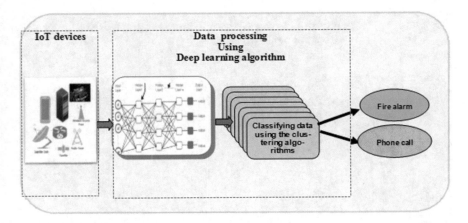

Fig. 7 Proposed deep fire monitoring system

part is linked to the distributor device as it acts as the brain and the main control center for smart home devices, to early control through a single interface.

After receiving a signal from the processing device, through the distributed device, it collects and sends data by linking to the cloud through a variety of methods including cellular devices, satellites, Wi-Fi, and Bluetooth, which use DL to process the information and check for true or false fire alarm. Then the output is reflected by making a phone call to a phone number activated for this purpose, or by applying the alarm as shown in Fig. 7.

5.1 Methodology

The number of groups in this algorithm is chosen into two groups. The algorithm will process the information and compare it to a database that contains all the data with the data from the sensors via the Arduino device and then make comparisons through the deep learning algorithm. Figure 8 shows the layers of deep learning algorithm, and Fig. 9 shows the proposed system.

6 Conclusion and Future Work

The smart home is one of the primary concerns for manufacturers, builders, and real estate residents to keep pace with the change of time. In this research, a home-based fire alarm system was proposed, where sensors connected to an Arduino device by using an artificial intelligence system have been used to detect a fire in the event of any change in the proportion and values of the sensors.

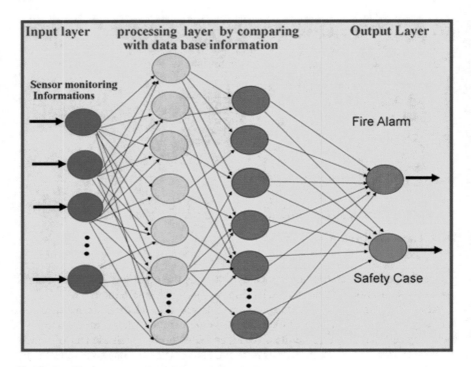

Fig. 8 Graphical representation of deep learning layers

The proposed research combines the monitoring system by linking it to tempera-ture sensors and carbon dioxide sensors to collect system information and save it to a database for performing data analysis.

The proposed system can obtain real-time working status information and send alarm to firefighting facilities such as temperature flow and carbon dioxide in the medium based on the analysis and control of the intelligent system for all the sensi-tive devices associated within the home. For future works, other different machine learning algorithms can be combined to achieve intelligent controlling system.

Conflict of Interest

The authors confirm that this article's content has no conflict of interest.

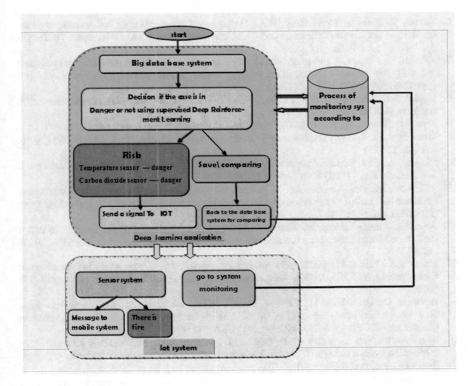

Fig. 9 Proposed system

Acknowledgements Declared none.

References

1. Shokouhi, M., Nasiriani, K., Khankeh, H., Fallahzadeh, H., Khorasani-Zavareh, D.: Exploring barriers and challenges in protecting residential fire-related injuries: a qualitative study. J. Inj. Violence Res. **11**(1), 81–92 (2019)
2. Kodur, V., Kumar, P., Rafi, M.M.: Fire hazard in buildings: review, assessment and strategies for improving fire safety. PSU Res. Rev. **4**(1), 1–23 (2019)
3. Salhi, L., Silverston, T., Yamazaki, T., Miyoshi, T.: Early detection system for gas leakage and fire in smart home using machine learning. In: 2019 IEEE International Conference on Consumer Electronics (ICCE), pp. 1–6. Las Vegas, NV, USA (2019)
4. Zhang, B., Sun, L., Song, Y., Shao, W., Guo, Y., Yuan, F.: DeepFireNet: a real-time video fire detection method based on multi-feature fusion. Math. Biosci. Eng.: MBE **17**(6), 7804–7818 (2020)
5. Mahzan, N.N., Enzai, N.M., Zin, N.M., Noh, K.S.S.K.M.: Design of an Arduino-based home fire alarm system with GSM module. J. Phys.: Conf. Ser. **1019**(1) (2018). Art. No. 012079

6. Suresh, S., Yuthika, S., Vardhini, G.A.: Home based fire monitoring and warning system. In: 2016 International Conference on ICT in Business Industry and Government (ICTBIG), pp. 1–6. Indore (2016)

7. Kanwal, K., Liaquat, A., Mughal, M., Abbasi, A.R., Aamir, M.: Towards development of a low cost early fire detection system using wireless sensor network and machine vision. Wireless Pers. Commun. **95**(2), 475–489 (2017)

8. Khalaf, O.I., Abdulsahib, G.M., Zghair, N.A.K.: IoT fire detection system using sensor with Arduino. Revista AUS 74–78 (2019)

9. Faraci, G., Raciti, A., Rizzo, S.A., Schembra, G.: Green wireless power transfer system for a drone fleet managed by reinforcement learning in smart industry. Appl. Energy **259** (2020). Art. No. 114204

10. Muneer, A., Fati, S.M.: Automated health monitoring system using advanced technology. J. Inf. Technol. Res. (JITR) **12**(3), 104–132 (2019)

11. Fati, S.M., Muneer, A., Mungur, D., Badawi, A.: Integrated health monitoring system using GSM and IoT. In: 2018 International Conference on Smart Computing and Electronic Enterprise (ICSCEE), pp. 1–7 (2018)

12. Ayaz, M., Ammad-Uddin, M., Sharif, Z., Mansour, A., Aggoune, E.H.M.: Internet-of-things (IoT)-based smart agriculture: toward making the fields talk. IEEE Access **7**, 129551–129583 (2019)

13. Muneer, A., Fati, S.M., Fuddah, S.: Smart health monitoring system using IoT based smart fitness mirror. TELKOMNIKA Telecommun. Comput. Electron. Control **18**(1), 317–331 (2020)

14. Hamdan, O., Shanableh, H., Zaki, I., Al-Ali, A.R., Shanableh, T.: IoT-based interactive dual mode smart home automation. In: 2019 IEEE International Conference on Consumer Electronics (ICCE), pp. 1–2. Las Vegas, NV, USA (2019)

15. Vishwakarma, S.K., Upadhyaya, P., Kumari, B., Mishra, A.K.: Smart energy efficient home automation system using IoT. In: 2019 4th International Conference on Internet of Things: Smart Innovation and Usages (IoT-SIU), pp. 1–4. Ghaziabad, India (2019)

16. Alqourabah, H., Muneer, A., Fati, S.M.: A smart fire detection system using IoT technology with automatic water sprinkler. Int. J. Electr. Comput. Eng. (IJECE) **11**(4), 2994–3002 (2021)

17. Li, Y., Yi, J., Zhu, X., Wang, Z., Xu, F.: Developing a fire monitoring and control system based on IoT. In: 2nd International Conference on Artificial Intelligence and Industrial Engineering (AIIE 2016), Advances in Intelligent Systems Research, vol. 133 (2016)

18. Hariveena, Ch., Anitha, K., Ramesh, P.: IoT-based fire detection and prevention system. IOP Conf. Ser.: Mater. Sci. Eng. ICRAEM (2020)

19. Saeed, F., Paul, A., Rehman, A., Hong, W.H., Seo, H.: IoT-based intelligent modeling of smart home environment for fire prevention and safety (2018)

20. Yadav, R., Rani, P.: Sensor based smart fire detection and fire alarm system. J. Sens. Actuator Netw. **7**(11) (2018)

21. Chu, H.-M., Lee, C.-T., Chen, L.-B., Lee, Y.-Y.: An expandable modular internet of things (IoT)-based temperature control power extender. Electronics **10** (2021)

22. https://arabic.alibaba.com/product-detail/mg811-carbon-dioxide-sensor-co2-sensor-gas-sensor-module-60609311308.html

23. https://rpmanetworks.com/ar/%D9%86%D8%B8%D8%A7%D9%85-%D8%A5%D9%86%D8%B0%D8%A7%D8%B1-%D8%A7%D9%84%D8%AD%D8%B1%D9%8A%D9%82-%D8%A7%D9%84%D8%B0%D9%83%D9%8A/

24. Elgendi, M., Al-Ali, A., Mohamed, A., Ward, R.: Improving remote health monitoring: a low-complexity ECG compression approach. Diagnostics **8**(10) (2018)

25. Ibraheem, N., Hasan, M.: Combining several substitution cipher algorithms using circular queue data structure. Baghdad Sci. J. **17**(4) (2020)

26. Pandey, S., Jaiswal, S., Yadav, N., Sonawane, J.: IOT based home automation and analysis using machine learning (2019)

27. Chen, J.I.-Z., Chang, J.-T.: Route choice behaviour modelling using IoT integrated artificial intelligence. J. Artif. Intell. **2**(04), 232–237 (2020)

28. Smys, S., Basar, A., Wang, H.: Artificial neural network based power management for smart street lighting systems. J. Artif. Intell. **2**(01), 42–52 (2020)
29. Li, H., Ota, K., Dong, M.: Learning IoT in edge: deep learning for the internet of things with edge computing. IEEE Network **32**(1), 96–101 (2018)
30. Al-Garadi, M.A., Mohamed, A., Al-Ali, A., Du, X., Guizani, M.: A survey of machine and deep learning methods for internet of things (IoT) security. Novel Deep Learning Architecture for Physical Activities assessment, mental Resilience and Emotion Detection (2018)
31. Minaee, S., Kafieh, R., Sonka, M., Yazdani, S., Soufi, J.: Deep-COVID: predicting covid-19 from chest X-ray images using deep transfer learning. Med. Image Anal. **65** (2020)
32. Vijayakumar, T., Vinothkanna, R., Duraipandian, M.: Fuzzy logic based aeration control system for contaminated water. J. Electron. **2**(01), 10–17 (2020)
33. Sungheetha, A., Sharma, R.: Fuzzy chaos whale optimization and bat integrated algorithm for parameter estimation in sewage treatment. J. Soft Comput. Paradigm (JSCP) **3**(01), 10–18 (2021)

Review on Methods to Predict Metastasis of Breast Cancer Using Artificial Intelligence

Sunitha Munappa, J. Subhashini, and Pallikonda Sarah Suhasini

Abstract Breast cancer is the most prevalent invasive malignancy in women world-wide. Around 70% of breast cancer patients will have cancer recurrence, which can emerge as a local, regional, or distant tumor recurrence. Breast cancer recurrence is normally observed after chemotherapy. Breast cancer metastasis is primarily observed in bone, liver, lung, and brain. PET scans are available to identify metastasis, but for saving lives, this research study is focusing on prediction approaches. If metastasis is anticipated before tumor development in a secondary part of the body, the patient's life expectancy can be extended. This paper explores the reasons for metastasis and methods to predict the occurrence of metastasis before the formation of tumor in other parts of the human body. Two vital prediction methods viz., lymph node analysis and detection of circulating tumor cells (CTCs) using deep learning techniques like convolution neural networks (CNNs), cycle-consistent generative adversarial community (CycleGAN), lymph node assistant (LYNA) based on inception V3 deep learning structure are illustrated in this paper.

Keywords Breast cancer · Metastasis · Lymph node analysis · Circulating tumor cells · Deep learning · CNN · CycleGAN · LYNA · Inception V3

1 Introduction

Breast cancer is a fatal disease that affects women and is one of the world's fastest growing diseases. Since identifying tumors in the breast by palpation is not practicable in tiny sizes, a greater proportion of patients detects them when they pass

S. Munappa · J. Subhashini (✉)
S R M Institute of Science and Technology, Chennai, India
e-mail: subhashj@srmist.edu.in

S. Munappa
e-mail: sm4461@srmist.edu.in

P. S. Suhasini
V R Siddhartha Engineering College, Vijayawada, India

© The Author(s), under exclusive license to Springer Nature Singapore Pte Ltd. 2022   475
V. Suma et al. (eds.), *Evolutionary Computing and Mobile Sustainable Networks*,
Lecture Notes on Data Engineering and Communications Technologies 116,
https://doi.org/10.1007/978-981-16-9605-3_32

the initial stage. The survival rate of breast cancer is determined by the type and stage of the disease. Treatment may also change based on the type and stage of the disease [1]. Treatment for breast cancer may be chemotherapy followed by surgery and radiation. Order depends upon stage and type of cancer. Early detection of tumor may only need surgery or chemotherapy. Chemotherapy is a drug that directly kill cancer cells. Chemotherapy is used to control the tumor growth and shrink the size of tumors before surgery or radiation therapy.

Chemotherapy given prior to surgery is called neoadjuvant therapy, and post surgery is called adjuvant therapy. When cancer cells from the original tumor develop cancers in other parts of the body, this is referred to as metastasis. In metastasis, after chemotherapy, cancer cells detached from tumor and travel through the blood and lymph system and form new tumors (metastatic tumors) in other parts of the body. This secondary tumor is the same type of first one.

Investigations done on mice suggests that chemotherapy given before surgery for breast cancer can cause the metastasis. By neoadjuvant therapy changes occur in tumor cells in and around the tumor, these changes lead to an increased risk of the cancer that spread to other areas of the body. Researchers used mouse models of breast cancer to demonstrate that chemotherapy drugs given to different types and stages of breast cancer increase the number of microscopic structures in breast tumors known as tumor microenvironment of metastasis (TMEM) and the number of tumor cells circulating in the blood. TMEM acts as way that help invasive cancer cells jump from a tumor and move into the circulation, this is the main cause for cancer metastasis.

Bone, liver, lung, and brain are the general primary spreading sites of breast cancer metastasis [3].

Mostly, metastasis is the cause for the increase in mortality rate in breast cancer patients [2]. The formation of breast cancer metastasis [3] is as shown in Fig. 1. If possibility of metastasis is identified before the formation of tumor in other parts of body, then life span of patients can increase and it will helpful for prognosis of treatment.

Breast cancer metastasis process is as shown in Fig. 2.

In the recent years, with the advancement of artificial intelligence, machine learning and deep learning are used widely in medical research. Many algorithms are developed for selecting the best weights for features in image recognition, including neural networks (NNs), decision trees (DTs), support vector machines (SVMs), Naive Bayes, and K-nearest neighbors. The CNNs outperform the machine learning methods and are successful in diagnosing the cancer [19]. The recent research works are more focused on applying different deep learning methods to predict breast cancer metastasis.

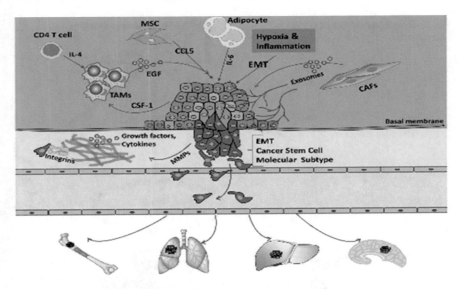

Fig. 1 Formation of breast cancer metastasis [3]

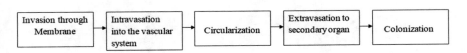

Fig. 2 Breast cancer metastasis process

2 Literature Review

The broad types of detection method for breast cancer are PET scan, lymph node analysis, and by the detection of circulating tumor cells (CTCs) by using different images that are obtained through imaging procedures. These methods are reviewed in this paper (Fig. 3).

2.1 Detection of Metastasis by PET Scan

A positron emission tomography [PET] is used to detect the tumors presence in the body. PET sweep can deliver multidimensional, shading pictures of functions of the human body. In a PET sweep, radiotracer is infused into the body, where it goes to cells that utilization glucose for energy. The more energy a group of cells need, the more the radiotracer develops in that region. This will be seen on images returned by a PCPET scan, which detects the tumor after it has formed [4] in the bones, lungs, liver, and other areas of the body (Fig. 4).

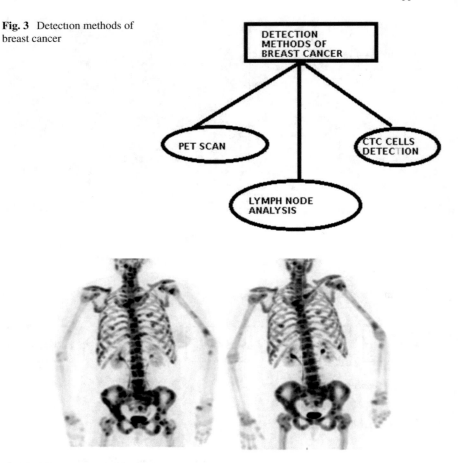

Fig. 3 Detection methods of breast cancer

Fig. 4 PET scan image [4]

The research is not very well focused for the detection of metastasis using PET scan, as this method can detect the tumor, i.e., after formation of tumor in other parts of the body. But if we detect the metastasis of breast cancer before formation of tumor, the patient's lifespan may be prolonged.

So, the research on the detection of metastasis is focused on two methods viz., lymph node analysis and CTC detection, which detect the metastasis before the formation of tumor, lymph node analysis, and detect the circulating tumor cells in blood.

2.2 Prediction of Metastasis by Lymph Node Analysis

Metastasis prediction by lymph node analysis done by taking biopsy images, i.e., it performs a dominant function in breast cancer analysis and prognosis. These images are non-uniform in nature, where training the model with rotated images will give an improved accuracy [20].

Wollmann et al. [5] in their research paper presented the detection process for metastasis of breast cancer by using lymph nodes. They used cycle-consistent generative adversarial community (CycleGAN) along-side a densely linked deep neural network. CAMELYON17 dataset is used to include whole slide images from hundred patients, who are gathered from five clinical centers within Netherlands. The whole slide images within the information show five lymph nodes according to a patient and labeled with a patient. The classification is performed by taking small image patches and uses model averaging for enhancement.

A new deep learning approach for sparse classification of WSIs and automated breast cancer grading is carried out. This approach uses end to end learning, and the model requires simplest side-level annotations in preference to pixel-degree annotations.

Lin et al. [8] have done object detection by patch-based frameworks. Here, data augmentation could be very flexible while training a deep neural network. While a window scans densely, a massive amount of redundant computations exist inside the overlapping area. To address this issue, a modified fully convolutional network (FCN) takes arbitrary-sized images as input for performing immediate prediction.

Modified FCN [9] eliminates the upsampling path that is necessary for segmentation, however, it is not essential for performing detection procedure. The upsampling direction consumes more time for the detection as by taking the large size of whole slide images but does not provide any extra advantage to the high-quality heat map for the detection procedure.

Modified FCN based on VGG16 network [10] is used by supplanting the 3 fully linked layers with fully convolutional layers $1024 \times 1024 \times 2$ with kernel size 1×1. The padding operations are taken out from the structure to avoid the downside of boundary impact and protect the inference equivalence [7]. Based on these changes in the architecture, FCN can recognize the transferred functions learned from an enormous set of images with those verified constant upgrades without transfer learning. Natural language processing (NLP) algorithm is applied to extract the important features of breast cancer metastasis [22].

The Camelyon16 challenge dataset is used in this work [8]. This contains a total of 400 whole slide images. These WSI images are classified as micro metastasis and macro metastasis. Fast ScanNet with dense analysis gives more metastasis detection sensitivity, less false positive rate, and advanced performance with a much quicker pace of less than 1 min localization of tumor (Fig. 5).

Steiner et al. [11] have evaluated the capability of interpretation in digitized slides. As a result, six pathologists have assessed 70 digitized slides from lymph node areas in 2 reader modes, namely unassisted and assisted.

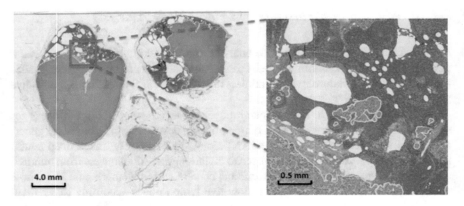

Fig. 5 View of WSI of lymph node [8]

For the assisted mode, deep gaining knowledge of the set of rules with lymph node assistant (LYNA) was utilized to detect the outline areas with a high probability of containing the tumor. A small version of the inception V3 deep learning structure is trained on whole slide images of digitized pathology slides, divided as $32 \times 32 \mu m$ tiles, and tiles-related marking determines whether the tissue region contains tumor or not (1 or 0). By changing the weights of the model, the similarities among its predicted and the labeled output are reduced. Thus, the model has been progressively learned to identify the irregularities from image tiles [12]. Then, applied the model to predict the reach of the tissue place and the probability of containing the tumor.

If no tumors are classified as negative, the tumors are classified as micrometastases if the tumor length is greater than 0.2 mm and equal to 0.2 mm, as macrometastases if the tumor length is greater than 2 mm, and it will be classified as remote tumor cellular clusters if scattered tumor cells or tumor cluster equal to 0.2 mm (Fig. 6).

LYNA-assisted pathologists have higher accuracy than unassisted pathologists. Specifically, the algorithm has significantly elevated the sensitivity of detection for

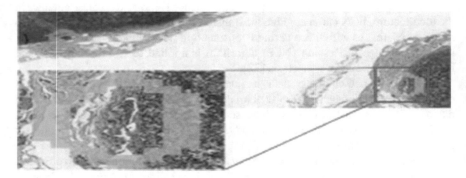

Fig. 6 LYNA-assisted image [11]

micrometastases, average evaluation time consistent with image, which has become significantly shorter without assistance. After predicting the metastasis, the result can be intimate to patient through IoT [21].

2.3 Prediction of Metastasis by CTC Cells

Circulating tumor cells known as CTCs are migratory cells inside the blood that causes to spreading of tumor cells throughout the body. CTCs are these days one among the predominant promising tumor biomarkers. CTC distinguish can be accomplished via liquid biopsy which is a non-invasive approach for taking certain tumor gadgets in blood. The popularity of CTCs is tough because of the presence of a very small range and no longer continuously disseminated into the bloodstream. CTCs may have a mean density of much less than one cell in line with ml, up to extra than 10 cells in line with ml of blood in patients having metastatic malignancy, while white blood platelets are inside the thousands and thousands and red blood cells are in billion.

Ciurte et al. [13] proposed an approach that depends on the assessment of image traits and boosting strategies, some of the absolute and well-known label-free microscopic techniques utilized for residing cells are phase contrast (PC), differential interference contrast (DIC), darkfield (DF), and pretty advanced virtual holographic microscopy. The images are obtained by the DF, the method is utilized for its simplicity and the awesome nature of live cell pictures, with numerous subtleties of the cellular shape. This process makes the identification CTCs from different blood cells.

Detection is done by the concepts of integral channel features and CNN [6]. For classifying CTCs, a few picture qualities, such as histogram statistics, grayscale co-occurrence matrix and color statistics [14], were used here. The values of the feature vector are the ones of the coordinates of the present pixels in all of the convolution channels and moreover the ones of a few other adjoining pixels. Characterized community as the association of pixels covering the space of the tumor cellular. In doing as such, ready to give a portrayal that models the texture of the cell (Figs. 7 and 8).

Binsheng et al. [15] carried out on actual-global samples for the detection of CTC cells, rather than the usage of public databases. By detecting the nuclei of CTC cells, CTC cells were detected and counted. AlexNet [16] was used, the architecture includes 8 weighted layers, the first five layers are convolution layers, and the final 3 layers from the 8 layers, i.e., the ultimate 3 layers are full connection layers. The five-fold move-validation was applied to prevent overfitting.

Six hundred patient's data are taken with 2300 images of circulatory tumor cells. Every image was divided into three or four channels with different shades. The acquired sensitivity and specificity are 93.7% and 92.1%, respectively.

DF microscopic images of impeccable blood are utilized by Ciurte et al. [17] in their work, to detect the existence of CTC cells. The images were procured with a

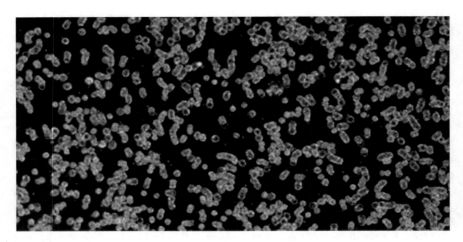

Fig. 7 DF microscopic image of BLOOD BIOPSY with high density of blood cells [13]

Fig. 8 Location of adjacent pixels indicated red (left), multi-resolution convolution channels (right) [13]

magnifying lens in DF mode at 10X and 20X optical magnification. Detection of red cells, white cells, and CTCs from blood samples of breast cancer patients is done.

The image was examined in several illustrations which include RGB color space, HSV coloration space and the radial profile of the cells, and computed number of capabilities on them [18], particularly histogram measurements, gray tone difference matrix, gray level co-occurrence matrix, and a few proposed radial functions. The classification is performed by using support vector machine (SVM) classifier, yielded 98% precision (Fig. 9).

2.4 Comparison of Methods

From the detailed aforementioned review, PET scan is affordable and it cannot be used for the prediction of breast cancer metastasis. While the other two methods: lymph node analysis and CTC cells detections are capable of predicting the metastasis.

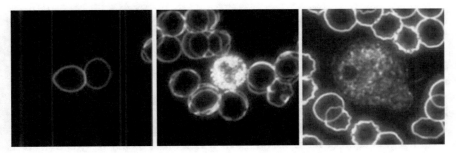

Fig. 9 Red (left), white (middle), and CTC cells (right) in darkfield microscopy [17]

Table 1 Comparison of metastasis detection/prediction methods

Parameter	PET scan	Lymph node analysis	CTC cells detection
Detection/prediction of metastasis	Detection	Prediction	Prediction
Possibility of detection	After formation of tumor	Before formation of tumor	Before formation of tumor
Type	Non-invasive	Invasive	Non-invasive
Cost	Expensive	Less compared to the other two methods	Expensive
Possibility of getting effected	Radiation effect will be more	No radiation effect	No radiation effect
Type of diagnosis	Scan	Biopsy	Liquid biopsy
Probability of detection	Less	More	More
Availability of test	In cities	In cities with well-equipped labs	Only one or two labs across India

However, they are having their own limitations in terms of cost, lab facility, etc. Table 1 illustrates the comparison of the three methods reviewed in this paper.

3 Conclusion

Neoadjuvant chemotherapy is a treatment given to breast cancer patients before proceeding to surgery. This causes cancer cells to migrate away from the site where they were previously observed. They spread through the blood or lymph system and develop secondary tumors, or metastatic tumors, in other parts of the body. In metastasis, the mortality rate is quite high. Detecting metastasis is critical for increasing the patient's life duration or prognosis of the treatment strategy. Methods for predicting metastasis using image processing and artificial intelligence are given in this study. The two ways to predict through image processing techniques are lymph node analysis and CTC cells detection.

The detection method used here is a PET scan, which is taken by injecting radio-tracer into a vein before the scan, where whole body scan will be displayed in reports, and here, we may identify the tumor after the formation in secondary location. However, this approach is costly. Instead of detecting the tumor after metastasis, it is better to predict metastasis to save or prolong the life of patient. For that, the lymph node analysis and CTC cells detection are considered.

In the prediction of metastasis through lymph node, biopsy of lymph nodes is taken near the breast or neck. By analyzing the presence of cancer cells, the chance of getting metastasis could be estimated. This method is not expensive and biopsy will be done in pathology labs.

In the prediction of metastasis through CTC cells identification, by taking blood biopsy (8–10 ml) of chemotherapy patients (one week after chemotherapy treatment), we analyze the WSI for identification of presence of CTCs and count the number of CTCs. Detection and counting are done by using image processing techniques with the help of artificial intelligence in order to avoid false detection and improve speed in both the methods. If CTCs are present and count is high, the doctor may change the treatment plan to prevent the metastasis. The data in blood biopsy images are not publicly available but we can get data in hospitals, where blood biopsy has been done. It is costly and availability of blood biopsy facility is very rare. Further, the research could also be focused on the identification of CTC cells in alternate ways, which would be helpful to increase the life span of cancer patients.

References

1. Boyle, P., Levin, B.: World Cancer Report. World Health Organization, Geneva, Switzerland (2008)
2. Allemani, C., Matsuda, T., Di Carlo, V., Harewood, R., Matz, M., Niksic, M., et al.: Global surveillance of trends in cancer survival 2000–14 (CONCORD-3): analysis of individual records for 37 513 025 patients diagnosed with one of 18 cancers from 322 population-based registries in 71 countries. Lancet **391**, 1023–1075 (2018)
3. Liang, Y., Zhang, H., Song, X., Yang, Q.: Metastatic heterogeneity of breast cancer: molecular mechanism and potential therapeutic targets. Semin. Cancer Biol. **60**, 14–27 (2020)
4. Azad, G.K., Taylor, B.P., Green, A., Sandri, I., et al.: Prediction of therapy response in bone-predominant metastatic breast cancer: comparison of [18F] fluorodeoxyglucose and [18F]-fluoride PET/CT with whole-body MRI with diffusion weighted imaging. Eur. J. Nucl. Med. Mol. Imag. 1–10 (2018)
5. Wollmann, T., Eijkman, C.S., Rohr, K.: Adversarial domain adaptation to improve automatic breast cancer grading in lymph nodes. In: IEEE 15th International Symposium on Biomedical Imaging (ISBI 2018), pp. 582–585 (2018)
6. Huang, G., Liu, Z., Weinberger, K.Q., van der Maaten, L.: Densely connected convolutional networks. In: iProc. CVPR 2017 (2017)
7. Zhu, J.Y., Park, T., Isola, P., Efros, A.A.: Unpaired image-to-image translation using cycle-consistent adversarial networks. arXiv:1703.10593 (2017)
8. Lin, H., Chen, H., Graham, S., et al.: Fast scanNet-fast and dense analysis of multi-gigapixel whole-slide images for cancer metastasis detection. IEEE Trans. Med. Imag. 1–11 (2018)

9. Chen, H., Qi, X., Yu, L., Dou, Q., Qin, J., Heng, P.-A.: Dcan: deep contour-aware networks for object instance segmentation from histology images. Med. Image Anal. **36**, 135–146 (2017)
10. Simonyan, K., Zisserman, A.: Very deep convolutional networks for large-scale image recognition, arXiv preprint arXiv:1409.1556 (2014)
11. Steiner, D.F., MacDonald, R., Liu, Y., et al.: Impact of deep learning assistance on the histopathology review of lymph nodes for metastatic breast cancer, impact of deep learning assistance. Am. J. Surg. Pathol. **42**, 12 (2018)
12. Bejnordi, B.E., Veta M., Johannes van Diest, P., et al.: Diagnostic assessment of deep learning algorithms for detection of lymph node metastases in women with breast cancer. JAMA **318**, 2199–2210 (2017)
13. Ciurte, A., Selicean, C., Soritau, O., Buiga, R.: Automatic detection of circulating tumor cells in darkfield microscopic images of unstained blood using boosting techniques. PLoS One, Open Access J. (2018)
14. Ciurte, A., Marita, T., Buiga, R.: Circulating tumor cells classification and characterization in dark field microscopic images of unstained blood. In: IEEE International Conference on Intelligent Computer Communication and Processing, pp. 367–374 (2015)
15. Binsheng, Lang, J., Li, S., Zhou, Q., Liang, Y.: A new method for CTC images recognition based on Machine Learning. Front. Bioeng. Biotechnol. **8** (2020). Article 897
16. Wainberg, M., Merico, D., Delong, A., Frey, B.J.: Deep learning in biomedicine. Nat. Biotechnol. **36**, 829–838 (2018). https://doi.org/10.1038/nbt.4233
17. Ciurte, A., Marita, T., Buig, R.: Circulating tumor cells classification and characterization in dark field microscopic images of unstained blood, 978-1-4673-8200 7/15. In: IEEE Conference on Biomedical Image Processing, pp. 367–374 (2015)
18. Li, K., Miller, E., Chen, M., Kanade, T., Weiss, L., Campbell, P.: Cell population tracking and lineage construction with spatiotemporal context. Med. Image Anal. **12**, 546–566 (2008)
19. Binsheng, H., Qingqing, L., Jidong, L., Hai, Y., Chao, P., Pingping, B., Shijun, L., Qiliang, Z., Yuebin, L., Geng, T.: A new method for CTC images recognition based on machine learning. Front. Bioeng. Biotechnol. **8** (2020)
20. Sungheetha, A., Sharma, R.: 3D Image processing using machine learning based input processing for man-machine interaction. J. Innov. Image Process. (JIIP) **3**, 1–6 (2021)
21. Joe, M.C.V., Raj, J.S.: Location-based orientation context dependent recommender system for users. J. Trends Comput. Sci. Smart Technol. (TCSST) **3**, 14–23 (2021)
22. Alzubi, A., Najadat, H., Doulat, W., Al-Shari, O., Zhou, L.: Predicting the recurrence of breast cancer using machine learning algorithms. Multimedia Tools Appl. **80**, 13787–13800 (2021)

Epileptic Seizure Prediction Using Geometrical Features Extracted from HRV Signal

Neda Mahmoudi, Mohammad Karimi Moridani, Melika Khosroshahi, and Seyedali Tabatabai Moghadam

Abstract The prediction of epileptic seizures in patients can help prevent many unwanted risks and excessive suffering. In this research, electrocardiography (ECG) signals for 7 patients under supervision and aged 31–48 years old, both male and female, were recorded. To predict recurrent seizures, we aimed to differentiate heart rate variation (HRV) signals in both seizure and seizure-free states. The evaluation of ECG signals alone was not able to predict recurrent seizures. Furthermore, we extracted geometrical features from our recordings. The results showed distinct differences between the two states with a high percentage of performance evaluation. Therefore, the proposed algorithm was successful and can be used to predict these epileptic seizures, thus decreasing the pain and suffering in patients. In this study, we have presented a new approach based on the geometric feature from extracted HRV analysis. We achieved a sensitivity of 100%, an accuracy of 90%, and a specificity of 88.33%, making our study practical for predicting seizures based on the pre-ictal state. With this approach, high-risk patients will be recognized, under control, and monitored sooner.

Keywords Epilepsy · HRV · Seizure prediction · Geometrical feature · Machine learning

N. Mahmoudi · M. Khosroshahi · S. T. Moghadam
Department of Biomedical Engineering, Islamic Azad University, North
Tehran Branch, Tehran, Iran

M. K. Moridani (✉)
Department of Biomedical Engineering, Faculty of Health, Tehran Medical Sciences,
Islamic Azad University, Tehran, Iran
e-mail: karimi.m@iautmu.ac.ir

© The Author(s), under exclusive license to Springer Nature Singapore Pte Ltd. 2022
V. Suma et al. (eds.), *Evolutionary Computing and Mobile Sustainable Networks*,
Lecture Notes on Data Engineering and Communications Technologies 116,
https://doi.org/10.1007/978-981-16-9605-3_33

1 Introduction

Epilepsy is a common chronic neurological disorder that affects people of all ages, and according to World Health Organization (WHO), data have a prevalence of 50 million people globally. Detection of this disease has been found in 80% of people under the age of 20 [1]. Contrary statistics have shown that in recent years detection of this disease has increased in the elderly. According to WHO, about 80% of people with epilepsy live in particularly low- and middle-income countries with three-quarters of this population not obtaining the care they require. Furthermore, according to this organization, "epilepsy is characterized by recurrent seizures which are brief episodes of involuntary movement that may involve a part of the body (partial) or the entire body (generalized) and are sometimes accompanied by loss of consciousness and control of the bowel or bladder function" [2].

Seizures are controlled in roughly two-thirds of patients with medication, whereas in remaining cases despite receiving appropriate therapy, they remain intact. The term for this condition is drug-resistant epilepsy. Epilepsies that are resistant to treatment cause significant emotional, family, and social hardship [3]. Depending on the type of epilepsy, an epileptic patient may experience two types of seizures: unilateral and bilateral seizures [4]. Most epileptic seizures are self-limiting, but they can occasionally develop to a condition called status epilepticus, a potentially life-threatening, long-term disease. Despite the fact that most seizures, including generalized tonic–clonic seizures (GTCS), do not cause permanent brain damage, they are linked to a slight risk of death due to cardiac and respiratory issues, as well as sudden unexpected death in epilepsy (SUDEP). Individuals who sleep alone and experience GTCS during sleep are at a particularly high risk of SUDEP [5]. SUDEP is defined as the sudden and unexpected death of a person with epilepsy for no apparent reason [6]. Another one of the episodes' damaging aspects is their unpredictability.

In the past few years, plenty of researchers have studied the prediction of epileptic seizures with the help of electroencephalography (EEG) signals. These signals obtained from video-electroencephalography were the gold standard for documenting epileptic episodes in hospitals [7]. Other studies have shown that research concerning epileptic seizure prediction and detection with the help of EEG recordings has been increasing. This incurs patients to have lower potential risks when taking action before the prevalence of the seizure [8]. Wet electrodes on the scalp are required for this type of EEG recording, which is bothersome and requires the assistance of a qualified nurse to position them.

In addition, the detection is dependent on manual human evaluation. As a result, the EEG must be analyzed by a skilled EEG analyst, which takes time. Consequently, EEG is not currently suited as a home-based, automated long-term seizure detection system [9]. It detects voltage fluctuations in the brain caused by ionic current flows. The EEG is assumed to record a time series of such voltage fluctuations (signals) that correlate to neural activity. The EEG can aid in the discovery and characterization of irregular brain activity by comparing and contrasting EEG records from numerous patients [10]. Some studies have shown an accuracy of 100% for seizure detection. One of the most effective methods for prediction had a sensitivity of 97.5% and a false positive rate of 0.27 per hour [11–18].

Past research has also observed that during these recurrent seizures, the nervous system functions (both voluntary and involuntary) are affected. More importantly, these seizures are accompanied by significant autonomic nervous system (ANS) modifications where heart rate and rhythm are frequently affected [12]. As a result, electrocardiography (ECG) signals have been utilized to detect and forecast seizure activity based on ECG patterns linked with known seizures. According to one study from 2009, this approach had a sensitivity of 85.7% and a specificity of 84.6% [13]. Notably, in all ECG signal-based prediction studies, the prediction is based on classical heart rate variability (HRV) indices retrieved in the time or frequency domain, which indicate the regulation of sympathetic and parasympathetic ANS on heart rate [14].

The ANS or the autonomic nervous system (also known as the visceral or involuntary nervous system) also has a vast effect on the heart without intentional control. The autonomic nervous system plays a big part in emotional experience and expression. The body tends to show certain reactions when feeling excitement, such as blood pressure and heartbeat increase, "butterflies," and dry mouth, all controlled by the ANS [15]. Autonomic changes might even result in sudden death in patients with epilepsy [16].

In this paper, we aim to explore the feasibility prediction of epileptic episodes with the help of ECG recordings of the pre-ictal state. However, we are incapable of diagnosis based on just these signals. A rather eye-opening problem is the incapability to find the triggering factor for heart rate oscillation in this state due to anxiety, epilepsy, and/or high heart rate. Part of our aim is to look for non-linear methods capable of showing differentiation between seizure and seizure-free status of the patients, such as the return map algorithm. Now, what grabs attention is how we plan on doing so using these ECG signals. The detection of epileptic seizures is not possible only based on the study of ECG signals because of the numerous triggering factors of oscillations in the recordings, thus making it hard for doctors to analyze. This is another reason why we cannot analyze just based on these signals. Instead, we extract the RR interval from QRS waves of each patient and seizure. HRV is known to be characterized by the differentiation of RR intervals [17]. To do so, we will use different phase spaces to extract more discriminant features to differentiate seizure from seizure-free states.

The rest of this paper is written out as follows: Sect. 2 covers the materials and methods employed in this study, which include data acquisition and ECG signals retrieved from PhysioNet. Section 3 is a description of our experimental findings, and lastly, Sect. 4 is a discussion and conclusion of our analyzes.

2   Materials and Method

2.1   Database

In order to achieve the study's goal, data of continuous ECG recordings during seizures were obtained. The study consisted of 7 patients aged 31–48 years old, both male and female, with partial epilepsy and no clinical evidence of cardiac disease. These patients had partial seizures with or without secondary generalization from frontal or temporal foci. All of the ECG recordings were collected while the patients were hospitalized. The seizure dataset used in this study has been obtained from Physiobank from the "post-ictal heart rate oscillations in partial epilepsy" database [18].

2.2   Proposed Method

Many different methods have been presented for the prediction of epileptic seizures in the past, but the methodology implemented in this paper is with the proposed approach of non-linear return maps and employment of the selected geometric feature that will be discussed ahead. Figure 1 is the block diagram of our proposed algorithm. The main reasoning behind the choice of our approach was the noticeable difference

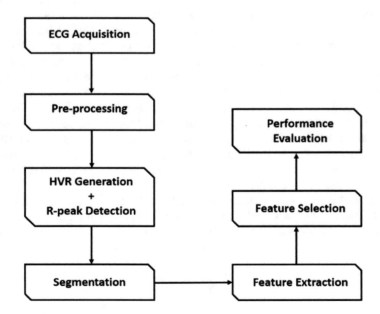

Fig. 1 Block diagram of proposed algorithm

between the amounts obtained for seizure and seizure-free periods. The first stage of the methodology was the extracting of the single-lead ECG recordings from our seven patients. Each patient and the time of their seizures have been recorded. Now, our primary goal is to distinguish the first indications of the onset of epileptic seizures. Because the pre-ictal interval has yet to be characterized, the duration of the interval varies between patients and from the same patient [19]. The table below is the time of seizure and its duration for each of the 7 patients. As presented in the table, patients 2, 3, and 6 each have had two seizures in the time hospitalized and while under supervision.

Preprocessing and R-peak Detection The ECG data were evaluated first, then detrended to remove significantly greater non-stationary heart rate changes that could disguise low-frequency oscillations [20].

The most commonly used algorithm for QRS detection is the Pan Tompkins algorithm [21]. After finding the R peaks and RR interval, they were filtered to eliminate any noise and false oscillations. In Fig. 2, a sample of patient HRV during seizure and seizure-free periods recorded is shown.

Feature Extraction and Selection We extracted each signal's features after the preprocessing, calculation, and filtration of the RR intervals.

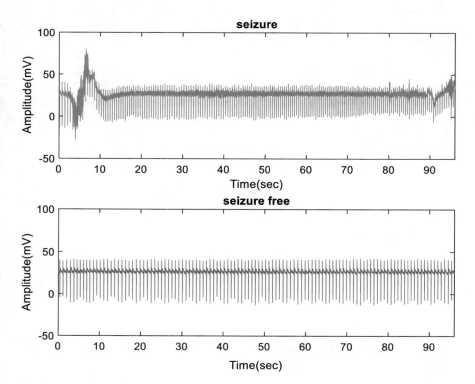

Fig. 2 Sample of patient HRV analysis

Poincare Plot. The Poincare plot, often known as the "return map," is a non-linear geometrical representation of a time series in the Cartesian plane. The RR time series on the phase space or Cartesian plane is represented by a two-dimensional plot built by plotting consecutive points [22].

The ECG signals consist of a time series where we find the RR intervals (the intervals between two successive R waves), and each R is drawn in the next R, which creates a graph where the density is shown as an ellipse and is where our features are extracted. We will need two short and long-term descriptors known as the standard deviation (SD) to extract features. SD1 and SD2, which reflect the minor and main axes of the plot's ellipse shape, are our two short- and long-term descriptors. Equations (1) and (2) can be used to describe the SD1 and SD2 equations.

$$SD_1^2 = \frac{1}{2}\sigma^2(S_n - S_{n+1}) = \frac{1}{2}SDSD^2 \tag{1}$$

$$SD_2^2 = 2SDS_n^2 - \frac{1}{2}SDSD^2 \tag{2}$$

Triangular Return Mapping (TRM). Because standard analysis of the Poincare plot tends to be mainly linear statistics, there are some limitations in extracting all physiological mechanisms in a time series [23]. This is where it is needed to bring in a new map to extract features to detect the other aspects of the HRV dynamics. In this new form of mapping, we used a geometric feature which was the sides of the assimilated triangle. This form of mapping has a high distinction which is very helpful for our study.

Construction of TRM. For this map, we will need the mean of the RR intervals. The Poincare plot points with respect to the mean of the RR intervals of the time series, as given in equation, form the basis for this new mapping (3). Figure 3 shows the distribution of points in TRM for seizure and seizure-free states.

$$\text{mean}(S_n) = \overline{S_n} = \frac{1}{n+1}\sum_{i=1}^{n+1} S_{ni} \tag{3}$$

Consisting of pairs:

$$\left(x_i, abc\left(\overline{S_n} - y_i^2\right)\right)$$

$$(S_{ni}, |S_n - S_{n+1}|)$$

And $i = 1, 2, 3, \ldots n$

By analyzing the distribution of points, we can form our new phase space as a triangle.

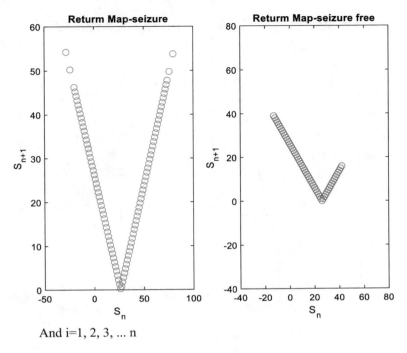

And i=1, 2, 3, ... n

Fig. 3 Distribution of points in TRM for seizure and seizure-free states

Extraction of Geometric Features in TRM: Initially, we have found the relation between all vertices and have extracted the sides (*D*1, *D*2, and *D*3) as our feature. Figure 4 shows proposed geometrical feature in the two states.

For the calculation of each side, we will be using Eq. (4) for the distance between the vertices:

$$
\begin{aligned}
DS1 &= \sqrt{(x_A - x_B)^2 + (y_A - y_C)^2} \\
DS2 &= \sqrt{(x_B - x_C)^2 + (y_B - y_C)^2} \quad [24] \\
DS3 &= \sqrt{(x_C - x_A)^2 + (y_C - y_A)^2} \quad [25]
\end{aligned}
\tag{4}
$$

Multilayer Perceptron (MLP): MLP is a type of artificial neural network that consists of three layers: input layer, hidden layer, and output layer. In this model, the state of seizure and seizure-free are labeled, respectively, with 1s and 0s. We have studied to see the distinction rate between both states. Figure 5 shows the structure of an MLP neural network. Because of the limited amount of samples, the MLP as well as the Poincare plot were not suitable for our study and were not able to differentiate well between the two states.

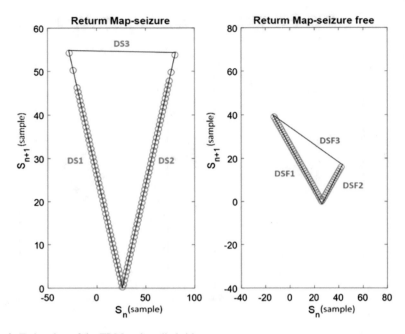

Fig. 4 Estimation of the TRM and studied sides

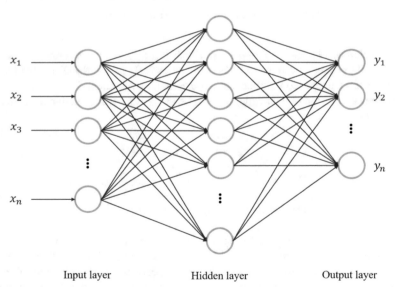

Fig. 5 Structure of an MLP

3   Results

In this research, 10 seizures were recorded for 7 patients, male and female, between 31–48 years old. In order to observe changes in both seizure and seizure-free states of patients, we have analyzed HRV signals of all patients during and before seizures. Table 1 is the recordings of the start and duration of each seizure. The duration of each seizure lasted from 25–110 s.

Figure 6 shows a significant difference between both seizure and seizure-free states with the help of the RR intervals derived from the ECG signals. Figure 6 is an example of patients' HRV recordings shown as Poincare plots. This first phase space was not ideal for the extracting of features for our study because the standard analysis of Poincare plots is linear which leads to limitations of extracting all physiological mechanisms in the time series. More time samples were also needed to create a more precise form where it would be viable to see a better difference in shapes and discriminate better between seizure and seizure-free periods. In Fig. 6, on the left is where we have our clustered form during the time of seizure, and on the right is our ellipse created from samples of our seizure-free time.

To convert our phase mapping from a non-linear approach of Poincare plots into a new phase space to extract features, it was required to convert into a new non-linear representation in which distinction between two states was evident. The new non-linear mapping constructed was the TRM. For the construction of this mapping, the mean of the RR intervals was calculated, which has been explained previously. By analyzing the point distribution, we obtained the triangular forms from the HRV signals. As shown in Table 2, each side of both TRMs for each seizure has been acquired. Subsequently, we calculated the mean and STD for each side in seizure and seizure-free states resulting in evident differences between the two states, confirming our approach was effective for prediction of epileptic seizures.

Table 1 Overview of the 7 patients under supervision and the time of seizures recorded

Subject ID	Seizure duration	
P1	00:14:36	00:16:12
P2	01:02:43	01:03:43
	02:55:51	02:56:16
P3	01:24:34	01:26:22
	02:34:27	02:36:17
P4	00:20:10	00:21:55
P5	00:24:07	00:25:30
P6	00:51:25	00:52:19
	02:04:45	02:06:10
P7	01:08:02	01:09:31

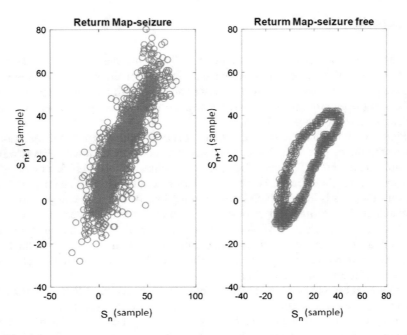

Fig. 6 Poincare plot (return map) for both seizure and seizure-free states

Table 2 Mean and STD calculated for each side of our TRM during both seizure and seizure-free states showing differentiation between the two

Seizure No.	DS1	DS2	DS3	DSF1	DSF2	DSF3
P1	76.3656	70.3843	104.0956	55.1571	22.5314	59.6696
P2	51.7042	89.0923	103.5497	8.0467	41.0128	42.2351
	19.1753	50.9139	54.9946	8.0482	39.5995	40.8470
P3	62.2240	85.8846	106.2823	8.4855	37.7607	38.7953
	175.3958	183.5984	254.0617	8.4851	37.7477	38.7859
P4	176.7754	183.8465	255.0490	11.0873	41.0111	42.7040
P5	73.2409	63.6378	97.2229	5.4804	18.3871	19.3570
P6	43.3235	62.2284	76.2482	17.8391	25.4531	31.5342
	22.2938	29.7017	37.4036	17.6750	25.4525	31.5791
P7	156.7135	179.6236	238.5644	74.7989	25.4568	79.0610
Mean	85.7212	99.8911	132.7472	21.5103	31.4413	42.4568
STD	61.0772	59.3055	83.5570	23.7285	8.7356	16.4552

Table 3 Performance evaluation of algorithm used

Feature	Sensitivity (%)	Accuracy (%)		Specificity (%)
DS1 DS2 DS3	100	90	83.33	

3.1 Performance Evaluation of the Proposed Algorithm

Results of each patient have been evaluated based on three factors of sensitivity (Sens), specificity (Spec), and accuracy (Acc), which determine the performance of this seizure prediction method. Equation (5) shows the evaluation criteria for the proposed method.

$$\text{Sens}\% = \frac{TP}{TP + FN} \times 100$$

$$\text{Spec}\% = \frac{TN}{TN + FP} \times 100 \tag{5}$$

$$\text{Acc}\% = \frac{TP + TN}{TP + TN + FP + FN} \times 100$$

TP, TN, FP, and FN were characterized as follows:
TP (true positives): accurately classified pre-ictal divisions.
TN (true negatives): accurately classified inter-ictal divisions.
FP (false positive): inaccurately classified inter-ictal divisions.
FN (false negative): inaccurately classified pre-ictal divisions [26]
In Table 3, we have calculated our proposed algorithm's performance evaluation (sensitivity, accuracy, and specificity).

4 Discussion and Conclusion

In this study, we proposed a method to predict seizures based on ECG signal extraction with the help of geometrical features. The dataset used was obtained from PhysioNet's "post-ictal heart rate oscillations in partial epilepsy" dataset. Our study consisted of both males and females from ages 31–48 who were under supervision and hospitalized. The main difference between our approach and the previous approaches made in the past is the optimal parameters we have obtained from our dataset. These parameters reflect the changes of patients around the time of onset seizures giving us better and more precise results.

The study of ECG signals was both during seizure and seizure-free states of the patients. As observed and known, seizures have vast effects on the ANS. As stated previously, changes in the ANS conclude with changes in heart rate and rhythm. As a result, ECG signals have been utilized to identify and predict seizures. The previous research has revolved mostly around EEG signals which were known as the golden standard. Both EEG and ECG signals have been widely used for the study of the pre-ictal state. EEG recordings have high-performance evaluations with high accuracy and a false positive rate, along with a series of limitations. We had predicted that we could obtain a high-performance assessment minus the cons of EEG recording signals with the help of ECG signals. The study of solely ECG recordings is inadequate and cannot distinguish the triggering factor of oscillations in the QRS waves of the patients' recordings. Our approach was to use the non-linear method of return maps. We had converted HRV signals of the patients into a different phase from the non-linear approach of Poincare plots into a new phase space for feature extraction. It was needed to convert into a new non-linear representation in which the distinction between seizure and seizure-free states was evident. The new non-linear mapping constructed was a phase space where we were able to assimilate into different geometrical figures giving us features to extract from to distinguish both states better. The discriminant feature extracted and used was the sides of the triangular forms of TRMs constructed with the help of standard deviation and means calculated.

Although the present study has obtained very good results, because of limitations, complicated and combined models are required to achieve higher accuracy for a more complete early prediction. One of the setbacks of this study was the limited database. By having a more complete dataset, results will also improve and become more precise. Future studies will include the extraction of other geometrical features for acquiring improved and beneficial features from the patients' ECG signals. The combining of EEG signals, ECG signals, MRI images, and electric impedance tomography (EIT) with the assistance of deep learning methods will also be put to use [27]. Small amounts of data from brain activity samples are used in neuro imaging developmental classification investigations. It promises to be engaging in terms of high-dimensional data analysis challenges [28]. Finally, the visualization of HRV signals and automated extraction of features from these images with deep learning can also be used in future studies for the better and faster detection of onset seizures.

In this paper, we have presented a new method using a geometric feature from extracted HRV signals. We have reached a sensitivity of 100%, accuracy of 90%, and a specificity of 88.33%, making our study practical for predicting seizures based on the pre-ictal state. With this approach, high-risk patients will be sooner recognized, under control, and monitored.

References

1. Sirven, J.I.: Epilepsy: a spectrum disorder. Cold Spring Harb. Perspect. Med. **5**(9), a022848 (2015)
2. Ogwang, R., Ningwa, A., Akun, P., Bangirana, P., Anguzu, R., Mazumder, R., Salamon, N., Henning, O.J., Newton, C.R., Abbo, C., Mwaka, A.D., Marsh, K., Idro, R.: Epilepsy in onchocerca volvulus sero-positive patients from northern uganda—clinical, EEG and brain imaging features. Front. Neurol. **12**, 687281 (2021)

3. Picot, M.C., Baldy-Moulinier, M., Daures, J.P., Dujols, P., Crespel, A.: The prevalence of epilepsy and pharmacoresistant epilepsy in adults: a population-based study in a Western European country. Epilepsia **49**, 1230–1238 (2008)
4. Moridani, M.K., Farhadi, H.: Heart rate variability as a biomarker for epilepsy seizure prediction. Bratislava Med. J. **118**(01), 3–8 (2017)
5. Devinsky, O., Hesdorffer, D.C., Thurman, D.J., Lhatoo, S., Richerson, G.: Sudden unexpected death in epilepsy: epidemiology, mechanisms, and prevention. Lancet Neurol. **15**(10), 1075–1088 (2020)
6. Devinsky, O., Bundock, E., Hesdorffer, D., Donner, E., Moseley, B., Cihan, E., Hussain, F., Friedman, D.: Resolving ambiguities in SUDEP classification. Epilepsia **59**(6), 1220–1233 (2018)
7. Van de Vel, A., Cuppens, K., Bonroy, B., Milosevic, M., Jansen, K., Van Huffel, S., Vanrumste, B., Cras, P., Lagae, L., Ceulemans, B.: Non-EEG seizure detection systems and potential SUDEP prevention: state of the art. Seizure **41**, 141–153 (2016)
8. Stirling, R.E., Cook, M.J., Grayden, D.B., Karoly, P.J.: Seizure forecasting and cyclic control of seizures. Epilepsia **62**(S1), S2–S14 (2020)
9. Van de Vel, A., Cuppens, K., Bonroy, B., Milosevic, M., Van Huffel, S., Vanrumste, B., Lagae, L., Ceulemans, B.: Long-term home monitoring of hypermotor seizures by patient-worn accelerometers. Epilepsy Behav. **26**, 118–125 (2013)
10. Vandecasteele, K., De Cooman, T., Gu, Y., Cleeren, E., Claes, K., Paesschen, W.V., Huffel, S.V., Hunyadi, B.: Automated epileptic seizure detection based on wearable and PPG in a hospital environment. Sensors **17**(10), 2338 (2017)
11. Park, Y., Luo, L., Parhi, K.K., Netoff, T.: Seizure prediction with spectral power of EEG using cost-sensitive support vector machines. Epilepsia **52**(10), 1761–1770 (2011)
12. Billeci, L., Marino, D., Insana, L., Vatti, G., Varanini, M.: Patient-specific seizure prediction based on heart rate variability and recurrence quantification analysis. PLOS ONE **13**(9), e0204339 (2018)
13. Malarvili, M.B., Mesbah, M.: Newborn seizure detection based on heart rate variability. IEEE Trans. Biomed. Eng. **56**(11), 2594–2603 (2009)
14. Ponnusamy, A., Marques, J.L., Reuber, M.: Heart rate variability measures as biomarkers in patients with psychogenic nonepileptic seizures: potential and limitations. Epilepsy Behav. **22**, 685–691 (2011)
15. Waxenbaum, J.A., Reddy, V., Varacallo, M.: Anatomy, Autonomic Nervous System. StatPearls Publishing [Internet] (2021)
16. Mueller, S.G., Nei, M., Bateman, L.M., Knowlton, R., Laxer, K.D., Friedman, D., Devinsky, O., Goldman, A.M.: Brainstem network disruption: a pathway to sudden unexplained death in epilepsy? Hum. Brain Mapp. **39**(12), 4820–4830 (2018)
17. Billman, G.E.: Heart rate variability—a historical perspective. Front. Physiol. **2**, 86 (2011)
18. Goldberger, A., Amaral, L., Glass, L., Hausdorff, J., Ivanov, P.C., Mark, R., Stanley, H.E.: PhysioBank, PhysioToolkit, and PhysioNet: components of a new research resource for complex physiologic signals. Circulation **101**(23), e215–e220 (2002)
19. Leal, A., Pinto, M.F., Lopes, F., et al.: Heart rate variability analysis for the identification of the preictal interval in patients with drug-resistant epilepsy. Sci. Rep. **11**, 5987 (2021)
20. Al-Aweel, I.C., Krishnamurthy, K.B., Hausdorff, J.M., Mietus, J.E., Ives, J.R., Blum, A.S., Schomer, D.L., Goldberger, A.L.: Postictal heart rate oscillations in partial epilepsy. Neurology **53**(7), 1590–1592 (1999)
21. Fariha, M.A.Z., Ikeura, R., Hayakawa, S., Tsutsumi, S.: Analysis of Pan-Tompkins algorithm performance with noisy ECG signals. J. Phys.: Conf. Ser. **1532** (2019)
22. Fishman, M., Jacono, F.J., Park, S., Jamasebi, R., Thungtong, A., Loparo, K.A., Dick, T.E.: A method for analyzing temporal patterns of variability of a time series from Poincare plots. J. Appl. Physiol. **113**(2), 297–306 (2012)
23. Karmakar, C.K., Khandoker, A.H., Gubbi, J., Palaniswami, M.: Complex correlation measure: a novel descriptor for Poincare plot. Biomed. Eng. Online **8**(1), 17 (2009)

24. Tang, G., Li, C., Wang, M.: Polarization discriminated time-bin phase-encoding measurement-device-independent quantum key distribution. Quantum Eng. e79 (2021)
25. Whitney, R., Donner, E.J.: Risk factors for sudden unexpected death in epilepsy (SUDEP) and their mitigation. Curr. Treat. Options Neurol. **21**(2), 7 (2019)
26. Baratloo, A., Hosseini, M., Negida, A., El Ashal, G.: Part 1: simple definition and calculation of accuracy, sensitivity and specificity. Emergency **3**(2), 48–49 (2015)
27. Adam, E.E.B., Ammayappan, S.: Survey on medical imaging of electrical impedance tomography (EIT) by variable current pattern methods. J. ISMAC **3**(2), 82–95 (2021)
28. Patcha, K., Ammayappan, S., Hamdan, Y.B.: Early prediction of autism spectrum disorder by computational approaches to fMRI analysis with early learning technique. J. Artif. Intell. Capsule Netw. **2**(4), 207–216 (2020)

Automatic Dent Detection in Automobile Using IR Sensor

Sudarshana S. Rao and Santosh R. Desai

Abstract In this era of automation, surface dent detection on cars is still manual and needs to be automated. Automobile surface irregularities like scratches or dents occur mainly due to accidents and in most worldwide automobile industries, the spotting of dents and other surface defects is performed by human vision, which is inefficient and faulty as minor dents maybe overlooked, so an automatic dent detector will obviate any human error in the detection of dents. An automatic dent detector is a vital component in any service garage as it makes the dent detection process simpler, faster, efficient and reliable. Automatic dent detection is accomplished by using Machine Learning (ML). Mask-R Convoluted Neural Network (CNN) algorithm is used which is accurate but very expensive and intricate. However, it is difficult to scan such surface irregularities like dents by a computer (machine) because of reflection, refraction, scattering or diffraction of light, imperfect illumination and limited surface defect features since dents are amorphous (neither perfectly round nor symmetric square). Therefore, the exact pixel location of the scratch or dent on the image will only help to identify the location and quantify the extent of damage accurately. So, IR sensor-based system presents an alternate cost-effective, simple and a hassle-free method of automatic dent detection. This paper talks about the detection of dents using IR sensors for distance measurement from defined threshold; interfaced with an Arduino Uno board and a computer which will display whether the dent is present on the surface of the car or not. Usage of IR sensors for dent detection makes this a unique solution as IR sensor is much reliable than any other sensor mainly because of its flexible range, and it can easily be integrated to any surface also can be interfaced with any of the micro-controllers.

Keyword Automatic dent detection using FC-51 IR sensors for distance measurement and interfacing with an Arduino Uno board (R3 CH340G ATmega328p)

S. S. Rao · S. R. Desai (✉)
Department of EIE, B.M.S. College of Engineering, Bangalore, India
e-mail: santoshdesai.intn@bmsce.ac.in

1 Introduction

Automated vehicle inspection is gaining importance these days. An important part of the inspection is examining the vehicle's body. This procedure involves checking for scratches, dents, patches and irregularities in painted parts. A vehicle's cost depends on this fact being discussed right now, whether it is new or used. Further, it is also agreed that a vehicle might need to be serviced. It is likely that a car owner who finds a dent on their car after it has been serviced at this garage will be enraged on the mechanic and won't return to the garage in the future. Also, the garage won't achieve 100% customer satisfaction and will lose a customer, which will result in liability. Generally, dent detection is done by human beings and human error (overlooking minor dents owing to poor eye sight) in this process is inherent. Therefore, this traditional manual detection process of surface irregularities needs to be automated in order to eliminate human error. When the surface defect is not clearly visible, the help of an automated dent scanner is needed to detect it. It is difficult to identify the surface defects only by the human eye. Moreover, it is impossible to achieve an efficient method of detecting surface dents by manual human inspection, resulting in time-consuming service provided by the garage [1], [6]. By integrating an infrared sensor, the above limitations are overcome, and this can be considered as a major contribution. The proposed automatic dent detector consists of IR sensors mounted on a hydraulic jack. This detector can be used to scan for a potential surface irregularity on the car. If the car does not have any surface dents, then this dent detector will certify that there are not any surface dents present. If the car consists of any dents on its surface, then this dent detector will detect dents efficiently, error-free and within a very short span of time. The dent detector can be used after a car's service to check for dents and ensure they are all fixed. IR sensor is used for distance measurement (range of 2–80 cm) over the ultrasonic sensor simply because IR radiation is faster than ultrasonic sound. Proximity sensor cannot be used for dent detection because a proximity sensor's range is very limited. IR sensor has a major advantage over all these sensors because it can be easily interfaced with the Arduino Uno board; it is compact and many IR sensors can be mounted on the hydraulic jack [2]. However, the FC-51 IR sensor is not reliable since its accuracy is not good and so LiDAR or SHARP IR sensor (GP2Y0A710K0F) can be an alternative but are very expensive and hence not feasible. Before interfacing the IR sensor with Arduino Uno, the Arduino Uno board has to be configured to the computer's serial COM3 port and the required libraries need to be downloaded in the Arduino IDE as shown in Fig. 1 [3]. The obtained distance measurement values can be ported into an excel file for further analysis [4]. The photometric method such as a three dimensional measurement using an 18 bit camera can be used as an alternate solution for dent detection [5]. Also, Convolution Neural Networks (CNN's) makes it technically feasible to recognize vehicle damages (surface dents) using deep convolution networks [7].

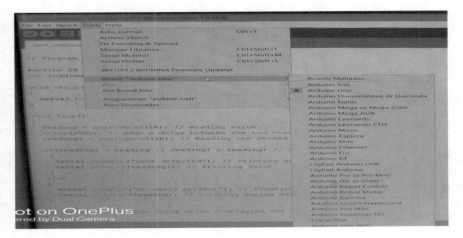

Fig. 1 Configuring Arduino Uno to computer

2 Proposed Methodology

IR sensors will be mounted on the stationary arms of the hydraulic jack and the IR sensors will be numbered. The car with no surface irregularities which is mounted on the hydraulic jack (in a fixed pre-defined position) will be moved back and forth for the IR sensors to measure the distance between the IR sensor and the car and this process is done in sections; for example, the IR sensors will first measure the distances in the front portion of the car (front bumper and bonnet), then the side of the car (front right and left doors, back left and right doors) and then the rear portion of the car like the rear bumper. These distance values will be stored in a database (the database stores the values under different sections like car's front portion, sides of the car and the rear end of the car under the rows—named by the IR sensors' number) and these values will be treated as threshold values and the IR sensors are calibrated to these threshold values using the potentiometer on the IR sensor. Before the car with surface defects is mounted on the jack, the car's model type is entered on the computer so that the IR sensors can be calibrated with the respective threshold values. Then a clean and dry car (with no potential dirt or mud on its surface which could affect the IR sensor's accuracy) with surface irregularities will be brought on the jack (in the same fixed position) and the IR sensors will once again measure the distance in the same sequence as mentioned above and these distance values will be compared to the threshold values and if they differ then the computer will display "Dent detected on < position > " and if the values do not differ, then "No dents present on < position > " will be displayed where < position > is car's front portion, side of the car or the rear end of the car. Figure 2 illustrates the block diagram of the system.

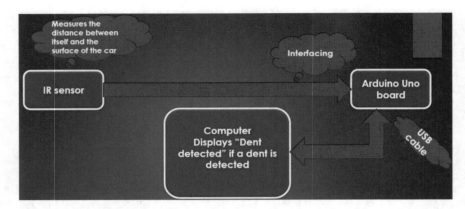

Fig. 2 System architectural block diagram

3 Working Mechanism

The overall setup consists of IR sensors mounted on the arms of the hydraulic jack with appropriate numbering. The hydraulic jack is mounted on the car surface having irregularities and moved back and forth through the process. Infrared sensors are used to measure distance between car surfaces and the IR sensors. To keep things in order, this process is divided into several sections, and the procedure adopted for distance measurement is outlined below. The details of the IR sensor being used in this setup is mentioned in Table 1. The model proposed for the dent detection is as shown in Fig. 3.

3.1 Distance Measurement Using IR Sensor

1. IR sensor's VCC pin is connected to the 5 V pin of the Arduino Uno, IR's GND pin is connected to Arduino Uno's ground pin and IR's OUT pin is connected to one of the analogue pins (A1) of the Arduino Uno as shown in Fig. 4. The internal circuit of IR sensor is as shown in Fig. 5.

Table 1 SHARP IR sensor specifications

IR sensor description	Values
Part number	SHARP GP2Y0A710K0F
Distance measuring range	100 to 550 cm
Package size	58 × 17.6 × 22.5 mm
Lag between subsequent measurements	5 ms

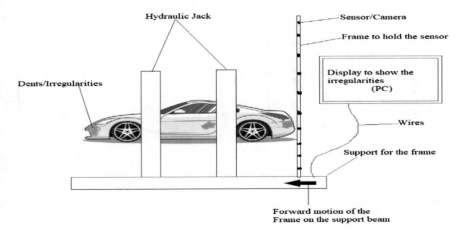

Fig. 3 Proposed model for automatic dent detection

Fig. 4 Interfacing of IR sensor with Arduino Uno

2. IR sensors mounted on the arms of the hydraulic jack will emit an infrared beam of light via its transmitter.
3. If an object is present in front of the sensor, the IR will be bounced back and it will be detected by the IR sensor's receiver.
4. The distance can be calculated using the formula:

$$Distance = \frac{Speed * Time}{2} \tag{1}$$

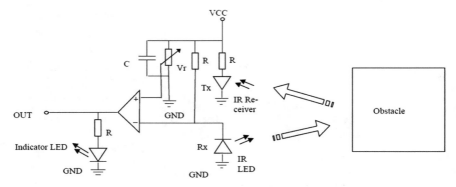

Fig.5 Internal circuit of IR sensor

5. The speed is 3×10^8 ms^{-1} (speed of light) and time is the interval between the transmission of IR beam and the detection.
6. The time is divided by 2 because only the time interval of the detection of IR beam is considered as the IR bounces from the surface of car to the IR receiver.
7. If the distance differs from the threshold value, then "Dent detected on < position > " will be displayed otherwise, "No dents present on < position > " will be displayed on the computer.

4 Results

The concept being discussed in this communication was tested on Hyundai i10 grand car. There are two dents on the left rear door near the tyre and one on the left front bumper above the tyre. The same is illustrated as shown in Fig. 6 and 7, respectively.

Fig. 6 Surface dent near the left front tyre

Fig.7 Surface dent near the back left tyre

As a first step, the distance between the stationary IR sensor and the area of the car without dents was measured. Next, the distance was measured between the stationary IR sensor and the area of the car with the surface dent. The same is illustrated in Figs. 8, 9, 10 and 11. To justify the concept behind the adopted methodology, Fig. 12 serves as a proof of concept. In this image, Arduino Board, Code and IR sensor output are clearly visible. IR sensor output value indicated in Fig. 13 clearly indicates that the test procedure was successful in detecting dent. IR distance measurements are shown between the car's surface and its IR sensor while the Arduino IDE is connected to a laptop.

Fig. 8 Distance measurement between the IR sensor and body of the car (back left tyre) with no surface defects

Fig. 9 Distance measurement between the IR sensor and body of the car (back left tyre) with surface dents

Fig. 10 Distance measurement between the IR sensor and body of the car (front left tyre) with no surface defects

5 Conclusions

The experiment results being discussed demonstrate that IR sensors can be effectively utilized to detect surface irregularities of a vehicle quite efficiently. Doing so will lead to automatic and fast dent detection, free of human errors, etc. Thus, one will be able to detect dents automatically and without human error. This detector is a cost-effective and an easy-to-use device as any lay person can operate this device with minimum computer training. Other features of the product include the following:

- It is a versatile and a compact device as it can be easily installed in any garage and can be used in any geographic location.

Fig. 11 Distance measurement between the IR sensor and body of the car (front left tyre) with surface dents

Fig. 12 Arduino circuit, Arduino code and IR sensor output

- The limitation of this detector is that it does not give the magnitude of the dent; it'll only return the general area of the dent on the car's surface. The FC-51 IR sensor has detected dents of magnitudes more than about 2 cm since it is not very accurate, and the FC-51 IR sensor can detect dents present on flat surface of the car like side doors, bonnet, etc.
- This proposed model does not distinguish between a design and a dent, as any deviation from the normal surface would be considered a dent, and the distances from the normal surface would not be displayed. To overcome the above, image processing can be used.

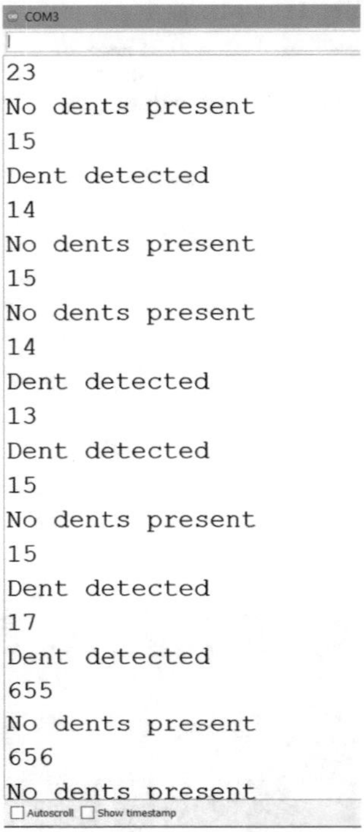

Fig. 13 Digital output
values of IR sensor

```
COM3
|
23
No dents present
15
Dent detected
14
No dents present
15
No dents present
14
Dent detected
13
Dent detected
15
No dents present
15
Dent detected
17
Dent detected
655
No dents present
656
No dents present
  Autoscroll   Show timestamp
```

6 Future Scope

- This scanner can be used to detect the surface irregularities on airplanes, ships, wind mills, etc.
- The output of this scanner can be used as an input to an automated dent fixer which will fix the dents.
- This scanner can be mounted on a dent fixing robot and hence one system can perform both the detection and the tinkering of the dents thereby increasing the efficiency and reducing the time consumption.
- Garages can use this device to provide dent detection as a regular service under proactive maintenance.
- For better accuracy, dent detection using image processing can be explored (mask RCNN).

References

1. Zhou, Q., Chen, R., Huang, B., Liu, C., Yu, J., Yu, X.: An automatic surface defect inspection system for automobiles using machine vision methods. Sensors 644–661 (2019)
2. Mohammad, T.: Using ultrasonic and infrared sensors for distance measurement." world academy of science. Eng. Technol. **3**, 272–278 (2009)
3. Kurniawan, A.: Arduino Uno: A Hands-on Guide for Beginner (Chapter 2-section 2.1: Installing Arduino software) (2015)
4. Mahaveer P., Raju R.S., Rai K., Thakur K.K., Desai S.R. Prognostic Monitoring and Analyzing System for Motors. In: Komanapalli V.L.N., Sivakumaran N., Hampannavar S. (eds) Advances in Automation, Signal Processing, Instrumentation, and Control. Lecture Notes in Electrical Engineering **700**. Springer, Singapore. (2021) https://doi.org/10.1007/978-981-15-8221-9_77
5. Hasebe, A., Kato, K., Tanahashi, H., Kubota, N.: Dent detection method by high gradation photometric stereo. In: Thirteenth International Conference on Quality Control by Artificial Vision 2017, vol. 10338, p. 103380P. International Society for Optics and Photonics (2017)
6. Tian, D., Zhang, C., Duan, X., Wang, X.: An automatic car accident detection method based on cooperative vehicle infrastructure systems. IEEE Access **7**, 127453–127463 (2019)
7. Malik, H.S., Dwivedi, M., Omakar, S.N., Samal, S.R., Rathi, A., Monis, E.B., Khanna, B., Tiwari, A.: Deep learning based car damage classification and detection. 3008. EasyChair (2020)

Applying ML on COVID-19 Data to Understand Significant Patterns

Amit Savyanavar, Tushar Ghumare, and Vijay Ghorpade

Abstract Corona viruses are a genus of viruses that infect vertebrate and birds, causing a variety of diseases. They induce a variety of respiratory problems in people. This study investigates COVID-19 infection rates and estimates the pandemic's scope, recovery rate, and death rate. We used Support Vector Machine (SVM), Random Forest, Decision Tree, K-nearest neighbor, and other well-known machine learning and mathematical modeling approaches. For disease diagnosis, study used three unique disease data sets (Asthma, Diabetes, and AIDS) provided there in UCI machine learning repository. After getting positive results, we applied these algorithms on COVID-19 data set. We used several categorization algorithms, each with its own set of benefits. The study's findings support the use of machine learning in disease early detection.

Keywords COVID-19 · Coronavirus · Machine learning · Support vector machine (SVM) · Random forest · Decision tree

1 Introduction

A germicide is a small infectious organism with a basic makeup and the ability to replicate only in alive cells of an organism, plants, and microorganisms. Some human pathogenic bacteria penetrate host cells and immediately begin making new duplicates of themselves, defeating the immune system's ability to produce antibodies. Coronaviruses are RNA viruses that infect animals and birds, causing illness.

In the twenty-first century, 5 epidemics have been recorded. The highly contagious Ebola virus epidemic between 2013 and 2016 in West Africa, before re-emerging

A. Savyanavar · T. Ghumare (✉)
Dr. Vishwanath Karad MIT World Peace University Pune, Pune, India

A. Savyanavar
e-mail: amit.savyanavar@mitwpu.edu.in

V. Ghorpade
Bharati Vidyapeeth's College of Engineering, Kolhapur, India

© The Author(s), under exclusive license to Springer Nature Singapore Pte Ltd. 2022
V. Suma et al. (eds.), *Evolutionary Computing and Mobile Sustainable Networks*,
Lecture Notes on Data Engineering and Communications Technologies 116,
https://doi.org/10.1007/978-981-16-9605-3_35

in 2018. To date, there are nearly 11,300 deaths. Influenza A (H1N1) In the early spring of 2009, a new virus developed in Mexico. This was first named swine flu, and in June of that year, it was proclaimed a pandemic. It turned out to be significantly less dangerous than expected, and various vaccine initiatives have been stymied as a result of its discovery. The World Health Organization believes that a virus cause 18,500 fatalities, but the Lancet medical magazine claimed 151,700 and 575,400 fatalities respectively. SARS, or severe acute respiratory syndrome, first appeared in the southern region of China in the fall of 2002. It is passed down from bats to civets, and subsequently from civets to humans. Avian influenza (bird flu) was the first to be eradicated from the poultry sector in Hong Kong and Mainland china, before spreading to the general public and resulting in the formation of global brain areas. Given that 400 individuals died, the number of victims turned out to be little. The COVID-19 pandemic, also known as the corona virus pandemic, is a global corona virus disease that is expected to become a pandemic in 2019 (COVID-19), and Corona infection causes severe acute respiratory illness.2. (SARS-CoV-2). The virus first became detected in December of this year in Wuhan, China. More than 172 million cases were confirmed in far more 3.71 thousand individuals on June 5, 2021, confirming COVID-19 mortality and calling it one of the worst epidemics in mankind.

Table 1 shows the major epidemics till date. The table consists the duration of the particular epidemic and the deaths caused. Swine flu caused 2 M deaths, HIV/AIDS caused 25–35 M deaths. Black death is on 1st rank which caused 200 M deaths till now (2019). Likewise COVID-19 is having higher death ratio and it caused 3 M deaths as of now 2019. That means after Black Death pandemic COVID-19 will be the largest epidemic of history.

In Fig. 1 we represented this statistical data in more suitable manner with the help of SPSS software tool.

The fight against pandemics and epidemics has been a never-ending adventure for humanity. Epidemics such as the bubonic plague, Ebola, Cholera, Encephalitis, and many others have struck humanity. They have, however, been successfully eradicated as a result of the development of vaccines and other preventative measures.

Table 1 Events their duration and fatalities

Event	Start	End	Deaths
Black death	1331	1353	200 M
Italian plague	1623	1632	1 M
Great plague of London	1665	1666	1 M
Ebola	2015	2016	11,000
Swine flu	2009	2010	2 M
SARS	2002	2003	770
MERS	2015	Present	850
HIV/AIDS	1981	Present	25–35 M
COVID-19	2019	Present	3 M

Fig. 1 Statistics of death
caused

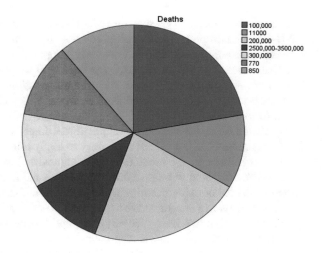

Traditional methods for forecasting epidemics include time series modeling and regression modeling. We used a combination of classic and cutting-edge machine learning approaches. K-Nearest Neighbor (KNN), Support Vector Machine (SVM), Random Forest, Nave Bayes and decision trees. We attempted to solve the problem of detecting the COVID-19 epidemic using the machine learning methodologies outlined above.

The content of the paper is described below. The work is broken into 5 parts: Sect. 2 discusses the review of literature, Sect. 3 discusses the machine learning techniques used, Sect. 4 discusses the results in further detail, and Sect. 5 concludes the paper.

2   Literature Survey

The death rates of heart, brain, and renal illnesses, according to our understanding, have been investigated on a regular basis, but the influenza virus has not. We'll look at how to provide future information on such efforts in this paper.

Chen et al. proposed [1] a new technique for forecasting CNN disease risk using both structured and unstructured hospital data. In various places, Chen et al. built an illness control system (Remove this, doesn't make sense). Three diseases, including diabetes, cerebral palsy, and heart problems, were predicted. The prognosis of a disease is based on organized data. Different machine learning algorithms, such as naive Bayes, Decision tree, and Regression algorithms, are used to predict heart problems, diabetes, and cerebral infarction. The Decision Tree algorithm outperforms both the Nave Bayes and the KNN algorithms. They also forecast whether the patient would have a higher potential repercussions of cerebral infarction. There's a chance you'll have a stroke. Use it to forecast the probability of cerebral infiltration by using text details to predict CNN disease risk. There are accurate comparisons between CNN-based multimodal illness risk prediction algorithm and CNN-based unimodal disease risk prediction algorithm. This study is applicable to both official and informal data databases. The author has experimented with a wide range of data. While the prior work was entirely based on structured information, no author has ever worked with unstructured or unstructured data. However, this work is based on both formal and informal data.

Kohli and Arora [2] used three unique illness databases (Heart, Breast cancer, and Diabetes) accessible in the UCI repository to apply multiple classification methods, with its own benefit, for prediction and diagnosis. Backward modeling using the *p*-value test was used to choose features within each data set. The study's findings support the idea of using machine learning to detect diseases early.

Arun and Iyer [3] examined the transmission of COVID-19 infection and anticipated the pandemic's size, recovery rate, and mortality rate. Among other well-known machine learning and mathematical modeling techniques. They used Rough Support Vector Machine (RS-SVM), Bayesian Ridge also Polynomial Regression, SIR model, and RNN.

The study by Prithivi et al. [4] reflects how a novel corona virus went from a simple influenza to an epidemic in a short amount of time. All factors, including the key countries responsible for spreading the epidemic owing to their size, the rate of human movement, and socioeconomic concerns, appear to be the reason, as per the evaluation. They've also drawn parallels.

With the ubiquitous availability of smartphones, they could be deployed to form a mobile grid for performing on-the-fly computing and quick decision-making. The subtasks of an application for COVID monitoring could be allotted [5, 6] to devices in the mobile grid to expedite the delivery of results. Remote monitoring of elderly and physically challenged patients is possible by using mobile grids. Also timely detection of complications could be determined for treatment purpose.

Chen [7] showed the working of the CNN classification method which is applied on the dataset contains the X-ray image records, and also explained the tenfold cross-validation along with the confusion matrix.

Sungheetha [8] compared the Data-Driven Intelligence Strategies and identified the purpose of this novel solution also proposed the SEIR model and collected the results. Also referred to one Report [9] on COVID-19 proposed by WHO. Chen et al. [10] proposed Data Visualization.

3 Methodology

This section delves into the specifics of our model. We'll start by going over the data set we obtained from the UCI repository. Then we'll go through how to estimate the pandemic's spread using machine learning.

Preparing data sets and pre-processing data—This study relied on real-time data from Kaggle. This is real-time patient information. The state of the country's regions and how the effected are all provided.

Following are the steps for data preprocessing:

- The non-null values are replaced with the mean of the column.
- The data was transformed using Python's StandardScaler object in order to create a Gaussian distribution in terms of predicting the pandemic's spread.
- A logarithmic function was used to standardize the data.

We start by obtaining a disease data sets from the Machine learning repository website or the Hospital Database, which usually consists of a disease list with characteristics and regions effected. Prior to cleaning, this data set has been pre-processed, which includes the removal of commas, punctuation, and whitespace. This can also be used as a data set for training. The object will then be picked after it has been deleted. The data is subsequently classified using classification algorithms such as K-NN, SVM, Random Forest, and CNN. We can properly anticipate the disease using artificial intelligence.

Observe Fig. 2 this is the system architecture we used in our research. Initially, we took data sets and we did preprocess steps on that which includes dealing with the missing tuples and much more. After Data set Preprocessing there sum two stages first one is the Training Phase and another one Testing Phase. We are going to train our model in Training Phase and with that model, we will test actual data set prediction in Testing Phase. Then we used some classification techniques and after that, we applied the following listed algorithms to understand the patterns and success rate of that particular data set.

3.1 Algorithms

We used some best Machine Learning algorithms including SVM, Decision Tree, K-nearest Neighbor and last but not the least Random Forest to understand the different patterns,

3.1.1 Nearest Neighbors (K-Nearest Neighbors) (KNN)

KNN makes predictions using only the training data. Predictions are made for a replacement instance (x) by scanning the whole training set for the K closest cases (neighbors) and summing the outcome variable for those Data instances.

$$\text{Euclidean Distance } (x, x_i) = \text{sqrt}\left(\text{sum}\left((x_j - x_{ij})^2\right)\right)$$

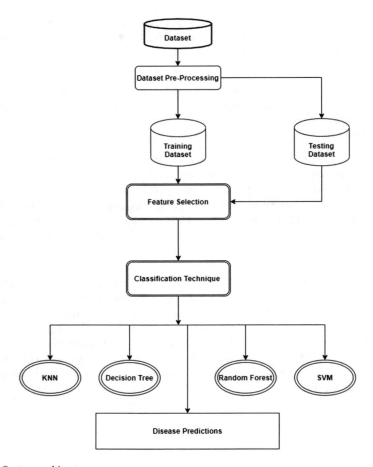

Fig. 2 System architecture

To determine which one of the K instances in the training collection are the most closest to a replacement input, a distance measure is utilized. The most extensively used distance measure for real-valued input is Euclidean distance.

3.1.2 Random Forest Algorithm

The random forest algorithm consists of several decision trees, all of which have the same components, but also makes use of a variety of information, which leads to different results.

Random Forest for Regression

$$MSE = \frac{1}{N} \sum_{i=1}^{N} (f_i - y_i)^2$$

where N is the number of data points,

f_i is the value returned by the model and

y_i is the actual value for data point i.

Entropy is used to determine how the node should branch based on the probability of a specific outcome.

$$Entropy = \sum_{i=1}^{C} -p_i * \log_2(p_i)$$

3.1.3   SVM

SVM (Support Vector Machine) is a supervised learning technique for solving regression and classification issues. Observe Fig. 3 if we encounter a strange cat with some dog-like characteristics, and we want a model to reliably determine whether it's a cat or a dog, the SVM method is frequently used to generate such a model.

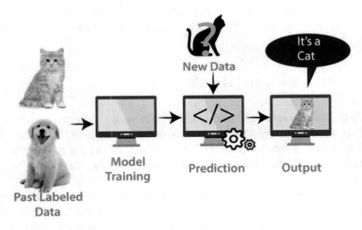

Fig. 3 SVM model example

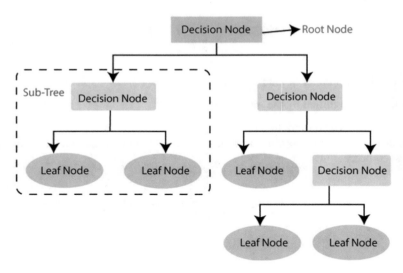

Fig. 4 Decision tree representation

3.1.4   Decision Tree

Decision Tree is a supervised classifier that could be used to solve both classification and regression problems, however it is most commonly employed to solve classification issues. In Fig. 4 internal nodes represent collection attributes, branches represent rule base, and each leaf node provides the conclusion in this tree-structured classifier.

4   Experimentation and Results

The analysis and outcomes of various machine learning algorithms on the data set and their comparisons are summarized below. For experimental purpose we have obtained data from Kaggle repository. The data consists of the count of confirmed cases, deaths, and the cured cases state wise. For that we have used the Weka machine learning tool.

Weka is a machine learning tool where we can apply various machine learning algorithms on different data sets for prediction. Also it has options for classification and clustering. The very first thing we need to do after entering into weka that is opening data set which can be in any format (ARFF, CSV). After loading data set preprocess step is important to delete unlabeled and missing tuples. And then we can choose any algorithm to process our data set.

Table 2 Algorithms with cross validation and accuraccy

Classification algorithm	Cross validation	Accuracy (%)
Decision tree	Fivefold	95.00
	Tenfold	95.66
SVM algorithm	Fivefold	93.08
	Tenfold	93.30
Random forest	Fivefold	84.08
	Tenfold	84.52
KNN algorithm	Fivefold	91.25
	Tenfold	91.86

4.1 Asthma Data Set Results

The data set is obtained from **kaggle repository** [11]. It contains 117 instances and 5 attributes which are nothing but Country, Year, Age Group, Number of Deaths, Mortality Rate. Missing values from the data set have been filled with the mean square values of them.

We are applying machine learning algorithms to predict the number of deaths and mortality rate in particular age group in various countries in particular time duration (in years) (Table 2).

Figure 5 showing the results obtained by applying Decision Tree algorithm. All the instances were classified correctly and received accuracy around 95%. Also confusion matrix is there showing the predicted class and its accuracy. While applying other algorithms SVM gives 93%, Random Forest gives 84%, and KNN gives 91% accuracy.

4.2 Diabetes Data Set Results

The diabetes data set which is obtained from **kaggle repository** [12] consists of 768 instances and 6 attributes which contains Pregnancies, Glucose, Blood-Pressure, Skin Thickness, Insulin, and BMI. İn the data set in the insulin column the 0 value indicates that it is non-diabetic and other values than 0 shows that it is diabetic. For attributes contain missing values we entered their mean square values. Table 3 shows that the results obtained after applying various machine learning models.

Figure 6 shows the results obtained by applying SVM algorithm on data set. Which has given the highest accuracy among all algorithms.

```
Number of Leaves  :        17

Size of the tree :        33


Time taken to build model: 0 seconds

=== Stratified cross-validation ===
=== Summary ===

Correctly Classified Instances          69              69    %
Incorrectly Classified Instances        31              31    %
Kappa statistic                         0.4173
Mean absolute error                     0.226
Root mean squared error                 0.4084
Relative absolute error                 61.9717 %
Root relative squared error             95.9294 %
Total Number of Instances               100

=== Detailed Accuracy By Class ===

            TP Rate  FP Rate  Precision  Recall  F-Measure  MCC    ROC Area  PRC Area  Class
            0.476    0.190    0.400      0.476   0.435      0.269  0.667     0.359     c0
            0.839    0.316    0.813      0.839   0.825      0.529  0.815     0.838     c1
            0.412    0.048    0.636      0.412   0.500      0.436  0.802     0.499     c2
Weighted Avg. 0.690  0.244    0.696      0.690   0.688      0.459  0.782     0.680

=== Confusion Matrix ===

 a  b  c   <-- classified as
10  8  3 |  a = c0
 9 52  1 |  b = c1
 6  4  7 |  c = c2
```

Fig. 5 Highest accuracy result (Decision Tree) 95% on Asthma data set

Table 3 .

Classification algorithm	Cross validation	Accuracy (%)
SVM algorithm	Fivefold	92.00
	Tenfold	92.56
KNN algorithm	Fivefold	91.01
	Tenfold	91.40
Decision tree	Fivefold	91.21
	Tenfold	91.68
Random forest	Fivefold	84.10
	Tenfold	84.57

4.3 COVID-19 Data Set Results

The COVID-19 data set is also obtained from **kaggle repository** [13] and it includes 1734 instances and 4 attributes. This datset contains the records collected from each state of India. Total confirmed cases, Deaths as well as cured/discharged records also included. Algorithms are applied on data set to predict the most affected state during

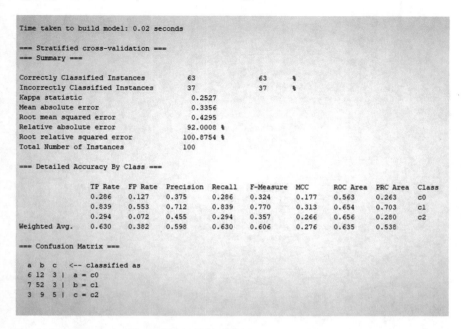

```
Time taken to build model: 0.02 seconds

=== Stratified cross-validation ===
=== Summary ===

Correctly Classified Instances        63          63    %
Incorrectly Classified Instances      37          37    %
Kappa statistic                        0.2527
Mean absolute error                    0.3356
Root mean squared error                0.4295
Relative absolute error               92.0008 %
Root relative squared error          100.8754 %
Total Number of Instances            100

=== Detailed Accuracy By Class ===

            TP Rate  FP Rate  Precision  Recall  F-Measure  MCC    ROC Area  PRC Area  Class
             0.286    0.127    0.375     0.286    0.324     0.177   0.563     0.263     c0
             0.839    0.553    0.712     0.839    0.770     0.313   0.654     0.703     c1
             0.294    0.072    0.455     0.294    0.357     0.266   0.656     0.280     c2
Weighted Avg. 0.630   0.382    0.598     0.630    0.606     0.276   0.635     0.538

=== Confusion Matrix ===

 a  b  c   <-- classified as
 6 12  3 |  a = c0
 7 52  3 |  b = c1
 3  9  5 |  c = c2
```

Fig. 6 Accuracy result (SVM) 92% on diabetes data set

the pandemic and the death toll. After applying machine learning algorithms on this data set we obtained some results as shown in Table 4. The SVM algorithm is given the highest accuracy i.e. 92%, K-NN got 89%, Decision Tree covered 90% accuracy, and Random Forest obtained 83% accuracy. Results contain all the data i.e. rate of infections, Death rate, and Recovery. With these results we can understand which state is affected most by this deadly virus. Also we can understand some patterns like which area is affected most Rural or Urban (Fig. 7).

Finally we determined the results on COVID-19 data set and we also derived the confusion matrix.

Comparing all the results we found that SVM algorithm is giving more accuracy than all other algorithms. For our COVID-19 data set SVM is working more efficiently and giving 92% of accuracy. Observe Table 4 for more detailed comparison.

Table 4 Comparison of each algorithm's accuracy with the data sets

Algorithm	Asthma (%)	Diabetes (%)	AIDS (%)	COVID-19 (%)
KNN	95	94	78	89
Random forest	84	84	76	83
Decision tree	94	91	71	90
SVM	96	92	83	92

```
Time taken to build model: 0.09 seconds

=== Stratified cross-validation ===
=== Summary ===

Correctly Classified Instances        63              63      %
Incorrectly Classified Instances      37              37      %
Kappa statistic                       0.2527
Mean absolute error                   0.3356
Root mean squared error               0.4295
Relative absolute error               92.0008 %
Root relative squared error           100.8754 %
Total Number of Instances             100

=== Detailed Accuracy By Class ===

              TP Rate  FP Rate  Precision  Recall  F-Measure  MCC    ROC Area  PRC Area  Class
              0.286    0.127    0.375      0.286   0.324      0.177  0.563     0.263     c0
              0.839    0.553    0.712      0.839   0.770      0.313  0.654     0.703     c1
              0.294    0.072    0.455      0.294   0.357      0.266  0.656     0.280     c2
Weighted Avg. 0.630    0.382    0.598      0.630   0.606      0.276  0.635     0.538

=== Confusion Matrix ===

 a  b  c   <-- classified as
 6 12  3 |  a = c0
 7 52  3 |  b = c1
 3  9  5 |  c = c2
```

Fig. 7 Highest accuracy result (SVM) 92% on COVID-19 data set

5 Conclusion

The findings of this study support the use of machine learning algorithms in disease prediction and understanding its patterns. To the best of our knowledge, the model created using the proposed method is more accurate than existing models. Our proposed approach has a prediction accuracy of 96% for Asthma Disease detection using Support Vector Machine, 94% for Diabetes prediction using K-Nearest Neighbor, 83% for AIDS disease prediction again using SVM, and finally for COVID-19 we got the highest accuracy using SVM as 92%. The analysis in this research was done using the data set given by the Kaggle website. We used some of the most widely used machine learning techniques and were able to get impressive results. Among all we found that SVM will give the best accuracy. Also we understood some patterns of COVID-19 after using these algorithms. The findings can be utilized to predict and control pandemics in any country or around the world.

References

1. Chen, M., Hao, Y., Hwang, K., Wang, L., Wang, L.: Disease prediction by machine dominating healthcare communities' huge data. IEEE Access 5(1), 8869–8879 (2017)

2. Kohli, P.S., Arora, S.: Application of machine learning in disease prediction. pp. 2119–2133 (2018)
3. Arun, S.S., Iyer, G.N.: Novel corona viral disease pandemic spread data using machine learning techniques (2020). ISBN: 978-1-7281-4876-2
4. Prithivi, P.P.R., Srija, K., Vaishnavi, P., et al.: Artificial intelligence would be used to analyse and visualise pandemics (2020). https://doi.org/10.1088/1757-899X/1022/1/012049
5. Savyanavar, A.S., Ghorpade, V.R.: Efficient resource allocation technique for mobile grids based on on-the-fly computing. Int. J. Inf. Technol. (2018). https://doi.org/10.1007/s41870-018-0269-y
6. Savyanavar, A.S., Ghorpade, V.R.: An exploit in mobile distributed systems checkpointing method for self-healing after mistakes is used. Int. J. Grid High Perform. Comput. **11**(2) (2019)
7. Chen, J.-Z.: Creation of a computer-assisted diagnostic Test-mVision for diagnostics COVID-19 disease in X-ray images. J. ISMAC **3**(02), 132–148 (2021)
8. Sungheetha, A.: COVID-19 risk aversion data selection method. J. Inf. Technol. **3**(01), 57–66 (2021)
9. Imai, N., et al.: Report 3: WHO working centre on 2019-nCoV disease transmission. Infect. Dis. Model., MRC Centre Global Infect. Dis. Anal. J-IDEA, Imperial Coll. London, UK (2020). [Online]. Available: https://doi.org/10.25561/77148
10. Chen, B.: COVID-19 information visualization evaluation and modeling forecast. Neural Computing and Applications (2021)
11. Asthma disease Data set-www.kaggle.com
12. Diebetas disease Data set-www.kaggle.com
13. COVID-19 disease Data set-www.kaggle.com

A Multi-Class Skin Cancer Classification Through Deep Learning

Naresh Kumar Sripada and B. Mohammed Ismail

Abstract Skin infection is one of the most frequent diseases all over the world, and persons under the age of 40–60 have a lot of skin problems. This paper shows how to use a very efficient deep-convolution neural network to analyze and predict skin lesions as accurately as possible. The dataset was acquired from the public domain and contains over 22,900 photos that include different categories of skin lesions and of which 2726 images related to squamous cell carcinoma, malignant melanoma, and basal cell carcinoma are extracted, and the remaining images are ignored. Some of the obtained photos may contain noise; filters are used to reduce the noise in the photographs. The suggested deep-convolution neural network technique encompasses six convolution layers, to reduce the size max-pooling layer applied wherever possible, and the model was used to categorize and forecast skin cancer once the data were cleaned. The model was able to distinguish skin lesions such as squamous cell carcinoma, malignant melanoma, and basal cell carcinoma generated an average accuracy of 97.156%. The paper's major goal is to predict skin cancer in its early stages and give the best accuracy with the least amount of error possible.

Keywords Convolution neural network (CNN) · Skin cancer · Melanoma · Machine learning

N. K. Sripada (✉)
Department of Computer Science and Engineering, Koneru Lakshmaiah Education Foundation, Vaddeswaram, Andhra Pradesh, India
e-mail: s.nareshkumar@sru.edu.in

B. Mohammed Ismail
Department of Information Technology, Kannur University Campus, Mangattuparamba, Kannur, Kerala, India

N. K. Sripada
School of Computer Science and Artificial Intelligence, SR University, Warangal, Telangana, India

V. Suma et al. (eds.), *Evolutionary Computing and Mobile Sustainable Networks*, Lecture Notes on Data Engineering and Communications Technologies 116, https://doi.org/10.1007/978-981-16-9605-3_36

1 Introduction

In today's world [1], computer-aided diagnostic (CAD) technologies are required to analyze and assess medical images to detect and anticipate many hazardous diseases that can lead to death if not detected early. According to a recent survey, there has been an increase in the number of cases of skin cancer growth due to changes in the environment, general propensities for human life [2] such as liquor use, smoking, diet, and radiation caused by the sun. The most well-known type of malignant growth includes skin disease and cancer known as melanoma.

In this article, we aim to diagnose the three skin cancer as squamous cell carcinoma, basal cell carcinoma, and malignant melanoma shown in Fig. 2. Cancer [3] can be treated more effectively if it is detected early [4]. Late diagnosis of sickness causes the tumor disease to spread toward other regions of the body, making treatment difficult and, in some cases, fatal, as well as disrupting people's social lives. The related work in recognizing, segmenting, and describing skin malignant growths discussed in [5, 6] using different machine learning algorithms visualization techniques [7], and by merging computer vision [8], neural organization, and characterization approaches. Table 1 shows the categories of cutaneous melanoma discussed in this paper.

When skin malignant development is detected and treated early, it has a greater cure rate; physicians routinely examine these skin abnormalities visually, followed by biopsy and obsessive investigation. If a dermatologist suspects that the appearance is related to any skin injuries, the visual evaluation approach is usually employed to make a determination; nonetheless, the visual examination does not detect all severe melanoma lesions. As a result, dermatologists want a tool that can accurately assess, detect, and predict skin lesions. The suggested algorithm illustrated in Fig. 1 can assist dermatologists in detecting skin disease at an early stage, lowering the chance of death. In the proposed model, before feeding the input image into the model, we need to customize the image size to 28 * 28, and a Gaussian filter is applied to reduce

Table 1 Types of skin cancer

Skin cancer	Description
Basal cell carcinoma	• Malignancy of basal keratinocytes of the epidermis • Very rarely metastasizes. Locally grows • Head, neck, and trunk
Squamous cell carcinoma	• Possibility of metastasis, or the spread of cancer to other parts of the body, causing significant health issues • Any region of the skin • Appearance of a sore or pimple that does not heal the wound might bleed or leak liquids
Malignant melanoma	• Malignancy of the pigment-forming cells or melanocytes within the epidermis • Melanomas tend to metastasize to lymph nodes, lungs, and brain • Most common location is back for men and lower legs for women

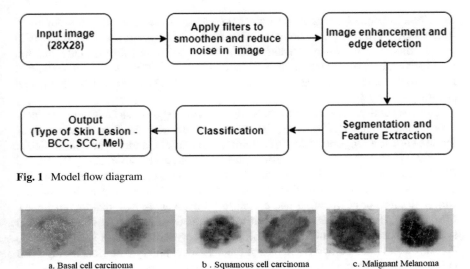

Fig. 1 Model flow diagram

a. Basal cell carcinoma   b . Squamous cell carcinoma   c. Malignant Melanoma

Fig. 2 Sample images of skin cancer [Codella, N., et al 2017]

the distortion [9], and to augment the picture quality [10], later skin lesion area is segmented [11]; feature extraction and prediction algorithm are deployed into the model to generate the appropriate results, and rest of the paper cover literature survey in Sect. 2, Sect. 3 focus on dataset preparation, whereas Sect. 4 demonstrates model designing and implementation; Sect. 5 exhibits results and discussion, and Sect. 6 provides the conclusion and future scope.

2 State-Of-Art: Overview

Cancer is a type of skin melanoma that generates pigmented spots on moles on the skin. It is the most prevalent type of melanoma in humans [12]; melanoma [13] is caused by abnormalities in the skin cells that produce melanin. Esteva et al. introduced convolution neural network-based prediction [14] for categorization of skin cancer, and the model performed better than the expert dermatologist, whereas Iyatomi et al. [15] introduced an automatic approach for classifying different skin lesions, initially model trained with 37 deadliest malignant melanoma and 176 nevi images and the detector able to identify patterns such as fibrillar, parallel furrow, and parallel ridge. AdaBoost and decision tree algorithms were used to classify dermoscopic images, and the model trained on 655 images that contain different skin lesions such as melanoma's 111 images and benign lesions 544 images [16, 17]. Isasi et al. developed a system for automatically detecting skin melanoma based on pattern recognition technique [18] with an accuracy of 85%, whereas Blum et al. [19]

described a method for diagnosing cutaneal melanoma and generated an accuracy of 81% using seven features melanoma, 83% for Menzie's score and 78% for ABCD rule. Ramlakhan and Shang created mobile-based software [20] for analyzing and classifying melanoma where images captured from a handheld device are converted into monochrome images and feed as input into KNN classifier which obtained an accuracy of 66.7%.

To detect and categorize diseases, Sardogan et al. suggested a CNN and the linear vector quantization (LVQ). The LVQ was supplied with the convolution part of the network training output feature vector in this article, and it was able to detect illnesses. [21], to classify skin lesions, a novel regularizes approach [22], ECOC SVM, and deep CNN [23] were used. Electrical impedance techniques were used by Manoorkar et al. [24] to distinguish normal skin tissues from malignant skin cancer tissue. For segmenting and classifying image areas with distinct attributes, Mim et al. [25] proposed usage of a texture values analysis tool such as histogram and statistical techniques. In this paper [26], author suggested a support vector machine algorithm and adapted two types of text features local binary pattern and gray-level co-occurrence matrix techniques to categorize non-melanoma and melanoma skin disease (Fig. 2).

Raju et al. [27] implemented a novel fuzzy set segmentation and SVM-black window optimization technique to classify skin diseases into five categories such as herpes, melanoma, psoriasis, benign, and paederus. On dermoscopy images, Karki et al. [28] implemented Ensemble and EfficientNets networks to identify melanoma cases, and the model attained a 0.94 ROC curve, whereas Dildar et al. [29] discussed and summarized the availability of numerous datasets as well as the diverse techniques utilized by various studies.

3   Dataset Preparation

Skin disease is the most prevalent malignancy in the world; doctors should have a great degree of competence and precision while diagnosing it; as a result, a computer-aided dermatological diagnosing model is required as a more valid and accurate method. We created a model that can predict the kind of skin lesion, and we used photos from the ISIC dataset [30] available in the public domain [31], which contains 22,900 images comprises different categories of disease such as actinic keratosis, vascular lesions, basal cell carcinoma (bcc), melanoma, seborrheic keratosis (SCC), dermatofibroma, and melanocytic nevi. 2726 images are gathered related to the proposed work, and rest of the images are ignored; a gathered image encompasses skin disease like basal cell carcinoma (bcc), squamous cell carcinoma (SCC), and malignant melanoma (mel).

The customized dataset is divided into two sets, such as a test set that contains images to test the model and a training set to prepare the proposed model (see Table 2). In this case, we will utilize an 80–20% ratio for testing and training the model, which implies that from the total images; 2182 images accompanied by three categories 412

Table 2 Training and test samples count

Class	Skin cancer name	Training images	Testing images	Total
1	Basal cell carcinoma (bcc)	412	102	514
2	Squamous cell carcinoma (sqq)	880	219	1099
3	Malignant melanoma (mel)	890	223	1113
Total		2182	544	2726

	lesion_id	image_id	dx	dx_type	age	sex	localization
0	HAM_0000118	ISIC_0027419	scc	histo	80.0	male	scalp
1	HAM_0000118	ISIC_0025030	scc	histo	80.0	male	scalp
2	HAM_0002730	ISIC_0026769	scc	histo	80.0	male	scalp
3	HAM_0002730	ISIC_0025661	scc	histo	80.0	male	scalp
4	HAM_0001466	ISIC_0031633	scc	histo	75.0	male	ear

Fig. 3 Top five rows of the given dataset

basal cell carcinoma images, 880 images of squamous cell carcinoma, 890 images of deadliest malignant melanoma are used to train the model, whereas 544 numbers of images are used to validate the model which comprises 102 images related to skin lesion bcc, 219 images related to SCC skin lesion, and 223 images related to the melanoma skin lesion. Figure 3 depicts the various parameters such as lesion_id, image_id, gender, age, localization, and type of skin lesion can help to analyze the data, whereas Fig. 4 depicts the data distribution by gender, age group, skin cancer type, and location (scalp, back, face, and so on).

Image preprocessing techniques [32] such as noise reduction [33], image enhancement and augmentation, rotation, and others are applied to the collected images. One of the best optimizer algorithms, an Adam optimizer used in the suggested model with filters applied to three RGB-based channels. For automatic feature extraction and categorization, we developed a six Conv2D-layered convolution neural network.

4 Proposed Methodology

A deep convolution neural network [34], a form of neural network commonly used for picture categorization and processing, is described. CNN is made up of neurons with learnable weights and biases, just like other neural networks. Each neuron receives many raw picture matrices as inputs, calculates their weighted sum, passes them through an activation function, and finally reacts with a result. The loss function of the network as a whole reduces with each iteration, and the network eventually

Distribution of Data

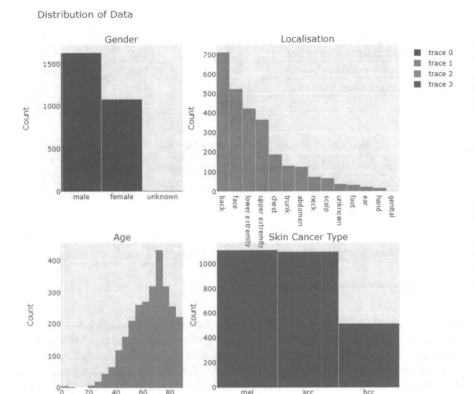

Fig. 4 Categorization of data based on gender, age, localization, skin cancer type

learns a set of parameters. A vector of final probabilities for each class is obtained as a consequence.

4.1 Convolution Layer

Convolution is a method for combining all of the pixels in an absorbent region into a single object (see Fig. 5). Before transmitting the input to the next layer, convolution layers convolution it. When you apply convolution to an image, for example, you reduce the image size, condensing all or most of the information in the field into a single pixel, and the convolution layer produces a vector as a result.

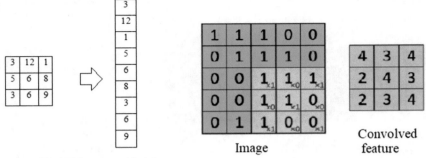

a. Flattening 3x3 image matrix into a 9x1 vector

b. Convoluting 5x5x1 image with kernel 3x3x1 to get 3x3x1 convolved feature

Fig. 5 Convolution of a given image

4.2 Batch Normalization Layer

Batch normalization is used to normalize the given inputs. To keep the mean output at zero and the standard deviation near one, it uses a transformation. Batch normalization works differently during training and inference, which is worth noting.

4.3 ReLU

The ReLU is the most often utilized activation function in deep neural networks. ReLU returns a positive value if the input value is positive; else, it returns 0. The main purpose of activation functions is to aid in the model's interaction impact accounting as well as its nonlinear effects.

4.4 Pooling

The pooling section is used to remove unnecessary arguments as needed. Incoming data are pooled to minimize its dimensionality while preserving its important features. There are numerous ways to say.

- Max pooling: selects the biggest element in the feature map.
- Average pooling: Compute the average for each function map patch. This means that each function map's average value is downsampled.
- Sum pooling: Each function map element's sum is calculated and utilized.

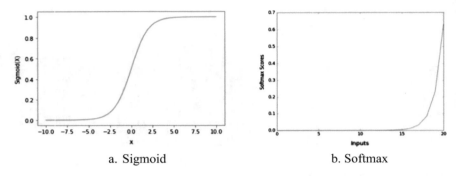

a. Sigmoid   b. Softmax

Fig. 6 S-shaped curve and a probability distribution over n

4.5 Fully Connected Layer

The fully connected layer comes next, which consumes the grid once it has been converted to vectors. Figure 5 shows how the matrix format of the feature map is translated into the vector $(v_1, v_2, v_3 \ldots v_n)$ format and fed into the network. In the fully connected layer, these vectored attributes are combined to create a model. Finally, for grouping, another activation mechanism is used, commonly sigmoid or softmax activation.

4.6 Sigmoid Function

A mathematical function having an "S" curve, commonly known as a sigmoid curve, mostly used in binary classifications. The formula describes it, and Fig. 6 visually shows it. The fact that this function's derivative is simple to compute is one of its main advantages. Based on the convention, we may predict an output ranging from -1 to 1.

4.7 Softmax Function

Over a collection of n-events, the softmax approximates a probability distribution event; softmax applies this notion to multiple classes. In other words, in a multi-class scenario, softmax assigns a decimal probability to each class. Their decimal probability should add up to 1.0.

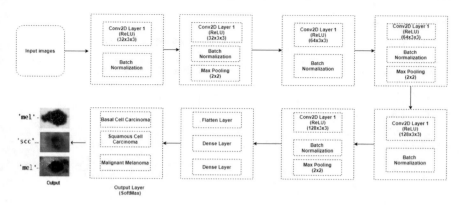

Fig. 7 Architecture of proposed model

4.8 Proposed Model

In contrast to neural networks, CNNs take a multi-channeled picture as input instead of a vector. We glide a filter of size (say, $3 \times 3 \times 1$) across the entire picture in a convolution approach, obtaining the dot product between the filter and the source image segment. In the proposed model, six convolutions-layered CNN architecture is implemented to identify three categories of skin disease such as squamous cell carcinoma, basal cell carcinoma, and melanoma. The CNN architecture is composed of blocks such as input image and feature extraction can be done by convolution layers; ReLU provides a nonlinearity operation which will be done, pooling layer to reduce the size, and then followed by flattening and dense layer as shown in Fig. 7.

To extract the features color, edge, and gradient, the proposed model employs convolution layers, followed by a batch normalization layer, a max pool layer with a 2×2 filter size to reduce the size, and to prevent model overfitting, a drop out of 0.2 is utilized. Deep learning CNN model parameters are shown in Table 3. A 28×28 image was used as the input for the suggested model; the image is examined using a succession of convolution layers, with the filters utilized having a 3×3 field of view.

A stochastic gradient descent Adam optimization approach was used, based on an adaptive estimate of moments first-order, second-order, and a default learning rate of 0.001 was used. The output result must belong to exactly one class; we recommend sparse categorical cross-entropy in place of categorical cross-entropy where the resulting sample can be categorized into multi-classes or labels based on probabilities ([0.1, 0.8, 0.1]). Softmax classifier is used in the output layer to predict the type of skin lesion.

Table 3 Proposed model parameters

Layer	Layer type	Filters and pooling size	Input image size	Activation function
Layer-1	Conv2D layer Batch normalization	$32 \times 3 \times 3$	$28 \times 28 \times 3$	ReLU
Layer-2	Conv2D layer Batch normalization Max pooling	$32 \times 3 \times 3$ $2 \times 2 \times 1$	$28 \times 28 \times 32$	ReLU
Layer-3	Conv2D layer Batch normalization	$64 \times 3 \times 3$	$14 \times 14 \times 32$	ReLU
Layer-4	Conv2D layer Batch normalization Max pooling	$64 \times 3 \times 3$ $2 \times 2 \times 1$	$14 \times 14 \times 64$	ReLU
Layer-5	Conv2D layer Batch normalization	$128 \times 3 \times 3$	$7 \times 7 \times 64$	ReLU
Layer-6	Conv2D layer Batch normalization Pooling layer Dense layer	$128 \times 3 \times 3$ $2 \times 2 \times 1$	$7 \times 7 \times 128$	Softmax

5 Results and Discussions

Deep-CNN-based architecture is used to detect skin cancer in its early stages. The model is trained with three types of skin cancers images including basal cell carcinoma, squamous cell carcinoma, and melanoma, and the model can generate very accurate results. Initially, the model was trained for 20 epochs; the model achieved training accuracy of 97.23% and validation accuracy of 96.75%, with a training loss of 0.0780 and a validation loss of 0.1676; however, to improve accuracy, epochs count increased to 50, and the model achieved training accuracy of 97.84% and validation accuracy of 96.82%, with a training loss of 0.0621 and a validation loss of 0.1810 as shown in Fig. 8.

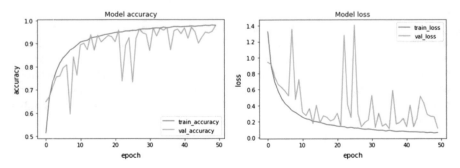

Fig. 8 Accuracy and loss generated by the proposed algorithm

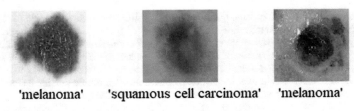

'melanoma'   'squamous cell carcinoma'   'melanoma'

Fig. 9 Different categories of skin lesions identified by the model

The model can achieve an average accuracy of 97.156% for the customized dataset, and be able to identify and categorize the test images to any one of the categories such as melanoma, basal cell carcinoma, or squamous cell carcinoma as shown in Fig. 9.

6 Conclusions and Future Work

Our proposed method would aid in the early detection and classification of skin lesions, decreasing death, and enabling the treatment of skin diseases. We employed a dataset from the public domain comprising 2726 images are used to train the proposed deep-CNN model, and model able to produce a good accuracy of 97.156%, indicating that the proposed model may be useful to dermatologists and persons suffering from skin lesions by diagnosing illnesses at an early stage; to obtain high accuracy, we may strive to enhance the architecture and train the model by increasing the number of sample images related to three categories of skin lesion: basal cell carcinoma, squamous cell carcinoma, and melanoma.

References

1. Sripada, N.K., et al.: Classification and clustering of gene expression in the form of microarray and prediction of cancersusceptibilit, cancerrecurrence and cancersurvival. J. Mech. Continua Math. Sci. **15** (2020)
2. Harshavardhan, A., et al.: Lifting wheelchair for limbless people. IOP Conf. Ser.: Mater. Sci. Eng. **981**(2) (2020)
3. Shahane, R., Ismail, M., Prabhu, C.: A survey on deep learning techniques for prognosis and diagnosis of cancer from microarray gene expression data. J. Comput. Theor. Nanosci. **16**, 5078–5088 (2019)
4. Subramanian, R.R., et al.: Skin cancer classification using convolutional neural networks. In: 2021 11th International Conference on Cloud Computing, Data Science and Engineering (Confluence), pp. 13–19 (2021)
5. Ismail, M., Vardhan, V.H., Mounika, V.A., Padmini, K.S.: An effective heart disease prediction method using artificial neural network. Int. J. Innov. Technol. Explor. Eng. 1529–1532 (2019)
6. Moulana, M., Kolapalli, R., Golla, N., Maturi, S.S.: Prediction of rainfall using machine learning techniques. Int. J. Sci. Technol. Res. **9**, 3236–3240 (2020)

7. Rajendra Prasad, K., Mohammed, M., Noorullah, R.M.: Visual topic models for healthcare data clustering. Evol. Intel. **14**, 545–562 (2021)
8. Kumar, S.N., et al.: A novel approach for detection of counterfeit indian currency notes using deep convolutional neural network. IOP Conf. Ser.: Mater. Sci. Eng. **981**(2), 022018 (2020)
9. Shaik, G.A., Reddy, T.B., Mohammed Ismail, B., Alam, M., Tahernezhadi, M.: Variable block size hybrid fractal technique for image compression. In: 2020 6th International Conference on Advanced Computing and Communication Systems (ICACCS) (2020)
10. Ismail, B.M., Reddy, T.B., Reddy, B.E.: Spiral architecture based hybrid fractal image compression. In: 2016 International Conference on Electrical, Electronics, Communication, Computer and Optimization Techniques (ICEECCOT) (2016)
11. Ismail, B.M., Basha, S.M., Reddy, B.E.: Improved fractal image compression using range block size. In: 2015 IEEE International Conference on Computer Graphics, Vision and Information Security (CGVIS) (2015)
12. Goldstein, B.G., Goldstein, A.O.: Diagnosis and management of malignant melanoma. Am. Fam. Phys. **63**(7), 1359–**136**8, 1374 (2001); Erratum in: Am. Fam. Phys. **64**(10), 1682 (2001)
13. Johnson, T.M., Headington, J.T., Baker, S.R., Lowe, L.: Usefulness of the staged excision for lentigo maligna and lentigo maligna melanoma: the "square" procedure. J. Am. Acad. Dermatol. **37**(5 Pt 1), 758–764 (1997). https://doi.org/10.1016/s0190-9622(97)70114-2.(1997)
14. Esteva, A., Kuprel, B., Novoa, R., et al.: Dermatologist-level classification of skin cancer with deep neural networks. Nature **542**, 115–118 (2017)
15. Iyatomi, H., Oka, H., Celebi, M.E., Ogawa, K., Argenziano, G., Soyer, H.P., Koga, H., Saida, T., Ohara, K., Tanaka, M.: Computer-based classification of dermoscopy images of melanocytic lesions on acral volar skin. J. Invest. Dermatol. (2008)
16. Capdehourat, G., Corez, A., Bazzano, A., Alonso, R., Musé, P.: Toward a combined tool to assist dermatologists in melanoma detection from dermoscopic images of pigmented skin lesions. Pattern Recogn. Lett. **32**, 16 (2011)
17. Abbas, Q., Celebi, M.E., Serrano, C., García, I.F., Ma, G.: Pattern classification of dermoscopy images: a perceptually uniform model. Pattern Recogn. **46**(1) (2013)
18. Isasi, A.G., Zapirain, B.G., Zorrilla, A.M.: Melanomas non-invasive diagnosis application based on the ABCD rule and pattern recognition image processing algorithms. Comput. Biol. Med. (2011)
19. Blum, A., Luedtke, H., Ellwanger, U., Schwabe, R., Rassner, G.: Digital image analysis for diagnosis of cutaneous melanoma. Development of a highly effective computer algorithm based on analysis of 837 melanocytic lesions. Br. J. Dermatol. (2004)
20. Ramlakhan, K., Shang, Y.: A mobile automated skin lesion classification system. In: IEEE 23rd International Conference on Tools with Artificial Intelligence, pp. 138–141 (2011)
21. Sardogan, M., Tuncer, A., Ozen, Y.: Plant leaf disease detection and classification based on CNN with LVQ algorithm. In: 2018 3rd International Conference on Computer Science and Engineering (UBMK), IEEE (2018)
22. Albahar, M.A.: Skin lesion classification using convolutional neural network with novel regularizer. IEEE Access **7**, 38306–38313 (2019)
23. Dorj, U.O., Lee, K.K., Choi, J.Y., et al.: The skin cancer classification using deep convolutional neural network. Multimed. Tools Appl. **77**, 9909–9924 (2018)
24. Manoorkar, P.B., Kamat, D.K., Patil, P.M.: Analysis and classification of human skin diseases. In: 2016 International Conference on Automatic Control and Dynamic Optimization Techniques (ICACDOT), pp. 1067–1071 (2016)
25. Mim, M.S., Das, M., Kiber, M.A.: Feature based skin disease estimation using image processing for teledermatology. In: International Conference on Computer, Communication, Chemical, Material and Electronic Engineering (IC4ME2) (2018)
26. Ebtihal, A., Arfan, J.M.: Classification of dermoscopic skin cancer images using color and hybrid texture features. Int. J. Comput. Sci. Netw. Secur. **16**(4), 135–139 (2016)
27. Raju, D.N., Shanmugasundaram, H., Sasikumar, R.: Fuzzy segmentation and black widow–based optimal SVM for skin disease classification. Med. Biol. Eng. Comput. (2021)

28. Karki, S., et al.: Melanoma classification using EfficientNets and ensemble of models with different input resolution. In: 2021 Australasian Computer Science Week Multiconference (ACSW '21). Association for Computing Machinery, New York, USA (2021). Article 17, 1–5. https://doi.org/10.1145/3437378.3437396
29. Dildar, M., et al.: Skin cancer detection: a review using deep learning techniques. Int. J. Environ. Res. Public Health **18**(10), 5479 (2021). https://doi.org/10.3390/ijerph18105479
30. Codella, N., et al.: Skin lesion analysis toward melanoma detection: a challenge. In: 2017 International Symposium on Biomedical Imaging (ISBI), Hosted by the International Skin Imaging Collaboration (ISIC) (2017)
31. Tschandl, P., Rosendahl, C., Kittler, H.: The HAM10000 dataset, a large collection of multi-source dermatoscopic images of common pigmented skin lesions. Sci. Data **5**, 180161 (2018)
32. Nethravathi, R., et al.: IOP Conf. Ser.: Mater. Sci. Eng. **981**, 022046 (2020)
33. Pisupati, S., Mohammed Ismail, B.: Image registration method for satellite image sensing using feature based techniques. Int. J. Adv. Trends Comput. Sci. Eng. 490–593 (2020)
34. Goceri, E.: Analysis of deep networks with residual blocks and different activation functions: classification of skin diseases. In: Ninth International Conference on Image Processing Theory, Tools and Applications (IPTA), pp. 1–6 (2019)

Preliminary Analysis and Design of a Customized Tourism Recommender System

Deepanjal Shrestha, Tan Wenan, Bijay Gaudel, Deepmala Shrestha, Neesha Rajkarnikar, and Seung Ryul Jeong

Abstract This work presents preliminary analysis and design of the tourism recommender system. The work considers Pokhara city of Nepal as the domain of study as it is the second-largest city and tourism capital of Nepal. The work is built on data from travel websites, published and unpublished reports, literature, and a pilot study conducted in the same city. Tourist decision factors and motivational factors along with seven demographic data are taken into consideration to build an open-end questionnaire for the study. A total sample of 250 respondents is taken as part of the pilot study. The study uses kNN as the classification algorithm to form clusters of data and calculate silhouette to confirm the consistency and validity of clusters. Scatter plots with different combinations of data, UML-based diagrams, and decision tree are used to come up with an initial recommender system design. The work is vital for the design and development of recommendation systems based on a model, mixed, and customized approach. As the requirements directly come from the user end, the system is more specific in terms of recommendations especially for tourists for a specific destination that can be further elaborated to build a generalized model. This work produces an initial system, a knowledge base, different design diagrams, and a fully functional algorithm for the recommender system that can be used by

D. Shrestha (✉) · T. Wenan · B. Gaudel
School of Computer Science and Technology, Nanjing University of Aeronautics and
Astronautics, Nanjing, China

T. Wenan
e-mail: wtan@foxmail.com

D. Shrestha · D. Shrestha
School of Business, Pokhara University, Pokhara, Nepal
e-mail: deepmala@pusob.edu.np

N. Rajkarnikar
School of Energy Science and Engineering, Nanjing Tech University, Pukou, Nanjing, China

S. R. Jeong
Graduate School of Business IT, Kookmin University, Seoul, South Korea
e-mail: srjeong@kookmin.ac.kr

© The Author(s), under exclusive license to Springer Nature Singapore Pte Ltd. 2022
V. Suma et al. (eds.), *Evolutionary Computing and Mobile Sustainable Networks*,
Lecture Notes on Data Engineering and Communications Technologies 116,
https://doi.org/10.1007/978-981-16-9605-3_37

individuals, business entities, and tourism organizations as a tool and knowledge base.

Keywords Recommender system · Tourism · Pokhara · Nepal · Clusters

1   Introduction

The advent of Information and Communication Technology (ICT) has brought massive changes to almost all sectors of human endeavor in the last 20 years [1]. There is no field of research or application that has not been affected by ICT. Business organizations have experienced a massive change with ICT and this has given rise to a new form of business, new boundaries, and new opportunities extending the spheres of business from domestic to global markets [2]. The technology has empowered big enterprises, SMEs, and independent ones by integrating them in a chain of inter-related and inter-dependent components [2]. The tourism industry is one of the leading examples that depicts the true impact of ICT on business. The unprecedented role of technology in the tourism industry in the last decade has challenged the way information is stored, consumed, and utilized in modern times [3]. Tourists, tourism organizations, and governments of different countries use technology extensively for managing their tourism products and services. Tourist is the central entity that serves as the core of this business and information sources play a very important role in the life of a tourist from tour planning to consuming tourism products and services. The rise of different technologies, tools, social applications, sensor networks, mobility systems, augmented reality, and computing paradigms has reinforced a new digital value [4] in the business scenario and created a huge volume of data [5]. The tremendous growth of data, on the one hand, has enriched a tourist but on the other hand, it has created difficulties in terms of processing and consuming data [4, 5]. A common search can aggregate thousands of information clusters and reviews that may pose a further challenge for a tourist to choose the best-desired product or service. Common recommender systems work on a generalized model of computing and do not specifically capture the specialties of a tourism destination. Every destination has its peculiarities and gets tourist based on it, who utilizes these products and services to form opinions, outlooks and shares experiences about them. A generalized model is unable to capture these concerns and does not exactly portray the best information about these tourism destinations. Thus, to solve all these problems a customized and well-researched tourism model is required for each specific tourism destination that can help a tourist to get processed, concise, and vital information that can save time and prove as a valuable asset to tourists in recommending tourism products and services.

2   Literature and Background

Information is a vital component of tourism business and governance, today. Internet and WWW have produced huge volumes of data, which put a tourist in a hard and complex situation of selecting the information that a user requires. The tourism recommender system serves as an important mode of application to help a tourist plan, consume and utilize tourism products and services. The term recommender system refers to an adaptive system that works on different models and principles to process and filter information to provide concise and value-based information to the user [6, 7]. A tourism recommender system is a specialized extension of a recommender system that particularly works for tourism products and services following the principles of a recommender system. There are different types of recommender systems that use different bases to provide a recommendation to the user as shown in Fig. 1. The advancement in technology and application areas of computer science has changed the way recommendation were made earlier in a recommender system. Today, AI, machine learning, deep learning, cluster-based methods, IoT, and similar techniques serve at the heart of these systems. Similarly, dimensions of the recommender system have also changed from content and context to enormous data obtained from different dimensions of technology. The IoT-based systems, Social networks, Big data, Mobile sensor networks, Augmented reality, Interactive web, Semantic web, and similar technologies have a huge potential to produce, process, and analyze data as shown in Fig. 2. The new recommender system must consider these advances to come up with a comprehensive recommender system.

Recommender system in the tourism industry is an active field of study and many scholars have worked on different aspects of these systems [7]. Tourism recommender system studies carried by different authors have focused on different attributes which

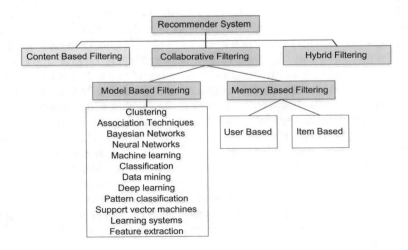

Fig. 1 Classification of recommender system [5–7]

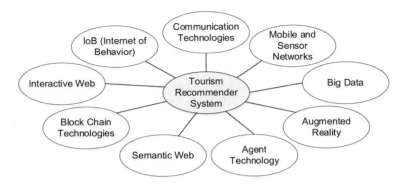

Fig. 2 Dimensions of recommender system [5, 6, 8]

include attractions, estimations, activities, tour planning assistance, tourism products, and services. Scholars have also studied technical aspects of the system which include sorting, filtering, association, matching, and also worked on different algorithms and techniques. Table 1 shows comparative information of different studies carried outduring different times of the year.

A tourism recommender system must be intelligent and work with different information parameters such as scalability, accuracy, information optimization, visualization, and data validity. Besides these, user acceptance and user requirement assimilation are one of the primary needs for a good recommendation system. Scholars have worked on the local, customized, and global nature of recommender systems. Different tools, methods, and technologies were used by scholars to come up with comprehensive, detailed, and realistic tourism recommender systems designs. To understand the recommendation needs of tourists more realistically and work closely with tourist needs, this work proposes a customized approach for the analysis and design of the recommendation system. The work considers Pokhara city of Nepal as a case study to understand tourist requirements and build customized rules and algorithms for the recommendation. Pokhara as a tourism center is one of the most significant areas popular with both national and international tourists. It is rich in a natural habitat with varied tourism activities and a center of attraction. Tourism research in and around Pokhara is always covered by Nepalese scholars and researchers whenever they work with the tourism sector of Nepal. The study of literature in terms of ICT and its application in Pokhara city for tourism is limited. There are only a few studies that are done concerning tourism, ICT, and tourism destinations of Nepal. Some of the notable work includes work on geotagged tweets and tourism points of interest [17, 18], Digital Tourism Security System [19], Study on the Role of ICT Tools and Technologies in Tourism Industry of Nepal [20] Digital Tourism Security System [21], etc. This work specializes to work with recommendations for tourists taking a pilot study as the base and build a recommendation system for the tourism industry of Nepal.

Table 1 Comparative study of tourism recommender system

Publisher and date	Title	Major component	References
2021, Complex and Intelligent System	Artificial intelligence in recommender systems	Role of AI in tourism recommendations	[7]
2020, IEEE Conference Proceedings of ICCC	Recommendation systems with machine learning	Demonstrating role of ML in recommendation	[9]
2020, Procedia Computer Science, ScienceDirect	A multi-level tourism destination recommender system	Destination recommender	[8]
2020, IEEE/ACS	A personalized hybrid tourism recommender system	Personalized recommender system	[10]
2021, Science Direct	A deep learning-based algorithm for multi-criteria recommender systems	Deep learning, multi-criteria, Items rating for multi-datasets	[11]
2021, Science Direct	Intelligent recommender system based on unsupervised ML	Demography, ML and recommendations	[12]
2019, Sustainability	Web-based recommendation system for smart tourism: multi-agent technology	Web-based multi-agent technology	[13]
2021, IEEE Digital Explore	Digital reference model system for religious tourism and its safeties	Religious recomm., safety, tourist	[14]
2019, IEEE Digital Explore	Tourism contextual information for tourism recommendation system	Contextual information tourism recommender	[15]
2016, IEEE Transactions on Big Data	Efficient image geotagging using large databases	Geotagged information for tourism	[16]

3 Research Framework

This work uses an exploratory research approach where data is collected from different sources to analyze, design, and verify a preliminary recommender system. The tourism rich data is obtained from tourism websites (TripAdvisor and google) in this case, which is pre-processed, cleaned, and enriched to come up with a tourism destination database. The method employed maxcopell TripAdvisor and google scrapper API to collect data as shown in Table 2.

The study also considers tourism survey data and other published reports by the Ministry of Tourism, Culture and Civil Aviation and Nepal tourism board to extract important information on tourist choices, expenditure, stay duration, and

Table 2 A sample data table for important tourism destinations [22]

POI	Type	Popularity	Address	Geo Loc	Time	Feature
Pokhara Grande	Hotel	4.5 star	Pardi Birauta Pokhara	28.1923, 83.9747	24 h. 365 days	Major tourism destinations within 2–3 km area
Peace temple	Tourist Point	3rd popular	Lakeside, Pokhara	28.2011, 83.9446	8 am–7 pm	Near major Tourism points
Lake side	Tourist Point	1st popular	Lakeside, Pokhara	28.2053, 83.9616	24 h. 365 days	90% restaurants and hotels
Mahendra cave	Tourist Point	Among top 10	Batulechaur, Pokhara	28.2715, 83.9797	8 am–7:00 pm	Natural cave. It is also near bat cave
Bindabasini temple	Religious Point	Top religious	Bagar, Pokhara	28.2379, 83.9841	5 am–6:30 pm	Near city center
Pame, Pokhara	Free Wandering Location	Top 5 wandering places	Pame, Lakeside Pokhara	28.2255, 83.9463	24 h. 365 days	Free walking, with good restaurants and hotels
Devis fall	Romantic Point	Top 10 Tourist Locations	Damside, Pokhara	28.1903, 83.9591	8 am–6:30 pm	Gupteshwor temple and cave

other important attributes. A pilot study in the field is carried out in Pokhara valley (Tourism capital) of Nepal for international and domestic tourists to further uncover the tourist needs. The field survey data is then classified and visualized using kNN classification and clustering algorithm. The cluster consistency, correlation, and other attributes are visualized using scatter plots and silhouette scores. Finally, data from two different sources is then analyzed to extract detailed information and structure for identifying dimensions, artifacts, and attributes of a customized recommender system in the form of system artifacts, decision rules, tourism taxonomy, and classes of products and services as shown in Fig. 3.

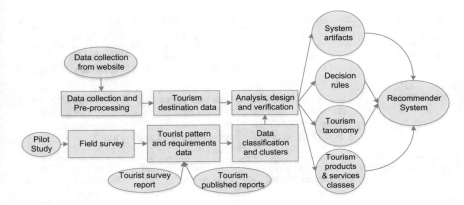

Fig. 3 The research framework of the study

4 Tourism Recommender System Analysis and Design

4.1 System Requirements and Analysis

User needs and tourism recommendations must correlate with each other to find good use of the tourism recommendation system. To understand the basic blocks of the recommender system, the requirements from various sources were identified. An analysis was carried out to design a recommender system in the context of Pokhara, Nepal, and study lapses in any existing system being used. The study of these aspects brought some general systems under investigation with their strength and weaknesses, as shown in Table 3.

As we can see the table depicts some of the most popular systems used for recommendations in tourism destinations with their information content, model, and shortcomings. It is noted that all the above-mentioned systems are either generalized or have static information content that is not suitable for a good recommender system. Further, a user recommender system must consider the local situations, information clusters, and local scenarios to build a comprehensive information recommender system. It was analyzed that the advantages of the existing general systems must be combined, removing the disadvantages to design a comprehensive tourism recommender system.

Table 3 Sources for data and recommendations for tourism destinations in Pokhara, Nepal

Source	Data provided	Model in use	Shortcomings
Trip advisor	Recommendation on hotels, restaurants, attractions, etc.	User rating and feedback	Dependent on user data, crowdsourced model
Google maps	Geotagged locations, distances, recommendations on product and services	User tagging, user information, distance-based, popularity-based, and others	User dependent and generalized for all countries and locations
Nepal tourism portal	Static web information inefficient chatbots, no real-time updates	Static web information system	No real-time updates, Static and fixed type of information
Social Sites Facebook, Twitter, etc.	Crowdsourced data from users	Social network model	User-based. Problems of accuracy and preciseness. Generalized model
Websites of Wikipedia, private tourism companies, etc.	Static, User-based, blogs, structured, etc.	Web content, static and dynamic model	Little data, biased data, static with no real-time updates. Static with very few updates

4.2 Tourism Governance and Business Management Scenario

A good tourism recommender system must consider all the tourism business entities to design and develop a good recommender system. It is important to start the analysis at the governance level and integrate the other entities by identifying the vital information required by the tourist. In this study, it was identified that the Government of Nepal (GoN) is the apex body that is linked to the Ministry of home affairs to check Visas, issue visas at the airport, and look for all immigration-related issues like a lawsuit, permit detention, and other issues in Nepal as shown in Fig. 4. The Ministry of Culture Tourism and Civil Aviation is the main body that is dedicated to all of the tourism-related management and promotion activities in Nepal. It regulates the different components of tourism through different subdivisions and departments and assures the quality, security, and stability of the related entities in the country. VISA information, security information, tourism assistance, destination regulations, accommodation information, etc. are all vital pieces of information that are needed by tourists at the beginning of their journey. The recommender system needs to identify these components and design them as a part of its recommendation. Moreover, as this body is responsible for any change in the tourism policy, its impact on other related entities must be communicated and updated in a tourism management system accordingly so that all necessary information content is updated and available in real-time.

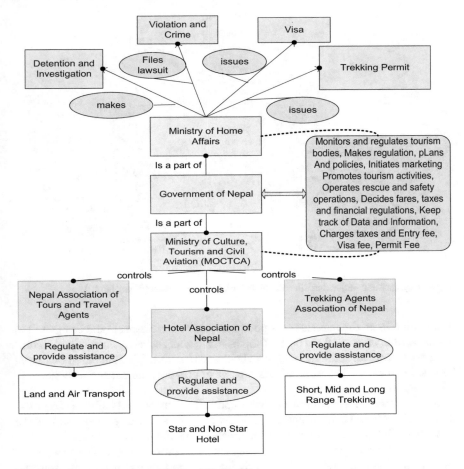

Fig. 4 Tourism governance architecture [20, 21, 23]

4.3 Analysis of Public and Private Tourism Bodies

The public and private tourism organizations are the main part of the tourism business. These include product development and service sectors like the hotel, restaurants, trekking agencies, local tour operators, destination management groups, transportations, health and beauty, shopping sites, etc. as shown in Fig. 5. These agencies have products and services that are consumed by tourists and in return provide their feedbacks and recommendations that help them to be in the business. The pool of information for these enormous ranges of products and services has increased posing challenges for tourists to select the best. User feedback, ranking, rating, and one-to-one recommendations have become a part for these business houses to grow and prosper. The technological growth has empowered users to access all the related

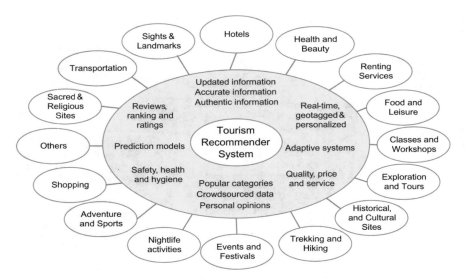

Fig. 5 Representing major tourism categories based on crowdsourced data [6–10]

information in no time and hence a good analysis of these aspects and design of the best algorithms to extract the best information is a vital aspect. A customized recommender system that combines all the technologies in use and is built based on tourist knowledge for specific destinations is capable of proving efficient and well-suited in this scenario.

A recommender system must capture updated, accurate, and authentic information of all tourism business entities in real-time, marked with geotags, and works in a personalized manner. The recommendations should consider reviews, ratings, rankings, personal opinions, popular categories, and crowdsourced data as a part of the information extraction process to finally provide the best from these sources. The system needs to capture information regarding safety, health, and hygiene and should be adaptive in learning from changing scenarios as shown in Fig. 5.

4.4 A Pilot Study of Tourist Decision and Motivation Analysis

Tourist decision factors and motivations to visit a particular place serve as the basis to build a recommender system. To understand these factors a pilot study was carried out with 250 tourist respondents who have visited Pokhara, Nepal in recent years. The respondent's data was collected from a set of open-ended questions that gave an understanding of the decision and motivation factors of a tourist. The data consisted of 7 demographic attributes that were combined with 14 decision factors, 11 motivation factors, and factors related to cost, spending behavior, choice of information, and sources of information to come up with a preliminary decision support system for

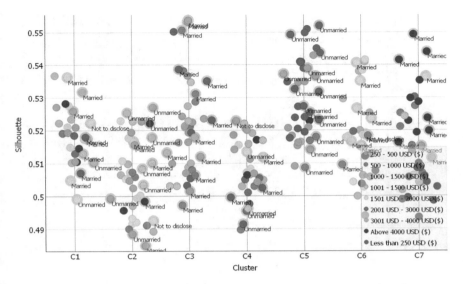

Fig. 6 Represents silhouette score of cluster based on income, marital status, and gender

the tourism industry of Pokhara, Nepal. The collected data is then classified using the k-Nearest Neighbor (kNN) algorithm to find out related clusters and silhouette scores to validate consistency and clusters of data. The data depicted two prominent groups of respondents that include inbound and domestic tourists, which were studied in combination with sex, marital status, monthly income, educational qualification, and profession to identify the specific recommendation classes and rules.

It can be seen from Fig. 6 that the silhouette score of clusters based on income, marital status, and gender is consistent as the majority of them have a score of above 0.5. We can also see that all clusters are evenly distributed based on income. Thus, we can conclude that Pokhara city has tourists that have an income from 250 USD minimum to above 4000 USD at maximum. Further Fig. 7 represents silhouette score for country of residence of a tourist, his average spending (USD), visit motive, and tourism interest. The domestic tourists form a close cluster and have a spending range of 100–400 USD whereas Australia, UK, China, and South Korea have a range of 400–1500 USD. The scatter plot also depicts that majority of the tourist are visiting the city for vacations.

The data of the pilot study was further analyzed to find the consistency between tourism interest, visit motive, accommodation type, and country of residence. The silhouette score for these factors was seen as consistent as shown in Fig. 8. The tourist had an interest in nature, entertainment, health, beauty and sports, and some had multiple interests. Hotels and homestay were the top two categories preferred by the tourists visiting Pokhara. Further, we can see that from the data of the pilot study, the ten most important factors were discovered. The cost was the most important factor, followed by transportation, safety and security, tourism activities, people and

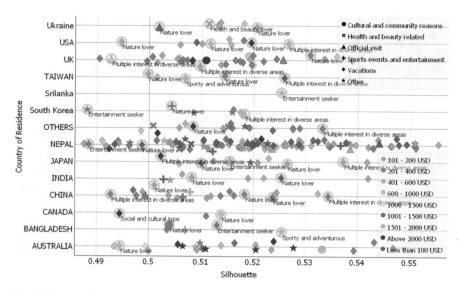

Fig. 7 Representing silhouette score for average spending (USD), visit motive, tourism interest, and country of tourist

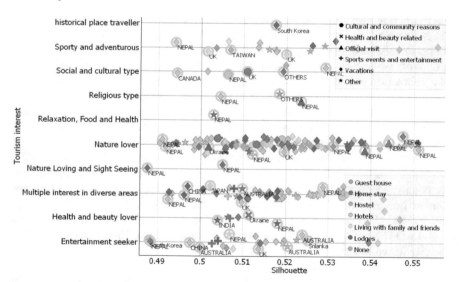

Fig. 8 Representing silhouette score for tourism interest, visit motive, accommodation type, and country of residence

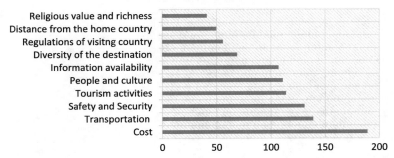

Fig. 9 Representing tourist decision factors and tourist visiting motives based on pilot study and literature review

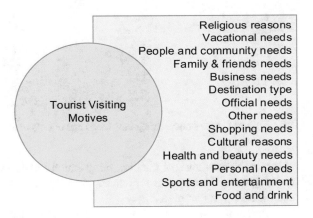

Fig. 10 Representing tourist visiting motives based on the pilot study. *Source* The World Travel & Tourism Council [24], pilot survey

culture, and information availability as shown in Fig. 9. The recommendation system must consider these factors to build a customized system for the tourist in Pokhara, Nepal. The analysis regarding the motivation depicts 14 important motives, out of which vacations were the first, followed by culture, health and beauty, and sports, conferences, and business needs as next important motivation factors as shown in Fig. 10.

4.5 The System Design Process

The study after identifying the requirements of the system proceeded with the system design and implementation. The next step in the design included the construction of the class diagram, taxonomy, use-case diagram, and object model. Figure 11 represents a Class diagram of tourism points of interest with three major sub-components, restaurants, hotels, and tourism attractions. These classes are further detailed to

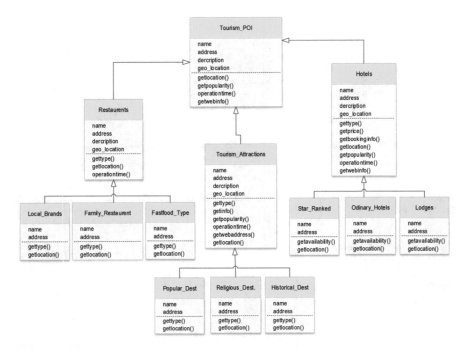

Fig. 11 Class diagram representing the classes and objects of a recommender system

represent the next level of the hierarchy, which finally represents the building block of the system. These class structures are bonded with attributes that include the recommendation parameters.

A taxonomy model was constructed based on tourism concepts and artifacts analyzed from the different sources of information and data. The model depicts an overall picture of tourism products and services that are shown as connected nodes in a graph with named links as seen in Fig. 12. It can be visualized that tourism activity has hiking and entertainment as its upper branches which further decompose into travel routes and theaters, which are further connected to related entities in a branching hierarchy. Similarly, all the required entities are detailed as bonded nodes with one another in a taxonomy of tourism hierarchy.

Use case models are an important aspect of any system design as they represent the processes and functions of a system as shown in Fig. 13. Altogether 14 important use cases are shown in the figure that connects to the actor tourist. A decision tree consists of nodes and leaves, where a node is called the root from where the condition test starts. This study applies the DT as the classifier to test the condition in the system serving as the base of the decision in the system. Figure 14 represents a snapshot of decision rules and their related parameters. It can be seen hostel is recommended to a tourist of age group 20–30, with a monthly income less than 250 USD, and is unmarried based on pilot survey data and rules created. An object diagram in Fig. 15

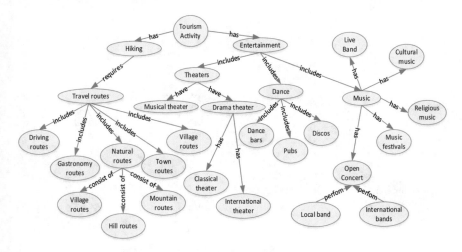

Fig. 12 A taxonomy of tourism concepts and artifacts

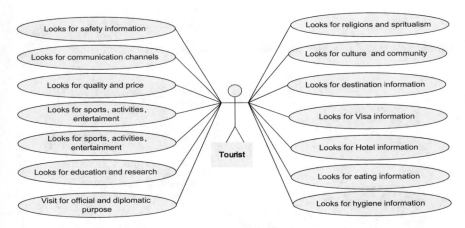

Fig. 13 A use case diagram of tourism concepts and artifacts

shows how a recommender matching module exactly works by calling instances of a class. All these diagrams depict the system design and modeling aspects that capture the essential activities and components of the system.

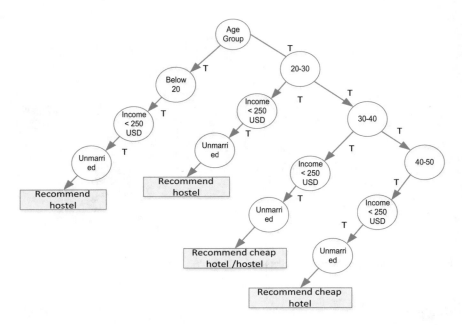

Fig. 14 Representing a decision tree in the decision-making process

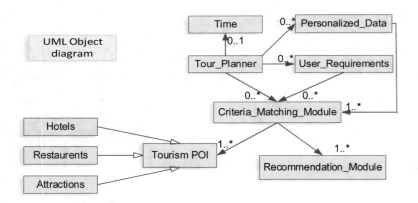

Fig. 15 An object diagram for tourism concepts and artifacts

4.6   The Recommender Algorithm

Algorithm   1. Recommender   system   algorithm   for   tourism destinations

---

Initialize real-time monitor , server system and fetch updates from weather server and security server

lookup = Initialize lookup table (sample content of lookup table = {"name": {"type": tn, "ranking":rn, "popularity": pn, "location": ln, "cost": cn, "rating": rt}})

user_location = get_location()

d = Ask user preferred distance

pt = Ask user preferred location type

(a, b) = Ask user spending range (lower and upper range)

rn = Ask user preferred ranking of choice

rt = User preferred rating of choice

possible_destination = [k for k, v in lookup if v["type"] == pt]

lookup_filter = []

for pd in possible_destination:

    if ((lookup[pd]["cost"]>= a && lookup[pd]["cost"] <=b) && (|user_location-lookup[pd]["location"] |<=d) && (rn==lookup[pd]["ranking"]), rt == lookup[pd]["rating"]):

        lookup_filter.append(pd)

    else if ((lookup[pd]["cost"]>= a && pd["cost"] <=b) && (|user_location-lookup[pd]["location"] |<=d) && (rn==lookup[pd]["ranking"])):

        lookup_filter.append(pd)

    else if ((lookup[pd]["cost"]>= a && pd["cost"] <=b) && (|user_location-lookup[pd]["location"] |<=d)):

        lookup_filter_append(pd)

    else if ((|user_location-lookup[pd]["location"] |<=d) && (rn==lookup[pd]["ranking"]) && rt == lookup[pd]["rating"])):

        lookup_filter.append(pd)

    else if ((|user_location-lookup[pd]["location"] |<=d) && (rn==lookup[pd]["ranking"])):

        lookup_filter.append(pd)

    else if ((|user_location-lookup[pd]["location"] |<=d) && (rt == lookup[pd]["rating"])):

        lookup_filter.append(pd)

    else if ((rn==lookup[pd]["ranking"]) && (rt == lookup[pd]["rating"])):

        lookup_filter.append(pd)

    else if (lookup[pd]["cost"]>= a && pd["cost"] <=b):

        lookup_filter.append(pd)

    else if (|user_location-lookup[pd]["location"] |<=d):

        lookup_filter.append(pd)

    else if (rn==lookup[pd]["ranking"]):

        lookup_filter.append(pd)

    else if (pr == lookup[pd]["rating"]):

        lookup_filter.append(pd)

if lookup_filter is None:

    lookup_filter = possible_destination

lookup_filter = sort(lookup_filter, argument = "Created Rules" and "Popularity")

return lookup_filter

The recommender system algorithm is built on customized attributes based on user requirements and scraped data from the web. The system first gets initialized and runs some early modules that capture real-time data from security servers, weather monitoring servers and makes a real-time update for changed information. In the second step, the updates are done in the database server based on any updates triggered by the other servers. After finishing the initial process of checking and updating, the system switches to user mode and captures its location. This location serves as the center point for the system to make any further analysis and recommendation to the tourist. Next, it asks the user for his/her preferences based on type, cost, rating, distance, and preferred ranking. The system then processes each input carefully entered by the user, checking each criterion and displays the processed data based on the popularity index in an ascending order. In case the criteria are not matched or users enter no choices, the system works based on rules created from pilot survey and popularity index of the destination type as shown in algorithm 1. The created rule takes demographic data along with other information clusters to make recommendations. Further at regular intervals, the system keeps scanning security servers, weather servers, and other important servers to identify any serious changes in the tourism destination and prepare warning messages to make tourists aware of any serious concerns in their journey. Table 4 represents the verification of all the logical paths executed in the algorithm based on user criteria and system updates. The algorithm takes each criterion and weight associated with it to provide a recommendation to the user. In case no single criteria are matched, the system makes recommendations based on created

Table 4 Representing logical path verification based on conditions in the algorithm

Assuming: Chosen type = *True*, and warning real-time messages = *Null*

User I/P	Check cost range (a, b)	Check distance (d)	Check ranking (Rm)	Check rating (Rt)	O/P (based on popularity in ascending order)
I/P1	Cost = T	d = T	Ranking = T	Rating = T	List with all T
I/P2	Cost = T	d = T	Ranking = T	Rating = F	List without Rt
I/P3	Cost = T	d = T	Ranking = F	Rating = F	List without Rt & Rn
I/P4	Cost = T	d = F	Ranking = F	Rating = F	List without d, Rt & Rn
I/P5	Cost = F	d = T	Ranking = T	Rating = T	List without (a, b)
I/P6	Cost = F	d = T	Ranking = T	Rating = F	List without (a, b) & Rt
I/P7	Cost = F	d = T	Ranking = F	Rating = F	List without (a, b), Rt & Rn
I/P8	Cost = F	d = F	Ranking = T	Rating = T	List without (a, b) & d
I/P9	Cost = F	d = F	Ranking = T	Rating = F	List without (a, b), d & RT
I/P10	Cost = F	d = F	Ranking = F	Rating = T	List without (a, b), d & Rn
I/P11	Cost = F	d = F	Ranking = F	Rating = F	List based on created rules and popularity index

rules and popularity index. It must be noted that user requirements form the basis of the recommendation rules and the popularity index is bonded to crowdsourced data that keeps changing with each review and ratings of the tourist.

5 Conclusion

The proposed work preliminary analysis and design of a customized tourism recommender system is carried out to design a customized recommendation system for a tourism destination. The work is undertaken in Pokhara city of Nepal as a domain to understand user requirements and build a customized recommender system that caters needs of any peculiar tourism destination by providing the best and optimal recommendations. This work uses crowdsourced data from the web by scrapping important tourism sites like TripAdvisor and Google, which help to build a database of tourism products and services for Pokhara city based on user reviews, opinions, feedback, and ratings. Further, the pilot study carried out as fieldwork helps to identify an important aspect of tourist requirements in terms of their demography, patterns, preferences, choices, and other related attributes. The parameters identified in a survey serve as the basis to design system artifacts, class diagrams, use-case models, taxonomy, and decision rules. The use of kNN clustering algorithm helps to visualize tourist clusters and their related attributes. The silhouette score depicts that clusters are well marked and are consistent. The visualization with scattered plots and a combination of attributes depicts the tourist scenario in the Pokhara city in terms of their choices, budget, average spending, and other tourism attributes providing a vital insight for tourism business entities along with the government of Nepal. The study being the first of its kind in Nepal is vital to research that brings the core tourist requirements of tourists for Pokhara city of Nepal. This study is well analyzed, designed, and proposes a good solution for a customized recommender system that is vital for tourists, business entities, government, and cities like Pokhara.

6 Limitation and Future Work

The current work is a preliminary study that has worked with the first cycle of recommendations for basic products and services. It does not include recommendations for all product types and services extensively, including parameters like safety and security. The testing of the system is limited and this works need to be extended considering factors related to behavior, satisfaction, and tour planning, combined with crowd-sourced data to design a comprehensive recommender system.

Acknowledgements The paper is supported in part by the National Natural Science Foundation of China under Grant (No. 61672022 and No. U1904186), and the Collaborative Innovation Platform of Electronic Information Master under Grant No. A10GY21F015 of Shanghai Polytechnic University.

References

1. William, J., Kramer, B.J., Ktaz, R.S.: The Role of Information and Communication Technology Sector in Expanding Economic Opportunity. Corporate Social Responsibility Initiative Report No. 22. Cambridge, MA, Kennedy, Harvard University (2007)
2. Cascio, W.F., Montealegre, R.: How technology is changing work and organizations. Annu. Rev. Organ. Psych. Organ. Behav. **3**, 349–375 (2016). https://doi.org/10.1146/annurev-orgpsych-041015-062352
3. Gössling, S.: Technology, ICT and tourism: from big data to the big picture. J. Sustain. Tour. **29**(5), 849–858 (2021). https://doi.org/10.1080/09669582.2020.1865387
4. Foris, D., Popescu, M., Foris, T.: A Comprehensive Review of the Quality Approach in Tourism. December 20th 2017. https://doi.org/10.5772/intechopen.70494
5. Badaro, G., Hajj, H., El-Hajj, W., Nachman, L.: A hybrid approach with collaborative filtering for recommender systems. In: 9th International Wireless Communications and Mobile Computing Conference (IWCMC), pp. 349–354 (2013). https://doi.org/10.1109/IWCMC.2013.6583584
6. Burke, R.: Hybrid recommender systems: survey and experiments. User Model User-Adap. Inter. **12**, 331–370 (2002). https://doi.org/10.1023/A:1021240730564
7. Zhang, Q., Lu, J., Jin, Y.: Artificial intelligence in recommender systems. Complex Intel. Syst. **7**, 439–457 (2021). https://doi.org/10.1007/s40747-020-00212-w
8. Alrasheed, H., Alzeer, A., Alhowimel, A., Shameri, N., Althyabi, A.: A multi-level tourism destination recommender system. In: The 11th International Conference on Ambient Systems, Networks and Technologies (ANT), Procedia Computer Science, vol. 170, pp. 333–340 (2020). https://doi.org/10.1016/j.procs.2020.03.047
9. Fanca, A., Puscasiu, A., Gota, D.-I., Valean, H.: Recommendation systems with machine learning. In: 21th International Carpathian Control Conference, pp. 1–6 (2020). https://doi.org/10.1109/ICCC49264.2020.9257290
10. Kbaier, M.E.B.H., Masri H., Krichen, S.: A personalized hybrid tourism recommender system. In: 2017 IEEE/ACS 14th International Conference on Computer Systems and Applications (AICCSA), 2017, pp. 244–250. https://doi.org/10.1109/AICCSA.2017.12
11. Shambour, Q., A deep learning-based algorithm for multi-criteria recommender systems. Knowl.-Based Syst. **211**, 106545 (2021). https://doi.org/10.1016/j.knosys.2020.106545
12. Yassine, A., Mohamed, L., Al Achhab, M.: Intelligent recommender system based on unsupervised machine learning and demographic attributes. Simul. Model. Pract. Theor. **107**, 102198 (2021). https://doi.org/10.1016/j.simpat.2020.102198
13. Hassannia, R., Barenji, A.V., Li, Z., Alipour, H.: Web-based recommendation system for smart tourism: multiagent technology. Sustainability **11**, 323 (2019). https://doi.org/10.3390/su11020323
14. Wenan, T., Shrestha, D., Rajkarnikar, N., Adhikari, B., Jeong, S.R.: Digital reference model system for religious tourism & its safeties. In: 2020 IEEE 7th ICETAS, pp. 1–6. https://doi.org/10.1109/ICETAS51660.2020.9484189
15. Saputra, R.Y., Nugroho, L.E., Kusumawardani, S.S.: Collecting the tourism contextual information data to support the tourism recommendation system. In: International Conference on Info. and Comm. (ICOIACT), pp. 79–84 (2019). https://doi.org/10.1109/ICOIACT46704.2019.8938546
16. Kit, D., Kong, Y., Fu, Y.: Efficient image geotagging using large databases. IEEE Trans. Big Data **2**(4), 325–338 (2016). https://doi.org/10.1109/TBDATA.2016.2600564
17. Devkota, B., Miyazaki, H., Witayangkurn, A., Kim, S.M.: Using volunteered geographic information and nighttime light remote sensing data to identify tourism areas of interest. Sustainability **11**, 4718 (2019). https://doi.org/10.3390/su11174718
18. Devkota, B., Miyazaki, H.: An exploratory study on the generation and distribution of geotagged tweets in Nepal. In: 2018 IEEE 3rd International Conference on Computing, Communication and Security (ICCCS), pp. 70–76 (2018). https://doi.org/10.1109/CCCS.2018.8586827

19. D. Shrestha, T. Wenan, B. Adhikari, D. Shrestha, A. Khadka, S.R. Jeong, Design and analysis of mobile-based tourism security application: concepts, artifacts and challenges. In: Lecture Notes in Electrical Engineering, vol. 733 (2021). Springer, Singapore. https://doi.org/10.1007/978-981-33-4909-4_15

20. Shrestha, D., Wenan, T., Gaudel, B., Maharjan, S., Jeong, S.R.: An exploratory study on the role of ICT tools and technologies in tourism industry of Nepal. ICMCSI 2020. EAI/Springer Innovations in Communication and Computing. Springer, Cham (2021). https://doi.org/10.1007/978-3-030-49795-8_9

21. Shrestha, D., Wenan, T., Khadka, A., Jeong, S.R.: Digital tourism security system for Nepal. KSII Trans. Internet Inf. Syst. **14**(11), 4331–4354 (2020). https://doi.org/10.3837/tiis.2020.11.005

22. Tripadvisor: Explore Pokhara. Access date: 20 Oct 2020. https://www.tripadvisor.com/Tourism-g293891-Pokhara_Gandaki_Zone_Western_Region-Vacations.html

23. Ministry of Culture, Tourism & Civil Aviation: NEPAL TOURISM STATISTICS 2019, Government of Nepal, Ministry of Culture, Tourism & Civil Aviation, Singhadurbar, Kathmandu. Singha Durbar, Kathmandu (2020). www.tourism.gov.np

24. The World Travel & Tourism Council: WTTC Travel & Tourism Economic Impact 2018 Nepal. The World Travel & Tourism Council, London (2018)

A Deep Learning-Based Face Mask Detection

Rushi Patel, Yash Patel, Nehal Patel, and Sandip Patel

Abstract The unprecedented spread of coronavirus disease 2019 has resulted in a global pandemic. During the lockdown time of the COVID-19 pandemic, face mask detection is the popular topic in machine learning paradigm. Face mask detection uses deep learning algorithms to determine whether or not a person is wearing a mask. Face detection and recognition have become a highly prevalent challenge in image processing and computer vision domain. Usually, the face detection is achieved with the help of facial landmarks with the help of the popular library OpenCV for real-time computer vision. There are several convolution network methods available to build a model as accurate as possible. Convolutional neural networks are commonly employed for tasks like image processing, recognition, and detection, which demand greater computational power and time to train any model or program from the initial concept. This paper uses the concept of pre-trained network and advanced techniques like fine-tuning and feature extraction with the base pre-trained network model. Transfer learning technique (fine-tuning) with the base model of a pre-trained network makes the program problem much faster, easier, and feasible rather than training any model from the initial stage.

Keywords Deep learning · Face mask detection · Transfer learning · COVID-19

1 Introduction

The COVID-19 pandemic is known as the global catastrophe of our time, and it was the biggest threat faced by humans after World War II. The effect of COVID-19 on almost all growth sectors has declined. Worldwide, the count of death cases is about to cross the tragic figure of two million. Many people have lost their loved ones,

R. Patel · Y. Patel · N. Patel (✉) · S. Patel
K D Patel Department of Information Technology, Chandubhai S. Patel Institute of Technology (CSPIT), FTE, CHARUSAT, Changa, Gujarat, India
e-mail: nehalpatel.it@charusat.ac.in

S. Patel
e-mail: sandippatel.it@charusat.ac.in

© The Author(s), under exclusive license to Springer Nature Singapore Pte Ltd. 2022
V. Suma et al. (eds.), *Evolutionary Computing and Mobile Sustainable Networks*,
Lecture Notes on Data Engineering and Communications Technologies 116,
https://doi.org/10.1007/978-981-16-9605-3_38

and many families are deeply grieving of their lost ones. In any case, the pandemic is significantly more than a well-being emergency; it is likewise an exceptional financial emergency. Focusing on all of the nations it contacts, it can possibly destroy social, financial, and political impacts that will leave profound and long-standing scars. UNDP is the specialized lead in the UN's financial recuperation close to the well-being reaction driven by WHO, and the global humanitarian response plan and working under the administration of UN Resident Coordinators [1].

According to the World Bank, remittances will drop by \$110 billion this year, leaving 800 million people unable to meet their basic needs. The main coronavirus symptoms identified by the World Health Organization [WHO] are fever, dry cough, weakness, diarrhea, loss of taste, and odor [2]. Many steps for preventing good size of this virus have been achieved in numerous methods like social distancing, wearing mask, sanitizing all the surfaces, and boosting up immunity. Every prevention measure has a critical role in terminating this virus. One of the essential preventive measures among them is sporting a mask. COVID's droplets are stronger than other viruses, and as a consequence, they remain in the environment for long duration, and thus, wearing a mask should help you avoid breathing it in our frame. Even in a crowded area, every other person's breath should be a stimulant for the spread of COVID-19 [3].

The rest of this paper is structured as follows. In Sect. 2, we analysis related works on face mask detection and neural network. A brief description about deep learning is given in Sect. 3. We have used the concepts like pre-trained network which are discussed in detail in the subsection of deep learning. The advance techniques like fine-tuning and feature extraction are well-known when the user is working with pre-network and keras. The proposed model is presented in Sect. 4. Section 5 describes results and comparison of approaches. Lastly, Sect. 6 accomplishes the paper.

2 Literature Review

Asmit et al. [4] found two datasets (i) Variety Masks—Segmentation (VAMA-S) (ii) Variety Masks—Classification (VAMA-C) for face mask detection and mask fit analysis and achieve 98% accuracy. In these, they have used social media images from Instagram which are used for classification using pre-trained network and semantic segmentation.

Yuzhen et al. [5] published face mask service stage detection to obtain stage of face mask with the use of gray-level co-occurrence matrix (GLCM) and K-nearest neighbor (KNN) algorithm. They obtained good results with a precision of 82.87 ± 8.50% on the testing dataset and also concluded that experiments give an idea of using photographs of masks to detect the use time of face masks to distinguish whether the face mask can be used sequentially or not.

Jignesh et al. [6] have proposed face mask detection using transfer learning of inception v3. This model is trained and applied on the Simulated Masked Face Dataset (SMFD) achieving 99% accuracy. Amit et al. [7] used a multi-stage

Table 1 Summary of the machine learning algorithms

References	Techniques	Accuracy (%)	Dataset
[9]	Fully convolutional networks (FCN)	93.85	Multi parsing human dataset
[10]	Convolution neural network (CNN)	98.7	Kaggle dataset
[11]	Keras, OpenCV, TensorFlow	98.55	Kaggle dataset
[12]	Convolutional neural network (CNN), classifiers	96.19	Dataset provided by Prajna Bhandary (1376 images)
[13]	YOLOv3, DNN, GPU	96	Web-scrapping tool

CNN architecture for face mask detection in which first stage uses Dlib, MTCNN, RetinaFace for face detection and second stage uses mobilenetv2, Densenet121, Nasnet for classification. Aqeel et al. [8] have proposed masked face recognition for performing secured authentication using facenet. Dataset used are VGGFace2-mini VGGFace2-mini-SM1, LFW-SM (combined) MFR2 achieving accuracy up to 95–97%.

Wang et al. [19] proposed an in-browser serverless edge computing-based face mask detection solution called web-based effective artificial intelligence recognition of masks, which can be deployed on any common device with Internet connection via web browsers, without the need for any software installation. The serverless edge-computing design reduces the need for additional hardware, where the proposed method makes a contribution by providing a holistic edge-computing framework that integrates deep learning models (YOLO), a high-performance neural network inference computing framework (NCNN), and a stack-based virtual machine.

Joshi et al. [20] to identify the faces and their accompanying facial landmarks contained in the video frame, the proposed framework uses the MTCNN face detection model. These facial images and cues are then analyzed by a neoteric classifier that employs the MobileNetV2 architecture as an object detector for masked region detection. The suggested framework was put to the test using a dataset consisting of films documenting people's movement in public spaces while adhering to COVID-19 safety protocols. By achieving excellent precision, recall, and accuracy, the proposed methodology established its effectiveness in detecting facial masks. Table 1 shows the popular techniques of machine learning with accuracy and dataset.

3 Deep Learning

Deep learning can be deliberated as a subcategory of machine learning (ML) which is based on the neural network (NN) inspired by animal brain anatomy. Deep learning is a machine model which mimics how the human brain is trained and learns things. Nowadays, deep learning is widely used because of its broader domain. Remote sensing, text documentation and summarization, fraud detection, autonomous vehicles, virtual assistants, supercomputing, customer relationship management (CRM) systems, investment modeling, and facial recognition systems are all examples of deep learning applications [14].

3.1 Neural Networks

To comprehend how an artificial neuron functions, we must first comprehend how a biological neuron functions. Each neuron in the mammalian nervous structure behaviors the electrical signal in three diverse ways, as shown in Fig. 1. Dendrites, which receives information or signals from other neurons that get connected to it. In the Soma, a pooling process of the operated input signals is performed through a spatial and temporal signal integrator. For information flow, axon transfers the output signal to another neuron. Each of the flanges connects to the dendrite or hairs of the one after it [15]. The different categories of neural networks are artificial neural network, radial basis functions neural network, Kohonen self-organizing neural network, recurrent neural network, convolution neural network, and modular neural network. This paper focuses on three essential forms of neural networks that are the foundation for most deep learning pre-trained models.

3.1.1 Recurrent Neural Network (RNN)

A recurrent neural network is based on the concept of saving a layer's output and feeding it back to the input in order to predict the layer's output. RNN uses the previous input memory to model the problems in real time. With normal back-propagation over time, these networks can be improved, trained, and drawn-out [16]. The graphic depiction is shown in Fig. 2.

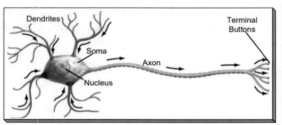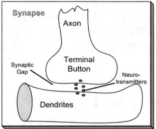

Fig. 1 A biological neuron (left) with the direction of the signal flow and a synapse (right) [15]

Fig. 2 Structure of basic
recurrent neural network [16]

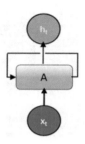

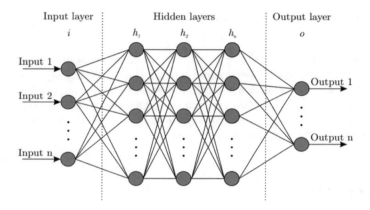

Fig. 3 Artificial neural network architecture [18]

3.1.2 Artificial Neural Network (ANN)

Multilayer perceptron (MLP) is of one or further hidden layers, and several hidden
units make up ANNs or general NNs. The general ANN architecture is depicted in
Fig. 3, which includes an input layer, a number of hidden layers, and an output layer.
Artificial neurons are interconnected through adaptive weights in individually hidden
and output layers. These weights are measured using input–output data in a training
phase. There is an activation function for each artificial neuron, which can be any
function with the range [−1, 1]; the tangent sigmoid and the logarithmic sigmoid are
the most common activation functions [17].

3.1.3 Convolutional Neural Network (CNN)

CNN is a multilayer neural network built on the visual cortex of animals. The most
advantageous feature of CNNs is that they reduce the number of parameters in ANNs.
Figure 4 shows the CNN architecture which contains two parts: feature extraction
and classification. Each layer of the network in the feature extraction layers takes the
output from the layer before it as input and passes it on to the next layer as output.

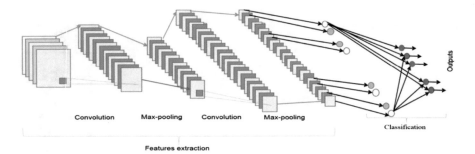

Fig. 4 Convolutional neural network architecture [16]

Convolution, max-pooling, and grouping are the three types of layers that make up the CNN architecture. Convolutional layers and max-pooling layers are the two types of layers in the network's low and middle levels [16]. Convolutions are done on even-numbered layers, while max-pooling operations are done on odd-numbered layers. The convolution and max-pooling layers' output nodes are grouped into a two-dimensional plane called feature mapping.

3.2 Pre-trained Networks

A pre-trained network/model is previously trained from very scratched on a very huge dataset, especially it is highly trained for image classification task. Either user can use a pre-trained model directly or use concepts like fine-tuning or transfer learning to customize the model for a particular task. Pre-trained networks for image classification have already learned to extract insightful features from a collection of images and use them as a starting point for learning every new task from scratch. Many pre-trained networks are trained on a subset of the ImageNet. These pre-trained networks have been trained on more than a million images and can classify into thousands of object class categories, for instance, keyboard, coffee mug, cat, dog, pencil, etc., many more. Pre-trained networks are highly used for the following tasks: 1. Classification 2. Feature Extraction 3. Fine-Tuning.

4 Proposed Model

A. COVID Face Mask Detection Dataset

A part of the dataset of COVID face mask detection was obtained from kaggle website which contains 1006 images in total. It has three sub-directory, and every directory has two different sub-directory (classes having label), namely mask and non-mask.

B.   Feature Extraction

It is an algorithm that extracts features from raw data that can be used for input to layers of the network. It may be color of pixel, shapes like edge, blog, and line or image segment. For instance, if x1 has the highest value then such a feature is a good feature to distinguish objects. In short feature or pattern with the highest appearance has the highest value (importance). If x 4 has lower value, then such a feature has the lowest appearance. Suppose we are training data of motorcycles and x1 is a feature which represents the wheel so throughout our dataset if more number of images have motorcycle then frequency of x1(Appearance of wheel) is higher because every motorcycle has wheels. In this case, the weight of x1 is increased during training to represent its relevance. Similarly, if x4 is a feature extracted from the same image which is not useful to distinguish motorcycle so possibility of appearance of that feature in every motorcycle image is less so during training the weight of x4 is adjusted to lower value indicating useless feature which is going to be removed in further training. So, when we use a pre-trained network as a feature extractor using representation or learning of that model, we are able to extract meaningful features from our dataset instead of extracting and training them from scratch. In short, we don't need to train the entire model we just have to train our custom model to extract high-level features.

Base_model.trainable = False #freeze

In feature extraction, user can freeze convolutional the basic layer of network which is going to use as a feature extractor. It is best practice to add a classifier-head on top of the model and train only these top-level classifier-head, which is as shown above whereby writing false, anyone can freeze the base layers.

C.   Fine-Tuning

Fine-tuning is a transfer learning approach through which we can reuse a pre-trained network. Transfer learning name suggests there is a transfer of learning between pre-trained layers as well as custom layers, which was trivial in feature extraction. Fine-tuning is an optional step in which some or all of the top layers of the base model are unfreezing and allowed learning. It is a process, which may or may not be performed after some feature extractions.

After training a custom classifier on the top of the base model, it has been initiated to extract some average-level features with the help of representations (features) made by pre-trained model. But using fine-tuning on that model some top-layers of base model and newly added layers are trained combined to get high-level features.

It is necessary to keep in mind that it will also lead to overfitting if we directly fine-tuned our network. It is like first newly added layers learning something for our dataset with the help of base model (feature extraction) and after some steps, both learn combined to extract high-level features (fine-tuning). So, from the above-mentioned workflow of feature extraction after completion of that we may try to unfreeze some layers to get some better accuracy. In this task, we have unfrozen some top layers for fine-tuning their weights which adjust according to our dataset for meaningful learning.

$$\text{Base\_model.trainable} = \text{True Unfreeze}$$

In fine-tuning, unfreeze a few layers from the convolutional frozen model base. Thus, it includes some of the base-level model layers as well as custom classifier-head, so it will train all these together. Therefore, it will give more accuracy from the feature extraction. As shown above by writing true, it will include some of layers of basic pre-trained network.

Figure 5 shows that proposed work can be divided into two parts: one is feature extraction and second is fine-tuning of model. So, for implementing the model-based upon flowchart the first part is loading a dataset of face mask detector which has different images with mask and unmask images. For every deep learning task, the most important part is to preprocess our raw data. Hence, preprocessing involves building input pipeline, data augmentation, rescaling, configure dataset for a performance. The next step is to compose the model. So, according to our proposed model we need to load a pre-trained network (having pre-trained weights) into a base model. And then we have to stack a custom classifier on top of the base model known as head. After that, only the head of the model is going to be trained because the base model is pre-trained and the head model uses its learning to extract some features from the dataset. Thus, this is the first part which we conclude as feature extraction. But these features are based on pre-trained models not almost according to our dataset so we need to fine-tune the model by unfreezing some of the top layers of pre-trained models. And training those layers too along with our custom classifier and therefore features extracted from dataset are based on our dataset, which leads

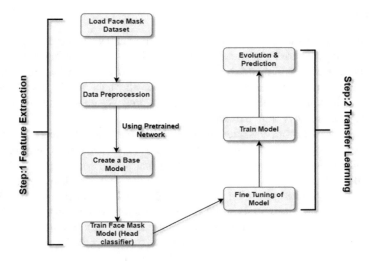

Fig. 5 Flowchart

to a better accuracy. After fine-tuning, the model is trained again and lastly assessed, and after evaluation, it is used for prediction purposes.

5 Results

According to the proposed model initially, when base model is created from mobilenetv2 without including top layers and using approach of feature extraction model had total 2,257,984 parameters from which none of them are trainable parameters; after adding our custom classifier, the total parameters are 2,259,265 = 2257,984 + 1281 in which 1281 are parameters of custom classifier, which are trainable.

Using transfer learning approach (fine-tuning), total parameters remain same but trainable parameters increase from 1281 to 1,862,721, which means the model is going to learn more compared to feature extraction. Table 2 demonstrates the accuracy of feature extraction after the completion of last epoch to be 81.83% and simultaneously accuracy of fine-tuning is 99.5% so, it is clearly concluded that the transfer learning is performing better than feature extraction as we can see there is a vast difference in trainable parameters. By creating the neural network (CNN) from the scratch the total parameters get increased which is around 24,948,161 which means it has more training parameters from the pre-trained network. It requires training from the very beginning-level layer. Total number of parameters gets increased for more time when compared to the two aforementioned techniques, which is clearly observed from Table 2. However, the basic model gives medium accuracy rate when compared to feature extractor but it consumes more computational power and CPU processing time, as it has more trainable parameters when compared to pre-trained network, which is clearly visible from Table 2.

Table 2 shows that the accuracy remains epochwise. The initial epoch accuracy of feature extraction is 48.17, 86.40% accuracy for fine-tuning, whereas the basic model gives the starting accuracy around 78.6%.

Figure 6 shows the outcome of proposed approach. Image dataset from directory generates the dataset and gives label to sub-directory. Sub-directory without mask data is labeled as 1 and with mask data is labeled as 0. Also, feed the same to the model so that the prediction will be based on 0 or 1, which means 0 indicates image with mask and 1 indicates image without mask. Figure 3 visualizes two list predictions

Table 2 Accuracy epochwise

Epoch	Feature extraction (%)	Fine-tuning (%)	Basic model (CNN) (%)
1	48.17	86.40	78.67
3	58.67	96.10	85.67
5	66.33	97.26	88.17
7	73.83	98.56	88.17
9	79.17	99.42	87.5

Fig. 6 With and without facemask predictions

and a label. So, in the prediction list, first 5 values are 1, which means the model predicts images where people have no mask. And also in the labeled (original) list, first 5 values are 1, and from same, we can see that the first five images in picture matrix are without mask.

6 Conclusion and Future Work

This paper utilizes the concept of pre-trained network and advanced techniques like fine-tuning and feature extraction with the pre-trained network model. A pre-trained model for feature extraction is the best practice to take advantage of the features extracted by a model, which is obviously trained on a larger dataset. The feature extraction is done by instantiating the pre-trained model and adding the classifier (Classification Head) at the top layer. The pre-trained model is "frozen" and only the weights of the classifier (Classification Head) get updated during training. Transfer learning is considered as a pre-trained model to increase the accuracy and performance of the model where one might use techniques like fine-tuning. In which the reuse of top layers (top specific layers) is possible for the pre-trained model. Fine-tune can tune weights along with the model and learn high-level features (unfreeze) to the dataset. This technique is usually recommended when working on a large dataset and very similar to the original dataset that the pre-trained model was trained before. The research opens up new avenues with the proposed technique, which can be combined into any high-resolution video surveillance device. Also, the model can be expanded to detect facial benchmarks while wearing a facemask for biometric determination.

References

1. Asia and the Pacific | About COVID-19 Pandemic UNDP. https://www.asia-pacific.undp.org/content/rbap/en/home/coronavirus.html
2. Li, L.-Q., Huang, T., Wang, Y.-Q., Wang, Z.-P., Liang, Y., Huang, T.-b, Zhang, H.-Y., Sun, W., Wang, Y.: COVID-19 patients' clinical characteristics, discharge rate, and fatality rate of meta-analysis. J. Med. Virol. **92**(6), 577–583 (2020)
3. https://www.health.harvard.edu/diseases-and-conditions/preventing-the-spread-of-the-corona virus
4. Singh, A.K., Mehan, P., Sharma, D., Pandey, R., Sethi, T., Kumaraguru, P.: (Un) Masked COVID-19 Trends from Social Media. arXiv preprint arXiv:2011.00052 (2020)
5. Chen, Y., Hu, M., Hua, C., Zhai, G., Zhang, J., Li, Q., Yang, S.X.: Face Mask Assistant: Detection of Face Mask Service Stage Based on Mobile Phone. arXiv preprint arXiv:2010.06421 (2020)
6. Chowdary, G.J., Punn, N.S., Sonbhadra, S.K., Agarwal, S.: Face mask detection using transfer learning of inceptionv3. In: International Conference on Big Data Analytics, pp. 81–90. Springer, Cham (2020)
7. Chavda, A., Dsouza, J., Badgujar, S., Damani, A.: Multi-Stage CNN Architecture for Face Mask Detection. arXiv preprint arXiv:2009.07627 (2020)
8. Anwar, A., Raychowdhury, A.: Masked Face Recognition for Secure Authentication. arXiv preprint arXiv:2008.11104 (2020)
9. Meenpal, T., Balakrishnan, A., Verma, A.: Facial mask detection using semantic segmentation. In: 2019 4th International Conference on Computing, Communications and Security (ICCCS), pp. 1–5. IEEE (2019)
10. Rahman, M.M., Manik, M.M.H., Islam, M.M., Mahmud, S., Kim, J.-H.: An Automated System to Limit COVID-19 Using Facial Mask Detection in Smart City Network. In: 2020 IEEE International IOT, Electronics and Mechatronics Conference (IEMTRONICS), pp. 1–5. IEEE (2020)
11. Rosebrock, A.: Covid:19-Face Mask Detector. https://www.pyimagesearch.com/2020/05/04/covid-19-face-mask-detector-with-opencv-keras-tensorflow-and-deep-learning
12. Gurucharan, M.K.: Face Mask Detection. https://towardsdatascience.com/covid-19-face-mask-detection-using-tensorflow-and-opencv-702dd833515b
13. Bhuiyan, M.R., Khushbu, S.A., Islam, M.S.: A Deep Learning Based Assistive System to Classify COVID-19 Face Mask for Human Safety with YOLOv3. In: 2020 11th International Conference on Computing, Communication and Networking Technologies (ICCCNT), pp. 1–5. IEEE (2020)
14. Dargan, S., Kumar, M., Ayyagari, M.R., Kumar, G.: A survey of deep learning and its applications: a new paradigm to machine learning. Arch. Comput. Methods Eng. 1–22 (2019)
15. Kiranyaz, S., Ince, T., Iosifidis, A., Gabbouj, M.: Progressive operational perceptrons. Neurocomputing **224**, 142–154 (2017)

16. Alom, M.Z., Taha, T.M., Yakopcic, C., Westberg, S., Sidike, P., Nasrin, M.S., Hasan, M., Van Essen, B.C., Awwal, A.A.S., Asari, V.K.: A state-of-the-art survey on deep learning theory and architectures. Electronics **8**(3), 292 (2019)
17. Haykin, S.: Neural Networks and Learning Machines, 3/E. Pearson Education India (2010)
18. Bre, F., Gimenez, J.M., Fachinotti, V.D.: Prediction of wind pressure coefficients on building surfaces using artificial neural networks. Energy and Buildings **158**, 1429–1441 (2018)
19. Wang, Z., Wang, P., Louis, P.C., Wheless, L.E., Huo, Y.: Wearmask: Fast In-Browser Face Mask Detection with Serverless Edge Computing for Covid-19. arXiv preprint arXiv:2101. 00784 (2021)
20. Joshi, A.S., Srinivas Joshi, S., Kanahasabai, G., Kapil, R., Gupta, S.: Deep learning framework to detect face masks from video footage. In: 2020 12th International Conference on Computational Intelligence and Communication Networks (CICN), pp. 435–440. IEEE (2020)

Anomaly in the Behavior of Silicon from Free Energy Analysis: A Computational Study

Chandan K. Das

Abstract Silicon shows a very different trend while melting. Melting has remained a challenging subject from a long time. Especially, predicting the melting temperature of any solid substance still exists as a problem in many cases. Recently, various studies and new rules and set of parameters have simplified things, but its mechanism is yet to be studied properly and there still does not exist any generalized concept regarding this. Also, there are certain anomalies in silicon, which makes its melting and phase transition mechanism more difficult to understand and predict. It is essential to study about its behavior, which also gives us an insight into melting behaviors of nanowires, clusters, defect, and surfaces. This project is an attempt to study the mechanism of melting using classical molecular dynamics simulation, to define a set of parameters that could help us predict the behavior of silicon at any temperature and also its phase transition mechanism. In order to understand the phenomenon, it is important to know the interaction potential governing the silicon system. Stillinger–Weber potential is a good model for Si atoms which takes into account two and three-particle interactions. Melting of silicon atoms is studied using molecular dynamics simulation with the help of LAMMPS software. Heating and quenching processes are implemented on a system of Si atoms. Variations of various parameters like density, volume per atom, and potential energy have been studied. Free energy gap connecting phases are estimated with the help reversible thermodynamic route. Supercritical path is constructed with the help of more than one reversible thermodynamic path. The best of my knowledge, this is first attempt to implement pseudo-supercritical reversible thermodynamic path for a system whose solid volume is higher than liquid volume at phase transition point. It has been found that melting in Si occurs in three stages. It involves pre-melting, melting, and relaxation. The potential energy and density appear to be steady, indicating pre-melting. Melting temperature can be predicted using Gibbs free energy. Gibbs free energy calculation involves with thermodynamics integration and multiple histogram reweighting (MHR) method. The density and

C. K. Das (✉)
Department of Chemical Engineering, National Institute of Technology Rourkela, Rourkela 769008, India
e-mail: dasck@nitrkl.ac.in

© The Author(s), under exclusive license to Springer Nature Singapore Pte Ltd. 2022 575
V. Suma et al. (eds.), *Evolutionary Computing and Mobile Sustainable Networks*,
Lecture Notes on Data Engineering and Communications Technologies 116,
https://doi.org/10.1007/978-981-16-9605-3_39

potential energy drop abruptly, for temperature more than 1750 K indicating complete loss of crystallinity.

Keywords Molecular dynamics · LAMMPS · Hysteresis loop ·
Pseudo-supercritical path · Thermodynamic integration

1 Introduction

Silicon, the fourteenth element of the periodic table, having symbol "Si", is a brittle and hard solid crystal having bluish gray color luster, a tetravalent metalloid that behaves as a semiconductor. In pure state, at room temperature, its behavior is that of an insulator, so doping is usually done with a group 13 element to increase its conductivity. Its resistivity drops on increasing temperature. This usually unreactive element has a high chemical affinity for oxygen, due to which it is difficult to be obtained in its pure form. The melting point of silicon is 1414 °C. Silicon boils at 3265 °C, which is the 2nd highest among non-metals, falling just after boron. It is also the 2nd most adequate element in the earth's crust.

Silicon is the very important element for the advancement of technology, due to widely used of silicon and its products in various fields of engineering and technology. It has been given/attracts greater attention due to widespread application of silicon in our day-to-day life. It is abundant in earth's crust. Silicon is widely used as semiconductor. Silicon is principal component of synthetic polymers silicones. Silicon is mostly found in compound form. It is rarely found in pure form and can be used with little processing of natural minerals. During phase transformation, it shows anomalous behaviors compare to conventional elements and compounds. Exact mechanism of phase transformation of silicon becomes a challenging problem and remains unanswered [1–5].

Phase transition is reported for many pure materials including silica and silicon [6–8]. Transition point is obtained either pressure swinging or temperature swinging method [8–10]. Transition temperature can also be evaluated using specific heat capacity information [11, 12]. Another robust technique for determination of transition point is calculation of entropy [13, 14]. Conventional methods like density hysteresis plot, Lindemann parameter [5], non-Gaussian parameter [15], radial distribution function, structure factor, orientation order parameter, etc., are employed to predict the transition point of a material.

Solid to liquid transformation of Lennard–Jones (LJ) system under confinement is reported [16]. Transition point is determined on the basis of density hysteresis plot, Lindemann parameter, non-Gaussian parameter, radial distribution function, structure factor, orientation order parameter [16]. From sudden jump in density, one can determine the transition point. Similar kind of phenomena is also observed in potential energy. However, density changes observed after complete phase transformation. So determined transition point is not accurate. Other parameters are Lindemann parameter and non-Gaussian parameter. From the Lindemann parameter value, one can estimate the melting and freezing transition point. For determination of melting transition, change in first and second coordination number is important too [15].

Most of the above-mentioned methods are not accurate to predict the melting transition [17]. Estimated melting temperature is often higher compared to true melting temperature. Melting transition can be predicted more precisely using the knowledge of free energy. Transition temperature of Lennard–Jones (LJ) and sodium chloride (NaCl) is reported from free energy information [18]. Free energy is evaluated employing thermodynamic integration. The thermodynamic route connecting solid–liquid is constructed employing reversible thermodynamic route [17, 18]. Phase transformation from solid to liquid under slit [19, 20] and cylindrical confinement is studied using free energy analyzes [21]. Machine learning techniques are perfect tools to predict the properties of complex systems [22–24].

Gibbs free energy is the extensive property to analysis a system. The only extensive property that remains constant during phase transformation. Using this beautiful property, one can estimate true thermodynamics melting temperature. Free energy gap between states is estimated deploying pseudo-supercritical path [18]. This work is mainly focusing on evaluation of free energy barrier connecting solid–liquid during phase transitions. Melting temperature can be predicted using Gibbs free energy. Gibbs free energy calculation involves with thermodynamics integration and multiple histogram reweighting (MHR) method [18]. Estimation of true thermodynamic temperature is evaluated with the help of pseudo-supercritical reversible thermodynamic cycle along the help of multiple histogram reweighting diagrams. Supercritical path is constructed with the help of more than one reversible thermodynamic path. Thermodynamic integration is performed using Gauss quadrature integration scheme.

In this work, I evaluate free energy gap connecting solid–liquid transitions. The best of my knowledge, this is first attempt to implement pseudo-supercritical reversible thermodynamic path for a system whose solid volume is higher than liquid volume at phase transition point. Moreover, due to very small density, difference between two phases make the simulations more complicated. I present briefly the technique. (a) The liquid state is transformed into a poorly interacting liquid with the help of slowly decreasing the interatomic attractions. (b) Gaussian wells are located to the corresponding particles; simultaneously, the volume is enlarged to obtain a poorly interacting oriented state. (c) Gaussian wells are removed gradually, and simultaneously, interatomic attractions are slowly brought back to its whole strength to obtain a crystalline state.

2 Methodology

In this work, I evaluate the free energy connecting solid–liquid state transition. The inclusive technique is described elsewhere [17]. However, I present very shortly the technique for the comprehensive understanding of the reader. The estimation of phase transition point from free energy analysis is combination of four stages. First step is evaluation of an approximate transition point from quenching and heating method. Second is estimation of free energy for the solid state with respect solid reference state. Similarly, for liquid phase is also determined with respect to liquid reference state. Equation of states is generated using multiple histogram reweighting technique [21]. Third stage is the computation of gap in free energy connecting two states at an estimated transition point [21]. Free energy computation is performed with the help of pseudo-supercritical transformation path. Then, ultimately with the help of second and third steps, evaluation of the transition point is done at zero energy difference [21]. That point is considered as true thermodynamic transition point. Each step is exclusively elaborated below. Interaction potential of silicon is as follows:

$$E = U_{\text{inter}}\left(r^N\right) = \sum_i \sum_{j>i} \varphi_2\left(r_{ij}\right) + \sum_i \sum_{j\neq i} \sum_{k>j} \varphi_3\left(r_{ij}, r_{ik}, \theta_{ijk}\right) \tag{1}$$

$$\varphi_2\left(r_{ij}\right) = A_{ij}\epsilon_{ij}\left[B_{ij}\left(\frac{\sigma_{ij}}{r_{ij}}\right)^{p_{ij}} - \left(\frac{\sigma_{ij}}{r_{ij}}\right)^{q_{ij}}\right]\exp\left(\frac{\sigma_{ij}}{r_{ij} - a_{ij}\sigma_{ij}}\right) \tag{2}$$

$$\varphi_3\left(r_{ij}, r_{ik}, \theta_{ijk}\right) = \lambda_{ijk}\epsilon_{ijk}\left[\cos\theta_{ijk} - \cos\theta_{0ijk}\right]^2$$
$$\exp\left(\frac{\gamma_{ij}\sigma_{ij}}{r_{ij} - a_{ij}\sigma_{ij}}\right)\exp\left(\frac{\gamma_{ik}\sigma_{ik}}{r_{ik} - a_{ik}\sigma_{ik}}\right) \tag{3}$$

The φ_2 represents two-particles interaction term. The φ_3 presents three-particles attraction expression. The summation in the expression is overall neighbors J and K of atom I within a truncated length a [25]. The A, B, p, and q parameters employed for two-particles attractions. The λ and $\cos\theta_0$ parameters are used only for three-particles attractions. The ϵ, σ, and a parameters employed for both cases. γ is employed for three-particles attraction. However, this is classified for pairs of atoms. The others extra parameters are dimensionless [25].

2.1 Estimation of an Estimated Transition Temperature

To detect an approximate transition temperature, gradually heating and quenching simulations are performed for solid and liquid states, respectively [21], by employing *NPT* simulation at $P = 1.0$ bar. Afterward, the estimated transition temperature is chosen within the metastable region at where a sudden change in the density is noticed [21].

2.2 Solid and Liquid Free Energy Curve with Respect to Their Corresponding Reference States

The next stage is the formation of the Gibbs energy. The Gibbs energy is expressed in terms of temperature. They are presented for both states with regard to their corresponding standard phase temperature. This free energy curves are obtained over a small region around the estimated transition point at the constant pressure [21]. Using the free energy, connecting the two states at the estimated transition point, the pure state corresponding free energy plots are altered to the solid–liquid free energy difference in the terms of temperature, which is required to determine the transition point where the gap in energy connecting solid and liquid states is zero [21]. This is carried out using multiple histogram reweighting (MHR) technique.

2.3 Computation of Solid–Liquid Free Energy Gap at an Estimated Transition Point

The Helmholtz free energy gap connecting the solid and liquid states at an estimated transition point is estimated by forming a reversible way connecting the solid and liquid states with the help others reversible stages [21]. The free energy throughout the connecting route is evaluated using a known integration scheme:

$$\Delta A^{\text{ex}} = \int \langle \frac{dU}{d\lambda} \rangle_{\text{NVT}\lambda} \, d\lambda \tag{4}$$

while ΔA^{ex} is the gap in Helmholtz free energy. Kirkwood's coupling parameter is used by the symbol λ. Generally, λ changes in between 0 to 1. The value of $\lambda = 0$ system act as an ideal state [18]. The angled bracket is indication of ensemble average for a specific λ parameter [18]. The three stages pseudo-supercritical conversion method is represented in Fig. 1. Very short explanation of the stages is presented below.

1. **Stage-a**

Initially, strongly attracted liquid state is transformed into a poorly interacting liquid using a coupling parameter λ, which controls interatomic potential [18] in the mentioned way:

$$U_a(\lambda) = [1 - \lambda(1 - \eta)]U_{\text{inter}}(r^N) \tag{5}$$

where $U_{\text{inter}}(\mathbf{r}^N)$ is the interatomic interaction energy due to location of all N particles [18]. The η is a scaling parameter. The value varies $0 < \eta < 1$. The first derivative of intermolecular interaction relation produces

Fig. 1 Presents the three stages pseudo-supercritical conversion route. **a** The liquid state is transformed into a poorly interacting liquid by slowly increasing the coupling parameter [18]. **b** Gaussian wells are located to the corresponding particles; simultaneously, the volume is enlarged to obtain a poorly interacting oriented state. **c** Gaussian wells are removed gradually while coupling parameter is slowly increasing to bring back its full strength to obtain a crystalline state

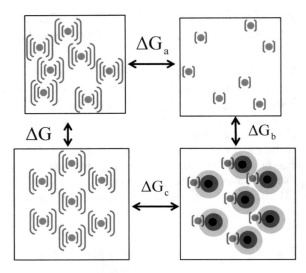

$$\frac{\partial U_a}{\partial \lambda} = -(1 - \eta)U_{\text{inter}}(r^N) \tag{6}$$

2.   Stage-b

During second stage, volume of liquid state is enlarged to the volume of solid state unlike other conventional substances. Enlarge volume is clearly visible in Fig. 1. This stage is most complicated among the three stages. The minute densities difference between two phases become challenging in this stage. Hence, length of the simulation box (L_x, L_y and L_z) for a particular system dimension must be predetermined at the estimated transition point, either from the MHR results or hysteresis diagram [18]. Liquid box dimension is 21.28163 Å(H_l), and solid phase dimension is 21.81412 Å (H_s). Greater solid phase dimension compares to liquid phase dimension make the simulation more complicated. The change in simulations' box dimension confirms that pressures remain unaltered at the start of thermodynamic path and at the completion of stage-c, which is presented in Fig. 7. The interatomic interaction on the basis of λ in this stage is represented following way:

$$U_b(\lambda) = \eta U_{\text{inter}}\left[r^N(\lambda)\right] + \lambda U_{\text{Gauss}}\left[r^N(\lambda), r^N_{\text{well}}(\lambda)\right] \tag{7}$$

where $\mathbf{r}^N(\lambda)$ and $\mathbf{r}_{\text{well}}{}^N(\lambda)$ are the representation of the positions of atoms and Gaussian wells, respectively [18]. These coordinates are purely function of *coupling parameter* due to the change in simulation dimension. U_{Gauss} presents interatomic potential because of the attraction in between the wells and corresponding particles (Eq. 9). The values of parameters "a" and "b" are taken from Gochola's works [17]. Particles coordinate due to enlarge volume from liquid to solid conversion follow the same manner as did in literature [18]. $H(\lambda)$ denotes box dimension at any value of coupling parameter λ. Equation 8 represents change in box dimension for coupling parameter values.

$$H(\lambda) = (1 - \lambda)H_l + \lambda H_s \tag{8}$$

$$U_{\text{Gauss}}\big[r^N(\lambda), r^N_{\text{well}}(\lambda)\big] = \sum_{i=1}^{N} \sum_{k=1}^{N_{\text{wells}}} a_{ik}\exp\big[-b_{ik}r_{ik}^2(\lambda)\big] \tag{9}$$

$$-\frac{\delta U_{\text{inter}}}{\delta H_{xz}} = \sum_{y} P_{xy}^{\text{ex}} V H_{zy}^{-1} H(\lambda) = (1 - \lambda)H_l + \lambda H_s \tag{10}$$

Derived form of potential expression with respect to λ is

$$\frac{\partial U_b}{\partial \lambda} = -\sum_{x,y,z} V(\lambda) H_{zy}^{-1}(\lambda) \Delta H_{xz}\big(\eta P_{xy}^{ex} + \lambda P_{Gauss,xy}^{ex}\big) + U_{Gauss}\big[r^N(\lambda), r^N_{well}(\lambda)\big] \tag{11}$$

3. Stage-c

Stage-c is ultimate step of the pseudo-supercritical conversion method [18]. In this stage, fully interacting solid configurationally phase is obtained. The interaction potential is presented of this final step in terms of λ

$$U_c(\lambda) = [\eta + (1 - \eta)\lambda]U_{inter}(r^N) + (1 - \lambda)U_{Gauss}\big[r^N(\lambda), r^N_{well}(\lambda)\big] \tag{12}$$

And the derivative terms can be rewritten:

$$\frac{\partial U_c}{\partial \lambda} = (1 - \eta)U_{inter}(r^N) + U_{Gauss}\big[r^N(\lambda), r^N_{well}(\lambda)\big] \tag{13}$$

2.4 Determination of Transition Point Where ΔG is Zero

The free energy ΔA^{ex} between phases at the estimated transition point is evaluated by thermodynamic integration [18]. Now, transfer of the Helmholtz energy into the Gibbs energy is important. It is obtained using the formulae given, $\Delta G = \Delta A^{\text{ex}} + \Delta A^{\text{id}} + P\Delta V$. The expression ΔA^{ex} is computed using reversible transformation path method. The second term ΔA^{id} is the contribution from ideality. Final term in the expression is work term due to change in volume from solid to liquid phase. Additionally, the histogram reweighting method produces two equations of states. Expression of the liquid state $\big[(\beta G)_{T_1,l} - (\beta G)_{T_l,l}\big]$ is familiar and for the solid state expression is $\big[(\beta G)_{T_1,s} - (\beta G)_{T_l,s}\big]$ [18], the term $(\beta G)_{Tm,n}$ expresses (βG) for hysteresis loop region at the estimated temperature T_{em} [21], provided that T_{em} is estimated transition temperature, is the transition temperature st which the reversible thermodynamic path [21] is performed [18], achieved the following:

$$\left[(\beta G)_{T_1,l} - (\beta G)_{T_{em},s}\right] + \left[\beta\left(G_{T_{em},s} - G_{T_{em},l}\right)\right]$$
$$- \left[(\beta G)_{T_i,l} - (\beta G)_{T_{em},l}\right] = \left[(\beta G)_{T_1,s} - (\beta G)_{T_i,l}\right] \tag{14}$$

Equation 10 further can be rearranged as

$$\left[(\beta G)_{T_1,l} - (\beta G)_{T_{em},s}\right] + \left[\beta\left(G_{T_{em},s} - G_{T_{em},l}\right)\right] + \left[(\beta G)_{T_1,l} - (\beta G)_{T_i,l}\right]$$
$$- \left[(\beta G)_{T_1,l} - (\beta G)_{T_{em},l}\right] = \left[(\beta G)_{T_1,s} - (\beta G)_{T_i,l}\right] \tag{15}$$

where all the terms except second term of Eq. 11 is achieved using the multiple histograms reweighting method, Whereas the second term is determined using three stages reversible thermodynamic path using thermodynamic integration.

3 Simulation Details and Software Work

3.1 Molecular Dynamics Simulation

Molecular dynamics simulation is a technique used to get insights about the movements and properties of a system of atoms and molecules. It is basically a computer simulation where the system of atoms and molecules is given to make interaction among themselves for a certain amount of period. As a result, we can predict the outcome of various real-world systems by computer simulations without the need for experiments. MD helps to do it in different length and time scales. It is said to bridge the gap between theory and experiments. Various parameters involving in a system's interaction are calculated in molecular dynamics simulation.

3.2 LAMMPS Software

It uses message passing interface (MPI) for parallel communication. Sandia National Laboratories and Temple University researchers maintain and distribute it. It was created as part of a Cooperative Research and Development Agreement among laboratories from the US Department of Energy and few other private sector laboratories [25].

Table 1 Values of parameters used in SW potential (in metals unit)

A	B	P	Q	A	Λ	Γ	ε (eV)	σ (Å)
7.0495562	0.6022245	4	0	1.80	21.0	1.20	2.1672	2.0951

3.3 Atomic Potential Used

Stillinger–Weber potential is a good model for Si. It considers both two-particle and three-particle interactions. The values of following parameters in metal units have been used. The potential of the silicon is provided in Eqs. 1, 2, and 3. Parameters values are listed in Table 1.

3.4 Simulation Details and Potential Model

The NPT MD simulations are conducted with the help of LAMMPS [25]. Unit system of the simulation process is metal. The integration of equation is performed employing velocity-verlet algorithm. Integration time step (Δt) is 5 fs. The temperature is monitored using a Nosé–Hoover thermostat. The pressure is monitored using Nosé–Hoover barostat. The time relaxation is of 100 ps. The pressure relaxation is of 500 ps. Number of particles are simulated around 512. The periodic boundary condition is applied for simulations. The pressure is varied from $P = 1$ atm to $P = 70$ atm to observe density dependence with pressure. During quenching, the initial liquid configurations are taken as ideal diamond structures at 3000 k. Cooling process is carried out gradually after each 5,000,000 MD time steps. Change of temperature T is 25 k for each NPT simulation. Temperature is dropped from 3000 to 500 k with a decrement of 25 K. At the time of heating process, the last configuration of the quenching simulation is initial coordinate of the heating system. Heating is also conducted same way as the quenching. The increment of temperature T is 25 K for each NPT simulation. Process of heat supply is done until the solid has completely converted into liquid. Temperature range of heating is from 500 to 4000 k. The density is determined at every interval of temperature from the knowledge of simulation volume, number of particles, and molecular weight of the silicon. The Gibbs energy gap for connecting states is estimated at transition temperature using pseudo-supercritical path. The true transition point is estimated at a point, Gibbs free energy gap connecting two states reach zero. Initially, I select an estimated transition point, T_{em}. The two different types of NPT simulations are carried out by LAMMPS [25]. The velocity-verlet algorithm is employed throughout all simulations. Integration time step is $\Delta t = 5$ fs. The temperature is monitored using a Nosé–Hoover thermostat. The pressure is monitored using Nosé–Hoover barostat. The time relaxation is of 100 ps for temperature. The pressure relaxation is of 500 ps. 512 number of silicon particles. Throughout the simulations process, applied boundary conditions for in all the three dimensions are periodic. The constructions of equations of states for both phases are done employing multiple histogram reweighting diagrams. Histograms are generated from NPT simulations based on volume and potential energy of the system. Total 41 simulations are carried out for individual phase. The temperature is selected in accordance with the given formulae

$$T_i = T_{em} + \sum_{n=-20}^{20} n\Delta T$$

where T_{em} is the expected transition point computed from the density vs temperature plot hysteresis data; $\Delta T(=10° K)$ is determined based on size of the with metastable region.

The initial configurations for both phases are used from the previous simulation run. The sufficient equilibration, run is performed duration of around 200 ps. The total simulation run time is for 10 ns. The standard reference temperatures are chosen at the minimum point, $T_i = T_{em} - 20\Delta T$. Histograms are generated base potential energy and volume of system.

For the reversible path evaluation (for the three steps of pseudo-supercritical path) as shown in Fig. 1, simulations are carried with NVT ensemble. The Nosé–Hoover thermostat algorithm is employed for temperature and pressure. The value of Gaussian parameters is selected in accordance with Grochola [17]. The scaling parameter is constant at $\eta = 0.1$ [18].

Simulations of Stage-a of the reversible thermodynamic path are initialized from a random initial coordinates of the particles. These coordinates are achieved during heating quenching simulations. After that for every coupling parameter value, initial configuration is obtained from simulation [18]. Total run time for each simulation of three stages time step of integration is 5 fs. Total run time for each coupling parameter value is 20 ns [21]. During the stage-b of three stages, final coordinates of step-1 are the starting point. But, to achieve the Gaussian potential wells, another 512 atoms are situated on its corresponding lattice point [18]. Now, for final stage, last configuration of stage-b for the third stage is taken as initial configuration. The way the three stages is performed the pressures of the system remains constant before and after of the transformation path as shown in Fig. 7. Integration for three stages is performed with ten, fifteen, and twenty points Gauss–Legendre integration techniques [18].

4 Results and Discussion

In this portion, I try to describe output results of various parameters like density, potential energy, and free energy with the change in temperature and coupling parameter (λ). Based on the variation of these parameters, we will try to formulate a basic idea about the melting of Si atoms.

4.1 Density

In this part, we describe the nature of density of the Si system as we perform heating and quenching. Sharp density changed is observed for both the heat and quenching case. Quenching and heating path are not reversible, for this reason hysteresis loop is formed. That indicates first order phase transition. Density of silicon for different temperature is shown in Fig. 2. It can be observed that in the heating process, there is a sudden increase in the density of the system around 1700 k. Also, in the quenching process, sudden decrease in the density can be seen around 1200 k. The plot clearly forms a hysteresis. This is an indication of first order phase change. Density plot shows anomaly behavior toward the phase transition. Metastable region is observed in middle portion of the hysteresis curve. True melting temperature lies in this loop. Vertical line indicates an approximate estimated transition point; corresponding horizontal lines indicate density of liquid and solid phase. These densities determine the solid phase and liquid phase box dimension for the reversible thermodynamic route.

If we perform the same NPT simulation by varying pressure, we can find a trend of the density as a function of the temperature for different pressure values. Density can be presented using the following equation during the heating processes, where temperature range is $500 \geq T \leq 2000$, other parameter values are presented in Table 2 (Fig. 3).

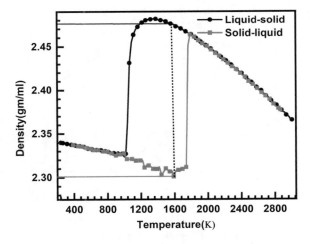

Fig. 2 Figure represents density vs temperature plot. Filled circle black in color for quenching the system whereas filled square for heating the system. Quenching and heating curves do not follow the same path which indicates first order transition. Hysteresis loop is clearly observed in density temperature plot. Metastable region is noticed middle of the curve and true transition point lies in this metastable region. Vertical dotted line indicates an estimated estimate transition point (T_{em}). Horizontal black line and red line indicate corresponding liquid density (2.477) and solid density (2.30), respectively. From where one can determine liquid phase as well as solid phase box dimension

Table 2 Before melting, the density of silicon can be predicted at different temperature with different constant pressure

Pressure (atm)	A	B	C	D	R^2
1	$-2*10^{-12}$	10^{-8}	$-3*10^{-5}$	2.3301	0.9999
5	$-2*10^{-12}$	10^{-8}	$-4*10^{-5}$	2.3302	1.0000
10	$-2*10^{-12}$	10^{-8}	$-4*10^{-5}$	2.3301	1.0000
30	$-2*10^{-12}$	10^{-8}	$-3*10^{-5}$	2.3301	1.0000
35	$-2*10^{-12}$	10^{-8}	$-4*10^{-5}$	2.3303	1.0000
40	$-2*10^{-12}$	10^{-8}	$-4*10^{-5}$	2.3303	1.0000
45	$-2*10^{-12}$	10^{-8}	$-4*10^{-5}$	2.3302	1.0000
50	$-2*10^{-12}$	10^{-8}	$-4*10^{-5}$	2.3303	1.0000
55	$-2*10^{-12}$	10^{-8}	$-4*10^{-5}$	2.3304	1.0000
60	$-2*10^{-12}$	10^{-8}	$-4*10^{-5}$	2.3305	1.0000
65	$-2*10^{-12}$	10^{-8}	$-4*10^{-5}$	2.3303	1.0000
70	$-2*10^{-12}$	10^{-8}	$-3*10^{-5}$	2.3303	1.0000

R^2 is regression coefficient of the fittings

Fig. 3 Presents the fitting curve before melting

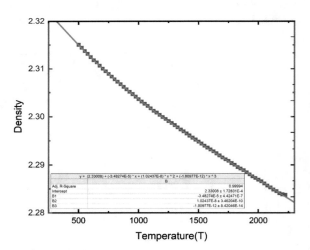

$$\rho = A*T^3 + B*T^2 + C*T + D$$

Similarly, we also predict the density of liquid silicon as an expression in terms of temperature for difference pressures. Where temperature range is $2200 \geq T \leq 4000$. Expression for density is (Table 3, Fig. 4)

$$\rho = \alpha*T^2 + \beta*T + \gamma$$

Table 3 After melting change of the density as a function of temperature with different constant at different pressure

Pressure (atm)	α	β	γ	R^2
1	-10^{-8}	$-3 * 10^{-5}$	2.5559	1.000
5	-10^{-8}	$-3 * 10^{-5}$	2.5559	1.0000
10	-10^{-8}	$-3 * 10^{-5}$	2.5573	1.0000
30	-10^{-8}	$-3 * 10^{-5}$	2.556 3	1.0000
35	-10^{-8}	$-3 * 10^{-5}$	2.2556	1.0000
40	-10^{-8}	$-3 * 10^{-5}$	2.5573	1.0000
45	-10^{-8}	$-3 * 10^{-5}$	2.5512	0.9999
50	-10^{-8}	$-3 * 10^{-5}$	2.5568	1.0000
55	-10^{-8}	$-3 * 10^{-5}$	2.5551	1.0000
60	-10^{-8}	$-3 * 10^{-5}$	2.5566	1.0000
65	-10^{-8}	$-3 * 10^{-5}$	2.5569	1.0000
70	-10^{-8}	$-3 * 10^{-5}$	2.5571	1.0000

R^2 is regression coefficient of the fittings. Others parameters values are given above

Fig. 4 Represents fitting curve after melting region

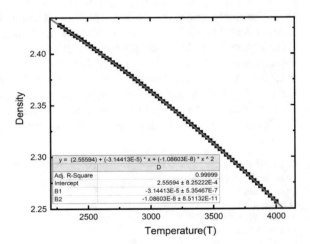

Potential energy also shows the similar kind of behavior like density which is shown in Fig. 5. The red rectangle represents heating process. The black circle represents quenching process.

Fig. 5 Potential energy in terms of temperature similar kind of nature is obtained as density vs temperature curve

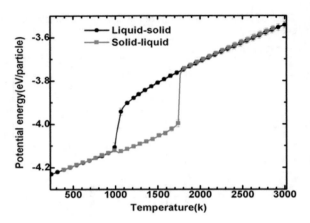

4.2 Free Energy

Helmholtz free energy difference between liquid and solid phase is determined using pseudo-supercritical path by constructing reversible thermodynamic paths [17]. Thermodynamics integration is performed using Gauss quadrature integration scheme. At the beginning of the reversible path, the interaction potential is changing according to Eq. 5. Integration is carried out using 10, 15, and 20 points. No significance difference is observed due to different data points. Derivative of interaction potential energy with respect to λ presents in Fig. 6. For the λ values, they coincide as shown in Fig. 6. Figures 8 and 9 represent for stage-b and stage-c, respectively. During stage-b, interaction potential and box dimensions are changing in accordance with Eqs. 7 and 8, respectively. Simultaneously, Gaussian potential wells are interacting

Fig. 6 $\langle \partial U / \partial \lambda \rangle_{NVT\lambda}$ as a variable of λ for three λ types values (10, 15, and 20) of stage-a for pseudo-supercritical path. Thermodynamic path is smooth and reversible, hence integrable. Error is so small, it submerges with symbol. There is no significant difference among them for stage-a of pseudo-supercritical transformation path

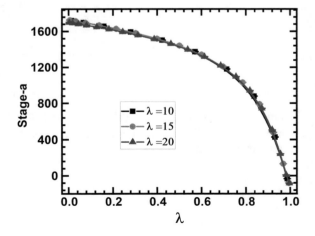

between particles and its corresponding lattice points as per Eq. 9. In the conclusion of the pseudo-supercritical route, the interaction potential follows the Eq. 11. The derivative of potential energy with respect to coupling parameter in stage-b is represented in Fig. 8. The path is reversible and smooth. Similarly, the derivative of potential energy with respect to coupling parameter in stage-c is represented in Fig. 9. Figures for all the stages are smooth and reversible, so we can easily integrate it. The free energy difference connecting solid–liquid is around $-42,731.2000 \pm 100$ eV. Results are reported in Table 4. Using this free energy difference along with equation of state which is obtained from multiple histogram reweighting (MHR) method, the gap in free energy connecting two phases is converted into single reference state and presented in Fig. 10. True thermodynamic melting temperature is the point where ΔG is zero. From Fig. 10, it is clear that true thermodynamic transition point is around $1684°$ K.

Fig. 7 Pressure at the start of stage-a and at the end of stage-c is constant. This is essential and the sufficient criteria for construction of the thermodynamic reversible paths

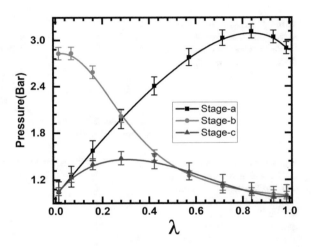

Fig. 8 $\langle \partial U/\partial \lambda \rangle_{NVT\lambda}$ as a variable of λ for ten λ of stage-b values. Thermodynamic path is smooth and reversible, hence integrable. Error is so small, it submerges with symbol

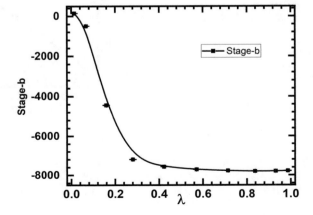

Fig. 9 $\langle\partial U/\partial\lambda\rangle_{\text{NVT}\lambda}$ as a variable of λ for three λ values for stage-c. Thermodynamic path is smooth and reversible, hence integrable

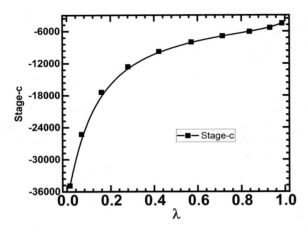

Table 4 Separation of the subscriptions to the gap in Gibbs free energy connecting the two states $T = 1600°$ K

Free energy terms (eV)	
$A_s^*\text{ex} - A_l^*\text{ex}$	$16,600 \pm 100$
$A_s^*\text{id} - A_l^*\text{id}$	$-60,073$
$P^*\Delta V^*$	741.7546
$G_s^* - G_l^*$	$-42,731.2000 \pm 100$

The pressure is maintained at $P = 1$ Bar for the silicon (Stillinger–Weber potential)

Fig. 10 ΔG as a function of T. Vertical arrow line blue in color indicates solid–liquid transition point temperature or true thermodynamic transition temperature temperature (T_m) of solid where Gibbs free energy difference, ΔG, between solid and liquid is zero

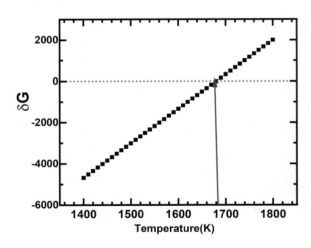

5　Conclusion

Various methods have been employed and we have been successful in observing the phase transition of silicon, depending on various parameters. The density of silicon is known to be 2.32 g/cc. While simulating with decreasing temperature, this is the value for density after the melting occurred at approximately 1880 K, which is slightly more than the reported melting point for silicon which is 1687 K.

Also, keeping in mind the traditional meaning of solid to liquid phase transition, the jump in potential energy is also used to indicate the melting stage of any substance. This jump is seen at 1700–1800 K for decreasing temperature and 2400–2500 K for increasing temperature. Anomaly behavior is observed in density for silicon system, which makes more complicated to implement pseudo-supercritical thermodynamic path.

Phase transition point is determined based on Gibbs free energy calculation. The only extensive property that remains constant during phase transformation. Using this beautiful property, one can estimate true thermodynamics melting temperature. Estimation of Gibbs free energy is performed with the help of pseudo-supercritical reversible thermodynamic cycle along the help of multiple histogram reweighting diagrams. The construction of supercritical path is combination of three stages. For each step of thermodynamics integration is applied using 10, 15, and 20 points. The thermodynamics integration is insignificance with respect to number of data point, which is shown in stage-a. Estimated true thermodynamics melting temperature is around 1684° K ± 2° K, which is in good precision with experimental results. Accuracy of determined melting temperature from free energy analysis is better comparing any other methods. This is a very clear indication of how the break-down of lattice occurs on heating a substance.

I have been successful in implemented pseudo-supercritical route to evaluate gap in free energy connecting two phase for complicated interaction potential model. Though; simulation is performed with a smaller number of particles. Simulation with larger system size is important to observe the system size effect. In future, I shall check the system size effect by varying number of simulated particles. In future, one can study the confinement effect on anomalous phase transition of silicon.

Acknowledgements This work is supported by National Institute of Technology Rourkela, Government of India

References

1. Cahn, R.W.: Melting from within. Nature **413**, 582–583 (2001)
2. Stillinger, F.H.: A topographic view of supercooled liquids and glass formation. Science **267**(5206), 1935–1939 (1995)
3. Stillinger, F.H., Weber, T.A.: Computer simulation of local order in condensed phases of silicon. Phys. Rev. B **31**(8), 5262–5271 (1985)

4. Stillinger, F.H., Weber, T.A.: Point defects in bcc crystals: structures, transition kinetics, and melting implications. J. Chem. Phys. **81**(11), 5095–5102 (1984)
5. Stillinger, F.H., Weber, T.A.: Lindemann melting criterion and the Gaussian core model. Phys. Rev. B **22**(8), 3790–3794 (1980)
6. Van Haong, H.: Atomic mechanism of glass-to-liquid transition in simple monatomic glasses. Philos. Mag. **91**(26), 3443–3455 (2011)
7. Löwen, H., Palberg, T., Simon, R.: Dynamical criterion for freezing of colloidal liquids. Phys. Rev. Lett. **70**, 1557 (1993)
8. Vasisht, V.V., Saw, S., Sastry, S.: *Liquid–liquid critical point in supercooled silicon*. Nat. Phys. **7**, 549–553 (2011)
9. Lascaris, E., et al.: Search for a liquid-liquid critical point in models of silica. J. Chem. Phys. **140**, 224502 (2014)
10. Lascaris, E.: Tunable liquid-liquid critical point in an ionic model of silica. Phys. Rev. Lett. **116**, 125701-1–125701-5 (2016)
11. Chun, Y., Min, C., Zengyuan, G.: Molecular dynamics simulations on specific heat capacity and glass transition temperature of liquid silver. Chin. Sci. Bull. **46**(12), 1051–1053 (2001)
12. Hong, Y., et al.: A molecular dynamics study on melting point and specific heat of Ni_3Al alloy. Sci China-Phys Mech Astron **50**(4), 407–413 (2007)
13. Piaggi, P.M., Parrinello, M.: Entropy based fingerprint for local crystalline order. J. Chem. Phys. **147**(11), 114112 (2017)
14. Nettleton, R.E., Green, M.S.: Expression in terms of molecular distribution functions for the entropy density in an infinite system. J. Chem. Phys. **29**(6), 1365 (1958)
15. Zhang, S.L., et al.: The study of melting stage of bulk silicon using molecular dynamics simulation. Physica B **406**, 2637–2641 (2011)
16. Das, C.K., Singh, J.K.: Melting transition of confined Lennard-Jones solids in slit pores. Theor. Chem. Acc. **132**, 1351 (2013)
17. Grochola, G.: Constrained fluid λ-integration: constructing a reversible thermodynamic path between the solid and liquid state. J. Chem. Phys. **120**, 2122 (2004)
18. Eike, D.M., Brennecke, J.F., Maginn, E.J.: Toward a robust and general molecular simulation method for computing solid-liquid coexistence. J. Chem. Phys. **122**, 014115 (2005)
19. Das, C.K., Singh, J.K.: Effect of confinement on the solid-liquid coexistence of Lennard-Jones Fluid. J. Chem. Phys. **139**(17), 174706 (2013)
20. Das, C.K., Singh, J.K.: Oscillatory melting temperature of stockmayer fluid in slit pores. J. Phys. Chem. C. **118**(36), 20848–20857 (2014)
21. Das, C.K., Singh, J.K.: Melting transition of Lennard-Jones fluid in cylindrical pores. J. Chem. Phys., 2014. **140**(20), 204703-1–204703-9.
22. Manoharan, S.: Early diagnosis of lung cancer with probability of malignancy calculation and automatic segmentation of lung CT scan Images. J. Innov. Image Proces. (JIIP) **2**(4), 175–186 (2020)
23. Shakya, S., Nepal, L.: Computational enhancements of wearable healthcare devices on pervasive computing system. J. Ubiquitous Comput. Commun. Technol. (UCCT) **2**(02), 98–108 (2020)
24. Smys, S., Basar, A., Wang, H.: Artificial neural network based power management for smart street lighting system. J. Artif. Intell. **2**(1), 42–52 (2020)
25. Plimpton, S.: Fast parallel algorithms for short–range molecular dynamics. J. Comp. Phys. **117**, 1–9 (1995)

An Extensive Survey on Outlier Prediction Using Mining and Learning Approaches

Swaroop Chigurupati, K. Raja, and M. S. Babu

Abstract The predictions of outliers are considered the foremost issue in various real-time applications. This work provides an extensive survey on outlier prediction that captures the researchers' attention to handle the issue substantially. The investigators intend to model various feasible and robust models to formulate the solution and generically predict the outliers. Some of the foremost baseline approaches are discussed, and the outlier's prediction process classification is discussed comprehensively. In every section, various outlier prediction processes are analyzed, and their prediction performance is validated. The associated merits and demerits of the models and the research challenges are highlighted to extend future research works and provide future research ideas for young researchers. The solution for specific research questions is consolidated provided for the experts to gain knowledge with the various prediction approaches. The open research problems and extension ideas are provided for the upcoming researchers to make a clear outlier for the prediction process.

Keywords Outliers · Detection process · Research questions · Expert knowledge · Solutions

1 Introduction

Outlier prediction acts as a substantial part that needs to be identified and remains an essential research field based on its widespread occurrence in various ranges of applications. With the prediction of outliers, investigators can attain essential expert knowledge, which helps provide superior detection regarding the data [1]. The outlier's prediction process also helps provide actionable information in various

S. Chigurupati (✉) · K. Raja
Department of Information Technology, FEAT, Annamalai University, Chidambaram, Tamil Nadu, India

M. S. Babu
SRKR Engineering College, Chinnamiram, Bhimavaram, Andhra Pradesh, India

© The Author(s), under exclusive license to Springer Nature Singapore Pte Ltd. 2022   593
V. Suma et al. (eds.), *Evolutionary Computing and Mobile Sustainable Networks*,
Lecture Notes on Data Engineering and Communications Technologies 116,
https://doi.org/10.1007/978-981-16-9605-3_40

applications like health diagnosis, fraud detection, and intrusion detection in cyber-security. Indeed of providing an uncertain definition of outliers, the baseline description regarding outliers is the data point that differs from other data points and does not imitate the average expected data points of the specified phenomenon [2]. The outlier prediction process strives to handle all the issues in predicting the patterns, and it does not adopt the expected nature of the data points. Assume a scenario that defines the general behavior of the data points and the specific regions. This assumption is highly complicated due to certain factors provided below: (1) noisy data that mimics real-time outliers; (2) various conflicting notations and applications make it complex to adopt approaches modeled in various fields; (3) the probability of standard data points behavior intends to evolve, and perhaps it is not a suitable specification, and (4) inappropriate boundaries among the normal and outlier behaviors is confronting to different and remove the outliers [3]. Some of the generally identified complexities are connected with the input data, data labels, outlier type, computational complexity, and accuracy associated with memory consumption and CPU time. Some investigators intend to predict superior solutions to address these issues and the issues related to efficiently predicting outliers in data cleaning and quality, efficient trajectories, multi-dimension data, wireless sensor networks, RFID reading streams, and distributed data streams [4]. However, the outlier prediction process faces enormous challenges, and various outlier prediction approaches are anticipated with various algorithms and methodologies to handle these issues [5].

For instance, assume an issue over the extensive multi-dimensional data where it is extremely large or relatively large as it always contains certain outliers. In some cases, when the data size increases, the number of outliers also increases substantially [6]. Thus, it is necessary to model a scalable outlier prediction process to deal with massive datasets with a vast amount of data. When the data size increases, it influences the computational cost proportionally by making the process expensive and slow. The outliers must be predicted periodically to reduce the influencing data, eliminate data infection, and offer well-known value [7]. Subsequently, with substantial data variants, some of the data are unstructured, mixed-value, structured, and semi-structured, where the prediction of this type of data is complicated and daunting. Some applications more confronted with the challenges include security surveillance, traffic management, trajectory streams, and mobile social networks in various application regions [8]. These application regions require constant prediction of abnormal objects for promptly delivering crucial information. Various other outlier prediction processes show similar, and novel challenges are discussed in the section given below.

Even though there is an enormous amount of surveys over the outlier prediction process, it remains a hot top in the field of research. As an outcome of the intrinsic significance of outlier detection in diverse regions, some big research ideas are discussed in the survey to predict outliers. Similarly, some novel ideas are designed by the researchers to address the prediction issues [9]. Thus, this review plays a significant role in updating the investigators with the recent and trending predictive models in the outlier prediction process. Various existing research carried out in the outlier prediction process concentrates only on certain regions indeed of providing in-depth and updated research ideas to the best of our knowledge. For instance, the comprehensive analysis concentrates on the high-dimensional numeric data, data streams, dynamic network model and emergent deep learning approaches. Most of the existing works like [10–12] provide enormous research ideas, and it needs to be published for future reference.

For the past few years, various contemporary investigators have explicitly performed deep learning approaches and ensemble models. Thus, most of the recent research ideas and predictions need to be analyzed [13]. This survey offers a comprehensive analysis of the prominent outlier prediction approaches, including the challenges of emerging and traditional approaches. This review shows some differences from the existing reviews as it acquires and provides a comprehensive analysis of diverse literature. Also, it complements and consolidates prevailing ideas in the outlier prediction process [14]. It is made to project the essential categories of the outlier prediction process and extensively evaluates the prediction process [15]. The well-established evaluation criteria, corresponding simulation tools, and available databases used for the outlier prediction process are discussed. This survey helps the young researchers and industrialists to acquire a thorough knowledge of various advantages, disadvantages, research gaps, and open research challenges with the overall detection approach. This analysis offers extensive insights into the future analysis process. The significant contributions with this survey are listed below:

- This survey presents various outlier definitions, types of outliers, causes, identification and prevention processes and research challenges encountered in various applications. This work includes the corresponding applications to grab the attention of the researchers.
- Here, various categories of the outlier prediction process with various other approaches are considered in this survey. Some general approaches are discussed, and the drawbacks are highlighted to improve the methodology with future modifications. An extensive analysis is done with the various prevailing approaches, and the corresponding citations are included.
- The research ideas are significantly enlarged based on the extensive analysis with the categories and the comparison with previous approaches and the advantages, disadvantages and shortfalls of the provided methods. The performance summary is also listed by addressing the solutions, complications and so on.
- Some of the standard or benchmark prediction process with corresponding datasets are analyzed for the presence of outliers.
- The contemporary open challenges are presented while evaluating the outlier detection approaches. The survey is extended with the broader discussion of the outlier prediction tools and the challenges related to selecting an appropriate dataset.
- Some of the research challenges are identified, and some solutions are recommended for the research questions to improve future research directions.

The remainder of the work is organized as: Sect. 2 shows outlier prediction background analysis; Sect. 3 discusses various existing outlier prediction approaches.

Finally, Sect. 3 shows the summary of the extensive survey and ideas for future research.

2 Background Analysis

This section discusses the commonly used outlier definitions, the rise of outliers, methodologies to predict outliers occurrence, etc. The applications which generally encounter the presence of outliers are also discussed to provide widespread knowledge.

2.1 General Outlier Definition

Generally, outliers are provided with enormous definitions where some of them are most generic and used for real-time interpretations. Patcha et al. [16] discuss various forms of outlier interpretations from various researchers' perspectives. It is complex to provide a generic flow of the outlier definition due to its complex definitions and vagueness. Outliers are the data points that are different from other points or data points that do not reflect the typical data points nature. The contrary form of outliers is known and described as inliers.

2.2 Emergence of Outliers

There are diverse factors that are directly or indirectly connected with the Emergence of outliers. The common cause of outliers is environmental variations, sampling errors, data entry errors, setup errors, human error, instrumental error, fraudulent nature, system misbehavior, mechanical fault, and malicious activities [17]. For example, outliers due to data errors generally outcomes in human error like data recording and entry. The preliminary research question of researchers is: (1) How to predict the occurrence of the outliers? (2) The second research question that hits the researchers is necessary to predict the outlier features and tests to be taken. There are various innovative concepts coined to address these questions. Researchers coins various names to the prediction process with mining or learning approaches like anomaly detection, outlier mining, modeling outliers, novelty prediction and so on [18]. The prediction process needs to be performed generically as outlier avoidance leads to the loss of essential information (hidden). The feature prediction can be made in univariate or multivariate models. There are diverse methods designed for predicting outliers. The categories of these prediction models are given below:

i. Statistical approaches help predict and label the outliers based on the relationship established among the distributive models. It is further classified as non-parametric and parametric approaches.

ii. The distance-based approach concentrates on evaluating the distance among the observations. Here, the data point is considered an outlier placed far away from the neighborhood outliers.

iii. Density-based model possess lesser density regions where the inliers are depicted as the dense neighborhood.

iv. The cluster-based model is used for predicting the outliers from the provided data, and it is measured as the observations that are away from the dense cluster model.

v. Graph-based models is used for capturing the interdependencies of entities to predict the outliers.

vi. Ensemble model concentrates on integrating the outcomes to acquire robust model to acquire the outliers.

vii. Learning-based approaches like Deep Learning (DL) assists in predicting the outliers by learning the features [18].

Huang et al. [19] proposed a novel idea for handling the outliers based on the Emergence of the model development. The visual analysis of data provides a clear idea to examine the degree of data point outliers. Some univariate approaches are adopted for searching the data points, and it holds the extreme values of the provided single variables. Similarly, other models like multivariate approaches are used for evaluating the entire variables, and Minkowski error diminishes the outliers during the training process. Weng et al. [20] illegally induce outliers, a part of data with high-dimensional data computation. The investigator uses appropriate data that are uncontaminated and more appropriate for the prediction process. Lei et al. [21] deal with independent outliers, which is domain-specific. The occurrence of outliers causes some crucial issues like errors, safety measures, and real-time situations trigger alarm set with no cause. Yu et al. [22] discuss the outlier prediction process over various applications. It is not probable to deal with the space limitations. An extensive outline and in-depth knowledge of the outlier detection process helps predict intrusion detection, unusual patterns in a time-series manner, e-commerce, trajectories, industrial machine-based behavioral patterns, cleaning, energy consumption and data quality, and extensive data analysis and textural outliers. Figure 1 depicts the local and global outliers.

Abid et al. [23] provide a solution for outlier detection with gaining concealed insights with added cost and efforts. While processing logs, specific automated data mining approaches are required to search unusual patterns with huge log volumes. The provided logs acquire the most acceptable source of information for monitoring the outliers. Shahid et al. [24] predict fault during card theft and the nature of the card user changes while purchasing. Thus, an abnormality is noted during pattern purchase. This process is valid during unauthorized access over the networking environment and outcomes in unusual patterns. The prediction of abnormal patterns is highly essential for security purposes. Shukla et al. [25] consider surveillance and

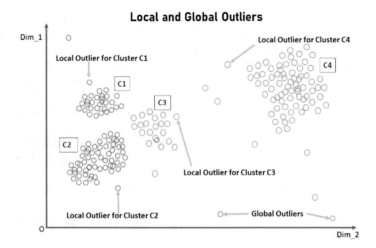

Fig. 1 Local versus global outliers

safety in the cyber-security field, and the process ensures log administration and safe logging to enhance the security and authentication intelligence. This prediction process is practical and considered a hot research topic. Tran et al. [26] discuss the fake news platform (social networks) and examine the complexity in differentiating the fake and real news. Moreover, some false reports are considered outliers from the constant source, and these fake news spread negative images over society and people. Thus, it is crucial during the identification process. Zhang et al. [27] discuss outliers in medical applications and the health care system. Here, some unusual patterns are identified, and the disease conditions are diagnosed. The understanding of the abnormal patterns assist in predicting the disease and its consequences, thus, provisioning the medical practitioners to take necessary actions. Schubert et al. [28] discuss audit logs during financial transactions where the essential information is stored in the database. It helps validate the legality, accuracy, and reporting risks and the constant monitoring of audit logs to report the abnormal behaviors. Kriegel et al. [29] predict outliers in the sensor network for target tracking to ensure network routing quality and provide appropriate outcomes. It assists in having a more comprehensive analysis with computer network performance for predicting the bottleneck. Keller et al. [30] analyses the data quality in various applications to measure the dirty and error data. Therefore, the outlier detection process helps to improve the data cleaning and quality. The process of correcting and cleaning the data is necessary for faster computation, training high-quality data and predicting appropriate outcomes. Momtaz et al. [31] examine predicting the outliers over the time-series data and abnormal patterns during data streaming. It is because the abnormal pattern may influence the prediction of appropriate outcomes and faster computation. Wu et al. [32] examine the sensors over the unceasing environmental condition. Here, sensors are merged to acquire specific information based on the desirable tasks. It is essential to validate the data quality as the data are contaminated due to the polluted outliers. It is highly essential to predict the outliers to restrict the overall efficiency.

2.3 Outliers Prediction Methodologies

Vazquez et al. [33] discuss outlier prediction approaches using statistical, distance, graphical, geometric, depth, profiling, and density-based model. Su et al. [34] used a density-based model for predicting outliers over the lower density regions, and inliers are noted in dense regions. The object considerably differs from neighborhood regions are flagged as outliers. The local point densities are compared with local neighborhood densities for modeling the outliers. The efficiency and simplicity of density-based techniques are extensively used for predicting outliers. Hido et al. [35] anticipated local outlier factors, loosely associated with a density-based prediction model. Here, k-NN is used for measuring local reachability density and evaluating the neighborhood of the k-NN model. It is expressed as in Eq. (1):

$$\text{Local outlier factor } (p) = \frac{1}{|\text{kNN}(p)|} \sum_{o \in \text{kNN}(p)} \frac{\text{lrd}_k(o)}{\text{lrd}_k(p)} \tag{1}$$

Here, $\text{lrd}_k(p)$ and $\text{lrd}_k(o)$ are specified as local reachability density. The significant concept measures that the degree of observations of the outliers. The lrd and k-NN values are maintained while evaluating the local outlier factors (See Fig. 2). Eskin et al. [36] discuss the density-based model and acts as a fundamental approach that outperforms distance and statistical model. It analyses the neighborhood object

Fig. 2 Representation of local outlier factor (LOF)

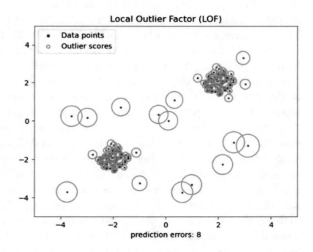

density and is beneficial for predicting the crucial outliers, and it has the competency to evaluate the local outliers. It is computationally expensive and complicated compared to the statistical approach as it has sensitive parameters, i.e., neighborhood size determination. The model gave a poor performance and was complicated in predicting the outlier.

Tang et al. [37] discuss a statistical model which may be supervised, unsupervised or semi-supervised. The data points are modeled in a stochastic manner, labeled as outliers based on the relationship. Here, inliers and outliers are provided based on data distribution. It is further divided into a non-parametric and parametric model. The significant differences among the model are based on the known data distribution. Satman et al. [38] discussed parametric models using regression, and Gaussian Mixture Model (GMM) approaches. The outlier prediction with the regression model is used for handling various issues. It may be either non-linear or linear based on the problem provided. Initially, data is trained with constructing a regression model, and it fits with the data. In the testing stage, the regression model evaluates the data instances where the outliers are labeled with notable deviations over the anticipated and actual value using a regression model. Some standard models with regression approaches used thresholding with Mahalanobis distance, mixture model and vibrational Bayesian approach. Dalatu et al. [39] predict outlier in a linear regressive manner. It relies on the non-interactive covariance matrix and shows huge advantages in predicting multiple outliers where the computational time is cost-efficient. The outcome is to diminish the variance and bias of the intercept estimator using the regression model. Latecki et al. [40] anticipated another regression model for predicting outliers over the sensor measures. It adopts weighted summation to construct independent variable synthesizes, and it is tested over the nominal environment. The author intends to compare the non-linear and linear regression model for predicting outliers and examines the receiver operating characteristics (ROC) for evaluating prediction accuracy and misclassification rate. The anticipated model provides an extensive insight toward the predictive outcomes for outlier prediction. The non-linear model intends to fit with the linear model for predicting outliers.

Samparthi et al. [41] discuss the kernel density estimation model, an unsupervised model for predicting the outliers. The evaluation is done with the comparison of local density points to the neighborhood local density. The comparison of the anticipated model is made with the various other standard models. However, the model gives superior detection performance but lacks high dimensional and considerable scale data adaptability. Boedi et al. [42] anticipated a superior model to overcome the shortcomings mentioned above. It shows superior scalability and enhanced performance using kernel-based approaches with only computational time. The author intends to build an approach for addressing inappropriate outlier detection. Here, variable kernel density evaluation is performed to handle the issues. The dependency is to measure the k-parameter with the local neighborhood. Thus, the overall performance of the model gives superior performance with better scalability with reduced computational complexity. Uddin et al. [43] use to evaluate the data distribution to predict malicious nodes. The data stream over an environment induces the challenge of adopting KDE directly and shows improved performance. It shows superior performance with reduced computational complexity. The order of evaluation is provided as $O(n^2)$. Zhang et al. [44] uses local semantic outliers to offer superior KDE and predicts the data streams locally. This work intends to handle the shortcomings of the prevailing models. It deals with the high-velocity data streams and unpredictability in data updation. The model unceasingly predicts the local outliers over the data streams. However, it suffers from the curse of dimensionality and high computational cost. It works effectually than the non-parametric outlier prediction process. Various other statistical approaches are used for the prediction purpose but lack in achieving better performance and reduced computational cost. Table 1 depicts the comparison of variants of the density-based algorithm.

The significant advantages identified in the approaches mentioned above are discussed below [46]:

i.   These approaches are accepted mathematically and perform faster evaluation after the construction of the model. Thus, they show enhanced performance with the probabilistic model.

ii.   Generally, they are fitted with quantitative real-world data or data distributions. The original form of data is transformed to the suitable processing value and outcomes in enhanced processing time over the complex data.

iii. It shows a more straightforward implementation even in case of restricted problems.

Similarly, the drawbacks identified with these problems, research challenges and gaps are discussed below:

i.   The distributive model assumption and their dependency are given in a parametric manner. Some quantitative outcomes are provided inconsistently for practical execution owing to the lack of knowledge. However, non-parametric models are determined as a poor choice for executing real-time applications.

ii.   The model adopts univariate feature space, which is not applicable for the multi-dimensional environment. It leads to higher computational costs while handling multivariate data.

iii. The preliminary shortcomings of multivariate data are to capture the interactions among various attributes. It cannot examine the feature simultaneously. Generally, specific statistical approaches are not applicable to deal with high-dimensional data. Thus, there are preliminary requirements to model statistical approaches to analyze high-dimensional data and examine multiple features.

iv.   These models face huge issues on incremental dimensionality where the statistical approaches adopt various approaches, and it outcomes in total processing time and misrepresentation of data distribution. Thus, statistical approaches give huge advantages and disadvantages. The significance of outlier accessible data is used for constructing reliable systems. The outliers show a drastic effect on the system efficiency; therefore, it is essential to predict or eliminate the system accuracy. The searching approaches like k-NN and neighborhood in high-dimensionality is considered expensive tasks.

Table 1 Comparison of variants of density-based algorithms

Methods	Performance	Issues	Disadvantages
Local outlier factor [35]	Finest index $O(n)$ Without index term $O(n^2)$ Average order $O(n\log n)$ Higher-dimensional order $O(n^2)$	• It handles the problem related to the binary property of outliers and intends to prove that it is not a binary property • It gives a solution to every local density outlier prediction problem	It fails to handle the granularity issues and is sensitive toward Minimum points It needs expensive object computation and eliminates the probable outliers where the local neighborhood is extremely nearer to the neighbor
COF [45]	Lower order $O(n)$ Medium order $O(n\log n)$ Higher-order $O(n^2)$	• It intends to handle the issues in local outlier factor (non-outliers patterns with lower density) • It predicts the dataset independently with pattern density	It requires complex computation than that of local outlier model
LOC [45]	Computational complexity $O(n^3)$ The computational complexity of memory $O(n^2)$	• Sensitive toward specific parameters while handling multi-granularity and local density • Handles huge sampling neighborhood	The added computational cost for evaluating standard deviation
Radial density function [34]	It is highly scalable and consumes lesser time compared to LOF and COF	It prunes the deeper data clusters and predicts the outliers with a subset of data	The running time is lesser compared to prevailing methods
INFLO	For smaller dimensions, the running time of the model is lesser and expensive computational time	• It uses data points for evaluating the density ratio, which makes outliers extremely meaningful • It identifies meaningful outliers	It predicts local outliers using symmetric neighborhood
DWARF	It shows enhanced performance (sensitivity and accuracy)	It overcomes the constraints of lesser accuracy and superior sensitivity to *the k* parameter	Unable to evaluate the computational cost. Concentrates only on prediction accuracy
GBP	It shows better performance compared to the existing model	• It resolves various challenges encountered in the centralized region with the massive amount of data • The processing efficiency is lower	Higher running time and it is not highly scalable

(continued)

Table 1 (continued)

Methods	Performance	Issues	Disadvantages
RODS	AOC shows superior performance than local outlier prediction	• It shows effectual relative density computation based on kernel density evaluation • It shows robust computation against outliers	It lacks in the utilization of distance approaches

v. Most prevailing models cannot handle the data streams as it is complex to preserve data distribution and predict data streams using k-NN. The scalability does not cost-effective.

vi. There are some disadvantages connected with high-dimensional space, and the performance is avoided owing to the dimensionality curse. The data object possess discrete attributes, which is challenging to determine the distance among the objects.

It is known that statistical approaches are effective in the prediction process. While the parametric method is not suitable for big data; however non-parametric approaches are not suitable. Additionally, the threshold definition of the SD is to distinguish the outliers with the superior probability of an inappropriate labeling process. Table 2 depicts the comparison of the diverse outlier prediction process.

2.4 Reviews on Evaluation Metrics

The researchers use various evaluation metrics to examine the significance of the anticipated model. Based on outlier prediction, the computational time and classification accuracy are considered the essential part of the prediction process [47]. Various measurements are considered for measuring outlier performance. They are discussed below:

i. Precision is depicted as the ratio of several appropriately predicted outliers to the total number of outliers. In concern to specific applications, the evaluation of a total number of outliers is highly complex. Thus, it is measured as the number of outliers over the ground truth;

$$\text{Precision} = \frac{TP}{TP + FP} \tag{2}$$

Table 2 Comparison of various outlier prediction processes

Approaches	Performance	Problems	Disadvantages
Gaussian mixture model (GMM)	Single iteration process $O(n^3)$ Successive iteration process $O(Nn^3)$	It is in contrast to general approaches and concentrates on local properties indeed of global properties It handles both issues	Shows higher complexity
	It gives enhanced prediction accuracy for TP and reduced accuracy while considering FP	The model fails in predicting the outliers in its multiple and multi-state Gaussian model	
Regression models	Computational time (Cost-efficient)	Predicts multiple outliers in the shorter period	Variance and bias of the estimator are not reduced
	It gives higher prediction accuracy for non-linear models	It evaluates the non-linear and linear models for predicting outliers	It is not provided in a detailed manner
Kernel density models	It shows superior prediction performance	Performs prediction with local density detection	It is not suitable for high-dimensional and large scale real-time data
	Shows superior performance and better scalability for the larger dataset with lesser computational time	Inappropriate in predicting outliers when the dataset is large and complex It entirely depends on $the k-$ parameter for evaluating local neighborhood weight	Considered as a complex method
	Computational cost is given in $O(n^2)$	Uses KDE model over data stream environment	Needs extensive computational requirements
Statistical models	Enhanced performance than prevailing approaches (accuracy)	Some statistical models are required for outlier prediction	Accuracy in predicting density ratio evaluation
	The model proves the significance compared to other models	It assists in predicting both the global and local outliers	Effectual method

ii   .R-precision value is depicted as the ratio of appropriately predicted outliers over the top layer of the potential outliers. It does not hold appropriate information as the number of outliers are minimal compared to the total data size.

iii. Average precision—It is depicted as the average precision score toward the outlier point rank. It merges precision and recall.

$$\text{Average Precision} = \sum_{k=0}^{k=n-1} [\text{Recall}(k) - \text{recalls}(k+1)] * \text{precisions}(k) \quad (3)$$

iv. Area Under Curve—It is depicted as the graphical representation that plots True Positive Rate (TPR) against False Positive Rate (FPR). The TPR or FPR specifies the number of inliners or outliers ranked between the potential outliers (maximum amount of outliers) in the ground truth.

v. Rank power—It needs to rank the true outliers (top layers) and average outliers (normal ones). It measures the accurate outlier ranking model.

vi. Correlation coefficient is depicted as the numerical measure of correlation, that is, the statistical relationship among the variables, such as Pearson correlation or spearman's rank similarity.

$$\text{Pearson correlation coefficient } 'r' = \frac{n\left(\sum xy\right) - \left(\sum x\right)\left(\sum y\right)}{\sqrt{[n\sum x^2 - \left(\sum (x)^2\right)][n\sum y^2 - \left(\sum (y)^2\right)]}} \quad (4)$$

$$\text{Spearman's rank similarity } '\rho' = \frac{\frac{1}{n}\sum_{i=1}^{n}\left(R(x_i) - \bar{R(x)}\right) \cdot \left(R(y_i) - \bar{R(y)}\right)}{\sqrt{\frac{1}{n}\sum_{i=1}^{n}\left(R(x_i) - \bar{R(x)}\right)^2 \cdot \left(\frac{1}{n}\sum_{i=1}^{n}\left(R(y_i) - \bar{R(y)}\right)^2\right)}} \quad (5)$$

Most of the general approaches concentrate on evaluating Area Under Curve (AUC), and the drawback of the evaluation process is that is no provision for similarity checking. AUC eliminates the slight variations encountered in the ranking scores. It is not suitable for unbalanced datasets compared to AUC and precision-recall curves.

$$\text{Recall} = \frac{\text{TP}}{\text{TP} + \text{FN}} \quad (6)$$

Moreover, ROC and AUC are determined as the standard way for measuring the outlier problem. Ranking based on correlations is a highly essential step for the modeling of a superior outlier detection process. Some proposed methods consider the class imbalance problem and provide a superior understanding of the redundancy and similarity of the outlier prediction process. An appropriate correlation measure is needed to attain better evaluation for both outlier scores and ranking.

2.5 Reviews on Datasets and Tools Used for Outlier Prediction

The outlier prediction process can be applied in various types of data in a high-dimensional and regular manner. Two diverse kinds of datasets are considered for the validation purpose, i.e., synthetic datasets and real-world datasets. Dang et al. [48] discuss the UCI machine learning repository, an online source used for the outlier prediction process. This dataset is constructed explicitly for classification purposes. Ghoting et al. [49] discuss outlier detection datasets with open access provisioning, which is highly appropriate for the outlier prediction process. It is grouped into various types like time-series graphs, univariate and multivariate datasets, and multi-dimensional datasets. Bhattacharya et al. [50] discuss the ELKI outlier dataset for evaluating the outlier prediction process. This dataset is used for validating the performance of various algorithms. Rado et al. [51] discuss an unsupervised anomaly detection dataset for measuring the unsupervised algorithm. It is highly suited for real-world datasets where it is not publicly accessible for security and privacy concerns. Finally, the synthetic dataset is constructed in a constraint manner which is less complex and eccentric. It provides better validation for supervised data classification, and it is extensively adopted and manipulated.

Similarly, there are various tools used for outlier detection. Ha et al. [52] discuss a scikit learning tool for predicting outliers. It works effectually for specific algorithms like isolation forest and LOF. Angiulli et al. [53] adopt python outlier detection for multivariate data. It is considered a scalable tool, and it is extensively utilized in various researches, including learning and mining approaches. Vu et al. [54] discuss the ELKI tool, best suited for mining algorithms and facilitates more accessible and fair evaluation with benchmark algorithms. Angiulli et al. [55] adopt rapid miner, which is best suited for unsupervised prediction algorithms like COF, LOF, and so on. Bhaduri et al. [45] uses MATLAB for outlier prediction and act as a user-friendly model for execution. Finally, Salehi et al. [56] adopt the MOA tool, an open framework suited for mining algorithms. Various researchers generally use these tools to perform validation and acquire superior results.

2.6 Research Gaps

Based on the extensive analysis, it is proven that predicting outliers is a complex task; however, various investigators intend to handle the research constraints and acquire superior outcomes [57]. But, enormous research questions and open issues occur due to the consideration of millions of data. Some of the research constraints and the gaps are discussed in the section below [58–60]:

i.   In recent days, enormous data is acquired from the data stream and reduces the efficiency while predicting the outliers. Thus, the data stream patterns need to be examined in a contemporary manner. It should be analyzed for the high-dimensional data without losing the efficiency;

ii.   Researchers need to categorize the data based on real-time applications, specifically for the high dimensional and large datasets. Some prevailing approaches are adopted for evaluating the data density and lead to the curse of dimensionality issues;

iii. Generally, outliers show some unusual local data behavior and intend to deal with high-dimensional data. Various researchers lack in predicting the correlation among the data, which is determined to be a confronting issue and triggered as an open, challenging environment.

iv.   While considering the evaluation process, researchers find it challenging to measure the broader and efficient performance metrics other than standard measures like prediction accuracy, precision, recall, F-measure, etc.

v.   Although deep learning approaches give superior prediction accuracy in various real-time applications, they show some relative shortcomings in the outlier prediction process. It is still measured as an open research challenge.

vi.   In learning approaches, the adoption of the baseline classification model provides some good performance. However, the prediction of appropriate base learners and subspaces is highly essential. The selections of appropriate combinations of learners are determined as a research gap.

vii. It is highly essential to measure the influence of the available features in the prediction process. But, the feature analysis process is time-consuming, and an efficient approach needs to be designed for handling this issue.

viii. The adoption of the massive dataset and the corresponding data analysis is an open research challenge. The data patterns give unpredictable insights toward the prediction process. It is essential to model a robust outlier prediction process in a scalable manner and higher dimensionality dataset. But, various works fail in this process with minimal run time.

ix.   The distance evaluation metrics are highly essential for the high-dimensional data, but the available algorithms are sensitive to determining neighborhood evaluation. However, various works fail to handle both the parametric and non-parametric approaches in outlier prediction. Thus, it is mandatory for the researchers to analyze these drawbacks and to draw a better outcome.

x.   Finally, the construction of a robust model is highly solicited for statistical analysis.

3   Conclusion

This survey provides an extensive analysis of the various outlier prediction processes. The advantages and disadvantages of these methods are analyzed, and some research ideas are drawn in a superior manner. However, some research challenges are directly or indirectly connected with the available approaches. Thus, mining and learning approaches are adopted to handle the issues connected with the massive data. The available approaches still encounter some metrics like time complexity, computational cost, and complexity during the analysis with the enormous datasets. Some researchers show a baseline idea for boosting the prediction performance and some lack in analyzing various performance metrics. Thus, some common research ideas are triggered and listed in the research gap section. In future, the young and the upcoming researchers need to concentrate on these points to design a superior outlier prediction method. The analysis of various deep learning approaches is also highly solicited to improve prediction accuracy. Based on various real-time applications, DL approaches work effectually to meet investigators' research objectives and goals in a superior manner.

References

1. Gebremeskel, C.Y., He, Z., Haile, D.: Combined data mining techniques based patient data outlier detection for healthcare safety. Int. J. Intell. Comput. Cybern. **9**(1), 42–68 (2016)
2. Angiulli, S.B., Pizzuti, C.: Distance-based detection and prediction of outliers. IEEE Trans. Knowl. Data Eng. **18**(2), 145–160 (2006)
3. Cao, M.E., Qian, W., Zhou, A.: Density-based clustering over an evolving data stream with noise. In: Proceedings of SIAM Conference on Data Mining, pp. 328–339 (2006)
4. Ayadi, O.G., Obeid, A.M., Abid, M.: Outlier detection approaches for wireless sensor networks: a survey. Comput. Netw. **129**, 319–333 (2017)
5. Wang, D.R.S., Bai, M., Nie, T.Z., Kou, Y., Yu, G.: An efficient algorithm for distributed outlier detection in large multi-dimensional datasets. J. Comput. Sci. Technol. **30**(6), 1233–1248 (2015)
6. Kriegel, P.K., Schubert, E., Zimek, A.: Outlier detection in axis-parallel subspaces of high dimensional data. In: Proceedings of Pacific-Asia Conference on Knowledge Discovery Data Mining, Berlin. Springer, Germany, pp. 831–838 (2009)
7. Yu, L.C., Rundensteiner, E.A., Wang, Q.: Outlier detection over massive-scale trajectory streams. ACM Trans. Database Syst. **42**(2), 10:1–10:33 (2017)
8. Djenouri, A.Z., Chiarandini, M.: Outlier detection in urban traffic flow distributions. In: Proceedings of IEEE International Conference on Data Mining (ICDM), pp. 935–940 (2018)
9. Ranshous, S.S., Koutra, D., Harenberg, S., Faloutsos, C., Samatova, N.F.: Anomaly detection in dynamic networks: a survey. Wiley Interdiscipl. Rev. Comput. Stat. **7**(3), 223–247 (2015)
10. Tamboli, Shukla, M.: A survey of outlier detection algorithms for data streams. In: Proceedings of 3rd International Conference on Computing for Sustainable Global Development, pp. 3535–3540 (2016)
11. Zimek, E.S., Kriegel, H.-P.: A survey on unsupervised outlier detection in high-dimensional numerical data. Stat. Anal. Data Mining **5**(5), 363–387 (2012)
12. Kwon, H.K, Kim, J., Suh, S.C., Kim, I., Kim, K.J.: A survey of deep learning-based network anomaly detection. Cluster Comput. **10**, 1–13 (2017)
13. Chalapathy, Chawla, S.: Deep learning for anomaly detection: a survey (2019), arXiv:1901.03407. [Online]. Available: https://arxiv.org/abs/1901.03407
14. Akoglu, H.T., Koutra, D.: Graph based anomaly detection and description: a survey. Data Mining Knowl. Discov. **29**(3), 626–688 (2015)
15. Nguyen, H.H.A., Gopalkrishnan, V.: Mining outliers with ensemble of heterogeneous detectors on random subspaces. In: Database Systems for Advanced Applications. Springer, Berlin, Germany, pp. 368–383 (2010)
16. Patcha, Park, J.-M.: An overview of anomaly detection techniques: existing solutions and latest technological trends. Comput. Netw. **51**(12), 3448–3470 (2007)

17. Achtert, H.P.K., Reichert, L., Schubert, E., Wojdanowski, R., Zimek, A.: Visual evaluation of outlier detection models. In: Proceedings of 15th International Conference on Database Systems for Advanced Applications (DASFAA), pp. 396–399 (2010)

18. Liu, X.L., Li, J., Zhang, S.: Efficient outlier detection for highdimensional data. IEEE Trans. Syst., Man, Cybern. Syst. **48**(12), 2451–2461 (2018)

19. Huang, D.M., Yang, L., Cai, X.: CoDetect: financial fraud detection with anomaly feature detection. IEEE Access **6**, 19161–19174 (2018)

20. Weng, N.Z., Xia, C.: Multi-agent-based unsupervised detection of energy consumption anomalies on smart campus. IEEE Access **7**, 2169–2178 (2019)

21. Lei, T.J., Wu, K., Du, H., Zhu, L.: Robust local outlier detection with statistical parameters for big data. Comput. Syst. Sci. Eng. **30**(5), 411–419 (2015)

22. Yu, X., Liu, Y.: Glad: group anomaly detection in social media analysis. ACM Trans. Knowl. Discov. Data (TKDD) **10**(2), 18 (2015)

23. Abid, A.K., Mahfoudhi, A.: Outlier detection for wireless sensor networks using density-based clustering approach. IET Wirel. Sens. Syst. **7**(4), 83–90 (2017)

24. Shahid, I.H.N., Qaisar, S.B.: Characteristics and classification of outlier detection techniques for wireless sensor networks in harsh environments: a survey. Artif. Intell. Rev. **43**(2), 193–228 (2015)

25. Shukla, Y.P.K., Chauhan, P.: Analysis and evaluation of outlier detection algorithms in data streams. In: Proceedings of IEEE International Conference on Computer Communications Control (IC4), pp. 1–8 (2015)

26. Tran, L.F., Shahabi, C.: Distance-based outlier detection in data streams. Proc. VLDB Endowment (PVLDB) **9**(12), 1089–1100 (2016)

27. Zhang, M.H., Jin, H.: A new local distance-based outlier detection approach for scattered real-world data. In: Proceedings of Pacific-Asia Conference on Knowl. Discovery Data Mining, pp. 813–822 (2009)

28. Schubert, A.Z., Kriegel, H.-P.: Local outlier detection reconsidered: a generalized view on locality with applications to spatial, video, and network outlier detection. Data Mining Knowl. Discov. **28**(1), 190–237 (2014)

29. Kriegel, P.K, Schubert, E., Zimek, A.: LoOP: local outlier probabilities. In: Proceedings of the 18th ACM Conference on Information and Knowledge Management, pp. 1649–1652 (2009)

30. Keller, E.M., Bohm, K.: HiCS: high contrast subspaces for density-based outlier ranking. In: Proceedings of IEEE 28th International Conference on Data Engineering (ICDE), pp. 1037–1048 (2012)

31. Momtaz, N.M., Gowayyed, M.A.: DWOF: a robust density-based outlier detection approach. In: Proceedings of Iberian Conference on Pattern Recognition and Image Analysis, pp. 517–525 (2013)

32. Wu, K.Z, Fan, W., Edwards, A., Yu, P.S.: RS-forest: a rapid density estimator for streaming anomaly detection. In: Proceedings of IEEE International Conference on Data Mining, pp. 600–609 (2014)

33. Vázquez, T.Z., Zimek, A.: Outlier detection based on low density models. In: Proceedings of ICDM Workshops, pp. 970–979 (2018)

34. Su, L.X., Ruan, L., Gu, F., Li, S., Wang, Z., Xu, R.: An efficient density-based local outlier detection approach for scattered data. IEEE Access **7**, 1006–1020 (2019)

35. Hido, Y.T., Kashima, H., Sugiyama, M., Kanamori, T.: Statistical outlier detection using direct density ratio estimation. Knowl. Inf. Syst. **26**(2), 309–336 (2011)

36. Eskin: Anomaly detection over noisy data using learned probability distributions. In: Proceedings of 17th International Conference on Machine Learning (ICML), pp. 255–262 (2000)

37. Tang, X., Yuan, R., Chen, J.: Outlier detection in energy disaggregation using subspace learning and Gaussian mixture model. Int. J. Control Autom. **8**(8), 161–170 (2015)

38. Satman, M.H.: A new algorithm for detecting outliers in linear regression. Int. J. Statist. Probab. **2**(3), 101–109 (2013)

39. Dalatu, A.F., Mustapha, A.: A comparative study of linear and nonlinear regression models for outlier detection. In: Proceedings of International Conference on Soft Computing and Data Mining, vol. 549, pp. 316–327 (2017)

40. Latecki, A.L., Pokrajac, D.: Outlier detection with kernel density functions. In: Proceedings of 5th International Conference on Machine Learning and Data Mining in Pattern Recognition, pp. 61–75 (2007)
41. Samparthi, V.S., Verma, H.K.: Outlier detection of data in wireless sensor networks using kernel density estimation. Int. J. Comput. Appl. **5**(7), 28–32 (2010)
42. Boedihardjo, C.-T.L., Chen, F.: Fast adaptive kernel density estimator for data streams. Knowl. Inf. Syst. **42**(2), 285–317 (2015)
43. Uddin, A.K., Weng, Y.: Online bad data detection using kernel density estimation. In: Proceedings of IEEE Power Energy Society General Meeting, pp. 1–5 (2015)
44. Zhang, J.L., Karim, R.: Adaptive kernel density-based anomaly detection for nonlinear systems. Knowl.-Based Syst. **139**, 50–63 (2018)
45. Sungheetha, A., Sharma, R.: 3D image processing using machine learning based input processing for man-machine interaction. J. Innov. Image Proces. (JIIP) **3**(01), 1–6 (2021)
46. Rousseeuw, P.J. Hubert, M.: Robust statistics for outlier detection. Data Mining Knowl. Discovery **1**(1), 73–79 (2011)
47. Angiulli, F., Pizzuti, C.: Fast outlier detection in high dimensional spaces. In: Proceedings of European Conference on Principles of Data Mining and Knowledge Discovery, pp. 15–26 (2002)
48. Dang, H.Y.T.N, Liu, W.: Distance-based k-nearest neighbors outlier detection method in large-scale traffic data. In: Proceedings of IEEE International Conference on Digital Signal Processing, pp. 507–510 (2015)
49. Ghoting, S.P, Otey, M.E.: Fast mining of distance based outliers in high-dimensional datasets. Data Mining Knowl. Discov. **16**(3), 349–364 (2008)
50. Bhattacharya, K.G., Chowdhury, A.S.: Outlier detection using neighborhood rank difference. Pattern Recognit. Lett. **60**, 24–31 (2015)
51. Radovanović, A.N., Ivanović, M.: Reverse nearest neighbors in unsupervised distance-based outlier detection. IEEE Trans. Knowl. Data Eng. **27**(5), 1369–1382 (2015)
52. Ha, S.S., Lee, J.-S.: A precise ranking method for outlier detection. Inf. Sci. **324**, 88–107 (2015)
53. Angiulli, Fassetti, F.: Very efficient mining of distance-based outliers. In: Proceedings of 16th ACM Conference on Information and Knowledge Management, pp. 791–800 (2007)
54. Vu, N.H., Gopalkrishnan, V.: Efficient pruning schemes for distance-based outlier detection. In: Proceedings of European Conference on Machine Learning and Knowledge Discovery in Databases, pp. 160–175 (2009)
55. Angiulli, Fassetti, F.: Detecting distance-based outliers in streams of data. In: Proceedings of 16th ACM Conference on Information and Knowledge Management, pp. 811–820 (2007)
56. . Dube, T., Eck, R.V., Zuva, T.: Review of technology adoption models and theories to measure readiness and acceptable use of technology in a business organization. J. Inform. Technol. 2(4), 207–212 (2020)
57. Vivekanandam, B.: Evaluation of activity monitoring algorithm based on smart approaches. J. Electron. **2**(03), 175–181 (2020)
58. Pasillas-Díaz, J.R., Ratté, S.: Bagged subspaces for unsupervised outlier detection. Int. J. Comput. Intell. **33**(3), 507–523 (2017)
59. Aggarwal, C.C., Sathe, S.: Theoretical foundations and algorithms for outlier ensembles. ACM SIGKDD Explor. Newslett. **17**(1), 24–47 (2015)
60. Rayana, S., Akoglu, L.: Less is more: building selective anomaly ensembles. ACM Trans. Knowl. Discov. Data **10**(4), 1–33 (2016)

Analyzing Mental States, Perceptions, and Experiences of Teachers and Students: An Indian Case Study on Online Teaching–Learning

Priti Rai Jain⑩, S. M. K. Quadri, and Elaine Mary Rose⑩

Abstract This online survey-based case study is a novel attempt to juxtapose the entwined experiences of teachers and students, their perspectives, and the effects of online teaching–learning during the COVID-19 Pandemic. The current work is an attempt to study the impact of the Pandemic on the mental health and emotional well-being of students and teachers across various parts of India. It presents a unique comparison of student responses at the beginning of the pandemic (in a previous study) with the recent data collected exactly one year after that, which shows a significant rise in negative emotional states, lower levels of satisfaction with personal progress, and much higher average screen-times. The novel dataset gathered for the current study has 572 responses of students with 51 unique attributes and 390 responses of teachers with 64 unique attributes were called "Covid-19 Go Away 2021" abbreviated as "C-19GA21" and was published on an open online data repository.

Keywords Artificial intelligence · Machine learning · Online teaching–learning · Mental health · Emotion analysis · Sentiment analysis

1 Introduction

In December 2019, the Pandemic Covid-19 swept across the world and halted action in all walks of life [1]. The implementation of unprecedented lockdown, movement restrictions, associated social and physical distancing due to Covid-19 affected

P. R. Jain (✉) · S. M. K. Quadri
Department of Computer Science, Jamia Millia Islamia, New Delhi, India
e-mail: pritirai.jain@mirandahouse.ac.in

S. M. K. Quadri
e-mail: quadrismk@jmi.ac.in

E. Mary Rose
Department of Mathematics, Miranda House, Delhi University, New Delhi, India
e-mail: elaine.maryrose@mirandahouse.ac.in

almost everyone in the world and India alike. This has led to considerable changes in lifestyles, manner of day-to-day functioning, availability of resources, etc. which in turn took its toll on behavioral patterns of individuals [2, 3]. Most of these consequences were outcomes of decreased economic activity. These when combined with loneliness, infection fears, boredom, inadequate supplies, and financial insecurity can potentially disturb individuals emotionally and boost risks of mental health problems [4]. Though lockdowns, curfews, and physical distancing have been critical to mitigating the spread of COVID-19 they have certainly had both short-term and long-term consequences on mental health and well-being [1]. There is a high probability that it could have triggered a variety of adverse psychological and social problems such as panic attacks, Generalized Anxiety Disorder (GAD), poor sleep quality, etc. [5]. Medical professionals anticipate a considerable increase in anxiety and depressive symptoms among people who did not have pre-existing mental health issues due to the Pandemic and the trepidation is that some of these individuals may experience Post-Traumatic Stress Disorder (PTSD) as time passes [6]. Studies mandate that there is an urgent need for research that can address mental health consequences for susceptible groups that can be moderated under pandemic conditions, and the impact of repetitive information consumption especially via social media (SM) about COVID-19 [7].

Just like all other facets of life, COVID-19 related restrictions have jeopardized the academic calendars of educational institutes across the world. Learners were affected in 157 countries across the world due to the closure of educational institutes [8]. Most educational institutions were forced to suddenly switch to online teaching–learning (OTL) to continue academic activities. Teachers and students had to abruptly change their styles drastically due to this switch [9]. Switching to online mode, without much preparedness in terms of teaching skills, availability of resources, suitable devices, bandwidth, design of courses, the effectiveness of learning, the efficiency of evaluation, and also conditioning of mindsets were not appropriately understood [10, 11]. These struggles could impact the cognitive, emotional, social, and behavioral aspects of the mental health of the stakeholders. It can have significant consequences for India which is still a developing economy but at the same time has a huge population. Thus, it calls for immediately focused efforts for prevention and direct intervention of any potential outbreak of mental health problems [1].

The current study considers a variety of facets of the lives of teachers and students across India, involved in OTL activities for almost over a year in the pandemic. The study gathered information from them by conducting online surveys using SM platforms. The current study is focused on understanding students' and teachers' perceptions and their preferences toward the OTL, availability of infrastructure and resources required, experiences and challenges involved, and ascertaining the psychological effects of being confined to home due to Covid-19. The information collected has state/union territory an individual has been living in, institute type, internet access, availability of suitable device(s), subjects studied, online platforms used for classes, class duration, average screen-time per day (AST), the support provided by their institute specifically for online classes (OCs), percentage of classes attended in the first six months of the pandemic versus the percentage of classes attended in the last six months and reasons for the increase/decrease in classes attended, active participation, reaction to online evaluation, FAEAD these days, predominant emotional responses (PER) in these times, attitude toward OCs, the efficacy of OCs in maintaining a daily routine, personal progress and feeling of solidarity toward the peer group, reaction to information consumed from SM and the trust therein, and change in the thinking process and change in attitude toward life due to the Pandemic, etc.

The uniqueness of the current work is that it juxtaposes the entwined teachers' and students' experiences, perspectives, and effects of online teaching–learning during a year of the COVID-19 Pandemic. It is an attempt to study the impact of the Pandemic on the mental health and emotional well-being of participants across various parts of India. To the best of the knowledge of the researchers, all other studies in this area either focus on students or teachers but none deal with both at the same time. It also presents a novel comparison of student responses at the beginning of the pandemic (in a previous study) with the recent data collected exactly one year after that, which shows a significant rise in negative emotional states, lower levels of satisfaction with personal progress, and much higher average screen-times. The current study implements a variety of data analytics and visualization techniques (both statistical and AI-based) for data cleaning (handling missing values and outliers), feature engineering (conversion, scaling, and selection using principal component analysis), univariate analysis, bivariate and multivariate analysis, hypothesis testing, opinion mining, and clustering, etc., to arrive at significant insightful conclusions. A limitation of this study is that the survey questionnaire could reach only those participants who are well conversant with English.

The results of the study show that the mental well-being of participants has been adversely affected during the Pandemic. Though online learning has provided an opportunity to teach and learn in innovative ways, there is a unanimous response from participants that face-to-face interaction of a classroom cannot be substituted by OTL. With the predictions for more waves of the Pandemic, online education is likely to continue for a while. Thus, educational institutions should ensure that adequate resources and support systems are in place to protect the mental health and well-being of teachers and students. A positive takeaway from the study is the underlying sentiment of hope and concern for each other's health and safety in the messages of participants to the world.

This paper begins by stating the objective of this study in Sect. 2. Section 3 discusses the data collection strategy for the study. Section 4 introduces the various dimensions of the collected novel dataset named "Covid-19 Go Away 2021" and is abbreviated as C-19GA21. Section 5 discusses the related work and Sect. 6 discusses the design and methodology used for the current study. Section 7 discusses the results of the analysis and also compares it with a previous study [12]. Section 8 is devoted to the conclusion of this study. Finally, Sect. 9 lists the limitations of this study.

2 Objective

The objective behind the current study is to use AI and statistical techniques to ascertain the psychological effects of the restricted movements and social distancing due to Covid-19. The study aims at using these indicators to gain insights for identifying issues related to teaching–learning practices, environment, resources, their impact on work-life balance, and the mental health of participants.

3 Instrument Design and Data Collection

Data regarding OTL experiences of students and teachers during the pandemic was collected by conducting an online survey using Google Forms. The questionnaires—one for teachers' experiences and another for students' experiences were floated via SM platforms from 28 March 2021 to 28 April 2021 and 4 April 2021 to 26 April 2021 respectively.

Snowball sampling was used to gather the dataset. The participants were requested to circulate the questionnaires among their peer groups to ensure better generalizability. All participants were informed about the purpose of the survey and were assured of the confidentiality of their identity. Data of only those participants who consented that their responses be used for academic purposes were considered in the study.

The surveys were responded to by 709 students and 420 teachers from different states of India. After removing records of individuals who did not agree to participate and retaining only one record for each email-id, 572 responses of students with 51 unique attributes and 390 responses of teachers with 64 unique attributes were obtained. The data was then anonymized to maintain confidentiality by removing the only column for personal details, namely the email address. Subsequently, the timestamp column was also removed as it did not add to our research. The resulting novel dataset was named called "Covid-19 Go Away 2021" abbreviated as "C-19GA21" and was published on data.mendeley.com [13].

4 Dataset Covid-19 Go Away 2021 (C-19GA21)

The dataset C-19GA21 has a variety of information such as socio-demographic details, experiences regarding OTL, responses that depict changes in emotional, social, and cognitive behaviors, attitudes toward the Pandemic, and insights about the participants' mental health.

Basic information such as age/age group, gender, nationality, state/union territory of residence, type of educational institute and subject taught/learned and responses related to social, emotional, and behavioral aspects like PER, feeling at the end of an average day (FAEAD) of online teaching/learning, AST, level of stress due to OCs, the impact of online education on social interaction and maintaining a daily routine, the extent of faith in SM content and extent of satisfaction with personal progress were common to both student and teacher data. Other attributes were related to the availability of online teaching/learning resources, status of internet access, device(s) and platform(s) used for OCs, and whether sharing of a device(s) led to resource crunch, etc. Text responses regarding aspects of conventional teaching–learning that cannot be substituted by online mode, change in thought process due to the pandemic and one-word message to the world after a year of the pandemic were also sought from all participants.

Additionally, teacher data attributes relate to new pedagogies adopted, change in preparation-time, areas of online teaching that would require further assistance, facilities provided by the employer/institution during the pandemic such as E-books, ergonomic furniture, data packs, software resources, hardware resources (laptop, tablet, webcam, etc.), technological or IT support, etc. Questions were based on five-point Likert scale responses ranging from strongly agree to strongly-disagree regarding possible reasons for the decrease in students' attendance in online learning (10 attributes), obstacles faced in online teaching (9 attributes), and overall experience of online teaching (5 attributes). Teachers' estimates of average student attendance, attendance compared to the previous semester, and the extent of active participation by students in OCs. Teacher preferences regarding resuming offline classes in the upcoming semester were also sought.

Similarly, student data attributes relate to the duration of each online session, total time spent in OCs per day, breaks between classes, the percentage of attendance in the last six months, and its comparison with the previous semester. Type of learning support provided by the students' institute and the areas of online learning that they require help with are also sought. Five-point Likert scale responses regarding possible reasons for the decrease in attendance (13 attributes) and overall experience of online learning (5 attributes) are also available in the dataset. It also sought students' interpretation of active participation in OCs and how often they scrolled through SM content after joining them, the extent of belief in SM content, whether or not they verify content before forwarding it on SM platforms, etc. Figure 1 shows the states and union territories where the participants have been residing during the Pandemic. Figure 2 depicts the various disciplines of subjects they teach/ learn. These subjects could be practical-oriented, theoretical, or vocational. The current study focuses on a cross-section of students and teachers in C-19GA21 teaching–learning in Secondary (Class IX and X), Senior Secondary (Class XI and XII), undergraduate courses, and postgraduate courses. The aforesaid cross-section has a total of 50 attributes of 493 students and 63 attributes of 282 teachers (hereafter called C-19GA21(S3)).

Cronbach's alpha score [14–16], is a function of the number of test items and the average inter-correlation among these test items. It is widely used as a measure of reliability and internal consistency for test items/datasets. It determines how closely the items in the group relate to each other. Its value ranges between 0 and 1. A higher value indicates a greater internal consistency. The values for Cronbach's alpha for C-19GA21(S3) is 0.81549 for students' data and 0.85093 for teachers' data.

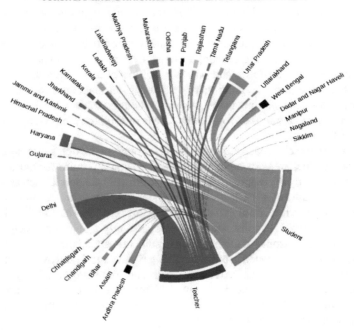

Fig. 1 States and Union Territories where the participants of C-19GA21 were residing during COVID-19

According to the recommendations in [15], these values can be interpreted to mean that the students' data is 'robust' and the teachers' data is 'reliable'. As per [16] it implies that the internal consistency of the dataset C-19GA21(S3) is 'good'. These values for Cronbach's alpha in combination with the sampling strategy suggest that it can be safely inferred that the analysis and results derived from C-19GA21(S3) reliably represent ground reality vis-a-vis the circumstances and mental state of Indian students and teachers during the Pandemic.

5 Related Work

The online survey-based study [17] found that learners and instructors both felt face-to-face interaction is important to build the confidence of a learner. It noted that online education lacks exposure to real-life implementations and practical knowledge. It also claims that students and teachers in rural areas face a scarcity of resources. It further states that poor internet connectivity and students' lack of concentration during OCs are some of the major issues faced. WhatsApp and email were found to be the best ways to circulate course contents and class notes to students. The study

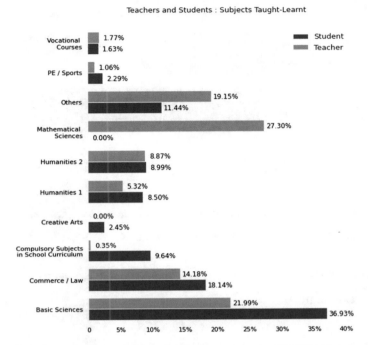

Fig. 2 Disciplines of subjects taught/learned online by participants of C-19GA21 during COVID-19

also points out that 51.43% of participants had not used ICT-based teaching–learning methodologies before the Pandemic. The accessibility to the internet is as shown in Fig. 3. The gender-wise distribution of age groups of the participants is as in Figs. 4 and 5. The types of institutes of participants of C-19GA21(S3) are as shown in Fig. 6. It suggests that nearly all participants had access to the internet for teaching–learning activities even though there could be limitations with bandwidth and connectivity. Figure 7 shows that a larger proportion of students used smartphones to attend classes compared to teachers who used phones to teach. It also highlights that the most frequently used devices are laptops/ desktops, followed by smartphones though there existed a small section of participants who used iPad/tablets for academic activities. Figure 8 shows that at times teachers as well as students faced device crunch and had to miss classes due to the non-availability of a shared device. The popularity of the various platforms used for OCs is summarized in Fig. 9. Figure 10 shows the average screen-time spent by the participants in academic and non-academic activities every day. Figure 11 shows the PER of C-19GA21(S3) participants after a year of OTL during the Pandemic. These clearly show that the participants experienced negative emotions more than positive emotions.

The study [8] by Word Bank, states that students around the world were having dissimilar experiences during the closure of schools due to the Pandemic. In most

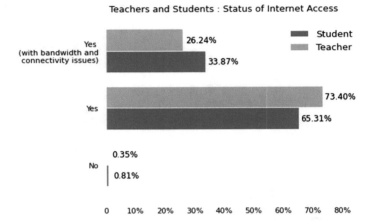

Fig. 3 Availability of Internet for OTL for participants of C-19GA21(S3)

Fig. 4 The gender-wise distribution of age groups of teachers in C-19GA21(S3)

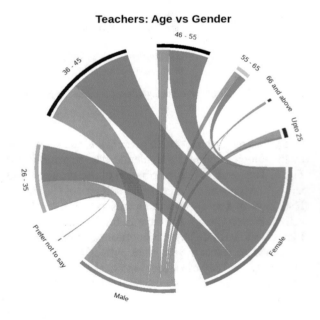

countries, the education systems tried to mitigate this by providing remote learning via OCs. It is important to note and recognize that on-ground what most educational institutions rolled out was an emergency response to teaching–learning. It was not a planned activity and thus it cannot and should not be expected to be perfect in quality or execution. The study underlines that there exists evidence to show that there exists a great deal of disparity both across and within countries vis-a-vis availability, access, and effectiveness of mitigation strategies needed to dampen the adverse effects of

Fig. 5 The gender-wise
distribution of age groups of
students in C-19GA21(S3)

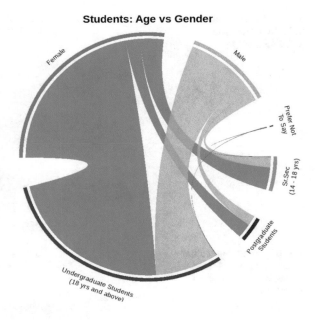

Students: Age vs Gender

Fig. 6 Types of institutes of
participants of
C-19GA21(S3)

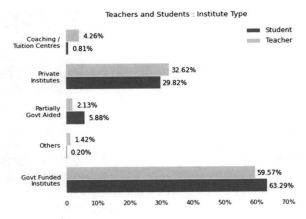

Teachers and Students : Institute Type

closures of schools on learning outcomes and mental health of students. The current
study is an effort to assess the situation regarding the same in India.

The study [18] states that digital literacy includes the ability to read, write,
and communicate aptly using digital tools and resources. It allows for building
connections with others for collaborative problem-solving and strengthens independent thinking. These skills have been an important asset for both teachers and
learners online during the Pandemic. Unfortunately, many teachers and learners
were not adequately digitally literate to adapt to this sudden shift in the mode of
teaching-leaning.

Fig. 7 Devices used for OTL by participants of C-19GA21(S3)

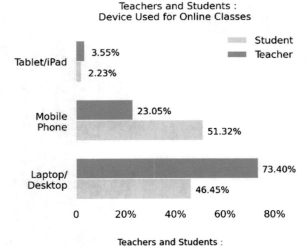

Fig. 8 Classes missed due to unavailability of a shared device by participants of C-19GA21(S3)

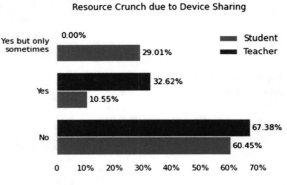

Fig. 9 Platforms used for OCs by participants of C-19GA21(S3)

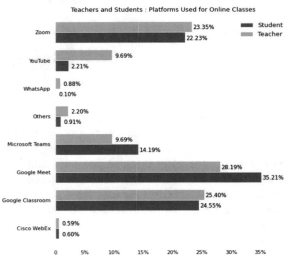

Fig. 10 AST spent on academic and non-academic activities by participants of C-19GA21(S3)

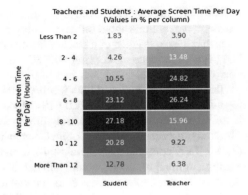

Fig. 11 PER of participants of C-19GA21(S3) after a year of OTL during the Pandemic

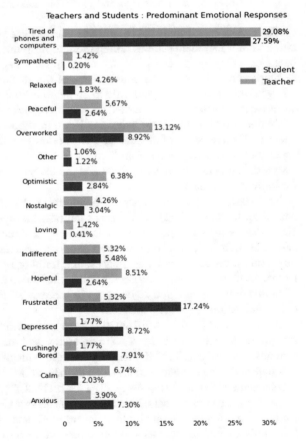

The main findings of the study [19] state that there exists a significant difference in teachers' perceptions about online teaching experiences and expectations. It emphasizes that the experience of a teacher plays a major role in this regard. Teachers who had prior experience of working with ICT tools found online teaching to be a more positive experience.

The study [20] explored how teachers who were new to the profession, adapted to online teaching during COVID-19. Most teachers reported that they used new learning content in addition to assigning tasks and providing feedback to students. The study discusses teachers' competence in Technological Pedagogical Knowledge (TPK). TPK comprises teachers' skills in using technologies for education. These are not confined to any particular subject. They are independent of specific subject domains. The situation that has emerged due to COVID-19 warrants that the teachers should not only have knowledge and skills but also have confidence for effective online teaching.

The review-based study [21] highlights that because students had to attend to their ailing family members such as taking them to hospitals, etc. by the time they were back home they had no energy for learning online. Thus, it became difficult for them to keep pace with OCs. As a consequence of the closure of businesses and offices, a large number of families had little to no income, so the data packs for continuous access to the internet for OCs burned holes in their pockets in these difficult times. The study finds that it is tough to design a uniform methodology that can fit the learning needs and convenience of all students. The review highlights that since all students' assessments are being done from home, teachers found it extremely challenging to assess the authenticity of work done and the extent of actual learning taking place for individual students.

The study [22] revealed issues being faced by teachers concerning online education include deficiency of basic facilities, family-related distractions during classes/assessments, issues related to insufficient technical support from institutions, and little clarity of objectives. Teachers experienced a dearth of technical infrastructure, had inadequate access, awareness, and training in using online teaching platforms. Teachers' problems during the Pandemic lead to negative emotions and a lack of motivation. These acted as dampers in the way of their fruitful engagement in online education.

A study conducted in April 2020 [12] to understand the effect of Covid-19 on the day-to-day living, activities, learning styles, and mental health of young Indian students in schools (class IX–XII) and those in colleges and universities in undergraduate and postgraduate courses when the pandemic began exploding in India. The dataset named "Covid-19 Go Away 2020 (C-19GA20)" [23] was created for and used in the aforesaid study. A set of attributes of the dataset C-19GA21(S3) of the present study when compared with similar relevant attributes of dataset C-19GA20 of the [12] study by taking them as a baseline present a very clear picture as to what has been the effect of the last one year of the Pandemic on the mental states of students. Some of these attributes comprise PER, feelings at the end of an average day, one-line message to the world, and things that online teaching mode cannot substitute when compared to a physical classroom. The comparison of student responses at the beginning of the pandemic in (C-19GA20) with the recent data (C-19GA21(S3)) that has been collected exactly one year after (C-19GA20), shows a significant rise in their screen-time, negative emotional states, and lower levels of satisfaction with personal progress. This could be suggestive of potential mental health issues that some participants could be inching toward.

These findings are in line with the findings of most studies included in the review [24]. These studies confirm that an increase in screen-time enhances the risk of behavioral, cognitive and, emotional disorders in adolescents and young adults. The review [24] suggests that an increase in screen-time harms brain development and memory. It is associated with lowered psychological well-being, lowered self-esteem, slackened learning and acquisition, increased risk of early cognitive decline, depression, anxiety, early neurodegeneration, and increases risk of early onset of dementia. The study [25] states that continuously viewing screens causes digital eye strain: the blurring of vision, dry/itchy eyes, headache and increases obesity. Late-night viewing of screens is associated with deteriorated sleep quality and duration. The study [26] concludes that excessive screen-time significantly deteriorates the regulation of emotions, the person is unable to stay calm, has little self-control, tends to leave tasks unfinished, has lower curiosity, and finds it difficult to make friends. Such individuals are doubly at risk of depression and anxiety. The study [27] suggests that individuals with high screen-time are prone to inadequate sleep, a wandering mind, higher stress levels, non-adaptive/negative thinking styles, and increased health risks. Excessive screen-time during childhood and adolescence is likely to habituate the mind to decreased satisfaction with life, depression, and anxiety.

6 Design and Methodology

As discussed in Sect. 3 two questionnaires were created as Google Forms (one each for students and teachers) were floated online via SM platforms. The cross-section C-19GA21(S3) was used for the work presented in this paper. Dataset C-19GA21(S3) has a total of 50 attributes of 493 students and 63 attributes of 282 teachers.

6.1 Tools and Techniques Used

The current study was implemented in Python and R in the virtual settings of the Google Collaboratory environment.

6.2   Data Cleaning and Preparation

The questions that appeared as headers in the response sheets downloaded from Forms were replaced with appropriate column names. The data was appropriately handled for missing/null values and outliers.

Special characters, undesirable words (manual list of unnecessary words), stop words (overly common words such as "and", "the", "a", "of", etc.), and words with less than three characters were removed from text in responses to the last three questions (L3Q) of questionnaires—"What are the things in classroom learning, if any, that you feel cannot be replaced in online classes?", "How has the pandemic COVID-19 changed you as a person? How has it changed your thinking process?" and "Your One-Line message to the World after ONE FULL year of this Pandemic!". This was done to ensure meaningful results on the sentiment analysis of these columns.

6.3   Feature Engineering

Nominal values were converted to ordinal values to compute correlation among features. Kaiser–Meyer–Olkin test was performed to check the data suitability for factor analysis. Principal Component Analysis (PCA) was done to identify features to include or exclude for dimensionality reduction. The conversion to ordinal values was also required to be able to use the Chi-square test of independence that is used to test the statistical association between two and more categorical variables.

6.4   Data Analysis

Exploratory data analysis (EDA) was done in the form of univariate analysis and results were visualized as over 100 charts: bars, pie, histograms, and density plots, etc. They revealed several insights as per discussions in this section. Students' perspective of online learning experiences is summarized in Fig. 12. Similarly, the online teaching experience is summarized in Fig. 13. Figures 14 and 15 summarize teachers' perception of reasons for the decline in attendance in OCs (October 2020 to March–April 2021) in the last six months in comparison to (April 2020 to September–October 2020) the initial six months when the Pandemic Covid-19 related restrictions were enforced 2020.

A major part of the bivariate analysis was done by creating more than 100 + contingency tables and heatmaps between the variables of interest. It was practically impossible to include all these results and visualizations, thus, the paper presents only some cherry-picked results and plots that the researchers found most interesting. The Chi-square test of independence was used to test the statistical association between two or more categorical variables [28]. This test can only assess associations among

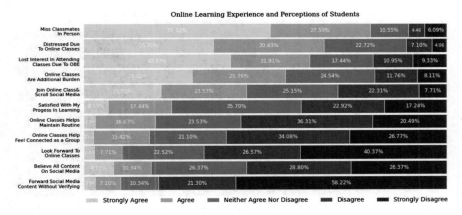

Fig. 12 Online learning experience: students' perspective in C-19GA21(S3)

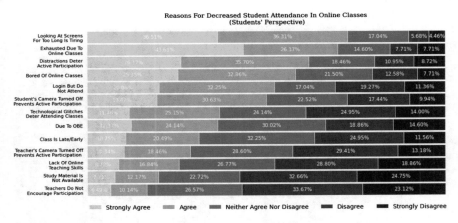

Fig. 13 Reasons for decline in attendance in OCs in Past 6 months as per students based on C-19GA21(S3)

variables, it cannot be used for inferencing causation in any manner. The test is applied using a contingency table where categories for one variable appear in the rows, and the categories for the other variable appear as columns. Every cell of the cross-tabulation gives the percentage of cases for the specific pair of categories that it intersects. To apply the Chi-square test of independence one must ensure that each variable must have at least two (or more) categories. Whenever the Chi-square value (statistic) is greater than the critical value it means that there is a significant difference. The critical value of a statistical test is that value at which, for any pre-determined probability (p), the test indicates a result that is less probable than p. Such a result is said to be statistically significant at that probability [28]. A rich and important set of 30 hypotheses were formulated and tested in the current study using the Chi-square

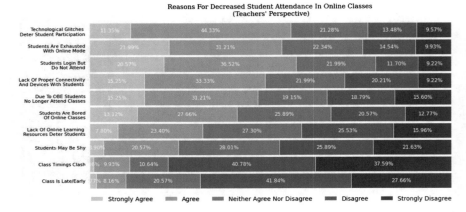

Fig. 14 Reasons for decline in attendance in OCs in Past 6 months as per teachers based on C-19GA21(S3)

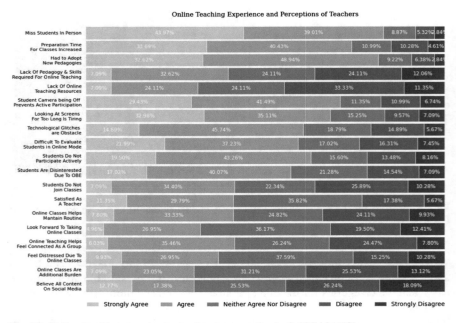

Fig. 15 Online teaching experience: teachers' perspective in C-19GA21(S3)

test of independence. Their detailed statements, values obtained and the results are summarized in Tables 1, 2, and 3.

A pirate plot was made to visualize the lexical diversity in PER of participants (refer Fig. 16). Another pirate plot was created to visualize diversity in general FAEADof the participants (refer Fig. 17). The representation was chosen as it can

Table 1 Hypothesis testing results on students and teachers data combined

Hypothesis testing results using chi-square test of independence for students and teachers in C-19GA21(S3) dataset (using α = 0.05)

S. No.	Combined: Attribute 1	Combined: Attribute 2	Null hypothesis (H0)	Test-statistic	p-value (up to six decimal places)	Outcome
1.	State/Union Territory	Status of internet access	Status of internet access has NO significant association on State/Union Territory of residence	Probability = 0.950, critical = 74.468, stat = 716.576	0.000000	Reject H0
2.	State/Union Territory	PER	PER have NO significant association on State/Union Territory of residence	Probability = 0.950, critical = 468.782, stat = 10,074.397	0.000000	Reject H0
3.	State/Union Territory	Feeling stressed due to OCs	Feeling stressed due to OCs has NO significant association on State/Union Territory of residence	Probability = 0.950, critical = 468.782, stat = 10,074.397	0.000000	Reject H0
4.	State/Union Territory	The extent of believing content on SM	Extent of believing content on SM has NO significant association on State/Union Territory of residence	Probability = 0.950, critical = 137.701, stat = 2528.521	0.000000	Reject H0

(continued)

Table 1 (continued)

Hypothesis testing results using chi-square test of independence for students and teachers in C-19GA21(S3) dataset (using $\alpha = 0.05$)

S. No.	Combined: Attribute 1	Combined: Attribute 2	Null hypothesis (H0)	Test-statistic	p-value (up to six decimal places)	Outcome
5.	The extent of believing content on SM	Gender	Extent of believing content on SM has NO significant association with gender	Probability = 0.950, critical = 15.507, stat = 131.447	0.000000	Reject H0
6.	Feeling connected as a group	Looking forward to attending OCs	Feeling connected as a group has NO significant association with looking forward to attending OCs	Probability = 0.950, critical = 26.296, stat = 36.844	0.002206	Reject H0
7.	Feeling stressed due to OCs	Experience too much burden due to OCs	Feeling stressed due to OCs has NO significant association with experiencing too much burden due to OCs	Probability = 0.950, critical = 26.296, stat = 37.812	0.001610	Reject H0
8.	Feeling stressed due to OCs	The extent of believing content on SM	Feeling stressed due to OCs has NO significant association with the extent of believing content on SM	Probability = 0.950, critical = 26.296, stat = 57.519	0.000001	Reject H0

(continued)

Table 1 (continued)

Hypothesis testing results using chi-square test of independence for students and teachers in C-19GA21(S3) dataset (using $\alpha = 0.05$)

S. No.	Combined: Attribute 1	Combined: Attribute 2	Null hypothesis (H0)	Test-statistic	p-value (up to six decimal places)	Outcome
9.	Feeling stressed due to OCs	Satisfaction with personal progress	Feeling stressed due to OCs has NO significant association with satisfaction with personal progress	Probability = 0.950, critical = 26.296, stat = 57.519	0.000000	Reject H0

simultaneously show three key aspects of the data: (a) the raw data in the form of individual points, (b) descriptive statistics in the form of lines (the median in Figs. 16 and 17 is shown as a red line), and (c) inferential statistics by way of Bayesian highest-density intervals and smoothed densities. The quartile markings and outliers can thus be easily identified in pirate plots.

For sentiment analysis of the L3Q, frequency plots for words, bigram term frequency plots, sentiment distribution plots, negation bigrams, negation trigrams, etc. were created to understand the emotions and sentiments in data using the colors of emotions in Robert Plutchik's Wheel of Emotions [29]. The chord diagram for emotions based on the responses in the attribute "Change in Thinking Process During a Year of Pandemic of the participants of C-19GA21(S3)" based on the NRC lexicon is shown in Fig. 18.

7 Discussion of Results

The results show that 48.6% of teachers and 35.1% of students who responded to the survey belong to Delhi NCR while the other participants were spread across parts of India. About 41% of teachers are in the age group of 36–45 years while 79.7% of students in C-19GA21(S3) are pursuing undergraduate courses (about 18–22 years of age). The fraction of females (60.3% of teachers and 70.6% of students) participating in C-19GA21(S3) is far more than males. In both categories, 0.4% of participants preferred not to disclose their gender.

Students' perspective of online learning experiences is depicted in Fig. 12. Results show that about 79% of students (51.32% strongly agree, 27.59% agree) miss seeing their classmates in person. Nearly 62% of students lost interest in classes due to online open-book exams being conducted by their institutes. About 56% of students

Table 2 Hypothesis testing results on students data only

Hypothesis testing results using chi-square test of independence for students data in C-19GA21(S3) Dataset (using α = 0.05)

S. No.	Students: Attribute 1	Students: Attribute 2	Null Hypothesis (H0)	Test-statistic	p-value (up to six decimal places)	Outcome
10.	Attendance of students during last 6 months	Attendance of students at the beginning of the pandemic	Attendance of students during the last 6 months has NO significant association with their attendance at the beginning of the pandemic	Probability = 0.950, critical = 26.296, stat = 208.680	0.000000	Reject H0
11.	Attendance of Students during last 6 Months	Scrolling SM after joining OCs	Attendance of students during the last 6 months has NO significant association with them scrolling SM after joining OCs	Probability = 0.950, critical = 26.296, stat = 58.368	0.000001	Reject H0
12.	AST	Satisfaction with personal progress	Student's perception about their personal progress has NO significant association with their AST	Probability = 0.950, critical = 36.415, stat = 106.193	0.000000	Reject H0
13.	The extent of believing content on SM	Scrolling SM after joining OCs	The extent of believing content on SM has NO significant association with their scrolling SM after joining OCs	Probability = 0.950, critical = 26.296, stat = 47.350	0.000060	Reject H0

(continued)

Table 2 (continued)

Hypothesis testing results using chi-square test of independence for students data in C-19GA21(S3) Dataset (using $\alpha = 0.05$)

S. No.	Students: Attribute 1	Students: Attribute 2	Null Hypothesis (H0)	Test-statistic	p-value (up to six decimal places)	Outcome
14.	The general feeling at the end of an average day	AST	AST has NO significant association with their general feeling at the end of an average day	Probability = 0.950, critical = 58.124, stat = 196.814	0.000000	Reject H0
15.	The general feeling at the end of an average day	Satisfaction with personal progress	Student's perception about their personal progress has NO significant association with their general feeling at the end of an average day	Probability = 0.950, critical = 41.337, stat = 196.725	0.000000	Reject H0
16.	The general feeling at the end of an average day	Scrolling SM after joining OCs	Scrolling SM after joining OCs has NO significant association with their general feeling at the end of an average day	Probability = 0.950, critical = 26.296, stat = 79.336	0.000000	Reject H0
17.	PER	Feeling stressed due to OCs	PER of students has NO significant association with stress experienced due to OCs	Probability = 0.950, critical = 79.082, stat = 253.436	0.000000	Reject H0

(continued)

Table 2 (continued)

Hypothesis testing results using chi-square test of independence for students data in C-19GA21(S3) Dataset (using $\alpha = 0.05$)

S. No.	Students: Attribute 1	Students: Attribute 2	Null Hypothesis (H0)	Test-statistic	p-value (up to six decimal places)	Outcome
18.	PER	AST	PER of students has NO significant association with their AST	Probability = 0.950, critical = 113.145, stat = 332.522	0.000000	Reject H0
19.	PER	Satisfaction with personal progress	PER of students has NO significant association with their perception about personal progress	Probability = 0.950, critical = 79.082, stat = 1025.630	0.000006	Reject H0
20.	PER	The general feeling at the end of an average day	PER of students have NO significant association with their general FAEAD	Probability = 0.950, critical = 129.918, stat = 1566.857	0.000000	Reject H0

find OCs a burden in these times of the pandemic and 66% of students were feeling distressed due to the same. A much smaller pie of students (6.69% strongly agree, 17.44% agree) feel satisfied with their learning progress in OCs. About 20% of students responded that OCs help them maintain a personal routine and about 10% of students look forward to these OCs. A little more than 18% of students feel connected as a group due to these OCs. Students' reasons for their decreased attendance in OCs are as described in Fig. 13. Nearly 73% of students felt that looking at the screen for a long time was tiring; 69% felt exhausted after attending OCs. About 61% of students felt that distractions around them deterred them from actively participating in OCs.

Teachers' estimation of the reasons for decreased student attendance in OCs is as described in Fig. 14. About 53% of teachers felt that students were exhausted with OCs while 48% of teachers felt that students do not have proper connectivity and devices which adversely affects their participation. Nearly 58% of teachers opined

Table 3 Hypothesis testing results on teachers data only

Hypothesis testing results using chi-square test of independence for teachers data in C-19GA21(S3) Dataset (using $\alpha = 0.05$)

S. No.	Teachers: Attribute 1	Teachers: Attribute 2	Null hypothesis (H0)	Test-statistic	p-value (up to six decimal places)	Outcome
21.	The age group of teachers	Platform used for communicating with students	Platform used by teachers for communicating with students has NO significant association with the age group of teachers	Probability = 0.950, critical = 37.652, stat = 83.114	0.000000	Reject H0
22.	The age group of teachers	Type of pedagogy adopted	Type of pedagogy adopted by teachers has NO significant association with their age group	Probability = 0.950, critical = 61.656, stat = 105.595	0.000001	Reject H0
23.	AST	Feeling satisfied with personal progress as a teacher	AST have NO significant association with feeling satisfied with personal progress as a teacher	Probability = 0.950, critical = 36.415, stat = 71.049	0.000007	Reject H0
24.	Feeling stressed due to OCs	Stance on resuming physical mode for the upcoming semester	The degree of stress experienced by teachers due to online teaching have NO significant association with their stance on resuming physical mode for the upcoming semester	Probability = 0.950, critical = 15.507, stat = 59.371	0.000000	Reject H0

(continued)

Table 3 (continued)

Hypothesis testing results using chi-square test of independence for teachers data in C-19GA21(S3) Dataset (using $\alpha = 0.05$)

S. No.	Teachers: Attribute 1	Teachers: Attribute 2	Null hypothesis (H0)	Test-statistic	p-value (up to six decimal places)	Outcome
25.	The general feeling at end of an average day	AST	General feeling at end of an average day has NO significant association with their AST	Probability = 0.950, critical = 58.124, stat = 89.999	0.000024	Reject H0
26.	Miss seeing students in person	Stance on resuming physical mode for the upcoming semester	Miss seeing students in person has NO significant association with the teacher's stance on resuming physical mode for the upcoming semester	Probability = 0.950, critical = 15.507, stat = 121.137	0.000000	Reject H0
27.	PER	Gender	PER of teachers have NO significant association with their gender	Probability = 0.950, critical = 24.996, stat = 109.221	0.007482	Reject H0
28.	PER	Type of teaching institute	PER of teachers have NO significant association with the type of teaching institute	Probability = 0.950, critical = 12.592, stat = 17.543	0.000009	Reject H0
29.	PER	Feeling satisfied with personal progress as a teacher	PER of teachers have NO significant association with feeling satisfied with their personal progress	Probability = 0.950, critical = 79.082, stat = 903.750	0.000000	Reject H0

(continued)

Table 3 (continued)

Hypothesis testing results using chi-square test of independence for teachers data in C-19GA21(S3) Dataset (using $\alpha = 0.05$)

S. No.	Teachers: Attribute 1	Teachers: Attribute 2	Null hypothesis (H0)	Test-statistic	p-value (up to six decimal places)	Outcome
30.	PER	The general feeling at the end of an average day	PER of teachers have NO significant association with their general FAEAD	Probability = 0.950, critical = 129.918, stat = 643.270	0.0000 00	Reject H0

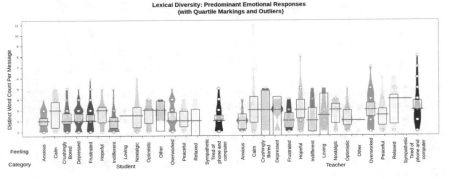

Fig. 16 Pirate plot for Lexical diversity in participants' message to the world based on their PER in C-19GA21(S3)

that students did not attend OCs after logging in, 46% felt that Open-Book Examination (OBE) was the reason for this attitude of students. About 41% felt that students are bored of OCs.

Teachers' perspective of online teaching experiences is as in Fig. 15. It shows that 83% of teachers miss meeting their students in person, 82% had adopted new pedagogies for OCs, 74% devoted more time to prepare for OCs. About 63% of teachers felt that students did not actively participate in OCs and 41% felt that students do not join classes. About 23% of teachers were not satisfied as a teacher, more than 25% felt distressed and about 32% of teachers did not look forward to taking OCs. Nearly 57% of teachers felt that students were not interested in OCs because of OBE. About 41% of teachers felt that OCs help maintain routine and make them feel connected. Results show that 30% of teachers believed in the content available on SM while 44% of teachers did not do so.

Bivariate analysis was done for over 100 + combinations of variables. On plotting PER of participants in C-19GA21(S3) against gender, (refer Fig. 19), it was found that 32.43% of females were tired of phones and computers as against 19.29% males,

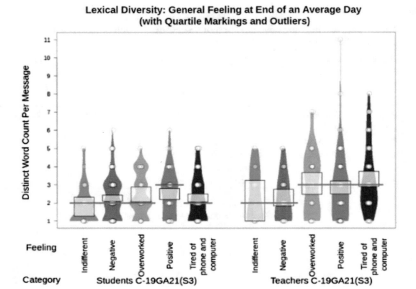

Fig. 17 Pirate plot for lexical diversity in participants' message to the world based on their general feeling at end of an average day in C-19GA21(S3)

Fig. 18 Emotions in the attribute—"change in thought process during a year of Covid-19" reported in C-19GA21(S3) based on the NRC lexicon depicted using colors of Robert Plutchik's Wheel of emotions

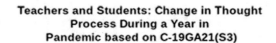

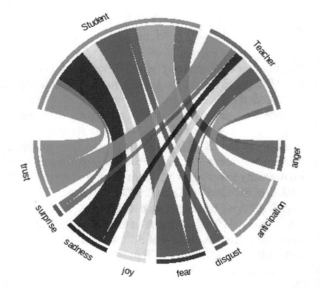

Fig. 19 PER versus gender
in C-19GA21(S3)

Teachers and Students: Predominant
Emotional Responses vs Gender
(Values in % per Column)

	Female	Male
Anxious	6.37	5.51
Calm	2.51	6.30
Crushingly bored	5.02	7.09
Depressed	5.21	7.87
Frustrated	13.32	12.20
Hopeful	4.83	4.72
Indifferent	5.02	6.30
Loving	0.58	1.18
Nostalgic	3.47	3.54
Optimistic	3.47	5.51
Other	1.16	1.18
Overworked	11.97	7.09
Peaceful	1.74	7.87
Relaxed	2.12	3.94
Sympathetic	0.77	0.39
Tired of phone and Computer	32.43	19.29

11.97% females felt overworked against 7.09% males. Males were calmer (6.30%) and more peaceful (7.87%) compared to females of whom only 2.51% were calm and 1.74% were peaceful, though frustration seemed to be manifesting almost equally in males (12.20%) and females (13.32%). This indicates that overall, more females were experiencing negative emotions as compared to their male counterparts. Plotting PER against geographies showed that irrespective of region/state, negative emotions in participants dominate positive emotions.

Analyzing the FAEAD (refer Fig. 20), shows that students are much more exhausted (33.27%) as compared with teachers (11.89%). Students are also more tired (19.88%) than teachers (13.31%) while the work goes on after classes for more teachers (23.30%) than students (11.16%). Though the students are more relieved (21.70%) than teachers (15.37%), the teachers are happier (19.65%) than students (5.48%). Figure 21 describes the feeling of connectedness of the participants due to OCs.

The general FAEAD of versus their AST is as in Fig. 22. It shows that students who spent less than 2 h per day on screen were not exhausted, whereas, 43 and 41.27% of students who had a screen-time of 10–12 h per day and more than 12 h per day respectively, felt exhausted by the end of the day. As AST increases the percentage of students feeling happy decreases appreciably. The Chi-square test of independence for AST against their general FAEAD (refer Table 2(14)) shows that the two attributes are not independent. AST of students per day versus their perception about personal progress during the pandemic in Fig. 23 shows that out of those students who spent

Teachers and Students: General Feeling at the end of an Average Day
(Values in % per column)
Confused 3.45 3.01
Exhausted 33.27 11.89
Happy 5.48 19.65
Joyful 2.23 9.67
Overwhelmed 2.84 3.80
Relieved 21.70 15.37
Study/Work Goes On Even After Classes 11.16 23.30
Tired 19.88 13.31
Student
Teacher

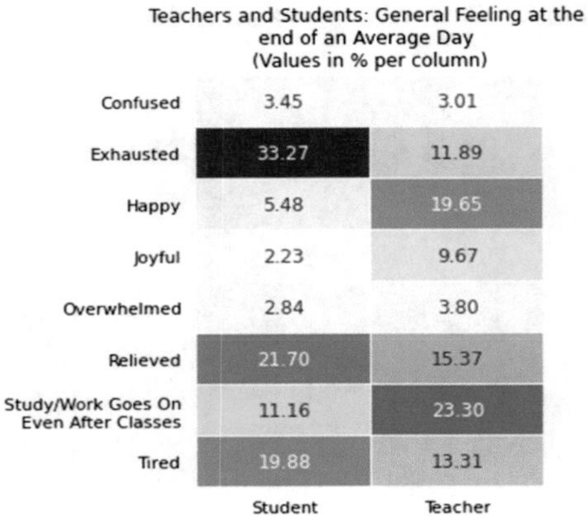

Fig. 20 FAEAD versus gender in C-19GA21(S3)

Teachers and Students:
Online Classes Help Feel Connected as a Group vs Look Forward to Online Classes
(Values in % per column)
Online Classes Help Feel Connected As A Group
Strongly Disagree 10.60 2.67 1.75 0.51 0.31
Disagree 7.30 9.57 6.07 1.44 0.00
Neither Agree Nor Disagree 3.50 3.70 8.13 2.37 0.62
Agree 2.16 3.09 5.25 6.79 0.82
Strongly Agree 0.51 0.10 0.72 0.62 1.13
Strongly Disagree
Disagree
Neither Agree Nor Disagree
Agree
Strongly Agree
Look Forward To Online Classes

Fig. 21 Feel connected as a group due to OCs and look forward to them in C-19GA21(S3)

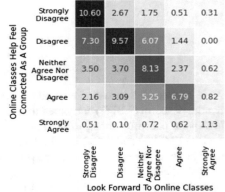

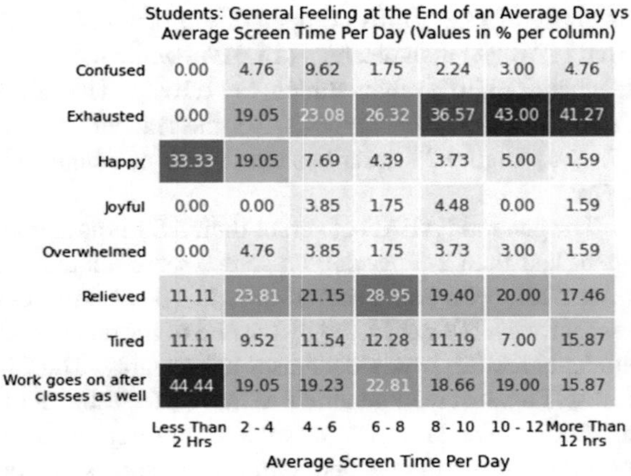

Students: General Feeling at the End of an Average Day vs Average Screen Time Per Day (Values in % per column)
Confused 0.00 4.76 9.62 1.75 2.24 3.00 4.76
Exhausted 0.00 19.05 23.08 26.32 36.57 43.00 41.27
Happy 33.33 19.05 7.69 4.39 3.73 5.00 1.59
Joyful 0.00 0.00 3.85 1.75 4.48 0.00 1.59
Overwhelmed 0.00 4.76 3.85 1.75 3.73 3.00 1.59
Relieved 11.11 23.81 21.15 28.95 19.40 20.00 17.46
Tired 11.11 9.52 11.54 12.28 11.19 7.00 15.87
Work goes on after classes as well 44.44 19.05 19.23 22.81 18.66 19.00 15.87
Less Than 2 Hrs
2 - 4
4 - 6
6 - 8
8 - 10
10 - 12
More Than 12 hrs
Average Screen Time Per Day

Fig. 22 FAEAD versus AST spent per day by students in C-19GA21(S3)

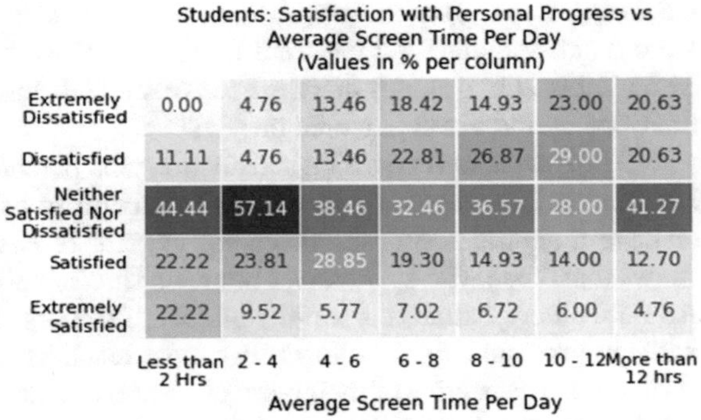

Fig. 23 AST of students per day versus their perception about personal progress as a student during the pandemic (C-19GA21(S3))

Students: Satisfaction with Personal Progress vs Average Screen Time Per Day (Values in % per column)						
Extremely Dissatisfied	0.00	4.76	13.46	18.42	14.93	23.00	20.63
Dissatisfied	11.11	4.76	13.46	22.81	26.87	29.00	20.63
Neither Satisfied Nor Dissatisfied	44.44	57.14	38.46	32.46	36.57	28.00	41.27
Satisfied	22.22	23.81	28.85	19.30	14.93	14.00	12.70
Extremely Satisfied	22.22	9.52	5.77	7.02	6.72	6.00	4.76
	Less than 2 Hrs	2 - 4	4 - 6	6 - 8	8 - 10	10 - 12	More than 12 hrs
	Average Screen Time Per Day						

less than 2 h of screen-time per day more than 44% were satisfied with their progress. Out of students who spent 10–12 h on screen per day, only 20% were satisfied, while 52% were dissatisfied. It shows that with an increase in AST, there is a decrease in students' satisfaction with their progress. Students' PER versus their satisfaction with learning is depicted in Fig. 24. Irrespective of their progress with learning, a large number of students were tired of using phones and computers. Out of students' who were extremely dissatisfied with their learning, 23.53% were depressed, 25.88% were

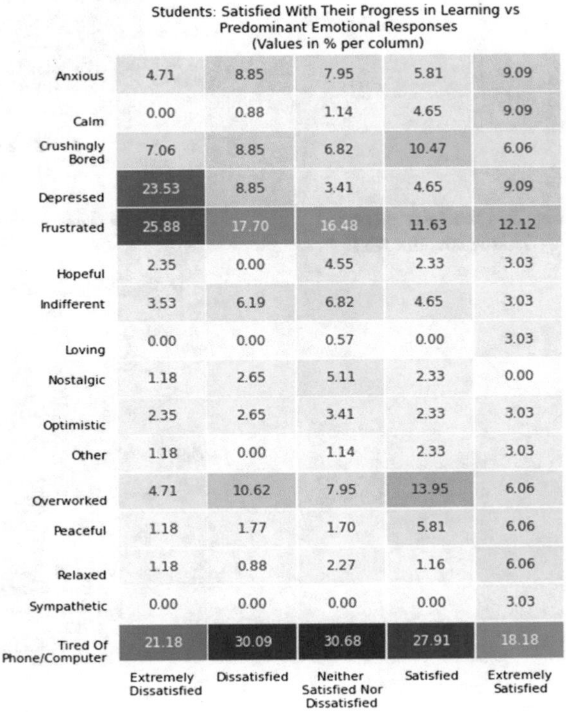

Fig. 24 PER of students and their perception about personal progress during the pandemic

Students: Satisfied With Their Progress in Learning vs Predominant Emotional Responses (Values in % per column)					
Anxious	4.71	8.85	7.95	5.81	9.09
Calm	0.00	0.88	1.14	4.65	9.09
Crushingly Bored	7.06	8.85	6.82	10.47	6.06
Depressed	23.53	8.85	3.41	4.65	9.09
Frustrated	25.88	17.70	16.48	11.63	12.12
Hopeful	2.35	0.00	4.55	2.33	3.03
Indifferent	3.53	6.19	6.82	4.65	3.03
Loving	0.00	0.00	0.57	0.00	3.03
Nostalgic	1.18	2.65	5.11	2.33	0.00
Optimistic	2.35	2.65	3.41	2.33	3.03
Other	1.18	0.00	1.14	2.33	3.03
Overworked	4.71	10.62	7.95	13.95	6.06
Peaceful	1.18	1.77	1.70	5.81	6.06
Relaxed	1.18	0.88	2.27	1.16	6.06
Sympathetic	0.00	0.00	0.00	0.00	3.03
Tired Of Phone/Computer	21.18	30.09	30.68	27.91	18.18
	Extremely Dissatisfied	Dissatisfied	Neither Satisfied Nor Dissatisfied	Satisfied	Extremely Satisfied

frustrated. Even out of those who were extremely satisfied with their progress 9.09% were depressed and 12.12% were frustrated. The Chi-square test of independence (refer Table 2(19)) shows that the PER of students has a significant association with their perception about personal progress.

AST of teachers versus their satisfaction with personal progress as a teacher during the pandemic is depicted in Fig. 25. The majority of teachers seem to be satisfied or have a neutral response. Teachers' general FAEAD versus their AST per day, shows that those who spent fewer hours on screen were happier. Those with a higher AST felt more exhausted (refer Fig. 26). Teachers' PER versus their satisfaction with progress in Fig. 27 shows that most teachers are tired of using phones and computers irrespective of their perception about personal progress. Out of extremely satisfied teachers, 21.88% were hopeful, 15.62% were peaceful and 9.38% were relaxed while 12.5% were anxious. Out of those who are extremely dissatisfied with personal progress 43.75% are tired of the screen, 18.75% feel overworked, 12.5% are frustrated and an equal percentage are predominantly anxious. The Chi-square

Fig. 25 AST of teachers per day versus their perception about personal progress as a teacher during the pandemic (C-19GA21(S3))

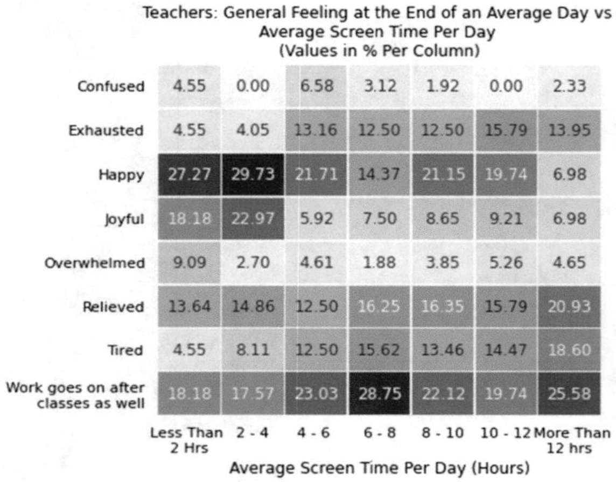

Teachers : Satisfaction with Personal Progress vs Average Screen Time Per Day (Values in % per column)

	Less Than 2 Hrs	2 - 4	4 - 6	6 - 8	8 - 10	10 - 12	More Than 12 hrs
Extremely Dissatisfied	0.00	0.00	1.43	6.76	11.11	15.38	5.56
Dissatisfied	18.18	7.89	17.14	24.32	13.33	15.38	22.22
Neither Satisfied Nor Dissatisfied	36.36	39.47	41.43	29.73	40.00	23.08	38.89
Satisfied	27.27	39.47	27.14	28.38	28.89	34.62	22.22
Extremely Satisfied	18.18	13.16	12.86	10.81	6.67	11.54	11.11

Average Screen Time Per Day (in Hours)

Fig. 26 Teachers' general FAEAD versus the AST

Teachers: General Feeling at the End of an Average Day vs Average Screen Time Per Day (Values in % Per Column)

	Less Than 2 Hrs	2 - 4	4 - 6	6 - 8	8 - 10	10 - 12	More Than 12 hrs
Confused	4.55	0.00	6.58	3.12	1.92	0.00	2.33
Exhausted	4.55	4.05	13.16	12.50	12.50	15.79	13.95
Happy	27.27	29.73	21.71	14.37	21.15	19.74	6.98
Joyful	18.18	22.97	5.92	7.50	8.65	9.21	6.98
Overwhelmed	9.09	2.70	4.61	1.88	3.85	5.26	4.65
Relieved	13.64	14.86	12.50	16.25	16.35	15.79	20.93
Tired	4.55	8.11	12.50	15.62	13.46	14.47	18.60
Work goes on after classes as well	18.18	17.57	23.03	28.75	22.12	19.74	25.58

Average Screen Time Per Day (Hours)

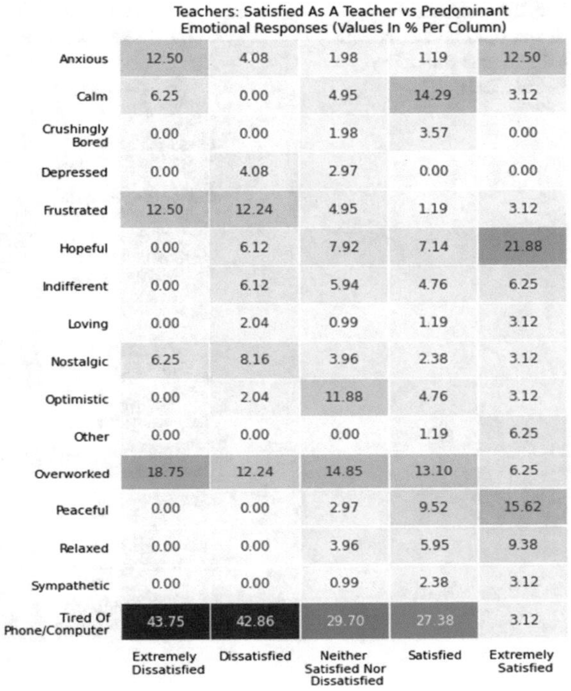

Fig. 27 Teachers' general FAEAD versus Their PER

Teachers: Satisfied As A Teacher vs Predominant Emotional Responses (Values In % Per Column)

	Extremely Dissatisfied	Dissatisfied	Neither Satisfied Nor Dissatisfied	Satisfied	Extremely Satisfied
Anxious	12.50	4.08	1.98	1.19	12.50
Calm	6.25	0.00	4.95	14.29	3.12
Crushingly Bored	0.00	0.00	1.98	3.57	0.00
Depressed	0.00	4.08	2.97	0.00	0.00
Frustrated	12.50	12.24	4.95	1.19	3.12
Hopeful	0.00	6.12	7.92	7.14	21.88
Indifferent	0.00	6.12	5.94	4.76	6.25
Loving	0.00	2.04	0.99	1.19	3.12
Nostalgic	6.25	8.16	3.96	2.38	3.12
Optimistic	0.00	2.04	11.88	4.76	3.12
Other	0.00	0.00	0.00	1.19	6.25
Overworked	18.75	12.24	14.85	13.10	6.25
Peaceful	0.00	0.00	2.97	9.52	15.62
Relaxed	0.00	0.00	3.96	5.95	9.38
Sympathetic	0.00	0.00	0.99	2.38	3.12
Tired Of Phone/Computer	43.75	42.86	29.70	27.38	3.12

test (refer Table 3(30)) shows that the PER of teachers has a significant association with their perception about personal progress.

Figure 28 depicts the responses of teachers as compared to students about the reasons they observe have led to decreased attendance in OCs in the past six months of the Pandemic. About 22.30% of students and 19.13% of teachers concur that students were exhausted with online learning. Over 22% of teachers felt that students did have adequate resources to attend OCs while only 1.6% of students reported the same. Nearly 9.41% of students were discontented with OCs and 15.24% did not enjoy participating in them. Technological glitches seemed to be an important factor on which students and teachers concur as a reason for reduced attendance.

Comparing the AST of students in the first six months of the Pandemic (C-19GA20) to the AST in the past six months (C-19GA21(S3)) shows that there is a major shift toward an increase in screen-time. While 27.85% of students in C-19GA20 spent 8 h or more on screen, the number more than doubled to 60.24% of students who were doing so in C-19GA21(S3) (refer Fig. 29). This unhealthy trend manifests as a sharp increase of predominant negative emotions (54.86% in C-19GA20 to 78.64% in C-19GA21(S3)) and a simultaneous sharp decrease in positive emotions of the same (38.67% in C-19GA20 to 15.81% in C-19GA21(S3)) as is evident in Fig. 30. This is an alarming signal and steps need to be put in place to increase awareness and mitigate these issues.

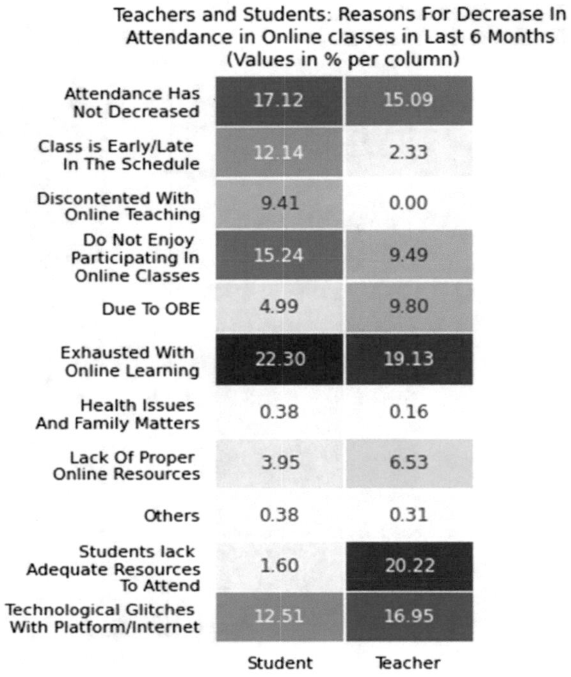

Fig. 28 Reasons for decrease in attendance in OCs in past 6 months C-19GA21(S3)

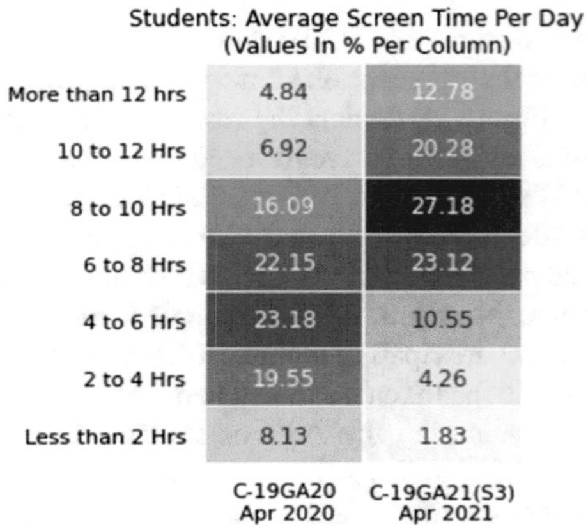

Teachers and Students: Reasons For Decrease In Attendance in Online classes in Last 6 Months (Values in % per column)

	Student	Teacher
Attendance Has Not Decreased	17.12	15.09
Class is Early/Late In The Schedule	12.14	2.33
Discontented With Online Teaching	9.41	0.00
Do Not Enjoy Participating In Online Classes	15.24	9.49
Due To OBE	4.99	9.80
Exhausted With Online Learning	22.30	19.13
Health Issues And Family Matters	0.38	0.16
Lack Of Proper Online Resources	3.95	6.53
Others	0.38	0.31
Students lack Adequate Resources To Attend	1.60	20.22
Technological Glitches With Platform/Internet	12.51	16.95

Fig. 29 AST of students in C-19GA20 and C-19GA21(S3)

Students: Average Screen Time Per Day (Values In % Per Column)

	C-19GA20 Apr 2020	C-19GA21(S3) Apr 2021
More than 12 hrs	4.84	12.78
10 to 12 Hrs	6.92	20.28
8 to 10 Hrs	16.09	27.18
6 to 8 Hrs	22.15	23.12
4 to 6 Hrs	23.18	10.55
2 to 4 Hrs	19.55	4.26
Less than 2 Hrs	8.13	1.83

7.1 Sentiment Analysis

Sentiment analysis is said to be a domain where AI meets the field of Psychology. The study explored sentiment lexicons and their word counts to determine the mood expressed in "L3Q"s in C-19GA21(S3). As the first task for sentiment analysis, the

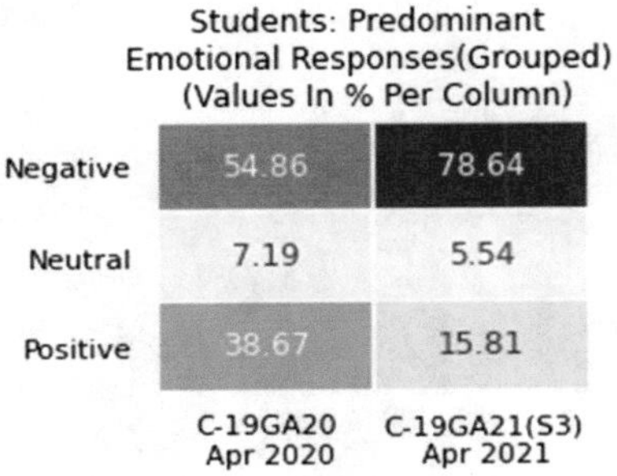

Fig. 30 Grouped PER of students in C-19GA20 and C-19GA21(S3)

Students: Predominant
Emotional Responses(Grouped)
(Values In % Per Column)

Negative	54.86	78.64
Neutral	7.19	5.54
Positive	38.67	15.81
	C-19GA20 Apr 2020	C-19GA21(S3) Apr 2021

researchers designed a customized list of stop words. Then, lemmatization was done while preserving word frequency. Lemmatization improves accuracy as lexicons contain only the root words [30].

In the third step, AFINN lexicon [31] based sentiment scores were assigned to the text responses in L3Q, based on the polarity strength. AFINN has over 3000 words with a preassigned polarity score $(-5, -4...0...4, 5)$. Data (i.e., each sentence) is compared against it and given a sentiment score. Negative scores indicate negative sentiment and positive scores indicate positive sentiment. AFINN score of zero is considered to be neutral. The scores can be used to carry out comparative sentiment analysis between different demographics [32], e.g., for teachers' and students' data and also for comparing results of the current study C-19GA21(S3) to the results of [12] in C-19GA20. The higher the value of the AFINN sentiment score the more positive is the emotional response. Further visualizations were done using negation bigram networks (refer Fig. 31) revealed a significant number of negating bigrams in the text attributes (L3Q). Simply applying a lexicon-based sentiment analysis approach to such data could result in false positives and false negatives. Therefore, the current study implements an algorithm to reduce such inaccuracies in AFINN scores.

The algorithm has the following steps:

i. Create a list of frequently occurring negating words in the dataset.
ii. Set initial value for sentiment_score $= 0$
iii. Search for the occurrence of negating words found in [i] in each response:
    (If found, go to step iii(a), else go to step iv)

    a. Check for words with non-zero AFINN scores preceded by a negation word

    b. Assign sentiment_score as the additive inverse of the AFINN score of the word being negated, e.g., in 'no hatred', since the AFINN _score of hatred is -3, the algorithm assigns the bigram a sentiment score of $+ 3$. Similarly, 'do not help' gets a score of -2 as 'help' has an AFINN score of $+ 2$.

    c. Remove the bigram from the original string/response so that the resulting string does not contain any negating bigrams.

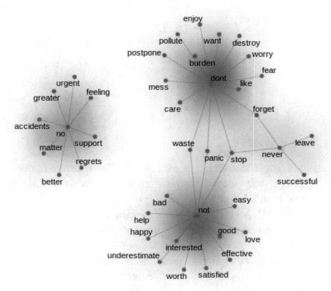

Fig. 31 Negation Bigram Network for the attribute 'change in thinking during the pandemic' (C-19GA21(S3))

iv.   sentiment _score = sentiment _score + sum of AFINN score of the remaining
        words in the string.

The AFINN sentiment scores obtained for C-19GA21(S3) and C-19GA20 are as shown in Tables 4 and 5. Higher positive scores in Table 4 for 'change in thinking during the Pandemic' and 'message to the world' from teachers versus the score for students is in agreement agrees with results of Students' versus Teachers' comparison

Table 4 AFINN sentiment scores for C-19GA21(S3)

Year	Change in thinking after a year in Pandemic	Message to the world
Students	−0.002028	0.906694
Teachers	0.801418	1.379433
Both	0.290323	1.078710

Table 5 AFINN sentiment scores for students in C-19GA21(S3) and C-19GA20

	Change in thinking due to pandemic	Message to the world
C-19GA20 in Apr 2020	0.704152	1.688581
C-19GA21(S3) in Apr 2021	−0.002028	0.906694

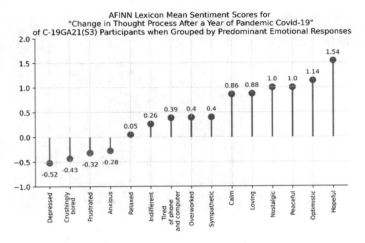

Fig. 32 Sentiment scores for "change in thought process" in C-19GA21(S3)

of the PER and FAEAD, which indicate that students experienced more negative emotions than teachers.

Table 5 shows students' messages to the world in both C-19GA21(S3) and C-19GA20 indicate hope for a relatively happier post-pandemic future despite their current adverse circumstances.

When grouped by PER, those participants in C-19GA21(S3) who chose anxious, crushingly bored, frustrated, and depressed were the ones with the most negative AFINN sentiment scores for "Change in Thinking after a year of Pandemic" (refer Fig. 32).

Further, when the sentiment scores for attribute Students' "Message one line to the World" in LQ3 in C-19GA21(S3) and C-19GA20 and sentiment scores for the attribute "Change in Thinking During the Pandemic" in C-19GA21(S3) and C-19GA20 were compared, it was found that C-19GA20 sentiment scores are more positive than C-19GA21(S3). This is in agreement with the results of PER of students in C-19GA21(S3) and C-19GA20. It confirms that there is a significant rise in negative emotions of students within a year, i.e., from April 2020 to April 2021 (refer Fig. 30). Sentiment classification was also done using the NRC lexicon. NRC classifies words into eight basic emotions (anger, fear, anticipation, trust, surprise, sadness, joy, and disgust) and two sentiments (negative and positive). The chord diagram for the emotions in attribute- 'Change in Thinking Process During a Year of Pandemic' for the participants of C-19GA21(S3) based on NRC lexicon is as shown in Fig. 18 in Sect. 6.

After pre-processing the text in L3Q, word clouds were rendered to visualize the sentiments and emotions. Figures 33 and 34 clearly show that teachers are missing in person student interaction and students are equally missing the same. This is in sync with the result obtained in [12] (refer Fig. 35). Also, it is interesting to note

Fig. 33 Teachers' response: cannot be replaced in online teaching C-19GA21(S3)

Fig. 34 Students' response: cannot be replaced in online teaching C-19GA21(S3)

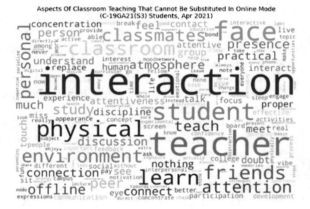

Fig. 35 Students' response: cannot be replaced in online teaching C-19GA20

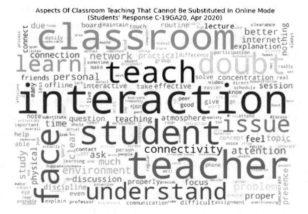

Fig. 36 Teachers' one line message in C-19GA21(S3)

Fig. 37 Students' one line message in C-19GA21(S3)

that the students and teachers are in concurrence on their message to the world in C-19GA21(S3) where "stay safe" predominates everything else (refer Figs. 36 and 37).

8 Conclusion

The findings of the study indicate that the mental well-being of both students and teachers has been adversely affected during the pandemic and many of them find OCs an additional burden during the Pandemic. Several studies substantiate evidence to support that longer screen-times can have consequences such as cognitive impairment, reduced attention span, poor quality of sleep, and addictive behaviors, etc. The significant rise in the AST of students and its high association with negative emotional states and high levels of stress is particularly alarming. For a developing economy like India, where the student community is the biggest asset, it is imperative to take the necessary steps to prioritize their mental health. The novel approach in the current work to juxtapose teachers' and students' responses conclude that there is unanimous agreement that face-to-face interaction of a classroom cannot be substituted by online teaching–learning. The unique comparison of student responses at the beginning of the pandemic (in a previous study) with the recent data collected exactly one year after that, shows that there is a significant rise in negative emotional states, decrease in satisfaction levels with personal progress, and much higher average screen-times per day.

Further, the negative state of mind reflected in their responses to 'change in thought process' needs to be viewed together with the increased risks of mental health issues among people who have endured prolonged periods of restricted social interaction. With the ongoing second wave of the Pandemic in India and predictions for a third wave, online education is likely to continue in the upcoming academic session. Educational institutions should therefore, ensure adequate resources and support systems are in place to protect the mental health and well-being of teachers and students. Amidst an unprecedented surge in fake news, it is heartening to see that the majority of respondents do not forward content without verifying nor do they blindly believe in SM content. Another positive takeaway from the study is the underlying sentiment of hope and concern for each other's health and safety in their message to the world. Online learning has provided the opportunity to teach and learn in innovative ways unlike the teaching–learning experiences in the normal classroom setting.

9 Limitations of the Study

A major limitation of the study is that the survey questionnaire was able to reach only students and teachers who are well conversant in the English language.

Acknowledgements The authors are grateful to all of the participants in this study, who generously gave their time to respond to the online survey. The authors are also grateful to Ms. Nitika Malhotra, Manager (Data Analytics Division), Ministry of Health (Singapore) for sparing time out of her busy schedule to share her valuable feedback about the findings of this study. Her inputs motivated the authors and ensured that findings in this study are presented as per the industry norms.

Funding The study did not receive any funding from anywhere.

Conflict of Interest Priti Rai Jain, S.M.K. Quadri and Elaine Mary Rose (an undergraduate student at Miranda House, University of Delhi) declare that there are no conflicts of interest.

Declaration

All procedures performed in this study involving human participants were done in online mode only. No physical in person interactions were done for this.

Informed Consent

The decision to participate in the survey was purely voluntary and informed consent was obtained from all individual participants included in the study.

Availability of Data And Material (Data Transparency)
The complete dataset C-19GA21 and the data collection formats are available at: https://data.mendeley.com/datasets/99hx2xg7gx/2

References

1. Galea, S., Merchant, R.M., Lurie, N.: The mental health consequences of COVID-19 and physical distancing. JAMA Intern. Med. **180**, 817 (2020). https://doi.org/10.1001/jamainternmed.2020.1562
2. Xiong, J., Lipsitz, O., Nasri, F., Lui, L.M.W., Gill, H., Phan, L., Chen-Li, D., Iacobucci, M., Ho, R., Majeed, A., McIntyre, R.S.: Impact of COVID-19 pandemic on mental health in the general population: a systematic review. J. Affect. Disord. **277**, 55–64 (2020). https://doi.org/10.1016/j.jad.2020.08.001
3. Tesar, M.: Towards a post-covid-19 'new normality?': physical and social distancing, the move to online and higher education. Policy Futur. Educ. **18**, 556–559 (2020). https://doi.org/10.1177/1478210320935671
4. Brooks, S.K., Webster, R.K., Smith, L.E., Woodland, L., Wessely, S., Greenberg, N., Rubin, G.J.: The psychological impact of quarantine and how to reduce it: rapid review of the evidence. Lancet **395**, 912–920 (2020). https://doi.org/10.1016/S0140-6736(20)30460-8
5. Wang, Y., Shi, L., Que, J., Lu, Q., Liu, L., Lu, Z., Xu, Y., Liu, J., Sun, Y., Meng, S., Yuan, K., Ran, M., Lu, L., Bao, Y., Shi, J.: The impact of quarantine on mental health status among general population in China during the COVID-19 pandemic. Mol. Psychiatry. (2021). https://doi.org/10.1038/s41380-021-01019-y
6. Cullen, W., Gulati, G., Kelly, B.D.: Mental health in the COVID-19 pandemic. QJM An Int. J. Med. **113**, 311–312 (2020). https://doi.org/10.1093/qjmed/hcaa110
7. Holmes, E.A., O'Connor, R.C., Perry, V.H., Tracey, I., Wessely, S., Arseneault, L., Ballard, C., Christensen, H., Cohen Silver, R., Everall, I., Ford, T., John, A., Kabir, T., King, K., Madan, I., Michie, S., Przybylski, A.K., Shafran, R., Sweeney, A., Worthman, C.M., Yardley, L., Cowan, K., Cope, C., Hotopf, M., Bullmore, E.: Multidisciplinary research priorities for the COVID-19 pandemic: a call for action for mental health science. The Lancet Psychiatry. **7**, 547–560 (2020). https://doi.org/10.1016/S2215-0366(20)30168-1
8. Azevedo, J.P., Hasan, A., Goldemberg, D., Iqbal, S.A., Geven, K.: Simulating the Potential Impacts of COVID-19 School Closures on Schooling and Learning Outcomes: A Set of Global Estimates. World Bank, Washington, DC (2020). https://doi.org/10.1596/1813-9450-9284
9. Rasmitadila, Aliyyah, R.R., Rachmadtullah, R., Samsudin, A., Syaodih, E., Nurtanto, M., Tambunan, A.R.S.: The perceptions of primary school teachers of online learning during the covid-19 pandemic period: a case study in Indonesia. J. Ethn. Cult. Stud. **7**, 90–109 (2020). https://doi.org/10.29333/ejecs/388
10. Muthuprasad, T., Aiswarya, S., Aditya, K.S., Jha, G.K.: Students' perception and preference for online education in India during COVID -19 pandemic. Soc. Sci. Humanit. Open. **3**, 100101 (2021). https://doi.org/10.1016/j.ssaho.2020.100101
11. Tyagi, H.: Online teaching in Delhi-NCR schools in India during Covid-19 pandemic. Indian J. Sci. Technol. **13**, 4036–4054 (2020). https://doi.org/10.17485/IJST/v13i38.1613
12. Khattar, A., Jain, P.R., Quadri, S.M.K.: Effects of the disastrous pandemic COVID 19 on learning styles, activities and mental health of young Indian students—a machine learning approach. In: 2020 4th International Conference on Intelligent Computing and Control Systems (ICICCS). pp. 1190–1195. IEEE (2020). https://doi.org/10.1109/ICICCS48265.2020.9120955
13. Jain, P.R., Quadri, S.M.K., Rose, E.M.: Covid-19 Go Away 2021 (C-19GA21) Dataset (2021). https://doi.org/10.17632/99hx2xg7gx.2
14. Taber, K.S.: The use of Cronbach's alpha when developing and reporting research instruments in science education. Res. Sci. Educ. **48**, 1273–1296 (2018). https://doi.org/10.1007/s11165-016-9602-2

15. Gardner, P.L.: Measuring attitudes to science: unidimensionality and internal consistency revisited. Res. Sci. Educ. **25**, 283–289 (1995). https://doi.org/10.1007/BF02357402
16. Glen, S.: Cronbach's alpha: simple definition, use and interpretation, From StatisticsHowTo.com: Elementary Statistics for the rest of us! https://www.statisticshowto.com/probability-and-statistics/statistics-definitions/cronbachs-alpha-spss/
17. Kumar, G., Singh, G., Bhatnagar, V., Gupta, R., Upadhyay, S.K.: Outcome of online teaching-learning over traditional education during covid-19 pandemic. Int. J. Adv. Trends Comput. Sci. Eng. **9**, 7704–7711 (2020). https://doi.org/10.30534/ijatcse/2020/113952020
18. Zimmer, W.K., McTigue, E.M., Matsuda, N.: Development and validation of the teachers' digital learning identity survey. Int. J. Educ. Res. **105**, 101717 (2021). https://doi.org/10.1016/j.ijer.2020.101717
19. van der Spoel, I., Noroozi, O., Schuurink, E., van Ginkel, S.: Teachers' online teaching expectations and experiences during the Covid19-pandemic in the Netherlands. Eur. J. Teach. Educ. **43**, 623–638 (2020). https://doi.org/10.1080/02619768.2020.1821185
20. König, J., Jäger-Biela, D.J., Glutsch, N.: Adapting to online teaching during COVID-19 school closure: teacher education and teacher competence effects among early career teachers in Germany. Eur. J. Teach. Educ. **43**, 608–622 (2020). https://doi.org/10.1080/02619768.2020.1809650
21. Pokhrel, S., Chhetri, R.: A literature review on impact of COVID-19 pandemic on teaching and learning. High. Educ. Futur. **8**, 133–141 (2021). https://doi.org/10.1177/2347631120983481
22. Joshi, A., Vinay, M., Bhaskar, P.: Impact of coronavirus pandemic on the Indian education sector: perspectives of teachers on online teaching and assessments. Interact. Technol. Smart Educ. (2020). https://doi.org/10.1108/ITSE-06-2020-0087
23. Jain, P.R., Khattar, A., Quadri, S.M.K.: Covid-19 Go Away 2020 (C-19GA20) Dataset. (2021). https://doi.org/10.17632/ncvfr2trwb.2
24. Neophytou, E., Manwell, L.A., Eikelboom, R.: Effects of excessive screen time on neurodevelopment, learning, memory, mental health, and neurodegeneration: a scoping review. Int. J. Ment. Health Addict. (2019). https://doi.org/10.1007/s11469-019-00182-2
25. Singh, S., Balhara, Y.S.: "Screen-time" for children and adolescents in COVID-19 times: need to have the contextually informed perspective. Indian J. Psychiatry. **63**, 192 (2021). https://doi.org/10.4103/psychiatry.IndianJPsychiatry_646_20
26. Twenge, J.M., Campbell, W.K.: Associations between screen time and lower psychological well-being among children and adolescents: evidence from a population-based study. Prev. Med. Reports. **12**, 271–283 (2018). https://doi.org/10.1016/j.pmedr.2018.10.003
27. Lissak, G.: Adverse physiological and psychological effects of screen time on children and adolescents: literature review and case study. Environ. Res. **164**, 149–157 (2018). https://doi.org/10.1016/j.envres.2018.01.015
28. Glen, S.: Chi-Square Statistic: How to Calculate It/Distribution. https://www.statisticshowto.com/probability-and-statistics/chi-square/. Last accessed 8 May 2021
29. Plutchik, R.: The Nature of Emotions: Human emotions have deep evolutionary roots, a fact that may explain their complexity and provide tools for clinical practice. http://www.jstor.org/stable/27857503 (2001). https://doi.org/10.1007/BF00354055
30. Manning, C.D., Raghavan, P., Schütze, H.: Introduction to Information Retrieval. Cambridge University Press (2008)
31. Nielsen, F.A.: A new ANEW: evaluation of a word list for sentiment analysis in microblogs. In: Proceedings of the ESWC2011 Workshop on "Making Sense of Microposts": Big Things Come in Small Packages, pp. 93–98 (2011)
32. Mohammad, S.M.: Practical and Ethical Considerations in the Effective use of Emotion and Sentiment Lexicons (2020)

Traffic Density Calculation Using PEGASIS and Traffic Light Control Using STLSD Algorithm

Ramya Srikanteswara⊕**, Aayesha Nomani, Rituraj Pandey, and Bhola Nath Sharma**

Abstract Road congestion and heavy traffic in cities are one of the most annoying problems of urbanization. Several projects and management systems have tried to overcome this problem, but no feasible solution has been devised yet. Modern traffic management systems are heavily dependent on traffic lights. But, traffic lights deployed nowadays work on obsolete logic of fixed time cycles. It is therefore extremely important to find an alternate approach for solving this problem. The proposed system makes use of sensors to detect traffic density at crossroads and reinforcement learning-based STLS algorithm to schedule traffic lights.

Keywords Power efficient gathering in sensor information system · Smart traffic light scheduling · Energy aware routing · Wireless sensor network

1 Introduction

Since the late 1800s, traffic control has been used to safely regulate the movement of vehicles. Problems such as traffic congestion and gridlock have risen exponentially as the number of cars on our roads has increased. Congestion in modern major cities such as Toronto and New York, for example, can add up to 60% to one's commuting time [1]. Congestion is also a significant contributor to global warming. The longer a vehicle is on the lane, the more CO_2 it emits. We will significantly minimize our pollution by reducing the number of traffic jams. As a result, increasing average traffic speed by about 20 miles per hour reduces CO_2 emissions by 21 t per month [2]. These issues illustrate the need for a more proficient traffic board framework. Conventional traffic signals presently do not address our issues, as they do not consider the present status of the traffic it is attempting to oversee. Along these lines, they can't satisfy the particular requirements of each crossing point.

R. Srikanteswara (✉) · A. Nomani · R. Pandey · B. N. Sharma
Nitte Meenakshi Institute of Technology, Bengaluru, India
e-mail: ramya.srikanteswara@nmit.ac.in

© The Author(s), under exclusive license to Springer Nature Singapore Pte Ltd. 2022
V. Suma et al. (eds.), *Evolutionary Computing and Mobile Sustainable Networks*,
Lecture Notes on Data Engineering and Communications Technologies 116,
https://doi.org/10.1007/978-981-16-9605-3_42

651

The use of STLS algorithms [3] as a replacement for conventional traffic light scheduling criteria is expected to maximize traffic flow at intersections and minimize congestion on roads. The STLS algorithms work on real-time data collected at crossroads using sensors operating on PEGASIS protocol as described in [4].

Many past researches have tried to determine this issue by utilizing a few algorithms that recognize the requirements for ideal traffic stream. With our proposed system, we have accomplished enhancements for the limitations that were observed in previous studies.

2 Related Work

Roads are the basic mode of transportation; therefore, they need the most attention. Recently, many algorithms have been developed to check the road congestion and to reduce the traffic density of the roads. To do so, several intelligent devices and techniques are used. Most widely used approach involves applying computer vision and object detection to record videos of real-time traffic and then applying scheduling techniques [5, 6]. Wireless sensors are used for one such method [7–9]. It is basically used to ensure safety of the driver and the vehicles. Similarly, VANETs are also developed to improve efficiency of traffic on the roads. Here, we will talk about the wireless sensor network (WSN) used for collecting traffic information [4]. Kanungo et al. [10] presents a dynamic approach toward management of traffic at any intersection. The main concept used in their approach is utilizing traffic density to make decisions for scheduling traffic lights. However, the approach has a major drawback. Their simulation allows only one traffic signal to be green at a particular time, whereas in the real world, two opposing lanes can simultaneously have a green signal without causing any problem in traffic flow. Another drawback with the approach is that the traffic densities for only lanes with red lights are taken into consideration. Ignoring traffic densities of lanes with green signal would result in shorter time cycles but would not lead to an efficient, smart traffic management system.

Similarly, Hawi et al. [9] discussed the use of smart grid network using mini-batch gradient descent method which gave each application its own quality of service, and provided different priorities to different emergency messages. The cognitive radio approach required 2 types of users, one having more priority than the other. The mini-batch gradient descent worked smartly when it came to traffic optimization, but the training period for it was so long that it could not be considered a fair trade.

Another approach for smart traffic management is described in [11]. This intelligent traffic light controlling algorithm is entirely based on traffic density. The problem with this approach is that it may lead to indefinite waiting time for a lane which has very less number of vehicles as compared to a major lane which has a constant flow of traffic. Also, the algorithm discussed in [8] does not include any specific strategy to give priority to emergency vehicles. Later, the deep reinforcement learning model with vehicle heterogeneity (DLVH) proposed in [12] reduced the waiting time for the emergency vehicles but took a lot of time in the training period of the model when the traffic increased.

Work is being done on including factors such as working days, holidays, and weather conditions and quantifying the effect of these conditions to construct multidimensional state vectors [13].

Jonnalagadda et al. [14] utilizes cloud-edge offloading to overcome the problems of resource management. The method suggested in the paper analyzes traffic flow as a set of inter-related tasks which are offloaded to the edge for computation and scheduling (Table 1).

Table 1 Various algorithms used for scheduling traffic lights

S. No.	References	Feature	Limitations	Ref. Year
1	[15]	Microcontroller: Arduino Uno ATMega 328	The model does not include mechanisms for high traffic density on two or more lanes	2020
2	[16, 17]	Fuzzy logic controller, WSN, Mamdami inference system	The model does not take into consideration the presence of emergency vehicles while making decisions. The model treats each intersection as independent and does not take into account neighboring intersections while making decisions	2017, 2020
3	[18]	Double dueling deep Q-network, Markov decision	With increase in traffic density, the training period increases which affects performance	2018
4	[19]	Mixed Q-network	Further work needs to be done in various traffic settings and for various network structures. In addition, the design of a pattern detector that suits the demand of the problem is another challenge	2019
5	[20]	DQN, DQRN	With increase in the traffic in urban areas, the training period of the model will increase, hence affecting the performance	2019
6	[21]	Model spatial arrangement	Heavily rely on the base. If a vehicle is not in the database, then the decision taken is completely random	2020

3   Methodology

This part is divided into two segments. The first one deals with getting information about traffic at an intersection of lanes, and the second one deals with the study of algorithms that can be used to schedule traffic lights for efficient traffic light management.

3.1   Collecting Traffic Information Using WSN

Sensor Network

Sensors are placed on different components of the roads, such as turns, junctions, and vehicles. When it comes to the types of sensor nodes, they are of three types based on where they are placed [22]:

1.   Roadside nodes
2.   Intersection nodes
3.   Vehicle nodes.

These three nodes are involved in all kinds of data sharing, so as to calculate the traffic density. To establish communication between the sensor nodes, certain protocols are used such as LEACH, PEGASIS, TEEN [23, 24]. The roadside nodes can be perceived as source, the nodes at intersections as intermediates and the vehicular nodes as sink. The diagram in Fig. 1 explains the same.

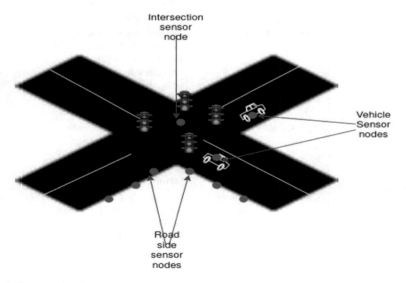

Fig. 1 Sensor network

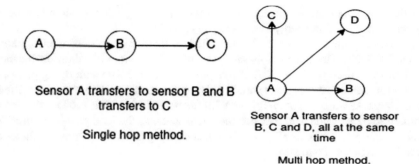

Fig. 2 Methods to transfer information

Collecting Information from Sensors: Road networks can be generated using SUMO. The nodes present on roadsides collect information and send it to the intermediate nodes present at the intersections, which in turn pass it to the vehicular nodes. Again, there are two methods to transfer such data as shown in Fig. 2.

1. Single hop method
2. Multi-hop method.

As the name suggests, the single hop method transfers the data one by one, i.e., one node transfers the data to another and that node further transfers the data to next node, whereas in multi-hop transfer, the data are transferred to many nodes at the same time, i.e., a single node transfers the data to all the nodes nearby.

This phase is called the information delivery phase. The routing can be done by fixing the nodes present roadside and on vehicles, i.e., source and sink nodes [25]. Data are transferred using the least hops possible. The routing schemes fetch the data such as cost, delay, number of hops, and efficiency [26]. In the end, we create routing tables and calculate the costs; then, the minimum cost path is selected [22, 27].

Routing Schemes To Be Used

Basically, two protocols are used for the given data:

1. EARQ
2. PEGASIS.

EARQ [28, 29] provides a reliable packet delivery between the sensors. This protocol selects a random path between two nodes and calculates the cost, efficiency, reliability, and energy of that path. It, then, repeats the process for all the possible paths in the area, with all the neighboring nodes. After a heavy calculation, the most efficient path is chosen, which obviously has the minimum cost.

EARQ constructs a routing table and maintains it using the data taken from the neighboring nodes [30]. In this protocol, there are two types of messages: beacon and data packets. The beacon messages are used to construct the routing table, while the data packets are sent using the routing table constructed. This protocol provides real-time communication between the sensors without any concession in the energy awareness of the basic energy aware routing (EAR) protocol.

PEGASIS which stands for "Power Efficient Gathering in Sensor Information System" [23, 27]. This is a chain-based system of sensor communication and the most efficient to be used till now. Why is this power efficient? The underlying principle of the protocol is that the nodes should only communicate to their nearest neighbor. Then, they take turns to communicate with the base station. This reduces the power required to transfer the data between the nodes and also the total power required for each cycle. Hence, the overall power is reduced giving the name—"power efficient." Now, PEGASIS has two aims:

1. Increase the overall life cycle of each node.
2. Allow communication between two closest nodes.

Working of PEGASIS Protocol

The basic concept of the PEGASIS protocol is to allow communication between the two closest nodes. It is an improved version of ant colony algorithm [31] which builds chains to communicate with each other. Each node present in the chain communicates with its closest node in the chain. For the chain to be formed, we use a greedy algorithm. There is no communication between one node and all other nodes; hence, it is power efficient too.

When a node receives certain data from its neighboring node, it then combines that data with the data collected by itself and passes the combined data to the next node. This is called data fusion. Every node performs data fusion except the end nodes. Also, each node takes turns to be the leader of the chain by transferring data to the base station as shown in Fig. 3 so that average energy spent by each node is reduced.

In PEGASIS, whenever a node dies (stops working due to some issue), the chain is reconstructed on its own, again, using the greedy algorithm. It reconstructs in the same manner so as to bypass the dead node [31]. Since the leader of the chain will be at a random position in each round, therefore, it is normal for nodes to die at random positions. This helps the system to learn and cope up with system failures.

We could also use the LEACH protocol, but it results in the formation of clusters, increasing the energy consumption.

Although percentage of node deaths is higher in PEGASIS when compared to LEACH, it is still considered the most suitable protocol for traffic density calculation.

Decision-making Technique: The values obtained, till now, are recorded in an Excel format.

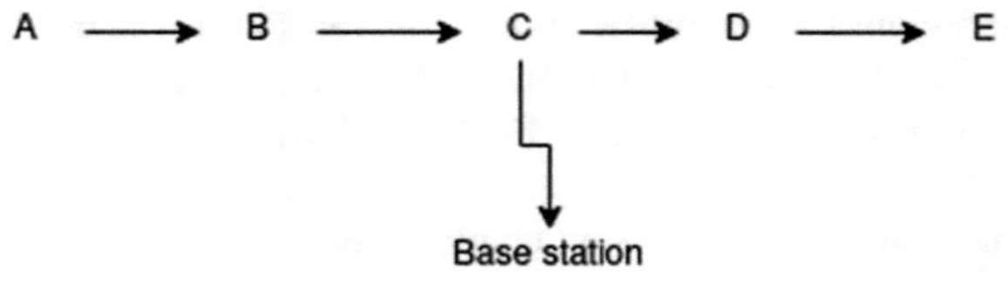

Fig. 3 Data transfer in PEGASIS

3.2   Smart Traffic Light Scheduling Algorithms

Conventional Approach: Conventionally, traffic lights operate on fixed time cycles to manage traffic. Determining when and how to change the flow of traffic is solely based on timed cycles.

STLSD: The authors in [3] proposed a smart traffic light scheduling algorithm based on traffic density, somewhat similar to the algorithm discussed in [11]. An intersection is perceived as a combination of eight lanes, each containing a "waiting zone." Vehicles located in any waiting zone are yet to pass the intersection and hence are waiting for green signal. Apart from these waiting zones, there are two lists: horizontal density list and vertical density list, which represent the traffic density in horizontal and vertical directions, respectively. Each list covers a total of four lanes. Based on where a vehicle is located, it is added to one of these lists. The lists are then compared by the system to verify to make sure that the list with the highest count has a green signal. If yes, no action needs to be taken. Otherwise, the algorithm transfers the flow of traffic to another lane which has the highest traffic density. After each round of calculation, both the lists are emptied before repeating the entire process. This algorithm, however, does not address the problem of abnormally long waiting time for a vehicle.

STLSDT: An improved version of the STLSD algorithm discussed in [3] is the STLSDT algorithm, which is based on traffic densities but also takes into account the waiting period of each vehicle. After determining which lane is to be awarded the green signal as discussed previously, the algorithm observes the waiting period of the vehicles in other lanes. If the waiting period of vehicles in a particular lane exceeds some predefined threshold, then a green signal is provided to that lane to allow the waiting cars to move. This approach helps in minimizing the waiting period of the vehicles at an intersection.

STLSDE: Further improvements in the initial STLSD algorithm discussed in [3] led to the STLSDE algorithm. This approach builds on the STLSDT approach but also takes into consideration the presence of emergency vehicles in any of the lanes. When an emergency vehicle is detected in any of the eight waiting zones, the algorithm provides a green signal to that lane if it already did not have a green signal. This helps in ensuring that an emergency vehicle has green signal at all times. In case there are emergency vehicles in two adjacent lanes, the lane with the green signal continues to have it till the emergency vehicle crosses the intersection. Once this is done, a green signal is provided to the other lane for the emergency vehicle to pass.

4 Conclusion

This paper presents a combination of PEGASIS and STLS to manage traffic lights in a way so as to yield better scheduling and lesser congestion at intersections of lanes. The proposed system overcomes the shortcomings of the models used previously and presents a better approach toward traffic management. This combination of PEGASIS protocol and STLS algorithms proves to be a promising solution for the problem. Both these techniques, though tested independently, were never combined together. This paper utilized the combined efficiency of the techniques and achieved better results than previously used methods. The algorithm has been tested on simulation software SUMO discussed in [32]. The simulation results affirmed the efficiency of the proposed method.

References

1. TomTomTrafficIndex:Toronto, TomTom.com (2019) [Online]. https://www.tomtom.com/en_gb/trafficindex/city/toronto. Accessed: 9 Aug 2019
2. Bsrth, M., Boriboonsomsin, K.: Traffic congestion and greenhouse gases. Access Magazine (2009)
3. Rezgui, J., Barri, M., Gayta, R.: Smart traffic light scheduling algorithms. In: 2019 International Conference on Smart Applications, Communications and Networking (SmartNets), Sharm El Sheikh, Egypt, pp. 1–7 (2019). https://doi.org/10.1109/SmartNets48225.2019.9069760
4. Vachan, B.R., Mishra, S.: A user monitoring road traffic information collection using SUMO and scheme for road surveillance with deep mind analytics and human behavior tracking. In: 2019 IEEE 4th International Conference on Cloud Computing and Big Data Analysis (ICCCBDA), Chengdu, China, pp. 274–278 (2019). https://doi.org/10.1109/ICCCBDA.2019.8725761
5. Perafan-Villota, J.C., Mondragon, O.H., Mayor-Toro, W.M.: Fast and precise: parallel processing of vehicle traffic videos using big data analytics. IEEE Trans. Intel. Transp. Syst. https://doi.org/10.1109/TITS.2021.3109625
6. Jonnalagadda, M., Taduri, S., Reddy, R.: Real time traffic management system using object detection based signal logic. IEEE Appl. Imagery Pattern Recogn. Workshop (AIPR) **2020**, 1–5 (2020). https://doi.org/10.1109/AIPR50011.2020.9425070
7. Zeeny, M.: Multiple Criteria Decision Making. McGraw-Hill, New York (1982)
8. Culler, D., Estrin, D., Srivastava, M.: Overview of sensor networks. IEEE Comput. **37**(8), 41–49 (2004)
9. Khan, M.W., Zeeshan, M., Usman, M.: Traffic scheduling optimization in cognitive radio based smart grid network using mini-batch gradient descent method. In: 2019 14th Iberian conference on information systems and technologies (CISTI), Coimbra, Portugal, pp. 1–5 (2019). https://doi.org/10.23919/CISTI.2019.8760693
10. Kanungo, A., Sharma, A., Singla,C.: Smart traffic lights switching and traffic density calculation using video processing. In: 2014 Recent Advances in Engineering and Computational Sciences (RAECS), Chandigarh, India, pp. 1–6 (2014). https://doi.org/10.1109/RAECS.2014.679954
11. Bani Younes, M., Boukerche, A.: An intelligent traffic light scheduling algorithm through VANETs. In: 39th Annual IEEE Conference on Local Computer Networks Workshops, Edmonton, AB, Canada, pp. 637–642 (2014). https://doi.org/10.1109/LCNW.2014.6927714
12. Kumar, N., Rahman, S.S.: Deep reinforcement learning with vehicle heterogeneity based traffic light control for intelligent transportation system. In: 2019 IEEE International Conference on

Industrial Internet (ICII), Orlando, FL, USA, pp. 28–33 (2019). https://doi.org/10.1109/ICII. 2019.00016

13. Yu, J., Yan, Y., Chen, X., Luo, T.: Short-term road traffic flow prediction based on multi-dimensional data. In: 2021 International Conference on Intelligent Transportation, Big Data & Smart City (ICITBS), pp. 43–46 (2021). https://doi.org/10.1109/ICITBS53129.2021.00019

14. Kimovski, D., et al.: Cloud—edge offloading model for vehicular traffic analysis. In: 2020 IEEE International Conference on Parallel & Distributed Processing with Applications, Big Data & Cloud Computing, Sustainable Computing & Communications, Social Computing & Networking (ISPA/BDCloud/SocialCom/SustainCom), pp. 746–753 (2020). https://doi.org/10. 1109/ISPA-BDCloud-SocialCom-SustainCom51426.2020.00119

15. Firdous, A., Indu, Niranjan, V.: Smart density based traffic light system. In: 2020 8th International Conference on Reliability, Infocom Technologies and Optimization (Trends and Future Directions) (ICRITO), Noida, India, pp. 497–500 (2020). https://doi.org/10.1109/ICRITO 48877.2020.9197940

16. Kumar, N., Rahman, S.S., Dhakad, N.: Fuzzy inference enabled deep reinforcement learning-based traffic light control for intelligent transportation system. IEEE Trans. Intel. Transp. Syst. https://doi.org/10.1109/TITS.2020.2984033

17. Hawi, R., Okeyo, G., Kimwele, M.: Smart traffic light control using fuzzy logic and wireless sensor network. In: 2017 Computing Conference, London, UK, pp. 450–460 (2017). https:// doi.org/10.1109/SAI.2017.8252137

18. Liang, X., Du, X., Wang, G., Han, Z.: A deep Q learning network for traffic lights' cycle control in vehicular networks. IEEE Trans. Veh. Technol. **68**, 1–1 (2019). https://doi.org/10.1109/TVT. 2018.2890726

19. Zeng, J., Hu, J., Zhang, Y.: Training reinforcement learning agent for traffic signal control under different traffic conditions. In: 2019 IEEE Intelligent Transportation Systems Conference (ITSC), Auckland, New Zealand, pp. 4248–4254 (2019). https://doi.org/10.1109/ITSC.2019. 8917342

20. Zhao, T., Wang, P., Li, S.: Traffic signal control with deep reinforcement learning. In: 2019 International Conference on Intelligent Computing, Automation and Systems (ICICAS), pp. 763–767 (2019). https://doi.org/10.1109/ICICAS48597.2019.00164

21. Manasi, P.S., Nishitha, N., Pratyusha, V., Ramesh, T.K.: Smart traffic light signaling strategy. In: 2020 international conference on communication and signal processing (ICCSP), 2020, pp. 1200–1203. https://doi.org/10.1109/ICCSP48568.2020.9182165

22. Wang, F.-Y., Agent-based control for networked traffic management systems. IEEE Intell. Syst. 20(5), 92–96 (2005). https://doi.org/10.1109/MIS.2005.80

23. Broustis, I., Faloutsos, M.: Routing in vehicular networks: feasibility, modeling, and security. Int. J. Veh. Technol. **2008**,(2008). https://doi.org/10.1155/2008/267513

24. Georgy, J., Noureldin, A., Korenberg, M., Bayoumi, M.: Modeling the stochastic drift of a MEMS-based gyroscope in Gyro/Odometer/GPS integrated navigation. IEEE Trans. Int. Transp. Syst. **11**, 856–872 (2011). https://doi.org/10.1109/TITS.2010.2052805

25. Blum, J.J., Eskandarian, A., Hoffman, L.J.: Challenges of intervehicle ad hoc networks. IEEE Trans. Intell. Transp. Syst. 5(4), 347–351 (2004). https://doi.org/10.1109/TITS.2004.838218

26. Jerbi, M., Senouci, S., Rasheed, T., Ghamri-Doudane, Y.: Towards efficient geographic routing in urban vehicular networks. IEEE Trans. Veh. Technol. **58**(9), 5048–5059 (2009). https://doi. org/10.1109/TVT.2009.2024341

27. Zhao, J., Cao, G.: VADD: vehicle-assisted data delivery in vehicular ad hoc networks. IEEE Trans. Veh. Technol. **57**(3), 1910–1922 (2008). https://doi.org/10.1109/TVT.2007.901869

28. M. Franceschinis, L. Gioanola, M. Messere, R. Tomasi, M.A. Spirito, P. Civera, Wireless sensor networks for intelligent transportation systems. In: VTC Spring 2009—IEEE 69th Vehicular Technology Conference, Barcelona, Spain, pp. 1–5 (2009). https://doi.org/10.1109/VETECS. 2009.5073915

29. Heo, J., Hong, J., Cho, Y.: EARQ: energy aware routing for real-time and reliable communication in wireless industrial sensor networks. IEEE Trans. Industr. Inf. **5**(1), 3–11 (2009). https:// doi.org/10.1109/TII.2008.2011052

30. Heo, J., Hong, J., Cho, Y.: EARQ: energy aware routing for real-time and reliable communication in wireless industrial sensor networks. 2009 IEEE Trans. Indus. Inform. **5**(1), 5–6. https://doi.org/10.1109/TII.2008.2011052
31. Guo, W., Zhang, W., Lu, G.: PEGASIS protocol in wireless sensor network based on an improved ant colony algorithm. In: 2010 Second International Workshop on Education Technology and Computer Science, pp. 2–4. https://doi.org/10.1109/ETCS.2010.285
32. Elidrissi, H.L., Tajer, A., Nait-Sidi-Moh, A., Dakkak, B.: A SUMO-based simulation for adaptive control of urban signalized intersection using petri nets. In: 2019 4th World Conference on Complex Systems (WCCS), Ouarzazate, Morocco, pp. 1–6 (2019). https://doi.org/10.1109/ICoCS.2019.8930774
33. Oliveira, L.F.P., Manera, L.T., Luz, P.D.G.: Smart traffic light controller system. In: 2019 Sixth International Conference on Internet of Things: Systems, Management and Security (IOTSMS), Granada, Spain, pp. 155–160 (2019). https://doi.org/10.1109/IOTSMS48152.2019.8939239

A Cluster-based Data Aggregation Framework for WSN using Blockchain

Arabind Kumar, Sanjay Yadav, Vinod Kumar, and Jangirala Srinivas

Abstract Data aggregation is an important process for wireless sensor networks (WSNs). The important aspects with WSNs are to increase the lifetime and to reserve the battery power of sensor nodes. The energy of nodes is mainly utilized in the process of sensing and transmitting the data. Clustering is one of the processes in which the total WSNs are divided into the form of groups and each group has a cluster head (CH) and advantage of doing this is decrease of energy consumption and reduce the data collision. The blockchain technology is utilized in WSN by so many ways for example security, data storing, node recovery etc. because the role of blocks in blockchain is same as the role of node in WSN. In this paper, we propose a cluster-based data aggregation framework for wireless sensor networks based on the blockchain technique. The proposed scheme is based on cluster head (CH) selection based on energy, and nodes communicate with each other via the shortest path. Further, the proposed scheme is also using the concept of blockchain technology for the purpose of maximum data storage.

Keywords Wireless sensor networks · Low energy adaptive clustering hierarchy (LEACH) · Data aggregation · Cluster head · Blockchain · And Energy consumption

A. Kumar · S. Yadav
Department of Applied Sciences, The NorthCap University, Gurugram 122017, India
e-mail: sanjayyadav@ncuindia.edu

V. Kumar (✉)
Department of Mathematics, PGDAV College, University of Delhi, New Delhi 110065, India
e-mail: vinod@pgdav.du.ac.in

J. Srinivas
Jindal Global Business School, O. P. Jindal Global University, Haryana 131001, India
e-mail: sjangirala@jgu.edu.in

© The Author(s), under exclusive license to Springer Nature Singapore Pte Ltd. 2022   661
V. Suma et al. (eds.), *Evolutionary Computing and Mobile Sustainable Networks*,
Lecture Notes on Data Engineering and Communications Technologies 116,
https://doi.org/10.1007/978-981-16-9605-3_43

1 Introduction

WSNs are self-configured and infrastructure-free wireless networks that monitor physical or environmental conditions such as vibration, pressure, motion, temperature, and sound pollutants and cooperatively transmit their data through the network to a central location or sink where the data can be observed and analyzed. A sink, also known as a base station, connects users to the network. One can retrieve information from the network by injecting queries and gathering results from the sink. Data transmission accounts for 70% of total energy consumption. As a result, the data transmission process should be optimized in order to maximize network lifetime. Data transmission can be optimized by using efficient routing protocols and effective data aggregation methods [1, 2].

Data aggregation is the method of gathering useful information from various sensor nodes and delivering it to the BS after removing redundant data. Since the working sensor nodes are totally dependent on energy, it is impossible for all the sensors to send the information in the form of data direct to the BS. Data sensed between neighboring sensors is some time trivial and highly similar. In integration, the total data engendered in sizably voluminous sensor networks is conventionally in huge amount for the BS for further analysis. As a result, the entire process necessitated approaches for coalescing data into excellent evidence at the sensor level or between nodes, thereby reducing the number of packets sent to the BS and saving energy and bandwidth [3]. In data centric routing, data aggregation is a strategy used to tackle the implosion and overlap problems. When data from many sensor nodes reaches the same routing node on the way back to the sink, they are aggregated as though they are about the same characteristic of the phenomena. In WSNs, data aggregation is a common strategy. When a sensor network is deployed in a hostile environment, security considerations such as data confidentiality and integrity in data aggregation become critical. The process of aggregating sensor data using aggregation algorithms is known as data aggregation [4].

In particular sectors of the Indian economy, like as banking and indemnification, voluntary retirement plans are scrutinized visually. It is desirable to establish the best time to withdraw a voluntary retirement scheme, while balancing the minimizing of high emolument for a section of employees, avoiding mass voluntary turnover from the grade, achieving maximum productivity, and assuring mundane many operations. An approach to derive an optimal policy to withdraw a voluntary retirement scheme is discussed in this paper, taking into account the probability of a retirement request being accepted, the cost of promulgating a voluntary retirement scheme, and the cost to the organization of one-time special payments to those who retire during the duration [5]. The device with the ability to work without the need of wires, as well as sensing units, processing, and decision-making capabilities, is combined with tangible world items to create the optimal accommodation distribution for the cyber world of things. Integrating a wide range of devices, from sensors to intelligent equipment, is the key to gaining access to the network and its vast resources. By extending communication via the contrivances of the cessation users, consumers were granted an unrestricted, global right to use resources everywhere in the world. By connecting clients to outside networks and accommodations through adaptable conveyance methodologies, the equipment facilitated a wide range of applications [6]. As computer technologies and the Internet of Things evolve, more data is captured from the environment, which requires proper maintenance and protection. However, the information acquired will only be beneficial if it is appropriately used to decision-making. With the use of data mining, the connection and impact of various data can be formed, paving the way for the proper construction of relationships between data to provide information to data consumers [7].

1.1 Problem Statement

The darning of battery power reduces the lifetime of a sensor node in a WSN. The main challenges are to maximize the working time of each sensor devices so that the sensing and transmitting of data is also maximized at the cluster level and reduce overall energy consumption for network. The secure data storage is also a main issue in WSN which can be improved by using blockchain technology and also sensor nodes are received too much redundant data so that data fusion and combing is required in the cluster.

1.2 Motivation

WSN is an integral part of coming technology era and plays a very significant role of data analysis. At present there are various approaches for aggregating the data in WSNs and hence there is a need for good algorithms for the same. Hence, this research involves clustering-based data aggregation algorithms and protocols on all possibilities. The required theoretical description has been identified after taking improved work of data aggregation in WSNs. We are proposing a scheme on collecting of data in which energy consumption will be less.

1.3 Contribution

The main contributions of this study include:

- The work performs data aggregation at bit level which produces aggregation data as significance of the entire bit are included in the data.
- We propose a new scheme to select the best CH in each cluster on the aspect of energy efficiency of nodes.

- To explore the trend of study of this arena, we discussed a theoretical overview of clustering methods and communication attributes reinforced by literature.
- We provide the performance analysis of the proposed scheme and comparisons with other algorithms.

1.4 Related Work

Low Adaptive Clustering Hierarchy is the first cluster-based algorithm for WSNs that has been proposed (LEACH). It shares the process into rounds and arbitrarily picks new CHs in every round to allocate the power burden among all nodes. Several algorithms improve power distribution by forming equal clusters. Hybrid Energy Efficient Distributed (HEED) recurrently chooses CHs and set up equal clusters just as a mixture of the remaining power of the nodes and subsequent variables. In general, there are two main phases of Clustering that are assembling nodes and allotting responsibilities [8]. Low Energy Adaptive Clustering Hierarchy (LEACH) imparts an incipient technique to explain the lifespan of the sensor that contains all the nodes consisting of the same amount of power and the possibility of setting off as a CH; subsequently after the first round, the leftover power will diverge and further the CH would supply the respective nodes along with more of the leftover power. Therefore, the node decides on their own about being CHs and further when it is not compulsory to attach it with the BS. The HEED protocol culls cluster heads predicated on their remaining power and node point. The node point avails maintaining the charge among the cluster heads. In the following protocol, the process of clustering is transferred out in label of succession and in every succession, the nodes are not further protected by any CH that duplicates their chances of becoming a CH. It has a reasonable projecting in terms of preparing cycles and text swapping. This agreement doesn't include any dissemination of nodes or position vigilance. Additionally, it involves equitably consistent CH dissemination all over the network and protracts the lifespan of the network apart from fortifying data collection [9]. The Tree—Clustered Data Accumulating Protocol (TCDGP) forms a mix up of tree and cluster predicted point of view. This protocol forms collection on the substructure of position and power details regarding the sensor nodes. This helps in developing a track in the minimal terming tree utilizing the Prim Algorithm. The sink helps in enumerating the extent between the CHs to form a tree in lieu of CHs that reduce the computed expenses of the CHs. This formation predicated data compilation protocol preoccupies paramount amplitude of power in the course of the establishment of network formation. The HEERTP is a hierarchical cluster predicated convey protocol that reduces power utilization by reducing observed unessential data transference along with the cooperation of the BS [10–12].

1.5 The Paper Organization

The following is a breakdown of the paper's structure. Section 2 defines some basic terms and explains some security requirements. Section 3 presents a detailed work of clustering methods for WSNs in terms of network properties and the background of blockchain in WSN. Section 4, describes the proposed scheme and in Sect. 5, performance analysis of algorithm is explained and followed by conclusion and future work.

2 Some Basic Definition and Security Requirement

2.1 Cluster-Based Approach in WSN

The role of data collection algorithms or protocols is to remove redundant data and collect only the WSN. Data transmission in WSNs is based on a multi-hop method in which every node directs its sensed data to a neighboring node that is closer to the trough. The clustering policy is repeated and forwarded by the adjacent node. However, because every node in the network is involved in the process of transferring and collecting data from many sinking sources, the process wastes more energy. Because randomly placed nodes may sense similar data, the above protocol cannot be considered energy efficient. Clusters will be an improvement over the above protocol in that each node forwards data to CH which accomplishes an aggregation procedure on the received metadata and sends it to sump. Performing the collect function on CH still causes significant energy waste. If there is a homogeneous sensor network CH it will be finished soon and the reassembly will have to be done again which again results in the power consumption. As a result, we presented an algorithm that aggregates data within a group and thus truncates the aggregate load at CH in order to maximize network life [4, 13] (Fig. 1).

In general, the research community widely applied clustering of nodes into clusters to achieve the above scalability goal as well as overall high energy efficiency and extended network life in astronomically massive WSN locations. The hierarchical transmitting and related data accumulation procedures indicate the block-based organization of sensor nodes in order to combine and aggregate data, resulting in significant energy savings. Each hierarchical network architecture has an amniotic cluster, called CH that usually performs the aforementioned special functions, as well as several regular sensor nodes as members [14].

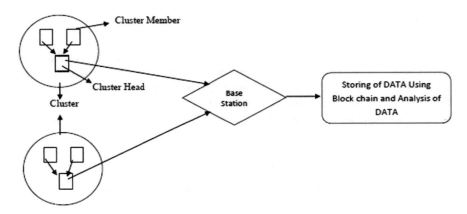

Fig. 1 Cluster approach in WSN using blockchain

2.2 Blockchain in WSN

In a WSN, node storage capacity is limited, and data storage capacity is likewise a valuable resource. In WSN, blockchain technology is employed to store data in the form of blocks. The data set that each node stores is termed a block of the blockchain in the network architecture, and nodes that keep more data get higher rewards. As a result, network nodes' storage space is considerably reduced [15].

Blockchain is observed to be one of the most primitive and ingenious transformations these days and it can also be applied to resolve the issues of double spending attack and to continue the structure of the agreement. It carries on with the progress in such a procedure that will help in developing the system we link, detects and marks payments among the members. Blockchain was originally established for electronic negotiations. The Blockchain innovation come up with clarity, distrustful yet assured agreement in the dispersed system that helps out in achieving vigorous and scrutinized documentation of entire agreement [16] (Fig. 2).

So accordingly, Blockchain is applicable to contrasting structure for Wireless Senor Networks (WSNs) like including bright commitment that consist of lines of rules of an agreement and implement the terms and conditions of the agreement among the two members. Blockchain in WSNs is really challenging as node in a sensor network possess restricted capacity of cells which in turn causes the restricted number of batteries thus, the entire network is bounded. The dependency of power utilization builds upon the group of transferred, collected, and prepared details, data accession from the nodes of the sensor network and some additional component. The major matter of WSN is the disclosing of secrecy. This secrecy disclosing issue results in reduced number of end users and incorrect details are uploaded to secure their privacy data. The nodes of WSN and digital phones contain a restricted amount of schemes. The utilization of Blockchain network in the network of sharp sensor is unfeasible because of the reason that sensors are system restricted devices and

Fig. 2 Flowchart for the
proposed technique

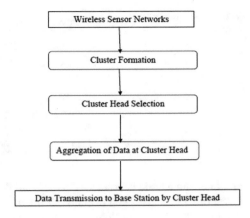

do not contain sufficient data processing energy to execute complex algorithms. A permanent relation between the nodes is required for the blockchain network, which is not power efficacious. Therefore, the blockchain without the power to perform is needed that permits the limited connectivity along with so as to combat inaccuracy and unfavorable actions [17, 18]

3 Proposed Scheme

The cluster formation methods in WSN have a substantial effect on increasing the network life time. The information in the working reason is collected by the cluster nodes and sent to the CH. The saved data is then sent to the BS by the CH. The energy calculation is used for the proposed scheme for the communication of data. Both the data transmitter and data aggregator are separated with each other by a distance d. The equations for the transmission and reception of data are given as [19]:

$$E_{\text{trans}}(k, d) = \begin{cases} kE_{\text{elec}} + k\varepsilon_{\text{efs}}d^2 & d < d_0 \\ kE_{\text{elec}} + k\varepsilon_{\text{amp}}d^4 & d \geq d_0 \end{cases} \tag{1}$$

$$E_{\text{receive}}(k) = kE_{\text{elec}} \tag{2}$$

$$d_0 = \sqrt{\frac{\varepsilon_{\text{efs}}}{\varepsilon_{\text{amp}}}} \tag{3}$$

Table 1 explains the symbols used in the above equation of proposed scheme along with the meaning. The proposed scheme is similar to LEACH in that it has two platforms: setup and steady; however, the setup stage differs in that during the setup stage, all regular nodes choose a random number. The setup stage differs in that during the setup stage, all regular nodes choose a random number between zero

Table 1 Parameter of equation

Symbol	Meaning
d	Distance between receiver and transmitter
k	Denotes the data bite
E_{trans}	Energy required for transmission
$E_{receive}$	Energy used in receiving data
E_{elec}	Data needed for sending a data bit
ε_{efs}	Amplification coefficient
ε_{amp}	Energy for amplification

and one, and a node becomes a cluster head if that number is less than or equal to the threshold $T(n)$ calculated using the following equation; otherwise, the node remains ordinary [8, 9] number between zero and one, and a node becomes a cluster head if that number is less than or equal to the threshold $T(n)$ calculated using the following equation; otherwise, the node remains ordinary [8, 9].

$$T(n) = \begin{cases} \frac{p}{1-p\left[r*\mathrm{mod}\left(\frac{1}{p}\right)\right]} & n \in G \\ 0 & \text{Otherwise} \end{cases} \tag{4}$$

where p denotes the percentage of nodes chosen as CHs, r denotes the current round, and G denotes the set of nodes that were not chosen as CHs in the previous $1/p$ rounds.

The blockchain is a peer-to-peer disseminated archive in which records called blocks are stored which are linked with node and secured by decentralizing data storage, we greatly improve the security of the data. Data storage on nodes works faster and secures access for the users. The distance between a node and a CH, as well as the distance between a CH and BS, is calculated using the equation below, and the shortest distance is chosen.

$$D = \sqrt{(x_1 - x_2)^2 + (y_1 - y_2)^2} \tag{5}$$

The proposed scheme using heterogeneous cluster-based WSNs. In the scheme of data aggregation in WSN, we follow three steps:

i.   Selection of CH with maximum energy node inside a cluster.
ii. Transmission of data by other nodes of cluster to CH by the shortest path.
iii. Data stored at CH in the form of block (packet) using blockchain and transmitted in the form of blocks to the base station by shortest path.

Proposed algorithm

```
Notation:
N: Total Number of Sensor Nodes Randomly Distributed
CH: Cluster Head
CM: Cluster Member
BS: Base Station
D: Euclidean Distance
E: Energy of Sensor Node
Declare node_id = {0, 1, 2, 3....N} for every sensor node
For Every Node [N]
        If (E = Maximum)
                N Selected as CH
                N changes its status to CH
                N send signal to all node about its CH status
        Else
                N become a ordinary node
                N receive a signal from CH
        End If
  For Every (CH)
        Calculate D from N to CH
        N choose the CH with minimum D
        N become CH
  End For
For Every Node [N]
        N stored sensed data
        N transmit data to CH
End For
For Every (CH)
        CH receives data from CH
        CH aggregate total data and compress it
        CH transmit data to BS via shortest path
End For
```

Step-(i) To cull the best CHs, clustering methods can be habituated to centralize and to distribute CH cull methods. In canalized techniques, the parameters used to cull CHs are collected in a central node (typically BS) and associated, analyzed, and processed for culling CHs. Though centralized approaches can yield macrocosmic results by comparing all nodes, they frequently have a high overhead for large and/or dynamic networks, especially when CHs are re-culled on a conventional substructure. In such networks, several organization packets must be exchanged, which can ingest a lot of possessions and decrease network productivity. Distributed methods, on the other hand, have lower overhead; however, due to their restricted system indication, culled CHs cannot always meet all network requirements.

The CH can be culled desultorily or predicated on some criteria. Cull of cluster head largely affects the life time of WSNs. Ideal cluster head is the one which has highest energy, the maximum number of neighbors' nodes and the most minute distance from the base station. Clustering is one of the most widely used topology management techniques for WSNs. The clustering procedure classified nodes into a number of cluster collections. Each cluster is led by one or more CHs, who gather data from the cluster's nodes. Cull the CH from among the cluster nodes with the most energy; if a node with the most energy is culled as Cluster head, the lifetime of the entire cluster is increased, and receiving and transmitting data is also more efficient than other methods.

Step-(ii) In previous step CH is selected, a node with maximum energy is CH in proposed scheme. The main role of CH is to obtain data after each sensor nodes inside a cluster for this process if in each cluster a sensor node will follow the minimum path algorithm or find the shortest path to send sensed data to cluster head by following this method nodes energy consumed and working time of a sensor is more.

Step-(iii) Node data is stored in the form of blocks and is transported either in its original configuration or as a merged value. To fuse data, various techniques are used, generally on CHs. The CH can send separate data matters to the BS or aggregate them and conduct the resulting data values to the BS. The final step in aggregation of data in WSNs is to send all the data transmitted to base station by CH. For this if CH sends stored data to BS by shortest path, applying this approach the CH performs more data transmission and number of rounds is also increased.

4   Performance Analysis

The following is the main feature of the proposed scheme:

(i)   Working capacity of sensor nodes will live longer in the proposed scheme as they will be performed a greater number of rounds. The time it takes for a WSN to operate from the start to the death of the first node. This indicates that the network is in a stable state.

(ii)   The data transmission will also be increased in the proposed scheme. The number of packets received by the BS while the network is stable.

(iii)   Energy consumption per round was less in the proposed scheme as node follows the minimum path for the transmission of data. The total amount of energy consumed in a stable network as well as the amount of energy used in a single round.

(iv)   Since the aggregated data was stored at CH in every cluster in the form of blocks so that in the proposed scheme the storage of data is maximized as well as secure.

5   Conclusion and Future Directions

In this study, we described a new scheme for gathering information from multiple sources, such as sensors that are based on cluster formation methods that reduce energy consumption by dividing work among nodes at different time intervals. This scheme has performed better as the selected CH is with high energy so that receiving and transmission process increase. At certain time periods, each cluster node has to transfer sense data to CH and at CH the information is combined and fused and finally sends BS for further analysis. When the data from source node is sent to sink through neighbors' nodes in a multi-hop fashion by reducing transmission and receiving power, the energy consumption is low as compared to that of sending data directly to sink that is aggregation reduces the data transmission then the without aggregation. Dividing total energy among network nodes is beneficial in positions of dropping global energy debauchery and maximizing system generation. In this paper, we have proposed an energy-efficient technique for data aggregation in WSNs. Our scheme has integrated energy-efficient and data storage mechanisms. The scheme can determine the optimal number of clusters for the WSN based on various factors such as network deployment process, environment, and relative cost.

The paper has shown that these techniques not only reduce power consumption but also prolong the lifetime of a network.

The research topic of maximizing the lifetime of sensor node and minimizing the energy consumption will be very useful in the era of technology and many more scope will be available. Furthermore, the proposed scheme will be improved and mathematical formulation can be done and compared with other existing algorithms.

References

1. Rajagopalan, R., Varshney, P.K.: Data Aggregation Techniques in Sensor Networks: A survey. Syracuse University (2006)
2. Liang, H., Yang, S., Li, L., Gao, J.: Research on routing optimization of WSNs based on improved LEACH protocol. EURASIP J. Wirel. Commun. Netw. (2019)
3. Randhawa, S., Jain, S.: Data aggregation in wireless sensor networks: previous research, current status and future directions. Wirel. Pers. Commun. 97, 3355–3425 (2017)
4. Varma, S., Shanker Tiwary, U.: Data aggregation in cluster based wireless sensor networks. Intel. Hum. Comput. Interact. 391–400 (2009)
5. Thilaka, B., Sivasankaran, J., Udayabaskaran, S.: Optimal time for withdrawal of voluntary retirement scheme with a probability of acceptance of retirement request. J. Inform. Technol. 2(4), 201–206 (2020)
6. Shakya, S., Nepal, L.: Computational enhancements of wearable healthcare devices on pervasive computing system. J. Ubiquitous Comput. Commun. Technol. (UCCT) 2(02), 98–108 (2020)
7. Wang, H., Smys, S.: Big data analysis and perturbation using data mining algorithm. J. Soft Comput. Paradigm (JSCP) 3(1), 19–28 (2021)
8. Heinzelman, W.R., Chandrakasan, A., Balakrishnan, H.: Energy-efficient communication protocol for wireless microsensor networks. In: Proceedings of the Proceedings of the 33rd Annual Hawaii International Conference on System Sciences, IEEE, pp. 1–10 (2000)
9. Salem, A.O., Shudifat, N.: Enhance LEACH protocol for increasing a lifetime of WSNs. In: Personal and Ubiquitous Computing, 13 Feb 2019
10. Yue, J., Zhang, W., Xiao, W., Tang, D., Jiuyang, J.T.: Energy efficient and balanced cluster-based data aggregation algorithm for wireless sensor networks. In: International Workshoap on Information and Electronic Engineering (2012)
11. Mohanty, P., Kabat, M.R.: Energy efficient structure-free data aggregation and delivery in WSN. Egypt. Inform. J. 17, 273–284 (2016)

12. Iqbal, A., Amir, M., Kumar, V., Alam, A., Umair, M.: Integration of next generation IIoT with blockchain for the development of smart industries. Emerg. Sci. J. **4**, 1–17 (2020)
13. Reza Farahzadi, H., Langarizadeh, M., Mirhosseini, M., Fatemi Aghda, S.A.: An improved cluster formation process in wireless sensor network do decrease energy consumption. Wirel. Netw. **27**, 1077–1877 (2020)
14. Shahraki, A., Taherkordi, A., Haugen, Q., Eliassen, F.: Clustering objectives in wireless sensor networks: a survey and research directions analysis. Comput. Netw. **180**, 107376 (2020)
15. Yang, J., He, S., Xu, Y., Chen, L., Ren, J.: A trusted routing scheme using blockchain and reinforcement learning for wireless sensor networks. Sensors (2019)
16. Mohanta, B.K., Jena, D., Panda, S.S., Sobhanayak, S.: Blockchain technology: a survey on applications and security privacy challenges. Internet of Things (2019)
17. Ren, Y., Liu, Y., Ji, S., Sangaiah, A.K., Wang, J.: Incentive mechanism of data storage based on blockchain for wireless sensor networks. In: Hindawi Mobile Information Systems, 29 Aug 2018
18. Khan, A.A., Kumar, V., Ahmad, M., Rana, S.: LAKAF: lightweight authentication and key agreement framework for smart grid network. J. Syst. Arch. **116**, 102053 (2021)
19. Handy, M.J., Haase, M., Timmermann, D.: Low energy adaptive clustering hierarchy with deterministic cluster-head selection. In: Mobile and Wireless Communications Network, 2002, 4th International Workshop on IEEE pp. 368–372 (2002))

Evaluating Hash-Based Post-Quantum Signature in Smart IoT Devices for Authentication

Purvi H. Tandel and Jitendra V. Nasriwala

Abstract The Quantum computing era has changed the idea of solving hard problems efficiently through its working mechanisms and algorithms. After the invention of Shor's and Grover's algorithms, existing public key cryptography will not be considered a quantum-safe cryptographic mechanism. Post-quantum cryptography is a promising way to replace vulnerable public key cryptographic primitives. This paper exhibits the implementation of hash-based post-quantum digital signatures on raspberry pi 0, comparing the performance achieved in raspberry pi 3, used in many popular IoT devices.

Keywords Signature scheme · Authentication mechanism · IoT applications · IoT devices · Cryptographic hash function · Raspberry Pi · Post-quantum cryptography · Cryptographic primitives

1 Introduction

Internet of Things has become an integral part of our day-to-day routines. Many of the smart devices are working on the principle of sensing, communicating, and processing architecture of IoT [1]. Applications of IoT device has been spread from industrial sectors to individual smart devices. IoT devices are used in almost all the sectors like agriculture, health care, smart appliances, logistics, energy, finance, manufacturing, and many more [1, 2]. In recent years due to pandemics, IoT applications in health care devices and logistics for such health care devices increased tremendously [3]. Medical devices have their special temperature requirements to be maintained and need to be transmitted from one place to another before their expiry. With the help of IoT devices, all such temperature requirements, locations, leakages

P. H. Tandel (✉)
Department of Information Technology, C G Patel Institute of Technology, Uka Tarsadia University, Bardoli, India

J. V. Nasriwala
Babu Madhav Institute of Information Technology, Uka Tarsadia University, Bardoli, India
e-mail: jvnasriwala@utu.ac.in

© The Author(s), under exclusive license to Springer Nature Singapore Pte Ltd. 2022
V. Suma et al. (eds.), *Evolutionary Computing and Mobile Sustainable Networks*,
Lecture Notes on Data Engineering and Communications Technologies 116,
https://doi.org/10.1007/978-981-16-9605-3_44

can be tracked for different purposes. IoT devices are of great help in many industries like medicine, transport, food, agriculture, smart gadgets.

Authentication is one of the very essential services required in IoT devices. There are chances of identity theft on such smaller smart IoT devices to break or to harm the application system. Taking the particular medicine into consideration which needs a particular temperature to be maintained while it is transported from one end to another end. It is very much essential that smart solutions used in medicine transport will not be compromised by any entities. Hence authentication of such small IoT devices needs to be strongly designed [1, 4, 5]. Earlier public key cryptography was widely used in all such applications to authenticate the devices. Public key cryptography relies on discrete mathematics and integer factorization concepts, which will be broken using quantum attacks [6, 7]. Therefore, researchers have worked on many post-quantum cryptographic techniques like hash-based, lattice-based, code-based, multivariate-based, and Super-singular Isogenies in the past few decades [8].

Among all post-quantum cryptographic techniques, hash-based approaches are more suitable as they are relying on the security of existing hash functions like SHA, BLAKE, etc. As existing hash functions are collision and pre-image resistant, this technique will protect against future quantum attacks [9]. Hash-based techniques use one-time/few-time/many-time signatures, existing hash functions, and tree constructions to provide authentication using digital signatures [10]. Security assumptions under this technique are well understood and standardized by researchers and security institutions [11, 12]. If at all any of the used hash functions is found vulnerable, it can be easily replaced with a different hash function. The hash-based cryptographic technique does not require additional infrastructure like quantum cryptography. On the other side, quantum cryptography requires a specific environment, transmission channels, protocols, quantum machines which are very costly to incorporate in such IoT-based systems.

There are two major categories under hash-based post-quantum signatures: stateful and stateless [12, 13]. Stateless hash-based signature schemes have very larger signature sizes, making the stateful hash-based signature schemes more suitable for constrained devices [13]. We have implemented a stateful signature scheme using XMSS with L-tree as a tree structure and WOTS+ as one-time signatures. We have conducted implementation on raspberry pi 0 and raspberry pi 3 development boards, as they are tiny in size and have enough computational power to perform the general task expected to be performed by IoT devices and any large IoT system as well.

2 Overview of Hash-Based Post-Quantum Signature Schemes

This section provides the necessary cryptographic background to implement a hash-based signature scheme. Initially, Merkle Signature Scheme became popular as the post-quantum hash-based approach. Merkle Signature Scheme (MSS) was using Lamport One-time Signature (LD-OTS) at leaf level to create the signature on a message. One-Time Signature Scheme produces key pair using which only one message can be signed [9]. If at all more than one message is signed using the same key pair attacker can retrieve and forge the signatures. As shown in Fig. 1c part which also exhibits a basic MSS scheme, where key generation and signature generation as been conducted at leaf level using OTS schemes [10, 11, 14]. Then public key of each node is hashed using the hash function and concatenated at their respective parent level, such a way that finally MSS public key is generated at the MSS root level [15]. For verification of the signature relevant node path and their direct hash concatenated values are given as shown as authentication path in Fig. 1c [9]. Rather than LD-OTS, another promising one-time signature scheme is Winternits One-Time Signature Scheme (WOTS). Winternitz one-time signature scheme gives performance time and signature size trade-off using winternitz parameter w which is a block size to be processed simultaneously [16, 17].

After successful MSS implementations, the study related to their security and performance was carried out by researchers. Later XMSS as an extended version of Merkle Signature Scheme (MSS) is proposed by Buchman et al. in 2011 and was published as an IETF draft [12]. eXtended Merkle Signature Scheme (XMSS) as a tree structure and XMSS is a stateful scheme. XMSS tree structure works almost similar to MSS tree structure, but it includes XORing bitmask at each node level and L-tree structures being used as a leaf node in XMSS tree to enhance the achieved security level before. In XMSS, L-tree structure initially at leaf node level key generation performed using OTS scheme and then public keys are being hashed but now before concatenating hashes of child node at parent node, additionally bitmask being XORed which enhance the security of the overall structure [12, 19, 20]. The final public key is generated at the root level of XMSS tree like the original MSS scheme. With the combination of WOTS as a signature scheme, XMSS provides smaller signature size achieved along with optimized key generation, signing, and verifying time to perform

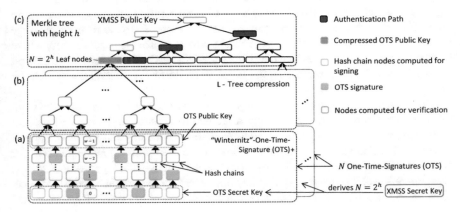

Fig. 1 XMSS tree construction using L-tree and WOTS+ as a one-time signature scheme [18]

the authentication. As these hash-based signature schemes follow tree structure key generation is the costlier process in terms of time parameters. Also, the signature size is being always a challenge in this category of signature schemes.

XMSS is proven forward secure and existentially unforgeable under chosen message attacks with minimal security requirements like pseudorandom function (PRF) and pre-image resistant hash function [12]. Due to minimal security assumptions and security is well understood and studied, XMSS has become the most promising post-quantum signature scheme [21]. One more important component of the scheme is a one-time signature scheme to sign individual messages. WOTS+ is a type of one-time signature that allows you to reduce the signature size more than previous versions of WOTS and also provides a higher level of security [16, 22, 23]. The detailed structure of XMSS with WOTS+ has been shown in Fig. 1.

WOTS+ is used at the leaf level. WOTS+ performs key generation and derives the 2^h singing and verification key pairs. On public key generated by the WOTS+, L-tree structure has been applied to construct the XMSS tree as shown in Fig. 1b. XMSS tree produces one public key, where there are 2^h verification keys. XMSS scheme is considered a stateful signature scheme because the signer needs to remember the state that which key has been already used as WOTS+ is a one-time signature scheme [22].

3   Implementation Scenario and Challenges in IoT Applications

Implementation Scenario: Here, as an experiment of a successful post-quantum signature scheme on IoT devices, XMSS with L-tree implemented as a tree structure, SHA-256 as a hash function, WOTS+ as a one-time signature scheme and pseudorandom function has been used. We have implemented the scheme for tree height 3 to 13, which means a total number of signatures are ranging from 8 to 8192 signatures. Although the scheme is capable of incorporating up to 65,536 and more users. Here experiments have been carried out with winternitz parameter 16, message output size 256 using SHA-256. As the XMSS structure provides b bit security where b is the output size of the used hash function, implemented scheme provides a 256-bit security level. Figure 2 represents the block diagram of the general authentication process in IoT applications.

The scenario is applicable to all large-scale user applications where users want to access their smart devices, initially a request to access the device will be sent over web application to the internet will reach to IoT gateway and then to an actual lower-level IoT device. Now small IoT device needs to perform authentication for which it will request the signing key from the IoT gateway. IoT gateway performs key generation and sends the signing key back securely to the requested IoT device. Now signature will be generated there and signature sent back to IoT gateway. IoT

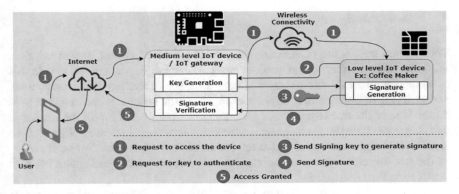

Fig. 2 Block diagram of the authentication scenario in IoT applications

gateway verifies the signature, if it matches then access is granted for the device else access denied will be returned to the user.

Implementation Setup: The experiment was carried out on both raspberry pi 0 and raspberry pi 3 processor boards as medium-level IoT devices. As explained in Fig. 2, medium-level IoT devices will perform most tedious task key generation of authentication and later on signature verification in the entire authentication process. Hence raspberry pi is the most favorable board as the IoT gateway or as a medium-level IoT server/controller. Raspberry pi 3 is a 1.2 GHz quad-core ARMv7(64 bit) with 1 GB RAM configurations [24]. On the other side, raspberry pi 0 is a 1 GHz single-core ARMv6 CPU with 512 MB RAM, Bluetooth, and wireless connectivity [25]. The majority of the IoT-like applications use processor boards that can effectively process their task associated with the application as well as such basic tasks like connection and authentication via signatures. The dimensions of raspberry pi 0 are 65.0 mm × 30.0 mm × 0.2 mm, which is the smallest processor board with wireless connectivity [25].

Implementation Challenges: When performance-driven authentication schemes like hash-based post-quantum signatures are implemented on constrained devices, challenges like limited memory, low processing capabilities affect the performance of the devices [26]. Such a signature scheme involves a lot of calculations to be performed and many library support to implement the scheme. Compatibility of such libraries and in-built functions is one of the challenges when implementing on such boards. The implemented signature scheme uses a tree structure, which utilizes more memory as the number of signatures increased in the system [26–28]. Due to limited RAM available on the boards, these processor takes more time in key generation, signature generation, and verification compared to other normal implementation setups. Also, signature size needs to be optimized to successfully implement the mentioned signature scheme.

4   Result Analysis

This section covers the observations and results derived during implementation. We have observed the time as well as size parameters. Parameters studied in this implementation are signature size, public key size, key generation time, signing time, and verifying of a signature time per user. As the entire scheme is constructed based on the idea of the tree structure for generating one public key for all the users at a leaf, key generation time is increased as the number of users increased. Signing time is individual sign time at the leaf level, it remains almost similar. For verification, as the tree structure grows as the number of users increased, authentication path verification time increased.

Figure 3 depicts the time taken for key generation for a different number of users in the scheme in both implemented environments. Key generation is carried out at leaf level, which represents all the users in the system and the phase generates the public key and private key pair for each user present in the system. Here in the key generation process, initial seeds are required which are generated using a pseudorandom function (PRF). PRF is one of the best candidates as it generates random output each time regardless of what input value is given. As the graph depicts in Fig. 3, generation time increases exponentially as the number of signatures increased. Because raspberry pi 0 is a single-core ARM processor it takes more time for key generation compared to raspberry pi with a quad-core processor. Due to the tree structure involved in the implemented signature scheme and key generation performed at the leaf node level, key generation time increases exponentially in practice when the number of users increased.

Figures 4 and 5 depict the time taken for signing and verifying individual signatures for a different number of users in the scheme. It is observed that signing time

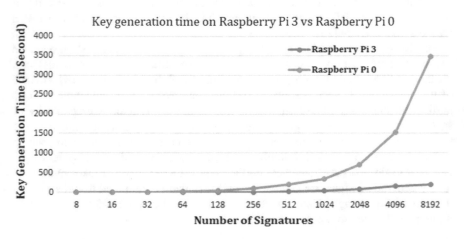

Fig. 3 Key generation time (sec) taken for different numbers of signatures

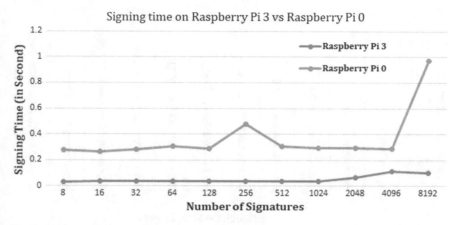

Fig. 4 Message signing time (sec) taken for different numbers of signatures

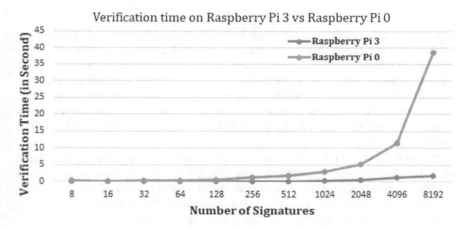

Fig. 5 Signature verification time (sec) taken for different numbers of signatures

per single user is almost constant, it does not have to variate much as the number of the signatures increased in the system.

The reason for the constant time taken for one signature is signature calculated at leaf level for the individual user. Hence the number of users in the system does not have an impact on signing time. But verification of signature requires the authentication path to be followed for verification, so as the number of signatures increased the height of the tree increased resulting in more time taken gradually for verification. Though 38 seconds to verify the signature with an 8192 user system, is perfectly fine for smart IoT applications having more users as displayed in Fig. 5 results for signature verification time as the number of users increased. In general raspberry pi 0 environment takes more time in all three main processes like key generation, encryption, and decryption for authentication. Here the verification time increases

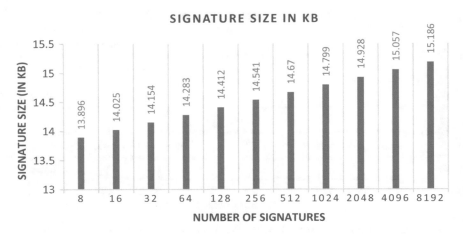

Fig. 6 Signature size derived for various number of signatures

exponentially, due to the authentication path verification procedure. As the users or height of the tree increases more nodes are to be verified for the authentication.

Another important parameter in the implemented scheme is signature size. XMSS with WOTS+ signature has a trade-off between signature size and performance time [21]. As winternitz parameter w increased performance time mentioned in the paper increased but the signature size is reduced. In this implementation winternitz parameter w has been considered 16, to achieve small signature sizes in practical applications in IoT.

Figure 6 shows the actual signature size retrieved for a different number of signatures. It is observed that signature size does not increase drastically or exponentially with the number of signatures. Also in both implementation environments mentioned signature size remains the same as the implemented scheme follows the same hash-based structure to authenticate. For 8 key pairs, the signature size is 13 KB and for 8192 key pairs, it is 15 KB. This depicts that signature size will be sufficiently small even for a large number of signatures, which is an expected performance parameter when we are working with memory-constrained devices for IoT applications.

In comparison with raspberry pi 3, raspberry pi 0 has low computational power, low RAM, tiny in size still, performance-wise raspberry pi 0 can incorporate such heavy computational hash-based post-quantum signature scheme for general authentication purposes. As raspberry pi 0 are tinier in the size many smart applications like supply chain systems, health care devices, automated controls in industries, agriculture, and many more can successfully incorporate them as client/server according to the size of the application which can perform advanced authentication scheme essential before performing the assigned tasks to the boards.

5   Conclusion

Strong authentication schemes for such IoT/embedded devices are of great concern as the need in real life increased tremendously. Hash-based signatures are one of the best options to authenticate tiny and smart IoT devices, as they are complex in nature as well as easiest to incorporate in existing IoT systems. Hash-based signature scheme like MSS with WOTS as one-time signature scheme provides n/2 bit security, while scheme involving XMSS with WOTS+ variant provides n bit security reducing the output size required for the used hash function. XMSS with WOTS+ performs key generation, signature generation, and verification tasks faster than MSS variants. Hash-based Post-quantum signature scheme XMSS with WOTS+ variant is performing well on such devices and has moderate performance parameters, which makes the scheme suitable for IoT applications.

References

1. Suhail, S., Hussain, R., Khan, A., Hong, C.S.: On the role of hash-based signatures in quantum-safe internet of things: Current solutions and future directions. IEEE Internet Things J. (2019)
2. Lee, I., Lee, K.: The Internet of Things (IoT): Applications, investments, and challenges for enterprises. Bus. Horiz. **58**(4), 431–440 (2015), ISSN 0007–6813, https://doi.org/10.1016/j.bushor.2015.03.008
3. Shakya, S., Nepal, L.: Computational enhancements of wearable healthcare devices on pervasive computing system. J. Ubiquit. Comput. Commun. Technol. (UCCT) **2**(02), 98–108 (2020)
4. Abdullah, G.M., Mehmood, Q., Khan, C.B.: Adoption of Lamport signature scheme to implement digital signatures in IoT. In: 2018 International Conference on Computing, Mathematics and Engineering Technologies (iCoMET), pp. 1–4 (2018)
5. Saldamli, G., et al.: Post-quantum cryptography on IoT: Merkle's tree authentication (2018)
6. Shor, P.W.: Polynomial-time algorithms for prime factorization and discrete logarithms on a quantum computer. In: SIAM Review **41**(2), 303–332, (1999)
7. Grover, L.K.: A fast quantum mechanical algorithm for database search. In: Proceedings of the Twenty-Eighth Annual ACM Symposium on Theory of Computing (STOC '96), pp. 212–219. Association for Computing Machinery, New York, NY, USA (1996). https://doi.org/10.1145/237814.237866
8. Niederhagen, R., Waidner, M.: Practical post-quantum cryptography. SIT-TR-2017–02 (2017)
9. Hulsing, A., Gazdag, S., Butin, D., Buchmann, J.: Hash-based Signatures: An outline for a new standard (2014)
10. Becker, G.: Merkle Signature Schemes, Merkle Trees and Their Cryptanalysis, seminar 'Post Quantum Cryptology' at the Ruhr-University Bochum, Germany (2008)
11. Coronado, C.: On the security and the efficiency of the Merkle signature scheme. IACR Cryptology ePrint Arch. **2005**, 192 (2005)
12. Buchmann, J., Dahmen, E., Hülsing, A.: XMSS—A practical forward secure signature scheme based on minimal security assumptions. In: Yang, B.Y. (eds.) Post-Quantum Cryptography. PQCrypto 2011. Lecture Notes in Computer Science, vol 7071. Springer, Berlin (2011). https://doi.org/10.1007/978-3-642-25405-5_8
13. Bernstein, D.J., Hopwood, D., Hulsing, A., Lange, T., Niederhagen, R., Papachristodoulou, L., Schneider, M., Schwabe, P., WilcoxO'Hearn, Z.: SPHINCS: practical stateless hashbased signatures. In: Fischlin, M., Oswald, E. (eds.) Advances in Cryptology, EUROCRYPT 2015, vol. 9056. LNCS. Springer, Berlin, pp. 368–397 (2015)

14. Kampanakis, P., Fluhrer, S.: LMS versus XMSS: Comparison of two Hash-Based Signature Standards (2017)
15. de Oliveira, A.K.D.S., López, J.: An efficient software implementation of the hash-based signature scheme MSS and its variants. In: Lauter, K., Rodríguez-Henríquez, F. (eds.) Progress in Cryptology—LATINCRYPT 2015. LATINCRYPT 2015. Lecture Notes in Computer Science, vol. 9230. Springer, Cham (2015). https://doi.org/10.1007/978-3-319-22174-8_20
16. Naor, D., Shenhav, A., Wool, A.: One-time signatures revisited: Have they become practical. Cryptology ePrint Archive, Report 2005/442 (2005). http://eprint.iacr.org/
17. Campos, F., Kohlstadt, T., Reith, S., Stoettinger, M.: LMS versus XMSS: Comparison of stateful hash-based signature schemes on ARM Cortex-M4, Cryptology ePrint Archive, Report 2020/470 (2020)
18. Pauls, F., Wittig, R., Fettweis, G.A.: Latency-optimized hash-based digital signature accelerator for the tactile internet. In: Pnevmatikatos, D., Pelcat, M., Jung, M. (eds.) Embedded Computer Systems: Architectures, Modeling, and Simulation. SAMOS 2019. Lecture Notes in Computer Science, vol. 11733. Springer, Cham (2019). https://doi.org/10.1007/978-3-030-27562-4_7
19. Wang, W., Jungk, B., Wälde, J., Deng, S., Gupta, N., Szefer, J. Niederhagen, R.: XMSS and Embedded Systems: XMSS Hardware Accelerators for RISC-V (2020). https://doi.org/10.1007/978-3-030-38471-5_21
20. Kannwischer, M.J., Genêt, A., Butin, D., Krämer, J., Buchmann, J.: Differential power analysis of XMSS and SPHINCS. In: Fan, J., Gierlichs, B. (eds.) Constructive Side-Channel Analysis and Secure Design. COSADE 2018. Lecture Notes in Computer Science, vol. 10815. Springer, Cham (2018). https://doi.org/10.1007/978-3-319-89641-0_10
21. Hulsing, A., Busold, C., Buchmann, J.: Forward secure signatures on smart cards. Sel. Areas Crypt (2012)
22. Hülsing, A.: WOTS+—Shorter signatures for hash-based signature schemes. IACR Cryptol. ePrint Arch. 2017, 965 (2017)
23. van der Linde, W.: Post-quantum blockchain using one-time signature chains (2018)
24. https://cdn.sparkfun.com/datasheets/Dev/RaspberryPi/2020826.pdf
25. https://cdn.sparkfun.com/assets/learn_tutorials/6/7/6/PiZero_1.pdf
26. Carneiro, J., Oliveira, L.B.: Evaluating post-quantum signatures for IoT devices. In: Miani, R., Camargos, L., Zarpel~ao, B., Rosas, E., Pasquini, R. (eds.) Green, Pervasive, and Cloud Computing. GPC 2019. Lecture Notes in Computer Science, vol 11484. Springer, Berlin (2019)
27. Pereira, G.C.C.F., Puodzius, C., Barreto, P.S.L.M.: Shorter hash-based signatures. J. Syst. Softw. 116, 95–100 (2016)
28. Rohde, S., Eisenbarth, T., Dahmen, E., Buchmann, J., Paar, C., Fast hash-based signatures on constrained devices. In: Grimaud, G., Standaert, F.X. (eds.) Smart Card Research and Advanced Applications. CARDIS 2008. Lecture Notes in Computer Science, vol. 5189. Springer, Berlin (2008)

Predictive Analysis of Clinical Outcomes Using an Enhanced Random Survival Forest for Heart Failure Patients

E. Laxmi Lydia, Karthikeyan Kaliyaperumal, and Jose Moses Gummadi

Abstract In the Intensive Care Unit cohorts, the monitoring of multiple risk factors and early estimation of mortality in heart failure patients was important to guide the proper decision-making. In this paper, we have established a complete risk model for estimating heart failure death with an enhanced random survival forest (eRSF) with a high degree of certainty. The proposed eRSF was used to classify predictors. Predictors that were more accurate for independent survivors and non-survivors, use a new split rule and stop guideline, and thus reinforce discrimination. 32 risk factors (including population, clinical knowledge, and medicines) were evaluated to develop a risk model for patients suffering from heart insufficiency, based on the available MIMIC II clinical database for 8059 cases. Further effective diagnostic forecasters, including glutamate plasma levels, alanine aminotransferases, complete bilirubin, blood urea nitrogen, serum creatine, and the results that could be shown difficult to control. These have been identified compared to previous studies and established as key indicators for forecasting cardiovascular mortality in eRSF. The observational effects indicate that the risk modeling approach is higher than those reported in previous studies and the classical random forest survival model instead of an out-of-bag C-statistical value of 0.821. The defined functional framework of the eRSF can however serve as a valuable tool in the prediction of mortality by clinicians with heart failure.

Keywords Heart disease · Survival · Risk estimation · Random forest survival · Forecaster

E. L. Lydia (✉)
Department of Computer Science and Engineering, Vignan's Institute of Information Technology (A), Visakhapatnam, Andhra Pradesh, India

K. Kaliyaperumal
IT @ IoT—HH Campus, Ambo University, Ambo, Ethiopia

J. M. Gummadi
Department of CSE, Malla Reddy Engineering College(A), Hyderabad, Telangana, India

© The Author(s), under exclusive license to Springer Nature Singapore Pte Ltd. 2022 683
V. Suma et al. (eds.), *Evolutionary Computing and Mobile Sustainable Networks*,
Lecture Notes on Data Engineering and Communications Technologies 116,
https://doi.org/10.1007/978-981-16-9605-3_45

1 Introduction

Heart failure, a significant cause of death, occurs if the heart is not able to pump the blood to sustain the flow of the body [1]. Around 2% of adolescents in developing countries have heart failure and the elderly aged over 65 have risen to 6–10% [2]. An important way of recognizing vital influences related to adverse results is the prediction of a patient's mortality by using a robust predictive model. This helps physicians to classify the patients needing extensive care and treatment or hospice services in ICU cohorts. For the prediction of cardiac insufficiency mortality [3–9], several risk models arise. The majority of them are focused on conventional risk factors including obesity, diabetes, and past cardiovascular conditions. There is also evolutionary research [10] that has identified areas in Brazil that are at risk for cardiovascular disease deaths. Clinical information including Leucocytes (WBC), blood urea nitrogen (BUN), Aspartate Amino Transferencease (AST), Serum Creatine (SCR), and total bilirubin signify that comorbid conditions are hard to process and therefore also result in high deaths. Such criteria were not extensively researched. Since most experimental findings are known, their standard test to such an extent is hard to classify. For an exact risk analysis that can classify important real variables, a good survival model to manage real variables is vital. Standard quantitative risks (i.e., Cox's maximum likelihood estimation model) in current risk models are low when manually transformed, to classify this type of predictor [11]. Machine learning methods continuously reconcile structural correlations within Big Data variables and response values, which offers a reliable method to enhance the efficiency and output of the identification of important predictors for conventional maximum likelihood estimation approaches [12]. Computer vision, like a minimum of squares, supports the vector machine, decision-making tree, the Bayesian Network, or the rules on associations, provide medical applications [13–17]. Survival trees represent versatility multivariate solutions to parametric or finite mixture modeling since certain interaction forms can be automatically defined and not specified for time-to-event data in advance. Survival forests merge survival trees to create a powerful prediction tool with the optimization technique. As an approximation of the random forest [18], an automated non-linear evaluation of non-linear impacts and potential differences of multi-variables was then formulated as an RSF [17] for non-parametric survival tests [19]. RSF was first reported in risk models on many disease forms, along with heart failure [7] and breast cancer [20]; but, as the observational results indicate, changes are being minimal. Besides, our prior studies have shown that determinants in comparatively fewer populations are poor to classify from our RSF [21, 22]. RSF is limited, for instance, in identifying the better new anti-arrhythmic agent class III. Although its prevention criterion is focused on limited individual deaths (their use was not common in large populations). Yet, since it has shown to be the ideal mechanism to minimize deaths in patients with heart disease, we cannot assume that it is not predictive [23, 24].

We use the latest enhanced RSF to generate a high sensitivity risk model with many more testing results to forecast clinical deaths in ICU patients and to establish appropriate risk factors. This paper consequently presents an enhanced RSF with just a new split rule which can define discriminatory factors that distinguish survivors from non-survivors in the sparse number.

2 Methodology

A. Propertiable Random Survival Forest

In 2008, RSF was suggested for random forest-based survival analysis. For each forest survival tree, a split rule, such as a log-rank test, was applied, which optimizes the difference in survival among daughter nodes. In conventional RSF, under the constraint that a terminal node should have no less than $d_0 > 0$ remarkable deaths every other survival tree is raised to the regular length. Conversely, together under proportional hazards substitute the log-rank test is exponential optimal due to an analogous endorsing sequence observation in the two classes. The stop measure is also arbitrary and shows bias toward determinants with a wider economy. This is because predictors with a relatively small population have trouble fulfilling the standard, particularly when d_0 is high.

Hence, we propose an efficient RSF (iRSF) with a unique split principle and a halt parameter to classify more reliable indicators that can distinguish survivors and non-survivors and therefore also increase the capacity for segregation. Firstly, a modified log-rank test had been used to break the node developed and can be extended to pro-proportional hazard circumstances to enhance the test for a variety of alternative possibilities. We, developers, $d(t)$ become the number of victims, and $Y(t)$ be the potential victims, and t be time. The hazard function calculation $H(t)$ can be described as at a time t with both the Nelson-Aalen assessment tool:

$$H(t) = \frac{d(t)}{Y(t)}, t <= \tau \tag{1}$$

where in our analysis $\tau_0 = 365$, In our analysis the proposed methodology will be used and presented as:

$$H_R(t) = \frac{\theta_1 \theta_2}{\theta_1 + (\theta_2 - \theta_1 Y_L(t))} H_L(t) \tag{2}$$

where $H_R(t)$ and $H_L(t)$ are indeed the corresponding hazard functions of both the right and left branches of the developed tree, $Y_R(t)$ and $Y_L(t)$ are indeed the corresponding survival functions of both the right and left branches, $\theta_1 = \lim_{t \downarrow 0} \frac{H_R(t)}{H_L(t)}$ and $\theta_2 = \lim_{t \uparrow 0} \frac{H_R(t)}{H_L(t)}$) are the corresponding survival mechanisms of both the right branches and the left branches. Right and left branch functions were used and years to test the significant gap hypothesis using χ^2.

We allow $t_{1;h} < t_{2;h} < \text{::::} < t_{N(h);h}$ be another separate occurrence times in a developed tree node. The representation of cumulative hazard function is presented as $CH(t) = \sum_{t_{l,h} \leq t} H(t)$.

$CH_R(t)$ and $CH_L(t)$ are accumulated hazard functions for branch right and branch left. The comparative cumulative hazard function could be estimated at every separate time of the occurrence utilizing:

$$rH(t) = abs\left(\log \frac{CH_R(t)}{CH_L(t)}\right), t_{1.h} \leq t \leq t_N(h) \tag{3}$$

To classify predictors that might distinguish between low risk (with a limited cumulative hazard function) and high risk (with a broad cumulative hazard function), in conjunction with the fact which high-risk groups have high mortality levels in the medium term, the split function has been identified as follows:

$$\text{Splitfun} = \sum_{t_{1.h} \leq t \leq t_N, h} \frac{rH(t)}{t} t_{N,h} \tag{4}$$

The stopping parameter is described as the decrease in the split function.

***Using the following methods based on the* Fig. 1 *mechanism, we have developed a prediction model based on the proposed iRSF*:**

1. There were 1000 random samples of bootstraps. The existing system has been specified (N samples, characteristics M). 37% of the preprocess data (i.e., the out-of-bag data (OOB)) was omitted for each study.

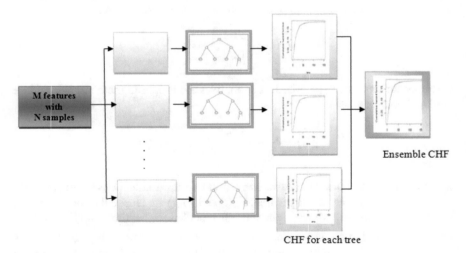

Fig. 1 Schematic representation for the implementation of proposed iRSF risk analysis cumulative hazard function (CHF)

2. A survival tree was cultivated with a growing testing set related to all variables. The Risk model was formulated by all survival trees. The previously discussed dividing rule had been used for dividing every node throughout the tree.
3. The constraining split function in formula (4) significantly reduced was that each survival tree was expanded to full size. The Nelson-Aalen estimators were then used for the accumulated hazard function.
4. On average, each tree in the forest has the cumulative hazard functions of each cumulative tree. In the future, the predictive error was calculated, while the b_{th} value was the error rate assessed with the first b trees in the forest. Probably, the most key parameters were determined by the split branches shut down to tree trunks. The probabilistic dependent variables were therefore based primarily upon the minimum depth of required substrates.

B. Identified Heart Failure Mortality Categorical Predictive Validity

In the cohort of 8059 ICU patients with heart failure, the comprehensive approach with iRSF indicated that ARB, ACE-I, and cardiac arrest were three categorical predictive factors of survival, whereas ARB was not identical to that with RSF. The survival role of subjects including, even without, associated variables can be further divided by ACE-I, ARB, and cardiac arrest while diuretics, particularly in the brief period, have reduced predictive potential. The observations of the iRSF model have been confirmed. For a person with a heart defect, increased levels of patient quality of life, ACE-I is the first line of therapy. Mortality in patients has been shown to decrease in numerous chosen random trials of ventricular dysfunction.

ARBs commonly utilized when ACE-I healthcare providers are prejudiced can be helpful because people avoid angiotensin II from acting on the AT1 receiver, choosing to leave the AT2 receptor unlocked. Consequently, they have been considered to be effective medicines in our research to reduce heart failure death. Furthermore, beta-blockers and diuretics had their efficacy constrained in their short-term survival, although a study showed that the absolute risk of mortality of beta-blockers had been decreased by 4.5% for the course of a period of 13 months. Evidencing their effectiveness and protection is limited, although diuretics are commonly used.

3 Result Analysis

Basic evidence of non-survivors and survivors who have suffered from heart problems throughout 1 year were collected as 1346 people who were monitored and 6713 people who were dead in hospital. e used to use a t-test with two samples to check if the parameter varied with $p < 0.05$ statistically relevant for survivors and non-survivors. We have considered age, cardiac murmur, cardiac arrest, and BMI, which were substantially different in demographic and socioeconomic risk factors as most clinical findings and medicinal products differed between patients who are dead. The comparison of the discrimination results for both frameworks in our sample was

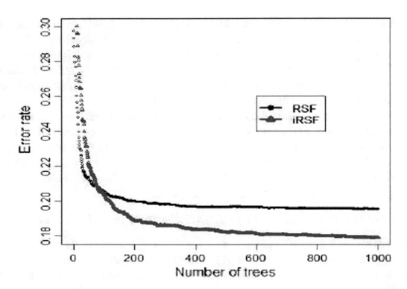

Fig. 2 Estimated analysis of error rate with independently grown trees for iRSF- and RSF-based models

presented (iRSF- vs RSF). The iRSF model thus increases the potential of discrimination to a certain extent, as opposed to 0.804 for RSF, with key characteristics of OOB C-statistics of 0.821.

Figure 2 represents the experimental requirements failure rates of contingency models based on iRSF and RSF, using 32 variables for various trees grown. Just as in Fig. 1, the minimum error rate of the iRSF model is 0.179. With another number of trees expanded and constant, the failure rate reduced as the number of trees expanded to more than 900. The minimum failure rate in the RSF-based model was 0.196. Although the iRSF-based model had comparatively higher rates of failure than the RSF method with less than 50 growing trees, it declined significantly with the rising number of growing trees. The one-sided t-test observed that the incidence of an error on the iRSF is substantially lower than most on the RSF, with less than a p-value of 0.01.

Figure 3 establishes correlations between the persistence role of the community and its estimator iRSF, RSF, and Kaplan–Meier. The survival function with iRSF is pretty close to the Kaplan–Meier estimator's curve compared to RSF. In terms of the Euclidean distance between the two curves, we measured their closeness.

The Euclidean gap is 0.00789, contrasted with 0.0687 for RSF, which shows a superior approximation of the survival function between the iRSF-estimated survival function and Kaplin-Meier.

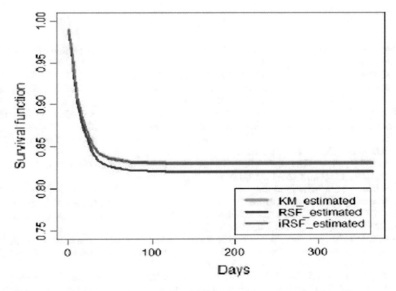

Fig. 3 Comparative analysis of the survival role of iRSF, RSF, and estimator Kaplan–Meier

4 Conclusion

In this current research, we have constructed a favor-prognostic model capable of productive treatment based on 32 risk factors, Demographic factors, clinicopathological details relevant data, and medicines besides physician prediction Creativity of heart failure ICU people just use an iRSF. The iRSF intended to identify the separating predictor's people who are at high risk and low risk, particularly real-life small samples of factors and variables. This analysis is similar to tested 11 independent heart failure mortality correlates. Most laboratory aspects that existing products can revert coorbidities into the risk model have been incorporated and Strong predictive ability, like AST, ALT, and complete bilirubin indicators of the liver, and SCR and BUN renal indicator are provided. In addition, the proposed iRSF presented a more precise estimation of the absolute risk feature compared to a conventional RSF in this report. We shall combine physiological signals like the electrocologic function that is only useful for small populations. ARB has shown that it is an independent predictor of the proposed iRSF. In contrast, a generalized risk model will be studied, featuring few variables that are widely applicable for the prediction of heart failure mortality, in various scenarios, for example, ICU cohorts or home control.

Funding Acknowledgments This Research Articles work does not have any funding support from any funding agencies.

Conflict of Interest Nil.

References

1. Denolin, H., Kuhn, H., Krayenbuehl, H., Loogen, F., Reale, A.: The definition of heart failure. Eur. Heart J. **4**(7), 445–448 (1983)
2. McMurray, J.J., Pfeffer, M.A.: Heart failure. Lancet **365**(9474), 1877–1889 (2005)
3. Pocock, S.J., et al.: Predicting survival in heart failure: a risk score based on 39372 patients from 30 studies. Eur. Heart J. **34**(9), 1404–1413 (2013)
4. Vazquez, R., et al.: The MUSIC risk score: a simple method for predicting mortality in ambulatory patients with chronic heart failure. Eur. Heart J. **30**(9), 1088–1096 (2009)
5. Manzano, L., et al.: Predictors of clinical outcomes in elderly patients with heart failure. Eur. J. Heart Fail. **13**(5), 528–536 (2011)
6. Levy, W.C., et al.: The seattle heart failure model: Prediction of survival in heart failure. Circulation **113**(11), 1424–1433 (2006)
7. Hsich, E., et al.: Identifying important risk factors for survival in patient with systolic heart failure using random survival forests. Circulat. Cardiovasc. Qual. Outcomes **4**(1), 39–45 (2011)
8. Senni, M., et al.: Predicting heart failure outcome from cardiac and comorbid conditions: The 3C-HF score. Int. J. Cardiol. **163**(2), 206–211 (2013)
9. Austin, P.C., Tu, J.V., Lee, D.S.: Logistic regression had superior performance compared with regression trees for predicting in-hospital mortality in patients hospitalized with heart failure. J. Clin. Epidemiol. **63**(10), 1145–1155 (2010)
10. Rodrigues, P.C.O., Santos, E.S., Ignotti, E., Hacon, S.D.: Space-time analysis to identify areas at risk of mortality from cardiovascular disease. BioMed Res. Int. **4**, (2015) Art. no. 841645
11. Breiman, L.: Heuristics of instability and stabilization in model selection. Ann Stat. **24**(6), 2350–2383 (1996)
12. Mitchell, T.M.: Machine Learning. McGraw-Hill, New York (2003)
13. Chen, M., Hao, Y., Hwang, K., Wang, L., Wang, L.: Disease prediction by machine learning over big data from healthcare communities. IEEE Access **5**, 8869–8879 (2017)
14. Lee, B.J., Kim, J.Y.: Identification of type 2 diabetes risk factors using phenotypes consisting of anthropometry and triglycerides based on machine learning. IEEE J. Biomed. Health Inform. **20**(1), 39–46 (2016)
15. Bandyopadhyay, S., et al.: Data mining for censored time-to-event data: A Bayesian network model for predicting cardiovascular risk from electronic health record data. Data Mining Knowl. Discov. **29**(4), 1033–1069 (2015)
16. Acharya, U.R. et al.: An integrated index for detection of sudden cardiac death using discrete wavelet transform and nonlinear features. Knowl.- Based Syst. **83**, 149–158 (2015)
17. Simon, G.J., Caraballo, P.J., Therneau, T.M., Cha, S.S., Castro, M.R., Li, P.W.: Extending association rule summarization techniques to assess risk of diabetes mellitus. IEEE Trans. Knowl. Data Eng. **27**(1), 130–141 (2015)
18. Breiman, L.: Random forests. Mach. Learn. **45**(1), 5–32 (2001)
19. Ishwaran, H., et al.: Random survival forests. Ann. Appl. Stat. **2**(3), 841–860 (2008)
20. Omurlu, I.K., Ture, M., Tokatli, F.: The comparisons of random survival forests and Cox regression analysis with simulation and an application related to breast cancer. Expert Syst. Appl. **36**(4), 8582–8588 (2009)
21. Miao, F., Cai, Y.P., Zhang, Y.T., Li, C.Y.: Is random survival forest an alternative to Cox proportional model on predicting cardiovascular disease? In: Proceeding of the 6TH European Conference of the International Federation for Medical and Biological Engineering, pp. 740–743 (2015)
22. Miao, F., Cai, Y.P., Zhang, Y.X., Li, Y., Zhang, Y.T.: Risk prediction of one-year mortality in patients with cardiac arrhythmias using random survival forest. Comput. Math. Methods Med. (2015) Art. no. 303250, https://doi.org/10.1155/2015/303250
23. Mason, J.W.: A comparison of seven antiarrhythmic drugs in patients with ventricular tachyarrhythmias. New England J. Med. **329**(7), 452–458 (1993)
24. Yang, S., Prentice, R.: Semiparametric analysis of short-term and long- term hazard ratios with two-sample survival data. Biometrika **92**, 1–17 (2005)

25. Yang, S., Prentice, R.: Improved logrank-type tests for survival data using adaptive weights. Biometrics **66**(1), 30–38 (2010)
26. Steyerberg, E.W., et al.: Internal validation of predictive models: Efficiency of some procedures for logistic regression analysis. J. Clin. Epidemiol. **54**(8), 774–781 (2001)
27. Uno, H., et al.: On the C-statistics for evaluating overall adequacy of risk prediction procedures with censored survival data. Stat. Med. **30**(10), 1105–1117 (2011)
28. Goldberger, A.L., et al.: PhysioBank, PhysioToolkit, and PhysioNet: Components of a new research resource for complex physiologic signals. Circulation **101**(23), E215–E220 (2000)
29. Saeed, M., et al.: Multiparameter intelligent monitoring in intensive care II (MIMIC-II): A public-access intensive care unit database. Crit. Care Med. **39**(5), 952–960 (2011)
30. Peterson, P.N., et al.: A validated risk score for in-hospital mortality in patients with heart failure from the American Heart Association get with the guidelines program. Circulat. Cardiovasc. Qual. Outcomes **3**(1), 25–32 (2010)
31. Lee, T.H., Kim, W., Benson, J.T., Therneau, T.M., Melton, L.J.: Serum aminotransferase activity and mortality risk in a United States community. Hepatology **47**(3), 880–887 (2008)
32. Kim, H.C., et al.: Elevated serum aminotransferase level as a predictor of intracerebral hemorrhage: Korea medical insurance corporation study. Stroke **36**(8), 1642–1647 (2005)
33. Lin, J.P., et al.: Association between the UGT1A1*28 allele, bilious- bin levels, and coronary heart disease in the Framingham Heart Study. Circulation **114**(14), 1476–1481 (2006)
34. Nguyen, A., et al.: Total bilirubin is an independent risk factor for the prevalence of coronary artery disease in men. Circulation **130**(Suppl 2), A16164 (2014)
35. Terg, R., et al.: 'Serum creatinine and bilirubin predict renal failure and mortality in patients with spontaneous bacterial peritonitis: A retrospective study.' Liver Int. **29**(3), 415–419 (2009)
36. Yokoyama, Y., et al.: Predictive power of prothrombin time and serum total bilirubin for postoperative mortality after major hepatectomy with extrahepatic bile duct resection. Surgery **155**(3), 504–511 (2014)
37. Investigators, T.S.: Effect of enalapril on survival in patients with reduced left ventricular ejection fractions and congestive heart failure. New England J. Med. **325**(5), 293–302 (1991)
38. Pritchett, A.M., Redfield, M.M.: β-blockers: New standard therapy for heart failure. Mayo Clin. Proc. **77**(8), 839–846 (2002)
39. Lueder, T.G., Atar, D., Krum, H.: Diuretic use in heart failure and outcomes. Clin. Pharmacol. Ther. **94**(4), 490–498 (2013)

Evolutionary Computerized Accounting Model of Colleges from the Perspective of ERP and Mobile Sustainable Networks

Zirui Gu

Abstract Evolutionary computerized accounting model of colleges from the perspective of ERP and mobile sustainable networks is designed in this paper. The upper limit of the forwarding threshold of the mobile network node directly limits the maximum forwarding bandwidth of the network node, due to the uncertain nature of mobile network traffic bandwidth. The absolute median difference game model is used to measure whether the communication data flow is stable, hence, in our cloud model, the complex data is sparse analyzed to obtain the comprehensive expressions. The ERP system is designed and implemented under the novel computerized accounting model with the integration of the evolutionary estimation. The proposed model is analyzed and simulated under different data sets.

Keywords Sustainable networks · ERP and mobile system · Evolutionary algorithm · Computerized accounting · Computer systems

1 Introduction and Background

In recent years, more and more companies have begun to apply ERP systems. Relying on its modern and high-level information management advantages, ERP system has been widely used in enterprise management, which helps to promote the reform and innovation of enterprise production, operation and management, and plays an important role in the improvement of enterprise economic efficiency and long-term development [1, 2]. As a new application system, SaaS ERP is, in essence, a reform and innovation of the ERP operating model. In the recent years, its application has received extensive attention from the academic community. As for the adoption of the SaaS ERP, the academic circles still focus on the research perspective of innovation diffusion. Researchers of innovation diffusion believe that adoption is a stage in the process of innovation diffusion [3]. The influencing factors of information system innovation diffusion mainly come from the three-dimensional framework

Z. Gu (✉)
Chongqing Chemical Industry Vocational College, Chongqing 400020, China
e-mail: 15025327699@163.com

of technology, organization and the environment. In the adoption of cloud ERP, trialability refers to the degree to which customers can trial cloud ERP and achieve the expected trial results. Because the current cloud ERP application environment is not mature enough and the application experience is lacking, it is very beneficial to adopters to choose to try cloud ERP before the formal adoption. If the testability is good, the possibility of customers adopting cloud ERP will be greatly improved [4, 5].

In addition, the applicability of cloud ERP can make customers familiar with cloud ERP in advance, which can effectively reduce the probability of errors in the formal use process, thereby reducing the customer's perceived risk. As the reflect, the Fig. 1 shows the general framework of the ERP systems.

Computerized accounting is a process of the obtaining, storing, transmitting, processing and analyzing accounting data by computer. Blockchain technology uses distributed data processing methods and is directly related to the process of data acquisition, storage, transmission, processing and analysis of the accounting computerization. We consider 2 core aspects as listed [6–8].

(1) The access to computerized accounting data has changed, and now the computerized accounting data is then stored in the centralized bookkeeping. Blockchain is decentralized, uses a distributed database system and uses the distributed accounting. In this way, data access changes, and the accounting computerization system should be adapted to the changes in the database system.

(2) Due to the decentralized, distributed storage technology and openness of the blockchain, except that the private information of both parties is encrypted, the data in the blockchain is public to someone, and anyone can query

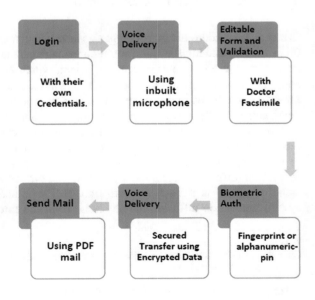

Fig. 1 The general framework of the ERP systems

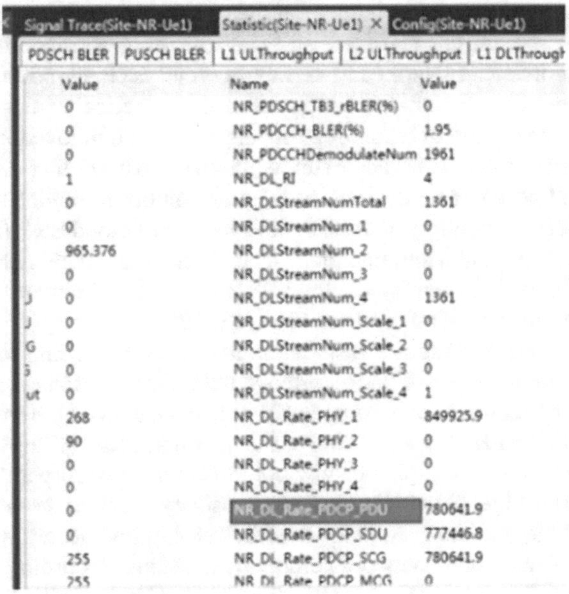

Fig. 2 Computerized accounting model UI

Signal Trace(Site-NR-Ue1)	Statistic(Site-NR-Ue1) ×	Config(Site-NR-Ue1)

POSCH BLER	PUSCH BLER	L1 ULThroughput	L2 ULThroughput	L1 DLThrough

Value	Name	Value
0	NR_PDSCH_TB3_rBLER(%)	0
0	NR_PDCCH_BLER(%)	1.95
0	NR_PDCCHDemodulateNum	1961
1	NR_DL_RI	4
0	NR_DLStreamNumTotal	1361
0	NR_DLStreamNum_1	0
965.376	NR_DLStreamNum_2	0
0	NR_DLStreamNum_3	0
0	NR_DLStreamNum_4	1361
0	NR_DLStreamNum_Scale_1	0
0	NR_DLStreamNum_Scale_2	0
0	NR_DLStreamNum_Scale_3	0
0	NR_DLStreamNum_Scale_4	1
268	NR_DL_Rate_PHY_1	849925.9
90	NR_DL_Rate_PHY_2	0
0	NR_DL_Rate_PHY_3	0
0	NR_DL_Rate_PHY_4	0
0	NR_DL_Rate_PDCP_PDU	780641.9
0	NR_DL_Rate_PDCP_SDU	777446.8
255	NR_DL_Rate_PDCP_SCG	780641.9
255	NR_DL_Rate_PDCP_MCG	0

the blockchain data and development and utilization data through the open interface.

In the Fig. 2, we demonstrate the computerized accounting model UI that shows the basic and general framework of the model and the special parts [9].

To conduct the systematic study, the rest of the paper is organized as the follows. In the Sect. 2, the literature review is provided. In the Sect. 3, the proposed model is provided. In the Sect. 4, the experiment is provided and in the final section, we summarize the work.

2 Literature Review

In this literature review section, we will focus on the ERP and the mobile sustainable networks. The forecasting tools embedded in the ERP system can allow managers to better understand the relationship between business and finance. This provides convenience for managers to manipulate data to realize personal gain. At the same time, although the ERP system embeds the internal control process, the premise for its function is that such a process is opened in the system. The management's selective opening of the embedded internal control process will reduce the effectiveness of the ERP system [10, 11].

Based on theoretical models such as TAM, TOE, DOI and other theoretical models, existing studies have conducted a preliminary analysis of the cloud ERP adoption related issues, and provided the certain reference value for subsequent research, but there are still the following shortcomings: ① Existing studies have influenced factors and factors in cloud ERP adoption. The influencing relationship between factors lacks the necessary theoretical exploration and empirical testing; ② The duality of technology-management and innovation of cloud ERP make its adoption process characteristic, while that most existing studies are based on the stateful IT adoption analysis of the existing conditions thoughts, lack of overall consideration of the adoption process. In the adoption of cloud ERP, the risks are mainly reflected in the potential and uncertain security risks and legal risks of cloud ERP. However, the lower the security of cloud ERP, the higher the relative risk, and the less conducive to the adoption of cloud ERP [12, 13].

For the mobile sustainable networks, according to the bandwidth utilization of mobile network nodes, remove the nodes whose load is closest to the upper limit of the set node forwarding threshold, and calculate the callable bandwidth of each node in the network according to the maximum forwarding bandwidth of the node and the currently occupied bandwidth. Select the best forwarding node; use the link evaluation algorithm to then comprehensively consider the criticality of the communication link, and then select the best forwarding link according to the remaining bandwidth of the mobile network communication link; according to the evaluation results of the nodes and links, adjust the dynamic load balance the flexible configuration threshold performs flexible scheduling of the general data transmitted by the mobile network, and realizes the load balancing configuration of the data transmission path. MEC is a technology based on the 5G evolution architecture, which deeply integrates base stations with various services. It has high bandwidth, low latency, close to users, real-time understanding of wireless network information and location recognition functions, and electronic health. Applications such as the connected cars, industrial automation and smart video acceleration will benefit from edge computing on mobile networks.

3   The Designed Methodology

3.1   The ERP System Details

The implementation and application of ERP system can improve the transmission efficiency of enterprise information and data and realize information sharing. ERP system can realize integrated management and effectively transmit and apply information and data. When the enterprise leadership makes major decisions and development plans, it can provide scientific and reliable reference information and data analysis, which is conducive to improving the accuracy and effectiveness of enterprise decision-making. At the same time, it can also reduce workload and enable the enterprise management to be based on the current situation of enterprise development to then improve the enterprise development strategic planning and management strategy, and make fast and accurate decisions on the overall and forward-looking issues. Although cloud ERP has the characteristics of on-demand use and strong scalability, it cannot be compatible with all the needs of all the customers. It often requires customers to increase investment in secondary development or adjustments to different degrees to improve the compatibility of cloud ERP. It can be seen that in the adoption of ERP innovation, compatibility is an important factor in the attitude of customers to adopt, and poor compatibility will hinder the adoption of cloud ERP [14, 15].

Generally speaking, there is a positive correlation between the completeness of internal resources and the decision-making group's perception of the value of SaaS ERP adoption. The more complete the internal resources, the higher the success of SaaS ERP implementation, and the higher the perception of the value of adoption by decision-making groups [16].

ERP system can provide information to help managers accurately predict the resources needed to expand capacity and increase resource input in time. With the support of ERP system, managers will no longer retain excessive idle resources due to lack of understanding of enterprise resource usage, resulting in high cost stickiness, but can timely adjust the resource holding amount according to market demand. When adjusting resource allocation, enterprise managers need to obtain high-quality relevant information to identify and locate idle resources within the enterprise. If the cost of obtaining high-quality information is too high, they may not be able to adequately respond to reduced resource requirements. Therefore, internal information cost is an important part of resource adjustment cost. The application effect of the ERP system can be effectively demonstrated through the application of the system, the improvement and upgrading process and other links. The enterprise must build a standardized and sound management system, carry out management work scientifically under the constraints of the management system, and formulate practical and feasible evaluation targets to enhance the awareness and ability of all employees to then apply the ERP system, promote the effective integration of the ERP system with specific tasks and then enhance the overall operational strength. Figure 3 shows the data patterns [17, 18].

Fig. 3 ERP system core data patterns

	CO1	0. 875			
COM	CO2	0. 951	0. 917	0. 946	0. 855
	CO3	0. 946			
	OB1	0. 788			
OBS	OB2	0. 964	0. 826	0. 869	0. 692
	OB3	0. 724			
	TR1	0. 824			
TRI	TR2	0. 912	0. 831	0. 898	0. 747
	TR3	0. 854			
	RI1	0. 847			

3.2   *Mobile Network Environment*

The physical network layer is obtained through the combination of wireless and core network, which provides routing access and selection functions for 5G networks; the middle environment layer is mainly composed of three modules, namely the QoS mapping module, address conversion module and completeness management module; The main function of the application network layer is to provide network application services for communication operators.

When data mining technology is used to solve the problem of the mobile communication network switching, it needs to then make prediction and statistical analysis. Simply process the data that needs to be entered into the system in terms of existing data samples collected. In general, the pretreatment operation includes two methods: ① Direct input; ② Direct analysis and processing of the original collected data, and then obtain preliminary conclusions, and then mining and analysis of the processed data, in the analysis will certainly generate a variety of statements. It is processed by support vector machine classifier, classifies the digital vector pattern according to the classification function and transmits the final classification result to the signal recognition control module, so as to control and process the mobile communication network signal. Data preprocessor is used to extract abnormal signal features of packet network to identify all sources and destinations. If the number of fixed connection IP addresses is the same as the number of fixed connection ports, it is a normal mobile communication network signal. If it is, transmit to the mobile communication network signal control module as a normal network signal. If it is no longer, mark it as an abnormal signal of the mobile communication network, so as to then realize the accurate identification of the abnormal signal [19, 20].

We assume that statistics of the number of data packets and bytes transmitted on each path of the mobile network at the same time point are expressed as the following forms:

$$S_p = \{p_1, p_2, ..., p_n\} \tag{1}$$

$$S_b = \{B_1, B_2, ..., B_n\} \tag{2}$$

We use Bayesian algorithm to classify domains based on network information characteristics to obtain the domain candidate information, and non-domain information is directly submitted to users to then ensure real-time filtering; the monitored network information is modally matched with the candidate network information. The Fig. 4 shows the dimension space model.

Introducing filtering discrimination threshold and pass Bayesian, the decision-making process to determine the type of network information and whether to filter is then defined as the formula 3.

$$\alpha = \frac{\lambda_{12} - \lambda_{22}}{(\lambda_{21} - \lambda_{11}) + (\lambda_{12} - \lambda_{22})} \tag{3}$$

Fig. 4 The dimension space model

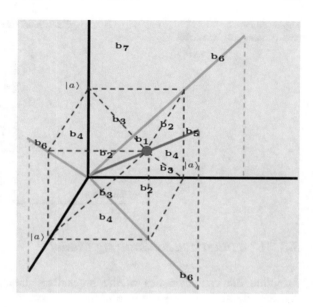

After obtaining the network information feature phenomenon, perform network information classification, while leaving only the necessary network information features use the Bayesian decision algorithm to then distinguish the network information category, and filter the bad information category to then achieve the minimum filtering cost to filter out the most as the purpose of harmful mobile network information. We consider 2 core aspects [21, 22].

(1) Network resource utilization, that is, the ratio of the network resources effectively used by data stream services to total network resources.
(2) The successful transmission rate of data flow services is the percentage of the number of data flow services received by the destination node to the total number of data flow services sent by the source node [23].

In mobile network information filtering, network information is divided into two categories: positive and negative. Positive network information is the bad information to be filtered, and also the negative network information is network information in other fields. Bayesian classification algorithm is used to classify the domain according to the weight of ontology elements, and the non-domain network information is directly submitted to users to ensure the real-time filtering. In the Fig. 5, the judging model is presented.

Fig. 5 Network
environment judging model

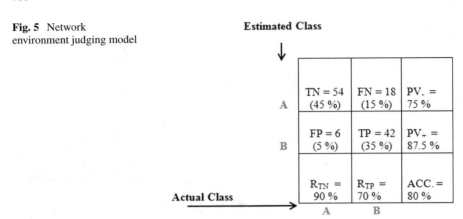

3.3 Computerized Accounting Model

Adopting the characteristics of the blockchain data recording and adopting a
distributed data storage system to ensure the accuracy and reliability of the data
obtained under computerized accounting. The use of the decentralized distributed
structure of the blockchain will change the way of computerized accounting data
storage and transmission, mainly for the acquisition, confirmation, and also the
authenticity of original documents in the computerized accounting fundamental
changes will occur [24, 25]. The computerized accounting system has introduced
a corporate financial information management model, which makes the corporate
account level and temporary payment management clearer. Enterprise financial information management adopts the financial voucher input method that conforms to the
working habits of corporate accountants. The voucher can take debit or credit details
and amount, which makes the operation of corporate accountants more convenient
and the filling of accounting vouchers is more humane.

The database in the computerized accounting system is a convenient tool for
managing a large number of reliable, shareable financial data within an enterprise.
The database in a complete computerized accounting system should include a financial information database, database applications and database management subsystems. The rationality of the database structure design of accounting computerization
system can improve the efficiency of financial data storage and ensure the integrity
of financial data. The design of accounting computerization system based on the
enterprise financial information management should fully understand the needs of
users, including the provision, storage, update and query of financial information.
The blockchain can use private keys to authorize private key encryption, so that the
blockchain can achieve data usage specifications, accurate authorization, and prevent
data abuse and also the illegal use; a credit investigation mechanism can be established in the data use process to achieve data traceability. This is conducive to more
the favorable data circulation, breaking through data and information islands, and

establishing a data horizontal circulation mechanism. This feature should be then used under computerized accounting to strengthen the control of data authorization.

4 The Simulation

In this section, the simulations are mainly focused on the communication systems. NR supports flexible setting of broadcast weights: including four dimensions of "azimuth offset, downtilt, horizontal lobe width, and vertical lobe width". The coverage adjustment and optimization of the front-end base station can be then implemented remotely through the background network management platform. The frequency of using tower workers to adjust industrial parameters has been greatly reduced. Due to the density difference between the normal data and the abnormal data of the data stream, similar to the cluster characteristics, this paper introduces microclusters to describe the evolution process of the data stream; since the nearest neighbor algorithm is essentially a clustering algorithm, it is suitable for searching for the optimal solution. And the algorithm overhead is small.

In the process of planning network sites, data mining technology is used to screen out a batch of sites that can be used. After repeated verification, multiple different initial results can be obtained. Then tabu search algorithm is used to select sites with large changes in the search process, and more appropriate web sites can be obtained after calculating in the corresponding order. The stability of data is also related to the accuracy of the final site selection. The application of data mining technology is also the basic condition to ensure the site. Meanwhile, the security and stability of the core network environment can be effectively guaranteed to further improve the communication service level and the efficiency of network operation. The Fig. 6 shows the results.

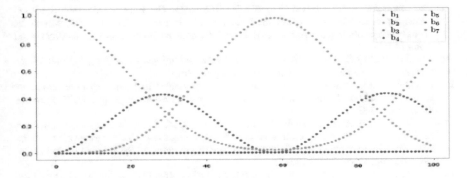

Fig. 6 The environment performance testing results

5 Conclusion and Future Study

Evolutionary computerized accounting model of colleges from the perspective of ERP and mobile sustainable networks is designed in this paper. Although cloud ERP has the characteristics of on-demand use and strong scalability, it cannot be compatible with all the needs of all the customers. It often requires customers to increase investment in secondary development or adjustments to different degrees to improve the compatibility of cloud ERP. Hence, the novel communication methods are combined for the systematic analysis. In our future study, the latest combinations with the modeling and simulation models will be combined.

References

1. Lanlan, Z., Ahmi, A., Popoola, O.M.J.: Perceived ease of use, perceived usefulness and the usage of computerized accounting systems: A performance of micro and small enterprises (mses) in china. Int. J. Recent Technol. Eng. (IJRTE) **8**(2), 324–331 (2019)
2. Gardi, B.: The effects of computerized accounting system on auditing process: a case study from northern Iraq. Available at SSRN 3838327 (2021)
3. Wang, Y.: The research on accounting information based on computerized accounting in the financial management. Development **4**, 5 (2017)
4. Jiang, M., Zhang, A., Li, T.: Distinct element analysis of the microstructure evolution in granular soils under cyclic loading. Granular Matter **21**(2), 1–16 (2019)
5. Hu, X., Xie, N., Zhu, Q., Chen, L., Li, P.C.: Modeling damage evolution in heterogeneous granite using digital image-based grain-based model. Rock Mech. Rock Eng. **53**(11), 4925–4945 (2020)
6. Chakraborty, B., Bhattacharjee, T.: A review on textual analysis of corporate disclosure according to the evolution of different automated methods. J. Financ. Reporting Acc. (2020)
7. Avdoulas, C., Bekiros, S.: Nonlinear forecasting of euro area industrial production using evolutionary approaches. Comput. Econ. **52**(2), 521–530 (2018)
8. Gerth, W.A., Srinivasan, R.S., Murphy, F.G., Gault, K.A.: A Probabilistic Model of Altitude Decompression Sickness Based on the 3RUT-MB Model of Gas Bubble Evolution in Perfused Tissue. A probabilistic model of altitude decompression sickness based on the 3RUT-MB model of gas bubble evolution in perfused tissue. Navy Experimental Diving Unit Panama City United States (2018)
9. Abouhawwash, M., Alessio, A.M.: Multi-objective evolutionary algorithm for PET image reconstruction: Concept. IEEE Trans. Med. Imaging (2021)
10. Purcinelli, L.M., Abreu, R., Vasconcelos, A.L.: Sustainable and smart system: rethinking accounting and taxation in Portugal. In: Governance and Sustainability, pp. 69–79. Springer, Singapore (2020)
11. Trawiński, B., Telec, Z., Krasnoborski, J., Piwowarczyk, M., Talaga, M., Lasota, T., Sawiłow, E.: Comparison of expert algorithms with machine learning models for real estate appraisal. In: 2017 IEEE International Conference on INnovations in Intelligent SysTems and Applications (INISTA), pp. 51–54. IEEE, (2017)
12. Dehaene-Lambertz, G., Monzalvo, K., Dehaene, S.: The emergence of the visual word form: Longitudinal evolution of category-specific ventral visual areas during reading acquisition. PLoS Biol. **16**(3), e2004103 (2018)
13. Baum, G.L., Cui, Z., Roalf, D.R., Ciric, R., Betzel, R.F., Larsen, B., Cieslak, M., et al.: Development of structure–function coupling in human brain networks during youth. In: Proceedings of the National Academy of Sciences, vol. 117, no. 1, pp. 771–778 (2020)

14. Park, J.Y., Cornillie, F., van der Maas, H.L.J., Van Den Noortgate, W.: A multidimensional IRT approach for dynamically monitoring ability growth in computerized practice environments. Front. Psychol. **10**, 620 (2019)
15. Alawaqleh, Q., Al-Sohaimat, M.: The relationship between accounting information systems and making investment decisions in the industrial companies listed in the Saudi Stock market. Int. Bus. Res. **10**(6), 199–211 (2017)
16. Naqi, M., AL-Hashimi, M., Hamdan, A.: Impact of innovative technologies in healthcare organization productivity with ERP. In: Applications of Artificial Intelligence in Business, Education and Healthcare, pp. 309–330. Springer, Cham (2021)
17. Noureldeen, A., Salaheldeen, M., Battour, M.: Critical success factors for ERP implementation: A study on mobile telecommunication companies in egypt. In: International Conference on Emerging Technologies and Intelligent Systems, pp. 691–701. Springer, Cham (2021)
18. Kumar, M., Garg, A., Kumar, A.: Critical factors of post implementation of ERP in higher education systems survey review. In: IOP Conference Series: Materials Science and Engineering, vol. 1149, no. 1, p. 012017. IOP Publishing, (2021)
19. Wang, X., Wang, Q., Zhao, Y.: Individual adoption of ERP system in intelligent manufacturing supply chain based on system dynamics. In: The Sixth International Conference on Information Management and Technology, pp. 1–5. (2021)
20. Hasan, M.S., Ebrahim, Z., Mahmood, W.H.W., Rahman, M.N.A.: An insight into enterprise resource planning system (ERP-S) research trend. J. ILMI **7**(1) (2017)
21. Choi, T.-M., Cai, Y.-J., Shen, B.: Sustainable fashion supply chain management: A system of systems analysis. IEEE Trans. Eng. Manage. **66**(4), 730–745 (2018)
22. Zakaria, M.A.M., Abdullah, R.I.R., Kasim, M.S., Ibrahim, M.H.: Enhancing the productivity of wire electrical discharge machining toward sustainable production by using artificial neural network modelling. EMITTER Int. J. Eng. Technol. **7**(1), 261–274 (2019)
23. Armand, M.T., Roger, A.E.: A comparative study of ERP types and their importance based on the African context. Acessed via ReserachGate **18** (2017)
24. Jones, P., Wynn, M.G.: The circular economy, natural capital and resilience in tourism and hospitality. Int. J. Contemp. Hospitality Manage. (2019)
25. Lee, M.J., Wong, W.Y., Hoo, M.H.: Next era of enterprise resource planning system review on traditional on-premise ERP versus cloud-based ERP: factors influence decision on migration to cloud-based ERP for Malaysian SMEs/SMIs. In: 2017 IEEE Conference on Systems, Process and Control (ICSPC), pp. 48–53. IEEE (2017)

Cloud Resource Hadoop Cluster Scheduling Algorithm Based on Evolutionary Artificial Bee Colony Model for Mobile Sustainable Networks

Haiyan Fan

Abstract Cloud resource hadoop cluster scheduling algorithm based on the evolutionary artificial bee colony model for mobile sustainable networks is designed and implemented in this manuscript. Artificial bee colony (ABC) algorithm is a swarm intelligence optimization algorithm proposed by simulating the honey-collecting behavior of bee colonies. It has potential parallelism and can be carried out at multiple points in the search process. In our designed model, the evolutionary model is combined to improve the traditional flowchart to strengthen the accuracy. Further, the Hadoop and the cloud clustering model are integrated to finalize the model. Then, we apply the proposed model into the mobile sustainable network's scenario. Through the experimental simulation, our designed model is proven to be effective.

Keywords Mobile sustainable networks · Artificial bee colony · Evolutionary model · Cloud resource hadoop · Cluster scheduling

1 Introduction

Hadoop is a widely used large-scale distributed data processing system, which can effectively expand data storage space, uses parallel computing to improve data computing and processing capabilities, and implements the MapReduce cloud computing programming model [1]. Cloud storage is an emerging network storage service. It is a new concept extended and developed from the concept of cloud computing [2]. Cloud storage uses traditional technologies such as cluster technology, grid technology, and distributed technology to integrate different types of networks, Storage devices from different manufacturers work together through software to provide users with storage services [3]. This means that cloud storage breaks through the barriers of incompatibility between traditional hardware, and is no longer a hardware device but provided to users as the service. The distributed data is clustered

H. Fan (✉)
Department of Information, Jiangsu College of Finance and Accounting, Lianyungang 222061, Jiangsu, China
e-mail: fanhaiyan410@163.com

© The Author(s), under exclusive license to Springer Nature Singapore Pte Ltd. 2022
V. Suma et al. (eds.), *Evolutionary Computing and Mobile Sustainable Networks*,
Lecture Notes on Data Engineering and Communications Technologies 116,
https://doi.org/10.1007/978-981-16-9605-3_47

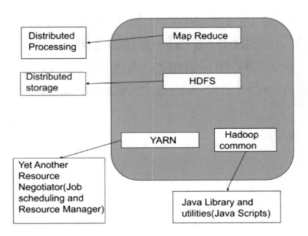

Fig. 1 Hadoop system structure

and clustered, and the data is abstracted, predicted, and analyzed, natural language processing, text processing, etc., and AT-IDS is implemented based on MapReduce to detect abnormal traffic. In Fig. 1, the Hadoop is presented [4, 5].

HDFS is used to store a large amount of traffic information and abnormal traffic information, and MapReduce is used for distributed processing, which improves the processing capacity and better realizes the monitoring of abnormal traffic in the cloud. For the goal of the HDFS, the listed aspects should be focused.

(1) HDFS pays attention to the throughput of data access and requires streaming access to its data sets. The design of HDFS considers batch processing of files more than user interaction processing [6].

(2) HDFS may be composed of thousands of ordinary and cheap servers. The components of the system are extremely large, and any component may fail. Therefore, the core architecture goal of HDFS is to quickly detect hardware faults and realize automatic data recovery, so as to ensure the integrity of data as much as possible [7, 8].

With the mentioned ideas, the mobile sustainable networks will be improved. As the basic technology in the cloud computing environment, virtualization technology can abstract the internal functions of the data storage process from programs, servers, and network resources, and then realize data management independent of applications and networks. Although various technologies can be used to prevent the emergence of a variety of illegal access processes in the cloud computing environment, due to the loopholes in many applications, illegal access still often occurs. Among them, passive routing is a traditional routing method based on table-driven, and its nodes update routing information by periodically exchanging routing tables; active routing is an on-demand routing based on the request/response. These routing methods are all single-path routing methods, and they have the advantages of simple routing and strong flexibility [9, 10].

In the following sections, the novel perspectives will be discussed clearly with the systematic simulations. The rest paper is organized as follows. In Sect. 2, we review the latest methods. In Sect. 3, the designed model is presented. In Sect. 4, the simulations are presented. In Sect. 5, the conclusion and the future work are discussed.

2 Literature Review

The mobile network environment based on cloud computing technology not only needs to guard against traditional network risk problems and risk threats in mobile communication technology but also should guard against risks in the cloud computing. In fact, the rerouting technology for the reliable transmission of information can also be regarded as a reconstruction method because it provides a way to bypass the network equipment or node that has failed or has lost its self-healing ability in the event of a failure, thereby ensuring network communication. Safe and reliable capability. Rerouting refers to bypassing the failed original relay node, recalculating to then generate the new route, and restoring the normal communication from the source node to the destination node [11, 12].

Each node saves a routing table, which contains a destination node list, next-hop address, and a destination sequence number for distinguishing routing updates to avoid loops. The destination sequence number is used to indicate the order of packets sent to a destination node. If the sequence number is higher, the routing information is considered newer; otherwise, if the destination sequence numbers are equal, the shorter distance route information is used. This solution selects one or several nodes with the best channel status from the source node as relay nodes, and collectively transmits data to the receiving node. Since the channel between the source node and the relay node is better, the error between the two is small, thus forming the virtual MISO structure with multiple transmissions and single reception [13]. For the optimization of the model, the artificial bee colony is essential. The ABC algorithm simulates the actual honey-gathering mechanism of bees to deal with function optimization problems. The artificial bee colony is divided into three categories: lead bees, follow bees, and scout bees. Lead bees and follow bees are used for the mining of nectar sources, and scout bees avoid too few types of nectar sources. According to the different roles in the process of searching for food sources and collecting nectar, individuals in the colony can be divided into employed bees and unemployed bees. The bees that find the food source are hired bees, also known as lead bees. Lead the bees to share the location information of the food source and the quality information of the food source with other bees with a certain probability. Unemployed bees are divided into follow bees and scout bees. The follow bee determines the rate of return of the food source according to the dance of the leading bee, and the level of the rate of return is the basis for determining whether to choose the food source for nectar collection and the mission of the scouting bee is to explore new sources of food. The algorithm combines the idea of the multi-parent hybridization and differential evolution operation. By adopting the multi-parent hybrid recombination strategy of the multi-parent hybrid algorithm, the subspace searched by the algorithm can cover the convex combination space of multiple parents, which enhances the overall situation of the ABC algorithm's search ability [14, 15].

3 The Proposed Methodology

3.1 Evolutionary Artificial Bee Colony Model

Although ABC algorithm has good performance in function optimization, it still has the problems of premature convergence and local optimal solution when solving complex functions with the high dimensions. The differential evolution operator is used as an acceleration operator. The differential evolution operator can make full use of the information contained in the optimal individual, and while fusing the idea of greedy algorithm, it also maintains a certain diversity, which not only accelerates the convergence speed of the algorithm also avoid premature convergence. In addition, for the same individual, a new individual is used to replace the original individual, which not only ensures the diversity of the group but also enables the differential evolution operator to always work [16].

The roles of the three types of bees can be interchanged. Collecting bees are bees that discover high-quality nectar sources and also conduct preliminary neighborhood searches. The number of them is equal to the number of nectar sources. Observing bees wait for the collecting bees to transmit nectar source quality information to determine which nectar source to conduct neighborhood searches for. Perform a global random search to find new nectar sources of higher quality. Each bee corresponds to a solution. Picking bees represents the existing solutions that constitute the current population; the observation bees represent potential neighborhood search solutions and have the opportunity to enter the population to become existing solutions; scouting bees represent global random search solutions, which can replace obsolescence existing solutions. Figure 2 shows the detailed framework.

We randomly generated 2 N positions, and took the superior N as nectar source positions, leading the bees to find the nectar sources and remember them, and search for new nectar sources near each nectar source according to formula (1):

$$V_{ij} = x_{ij} + R_{ij}(x_{ij} - x_{kj}) \tag{1}$$

We set the main parameters of the algorithm: the number of populations NP, the maximum number of cycles MCN, the dimension D of the problem, and the constant algebra limit of the picking bees. The number of leading bees, follow bees, and nectar sources are all NP/2, 1 scout bee. The formula 2 is the updating.

$$x_{ij} = (x_{ij})_{\min} + \text{rand}(0, 1) * ((x_{ij})_{\max} - (x_{ij})_{\min}) \tag{2}$$

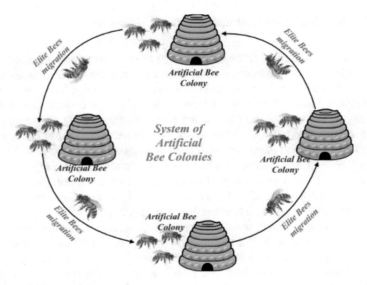

Fig. 2 The detailed framework of the ABC

Follow the bee to continue searching near the nectar source, calculate the fitness of the new nectar source, and compare it with the original nectar source, if the adaptability of the new nectar source is good, then mine the new nectar source; or otherwise, continue to mine the original nectar source. In the evolution of the algorithm, the best individual retention mechanism is used to then ensure the convergence of the algorithm. In addition, if there are duplicate individuals in each generation of nectar sources, individuals will be randomly initialized to replace the duplicate individuals, so that the diversity of nectar sources can be maintained and more information can be reserved for searching. In Fig. 3, the updating model is demonstrated for the analysis of the related further models [17, 18].

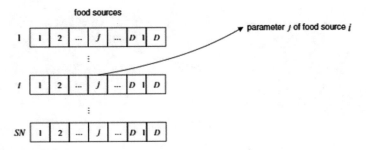

Fig. 3 The updating analytic framework

3.2   Mobile Network Communication Structure

Access control is the main technical means to allow external access. It can ensure that network resources are free from illegal operations. Access control is not only the subjective prevention but also conducive to resource isolation. There are currently seven types of access control in use, and the use of access control should be used reasonably in accordance with the actual situation. Since relay cooperation often involves the physical layer, the MAC layer, and the network layer at the same time, the original strict layer-based protocol does not support the relay and forwarding function. The original protocol must be improved, and the MAC layer protocol is the key. Because WLAN mostly adopts 802. 11 agreements at present, especially MAC agreement that adopts RTS/CTS handshake mechanism, must consider improving it to support the relaying and forwarding function [19, 20].

For the designed system, we should consider 3 aspects of the novelties. (1) The DSR protocol detects the link disconnection through the MAC layer and will determine whether the rescue position is set and whether there is a replacement path. If it is not set and there is a replacement path, the data packet will be resent through the replacement path. (2) AODV route discovery is similar to DSR. First, the source node initiates a routing request RREQ. This routing request packet contains the source node, destination node, last-hop address, hop count, and destination sequence number. The destination sequence number is used to indicate the update of the route, which is similar to the DSDV destination sequence number to avoid the loops. (3) Information encryption technology is the most common way of network security prevention. Encrypt some important information to ensure the safety of its operation. Commonly used information encryption technologies include link encryption, endpoint encryption, and node encryption. The use of encryption technology must be based on reality, not blindly.

In Fig. 4, the structure details are denoted. The back-end processing system imports the original data of the front-end test system and processes and outputs

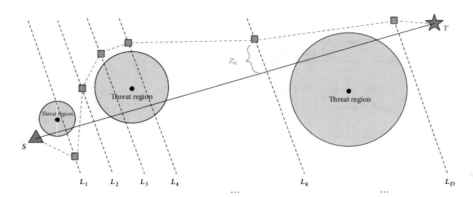

Fig. 4 The mobile network communication structure details

the final test results. It also has statistical analysis capabilities. The front-end test system and the back-end processing system can be located in different PCs. The GPS receiver is used to record the moving route of the test equipment. The external power supply is to provide a safe and stable power supply for the test system. Using this system, the designed model will be implemented.

3.3 Cloud Resource Hadoop Cluster Scheduling Algorithm

Hadoop clusters can be deployed on low-cost storage devices, and the number of DataNodes can be expanded at any time as needed. As long as ordinary computers supporting the Linux system can be deployed to the cluster. The MapReduce framework consists of a Master JobTracker running on the master node and Slave Task-Tracker running on each cluster slave node. The master node is responsible for scheduling all tasks that constitute a job distributed on different slave nodes.

The job set is divided into delayed and undelayed job subsets according to the average waiting time, and enters the double queue according to the degree of urgency, and obtains new subsets according to different strategies to form the largest subset under the condition of the available resources. In formula 3, the scheduling algorithm is defined [21, 22].

$$\text{Average}(WT) = \frac{1}{n} \sum_{i=1}^{n} WT(J_i) \tag{3}$$

In the execution process of the existing access control model of the Hadoop platform, a gated access control model is realized. That is, the user applies to the KDC for the identity card. Once the user obtains the authentication information and ticket, the user can apply for services to the master node of the cluster based on the information and can use the resources of the cluster after being authorized. Figure 5 shows the pattern.

At present, academia tends to monitor user behaviors, obtain their activity status in the cluster, and record them in the master node, and supervise users' permissions based on user behaviors, thereby changing the existing access control mode. There may be multiple job sets that meet the conditions. Choose the set with the largest number of jobs in each round. Finally, keep the first set with the largest number of jobs starting from the head of the team in each round of job set. The first one from the head of the team is chosen to comply with the fairness principle of first-come, first-served. The work set is closest to the head of the team and has the highest urgency value.

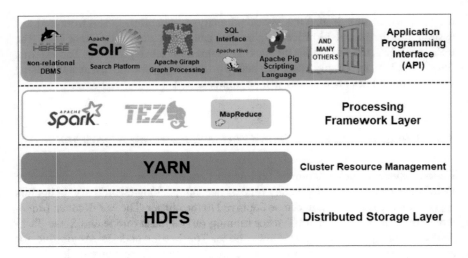

Fig. 5 Cloud resource hadoop cluster scheduling pattern

4 Experimental Analysis

To monitor user behavior in a Hadoop cluster and record the behavior data for quantification, it is necessary to define a reasonable data structure for storage. Call the copyFromLocalFile() method of DistributedFileSystem to upload the local file to HDFS according to the path of the local file and the path of the HDFS storage directory. After uploading the file, we can view the disk storage distribution status of each node through the Web. In Fig. 6, we show the parameters.

Multi-objective intelligent optimization algorithms are currently hot research field. Compared with other multi-objective intelligent optimization algorithms, the current research results of the multi-objective ABC algorithm are less and not systematic enough. The ABC algorithm has demonstrated excellent performance in solving single-objective problems. How to design an efficient multi-objective ABc algorithm will be a topic worthy of in-depth study. After experimentation, it can be concluded from the analysis of the chart that the multi-objective artificial bee colony algorithm is an effective optimization method, but in terms of operating efficiency, the ABC

Name	Type	Size	Replication	Block Size
jdk-7u51-linux-x64.tar.gz	file	131.80 MB	3	128 MB
mysql-5.5.35-win32.msi	file	33.67 MB	3	128 MB
ubuntu-12.04.2-desktop-x64.iso	file	695 MB	3	128 MB
windows-server_2008_x64.iso	file	3.14 GB	3	128 MB

Fig. 6 The parameters considered for the model

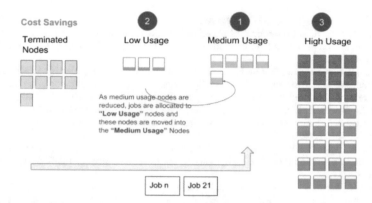

Fig. 7 The experiment on clustering structure

has more objectives and the artificial bee colony algorithm is better, and the multi-objective artificial bee colony algorithm is better. The running speed is relatively slow. In Fig. 7, we show the structure.

5 Conclusion and Future Work

Cloud resource hadoop cluster scheduling algorithm based on the evolutionary artificial bee colony model for mobile sustainable networks is designed and implemented in this manuscript. This article discusses the basic principles of the artificial bee colony algorithm and the specific process of the algorithm. Based on the artificial bee colony algorithm, a core multi-objective artificial bee colony algorithm is proposed and applied to the Hadoop structure. The simulation has proven the satisfactory performance. In the future study, we will consider some other application scenarios.

References

1. Radenkovic, M., Huynh, V.S.H.: Energy-aware opportunistic charging and energy distribution for sustainable vehicular edge and fog networks. In: 2020 Fifth international conference on fog and mobile edge computing (FMEC), pp. 5–12. IEEE (2020)
2. Han, Y., Jiao, J., Weissman, T.: Minimax estimation of divergences between discrete distributions. IEEE J. Sel. Areas Inf. Theor. **1**(3), 814–823 (2020)
3. Jain, M., Sharma, N., Gupta, A., Rawal, D., Garg, P.: Performance analysis of NOMA assisted mobile ad hoc networks for sustainable future radio access. IEEE Trans. Sustain. Comput. (2020)
4. Zhu, T., Li, J., Cai, Z., Li, Y., Gao, H.: Computation scheduling for wireless powered mobile edge computing networks. In: IEEE INFOCOM 2020-IEEE conference on computer

communications, pp. 596–605. IEEE (2020)

5. Mosharmovahed, B., Pourmina, M.A., Jabbehdari, S.: Providing a new way for sustainable communication on the VANET network on mountainous roads. Wirel. Pers. Commun. 113(4), 2243–2269 (2020)

6. Manman, L., Goswami, P., Mukherjee, P., Mukherjee, A., Yang, L., Ghosh, U., Menon, V.G., Nkenyereye, L.: Distributed artificial intelligence empowered sustainable cognitive radio sensor networks: A smart city on-demand perspective. Sustain. Cities Soc. 103265 (2021)

7. Mardia, J., Jiao, J., Tánczos, E., Nowak, R.D., Weissman, T.: Concentration inequalities for the empirical distribution of discrete distributions: Beyond the method of types. Inf. Infer. J. IMA 9(4), 813–850 (2020)

8. Yang, Y., Duan, Z.: An effective co-evolutionary algorithm based on artificial bee colony and differential evolution for time series predicting optimization. Complex Intell. Syst. 6, 299–308 (2020)

9. Wang, H., Wang, W., Xiao, S., Cui, Z., Minyang, X., Zhou, X.: Improving artificial bee colony algorithm using a new neighborhood selection mechanism. Inf. Sci. 527, 227–240 (2020)

10. Boudardara, F., Gorkemli, B.: Solving artificial ant problem using two artificial bee colony programming versions. Appl. Intell. 50(11), 3695–3717 (2020)

11. Zou, W.-Q., Pan, Q.-K., Meng, T., Gao, L., Wang, Y.-L.: An effective discrete artificial bee colony algorithm for multi-AGVs dispatching problem in a matrix manufacturing workshop. Expert Syst. Appl. 161, 113675 (2020)

12. Wu, Y., Qi, Z., Jiang, L., Dai, Z., Zhang, C., Zhang, C., Jian, X.: Weights optimization method of differential evolution based on artificial bee colony algorithm. In SimuTools 1, 626–635 (2020)

13. Lin, Y., Wang, J., Li, X., Zhang, Y., Huang, S.: An improved artificial bee colony for feature selection in QSAR. Algorithms 14(4), 120 (2021)

14. Ruan, X., Wang, J., Zhang, X., Liu, W., Fu, X.: A novel optimization algorithm combing gbest-guided artificial bee colony algorithm with variable gradients. Appl. Sci. 10(10), 3352 (2020)

15. Alquthami, T., Butt, S.E., Tahir, M.F., Mehmood, K.: Short-term optimal scheduling of hydro-thermal power plants using artificial bee colony algorithm. Energy Rep. 6, 984–992 (2020)

16. Yao, Y., Zhao, X., Ning, Q., Zhou, J.: ABC-Gly: Identifying protein lysine glycation sites with artificial bee colony algorithm. Curr. Proteomics 18(1), 18–26 (2021)

17. Liu, B., Li, J., Lin W., Bai, W., Li, P., Gao, Q.: K-PSO: An improved PSO-based container scheduling algorithm for big data applications. Int. J. Netw. Manag. 31(2), e2092 (2021)

18. Maurya, A.K.: Resource and task clustering based scheduling algorithm for workflow applications in cloud computing environment. In: 2020 Sixth International Conference on Parallel, Distributed and Grid Computing (PDGC), pp. 566–570. IEEE (2020)

19. Gill, S.S., Ouyang, X., Garraghan, P.: Tails in the cloud: a survey and taxonomy of straggler management within large-scale cloud data centres. J. Supercomputing (2020)

20. Nirmalan, R., Gokulakrishnan, K.: An intelligent surveillance video analytics framework using NACT-Hadoop/MapReduce on cloud services. Distrib. Parallel Databases 1–17 (2021)

21. Hussein, A.A.: Using hadoop technology to overcome big data problems by choosing proposed cost-efficient scheduler algorithm for heterogeneous hadoop system (BD3). J. Sci. Res. Rep. 58–84 (2020)

22. Wang, J., Li, X., Ruiz, R., Yang, J., Chu, D.: Energy utilization task scheduling for MapReduce in heterogeneous clusters. IEEE Trans. Serv. Comput. (2020)

Application of Evolutionary Big Data Statistical Analysis Method in Computer Guiding Management Under Mobile Sustainable Network Scenarios

Jikui Du

Abstract In this paper, we consider the application of evolutionary big data statistical analysis method in computer guiding management under mobile sustainable network scenarios. Based on the analysis of big data, we comprehensively consider the progress space of urban development and mobile communication and use wireless network transmission technology to provide 5G network access centered on areas with dense population distribution and large number of users. Under this access mode, diversified business requirements can be met, and the content of data business can be improved, the utilization of resources can be saved, and the program can be simplified. From the perspective of the choice of specific ways to improve the quality of big data statistical analysis, it should be carried out from two aspects: theoretical research, practical application, and training and learning. Hence, we understand the model from the different perspectives. The numerical simulation has been conducted to show the performance.

Keywords Network scenarios · Computer guiding management · Evolutionary big data · Statistical analysis method · Mobile systems

1 Introduction and Review

Theoretically, multiple variables are separated, one variable at a time or the relationship between two researches is studied. Although simple, its shortcomings are also obvious. It does not take into account the relationship between the variables [1]. Separate research will lead to the relationship between the variables. The relationship between is lost at the beginning of the process, which will greatly then affect the accuracy of the final model. Therefore, the multivariate statistical analysis method is very effective and reveals the internal relationship between the variables. After testing, the results of this analysis are usually valid and typical.

J. Du (✉)
Internet of Things Technology College, Wuxi Vocational Institute of Commerce, Wuxi 214153, Jiangsu, China
e-mail: djkxyz@163.com

© The Author(s), under exclusive license to Springer Nature Singapore Pte Ltd. 2022   715
V. Suma et al. (eds.), *Evolutionary Computing and Mobile Sustainable Networks*,
Lecture Notes on Data Engineering and Communications Technologies 116,
https://doi.org/10.1007/978-981-16-9605-3_48

From the perspective of the choice of specific ways to improve the quality of big data statistical analysis, it should be carried out from two aspects: theoretical research, practical application, and training and learning, so as to achieve the goals and requirements of improving the quality of the big data statistical analysis. (1) Through the study of traditional statistical theories and the new theories and new techniques of big data statistical analysis, the professional and technical personnel of statistical analysis can be comprehensively improved in terms of professional technical ability and comprehensive literacy. (2) In the process of research and analysis, from the perspective of statistics and data analysis of the information technology in two dimensions, in the perspective of statistics, the big data collection, data analysis, from the method of statistics, statistics of thinking, with the practical experience of working with large data analysis of the statistics of combination were studied. Figure 1 shows the model [2–4].

Because in the mobile network environment, nodes have the characteristics of high dynamics, heterogeneity, and also strong autonomy, a network without QoS guarantee will make it difficult to deploy and implement distributed file-sharing systems. The construction of a smart city requires the use of modern information technology to make the management and operation of the city smarter, so as to create a better city life and also show more harmonious and sustainable growth characteristics in the process of urban development. Today, the rapid increase in mobile terminals has brought about a rapid increase in mobile data, and its growth trend has shown

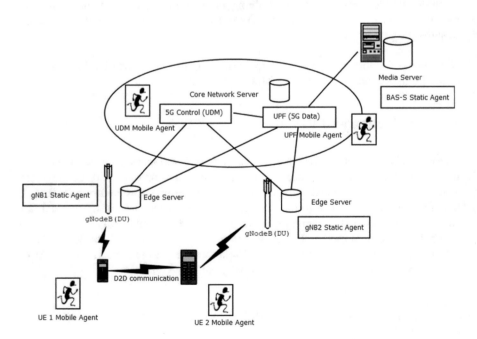

Fig. 1 Big data statistical analysis method

Fig. 2 Mobile sustainable
network scenarios

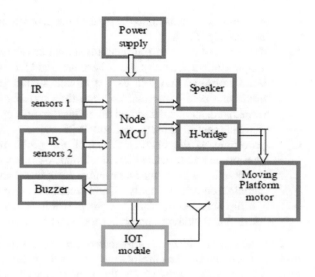

an exponential growth [5]. If there is no effective 5G wireless network architecture and effective resource allocation algorithm to then schedule system resources to process various user requests, the general consequences are immeasurable. It should include cloud access network architecture, heterogeneous network architecture, and mobile edge computing network architecture. Among these 5G wireless network architectures, heterogeneous network architecture is the research hotspot of experts and scholars.

It has been one of the candidate frameworks of 5 g wireless access network framework. Heterogeneous network architecture has the strong flexibility and scalability, and it also has many functional advantages of information processing. In Fig. 2, the model is discussed and in the next parts, details will be shown.

2   The Proposed Methodology

2.1   Big Data Statistical Analysis Method Review

The idea of statistical analysis is to discover and summarize. Finally, the statistical inference analysis theory is based on the principle of distribution theory, based on the probability in mathematics, and then infers the overall characteristics based on the characteristics of the sample. The analytic model has listed aspects.

(1)   Ask questions and set goals. Due to the complexity of the actual problem, the large number of variables and the complex relationship between variables, it

is necessary to determine the goal of the problem in the complex quantitative relationship [6, 7].

(2) The establishment of multivariate statistical analysis model is based on the sample statistical data of index variables, and the quality of the selected index data determines the role of the model. In the process of sorting out and processing data, abnormal data are then roughly eliminated, index data are normalized, standardized and other data processing, so as to then facilitate the establishment of subsequent models [8–10].

(3) According to the characteristics of the problem and data, we select the appropriate multivariate statistical analysis method to construct the theoretical model. Factor analysis, principal component analysis, regression analysis, canonical correlation analysis, and other methods can be used to study the relationship between multiple variables. If things are classified, cluster analysis, discriminant analysis, and other tools can be used.

The first stage divides the entire data set into several sub-samples so that each sub-sample is suitable for the current computer management capabilities, and estimates the parameters of each sub-sample; the second stage takes the average of the estimated results of each sub-sample. In Eq. 1, the matrix is defined.

$$
\begin{bmatrix}
x_{11} & \cdots & x_{1\alpha n} \\
\vdots & \ddots & \vdots \\
x_{\beta n1} & \cdots & x_{\beta n\alpha n}
\end{bmatrix}
\tag{1}
$$

Multiple regression analysis is mainly used for prediction and control analysis to find the relationship between multiple variables. This processing method can be seen in all fields. When constructing multiple regression model, we must first find the variables with strong correlation. The stepwise regression method is usually used to find the variables, and the data is dimensionless, that is, the data is standardized and the variables are selected to modify the parameters to meet the requirements, and use F test for the variable to determine whether the variable should be eliminated and then when using the least square method to estimate the parameters, it is assumed that the expression of the final parameters is unique. The basic idea of the sampling method is to extract sub-samples with the certain probability distribution from the initial data instead of the original big data to estimate, predict, and statistically infer the model. The difficulty of the sampling method lies in the design of the probability distribution of each sub-sample [11–13].

Figure 3 shows the method structure. Correlation analysis of a single chart often lacks persuasiveness and credibility, so it is necessary to measure the correlation relationship, so it is necessary to use covariance and covariance matrix analysis defined as the Eq. 2 [14, 15].

$$
\mathrm{cov}(X, Y) = \frac{\sum_{i=1}^{n} (X_i - \overline{X})(Y_i - \overline{Y})}{n - 1}
\tag{2}
$$

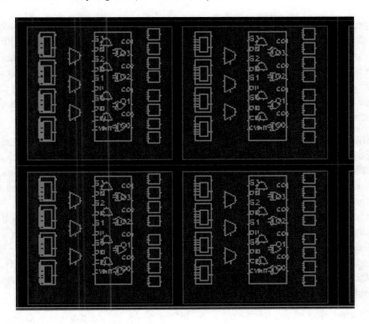

Fig. 3 Statistical analysis method structure

The X and Y are the parameters considered. In big data analysis, in order to obtain more accurate results, usually not only more complex models are required, but also more computing resources are required, which often results in extremely complicated calculations, high calculation costs, and low calculation efficiency. Then how to calculate accurately and the trade-off between performance and computational efficiency is a big problem we face. This requires the design of effective algorithms for simple models, that is, in the context of big data, simple models have better performance.

2.2 Evolutionary Big Data Statistical Analysis

For classification, irrelevant and redundant features are obviously meaningless, and even the increase of the search space will cause the classification performance to decrease. Through feature selection, the extraction of representative features in the data set is a common method to shorten the training time of the classifier and improve the classification ability. In many real-world application problems, there are multiple interrelated operation steps or the business processes, which can be described formally by the workflow model. The task of the cloud workflow scheduling problem is to allocate each work in the workflow computing task on the cloud platform to a set of virtual computing resources rented in a pay-as-you-go mode for execution, on the premise of meeting the business process constraints defined by the workflow and next, achieve optimization goals such as shortening the maximum completion time, saving costs, and improving resource utilization. Because there is the certain conflict between these goals, that is, there is no scheduling scheme that can achieve the optimum on each goal at the same time, the cloud workflow scheduling problem is essentially the general multi-objective optimization problem. The computation can be separated as follows [16, 17].

(1)   The researched online fuzzy clustering, the vector machine-based fuzzy classification, fast fuzzy correlation rule mining algorithm, etc., use fuzzy logic to soften the boundary of correlation rules and the conditions of rule matching, which are important for the correlation analysis of big data and the research significance and application value.

(2)   It is a heuristic search algorithm based on population behavior to optimize a given target. Its optimization process embodies randomness, parallelism, and distribution. It is suitable for solving complex distributed problems without centralized control and no global model. Swarm intelligence algorithms can also provide effective means for reduction of large-scale data, such as the text feature reduction [21, 22].

Multi-objective cloud workflow scheduling problem. At the same time, considering the two conflicting goals of minimizing the maximum completion time and then minimizing the calculation cost, the following multi-objective optimization model is established:

$$\min F(X) = \{S(X), C(X)\} \tag{3}$$

$$S(X) = T(W_{\text{end}}) \tag{4}$$

Although inspired by the optimal food source in the neighborhood, the iterative evolution of the bee colony has both global and local search capabilities. However, from the analysis of the algorithm's overall search capabilities, it can be seen that the algorithm's global search capabilities are relatively large. The range has been improved, but there are still insufficient in-depth search capabilities. Figure 4 gives the model trend [18–20].

Fig. 4 The model trend considered

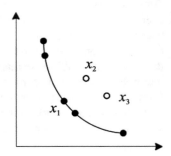

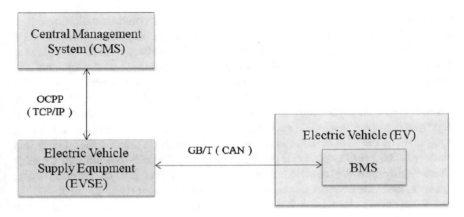

Fig. 5 Evolutionary analytic framework

Data reduction methods based on neural networks, multi-layer neural network learning algorithms, etc., all use artificial neural network algorithms to build big data analysis models, which can deeply reveal the rich and complex information in massive data and can make more accurate predictions for the future. In Fig. 5, the analytic framework is demonstrated.

2.3 Computer Guiding Management Under Mobile Sustainable Network Scenarios and Simulations

The most essential function of the mobile distributed file-sharing system is to share the resources and services provided by other nodes in the network. In addition to the characteristics of peer-to-peer and easy deployment, it must also meet the real-time requirements of multimedia data and services provided by new services with the quality and other issues. The reverse proxy first extracts the specific marking information in the packet header, and then ic. The Honeynet controller generates the corresponding SDN flow rules and checks the marking information on the SDN switch to implement the corresponding network flow control strategy [23]. The agent module has three operation modes, according to the decision of the controller, the corresponding operation mode is configured on demand. In order to prevent signature attacks, the proxy module will check the payload of the response message, focusing on checking whether it contains any signatures that may expose the deployed honeynet/honeypot. Once such an identification is found, the agent will immediately notify the controller to take corresponding actions, such as changing the agent mode or updating the network configuration. After the 5G mobile network has completed the test network layout, it has passed the operation and testing of the pilot area to the full commercial use of the 5G mobile network, which provides a guarantee for the informatization construction of smart cities. Mobile communication operators need to give full play to the role of leading operating companies, adhere to high standards and strict requirements, do research and data collection in advance, lay the foundation for forward-looking 5G mobile network planning and deployment, and ensure that high-speed communication networks can complete in one step [24, 25].

The MEC system is usually located between the wireless access point and the wired network. In the telecommunications cellular network, the MFC system can be deployed between the wireless access network and the mobile core network. The core equipment of the MEC system is the MEC server built on the IT general hardware platform. The MEC system can provide localized cloud services through the edge cloud-deployed inside the wireless base station or at the edge of the wireless access network, and can connect to other networks such as the private cloud inside the enterprise network to achieve hybrid cloud services. In Fig. 6, the estimation curve is presented.

Master the research and development technology of network application systems, including research and also development technologies related to new technologies such as distributed processing and cloud computing, and have the ability to design and develop new network application systems; master the design, implementation, testing, and verification of comprehensive network engineering schemes technology, with the plan design, demonstration, and implementation of network engineering. The ratio of the number of service response times to the total number of service requests in a cycle, and the extent to which the service is immediately available when the user initiates the request.

There are many factors that cause service unavailabilities, such as queuing, congestion, and system maintenance. The difference between availability and reliability is that availability only counts whether the service returns results, not whether the service is successful. In the context of big data, the security defense of mobile network information transmission requires active defense against information intrusion. Combined with the concept of the immune agent, a mobile network security model is constructed. According to the information on the mobile network host node under the distributed big data, the intrusion attack events are identified, the unknown

Fig. 6 The estimation curve for analysis

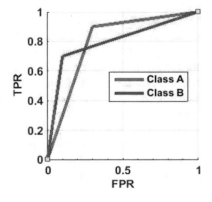

Fig. 7 The network
scenarios and simulations

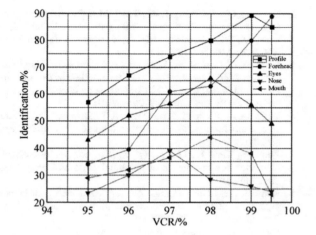

mobile network information intrusion attack is found by using the learning and memory mechanism, and the network information is transmitted to the destination node. At the same time, the identified intrusion attack vaccine is distributed to each node in the same domain to improve the intrusion attack defense capability of all nodes in the domain, and analyze the human intrusion attack information sent by each node in the domain. Figure 7 gives the simulation results for consideration.

3 Summary

In this paper, we consider the application of evolutionary big data statistical analysis method in computer guiding management under mobile sustainable network scenarios. The linear growth of cloud computing capabilities cannot match the explosive growth of massive network edge data, and the rapid increase in transmission network bandwidth load causes long network delays, and it is difficult to meet the transmission needs of the control data, real-time data, and streaming data. Under this perspective, the designed model is analyzed with the evolutionary model. In the future, we will test different parameters.

References

1. Sun, W., Liu, J.: Congestion-aware communication paradigm for sustainable dense mobile crowdsensing. IEEE Commun. Mag. **55**(3), 62–67 (2017)
2. Ordonez-Lucena, J., Chavarria, J.F., Contreras, L.M., Pastor, A.: The use of 5G non-public networks to support industry 4.0 scenarios. In: 2019 IEEE Conference on Standards for Communications and Networking (CSCN), pp. 1–7. IEEE (2019)

3. Lee, Y.L., Qin, D., Wang, L.-C., Sim, G.H.: 6G massive radio access networks: Key applications, requirements and challenges. IEEE Open J. Veh. Technol. **2**, 54–66 (2020)
4. Long, Q., Chen, Y., Zhang, H., Lei, X.: Software defined 5G and 6G networks: a survey. Mob. Networks Appl. 1–21 (2019)
5. Montori, F., Cortesi, E., Bedogni, L., Capponi, A., Fiandrino, C., Bononi, L.: Crowdsensim 2.0: A stateful simulation platform for mobile crowdsensing in smart cities. In: Proceedings of the 22nd International ACM Conference on Modeling, Analysis and Simulation of Wireless and Mobile Systems, pp. 289–296 (2019)
6. Malik, A.W., Rahman, A.U., Qayyum, T., Ravana, S.D.: Leveraging fog computing for sustainable smart farming using distributed simulation. IEEE Internet Things J. **7**(4) 3300–3309 (2020)
7. Zhou, Y., Tian, L., Liu, L., Qi, Y.: Fog computing enabled future mobile communication networks: A convergence of communication and computing. IEEE Commun. Mag. **57**(5), 20–27 (2019)
8. Kamble, S.S., Gunasekaran, A., Gawankar, S.A.: Sustainable Industry 4.0 framework: A systematic literature review identifying the current trends and future perspectives. Process Saf. Environ. Prot. **117**, 408–425 (2018)
9. Chaudhary, S., Johari, R., Bhatia, R., Gupta, K., Bhatnagar, A.: CRAIoT: concept, review and application (s) of IoT. In: 2019 4th İnternational Conference on İnternet of Things: Smart İnnovation and Usages (IoT-SIU), pp. 1–4. IEEE (2019)
10. Peng, K., Leung, V., Xu, X., Zheng, L., Wang, J., Huang, Q.: A survey on mobile edge computing: Focusing on service adoption and provision. Wireless Commun. Mob. Comput. 2018 (2018)
11. Zaunbrecher, B.S., Linzenich, A., Ziefle, M.: A mast is a mast is a mast…? Comparison of preferences for location-scenarios of electricity pylons and wind power plants using conjoint analysis. Energy Policy **105**, 429–439 (2017)
12. Yang, Z., Dehghanian, P., Nazemi, M.: Enhancing seismic resilience of electric power distribution systems with mobile power sources. In: 2019 IEEE Industry Applications Society Annual Meeting, pp. 1–7. IEEE (2019)
13. Beltagui, A., Kunz, N., Gold, S.: The role of 3D printing and open design on adoption of socially sustainable supply chain innovation. Int. J. Prod. Econ. **221**, 107462 (2020)
14. Li, X., Dang, Y., Aazam, M., Peng, X., Chen, T., Chen, C.: Energy-efficient computation offloading in vehicular edge cloud computing. IEEE Access **8**, 37632–37644 (2020)
15. Foukas, X., Marina, M.K., Kontovasilis, K.: Orion: RAN slicing for a flexible and cost-effective multi-service mobile network architecture. In: Proceedings of the 23rd Annual İnternational Conference on Mobile Computing and Networking, pp. 127–140. (2017)
16. Han, Y., Jiao, J., Weissman, T., Yihong, W.: Optimal rates of entropy estimation over Lipschitz balls. Ann. Stat. **48**(6), 3228–3250 (2020)
17. Elkhateeb, M., Shehab, A., El-Bakry, H.: Mobile learning system for egyptian higher education using agile-based approach. Educ. Res. Int. 2019 (2019)
18. Jo, J.H., Sharma, P.K., Sicato, J.C.S., Park, J.H.: Emerging technologies for sustainable smart city network security: Issues, challenges, and countermeasures. J. Inf. Process. Syst. **15**(4), 765–784 (2019)
19. Hui, Z., Li, J., Wang, X., Gao, X.: Learning the non-differentiable optimization for blind super-resolution. In: Proceedings of the IEEE/CVF conference on computer vision and pattern recognition, pp. 2093–2102 (2021)
20. Hui, Z., Li, J., Gao, X., Wang, X.: Progressive perception-oriented network for single image super-resolution. Inf. Sci. **546**, 769–786 (2021)
21. Wang, K., Machuca, C.M., Wosinska, L., Urban, P.J., Gavler, A., Brunnstrom, K., Chen, J.: Techno-economic analysis of active optical network migration toward next-generation optical access. IEEE/OSA J. Opt. Commun. Networking **9**(4), 327–341 (2017)
22. Glória, A., Dionisio, C., Simões, G., Cardoso, J., Sebastião, P.: Water management for sustainable irrigation systems using internet-of-things. Sensors **20**(5), 1402 (2020)

23. Papa, E., Ferreira, A.: Sustainable accessibility and the implementation of automated vehicles: Identifying critical decisions. Urban Sci. **2**(1), 5 (2018)
24. Soldani, D., Shore, M., Mitchell, J., Gregory, M.: The 4G to 5G network architecture evolution in Australia. J. Telecommun. Digit. Econ. **6**(4), 1–30 (2018)
25. Sjödin, H., Johansson, A.F., Brännström, Å., Farooq, Z., Kriit, H.K., Wilder-Smith, A., Åström, C., Thunberg, J., Söderquist, M., Rocklöv, J.: COVID-19 healthcare demand and mortality in Sweden in response to non-pharmaceutical mitigation and suppression scenarios. Int. J. Epidemiol. **49**(5), 1443–1453 (2020)

Governance Plan and Implementation Path of Community Informatization in the Big Data Era with Mobile Sustainable Networks

Guofeng Bian

Abstract Governance plan and the implementation path of community informatization in the big data era with the mobile sustainable networks is studied in this paper. Community informatization is the application of the modern communication technology, especially Internet technology, to build the information technology application platforms and channels. In the proposed model, the efficient mobile sustainable networks system is proposed to enhance the performance. The basic idea is to use pointer technology to update the distributed mobile database on demand. When the mobile node switches the subnet, it does not register with the mobile database but tracks the location change of the mobile node by creating and modifying the pointer between the mobile routers. Then, the efficiency of the model and robustness of the framework are both studied. Through the testing under different conditions, the performance of the model is shown.

Keywords Mobile sustainable networks · Big data · Governance plan · Community informatization · Smart cities · Computational intelligence

1 Introduction

Community informatization is the application of the modern communication technology, especially Internet technology, to build the information technology application platforms and channels for community government affairs, community management, community services, communities, and family life, and organically link them with the real community system [1]. To enable all members related to the community to share and use community information resources more fully and effectively when communicating information, and ultimately reach the goal of improving the quality of life of community members and promoting the overall progress of society. The so-called community service refers to an activity for residents to use and develop the community resources under the government's initiative and organization [2]. It

G. Bian (✉)
College of Public Management, Tianjin Vocational Institute, Tianjin 300410, China
e-mail: aningma1977@163.com

© The Author(s), under exclusive license to Springer Nature Singapore Pte Ltd. 2022
V. Suma et al. (eds.), *Evolutionary Computing and Mobile Sustainable Networks*,
Lecture Notes on Data Engineering and Communications Technologies 116,
https://doi.org/10.1007/978-981-16-9605-3_49

is a social service carried out by the community as a unit. It has the characteristics of welfare, service, and mutual assistance. It is a purpose. Comprehensive services to improve the quality of life of community residents and satisfy people's material and spiritual life [3].

Currently, there are two main ways of thinking about information integration in academia: One way of thinking can be summed up as the "retroducing and rein-venting" [4]. This idea is to overthrow the isolated information systems that were originally established in the process of urban informatization [5]. Unified planning, unified establishment of the data standards, unified hardware construction, and estab-lishment of a brand-new information system to eliminate the "information island" problem and other problems caused by it. This can completely solve the problem, but this method is not suitable for community-level applications, because doing so is tantamount to completely abandoning the previous construction results of urban informatization, and the cost of reconstruction is huge [6, 7]. Compared with the "retrofit and start over" solution, a "bonding" approach can be adopted to achieve the same effect at a lower cost. In Fig. 1, the model is presented [8].

The meaning of the community informatization can be summarized as follows.

(1) An indispensable foundation and important part of urban informatization construction. The construction of the community informatization is not only conducive to then improving the comprehensive competitiveness of the city, enhancing the level of city management, and improving the ability to deal with emergencies, but also conducive to then improving the city's cultural taste, improving the quality of life of the people, and building a harmonious society in an all-round way [9, 10].

(2) The fundamental task of domain analysis is to then identify common require-ments and individual characteristics among different application systems in

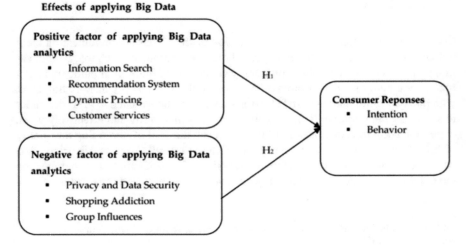

Fig. 1 Community informatization framework

Fig. 2 The initial
communication system

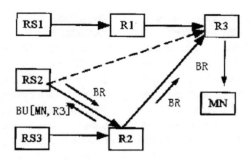

the domain, determine domain boundaries and information sources, find out which requirements are widely shared in the domain, and also summarize the reusable system requirements in the domain to form the domain model.

For the implementations, the network model should be considered. Optimal network design is to optimize the organizational structure of various networks and reduce network energy consumption. (1) Integrate scattered physical resources such as multiple processor cores, storage, and network broadband, reduce the construction and operation, and maintenance costs of network projects from all angles, and realize the resource optimization. Network optimization design can improve the flexibility and expansibility of mobile communication resources and improve work efficiency; (2) Simplify the topology and hierarchy, which can not only improve the resource integration of communication equipment to reduce energy consumption but also save the cost of network project construction [11–13].

The so-called core issue is green wireless mobile communication technology, and the pursuit of faster and higher data transmission capabilities will be realized. Transformation in pursuit of efficiency, effective transmission will be under the condition of limited spectrum resources. Our goal is to apply reasonable and effective requirements, Fig. 2 shows the initial framework.

Switching between different cells of a mobile node is regarded as a change in the membership of the multicast group. Instead of joining or exiting the multicast group, the mobile node only transfers the relationship, that is, the mobile node enters the cell to be switched with the flag of the multicast group. In the next parts, the proposed model will be discussed in detail.

2 The Proposed Methodology

2.1 The Big Data for Smart Cities

Data presentation is the use of visualization software to visually and quickly present the results of data analysis and mining to readers, providing a basis for scientific decision-making. At present, the visualization technology can quickly display data processing results on the web, mobile phones, smart terminals, and other terminals through charts, maps, animations, and more intuitive and vivid forms of physical phenomena or physical quantities that change with time and space, making it convenient people's understanding of data improves decision-making efficiency. Big data technology can process various types of data and present the results with real-time, high-efficiency, and the visual characteristics, changing the current operating mode of computers. Big data technology relies on cloud computing to provide users with computing resources, storage resources, network resources, and information services. On the basis of cloud computing technology, the processing and utilization of big data has become a reality. Big data technology is a series of processing technologies used to collect, transmit, store, process, analyze, and manage data [14, 15].

In our consideration, the core aspects should be analyzed in two ways. (1) On the one hand, it refers to the unprecedented speed of data update and growth; on the other hand, with the further development of informatization, the processing speed of data storage and transmission has reached unprecedented levels. (2) Diversity is a fact that cannot be transferred by the will. The density of data is different, and the processing cost is different. Compared with the traditional questionnaires, GPS and web logs can reflect the content and changes of residents' activities in a timely manner. With the continuous advancement of GIS technology and network map technology, the application of this type of method gradually shows its core advantages. The data layer adopts cloud computing technology, combining the four basic databases of economy, population, legal person, and geography, and the construction of professional databases such as transportation, public security, education, and medical care to provide data sharing, data mining, data analysis, decision support, statistical analysis, etc. "Smart" service effectively supports smart decision-making shown as follows [16–18] (Fig. 3).

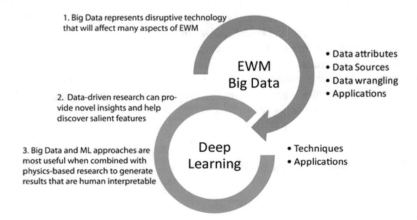

Fig. 3 Big data for smart cities: framework and detail

2.2 Mobile Sustainable Networks

The QoS model defines a structure by which certain types of services can be provided in the network. This is the system goal to be achieved. All other QoS components in the system, such as QoS signaling, QoS routing, and QoS MAC, must work together to achieve this goal. Simple port does not support seamless movement between ASNs, and the IP address needs to be then re-obtained when roaming across ASNs. Therefore, the mobile port came into being, its purpose is to make the mobile node change the network access point. Without changing its IP address, the continuity of communication can be maintained during the movement, thereby maintaining transparency to the upper layer protocol. It can communicate normally with other mobile nodes or nodes that do not have mobile IP functions.

Hence, ideas are separated into the following aspects. (1) The mobile agent of the local cell is responsible for establishing the current range reservation and the pre-range reservation for the cell that the mobile node will enter, and sending the result information to the mobile node [19]. (2) Now in the information age, the communications industry is a leader in the information age. It should actively take advantage of its information technology, adopt advanced information technology, and apply it to all aspects of society, which will be very beneficial to production and life. For example, the combination of the cloud computing and unified communications enables users to use various communication services anytime and anywhere. (3) The new mobility management scheme is called pointer strategy. The basic idea is to use pointer technology to update the distributed mobile database on demand. When the mobile node switches the subnet, it does not register with the mobile database but tracks the location change of the mobile node by creating and modifying the pointer between the mobile routers. When a connection request arrives, the mobile node is addressed according to the pointer, and the possible update of the mobile database is completed at the same time. Figure 4 gives the framework discussions [20, 21].

The local cellular mobile agent provides periodic location, speed, and direction information based on GPS, combined with cellular network and other information, combined with formula 1 for analysis.

$$\text{Dis} = \left(-(ab + cd) + \sqrt{(a^2 + b^2)r^2 - (ad - bc)} \right) / (a^2 + b^2) \qquad (1)$$

The parameter definition of the a, b, c, d can be referred to [12, 17], the cooperative distributed antenna system can effectively solve this problem. For cooperative distributed antenna systems, the base station has multiple distributed antenna transceiver points and a centralized base station processor. "Green" is a broad concept involving many fields. For wireless mobile communications, the meaning of "green" is to save resources and protect the environment. The most important resource in wireless mobile communications is spectrum resources, which is a major resource feature that distinguishes it from wired communications and the other industrial industries. In addition, there are energy resources, site resources, equipment and

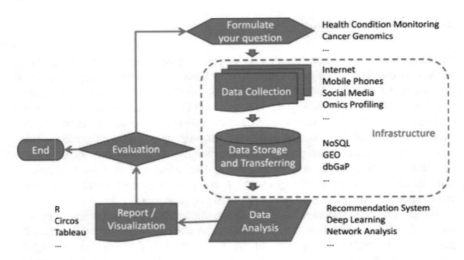

Fig. 4 Mobile sustainable framework details

materials resources, etc. In Fig. 5, the setting of the performance analysis is presented [22, 23].

Fig. 5 Sustainable performance setting

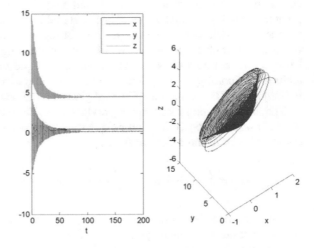

2.3　Governance and Implementation Path of Community Informatization

The construction of community informatization should proceed from the overall situation, break the boundaries between departments and units, mobilize the enthusiasm of all the parties, and take joint actions. The planning, layout, and construction must conform to the government's overall plan and requirements; in specific implementation, unified leadership must be strengthened., Strengthen the joint participation of relevant departments and enterprises involved in community informatization construction, clarify their specific responsibilities and tasks in informatization construction, and do a good job in departmental collaboration. It is necessary to effectively solve the problem of information islands and realize data sharing. Relying on the various e-government information systems, residents can apply to the government for services, consulting services, and online payments at home, and the government can also handle more official duties under the premise of streamlining institutions. The informatization and networking of the public management can replace a lot of tedious manual work. The establishment of the virtual organization simplifies the work process, overcomes a large number of unnecessary intermediate links, and promotes the improvement of work efficiency while reducing office costs and personnel.

The community information system is also constantly developing with the development of society and the needs of the people. It is a dynamically developing system. Using the idea of domain engineering to then analyze and design the community information system, effectively avoiding repetitive labor in the system development process, reducing the labor of the software developers, and saving production costs. Intelligent community requires the establishment of a public management service network system to digitize all kinds of community data and information, facilitate managers to view and process information, and effectively provide targeted services. The intelligent community also needs to rely on advanced network technology to build a reasonable and supporting community management network, connect various departments through the network, and the network various daily work such as community organization construction, public affairs management, cultural construction, urban management, and community service, so as to realize the office management of community affairs with flexible communication and rapid response. Figure 6 shows the estimation.

3　The Simulations

The simulation will be primarily focused on the communications. It is one of the most important tasks to study the energy-saving and environmentally friendly wireless mobile communication community network structure and to establish the community design parameters and the evaluation system under environmental protection. To this

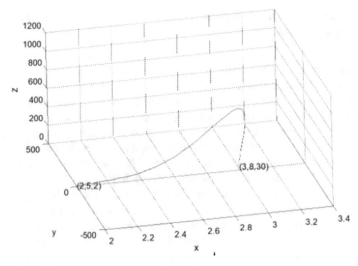

Fig. 6 Community informatization level estimation

end, we propose the comprehensive design parameters of the community network under environmental protection: the lowest terminal reception quality, the necessary access capability of the base station, the core maximum transmission power of the base station, and the network connection overhead per unit area. After receiving the binding request message, the mobile node directly returns a binding update message to the peer host to notify the peer node of the current mobile binding and the remaining registration validity period. Especially, if the mobile node is in the hometown subnet, the care-of address field in the binding update message is the hometown address of the mobile node, and the lifetime is 0. In this way, the peer node can obtain the current location information of the mobile node, and directly send data packets according to the optimized route. The design of the residential area under environmental protection is related to the energy consumption and resource cost of the residential base station and the number of the base station within the service area. So, the important constraint to realize the total cost of base station energy consumption under the communication quality assurance is the total energy cost of the total base stations required under the premise of given service area, given the total number of user communication alligators, Fig. 7 shows the results.

4 Conclusion and Summary

Governance plan and the implementation path of community informatization in the big data era with the mobile sustainable networks are studied in this paper. For wireless communication technology, "green" is an effective definition that covers

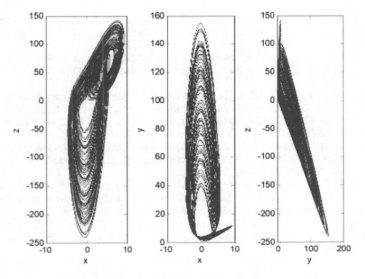

Fig. 7 The communication efficiency simulation result

multiple fields and covers a wide range. Looking at the wording, the word green is an obvious concept of the environmental protection and resource conservation. Spectrum resources are the most important resources in the wireless mobile communications. Spectrum resource is an industrial industry that is different from other wired communications, and it is undoubtedly the main resource feature. This paper starts from the analysis of the mobile sustainable networks and then applies to the path of community informatization. In the next stage, the comparison study will be considered to obtain the better performance.

Acknowledgements Tianjin 2020 Philosophy and Social Science Planning Project: Study on the Cultivation of Community Comprehensive Service Ability of Tianjin Community Social Organizations-Based on the perspective of project operation of civil administration system purchasing social services(TJSR20XSX_017).

References

1. Tian, X., He, J.: Research on informatization monitoring and evaluation index system of sports and medical integration demonstration community based on analytic hierarchy process (AHP). In: 2021 2nd International Conference on Education, Knowledge and Information Management (ICEKIM), pp. 787–791. IEEE (2021)
2. Tangirov, K.: Didactical possibilities of mobile applications in individualization and informatization of education. Mental Enlightenment Sci.-Methodol. J. **2020**(1), 76–84 (2020)
3. Kravtsov, Y.S.: Pedagogical innovation in the conditions of informatization and computerization of humanities education. Turk. J. Comput. Math. Educ. (TURCOMAT) **12**(10), 1573–1577 (2021)

4. Lin, J.: Customized agricultural informatization in big data era. In International Conference on Applications and Techniques in Cyber Security and Intelligence, pp. 497–503. Springer, Cham (2020)
5. Kravtsov, Y.S., Oleksiuk, M.P., Halahan, I.M., Lehin, V.B., Balbus, T.A.: Pedagogical innovation in the conditions of informatization of humanities education. Univers J. Educ. Res. **8**(11D), 117–121 (2020)
6. Wang, P., Lu, B., Yu, C., Zhang, S.: Research on the influence of internet development level on community education in China based on the panel data analysis of Beijing and Shanghai From 2009 to 2016. In: Third International Conference on Social Science, Public Health and Education (SSPHE 2019), pp. 197–200. Atlantis Press (2020)
7. Sang, Y., Xiao, Y.: Analysis and reflection on the current situation of informatization construction in tertiary TCM hospitals in China. In: Journal of Physics: Conference Series, vol. 1774, no. 1, p. 012004. IOP Publishing (2021)
8. Zhang, L., Zhou, Y.: Education informatization: An effective way to promote educational equity. In: 2020 International Conference on Data Processing Techniques and Applications for Cyber-Physical Systems, pp. 837–845. Springer, Singapore (2021)
9. Bag, S., Wood, L.C., Xu, L., Dhamija, P., Kayikci, Y.: Big data analytics as an operational excellence approach to enhance sustainable supply chain performance. Res. Conserv. Recycl. **153**, 104559 (2020)
10. Araz, O.M., Choi, T.-M., Olson, D.L., Salman, F.S.: Role of analytics for operational risk management in the era of big data. Decis. Sci. **51**(6), 1320–1346 (2020)
11. Sung, Y.-A., Kim, K.-W., Kwon, H.-J.: Big data analysis of korean travelers' behavior in the post-COVID-19 era. Sustainability **13**(1), 310 (2021)
12. Bibri, S.E., Krogstie, J.: A novel model for data-driven smart sustainable cities of the future: A strategic roadmap to transformational change in the era of big data. Future Cities Environ. **7**(1) (2021)
13. Zhuravleva, N.A., Wright, J., Michalkova, L., Musa, H.: Sustainable urban planning and internet of thingsenabled big data analytics: designing, implementing, and operating smart management systems. Geopolitics, Hist. Int. Relat. **12**(1), 59–65 (2020)
14. Ouyang, B., Cui, Y: Research on computer network security prevention in the Era of big data. In: Journal of Physics: Conference Series, vol. 1648, no. 2, p. 022011. IOP Publishing (2020)
15. Li, C., Niu, B.: Design of smart agriculture based on big data and Internet of things. Int. J. Distrib. Sens. Netw. **16**(5), 1550147720917065 (2020)
16. Ali Haidery, S., Ullah, H., Khan, N.U., Fatima, K., Rizvi, S.S., Kwon, S.J.: Role of big data in the development of smart city by analyzing the density of residents in shanghai. Electronics **9**(5), 837 (2020)
17. Huang, B., Wang, J.: Big spatial data for urban and environmental sustainability. Geo-spat. Inf. Sci. **23**(2), 125–140 (2020)
18. Ji, Y., Jiang, G.: Garment customization big data–processing and analysis in optimization design. J. Eng. Fibers Fabr. **15**, 1558925020925405 (2020)
19. Delanoy, N., Kasztelnik, K.: Business open big data analytics to support innovative leadership and management decision in Canada. Bus. Ethics Leadersh. **4**(2), 56–74 (2020)
20. Lepore, D., Micozzi, A., Spigarelli, F.: Industry 4.0 accelerating sustainable manufacturing in the COVID-19 Era: Assessing the readiness and responsiveness of Italian regions. Sustainability **13**(5), 2670 (2021)
21. Kim, J., Seo, D., Chung, Y.S.: An integrated methodological analysis for the highest best use of big data-based real estate development. Sustainability **12**(3), 1144 (2020)
22. Zhang, F., Zhang, Y.: A big data mining and blockchain-enabled security approach for agricultural based on Internet of Things. Wireless Commun. Mob. Comput. 2020 (2020)
23. De Las Heras, A., Luque-Sendra, A., Zamora-Polo, F.: Machine learning technologies for sustainability in smart cities in the post-covid era. Sustainability **12**(22), 9320 (2020)

Computer-Based Mathematical Algorithms and Conceptual Models of Complex Networks for Evolutionary Computing

Qian Liu

Abstract Computer-based mathematical algorithms and conceptual models of complex networks for the evolutionary computing are implemented in this paper. The development of modern science, especially system science and complexity science, that has increasingly revealed the interaction, operating mechanism, and internal laws of the objective world in the sense of a larger, deeper, and more complex relationship. A complex network is not a graph composed of nodes and their relationships, but a collection of graphs, or a random process of graphs. From this perspective, complex networks are an extension of graphs. We adopt this structure as the basis for the modeling, through the merger operator, the strong organization merges the weak organization to further strengthen the size and strength of the strong organization, and the modified algorithm is constructed. We applied the proposed model under different data sets and the experimental results compared with the other model has proven the performance.

Keywords Evolutionary computing · Complex networks · Mathematical algorithms · Computer system · Conceptual models

1 Introduction

At present, as a kind of optimization technology with adaptive adjustment function, evolutionary algorithm has been successfully applied in many fields, focusing on solving complex problems such as structural optimization, nonlinear optimization, and parallel computing [1]. When solving problems, its fundamental goal is to pursue group convergence and ensure that the algorithm tends to the global optimum [2, 3].

Generally, a location update operation occurs when a mobile user changes his location in the network, and the mobile terminal carried by the user notifies or updates his current location to the network. When the network needs to deliver the incoming call to a specific mobile user, the system implements a location query operation to

Q. Liu (✉)
College of China West Normal University, Nanchong 637000, Sichuan, China
e-mail: melody332568@163.com

© The Author(s), under exclusive license to Springer Nature Singapore Pte Ltd. 2022
V. Suma et al. (eds.), *Evolutionary Computing and Mobile Sustainable Networks*,
Lecture Notes on Data Engineering and Communications Technologies 116,
https://doi.org/10.1007/978-981-16-9605-3_50

determine the user's current location. In the actual network, both location update and location query need to be then completed through signal transmission, which leads to an increase in network load. Evolutionary algorithm is a classical solution in the field of global optimization and data mining. It can be used in the optimization field to search different method combinations globally and effectively and evaluate the results in each cycle [4, 5].

By adding heuristic guidance in the search process, each search cycle can get better search results than the previous cycle. As follows, details are presented.

(1) The diversity of antibodies. Through cell division and differentiation, the immune system can produce a large number of the antibodies to resist various antigens. This requires that when searching the solution space, the new algorithm should be based on a diverse population [6, 7].

(2) The immune system's response to different antigen invasion, the body's immune cells to respond to the different microbial infection is different, virus infection, usually white blood cell count showed higher percentage of general lymphocytes in, and bacterial infection, neutrophils ratio is higher, which means that the immune system to produce antibodies have very strong pertinence. In the Fig. 1, we show the process.

A complex network is not a graph composed of nodes and their relationships, but a collection of graphs, or a random process of graphs. From this perspective, complex networks are an extension of graphs. In the real society, due to the social environment and society itself, it is undergoing rapid development and changes. Therefore, there is no structure that has remained unchanged for thousands of years. The various nodes in the system and the relationships between them are constantly changing, thus forming a complex network. The laws that run through and are deeply embedded in the complex dynamic network system of this practical process and play a deep role is increasingly showing complex functions and rich phenomena. This requires people to grasp the essence through the phenomenon, reveal the trend and deep law of

Fig. 1 Evolutionary computing process

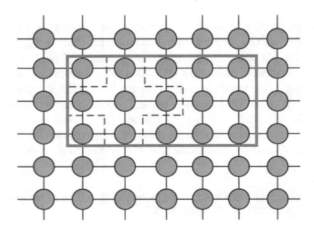

Fig. 2 Conceptual models of complex networks

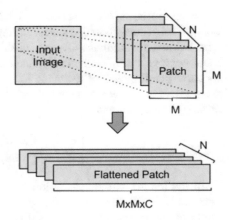

the complex dynamic network system inherent in this practical process, and explore the complex dynamic network system formed by the complex interaction of these laws and mechanism of action and internal composition [8]. The nodes in the network represent individuals, and the edges between nodes represent the interaction between individuals. The individual occupies the nodes in the network and only interacts with the directly connected neighbors. The individual adjusts his game strategy according to the income of the surrounding neighbors. The Fig. 2 shows the model [9].

2 The Proposed Methodology

2.1 Evolutionary Computing as Theoretical Basis

Organizational evolution algorithm is proposed by simulating the competition and cooperation among social organizations. Individuals are stored in organizations. Each organization has a leader, that is, the individual with the greatest fitness in the organization. The whole population is composed of several organizations. The population is evolved through competition and cooperation among organizations. In computer networks, the correct choice of routing is the key to ensuring its orderly operation, and quantum evolutionary algorithms often appear in routing selection. The so-called quantum evolutionary algorithm mainly refers to the product of the combination of evolutionary algorithm and corresponding quantum computing. Specifically, it is based on the expression of the quantum vector and uses the corresponding qubit code to represent the corresponding chromosome and then uses the quantum not gate and quantum rotating gate to update the corresponding chromosome [10, 11]. In turn, the corresponding target problem can be optimally solved. In terms of processes, quantum algorithms generally include the following steps [12] and first, the corresponding population needs to be initialized.

On this basis, the different bodies in the relevant initial population are measured to obtain a set of corresponding states. Chromosomes are also the best representation of correlation to maintain the diversity of solutions. In this way, it makes its expression more concise. At the same time, quantum revolving door evolution is needed for its corresponding way of the evolution. Then, to get the description of the specific method can be expressed in the following expression [13, 14]. We propose the model as the following lemma 1.

$$\begin{bmatrix} \alpha' \\ \beta' \end{bmatrix} = \begin{bmatrix} \cos \theta_i & -\sin \theta_i \\ \sin \theta_i & \cos \theta_i \end{bmatrix} \begin{bmatrix} \alpha_i \\ \beta_i \end{bmatrix} \tag{1}$$

In the lemma 1, the calculation is defined. The main function of the splitting operator is to split the oversized organization into two smaller organizations. In this algorithm, if the size of an organization exceeds the maximum allowable size, it will be split, and the resulting organization will be stored in the next generation population [15]. The merger operator is to merge two core organizations into one organization. The merger operator is designed to imitate the competition and mergers among social organizations. Through the merger operator, the strong organization merges the weak organization to further strengthen the size and strength of the strong organization. Before the merger, the individuals in the weak organization can increase their own strength by learning from the individuals in the strong organization and join a strong organization, so get a better organization defined as the following lemma 2 [16, 17].

$$\max\{f(x)|x \in X\} \tag{2}$$

To optimize and adjust the corresponding functions, the adjustment can adopt the combination optimization method to make the function in the optimal state and provide corresponding conditions for obtaining the optimal solution. By using the method of combinatorial optimization, it can be said that there is no strong correlation between individual genes [18].

Therefore, in the selection of computational network routing, the functions in quantum evolutionary algorithm can be adjusted and optimized accordingly. Based on the sorting method, the population is sorted by the target value. Fitness depends on the order of individuals in the population, rather than the size of the specific target value. It overcomes the scale problem of proportional fitness calculation and possible premature convergence. It has good robustness and can be used as an advantage allocation method for individual selection probability below. lemma 3

$$P_i = \frac{f_i}{\sum_{i=1}^{M} f_i} \tag{3}$$

The $P()$ is the defined function. Among the three operators mentioned above, the merge operator and the cooperative operator mainly play the role of searching for

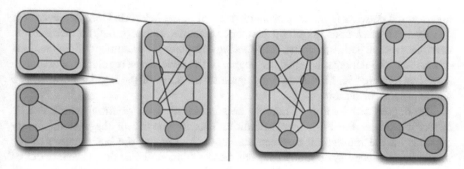

Fig. 3 Computing connection model

new solutions. The merge operator is equivalent to a real number variation operator, while the cooperative operator is equivalent to a crossover operator.

Although the split operator has no ability to search for new solutions, it can play the role of balancing organization scale. We consider listed aspects.

(1) Computing services allow users to securely perform application jobs independently or collaboratively on distributed computing resources. The grid that provides computing services is called a computing grid.

(2) Application services allow users to transparently and securely access remote applications and libraries. Application service will be realized by means of computing grid and data grid.

(3) Knowledge service can help the users achieve specific goals through knowledge mining, knowledge management, and knowledge maintenance. Data mining technology may be a main tool applied to knowledge services.

The first is the limitation of computing resources. The task of evolutionary computing is usually allocated to a small range of the computing resources. The second is the limitation of data sources. Distributed evolutionary computing can use distributed database technology to expand the range of its data sources.

However, distributed database technology can only be applied within an organization. Grid computing breaks through the limitations of the traditional distributed computing technology and expands the scope of computing from a single organization to cross organization computing. Therefore, we propose a grid-based evolutionary algorithm model. In the Fig. 3, the model is defined.

2.2 Complex Network Optimization

The clustering coefficient of a node refers to the ratio of the number of edges that actually exist between the nodes adjacent to the node and the maximum number of interconnected edges of these nodes. The average of the clustering coefficients of all nodes in the network is the clustering coefficient of the network. Complex networks mainly have three core important network measures: network degree distribution, network average path, and network clustering coefficient. The number of connections of a node in the network is called the degree of the node. For a regular network, each node has a degree K. Therefore, the degree distribution of the regular network is a single point distribution.

The degree distribution of random nets is binomial distribution, which can be approximated by the Poisson distribution. The complexity of the system mainly depends on the interaction between elements. As long as the basic nature of the interaction between system elements can be maintained, the basic characteristics of the core system will not change. Although the elements constituting various networks are different and various actual networks have their own complexity, there are some common properties representing universal laws.

Therefore, network model is the most effective model to describe complex systems. At present, a large number of recent studies have found that the network topology determines the characteristics of the network. The development of the inter-disciplinary science has revealed that there are some common laws of the different types and different levels in different fields, different processes, and different levels of intersecting belts and junctions. The common law is not of the same level and type, it is itself a complex dynamic network system.

Due to the connection and intermediary of the common law, a complex dynamic network system with a higher level of law is formed. In a directed network, the degree of a node includes out-degree and in-degree. There is no difference between out-degree and human degree in the undirected network. The degree of each node can describe the importance of the node. $P(k)$ is used to represent the number of nodes in the network or the proportion in the network. The $P(k)$ function of the network reflects the overall characteristics of the network. The Fig. 4 gives the model details.

The clustering algorithm classifies the data based on a certain distance, and the fuzzy algorithm transforms the problem of solving the exact solution into solving the satisfactory solution and solves the problem based on the membership function. The fuzzy clustering algorithm which combines fuzzy and clustering usually starts from constructing the node distance and then calculates the fuzzy membership degree from node to community to reveal the community relationship of node. In the following lemma 4, the model is defined.

$$C_i = \frac{E_i}{k_i(k_i - 1)/2} \tag{4}$$

As the scale of the network continues to increase, the network presents some phenomena that cannot be explained by the small world model. As items and edges join the network in a certain way, the network begins to grow in a certain way. In the process of comprehensive integration, the comprehensive effect and overall advantages of the cognitive network system composed of the cognitive subjects, cognitive methods, and cognitive means can be exerted, so as to grasp the complex dynamic

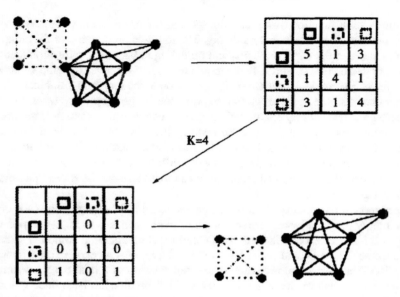

Fig. 4 Model structure details

network system of the law in the process of the gradual approach and dialectical development. The comprehensive integrated seminar hall system from qualitative to quantitative, through human–computer interaction, repeated comparisons, and successive approximations, realizes the transformation from perceptual to rational knowledge, from qualitative to some quantitative, and quantitative knowledge, thereby correcting empirical assumptions to make a clear conclusion or not. The idea based on the seed expansion algorithm is to take the node that meets the maximum selection strategy as the seed node through a specific selection strategy, then calculate the fitness of other nodes and seed nodes according to the fitness function set, and group the node with the maximum fitness and seed node into a community, and perform the above operations successively. Until the obtained community satisfies the evaluation function to obtain the maximum value. The next sections will discuss further.

2.3 Computer-Based Mathematical Algorithms and Conceptual Models

The emergence of computers is the product of mathematical modeling, and the development of computer technology has greatly promoted modeling activities. The computer's high-speed computing power is very suitable for numerical calculations in the mathematical modeling process, which can replace complex and tedious data processing. The computer's large-capacity storage capacity and computer network communication function make data storage and the retrieval convenient and quick in the process of mathematical modeling. The multimedia function of the computer enables some problems in mathematical modeling to be realistically simulated and demonstrated on the computer. In the loop, always first initialize the loop variable or the value of a variable dependent on the operation in the loop body, and then continuously calculate the value of certain expressions in the loop body when the loop variable performs subsequent operations; in the recursive algorithm, it is required to have export function and direct or indirect self-invoking, and also the first equation in these two equations is exactly an initialization or an exit function, and the second equation represents self-invoking. The only difference between them is that they have no parameters. Loop and recursion mean that there can be several parameters in loop and recursion.

In computational theory, the research on undecidability belongs to the category of basic theory. As mentioned earlier, computer mathematics is more concerned with how to achieve efficient calculations. Therefore, computational complexity theory has become the mainstream of computational theory. We can explore the essence of recursion. The essence of recursion is the transmission of a relationship. From the perspective of meta-mathematics, if a series of information has been linearly sorted, and if the latter information directly or indirectly depends on the previous value, there should be an inference rule acting on the known information, and a certain axiom can be obtained. This axiom is an algorithm, a theorem, or a specification in a programming language or database system.

If the transfer of a certain relationship acts on the axiom mode itself, and the conclusions obtained still conform to the form of the axiom mode, then the transfer of this relationship is recursive. The author applies this method to the database field, that is, first formalizes concepts in the real world, extracts axioms and logical rules, and then generates similar postulate tables. Among them, axioms are the set of specifications of database technology itself, and logical rules provide methods to solve related problems with the specifications of database itself. In the Fig. 5, the model is presented.

3   Systematic Verification

The community structure recognition of complex networks is a new research field. Although some community discovery algorithms have been proposed, there are still some open problems that need to be solved and are worthy of further research. Hence, our simulation will be focused on this. The node degree standard deviation and Pearson correlation coefficient of the growth network increase significantly, but the aggregation coefficient decreases and this also shows that the degree heterogeneity and degree correlation of growth networks are enhanced, and the aggregation is weakened. The network growth mechanism in the evolutionary model enhances the degree of network heterogeneity. After the network evolves to a stable state, the cooperative subject occupies the central node. The Fig. 6 defines the eventual results.

EX_IRQ

	t_0	t_1	t_2	t_3	t_4	t_5	t_6	t_7	t_8	t_9	t_{10}	t_{11}	t_{12}
CPU_A	0	1	2	3	4	5	6	7	8	9	10	11	12
CPU_B	2	3	4	5	6	7	8	9	10	11	12	13	14
CPU_C	3	4	5	6	7	8	9	10	11	12	13	14	15
IRQ_A_PENDING	0	0	0	0	1	1	1	1	1	1	1	1	1
IRQ_B_PENDING	0	0	1	1	1	1	1	1	1	1	1	1	1
IRQ_C_PENDING	0	1	1	1	1	1	1	1	1	1	1	1	1
IRQ_A	0	0	0	0	0	0	0	0	1	1	1	1	1
IRQ_B	0	0	0	0	0	0	1	1	1	1	1	1	1
IRQ_C	0	0	0	0	0	1	1	1	1	1	1	1	1

Fig. 5 Designed model framework

Fig. 6 Dynamic simulation result

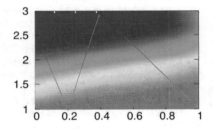

4 Conclusion

Computer-based mathematical algorithms and conceptual models of the complex networks for the evolutionary computing are implemented in this paper. In the network, where the network structure and the strategic behavior of the main body co-evolve, a network growth mechanism is added. Through the comparison and analysis of the simulation results, the level of cooperators in the network is lower than that of the fixed network scale. The decrease in the level of collaborators puts the network in a relatively stable state. Under this condition, the models are presented considering different parameters. In future study, the robustness test will be validated.

References

1. Herrera, M., Pérez-Hernández, M., Parlikad, A.K., Izquierdo, J.: Multi-agent systems and complex networks: Review and applications in systems engineering. Processes **8**(3), 312 (2020)

2. Shi, Y., Han, B., Zeng, Y.: Simulating policy interventions in the interfirm diffusion of low-carbon technologies: An agent-based evolutionary game model. J. Cleaner Prod. **250**, 119449 (2020)
3. Wen, T., Deng, Y.: Identification of influencers in complex networks by local information dimensionality. Inf. Sci. **512**, 549–562 (2020)
4. Milano, G., Pedretti, G., Fretto, M., Boarino, L., Benfenati, F., Ielmini, D., Valov, I., Ricciardi, C.: Brain-Inspired structural plasticity through reweighting and rewiring in multi-terminal self-organizing memristive nanowire networks. Adv. Intell. Syst. **2**(8), 2000096 (2020)
5. Estarellas, M.P., Osada, T., Bastidas, V.M., Renoust, B., Sanaka, K., Munro, W.J., Nemoto, K.: Simulating complex quantum networks with time crystals. Sci. Adv. **6**(42) eaay8892 (2020)
6. Iacobello, G., Ridolfi, L., Scarsoglio, S.: A review on turbulent and vortical flow analyses via complex networks. Phys. Stat. Mech. Appl. 125476 (2020)
7. Martino, A., Giuliani, A., Todde. V., Bizzarri, M., Rizzi, A.: Metabolic networks classification and knowledge discovery by information granulation. Computat. Biol. Chem. **84**, 107187 (2020)
8. Vlachas, P.R., Pathak, J., Hunt, B.R., Sapsis, T.P., Girvan, M., Ott, E., Koumoutsakos, P.: Backpropagation algorithms and reservoir computing in recurrent neural networks for the forecasting of complex spatiotemporal dynamics. Neural Netw. **126**, 191–217 (2020)
9. Will, M., Groeneveld, J., Frank, K., Müller, B.: Combining social network analysis and agent-based modelling to explore dynamics of human interaction: A review. Socio-Environ. Syst. Modell. **2**, 16325–16325 (2020)
10. Amamuddy, S., Olivier, W.V., Manyumwa, C., Khairallah, A., Agajanian, S., Oluyemi, O., Verkhivker, G.M., Bishop, Ö.T.: Integrated computational approaches and tools for allosteric drug discovery. Int. J. Mol. Sci. **21**(3) 847 (2020)
11. Zareie, A., Sheikhahmadi, A., Jalili, M.: Identification of influential users in social network using gray wolf optimization algorithm. Expert Syst. Appl. **142**, 112971 (2020)
12. Smys, S., Chen, J.I.Z., Shakya, S.: Survey on neural network architectures with deep learning. J. Soft Comput. Paradigm (JSCP) **2**(03), 186–194 (2020)
13. He, C., Tian, Y., Wang, H., Jin, Y.: A repository of real-world datasets for data-driven evolutionary multiobjective optimization. Complex Intell. Syst. **6**(1), 189–197 (2020)
14. Zuo, L.: Computer-based mathematical modeling method and application. In: Journal of Physics: Conference Series, vol. 1744, no. 3, p. 032145. IOP Publishing (2021)
15. Kovács, L., Czakó, B., Drexler, D.A., Eigner, G., Ferenci, T.: Integrative cybermedical systems for computer-based drug delivery: Research results of the physiological controls Research Center of Óbuda University. In: Automated Drug Delivery in Anesthesia, pp. 269–315. Academic Press (2020)
16. Driscoll, J.C.: Fractal Patterns as fitness criteria in genetic algorithms applied as a design tool in architecture. Nexus Netw. J. **23**, 21–37 (2021)
17. Florez, H., Cárdenas-Avendaño, A.: A computer-based approach to study the Gaussian moat problem. In: International Conference on Applied Informatics, pp. 481–492. Springer, Cham (2020)
18. Calandriello, L., Walsh, S.L.: The evolution of computer-based analysis of high-resolution CT of the chest in patients with IPF. Br. J. Radiol. **94**, 20200944 (2021)

Performance Comparison of Data Security Strategies in Fog Computing

S. Navya and R. Sumathi

Abstract Fog computing extends cloud computing by providing resources on the edges of a network. It offers a decentralized computing structure that lies between the cloud and data-generating devices. Fog computing is emerged as the remarkable technology that aims at providing various services in the form of resources like servers, databases, storage, networking, and intelligence-over the Internet ("the cloud") to customers. The demand for these services has drastically increased over the years due to flexibility and cost-effectiveness. As the demand increases for services, there is also the possibility of an increase in the threats. Security is critical in solving the challenges posed by these threats. In this paper, an attempt has been made to analyze the available security mechanisms to ensure security in the cloud and the fog environment such as decoy, advanced encryption standard, emoticons, blockchain-based security, attribute-based encryption, and identity-based encryption.

Keywords Fog computing · Cloud computing · Services · Security mechanism · Blockchain

1 Introduction

It is a well-known fact that the Internet changed the world of computing. Today, we have plenty of sources like social media, E-commerce Websites, IoTs which creates a huge amount of data every second. Since, the amount of data created by these sources is very huge, it is termed as big data. To handle the big data, cloud services are the major platform as it provides computation and storage service at a low cost. Cloud computing is facilitated through data centers, which are placed at different geographical location. Data center is a pool of resources in the form of servers. The cloud provides computing power, storage, software, and other services.

S. Navya (✉) · R. Sumathi
Siddaganga Institute of Technology, Tumakuru, Karnataka, India

R. Sumathi
e-mail: rsumathi@sit.ac.in

© The Author(s), under exclusive license to Springer Nature Singapore Pte Ltd. 2022
V. Suma et al. (eds.), *Evolutionary Computing and Mobile Sustainable Networks*,
Lecture Notes on Data Engineering and Communications Technologies 116,
https://doi.org/10.1007/978-981-16-9605-3_51

Cloud architecture is of centralized nature, so it is very difficult to respond to the requirements like high mobility and low latency. It takes more time to deliver the computing services carried out at data center. To handle the huge number of IoT devices and big data with low latency, fog computing emerged as the extension for cloud computing. As the volume of data increases, there is a greater probability of threat in computing world. The attacks will be more difficult to address and analyze with the growing technologies. Fog computing is an emerging technology that has been an interesting area for malicious attackers around the globe. To ensure the defense against the threats, efficient security mechanisms according to our needs should be implemented.

1.1 Fog Computing: An Overview

Cisco coined the phrase "fog computing." The definition of "fog computing" as per [1], "fog computing is a highly virtualized platform that connects end devices to standard cloud computing data centers via computation, storage, and networking services. It is often, but not always, positioned at the network's edge." Fog computing is not a replacement for the cloud; rather, it is an extension of the existing cloud architecture. The extension means that the cloud's services are brought one step closer to the client or data-generating device. The fog architecture acts as a bridge between the cloud and the end devices. By deploying distributed computing infrastructure, fog computing overcomes the cloud's centralized computing nature. According to Cisco [2], the fog computing environment will be made up of fog nodes that are connected to the end devices in the network. Fog nodes can be placed anywhere within the network. Any device that can compute, store, and has network compatibility can be a fog device. Fog infrastructure supports end devices, edge devices, access points, and switches. Hence, it is of heterogeneous in nature [3].

1.2 Characteristics of Fog Computing and Cloud Computing

NIST [3] has defined a set of essential characteristics to distinguish between fog computing and cloud computing and are listed in Table 1.

Architecture: The architecture of cloud computing is centralized, and it consists of data centers. The data centers are located around the globe and are away from client devices. But fog architecture is decentralized and consists of various small nodes that are close to client devices.

Latency: Delays are expected in the cloud computing and can be tolerated depending upon the application type. But when considered an emergency events like medical applications, these delays cannot be tolerated. Computing data through cloud may

Table 1 Cloud and fog computing characteristics

Characteristics	Fog computing	Cloud computing
Architecture	Decentralized	Centralized
Latency	Low	High
Mobility	Supported	Not supported
Location awareness	Yes	No
Geographic distribution	Yes	No
Scalability	High	Low

result in high latency. Hence, fog brings data to edge of network by reducing latency [4].

Mobility: Mobility feature is supported by fog computing. According to Cisco, any device that has the ability to compute, store, and connect to network can be called as fog node [3].

Location Awareness: Users demand applications that are aware of their device's location. Devices like as automobiles and smart traffic signals, for example. These applications require information that is specific to their geographic area. A centralized service is provided by the cloud. As a result, maintaining location awareness with mobile and geographically spread devices is problematic in cloud computing. Fog computing, on the other hand, is closer to the network edge, so it provides location awareness and allows apps to customize services to users based on device location.

Geographical Distribution: Fog nodes are geographically distributed, unlike cloud nodes, which are centralized. For example, to transmit a high-quality video while traveling, fog nodes are strategically placed along the roadway [5].

Scalability: With a significant growth in the number of edge devices, it is impossible to conduct calculations through a centralized structure like a cloud. Fog computing addresses this problem by utilizing dispersed fog nodes that minimize processing time while also allowing for hierarchical scalability [4].

1.3 Advantages of Fog Computing and Cloud Computing

When compared to traditional computer systems, cloud and fog computing offer additional benefits. While the cloud offers the distributed type of computing, fog enhances it by providing security, reducing bandwidth, etc. The advantages of fog computing on top of the cloud are discussed below.

Distance between cloud content and end device: In fog computing, the cloud contents are brought near the end devices, hence this method resolves the issues of latency and computation delays with respect to the time-sensitive applications like medical emergency, transportation, etc.

Network bandwidth: When fog computing is used as an additional layer between the cloud and the end device, there is no need to send all data to the cloud. As a result, network bandwidth is decreased. As a result, the operation's cost will be reduced as well.

Storage close to end device: Fog node provides temporary storage to the end devices. Hence, during the computation process, whole data will not be transferred to the cloud instead only the permanent data after the computation will be stored in the cloud.

Computation power brought close to end device: Whenever a process needs quick and small computations to be done, fog nodes can efficiently handle those computations without transmission of the work to the main cloud. This saves a lot of power and time. Big data computations still need the cloud to take care of it.

2   Security Challenges in Fog Computing

Fog infrastructure consists of large number of fog devices that can be deployed anywhere. The wireless medium is the only sort of communication between these fog devices. The continuous surveillance of these devices is also not possible; hence, they can be tampered or can be replaced with malicious nodes to disrupt the normal operations. The decentralized nature of the fog architecture also makes it difficult to locate and resolve the issue. Some of the challenges faced by the fog computing are described below:

Access Control Issues: Access control defines the data access policy that specifies the privilege granted to the user. Access control tries to avoid the events where a malicious entity can try to break into the node by assuring all the security policies defined are followed properly [6].

Data Breaches: Normally, data are sent from end devices to the nearest fog node for processing. During the transmission of data from the end device to the fog node, there may be a data breach [7]. This could lead to the disclosure of sensitive information.

Malicious Insider: The person who acts maliciously even by having an authentic entrance to the network can be treated as a malicious insider [8]. The usual attack with respect to the malicious insider is data stealing. The loss that occurs will damage both the service provider and the customers in this type of threat.

Unsafe APIs and Interfaces: The weak end points and APIs result in many security problems related to confidentiality, integrity, and availability (CIA) triad [9]. APIs are the only assets where an attacker can try to get into organization network. Even the service providers disclose many APIs to the clients used in their network. This makes an attacker easy to perform a security attack on the respective APIs.

Authentication: Authentication is the process where an entity verifies its identity. Fog nodes and end devices must register to network and authenticate themselves before performing any operation in the network [7]. Since, both cloud and fog nodes need access to servers, there can be specific authenticated identities with each other. Attackers can make use of these nodes and pose as the legitimate users to access the server.

Integrity: Integrity is the process of maintaining validity, firmness, and consistency of the data [10]. If an attacker is able to crack the encryption key, payload data can easily be altered. Hence, maintaining integrity plays an important role during data transmission. These attacks have the ability to alter data while it is being transmitted from the end device to the fog nodes and from the fog nodes to the cloud.

Trust: The level of trust lowers when end devices connect to the nearest fog node for real-time data processing. As a result, the fog nodes and the end devices trust levels must be assessed. The fog node that offers service to the end device must verify that it is a genuine node. As a result, trust between the fog node and the end device can be established.

Privacy: End devices provide data to fog nodes for compute purposes, which can result in a data breach. Because the data will be relayed to the nearest fog node, the end device's position will be revealed as well [10].

Shared Technology Issues: Usually, the fog node will be shared between hardware devices, storage, platforms, and applications. The infrastructure does not provide strong separation in case of multitenancy. Virtual supervisors are used to separate the computing resources between the clients. There are many flaws with respect these virtual supervisors. These issues can cause access of data to unauthorized entities [7].

Insufficient Due Diligence: Whenever an organization tries to deploy a network in a hurry, there may be insufficient due diligence. Many companies do not gather full information about the service providers [11]. For example, many fog service providers will not have their own data centers.

3 Data Security Strategies in Fog Computing

Fog computing is still in the evolving stage, many types of research have been carried out concerning security issues in fog computing. As discussed in the previous section, there are numerous security concerns with fog computing. Some of the strategies for dealing with these issues are presented below.

3.1 Decoy Technique

The decoy technique is the most efficient way to confuse an attacker. Whenever the server notices unauthorized access to any of the fog services, decoy information can be returned to the attacker [12]. The decoy information consists of fake data that looks like a legitimate one. Once the malicious activity is reported, the service provider can verify it with the customer and responsive measures can be taken.

In this security model [13], the decoy technique is used to preserve data in fog computing. The model mainly consists of 2 stages as shown in Fig. 1. In the first stage, all users can refer to decoy data. In the second stage, the legitimate users can access the original data only if they pass all security challenges. The user-behavior profile, decoy technology, and film production are the three key modules.

To detect anomalous data access, the fog's user-behavior profile module is used. This is used to evaluate whether a real user is accessing data in the fog by looking at things like search behavior, data downloaded, and so on. To determine anomalous access to the user's information in the fog, the user's behavior is observed. If unusual access is detected, a decoy file is created and given to the unauthorized user, which looks just like the actual material. The malicious user is identified using the behavioral profile identification algorithm. Only authorized users have access to the original file/data. If the user is found to be the malicious or unauthorized the bogus information is fed to the user without any notification or warning. The honeypot is made up of decoy data, which is made up of bogus information that protects original data from illegal access. The bogus information will represent the same type of original file, which makes it very hard to differentiate between the original and decoy files. Hence, an unauthorized user can be easily tricked using this method. In decoy technology, two security features are implemented.

- Validation-Once abnormal behavior of the user is detected, it validates whether the authorized or unauthorized user is accessing the data.
- To confuse the attacker, a fake amount of decoy information is provided.

When an anonymous user is detected, a fake file should be automatically generated. In file generation, the file created should look similar to the original data. It is difficult for an unauthorized user to differentiate between fake data and original.

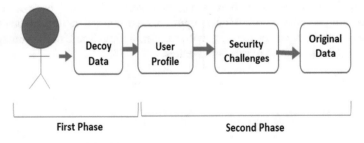

Fig. 1 Security model using decoy technique

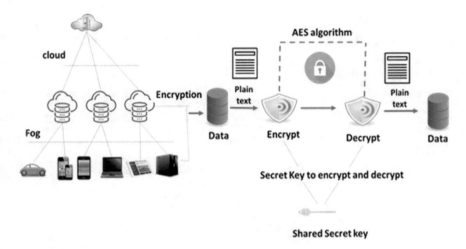

Fig. 2 AES security model

Hence, download fake data. An only authorized users will be able to access original data.

3.2 Advanced Encryption Standard Technique

Unauthorized access is detected in the decoy system via user-behavior profiling. However, there are various issues that can lead to data access and hacking in fog. As a result, advanced encryption standard [14] is used to encrypt data at the fog level in the cloud. The advanced encryption standard (AES), commonly known as Rijndael, is the most widely used encryption technique today. AES uses permutations and combinations in its design, making it faster on both hardware and software. To turn plain text into cipher text, AES can be repeated several times, although the typical configurations are 10 cycles for a 128-bit key, 12 cycles for a 192-bit key, and 14 cycles for a 256-bit key [14]. Since a result, including encryption technology is always a preferable solution, as it may help to prevent man in the middle (MITM) attacks. The AES can be utilized to secure the cloud system at the second level and deployed at mobile edge devices (Fig. 2).

3.3 Emoticon Technique

For fog computing, the advanced encryption standard (AES) method is regarded the most effective security strategy. It consists of encryption and decryption of data

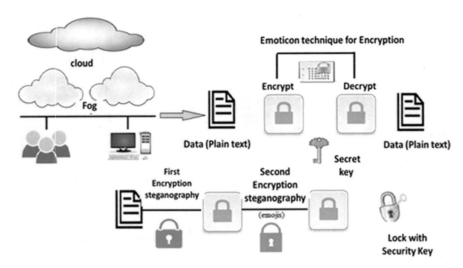

Fig. 3 Security model using emoticon technique

in fog engine to get valid data. AES algorithm requires more processing time and has security overhead. It does not provide protection and privacy to data. In order to overcome the problem, emoticon technique was proposed by Kumar et al. [15], which provides two-layer encryption as shown in Fig. 3.

The emoticons used in chats and comments are used as cover media to deliver data in hidden format. The emoticons are icons or pictorial representation of text that explains a user's perception and reflex it in text mode. As shown Fig. 3, at first stage data are collected by edge devices before transmitting it to fog engine and data are encrypted using emoticons. There are 2 phases in emoticon technique, in the first phase, encryption of data is carried using cryptography and data are transmitted. This encrypted data are input to second phase. In the second phase, encrypted data are hidden by emoticons. These emoticons are put into cover text that will generates stego text that is to be transmitted. The receiver will receive stego text and extract the emoticons from stego text, and then map each emoticon to get the encrypted message. Then, the authorized user will decrypt the message using private key to get original data.

3.4 Blockchain Technique

Blockchain technology was introduced in 2009, and fog computing was introduced in 2013, several efforts were made to integrate both. In recent years, blockchain technology has emerged as the preferred platform for developing secure apps. Blockchain facilitates the development of decentralized applications based on principles such as ledger, cryptography, consensus protocols, and smart contracts.

Ledgers: In blockchain, ledgers are databases that keep track of all transactions and keep them up to date. Because these ledgers are disseminated throughout all nodes in the network, there are several copies. When data in a node are updated in the network, other copies are updated as well. As a result, each data copy in a network is consistent. These ledgers are used to keep track of precious assets such as land records, gemstones, and so on. It is used to hold digital money in the Bitcoin implementation. For storage considerations, ledger records are organized in a chain pattern. A block is a collection of transactions in the asset [16]. The $(n + 1)$th block is linked to the nth block, and the nth block is linked to the $(n - 1)$th block, and so on [17]. Blockchain is the name given to ledgers because of this chain relationship.

Cryptography: To ensure confidence among user transactions in the blockchain network, blockchain technology employs a variety of cryptographic functions. Cryptographic functions are used for a variety of reasons, including maintaining privacy and verifying the identity of agents in a business transaction.

Consensus protocols: Because blockchain is a decentralized system, transactions are duplicated in a network. A unique state of transaction issue can be solved via a consensus process. There are three stages to the consensus process. A node is chosen as the leader node in the first phase. In the second phase, the transaction is validated. Transactions are applied in the third phase. Blockchain employs a variety of consensus algorithms. Some examples are (1) Bitcoin's proof-of-work (PoW) algorithm, in which the leader node is in charge of determining the global state by solving a cryptographic puzzle. (2) The proof-of-state (PoS) algorithm is a consensus process in which the network's leader node is chosen based on the network's greatest stakes.

Smart Contracts: Smart contracts can change the way the blockchain works. Blockchain programmers write smart contracts in a scripting language, which are then executed when a specific event occurs in the system. When additional signatures are validated in Bitcoin, a coin is released. Smart contracts are written in a variety of languages. For example, turing's complete native language, solidity, is used in Ethereum, while general-purpose languages such as Java/go are utilized in Hyperledger.

The centrally administered system in the cloud or fog network may ease the operations but concerning the security issues distributed approach is more effective [17]. The decentralized property of the blockchain along with immutability and transparency can be efficiently implemented in fog and cloud environments. The blockchains can establish trust in the trustless environment of the fog and cloud networks.

Blockchain technology provides the immutable repository to each of the participants in the network. To agree on a single copy of the data repository, all network

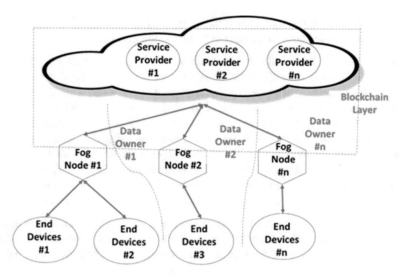

Fig. 4 Blockchain layer in fog computing

members will use the consensus process. The blockchain is practically unchangeable, and with some design assistance, it may also provide non-repudiation and authenticity.

Figure 4 shows the blockchain running over the fog architecture. Blockchain can be integrated into the fog architecture as a layer that guarantees data security. The blockchain's architecture is of distributed nature, hence single point of failure is not possible. Also, the confidentiality is achieved by encrypting the data on the blockchain. The network architecture of the fog nodes in cloud computing makes the blockchain approach more flexible and efficient.

Fog nodes are usually placed near the data sources. This decreases the bandwidth usage and improves the latency in the network. Fog nodes are usually used in IoT networks where data are generated by end devices. Hence, integrating the blockchain with these networks feasibly provide more security. The data gathered from the fog nodes are not always trustworthy, the security and trust play a main role in fog computing environments. When the gathered data are dishonest or when the data are spread without any authorization, it can cause many serious problems. The blockchain can be employed in the network to solve these problems more efficiently and in flexible manner. The blockchain consensus mechanisms, smart contracts, and transaction data mechanisms make it trustworthy and provide accountability in the network.

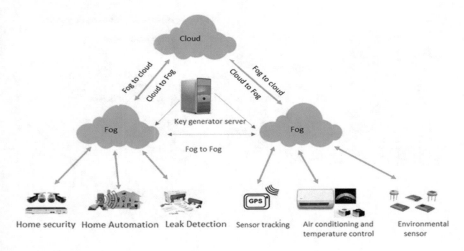

Fig. 5 ABE security model

3.5 Attribute-Based Encryption Technique

Attribute-based encryption (ABE) is an encrypted key exchange protocol. It focuses on providing data privacy along with security by establishing secure communication among a group of nodes in fog. Cipher text policy (CP-ABE) ensures security along with privacy and fine-grained access control [18]. It can be utilized well in the cloud environment because the user is having control over the access policy. The goals of the CP-ABE ensure confidentiality, authentication, access control, and verifiability. The network's fog nodes are linked to a set of attributes, and the encrypted texts are assigned to the access structures defined in these attributes. Decryption is also based on the properties of the fog node, as shown in Fig. 5. The node can only decrypt the cipher text and access the shared key if it has the set attributes. From the beginning, each node is assigned a private key based on their attributes. The cloud performs encryption that is based on a symmetric key. The fog nodes run the decryption process and extract the symmetric key. CP-ABE has high-security strength in collision attacks, message authentication, and unforgeability.

3.6 Identity-Based Encryption Technique

Identity-based cryptography (IBC) concept was introduced by Adi Samir in 1984. It ensures secure data transmission to authorized users. As a public key, users' unique identities such as their name, phone number, email address, and fingerprint are used. The scheme is very useful when it comes to resolving the burdens of the public key directories. It also tries to reduce the delay and focuses on local processing.

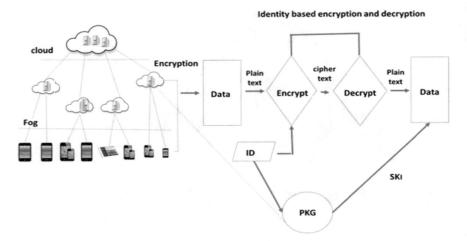

Fig. 6 IBC security model

Because it reduces the stress on the cloud and provides a faster response during data transmission, identity-based encryption (IBE) can be considered to reduce the strain on the cloud and provide efficient security [19].

When IBE is used in a four-layer system, each layer is dependent on the security of the previous level, as shown in Fig. 6. For each fog node in the network, the approach generates efficient unique keys. This ensures that the security of each node is improved at all levels. Knowing the secret keys of each node and generating their private keys is extremely tough for an attacker. Every time the secret code for nodes is generated randomly making it more difficult for an attacker to generate the keys. The secret element is chosen from the prime number which is known only to the private key generator and the fog node. Thus, the algorithm ensures efficient protection to the data theft attacks in the network.

4 Comparative Analysis

This section focuses on a comparative analysis of the security technologies which were discussed above. The comparative analysis provides us with a broad picture of understanding the techniques and their purposes. Most of the techniques discussed are being developed, and continuous research is taking place in their respective field. This section focuses on different security approaches along with their advantages and disadvantages as seen in the Table 2.

Table 2 Comparison of data security techniques

Security technique	Description	Advantage/s	Disadvantage/s
Decoy	Attackers are caught off guard by a false file in the decoy approach. The system is warned when an attacker attempts to download a file, and the information is safe [13]	• It verifies whether the information is accessed by authorized user • It confuses unauthorized user by providing decoy information	• When the attacker knows the authorized user, attacker can guess the security question and hack the data
Advanced encryption standard (AES)	AES is divided into two phases. The first phase consists of data encryption, after which the encrypted data are passed to the fog engine. In the next step, it will provide data decryption using the fog engine in order to obtain genuine data [14]	• It overcomes the man in the middle attack	• It requires more processing time • It has complex security overhead • It does not provide data privacy
Emoticons	Data are encrypted and covered with language, such as emoticons, in the first step, making it impossible to access data. As a result, when an attacker accesses data using this method, he can only see the encrypted data of the original message [15]	• It ensures confidentiality and reliability of communication	• It requires more memory
Blockchain	With blockchain, distributed approach in a network can be implemented instead of a centrally administered approach. Blockchain can establish trust between the fog nodes and enhance the security [17]	• It provides immutability and transparency in the network • Non-repudiation and authentication can also be achieved with additional support	• Difficult to implement • Cost is high because of ever-growing network

(continued)

Table 2 (continued)

Security technique	Description	Advantage/s	Disadvantage/s
Attribute-based encryption (ABE)	Both encryption and decryption rely on the network's fog nodes, which are linked to a set of properties [18]	• CP-ABE has high security • Strength in collision attacks • Message authentication and unforgeability	• Huge load on cloud • Unnecessary delay • More bandwidth consumption
Identity-based encryption (IBE)	It is a certificate free solution in which identifiable data can be utilized as the network's public key for encryption [19]	• It minimizes the load on the cloud while still providing effective security • It decreases the load on the cloud while also providing a faster response time during data transfer	• Key escrow problem is eliminated completely • It fully depends on third party public key generator

5 Conclusion

Nowadays, fog computing is emerging as a major part of the computing world. It plays a crucial role while connecting the end devices like IoTs to the network. The paper presents summary of different set of security mechanisms that can be employed on fog architecture for better security. The analysis was limited to the current trends and available security technologies. But the fog technology is evolving continuously and can expect many innovations in the field of security, which makes fog technology one of the future-oriented technologies. The paper intended to analyze some of the strategies like attribute-based encryption which provides high-security strength to the network, identity-based encryption which ensures security in a certificate free manner, blockchain-based security which ensures immutability and the transparency in the network, private data security using decoy, and AES which is a trivial mechanism for providing data security. Every algorithm has its own benefits and limitations. As per the requirements, one can use the set of algorithms or combine them to provide the efficient security in the network.

References

1. Bonomi F, Milito R, Zhu J, Addepalli S (2012) Fog computing and its role in the internet of things. Proc first Ed MCC Work Mob Cloud Comput 13–16
2. Computing F (2015) Fog computing and the Internet of Things: extend the cloud to where the things are. Available: https://www.cisco.com/c/dam/en_us/solutions/trends/iot/docs/computing-overview.pdf

3. Iorga M, Feldman L, Barton R, Martin MJ, Goren N, Mahmoudi C (2018) Fog computing conceptual model. National Institute of Standards and Technology, Gaithersburg, MD, NIST SP 500-325, Mar 2018
4. Guan, Y., Shao, J., Wei, G., Xie, M.: Data security and privacy in fog computing. IEEE Network **32**(5), 106–111 (2018). https://doi.org/10.1109/MNET.2018.1700250
5. Naha, R.K., et al.: Fog computing: survey of trends, architectures, requirements, and research directions. IEEE Access **6**, 47980–48009 (2018). https://doi.org/10.1109/ACCESS.2018.286 6491
6. Salonikias S, Mavridis I, Gritzalis D (2016) Access control issues in utilizing fog computing for transport infrastructure
7. Khan S, Parkinson S, Qin Y (2017) Fog computing security: a review of current applications and security solutions. J Cloud Comput: Adv Syst Appl
8. Stolfo SJ, Salem MB, Keromytis AD (2012) Fog computing: mitigating insider data theft attacks in the cloud
9. Cloud Security Alliance (2010) Top threats to cloud computing. Available: http://www.clouds ecurityalliance.org/topthreats/csathreats.v1.0.pdf, 15 Nov 2018
10. Yi S, Li C, Lie Q (2015) A survey of fog computing: concepts, applications and issues
11. Walsh J (2014) Lack of due diligence: how it can hurt your company. iCorps Technologies, 09 Nov 2014
12. Amrita Vishwa Vidyapeetham PP (2017) Fog computing: issues, challenges and future directions. Int J Electr Comput Eng (IJECE) 7(6)
13. Bindu Madavi KP, Vijayakarthick P, Decoy technique for preserving the privacy in fog computing
14. Vishwanath A, Peruri R, (Selena) He J (2016) Security in fog computing through encryption. Int J Inf Technol Comput Sci (IJITCS) 8(5):28–36
15. Kumar H, Shinde S, Talele P (2017) Secure Fog Computing System Using Emoticon Technique
16. Almadhoun R, Kadadha M, Alhemeiri M, Alshehhi M, Salah K (2018) A user authentication scheme of IoT devices using blockchain-enabled fog nodes. In: 2018 IEEE/ACS 15th international conference on computer systems and applications (AICCSA)
17. Farhadi M, U-Hopper (2019) Blockchain enabled fog structure to provide data security in IoT applications, 01/15/2019
18. Alrawais A, Alhothaily A, Hu C, Xing XS, Cheng X (2017) An attribute-based encryption scheme to secure fog communications. IEEE
19. Farjana N, Roy S, Mahi MJN, Whaiduzzaman M (2019) An identity-based encryption scheme for data security in fog computing. In: Algorithms for intelligent systems, pp 215–226. https://doi.org/10.1007/978-981-13-7564-4_19
20. Mouradian, C., et al.: A comprehensive survey on fog computing: state-of-the-art and research challenges. IEEE Commun Surv Tutor **20**, 09 (2017)
21. Shakya, S., Nepal, L.: Computational enhancements of wearable healthcare devices on pervasive computing system. J Ubiquitous Comput Commun Technol (UCCT) **2**(02), 98–108 (2020)

Monitoring the Elderly Using Time of Flight Kit

Arvind Vishnubhatla

Abstract It is seen that the physical function of elderly people decreases with age. Most of the problems are related to blood circulation. We need to monitor the state of life before a heart attack or stroke. The falling rate of elderly people also increases. The hustle and bustle of daily existence causes young couples to be bread earners. This causes the elderly people to be left alone without the supervision of others. Activities like frequency of breathing, position inside a room, heart attacks, and stroke are leading causes of mortality in elderly population. Activity monitoring of elderly people provides clues to rehabilitate them before disaster happens. In most cases of long-term attention, tracking of daily activities helps prevent injuries and guarantees safety. In the event of a crisis, late treatment can result in disastrous consequences. Time of flight sensors are used to acquire depth information about the environments. A time-of-flight camera emits modulated laser light and the reflected light from the point to measure distance. Thus, dense depth measurements can be made with low latency. From the time-of-flight data feature, information is extracted and judgmental decisions are made. From the camera, an image frame is obtained and after background subtraction, the target information for extracting the anomaly is obtained. Notable activities which signify important events like wake time, sleep time, and mealtime need to be estimated. An analog device AD-96TOF1-EBZ development kit and a dragon board 410c are employed for the detection and live streaming of depth and IR data is realized. A color map of the depth data with colors going from warm to cold as distance increases is obtained. The depth map is classified using a convolutional neural network.

Keywords Elderly · Problems · Supervision · Time of flight · Depth data · Color map

A. Vishnubhatla (✉)
Electronics and Communications Department, Gokaraju Rangaraju Institute of Engineering and Technology, Hyderabad, India

© The Author(s), under exclusive license to Springer Nature Singapore Pte Ltd. 2022   763
V. Suma et al. (eds.), *Evolutionary Computing and Mobile Sustainable Networks*,
Lecture Notes on Data Engineering and Communications Technologies 116,
https://doi.org/10.1007/978-981-16-9605-3_52

1   Introduction

It is seen that the physical function of elderly people decreases with age [1]. Most of the problems are related to blood circulation. We need to monitor the state of life before a heart attack or stroke. The falling rate of elderly people also increases. The hustle and bustle of daily existence causes young couples to be bread earners. This causes the elderly people to be left alone without the supervision of others [2]. Activities like frequency of breathing, position inside a room, heart attacks, and stroke are leading causes of mortality in elderly population [3]. Activity monitoring of elderly people provides clues to rehabilitate them before disaster happens. In most cases of long-term attention, tracking of daily activities helps prevent injuries and guarantees safety [4]. In the event of a crisis, late treatment can result in disastrous consequences [5].

Time-of-flight sensors are used to acquire depth information about the environments. A time-of-flight camera emits modulated laser light and the reflected light from the point to measure distance [6]. Thus, dense depth measurements can be made with low latency. From the time-of-flight data, feature information is extracted and judgmental decisions are made [7].

From the camera, an image frame is obtained and after background subtraction, the target information for extracting the anomaly is obtained [8]. Notable activities which signify important events like wake time, sleep time, and mealtime need to be estimated [9] (Fig. 1).

An analog device AD-96TOF1-EBZ development kit and a dragon board 410c are employed for the detection and live streaming of depth, and IR data is realized. A color map of the depth data with colors going from warm to cold as distance increases is obtained [10].

Fig. 1 Experimental Setup

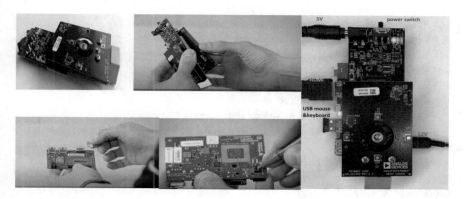

Fig. 2 AD-96TOF1-EBZ

2 About the Method

The AD-96TOF1-EBZ is a tested platform for depth perception. It has been used with 3D software for algorithm development. A VGA resolution helps objects with a higher level of granularity to be detected. This helps detect depth when ambient conditions are strong. A range of depth detection modes can be employed with increased accuracy. The system consists of two boards, a laser transmitter board and an AFE Receiver board.

A precision driver fires 4 individual lasers from a laser board. This is where a CCD sensor from an AFE board mates the laser board [11]. This minimizes the shadowing effects and gives optimum performance. The interface connector provides power and control I/O. A frame rate of 30fps with a resolution of 640 × 480 pixels is obtained with a 940 nm VCSEL. Three ranges, namely near, medium, and far are obtained [12] (Fig. 2).

To boot the system, an SD-card is put in the dragon board. The switch S6 is put in the SD-boot position. A 5 V supply is connected to the camera board, and a 12 V power supply is given to the dragon board. On booting a shortcut to the evaluation, application is visible on the HDMI terminal. Live streaming of the depth and IR data as well as recording option is visible. A color map of the depth data is now plotted.

3 The Software Flow

Algorithm 1: The TOF Algorithm Processing
• Initialise the system
  • Get the camera list
  • Initialise the camera
  • Get available frame types
  • Set the frame type to depth only

- Set range of depth modes
- Get camera details
- Set noise reduction threshold
- Request frame from camera
- Convert frame to depth map
- Calculate the distance factor
- Convert from raw values to values that opencv can understand
- Apply a rainbow color map to the mat to better visualize the depth data
- Display the image
- Use CNN to perform object detection and classification.

Algorithm 2: Classify image
- Import the necessary packages
  - Use SSD with MobileNet for classification
  - Random color choices for each class so its easy to distinguish
  - Load pre-trained model
  - Setup video stream from file
  - Loop over the frames from the video stream
    - While True:
      - Read from stream
      - Convert to blob
        - pass the blob through the network and obtain the predictions
        - loop over the detections
        - for i in np.arange(0, detections.shape[2]):
          - get the probability
          - thresholding
          - if probability > probability_treshold:
            - get the index of the detected class
            - get the location of the object
            - display the prediction
      - show the output frame
    - end

See Figs. 3, 4, 5, 6, 7 and 8.

Fig. 3 A color map to show depth information in near mod

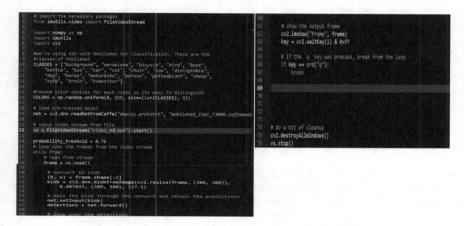

Fig. 4 Algorithm for identifying classes in Python

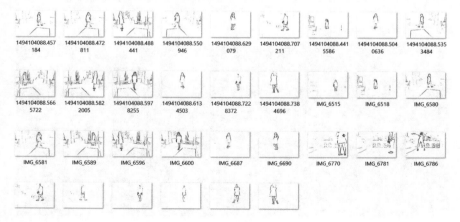

Fig. 5 Training data for standing

Fig. 6 Training data for sitting

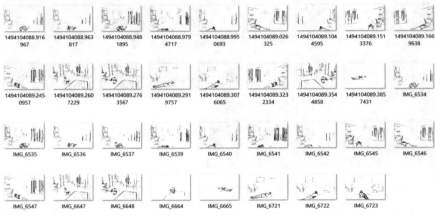

Fig. 7 Training data for falling

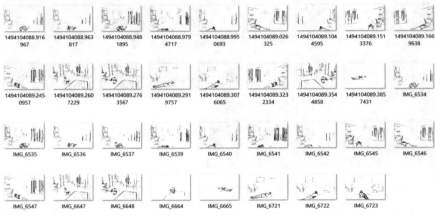

Fig. 8 Algorithm for fall detection in Python

----------------------- Fall Detection -----------------------

```
model.add(Conv2D(32, kernel_size = (3, 3), activation = 'tanh',input_shape = input_shape))

model.add(MaxPooling2D(pool_size = (2, 2)))

model.add(Conv2D(32, kernel_size = (3, 3), activation = 'tanh'))

model.add(MaxPooling2D(pool_size = (2, 2)))

model.add(Flatten())

model.add(Dense(num_classes, activation = 'softmax'))
```

Epoch 1/10

150/150 [==============================] - 69 s - loss: 1.1000 - acc: 0.3533

Epoch 2/10

150/150 [==============================] - 69 s - loss: 1.1038 - acc: 0.3800

Epoch 3/10

150/150 [==============================] - 65 s - loss: 1.1085 - acc: 0.3133

Epoch 4/10

150/150 [==============================] - 62 s - loss: 1.1363 - acc: 0.3267

Epoch 5/10

150/150 [==============================] - 62 s - loss: 1.1335 - acc: 0.3067

Epoch 6/10

150/150 [==============================] - 64 s - loss: 1.1233 - acc: 0.2600

Epoch 7/10

150/150 [==============================] - 64 s - loss: 1.0998 - acc: 0.3200

Epoch 8/10

150/150 [==============================] - 63 s - loss: 1.0991 - acc: 0.3800

Epoch 9/10

150/150 [==============================] - 61 s - loss: 1.0988 - acc: 0.4733

Epoch 10/10

150/150 [==============================] - 60 s - loss: 1.0975 - acc: 0.5467

Training time: 643.820997953

4   Results and Discussion

The color depth map is now used to perform various tasks such as object detection, classification and fall detection. Training data for 'falling,' 'sitting,' and 'standing' are used to train the model. Use is made of backpropagation algorithms to learn the weights. A system using python is designed which predicts 'fallen' or 'not fallen'.

5   Conclusion

The depth information has been successfully used with CNNs to improve classification accuracy. Thus time of flight based color map is an interesting method to track the elderly patients.

References

1. Dinh A, Teng D, Chen L, Shi Y, McCrosky C, Basran J, BelloHass VD (2009) Implementation of a physical activity monitoring system for the elderly people with built-in vital sign and fall detection. In: 2009 Sixth international conference on information technology: new generations
2. Du Y, Zhang B (2018) Research on family monitoring system of elderly solitaries based on embedded system. 978-1-5386-1243-9/18/$31.00 c 2018IEEE
3. Jia G, Zhou J, Yang P, Lin C, Cao X, Hu H, Ning G (2013) Integration of user centered design in the development of health monitoring system for elderly. In: 35th annual international conference of the IEEE EMBS Osaka, Japan, 3–7 July 2013
4. Watanabe S, Asano M, Nakazawa T, Ando R, Tasaki T, Aoki H (2017) Possibilities of simple IoT system for monitoring elderly people living alone. In: 2017 IEEE 6th global conference on consumer electronics (GCCE 2017)
5. Yu L, Chan WM, Zhao Y, Tsui K-L, Personalized health monitoring system of elderly wellness at the community level in Hong Kong. IEEE Access. https://doi.org/10.1109/ACCESS.2018. 2848936
6. Arvani F, Carusone TC, Rogers ES (2019) TDC sharing in SPAD-based direct time-of-flight 3D imaging applications. 978-1-7281-0397-6/19/$31.00 ©2019 IEEE
7. Beer M, Hosticka BJ, Schrey OM, Brockherde W, Kokozinski R (2017) Range accuracy of SPAD-based time-of-flight sensors. 978-1-5386-3974-0/17/$31.00 ©2017 IEEE

8. Bolsee Q, Munteanu A (2018) CNN-based denoising of time-of-flight depth images. 978-1-4799-7061-2/18/$31.00 ©2018 IEEE
9. Choi O, Lee S (2012) Wider angle stereo time-of-FLIGHTC AMERA. 978-1-4673-2533-2/12/$26.00 ©2012 IEEE
10. Durini D, Brockherde W, Member, IEEE, Ulfig W, Hosticka BJ (2020) Time-of-flight 3-D imaging pixel structures in standard CMOS processes. IEEE J Solid-State Circuits 43(7)
11. Mugunthan SR (2020) Concept of Li-Fi on smart communication between vehicles and traffic signals. J Ubiquitous Comput Commun Technol 2:59–69
12. Shrestha, S., Shakya, S.: A comparative performance analysis of fog-based smart surveillance system. J Trends Comput Sci Smart Technol (TCSST) 2(02), 78–88 (2020)

Corpus Creation and Annotation Framework for Sentiment Analysis of Hindi News Articles

Amey K. Shet Tilve, Gaurang Patkar, Leeroy Fernandes, Prathmesh Prabhudesai, Meghana Prakash Sawant, and Sakshi Maurya

Abstract This research is based on the application of sentiment analysis for Hindi news articles. When it comes to text and sentiment analysis, very less research has been done specifically in the domain of Hindi news articles. The system is targeted toward specific domain of defense news articles (Hindi) to understand and analyze emotion of articles from different news agencies. Custom scrapers were designed for data extraction and corpus creation which was later used for training the model. URLs of one of the selected news agencies are provided to the system to extract the content of the articles. Further Multinomial Naive Bayes is used to predict the polarity of a given sentence in the given article based on the previously trained corpus. The polarity count of individual sentences will be used to calculate the maximum value for polarity which will be considered as the overall sentiment of the article.

Keywords Natural language processing (NLP) · Dataset annotation (DA) · Text classification (TC) · Machine learning (ML) · Multinomial Naive Bayes (MNB)

1 Introduction

Sentiment Analysis helps to understand and categorize emotions within a given text, for example: tweets or user reviews. These emotions can be positive, negative, or neutral. A written piece can be conceptual based, facts or sentiments. To understand the dissimilarities among them is crucial. If we treat all of them in a same way, the result will not be prompt and accurate. So, it is mandatory that we first understand the context and identify to which category it belongs to. It concentrates on organizing at many levels of various natures. Today, it has a wide application, especially in the fields of marketing, customer services. It concentrates on sorting the content as

A. K. Shet Tilve (✉) · G. Patkar · L. Fernandes · P. Prabhudesai · M. P. Sawant · S. Maurya
Don Bosco College of Engineering, Fatorda, Goa, India
e-mail: amey.tilve@dbcegoa.ac.in

G. Patkar
e-mail: gaurang.patkar@dbcegoa.ac.in

© The Author(s), under exclusive license to Springer Nature Singapore Pte Ltd. 2022   773
V. Suma et al. (eds.), *Evolutionary Computing and Mobile Sustainable Networks*,
Lecture Notes on Data Engineering and Communications Technologies 116,
https://doi.org/10.1007/978-981-16-9605-3_53

per the required subject. Application of this is present across different domains like sociology, psychology, etc. Such algorithms also hold a dominant part in feedback and recommendation systems. In this approach, defense news headlines are scrapped and articles from different Hindi news agencies and then extracting sentiments from the body of the defense-based articles. Multinomial Naive Bayes is used to predict the polarity of given sentence in the given article based on the previously trained corpus. The polarity count of individual sentences will be used to calculate the maximum value for polarity which will be considered as the overall sentiment of the article [1, 2].

2 Literature Survey

2.1 Sentiment Analysis Model

Data Collection: Data collection is defined as the process of collecting and analyzing. The approach of data collection is different for different fields of study, depending on the required information. The articles which will be fed in the intelligent system are collected from different Hindi news websites and are stored in an excel file (Fig. 1).

Text Preprocessing: The data which has been collected is in raw form and is difficult to understand the motive or sentiment because the text contains lots of uninformative and unnecessary parts. If we try to classify all the data in the same way, the result will not be prompt and accurate, hence Text Preprocessing is required.

Fig. 1 Flowchart for sentiment analysis model

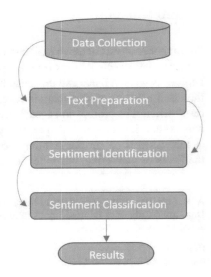

Sentiment Identification: Sentiment analysis is the task which associates itself with the identification and classification of sentiments in texts. Some tasks aim at getting the polarity for the text categorizing them into various labels. Whenever a sentiment is present in the text, it comprises of a source and the sentiment is with respect to some entity, object, event or person. Sentiment analysis is focused on finding the subject, the target, and the polarity for the data. Sentiment Classification [3].

2.2 Web Scraping

Web scraping is an approach, which finds its use in extracting huge amounts of data from websites in which the required data is drawn out and saved locally on a file or a database as .csv, json formats. A web scraping software automates the loading and extraction of website data based on your need. It can be built on your own for a particular website or can be designed to work for any website. Automated tools may also be used for web scraping.

Working—First, a set of URLs are given to the web scraper before scraping. Then the whole HTML code of that particular page is extracted by the scraper. More complicated scrapers will show the full website, containing CSS and JS parts. Next, the scraper can either extract all or specific data selected by the user. Preferably, the user will select the specific data that is needed from the page. Lastly, the web scraper displays all the scraped data in a more formatted way to the user. Most web scrapers will store data in a CSV or Excel spreadsheet.

There are various types of web scrapers that can be used. They can be self-built, an extension, or UI. The self-built scrapers are built using python and its inbuilt libraries such as Scrapy, BeautifulSoup. These can help increase number of features for your scraper. Browser extensions are app-like programs that can add additional functionality to your browser. It can include themes, ad blockers, messaging extensions, and more. Web scraping is simple and lightweight to run and provides easy and seamless integration. However, extensions can be limited by living in your browser. There is also actual web scraping software that can be downloaded and installed on your computer. They are more complicated to use than web scrapers but provide some additional features which are not limited to your browser.

2.3 Annotation

Annotations are certain comments, notes, observations, underlines of text used to highlight important areas of the given text. The use of human annotation is considered an important part of the process which is necessary for supervised learning approach to sentiment analysis. The annotation of a document is performed by the annotator with the polarity of sentiment that it contains with respect to a pre-defined topic. Annotating helps understand the texts better. When we annotate, we are forced to understand what the text means, hence painting a clear image. It results in more meaningful reading and helps in remembering important information. It can provide a set of functions for both private and public reading and communication practices. Simple and to the point instructions are important to improve quality of annotations. This holds true even for very simple annotation tasks, such as sentiment annotation [4–6].

Gold Standard Dataset Creation: It is considered the most accurate and reliable of its type, which can be useful to measure the quality of other datasets. It is created based on the opinion of the majority of annotators.

Below are the steps for the creation of a gold standard dataset:

1. We collect datasets from the website through web crawling.
2. The data collected from the website may include a lot of sentences/words from various sections (if included).
3. The data is annotated to sentence level.
4. The polarity for the sentences can be assigned using the words pos, neg, nu. (pos for positive, neg for negative, nu for neutral).
5. The sentences are annotated by more than two people who speak the Hindi native language. So, for each sentence, there would be many different annotations.
6. The final polarity is decided by the majority basis among the annotations.
7. If for the sentence, the polarity given by all annotators differs, then we discard that sentence [7, 8].

2.4 Text Preprocessing

Sentiment Analysis helps in understanding written pieces of data and categorizing emotions within a given text. This data can be conceptual based, facts or sentiments, sometimes it is difficult to understand the motive or sentiment because the text contains unnecessary parts. If we try to classify all the data in the same way, the result will not be prompt and accurate. It is necessary to transform text into a simpler form to increase performance of machine learning algorithms. Here comes text preprocessing into picture. It is a crucial step in Natural Language Processing tasks [9]. Text Preprocessing gets the text ready for classification by cleaning and preparing it. As we all know that many online text data contain lot of unwanted parts such as HTML tags, scripts, and many words in the text which do not play a huge role in classification.

General Outline of Text Preprocessing: So how do we go about doing text preprocessing?

After a text is obtained, we proceed with the following steps:

1. *Noise Removal*: Remove numbers, special characters, white space, and punctuation.

The following code removes this set of symbols [!"#$%&'()*+, −
./:;<=>?@[\]^_'{|}~]:

2. *Tokenization*: It is a method of forming smaller components by breaking text.
   These smaller components are called tokens.
3. *Removing stop words*: The most commonly used words in a language like "
   आपका", " किसी", " इसलिए", " मैं" are known as "Stop words". These words do not
   have any specific meaning and are normally removed from texts. We can remove
   stop words by using Natural Language Toolkit.
4. *Stemming/Lemmatization*: Stemming is a method by which words can be transformed to its source form (for example: जाएं(Jaen) [Go to], जाएंगे(jaenge) [Will
   go], " जाएँ(jaayen) [Go to], जा(jaen) [Go to], जाओगे(jaoge) [Will go], जाते(jaate)
   [Goes]," जाते(jaaten) [Go to], are different morphological variants of the word
   जा(ja) [Go]) [9].

2.5 Naive Bayes

Naive Bayes Classifiers are classification algorithms that are based on Bayes
Theorem. Naive Bayes classifier accepts that the presence of a specific feature in
a class is not identified with the presence of another feature.

Naive Bayes techniques are a bunch of supervised learning algorithms dependent
on applying Bayes' theorem with the "naive" suspicion of conditional autonomy
between each pair of features given the value of the class variable. Bayes' theorem
states the following relationship, given class variable y and dependent feature vector
$x1$ through xn:

$$P(y|x_1, \ldots, x_n) = \frac{P(y)P(x_1, \ldots, x_n|y)}{P(x_1, \ldots, x_n)}$$

Using the naive conditional independence assumption that

$$P(x_i|y, x_1, \ldots, x_{i-1}, x_{i+1}, \ldots, x_n) = P(x_i|y),$$

for all i, this relationship is simplified to

$$P(y|x_1, \ldots, x_n) = \frac{P(y) \prod_{i=1}^{n} P(x_i|y)}{P(x_1, \ldots, x_n)}$$

Multinomial Naïve Bayes Model: Automatic document sorting becomes increasingly important as handling and organizing documents manually is a time-consuming
and not a viable solution given the number of documents is very huge. The Naive
Bayes technique is a known method that is used for text classification because of its
efficient grating predictions, fast and simple implantation. This article consists of the

basic, probabilistic outcomes to the problems with Multinomial Naive Bayes (MNB) which is used for addressing both systemic problems and the problems which arise because the text is not the case produced based on multinomial model. An MNB classifier is a type of NB classifier which can be often used as a baseline for text classification but here it is applied for Sentiment Analysis [10].

3 Working of the Model

The Hindi dataset used was first split into 60–40 split to be used as a training and testing set. CountVectorizer() is used for transforming a provided text into a vector-based on the frequency of every word that is present in the entire text. The training data was first fitted to create a vocabulary which was then used to transform the training and testing sentences using transform(). TfidfTransformer() function was used to transform the count matrix to a normalized tf or tf-idf representation. Tf means term-frequency while tf-idf means term-frequency times inverse document-frequency.

MultinomialNB() was used to fit the transformed training sentences and their corresponding class labels. This data is used to train the model.

GridSearchCV() is a function that is provided in Scikit-learn's model_selection package. It takes pre-defined hyperparameters and loops through it and fits your model on the training set. So, in the end, we can select the best parameters from the listed hyperparameters. params variable was used to store all hyperparameters such as "alpha", "fit_prior", "class_prior", etc. used for tuning (Fig. 2).

The confusion matrix results after hyperparameter were as Fig. 3.

In the training set: 554 Hindi neg sentences were predicted correctly. 1919 Hindi neu sentences were predicted correctly. 247 Hindi pos sentences were predicted correctly.

In the testing set: 132 Hindi neg sentences were predicted correctly. 634 Hindi neu sentences were predicted correctly. 59 Hindi pos sentences were predicted correctly.

The newer version of article model with hyperparameter tuning showed higher accuracies due to re-attempting the annotations to check for errors and gaining an accurate solution by fine-tuning the parameters.

```
from sklearn.naive_bayes import MultinomialNB
from sklearn.svm import SVC
from sklearn.model_selection import GridSearchCV
alphas = [0.01, 0.1, 0.5, 1.0, 10.0, ]
params= {'alpha': alphas, 'fit_prior' : [True, False], 'class_prior' : [None, [.1,.9],[.2, .8]]}

grid = GridSearchCV(MultinomialNB(), param_grid = params, n_jobs=-1, cv=5, verbose=5)

model=grid.fit(title_tfidf,y_train)

Fitting 5 folds for each of 30 candidates, totalling 150 fits
```

Fig. 2 Implementation of multinomial Naive Bayes

Fig. 3 Confusion matrix for training and testing set

```
from sklearn.metrics import confusion_matrix
confusion_matrix(y_train, train_predictions)
```

```
array([[ 554,  203,   27],
       [ 144, 1919,   74],
       [  55,  145,  247]], dtype=int64)
```

```
confusion_matrix(y_test, test_predictions)
```

```
array([[132,  86,  14],
       [ 65, 634,  32],
       [ 29,  72,  59]], dtype=int64)
```

Table 1 F1 score on training and testing set

	F1 score for training set	F1 score for testing set
Negative sentences	0.72	0.58
Neutral sentences	0.87	0.83
Positive sentences	0.62	0.45

Accuracy on training data is 81%.

Accuracy on testing data is 73%.

Table 1 shows the f1 score for training and testing dataset for negative, positive, and neutral Hindi sentences [10].

4 Experimentation

Analysis for the first confusion matrix (Fig. 4 and Table 2).

In the training set confusion matrix: 554 neg sentences were predicted correctly. 1919 neu sentences were predicted correctly. 247 pos sentences were predicted correctly.

Figure 5 is the graphical representation of the confusion matrix of training set.

Analysis for second confusion matrix (Table 3).

In the testing set: 132 neg sentences were predicted correctly. 634 neu sentences were predicted correctly. 59 pos sentences were predicted correctly.

Figure 6 is the graphical representation of the confusion matrix of testing set.

The accuracy of the latest version of article model after hyperparameter tuning is shown in Table 4.

Figure 7 shows the graphical representation of the accuracy levels [10, 11].

```
from sklearn.metrics import confusion_matrix
confusion_matrix(y_train, train_predictions)
```

```
array([[ 554,  203,   27],
       [ 144, 1919,   74],
       [  55,  145,  247]], dtype=int64)
```

```
confusion_matrix(y_test, test_predictions)
```

```
array([[132,  86,  14],
       [ 65, 634,  32],
       [ 29,  72,  59]], dtype=int64)
```

Fig. 4 Analysis of confusion matrix

Table 2 First confusion matrix

No. of Hindi negative sentences predicted correctly	554
No. of Hindi neutral sentences predicted correctly	1919
No. of Hindi positive sentences predicted correctly	247

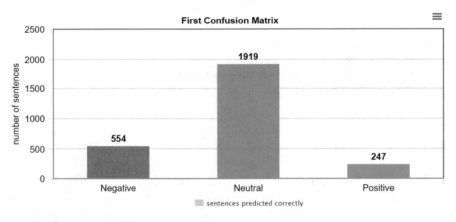

Fig. 5 First confusion matrix

Table 3 Second confusion matrix

No. of Hindi negative sentences tested correctly	132
No. of Hindi neutral sentences tested correctly	634
No. of Hindi positive sentences tested correctly	59

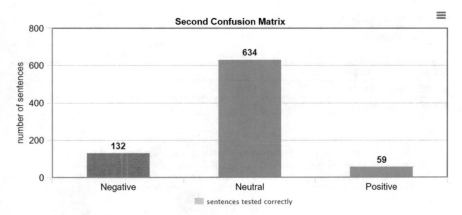

Fig. 6 Second confusion matrix

Table 4 Accuracy for the latest version of article model

Accuracy for training data (60%)	80.76%
Accuracy for testing data (40%)	73.46%

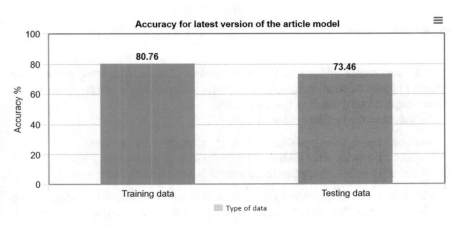

Fig. 7 Accuracy for latest version of the article model

5 Testing

Table 5 shows testing results on 29 URLs of Hindi defense articles of different news agencies which were collected and were analyzed by the annotators to determine the polarity of the articles as positive, negative, and neutral.

Table 5 Testing analysis

Total no. of articles	29	
	Annotator	System
No. of positive articles	7	4
No. of negative articles	11	9
No. of neutral articles	11	10

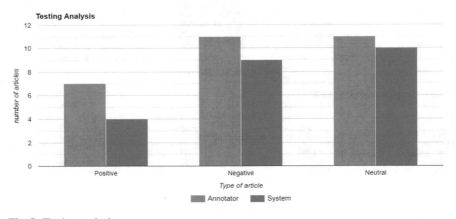

Fig. 8 Testing analysis

After annotation of the articles by annotators, Same URLs were given to the system to predict overall polarity of the article, and prediction results of the model were compared with result of the annotators.

As shown in the tabular representation:

4 out of 7 articles were predicted correctly as positive by the system.

9 out of 11 articles were predicted correctly as negative by the system.

10 out of 11 articles were predicted correctly as neutral by the system.

Figure 8 shows the graphical representation of testing results which displays the number of articles that were predicted correctly by the system based on annotators' predictions.

6 Conclusion

Sentiment Analysis is a technique in which a series of procedures are followed so as to accurately identify the sentiment of the speaker of a sentence, so as to make use of this sentiment information to develop and train prediction and decision-making machine learning models.

The project started with each of the annotators trying to get a better understanding of annotations and sentiment analysis by trying to analyze and annotate English sentences. Once a good understanding of the concepts of class by the basis of popularity and unbiased opinion, various news agencies were shortlisted for scraping of the news articles. Finally, news agencies: NDTV, INDIA TV, AAJTAK, TV9 were selected and web scrapers were designed for them.

A custom dataset was created by using sentences of the articles collected and assigning polarity labels as positive, negative or neutral to them by individual annotators. The annotations were done according to well-designed annotation guidelines based on a gold standard.

Several modifications were done to the dataset like using a different number of sentences or rechecking annotations and its impact on the performance of the models was carefully documented. Finally, the model version with best accuracy and performance was used for integration with the front end created in streamlit. The user interface was kept minimalistic so that it seamlessly performed the desired function of show the user a detailed analysis of the article which included the headline, article, individual sentences in the article along with their polarity. A graphical visualization is used to show the result of the analysis and the final overall sentiment of the article. The project helps identify the emotions of the author and understand what the author is trying to signify in more effective way which can prove to be beneficial especially in a sensitive domain like national security.

References

1. Joshi A, AR B, Bhattcharya P (2010) A fallback strategy for sentiment analysis in Hindi: a case study. In: International conference language processing
2. Mittal N, Agarwal B, Chouhan G, Pareek P, Bania N (2013) Discourse based sentiment analysis for Hindi reviews. In: International joint conference on natural language processing, Nagoya, Japan, 14–18 Oct 2013, pp 45–50
3. Bongirwar VK (2015) A survey on sentence level sentiment analysis. Department of Computer Science and Engineering Ramdeobaba College of Engineering & Management Nagpur, India. Int J Comput Sci Trends Technol (IJCST) 3(3)
4. Mohmad SM (2016) A practical guide to sentimennt annnotation: challenges and solutions. In: Proceedings of the 7th workshop on computational approaches to subjectivity, sentiment and social media analysis, June 2016
5. Liu, B.: Sentiment analysis and opinion mining. Synthesis Lect Hum Lang Technol 5(1), 1–167 (2012)
6. Shrivastava K, Kumar S (2020) A sentiment analysis system for the Hindi language by integrating gated recurrent unit with genetic algorithm. Department of Computer Science Engineering, Jaypee University of Engineering and Technology, India. Int Arab J Inf Technol 17(6)
7. de Arruda GD, Roman NT, Monteiro AM (2015) An annotated corpus for sentiment analysis in political news. In: Proceedings of the 10th Brazilian symposium in information and human language technology, Nov 2015

8. Arora P (2013) "Sentiment analysis for Hindi language" MS by Research in Computer Science Search and Information Extraction Lab (SIEL) Language Technology and Research Center (LTRC) International Institute of Information Technology, Hyderabad, 500032, India, Apr 2013
9. Yadav O, Patel R, Shah Y, Talim S (2020) Sentiment analysis on Hindi news articles. Int Res J Eng Technol (IRJET) 07(05)
10. Shamefulness S, Gopi SC, Viswanadapalli A, Nelapati PK (2020) A Naive Bayes algorithm of data mining method for clustering of data. Int J Adv Sci Technol
11. Singh K (2021) Lexicon based sentiment analysis for Hindi reviews. Linguistics and Language Technology Department. Int J Adv Res Comput Commun Eng 10(1)

Multi-agent-Driven Safety Information Dissemination in V2I Networks

Ramesh B. Koti🆔 **and Mahabaleshwar S. Kakkasageri**🆔

Abstract The intelligent information exchange techniques in vehicular adhoc networks (VANET) have gained huge attention in intelligent transport system (ITS). ITS connects people on the road, cars, and fixed infrastructure using wireless communication technologies to solve different traffic-related difficulties. Dissemination of information is the basis of communication that plays a major role in VANETs and has recently become an active field of research. Vehicle-to-infrastructure (V2I) communication is an essential study topic for enhanced data exchange for implementing security and integrity in communication. In this, we are proposing an effective data dissemination mechanism in V2I network scenario using multiple software agents. These agents are set up and used for the construction of high-quality communication paths between the nodes. The intermediary connections are picked based on each of the moving nodes radio signal strength and relative mobility. This protocol creates an environmentally beneficial system that also improves fuel economy by optimum traffic management techniques. To gather and distribute safety information, we employed a three-layer vehicle to infrastructure (V2I) network architecture with static and dynamic intelligent agents in the proposed study. The suggested algorithm performs better in terms of coverage area, lossless transmission, and decreased latency in a highway networks scenario with medium traffic density. Finally, qualitative comparison is made with present V2I system and found the significance improvement in its performance metrics.

Keywords V2I · Multi-agent · Vehicle manager agent (VMA) · Knowledge base (KB) · Routing node (RN) · High-quality link

R. B. Koti (✉)
Electronics and Communication Engineering Department, Gogte Institute of Technology (Autonomous), Belagavi 590008, Karnataka, India

M. S. Kakkasageri
Electronics and Communication Engineering Department, Basaveshwar Engineering College (Autonomous), Bagalkot 587102, Karnataka, India
e-mail: mskec@becbgk.edu

© The Author(s), under exclusive license to Springer Nature Singapore Pte Ltd. 2022
V. Suma et al. (eds.), *Evolutionary Computing and Mobile Sustainable Networks*,
Lecture Notes on Data Engineering and Communications Technologies 116,
https://doi.org/10.1007/978-981-16-9605-3_54

1   Introduction

Vehicular adhoc networks (VANETs) provide new ways to increase traffic safety and information distribution efficiency. V2I communication involves the broadcasting of recognized hazardous occurrences to neighboring vehicles using unicast, point to multipoint (PMP), or broadcast communication systems, in which a single vehicle's message is received by many receivers. When vehicles are traveling in difficult circumstances such as heavy fog, rain, or scenarios such as difficult to see cyclists or pedestrians running across the road, when these events are communicated in advance with all neighbors to overcome the damage together, this is referred to as a cooperative intelligent transportation system (C-ITS). A signal at a crosswalk, a digital signboard beside the road, and specialized equipment known as roadside units (RSU) are all examples of infrastructure-based network. The received signals from the source cars can be amplified and routed by RSU for onward transmission. The C-ITS used in connected vehicles focuses on digital technology for the information sharing via wireless communication channels.

1.1   Related Works

The vehicular adhoc network (VANET) is a viable answer to today's complicated transportation problems. Vehicles on the road that identify dangerous circumstances and are capable of sharing information collaboratively to prevent additional harm are an obvious example of ITS. The goal of ITS is to improve transportation safety and mobility by combining a variety of technologies and applications. It will enhance people's productivity while minimizing traffic's adverse effects. Road congestion and accidents are one of the most prominent causes for such an adverse impact on traffic. Road accidents have claimed the lives of millions of people and wounded millions more. The impact of these occurrences can be mitigated if information is disseminated quickly. We need to handle these challenges as a global concern. By employing appropriate distribution algorithms, ITS will assist us in improving road safety. Researchers, academia, and industry are now paying close attention to the dissemination of safety information in order to enhance vehicle traffic conditions, but they really make the transport industry more convenient and effective. Post-crash notification, proactive collision warning, parallel parking assistance, emergency electric brake light, left/right turn assistance, traffic signal violation warning, and other information broadcasts are all subject to strict time limitations.

Researchers have proposed a variety of works on information distribution to reduce packet transmission delay. In [1], it provides a comprehensive examination of dissemination focusing on the critical problems of privacy and data security, as well as data forwarding solutions that use fixed infrastructure or centralized administration. Vehicles in a priority-based traffic flow scheme are assigned a sequence ID depending on their speed, direction, and opposite side front. By combining a continuous monitored and disciplined style of dissemination protocol, the wireless communication module (NRF24L01) and Arduino module are utilized for sustainable traffic control. Using the GDSOM-P2P method, the data searching algorithm collects the critical data about the road characteristics and then distributes it to neighbors [2, 3]. In [4, 5], real-time traffic management can be achieved by monitoring V2I communication using an agent-based framework that comprises heavyweight static cognition (based on belief desire intention: BDI) and lightweight mobile agencies in a road works scenario. Based on the information relevancy, it does push (gather/store and disseminate) and pull (gather/store) actions. With a central traffic control database connected by multihop wireless technologies, the vehicle preemption approach improves the possibilities of emergency message distribution. The dataflow control is provided by the different dashboard controls on the logical platform architecture, which are based on DSRC connection [6].

The authors used Wi-Fi, DSRC, and LTE to demonstrate the potential of V2V and V2I in a Het-Net context, ensuring the best possible use of current communication options while reducing backhaul telecommunications network and taking into consideration connected vehicle application needs [7]. The proposed technique has important implications in [8], including overcoming the limitations of K-means-based clustering and increasing clustering accuracy as a crucial tool in data mining and expert systems.

A systematic reading and analysis of the relevant literature produced extremely valuable insights such as motivations, challenges, issues, and suggestions in connection to V2I systems and performance measures using traffic data and methodologies [9, 10]. In [11], it represents an adaptive load balancing method (RBO-EM) for effective data distribution with reduced end-to-end latency, whereas in [12], it talks about the secure cryptography-based cluster mechanism (SCCM) for MANET. Safe routing, encryption, signature creation, signature verification, and decryption were all steps in the development of the AOMDV routing protocol.

The global timeout method and the anti-packet broadcast scheme, where control signals are given in social-based end-to-end and local-based adhoc ways, are two common systems for lossless data transfer [13]. A trust evaluation approach is developed using the Dempster–Shafer theory. The method of evaluation is based on two forms of trust: direct and indirect trust, with the latter being more important [14]. Better results in terms of connection latency and transmission speed proved the reliability of the fragmentation technique. For effective traffic management, vibrant mapping of highway traffic with the collection of a digital map via swarm mapping with grouping is used to enable path optimization to reduce reaction time delays [15]. Our suggested work's limitations are limited to increased vehicle density, which is prevalent in highway scenarios in order to maintain consistent intermediate linkages. In the future scope, the issue of decreased vehicle density will be addressed.

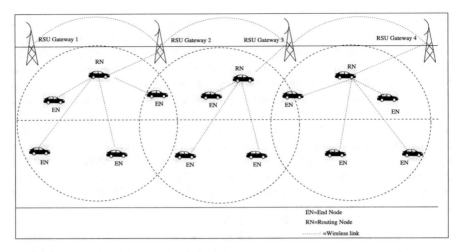

Fig. 1 V2I communication scenario in VANETs

1.2 Proposed Work

The suggested work uses several static and dynamic agents to fulfill two primary functions via the module's node agency and RSU agency. In a vehicle-to-infrastructure (V2I) architecture, the link quality network protocol (LQNP) is a multihop wireless protocol that allows for the fast broadcast of safety information. The nodes in the network under examination might fall into one of three categories, as described below. The communication method distributes traffic across its enabling devices in a tree-like topology. The architecture of the LQNP protocol is seen in Fig. 1. It comprises three tiers of nodes: road side units (RSU), routing nodes (RN), and end nodes.

RSU and RN will handle packet routing, while mesh equipped end nodes will link the RSU to the forwarding nodes for information dissemination. Hundreds of end nodes, numerous routing nodes (RN), and few RSUs along the road side are presented for redundancy and workload sharing to form flexible network environment. Based on their distance and radio signal strength (RSS) level, any end node can be set to operate as RN. A LQNP end node works in power sleep simplex mode for battery operation and does not execute information redistribution until it is awakened by external stimuli or a clock. Table 1 shows the predicted relationship between RSS levels and related connection possibilities for a vehicle density of 100 in a road length of 1000 m, based on current network scenarios.

Table 1 RSS thresholds with connection possibilities

Case	RSS levels (db)	Vehicle mobility (km/h)	Predicted connections (%)
1	30	80	70
2	40	80	60
3	50	80	50
4	60	80	40
5	70	80	30

1.3 Our Contributions

There are many methods for disseminating safety information that are currently in use. We proposing consistent connection-oriented safety information dissemination protocol in this part, which comprises the following set of actions: (1) development of static and dynamic cognitive agents for the gathering and dissemination of safety information. (2) The belief desire intention (BDI) model was used to identify critical and non-critical road danger occurrences. (3) Using the RSS levels of receiving cars, establish congestion-free wireless connections for information transfer. (4) Static agents are triggered to collect information on target cars, and mobile agents are activated to disseminate that information via error-free bandwidth-oriented networks. For instance, when a set of connections is available, the static agent always chooses the higher bandwidth link for dissemination. Rest of the paper is presented as follows. Safety information dissemination using cognitive agents is described in Sect. 2. Simulation model and performance parameters are described in Sect. 3, Sect. 4 includes the result analysis, and finally Sect. 5 concludes the paper.

2 Multi-Agent-Based Safety Data Dissemination Scheme

The roles of static and mobile agents used to improve the data dissemination process are described in this section. The techniques for node categorization and node search are described mathematically, and a network scenario is provided.

2.1 Network Scenario

The suggested network scenario includes a large number of densely packed cars on the road separated by a significant distance. Each vehicle has a GPS navigation system, event collection sensors, communication (TX and RX) capability, and a processing unit. All of the cars are expected to respect the road lane directions and be able to interact with their neighbors utilizing the 802.11 wireless communication standard. Clusters are produced by considering the city environment with curving road lanes and top structures, as well as the relative geographical area in the direction of vehicle moments. The equipment in the cars' onboard systems is expected to enable agent-based platforms. Platform-independent agent programmers launched on each vehicle resolve compatibility concerns. Each end node has a mobile agent, a static agent, and a knowledge base expert system, all of which help with information forwarding via a secure communication channel. The codes used by agents are used to interact with other agents in order to create a greater level of collaboration in order to achieve the given goal.

2.2   Software Agents

An intelligent agent is a self-contained entity that uses its knowledge base to achieve its objectives. Software agents are self-contained programmers that run on a host's agent platform. Agents employ their own knowledge base to accomplish certain tasks without interfering with the host's operations. Mobile agents are modular, adaptable entities that may be built, moved, deployed, and destroyed in real time. Mobile code should be platform agnostic, meaning it can run on any remote host in a mixed network environment. Node information retrieval agent (NIRA), vehicular manager agent (VMA), RSU vehicular manager agent (RVMA), and dissemination agent (DA) are some of the static and dynamic agents used in the proposed dissemination protocol.

2.3   Definitions

Belief Generation: Based on the installed sensors, beliefs are formed and compared to the node parameters. For the data collecting agent, the information tuple consists of source ID, destination ID, TTL, RSS level, mobility, and distance. Because the node with distance and mobility is a critical parameter for the RN selection for the node agency, beliefs are updated with these characteristics.
Desire Updation: The distance and mobility properties of nodes are critical for RN node selection; the desires are to identify nodes with distance and mobility parameters.
Intention Based on Desires: Based on the parametric values for mobility and distance for all the database collected in the interval t, the lowest value of k in KNN model generates the intention of lowest value of distance and mobility.
Cluster Mobility Pattern: It is defined as the collective pattern of vehicle movement in a cluster.
Lane Intersection Pattern: It is the collective segment of lane intersection points with one another.

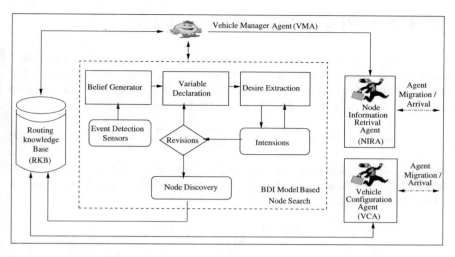

Fig. 2 Node agency functional model

2.4 Node Agency

Each vehicle will have a node agency and an RSU that will send invitation commands to the lower tier nodes; if the request is accepted, the end nodes will join the cluster. Figure 2 depicts the functional components of node agency.

- **Node Information Retrieval Agent (NIRA)**: It is a mobile agent that will be activated by VMA at regular intervals to update the knowledge base's node state. By saving a bit in the header for acknowledgment status, it hosts the HELLO packets in the cluster. The acknowledgment status of end nodes and RN in the coverage range is collected by this agent. As a result, it establishes communication across three layers of nodes, forming a network of high-quality links. NIRA updates the latest network topology dynamics by refreshing the HELLO packet broadcasting every 5ms. The low refresh rate aids the updated topology nodes in catching all essential events.
- **Routing Knowledge Base (RKB)**: This is a centralized data storage system that keeps track of the state of network nodes. NIRA keeps this up to date and uses it to help the VMA in detecting end nodes. RN gateways will use this information to conduct optimal routing. Every 10ms, the knowledge base will be updated. For fast uplink information distribution, RKB implements data gathering in a tree hierarchical structure. For data integrity, it employs the BDI software model.
- **Vehicle Manager Agent (VMA)**: VMA is installed on each end node and activates the mobile agent NIRA, which updates the knowledge base with vehicle status parameters. It manages and coordinates the BDI model's functions. It is aware of all node parameters, including node position, movement, RSS level, and road conditions. VMA, when used with a BDI model, generates beliefs based on the data obtained.

2.5 *K-Nearest (KNN) Neighbor Algorithm*

It is a nonparametric instance-based classification technique that looks for nodes utilizing computation approximation by majority parametric values and K value. The KNN algorithm determines the criticality of data. Select the number K of neighbors in a cluster and compute the Euclidean distance between them as follows:

$$d(p, q) = \sqrt{(q_1 - p_1)^2 + (q_2 - p_2)^2 + \cdots (q_n - p_n)^2} \tag{1}$$

In Eq. (1), take the K-nearest neighbors as per the calculated Euclidean distance and among these k neighbors find the RSS level for each node to satisfy the condition RSS>=threshold value.

$$d(p, q) = \sqrt{\sum_{i=1}^{n} (q_1 - p_1)^2} \tag{2}$$

Extract the safety data for each category by counting the number of nodes in each cluster. Assign the system ID and device ID to the node that has the highest RSS. We found the nearest neighbors as three nearest neighbors in category A and two nearest neighbors in category B by computing the Euclidean distance indicated in equation (2). This formula is used by KNN to calculate the distance between each RSS level of the node in question and the threshold RSS level. It then calculates the likelihood of a node having high-quality connections to RN among all available nodes in the cluster, as shown in Eq. (3)

$$p(Y = j | X = x) = \frac{1}{k} \sum_{i=A} I(Y^i = j) \tag{3}$$

where X is output class, k is the neighbor distance, and $I(Y)$ is the decision boundary. The node discovery by the mobile agent with respect to available information tuple is given by

$$E(d) = (X, X_i) = \sqrt{\sum_{i=1}^{n} (X_j - X_{ij})^2} \tag{4}$$

Since the usage of software static and dynamic agents works concurrently with the inherent routing functions in the V2I model's three-layer architecture, there is less end-to-end latency and congestion. The mobile agents may move between nodes and establish a database, which is accessed by the RSU for intelligent routing and forwarding information that is updated in real time. The current system, which employs an agent-based forwarding technique, can provide routing flexibility while also enabling heterogeneous communication for quick message delivery. The system is organized in a layered design that allows for distributed processing. For greater coverage, traffic signals and digital sign boards can participate as additional RSUs for location-based services and information dissemination in the neighborhood region.

2.6   RSU Agency

It uses the static and dynamic agency components, namely RVMA, KBES, IFA, and DA for its optimum functionality. Here the system finds the target identification and route discovery and establishes the error-free links approaching the target vehicle. The functions of RSU agency components are as follows.

- **Knowledge Base Expert System (KBES)**: It is the data center storage area where the routing node information and optimum hops to the target nodes are fetched and updated by the IFA. VMA uses this data during safety information dissemination after route discovery. The information includes RN ID, Node ID, TTL time stamp, and available bandwidth.
- **Information Fetch Agent (IFA)**: It is a mobile agent which will be triggered by RVMA at regular intervals to collect the information about RN nodes, RSU, and available bandwidth for error-free dissemination. IFA agent is a mobile agent that configures the end nodes on the air with RSS level threshold setting, data communication modes, and required packet TTL intervals. This provides the flexible operating modes of system function. It uses four byte addressing scheme for the header excluding the payload. IFA agent in connection with the KBES synchronizes all the events triggered by three-level network nodes that are associated with regional cluster.
- **RSU Vehicle Manager Agent (RVMA)**: It is an static agent deployed in RSU having the responsibilities of coordinating the activities of information dissemination. It triggers the IFA to collect the information about target vehicles based on the pre-stored route maps and finding the error-free route based on the probabilistic node searching algorithm. It uses the knowledge base data to take the appropriate decisions on route calculation and triggers the safety information dissemination. This unit is responsible for discovering an appropriate neighbor based on node geographic position, vector information, and predicted future direction. It uses HELLO messages to discover neighboring nodes. The next hop selection algorithm is used to find an appropriate neighbor for data forwarding.

2.7   Features of Proposed Model

The well-defined procedural steps in algorithms aid in the construction of programming and make the relationship between input and output parameters easier to comprehend. The following are some of the key aspects of the suggested model:

- **Node Hierarchy**: During node classification process, broadcasting higher level of RSS level for accepting the connections creates fever communication links in network which leads to error-free information dissemination.
- **Dynamic Connections**: Self-healing nature of end nodes in mesh networks continuously checks the alternate connections by comparing the hop level and signal

strength of received HELLO packets. In cases where the primary link becomes failed, the device will automatically change to the alternate routing if such routing is possibly available. If the alternate routing is also unresponsive, the device will enter a state where it searches for new routing possibilities.

- **Cognitive Agents**: Cognitive agents perspective in the dissemination process initiates the faster end to delay during packet transfer. Information is preserved in the knowledge base until the valid connection is established, thus this scheme ensures the guarantee of service.

2.8   Algorithm Properties

The characteristics of proposed algorithm are well defined in terms of sequence of actions performed, during information dissemination. Some of the useful properties of algorithm are as follows.

- **Property 1**: The dissemination process is secure and fast because of high performed link participation in the communication. The system has lower control overhead during connection setup due to the simple registration setup with finite size HELLO packets.
- **Property 2**: The knowledge base provides the nearby digital data centers along the road and traffic signal centers for enhanced target coverage which significantly increases the packet delivery.
- **Property 3**: This algorithm tends to minimize the delay by reducing the frequent handover using RSU wired communication which provides the stable connectivity for dissemination.
- **Property 4**: Automatic registration process of cluster members, RN and on air configuration of end nodes for RSS thresholds gives the better control over the speed and congestion in the network.

3   Simulation

We used C++ developer to evaluate our suggested method, the LQNP protocol, which takes use of the V2I naming scheme. To generate mobility situations, we randomly placed a set of vehicles (mainly 100) on a 5-km route. Different road incidents (e.g., accidents, poor roads, etc.) can have a significant impact on traffic conditions. The distribution of event notification to cars traveling to the event location may aid in making timely decisions, such as rerouting or slowing down. The LQNP uses the BDI machine learning algorithm to give a smart manner of broadcasting. Vehicle mobility patterns, as a result of variable traffic dynamics and road behavior, complicate vehicular information distribution. Broadcast storm and inconsistent connection are the two primary problems in a busy highway traffic scenario.

Table 2 Simulation parameters

Parameter	Considered values
Highway range	5 km
Number of nodes	100
File size	5 kb
Number of segments	4
Transmission range	75–100 m
Antenna model	Unidirectional
Vehicle density per segment	10
TTL of packet	500 ms
Traffic pype	Moderate
Connecting establishment time	2000 ms
Vehicle mobilty	10 m/s, 20 m/s, 30 m/s

3.1 Simulation Procedure

The simulation input parameters are summarized in Table 2. Simulation procedure for the proposed scheme is as follows: (1) Generate VANET network scenario in given road length of 10KM by deploying the vehicles based on geographical clusters. (2) Each vehicle maintains a data structure to store information as specified by scheme (RSS level, mobility, and distance from RN node.). (3) Generate the mobile and static agency to deliver the safety information and vehicle parameters to the RN node (agents are implemented as objects). (4) Apply mobility to vehicles. (5) Randomly generate the vehicle parameters at each vehicle and select RN members using the agency. (6) Use agency to identify RN and announce the intersection mobility pattern. (7) Compute the performance of system.

3.2 Performance Metrics

Some of the performance metrics evaluated are packet delivery ratio, energy consumption, dissemination delay, target coverage, handoff delay, and success rate.

- **Energy Consumption**: It is defined as the amount of energy necessary to establish a connection and transport packets. It is measured in millivolts (mV) and represented in terms of individual nodes.
- **Packet Delivery Ratio**: It is defined as the ratio of total packets received to total packets sent over a certain time span. The maximum time interval for recording the PDR was one TTL duration. It is expressed in percentage.
- **Dissemination Delay**: It is referred to as overhead delay because it contains the time component associated with queuing delay, which occurs when data switching phases in intermediary nodes retain data for a random length of time. It is measured in ms.

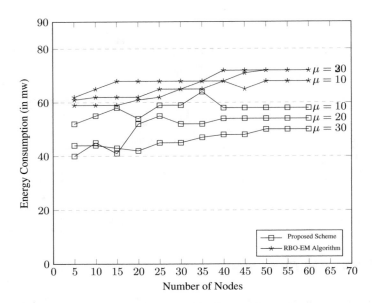

Fig. 3 Energy consumption in mw versus number of nodes

- **Handoff Delay**: It is the time it takes for packets to get from the end mobile agent to the RN nodes, the RN nodes to the RSU, and the RSU to the RSU. It is the sum total of all three delays. It is counted in milliseconds (ms).

4 Result Analysis

To design the simulation, we utilized developer C++ and Python scripting. Both are used to depict traffic flow on roadways that are connected to roadside infrastructure through a transmission control protocol (TCP) connection, with vehicle movement represented by changing vehicle mobility. Table 2 lists the additional parameters utilized in the simulation. The interactive user interface was utilized to deploy nodes randomly in the simulated scenario. Vehicle movements were planned from source to destination via several intermediary nodes, and vehicle routes were meant to last at least 10 s. The vehicles per kilometer were restricted to a minimum of 20 and a maximum of 25. The vehicle speed ranged from 10 to 30 m/s, and each vehicle's transmission range was set at 100 m. The simulation was run ten times, with the averages presented.

The suggested technique minimizes energy consumption in network infrastructure by limiting handover between nodes and selection of relaying nodes based on signal strength. Figure 3 shows the energy consumption simulation results for various node counts and mobility levels using the proposed method. Due to the availability of

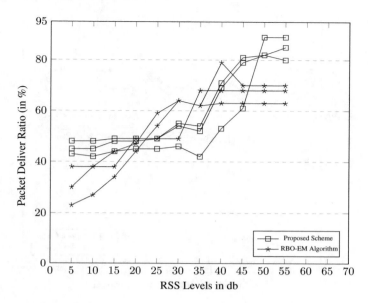

Fig. 4 PDR versus RSS Levels

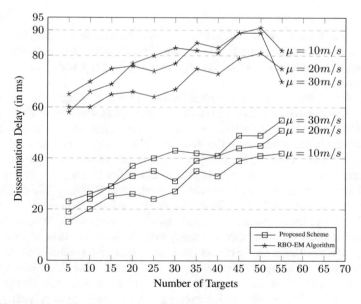

Fig. 5 Dissemination delay in ms versus number of targets

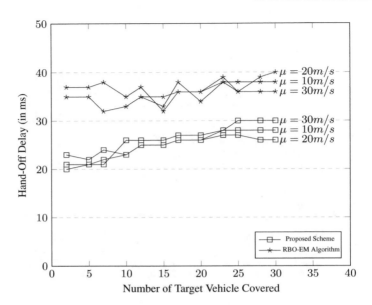

Fig. 6 Handoff delay in ms versus number of target vehicle covered

reliable connections between the nodes, the proposed approach improves outcomes with a reduced energy consumption of 45 mv when the number of nodes is approximately 50. When compared to the RBO-EM method, the proposed algorithm utilizes 10% less energy.

The packet delivery ratio in a communication is the percentage of received packets to transmitted packets. The effect of a greater RSS level on PDR is seen in Fig. 4. When the RSS level is low, certain messages fail to deliver under the suggested approach. However, when the RSS level rises, more nodes are likely to join in communication, lowering the overhead latency and therefore improving PDRs. The outcome depicts PDR with varying vehicle mobility levels. In comparison with the RBO-EM method, the suggested technique improves PDRs by 5%, 6%, and 7%, respectively. The nature of curves for other models has lower bent due to the absence of quality links in the communication path and delays involved in hand on process.

Figure 5 depicts the dissemination delay with various numbers of information receiving vehicles (targets). To obtain the best latency and reachability for a large number of targets, the cognitive multi-agents and probabilistic dispersion approach are used to establish a trustworthy path to the intended receivers. Because each target uses bandwidth for information exchange, the dissemination delay rises as the number of target nodes grows.

The handoff delay is made up of packet transmission notification (handshaking), path update information transmission (queuing), and mobile node processing delay. For a effectual analysis, the various mobility levels are taken into account while calculating the handoff latency. Figure 6 shows the simulated results, which show that the suggested system has a 2RBO-EM scheme. The better outcomes are attributable to the fever intermediary connections, which have larger bandwidth and latency capabilities for information exchange.

5 Conclusion

We proposed a V2I communication protocol to establish stable communication path with higher values of RSS levels to achieve lossless and secure information transport in this research project. The use of agents ensures that vehicle data are updated and disseminated on a regular basis. Furthermore, having a worldwide navigational system in OBU aids in time synchronization and dispersed distribution management. The three-tier design, along with the delegated communication technique, quickly generates a broad coverage area for rapid safety information distribution. The productivity of the V2I system is increased by better network performance characteristics such as end-to-end latency, energy consumption, and resource use. As a result, a 95 percent success rate is visible. The current innovative multilayer communication simulation was performed in developer C++ and produces excellent results with a 95 percent success rate in terms of coverage and broadcast efficiency. Working with reduced node density for single-lane road settings is part of the future scope. For timely synchronization among the V2I interface, the static and mobile agent maintains routing information and refreshes the knowledge base regularly for fast data delivery.

References

1. Ghebleh, R.: A comparative classification of information dissemination approaches in vehicular ad hoc networks from distinctive viewpoints: a survey. Comput. Netw. **131**, 15–37 (2018)
2. Azimi, R., Sajedi, H.: A decentralized gossip-based approach for data clustering in peer-to-peer networks. J. Parallel Dist. Comput. **119**, 64–80 (2018)
3. Markus, P., Matt, G., Sielen, N.: "Real-time traffic management analyzing V2I communication at the edge of the network. In: IEEE International smart cities conference (ISC2). Kansas City, MO **2018**, pp. 1–2 (2018). https://doi.org/10.1109/ISC2.2018.8656899
4. Kakkasageri, M., Manvi, Sunil: Safety information gathering and dissemination in vehicular ad hoc networks: cognitive agent based approach. Wireless Personal Commun. **69** (2011). https://doi.org/10.1007/978-3-642-17878-8-26
5. Manvi, S., Kakkasageri, M., Pitt, J.: Multiagent based information dissemination in vehicular Ad Hoc networks. Mobile Inf. Syst. **5**, 363–389 (2009). https://doi.org/10.1155/2009/518042
6. Chen, J., Mao, G., Li, C., Zhang, D.: A topological approach to secure message dissemination in vehicular networks. IEEE Trans. Intell. Transp. Syst. **21**(1), 135–148 (2020). https://doi.org/10.1109/TITS.2018.2889746
7. Dey, K.C., et al.: Vehicle-to-vehicle (V2V) and vehicle-to infrastructure (V2I) communication in a heterogeneous wireless network performance evaluation. Transp. Res. Part C Emerg. Technol. 68, 68–184 (2016)

8. Azimi, R., Ghayekhloo, M., Ghofrani, M., Sajedi, H.: A novel clustering algorithm based on data transformation approaches. Expert Syst. Appl. **76**, 59–70 (2017)
9. Malik, R., Alsattar, H., Ramli, K.N.B., Bahaa, B., Zaidan, A., Hashim, Z., Ameen, H., Garfan, S., Mohammed, A., Zaidan, R.: Mapping and deep analysis of vehicle-to-infrastructure communication systems: coherent taxonomy, datasets, evaluation and performance measurements, motivations, open challenges, recommendations, and methodological aspects. IEEE Access 1 (2019). 10.1109/ACCESS.2019.2927611
10. Li, T.-H., Khandaker, M.R.A., Tariq, F., Wong, K.-K., Khan, R.T.: Learning the wireless V2I channels using deep neural networks. arXiv e-prints (2019)
11. Ullah, S., Abbas, G., Abbas, Z.H., Waqas, M., Ahmed, M.: RBO-EM: reduced broadcast overhead scheme for emergency message dissemination in VANETs. IEEE Access **8**, 175205–175219 (2020). https://doi.org/10.1109/ACCESS.2020.3025212
12. Mohindra, A.R., Gandhi, C.: A secure cryptography based clustering mechanism for improving the data transmission in MANET. Walailak J. Sci. Tech. 18(6):8987 (2021). https://doi.org/10.48048/wjst.2021.8987
13. Chen, P.-Y., Cheng, S.-M., Sung, M.-H.: Analysis of data dissemination and control in social internet of vehicles. IEEE Internet of Tings J. 5(4):2466–2477 (2018)
14. Mudengudi, S., Kakkasageri, M.: Agent based trust establishment between vehicle in vehicular cloud networks. Int. J. Comput. Netw. Inf. Secur. **11**, 29–36 (2019). https://doi.org/10.5815/ijcnis.2019.07.05
15. Haoxiang, W., Smys, S.: Enhanced Vanet routing protocols for dynamic mapping in real time traffic. IRO J. Sustain. Wireless Syst. **01**, pp. 139–147 (2019). https://doi.org/10.36548/jsws.2019.3.001

Development of Test Pattern Generation for QCA-Based Circuits

Aishwary Tiwari and Vaishali Dhare

Abstract Quantum-D Dot cellular automata (QCA)-based circuits are nanometer-scale circuits. It is an emerging technology with many advantages over CMOS technology, for instance, it can operate in ultra-low power with improved speed and have high packaging density. Testing is an essential requirement or any circuit to detect the fault caused by physical defects that occur during the fabrication. In this paper, automatic test pattern generation (ATPG) algorithm is developed based on D-algorithm to test the stuck-at faults in the QCA circuits. The ATPG is developed using Python and tested on the QCA circuit set synthesized by standard Boolean functions.

Keywords Quantum dot cellular automata (QCA) · Majority voter (MV) · Stuck-at faults · ATPG

1 Introduction

With the advancement in technology, devices are getting replaced with faster, efficient, and more compact versions. QCA [1–3] is one of them. QCA circuits build-up using QCA primitives, namely majority voter (MV), inverter, and binary wires [4]. Being at the nanoscale proper device and circuit, functioning is one of the important aspects. Therefore, testing is very important for QCA-based circuits. Testing consists of two major parts, namely test generation and application. In this paper, test generation is developed. It consists of the generation of test vectors to detect the faults caused by a defect in the fabrication process. Test generation framework for QCA combinational circuits consists of MV, MV as AND and MV as OR gates are presented in [5–7].

In layman's terms, the basic algorithm for testing is to apply input vectors to the circuit and monitor its output followed by its comparison with the desired output, if they both mismatch then it can be declared that the applied input is capable of detecting error. But this simple algorithm would be very tedious if the input pins are

A. Tiwari · V. Dhare (✉)
Institute of Technology, Nirma University, Ahmedabad, India
e-mail: vaishali.dhare@nirmauni.ac.in

© The Author(s), under exclusive license to Springer Nature Singapore Pte Ltd. 2022
V. Suma et al. (eds.), *Evolutionary Computing and Mobile Sustainable Networks*,
Lecture Notes on Data Engineering and Communications Technologies 116,
https://doi.org/10.1007/978-981-16-9605-3_55

large in number. For solving such a dilemma associated with time complexity, other algorithms like D-algorithm and PODEM algorithm are adopted.

In this paper, a modified D-algorithm is developed to detect the single stuck-at fault caused by the missing cell defects in QC circuits [8]. The developed test generation algorithm is efficient and generates fewer test vectors. The test generation algorithm is developed using Python. It is tested on standard circuits synthesized by [9, 10].

The contents of the paper are Sect. 2 presents the background of the testing principle and QCA. The test generation algorithm is presented in Sect. 3. Results are discussed in Sect. 4, and the paper concludes in Sect. 5.

2 Background

In this section, the basics of testing and QCA cell is presented.

2.1 Testing Principle

To check whether the circuit design made by using any device works efficiently or not, testing is required. The basic algorithm of testing is to apply input vectors to the circuit and monitor its output followed by comparing it with the actual desired output, dual input XOR logic is used for comparing different input values (undesired output in our case) results in logic 1. Figure 1 represents this testing mechanism. In CMOS technology, for instance, if this mechanism is used, faulty nets be it with stuck at 1 or stuck at 0 fault are simulated using this basic mechanism and a test pattern is obtained for any particular net with faults or for multiple nets. In the later section, one more algorithm for test vector generation is discussed and both are implemented for circuits made of QCA.

The basic building block of QCA is majority voter (MV).

Any type of circuit except NOT gate can be implemented from MV.

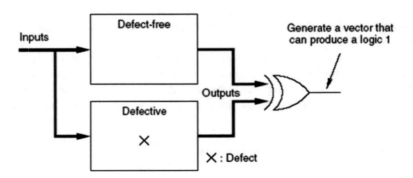

Fig. 1 The basic testing system mechanism

2.2 QCA

A QCA cell comprises a square structure with four quantum dots positioned at its vertices, out of four quantum dots, two diagonal ones consist of 2 electrons each which can move to any quantum dot through electron tunneling. A coulombic force of interaction exists between making the electrons to be placed diagonally, thus only two possible configurations as shown in Fig. 2 are possible. A polarization of -1 and $+1$ is assigned in these two configurations as shown in Fig. 2. Here, $P = -1$ is interpreted as logic 0, and $P = +1$ as logic 1 [1–3].

The advantage of QCA over CMOS is improved speed and reduction in area.

Wires can be formed using QCA cells by placing them next to each other either in the horizontal direction or in the vertical direction as shown in Fig. 3. Due to the existence of repulsive force between electrons inside a QCA cell, if a QCA cell is placed next to another QCA cell then the electrons in it attains the same position as the neighboring cell, if it is placed in opposite orientation then norms of Coulombic force of repulsion is violated. It can be seen in Figs. 3 and 4 that Coulombs's law of repulsion is not violated.

Similarly, NOT or invert logic can be implemented by QCA by placing it like it is done in Fig. 4d.

Fig. 2 Basic QCA cell with polarization

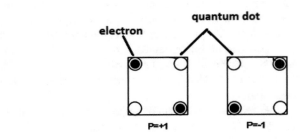

Fig. 3 Wire using QCA cells [4]

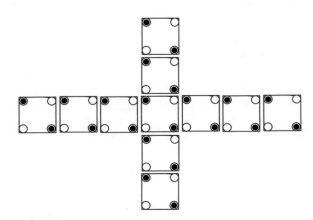

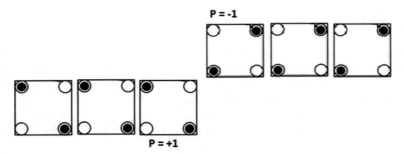

Fig. 4 Inverted logic using QCA cells [4]

Based on research performed on QCA circuits, there are three kinds of defects in QCA implementation based upon their placements. These are cell displacement, misalignment, and omission. Stuck-at faults (SA-0, SA-1) like in CMOS circuits are also possible in QCA and further sections, we have implemented algorithms for simulation and input test vector generation for such faults.

The majority voter gate (MV) is a digital logic circuit with an odd number of inputs. The output of MV is the majority of the inputs, for instance, the output will be logic 0, if the number of logic 0 s in input is more than the number of logic 1 s and vice versa. Figure 5 represents the gate level representation of a three-input MV and its truth table. Logically, it can be implemented by using 3 AND gates and one three-input OR gate. F represents the logic function for the same.

This can be implemented using QCA cells as shown in Fig. 6. The three outer cells act as the input and the fourth one as an output. MV is the fundamental building block of the QCA cell which can implement all the Boolean expressions except NOT gate [9].

ATPG algorithms can be implemented for test vector generation in circuits made from MVs which in turn are made from QCA cells. In further sections, ATPG algorithms implementation is represented for some of the small circuits made using MVs

Fig. 5 Majority voter representation

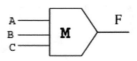

A	B	C	F
0	0	0	0
0	0	1	0
0	1	0	0
0	1	1	1
1	0	0	0
1	0	1	1
1	1	0	1
1	1	1	1

F = AB + BC+ CA

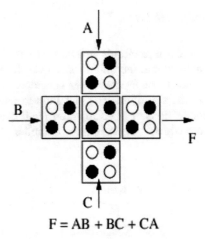

$$F = AB + BC + CA$$

Fig. 6 Majority voter using QCA cells [4]

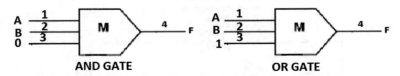

AND GATE   **OR GATE**

Fig. 7 Majority voter using QCA cells

and NOT gates. The minimum number of QCA cells for NOT gate implementation is two and for MV implementation is five. To implement an AND gate using MV just look at Fig. 7, one of the inputs of the MV is set to zero and the other two inputs will act as the inputs to the AND gate. Similarly, for the OR gate, set one of the inputs as logic 1.

3   Test Generation Algorithm

In this section, two ATPG algorithms, general, and D-algorithm [11] are discussed for circuits made of MVs and NOT gates, major aim is to determine all possible test vectors for some of the circuits with stuck-at faults. Like for example, in Fig. 7, if 4th net is at SA-1 fault, then the test vectors will be $(A, B) = (0, 0), (0, 1), (1, 0)$. As these inputs result in logic zero, but since the output net is stuck at 1 fault, so no matter what the input is, it will always result in output $F = 1$, In a similar way, test vectors for any type of circuit are determined.

3.1 Basic ATPG Algorithm

As shown in Fig. 1, a defect-free circuit will be first simulated and then a circuit with a fault will be simulated and the output of both is different for any particular input or multiple inputs then those inputs will be considered as test vectors. Figure 8 explains the flowchart of the algorithm. The first and most step is to read the netlist which tells about the connections of MVs and after reading the netlist, the circuit is converted into the format which is easier for the compiler to interpret and perform operations after that.

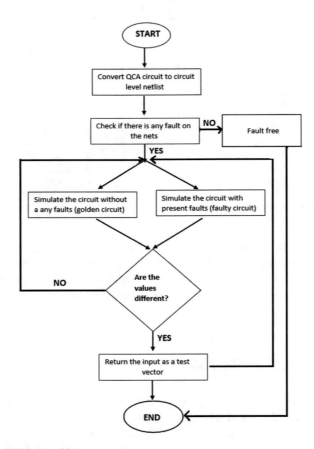

Fig. 8 Basic ATPG Algorithm

3.2 Modified D-Algorithm

In this algorithm, D algebra is used which comprises of 5 values 0, 1, ×, D, D'. Where D stands for discrepancy. It comprises of three steps:- fault propagation, where the site with fault is activated by justifying some of the signals associated with it, in the second step: fault propagation, D/D' is propagated to the POs by frequently simulating the forward gates until the PO is simulated. If for a given fault, D/D' cannot be reflected at the POs, then the fault is undetectable. In the third step, Backtracking is done for reaching back to the inputs. The values obtained at the inputs after back-tracking are the test vectors. All three steps are sequential and have some iterations at individual levels depending upon the complexity of the circuit. After execution of one step completely, the next step execution should start.

In the case of MV, if one of the inputs is at SA-0, SA-1, then to propagate that fault to its output, the other inputs should be either (1, 0) or (0, 1). And by putting all these combinations one at a time for multiple times. However, to simplify our codes efforts, we can put set both the values to "x" if these nets are not PIs as the nets will be iterating for both 0 and 1 possible combination, However, if it is in primary input, then the value cannot be set to "x." By considering this behavior of MV, the algorithm should be implemented. Just by defining the logic by taking into consideration the 5 possible values in our work is done, the rest of the scripting language will take care of the execution as per the instructions given for the three major steps of the D-algorithm (Fig. 9).

Similarly, the function for NOT gate is created. If the output of NOT is D/D', then input shall be 0/1.

4 Simulation Results

Python is used for the implementation of both the algorithms individually, the input to the ATPG code will be ISCAS 85 [12] netlist format which will define each net also faults them, note that the ISCAS netlist format is generally for logic gates like NAND, AND, etc. The idea behind choosing such a format is because it is well organized. However, with a few changes in the coding part, another format can also be made supported. Given below are examples of some of the basic small circuits (Fig. 10).

4.1 Circuit 1

The circuit is made from three MVs, and the fault is simulated for the SA-1 fault at its 8th net Fig. 11 shows the output returned by the compiler.

It returns 16 inputs out of 2^{6} total inputs. These 16 inputs will act as a test vector for SA-1 at net 9 (Fig. 12).

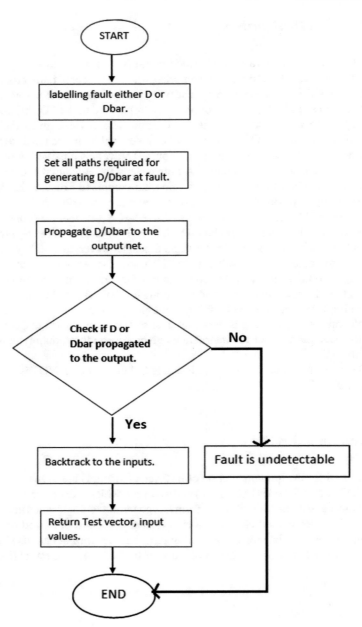

Fig. 9 D-algorithm flowchart

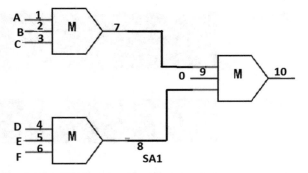

Fig. 10 Circuit with SA-1 fault on net number 8

Fig. 11 Test vector returned by compiler for SA-1 at 8th net

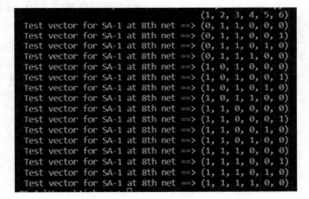

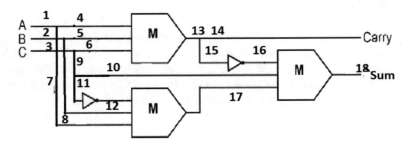

Fig. 12 Full adder implementation using MV and NOT gates [13]

4.2   Circuit 2

The full adder circuit is made by using three MVs and 2 NOT gates, Figs. 13, 14, and 15 show how our compiler returns test vectors for SA-1 fault at 17, SA-0 fault at 5, and SA-1 fault at 10.

Fig. 13 Test vector for SA-1 at net 17

```
                                                          (1, 2, 3)
Test vector for SA-1 at 17th net ==> (0, 0, 0)
Test vector for SA-1 at 17th net ==> (0, 1, 1)
Test vector for SA-1 at 17th net ==> (1, 0, 1)
```

Fig. 14 Test vector for SA-0 at net 5

```
                                                          (1, 2, 3)
Test vector for SA-0 at 5th net ==> (0, 1, 1)
Test vector for SA-0 at 5th net ==> (1, 1, 0)
```

Fig. 15 Test vector for SA-1 at net 10

```
                                                          (1, 2, 3)
Test vector for SA-1 at 10th net ==> (0, 0, 0)
Test vector for SA-1 at 10th net ==> (1, 1, 0)
PS C:\Users\Aishwary>
```

Figures 10 and 12 are symbolic representations of circuits made of MV. However, QCA designer is used to simulate such circuits made from QCA-based cells. Compared to CMOS, QCA cells offer a large reduction in the area with increased speed.

This was a demonstration for two of the circuits. Similarly, any type of circuit made of MVs and NOT gates can be simulated for test vector generation. The results of the algorithms are the same. However, D-algorithm is fast and is more convenient if the PIs to the circuit are large in number.

5 Conclusion

In this paper, the test vector generator is presented to test the stuck-at faults caused by missing cells in the QCA circuits. Python is used to develop the ATPG algorithm. Test vectors for the standard QCA synthesized circuits obtained from the Boolean functions are generated. The algorithm is suitable for QCA circuits, and a smaller number of test vectors are generated using the developed algorithm.

References

1. Lent, C.S., Tougaw, P.D., Porod, W., Bernstein, G.H.: Quantum cellular automata. Nanotechnology **4**(1), 49–57 (1993)
2. Lent, C.S., Tougaw, P.D., Porod, W.: Quantum cellular automata: the physics of computing with arrays of quantum dot molecules. In: Proceedings Workshop on Physics and Computation, pp. 5–13. IEEE (1994)
3. Dhare, V., Mehta, U.: Defect characterization and testing of QCA devices and circuits: a survey. In: 19th International Symposium on VLSI Design and Test, pp. 1–2. IEEE (2015)

4. Tougaw, P.D., Lent, C.S.: Logical devices implemented using quantum cellular automata. J. Appl. Phys. **75**(3), 1818–1825 (1994)
5. Gupta, P., Jha, N.K., Lingappan, L.: A test generation framework for quantum cellular automata circuits. IEEE Trans. Very Large Scale Integr. (VLSI) Syst. **15**(1), 24–36 (2007)
6. Karim, F., Walus, K., Ivanov, A.: Testing of combinational majority and minority logic networks. In: Mixed-Signals, Sensors, and Systems Test Workshop, pp. 1–6 (2008)
7. Dhare, V., Mehta, U.: Development of basic fault model and corresponding ATPG for single input missing cell deposition defects in majority voter of QCA. IEEE TENCON 2016, Singapore
8. Tahoori, M., Momenzadeh, M., Huang, J., Lombardi, F.: Testing of quantum cellular automata. IEEE Trans. Nanotechnol. **3**(4) (2004)
9. Dhare, V., Mehta, U.: A simple synthesis process for combinational QCA circuits: QSynthesizer. In: 2019 32nd International Conference on VLSI Design and 2019 18th International Conference on Embedded Systems (VLSID), pp. 498–499. IEEE (2019)
10. Dhare, V., Agarwal, D.: Implementation and defect analysis of QCA based reversible combinational circuit. In: Technologies for Sustainable Development, pp. 244–249. CRC Press (2020)
11. Jain S.K., Agrawal, V.D.: Test generation for MOS circuits using D-Algorithm. In: 20th Design Automation Conference Proceedings, pp. 64–70. (1983). https://doi.org/10.1109/DAC.1983.1585627
12. Hansen, M.C., Yalcin, H., Hayes, J.P.: Unveiling the ISCAS-85 benchmarks: a case study in reverse engineering. IEEE Design Test Comput. **16**(3), 72–80 (1999)
13. Zhang, R., Walus, K., Wang, W., Jullien, G.A.: A method of majority logic reduction for quantum cellular automata. IEEE Trans. Nanotechnol. **3**(4), 443–450 (2004)

A Comprehensive Review of Student Data Management System

Ozwin Dominic Dsouza, B. Tanvith Shenoy, Mrinal Singh, Pratiksha U. Kottary, and Shringar Agarwala

Abstract Student data management system deals with the storing of attendance and exam mark data. It gives the class attendance summary in the form of absence or presence in a particular class. The faculty will be provided with login credentials to maintain class data. When it comes to analytics, a consolidated student report will be generated typically once a month. Off late due to the advent of the accreditation process in education, it is crucial to maintain a reliable, user-friendly database of students to carry out further analysis. That is where this student data management system comes into the picture. In this paper, a comprehensive study is been made on the available tools/software available for student data management. An improved methodology is also been proposed in the end.

Keywords Application programming interface · Hypertext pre-processor · Structured query language · Operating system · Software development life cycle · Rapid application development

1 Introduction

The student data management system has been developed to eliminate the time and effort invested in taking and maintaining student data in academic institutions. It also helps in reducing the paper usage required for data management.

In a traditional system, what happens is that a teacher would take the attendance manually and records it in a book called the attendance register. At the end of an internal assessment, the marks scored by the students will be entered in the attendance register itself. Bajpai [1], a blogger by profession mentioned in one of his blogs that, manual entry is susceptible to human error and can have transparency issues. This is because, if a teacher wrongly marks a student absent, then the student will not get to know about this error. Also, the parents will not be able to know about their ward's attendance summary. Most of the state-affiliated colleges in Karnataka follow

O. D. Dsouza (✉) · B. Tanvith Shenoy · M. Singh · P. U. Kottary · S. Agarwala
BMS Institute of Technology and Management, Bangalore, India
e-mail: ozwindsouza@bmsit.in

© The Author(s), under exclusive license to Springer Nature Singapore Pte Ltd. 2022
V. Suma et al. (eds.), *Evolutionary Computing and Mobile Sustainable Networks*,
Lecture Notes on Data Engineering and Communications Technologies 116,
https://doi.org/10.1007/978-981-16-9605-3_56

a monthly basis attendance reveal practice. So it is evident that all the manual errors incurred during registering the attendance will come into the light only during this day. This can be considered as a drawback in the existing system.

To overcome this problem, institutions have started using cloud-based platforms such as Google spreadsheets. It defiantly helps in eliminating the data storage problem, but it is tedious to use. Article published in a micro blogging site, time doctor reveals the fact that entering the attendance in the online spreadsheets sheet is a hectic task [2]. This is attributed to the tiny cells present in the spreadsheet. Hence, the user needs to zoom in on the page to view the attendance status.

We all know that the student community is quite comfortable using a mobile phone, but the same cannot be considered for the parents. This could be due to a lack of awareness in accessing the sheets over messaging apps or email platforms.

These drawbacks made us look for a system that is user-friendly from the teacher, student, and parent's point of view. It would be very convenient if we could give an option to a faculty member to login using their credentials and then select the course and semester for reading and editing student data. After selecting the course and semester, they will get subcategories to show attendance and take attendance. Attendance of students based on dates will be listed in "show attendance," and in "take attendance," teacher will get the list of students where she/he must mark and submit the attendance. The marks of the students can be managed similarly.

Moreover, the survey conducted to know about smart phone usage in the parent community, revealed that more than 75% of the parents are using smart phones. Hence, a smart phone-based application can be built to facilitate the interaction between students, parents, and teachers. The application also provides an option for students and parents to view the managed data and also the analytics of the entire class. This makes the data management system transparent, user-friendly, efficient, and flexible, in addition to this, it avoids miscommunication and faulty management of the student data. Faculty can download the test data in the form of a spreadsheet to carry out further data analytics.

2   Related Work

2.1   College Android Chatbot

College app with attendance management system (where course instructor takes and checks scholar attendance, and students can login to check their status about the attendance), option to provide opinion about the application, and incorporated chatbot (using dialog flow API) for assistant and college inquiries with data security for guest users [3, 4].

College app using android studio, PHP, and MySQL. Used shared hosting for servers that contained PHP files and databases. It features an attendance management system, feedback system, and integrated chatbot for inquiry, notes, and assistant.

The cons are:

App uses PHP which is an outdated language
The app can be only used on Android phones; it is not supported on IOS
Does not display the analytics (representation of attendance and performance of class in graphical format) and chatbot provides limited solutions for queries.

2.2 Smart Student Attendance System

The student attendance system is a mechanism for registering student attendance daily in a specific course/subject. It is done on daily basis. All schools and universities embark on this mechanism to start the day. This process engages students in the learning journey by making sure that they will come to the class on regular basis. This is where the punctuality seed is sown in student minds that is why the attendance system is so important.

In this work, an insightful student attendance system is designed and implemented in the Android OS platform. In comparison with mundane turnout systems, the planned system provides a reliable, cost-effective, and accessible system for online scholar attendance and generates the turnout report without human intervention [5, 6].

The drawbacks are:

The system is not scalable (i.e., limited data entry)
No guest login facility
No provision for marks management
Supports only Android.

2.3 Simple Attendance System

For the very first when a person accesses the system, it prompts the user to create an administrator account without logging in. If this user is deleted, we will be prompted to create another one. This is the only time we can create a user without being logged in as a teacher or an administrator. Login email addresses are just used to determine our role. The register page enables us to create both teacher and student login. There is an option to create students in more than one class. Administrator login is used to add attendance, whereas student login is used to view the attendance report [5].

The cons are:

No security here, email addresses are not validated, and there are no passwords
No filtering available
No flexibility as new features cannot be added.

2.4 Attendance Management System

This work talks about the data handling of the student's attendance. It is used to record a daily basis of attendance. The teaching faculty is given unique login credentials to register student attendance. The faculty in charge of a particular course is made accountable to capture the turnout of all the students. If a learner is present on a particular date, the turnout gets punched in the system.

Major cons are:

Requires skilled developers
A complex Web application
The interface supports the English language only.

3 Software Development Methodology

3.1 Agile Model

Agile software development life cycle model is an amalgamation of the incessant and gradational scheme, with an aim on development, flexibility, and customer contentment by swift delivery of operational software products. Agile method splits the product into smaller gradational builds. These builds are given in iterations. The duration of every build is about one to three weeks typically.

The agile model is based on the belief that every project is distinctive and requires careful attention. The existing methods are to be customized that is fit well with the project necessities. In agile, the responsibilities are spitted into small boxes called time boxes to transport explicit features for a release.

Agile program principles are as follow:

Individuals and interactions: Stakeholders, business, product, and the business teams have to work collectively closely all the way through the project. Assemble projects around self-inspired individuals give them the atmosphere. The most effective and efficient method of conveying the information within the cross-functional teams are through face-to-face communication. The best architectures, requirements, and designs come out from self-organizing teams.

Working software: Functioning software is the crucial measure of evolution. The main concern is to satisfy the customer through early and constant handover of working software. Documentation is valuable and necessary, but we should make sure that the documentation is not comprehensive and is barely sufficient; the required/necessary details are covered.

Customer collaboration: We can only chase value if we are collaborating often. Accepting the changing requirements at the end of the development. Agile processes connect change for the customers' benefit. Agile processes encourage sustainable development. The stakeholders, team, and users should be able to maintain constant peace indefinitely. The top priority is to meet the customer through early and incessant delivery of functioning software. Contracts are a necessary part of doing business, but one should make sure the contracts support or hinder your ability to chase value.

Responding to change: A key benefit of agile is the ability to change quickly. Plans change and plans should account for that. Success is determined by how well you respond to change.

Unlike the conventional SDLC models such as the waterfall model which is mainly built upon a prognostic approach, an agile model is built on adaptive and flexible software development methods. In the conventional SDLC models, the traditional predictive teams usually work with comprehensive scheduling and have a total estimate of the accurate errands and characteristics to be delivered in the subsequent months or during the product life cycle.

The planning is done at the beginning of a cycle itself since the predictive methods depend on the requirement analysis. If any changes are to be included, then these changes must go through with stringent change control management and prioritization.

We can summarize that agile is all about chasing value, doing the next right thing. Therefore, scope is variable. If you are stuck to following a plan, you will risk missing the opportunities.

3.2 Rapid Application Development

Rapid application development is a software development methodology that makes use of limited planning in favor of rapid prototyping. A prototype is an operational model that is functionally comparable to a constituent of the product.

The functional modules are developed in parallel as prototypes in a typical RAD model scenario. And then, they are integrated to make the complete product for faster product delivery. Changes within the development process can be easily incorporated as there is no detailed preplanning.

RAD projects follow the incessant and gradational model and consist of minute teams with a mixture of developers, area experts, customer legislatures, and other IT resources working gradually on their component or prototype. The reusability of the prototype is the main important aspect of this model. So that with minor modifications, the system can be used in future tasks as well.

The customary SDLC follows a rigid procedure model with a high prominence on necessity analysis and congregation prior to the commencement of coding. It makes the customer quickly sign off from the process by giving way the requirements before the commencement of the project.

Once the software is shown to the customer, the customer may suggest some modifications to the software. But, the change process is quite inflexible and it may not be possible to include key changes in the product in the conventional SDLC.

In the RAD model, iterative and gradational handover to the customer are kept as the focal point. Customer participation during the entire development cycle reduces the threat of noncompliance with the actual user requirements.

3.3 Recent Trends

The work presented by Pandian [7] summarizes the role of different algorithms used in handling huge data set. This work opens up the avenue for data analytics in the field of education. Incorporation of maximization algorithms in understanding social network behavior is discussed in article [8]. This technical article throws light in the area of challenges posed by actual world social network size. Eigenvalue and CNN-based face detection-based attendance monitoring system are presented by Muthunagai [9] does not talk about the analytics part. But registering student attendance using face detection algorithm is quite challenging in scenarios were large student population exists, as it may take lot of time to register the attendance in every class. Few CNN-based algorithms have been proposed by Baba Shayeer in their experimental work [10]. Work presented by Kovelan [11] shows that time taken to register student attendance using manual method is about 15 min. To minimize this entry time, IoT-based attendance system is presented. This approach only talks about the attendance entry, it does not speak about the real-time attendance monitoring by the parent and student community. Devaprakash and team have explored the possibility of having a centralized attendance system using biometry [12]. In this work, the data segregation becomes the bottleneck. A SURF-based attendance system is discussed in [13]. In this work, face detection-based implementation has been done, but the data analytics part is missing. A network-enabled low-cost attendance system designed by Hashmi [14] is a low-cost system capable of integration of GUI.

4 Software Testing

Software testing is a process followed in SDLC to evaluate the system or sub-system performance to find whether it fulfills the desired task or not. The task is generally comprised of product requirements. Testing reveals the facts regarding drift from the expected value, errors, or lost requirements in contrast with the tangible requirements.

In the IT industry, testing itself is a separate team, where testers will extensively test the systems or sub-systems. Larger enterprises will have a team with responsibilities to assess the developed software in the framework of the given requirements. Furthermore, in few scenarios, developers assume the testing role which is called unit testing. The following cadre people are generally involved in the testing of a system contained by their capacities:

Software tester
Software developer
Project lead or project manager
End-user or user experience testing.

Many enterprises will have their way of allocating the designations for people who are assuming testing tasks. These designations are majorly given based on qualification and experience such as software tester, software quality assurance engineer, QA analyst, etc.

4.1 Applications of Software Testing

Cost-Effective Development—If testing is done at the beginning of development, then a lot of time, as well as revenue, can be saved. This is because the early bugs found in the software development can be fixed immediately. This avoids the backpropagation of work. If it is not done at the beginning stage, then in the latter half of or in the midway way, it can hinder the project flow.

Product Improvement—It is been observed that testing generally does not consume much time. Testing can be done quickly during the ongoing SDLC phases, provided a comprehensive test plan is built. But, discovering the cause and fixing the errors captured during the testing phase is a prolonged yet prolific activity.

Test Automation—Automating the testing can save a lot of time during SDLC. But test automation suits the scenarios in which the product requirements are fixed. However, test automation cannot be introduced during software development. The best time to introduce test automation is when the software is manually tested and found stable to some extent. Furthermore, test automation should not opt if the product requirements are changing or bound to change.

Quality Check—Software testing helps in identifying the following set of properties of any software such as.

Functionality
Reliability
Usability
Efficiency
Maintainability
Portability.

5 Entity-Relationship Diagram

See Fig. 1.

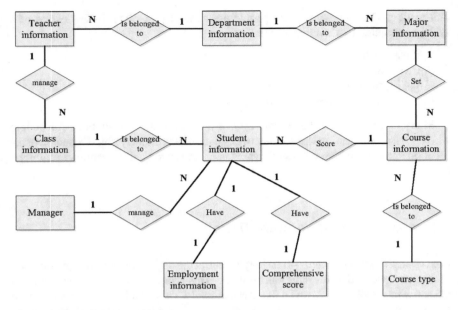

Fig. 1 Entity-relationship diagram

6 Hardware Requirements and Software Requirements

The hardware and software specifications have been discussed here. The following is the minimum requirement for the application to work in mobile and computing device.

6.1 Hardware Requirements

Laptop or PC with i3 processor-based computer or higher, 2 GB RAM, 20 GB hard disk.

Smartphone specification.

1.2 Quad-core processor or higher (Android or IOS), 1 GB RAM.

6.2 Software Requirements

Laptop or PC- Windows 7 or higher. Node js, Visual Studio Code, Npm or Yarn. *Android Phone or Tablet or iOS-* Android v5.0 or higher, iOS.

Testing- Expo, simulator.

Framework- Front End: React Native, Back End: Google Firebase Language: JavaScript.

7 Conclusion

The idea for this project came up because of the challenges faced by the teaching community while updating and reviewing the student report (marks and attendance) on the existing software. Also, there was no common platform for teachers and parents to discuss student analytics regularly. By developing this application, we aim to simplify the existing attendance recording mechanism as well as the transparency of data to all the stakeholders. The application will have separate spaces for teachers, students, and parents. The system administrator will assume the supervisor role. Presently, software implementation work is under progress. The software testing results with interpretations will be summarized in future articles.

8 Future Scope

The project has immense scope for improvement. A lot of feature addition can be brought in to meet the specific needs of the institute. The project may be implemented on the intranet platform. Data analytics part can be cascaded in the existing system to carry out the important analysis such as CO-PO analysis required for the accreditation process. A separate tab can be created for CO-PO analysis.

References

1. Bajpai, K.: Manual vs automated attendance system: comparison. Kiran Bajpai, March 2010. https://www.softwaresuggest.com/blog
2. Borja, C.: Google sheets timesheet templates: pros and cons. Blog article available at https://biz30.timedoctor.com/google-sheets-timesheet/2021
3. Gupta, P.: Available online: https://github.com/Pranjulcr7/College-android-chatbot
4. Chaudhry, W.: Available Onle https://github.com/waqasraza123/simple-attendance-system-in-php--mysql--bootstrap-
5. Kumar, M.: Attendance-management system U.G dissertation, Department of Computer Science Engineering, IFTM University Moradabad, 2019
6. Hameed, M.A.J.: Android based smart-attendance system. Int. Res. J. Eng. Technol. (IRJET), 1–5 (2017)
7. Pandian, A.P.: Review of machine learning techniques for voluminous information management. J. Soft Comput. Paradigm (JSCP) 1(02), 103–112 (2019)
8. Sivaganesan, D.: Novel Influence Maximization algorithm for social network behavior management. J. ISMAC 3(01), 60–68 (2021)

9. Muthunagai, R., Muruganandhan, D., Rajasekaran, P.: Classroom attendance monitoring using CCTV. In: 2020 International Conference on System, Computation, Automation and Networking (ICSCAN), 2020, pp. 1–4. https://doi.org/10.1109/ICSCAN49426.2020.9262436

10. Babu Naik, G., Prerit A., Baba Shayeer, N., Rakesh, B., Kavya Dravida, S.: Convolutional neural network-based on self-driving autonomous vehicle (CNN). In: 2021 International Conference on Innovative Data Communication Technologies and Application (ICIDCA 2021), August 2021, pp. 881–895, ISSN: 2367-4512

11. Kovelan, P., Thisenthira, N., Kartheeswaran, T.: Automated attendance monitoring system using IoT. In: 2019 International Conference on Advancements in Computing (ICAC), 2019, pp. 376-379. doi: https://doi.org/10.1109/ICAC49085.2019.9103412

12. Devaprakash, Gowtham, Murali, Muralidharan, Vijayalakshmi, V.J.: Centralized attendance monitoring system. In: 2020 6th International Conference on Advanced Computing and Communication Systems (ICACCS), 2020, pp. 1288–1291. https://doi.org/10.1109/ICACCS 48705.2020.9074162

13. Mohana, H.S., Mahanthesha, U.: Smart digital monitoring for attendance system. In: 2018 International Conference on Recent Innovations in Electrical, Electronics & Communication Engineering (ICRIEECE), 2018, pp. 612–616. https://doi.org/10.1109/ICRIEECE44171.2018. 9009166

14. Hashmi, A.: An inexpensive but smart MAC-address based attendance monitoring system. In: 2020 3rd International Conference on Advanced Communication Technologies and Networking (CommNet), 2020, pp. 1–5. https://doi.org/10.1109/CommNet49926.2020.9199613

E-commerce Website with Image Search and Price Prediction

Shambhavi Sudarsan, Atharva Shirode, Ninad Chavan, and Rizwana Shaikh

Abstract E-commerce has become a necessity for individuals when it comes to buying and selling products at the comfort of sitting at their homes. On conventional e-commerce platforms, if a vendor sets the price too high, they might not be able to attract consumers, and if they set their price too low, they might miss out on possible profits. Before pricing or selling their items, sellers generally look for similar listed products on other websites to get an estimate of current market prices. This can be quite monotonous and time-consuming. Also, buyers could sometimes have an image of a product but might not be able to search the product easily without knowing the brand of the product. The proposed system includes an image search feature which may help the user to search the specific image required by uploading the image on the website which would then display goods in decreasing order of visible similarity of the uploaded image. It also contains a price prediction feature which can help ease the sellers by providing a predicted price for used phones. The project E-Commerce Website with Price Prediction and Image Search is a website that enables buying and selling new and used products. This is a functional website which includes cart functionality, a payment gateway for processing financial transactions, image search functionality for finding similar products to the uploaded product image (phone and shoes), and price prediction feature which estimates the price of the used or new product (phone) being sold with the help of a machine learning model which accepts various features of the product as input.

S. Sudarsan (✉) · A. Shirode · N. Chavan · R. Shaikh
Department of Computer Engineering, SIES Graduate School of Technology, Navi Mumbai, India
e-mail: shambhavi.sudarsan17@siesgst.ac.in

A. Shirode
e-mail: atharva.nilesh17@siesgst.ac.in

N. Chavan
e-mail: chavan.ninad17@siesgst.ac.in

R. Shaikh
e-mail: rizwana.shaikh@siesgst.ac.in

© The Author(s), under exclusive license to Springer Nature Singapore Pte Ltd. 2022
V. Suma et al. (eds.), *Evolutionary Computing and Mobile Sustainable Networks*,
Lecture Notes on Data Engineering and Communications Technologies 116,
https://doi.org/10.1007/978-981-16-9605-3_57

823

Keywords Image search · Price prediction · Web scraping · Convolutional neural network · Residual neural network · Lasso regression · Feature extraction · Nearest neighbour · Electronic commerce

1 Introduction

The Indian e-commerce market is expected to grow to US $200 billion by 2026 [1]. Recent trends in online shopping in India show that people are spending high on apparels and mobile phones [2]. Also, COVID-19 pandemic [3] caused a surge in online shopping. People prefer ordering online because of convenience and comparatively cheaper prices online compared to buying products in store [4].

Existing e-commerce websites have text-based search systems. These methods lack efficiency especially in the apparel and accessories category where a better search system which can provide the user with similar existing products and may escalate the business in e-commerce. Through image search, the user can browse visually similar products to the ones they might have an image of. This feature can be used by the customers who are in search of a product of a particular brand that looks similar to the image of the product they have.

While selling used or new products, there are very few systems providing a service that predicts the prices of them. Having such a system can be beneficial for e-commerce systems as it can be helpful for sellers to get easy access to current prices and encourage more sellers to sell used products. The price prediction feature provides the seller with a uniform method of predicting the price of used phones which will help the seller to choose a price that can yield him/her profit and ensure that the price assigned to the product is neither too high that he/she may lose out on potential earnings nor too low to lose out on profits. Although this feature is provided to the seller, pricing decisions are left to the seller.

2 Related Work

2.1 *Image Search*

In [5], they are using indexing and searching to reach their goal. In indexing, they have used perceptual hashing to extract features vectors of the images. Once they get the feature vectors of the images, they store it in a database. Now, whenever a query image is given, its feature vectors are extracted and compared with all the feature vectors in the database. This is the searching phase, and here, they have used threshold base searching and K-NN (*K*-nearest neighbours). To evaluate the performance, they have calculated precision and recall for the sample dataset.

In [6], the ResNet50 model without top layers is used so as to get convolutional features as output instead of the image class probability. Keras Image data generator object is used for processing. Nearest neighbour algorithm is fitted to these extracted features.

In [7], they are performing reverse image search to retrieve similar images to the query image. They have used convolutional neural networks (CNN) to extract features from images. The image is input into the first layer, and each time it passes through a new layer, the most meaningful features are extracted and passed forward. After each layer, the semantic representation of the new image becomes more dense. Finally, the last layer gives a single dimensional array of features which is our feature vector. The feature vectors are compared with cosine distance to find the similar images from the database to our input image. Euclidean distance (the length of a line between two points), although simple, fails as a good metric in this context. One reason for this is that for our feature vectors, it is better to measure the correlation between features, rather than the features themselves.

2.2 Price Prediction

Used car price prediction has been studied in [8, 9]. Both the papers aim to predict used car prices with the help of the dataset provided in Kaggle [8]. In both papers, pre-processing steps involved filling the missing data. Enes [8] used average values to fill NaN data whereas [9] used the 'ffil' method. Other pre-processing steps involved in both these papers were normalization and label encoding. Enes [8] used models like multiple linear regression, lasso regression, ridge regression, decision tree, random forest and K-NN classifier to train the data. While [9] used models like random forest regressor, AdaBoost regressor and XGBoost to train the data. For data in [9], XGBoost gave the best results whereas for [8] K-NN and XGBoost gave similar results. The disadvantage of these papers were that the models were limited to the dataset which is relatively small and which has a lesser number of features. Also, both these papers used less accurate measures in the pre-processing step, i.e. they used average and 'fill' methods.

3 Dataset Description

The dataset is separated for two purposes, namely image search and price prediction. The dataset for image search has been retrieved with the help of the BeautifulSoup library of Python. This combined dataset has been collected from various e-commerce websites, namely amazon and Flipkart. This dataset contains links of images of the category 'Shoes' and 'Phones'. Total number of images collected for this category is 6000.

Two websites, namely 2gud and GSMArena were scrapped for the price prediction model. They were scrapped with the help of the BeautifulSoup library from Python. 2gud is a website which sells used and refurbished phones. Features contained in

product_image	product_ram	product_condition	product_cost	product_name	product_storage	product_color	original_cost
https://rukminim1.flixcart.com/image/612/612/j...	4 GB RAM	Refurbished - Good	₹ 7399.00	Nokia 6.1 Plus	64 GB	Black	10499
https://rukminim1.flixcart.com/image/612/612/k...	4 GB RAM	Refurbished - Superb	₹ 6790.00	lenovo A7	64 GB	Black	8893
https://rukminim1.flixcart.com/image/612/612/k...	4 GB RAM	Refurbished - Good	₹ 10520.00	SAMSUNG Galaxy M21	64 GB	Raven Black	13999
https://rukminim1.flixcart.com/image/612/612/k...	3 GB RAM	Refurbished - Superb	₹ 6890.00	Nokia 3.2	32 GB	Steel	9999
https://rukminim1.flixcart.com/image/612/612/j...	4 GB RAM	Refurbished - Superb	₹ 5490.00	InFocus Snap 4	64 GB	Platinum Gold	9999

Fig. 1 Merged dataset

this website are, namely image, RAM, condition, cost, name, brand and colour. GSMArena is a website that contains the current prices of all the used phones contained in the dataset collected from 2gud.

Both the datasets were compared with the help of the feature 'Product Name' and merged by excluding special characters, symbols and spaces. On merging both these datasets, a final dataset was generated. The merged dataset contains features namely 'product_image', 'product_ram', 'product_condition', 'product_cost', 'product_name', 'product_storage', 'product_color' and 'original_cost'. All features except 'original_cost' belong to the used and refurbished phones. The feature 'original_cost' determines the current market price of the identical phone.

This merging is done for the purpose of grasping the correlation between the features of a used phone with the corresponding market price of the same phone. The feature 'original_cost' is the market price for the used phone with refurbished price 'product_cost'. Figure 1 contains the merged dataset, and Fig. 2 contains the distribution of dataset correspondingly to the various phone brands. The various brands contained in the dataset are Nokia, Samsung, Lenovo, Vivo, OPPO, LG, Apple, etc.

The dataset contains data of 22 phone brands. Total number of phones contained in this dataset is 1000.

4 Methodology and System Architecture

4.1 Website Architecture

Figure 2 represents the proposed website architecture. Frontend technologies used in the construction of this website are HTML and CSS, and JavaScript has been used in the backend. React JS is used as a framework, and Node JS provides a local runtime environment for the website. MongoDB database has been used to store the data of users and all the items. Flask is used for interaction between the ML models and the

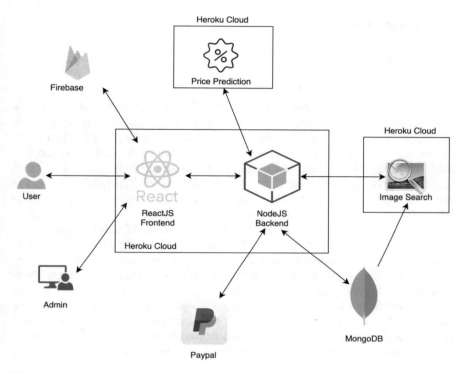

Fig. 2 Architecture of proposed work

website. Firebase is used as an image bucket to store and retrieve images. PayPal API is used for online transactions.

Price prediction model interacts with the website with the help of API's. Data sent from the website reaches the price prediction model and it sends the corresponding predicted price back to the website.

Image search model requires interaction with the database. Comparison with the uploaded image's feature vectors to the feature vectors of images present in the database requires connection and retrieval of feature vectors from the database. The uploaded image's feature vectors are sent to the ML model where the nearest neighbour Algorithm finds similarity between the two images and returns the Product_ID of the corresponding five similar products to the website. This is displayed by pressing the 'Search' button.

4.2 Image Search

4.2.1 Implementation

CNN Model

A convolutional neural network (CNN) is a deep learning algorithm that can take an input image, assign importance (learnable weights and biases) to various aspects in the image to be able to classify them. It consists of various layers like input layer, hidden layers and output layer. The hidden layers perform convolution. Convolution detects features of the images like edges of the image. This is followed by other layers like pooling, fully connected, normalization layers. The pooling layers are used to reduce the dimensions of the feature vectors. This feature vector is passed to an output layer which classifies the images into classes.

Firstly, a CNN model of 10 layers with different configurations of feature extraction and max pooling layers was used. From those configurations, the best configurations were picked which were feature extraction, feature extraction, max pooling, feature extraction, feature extraction, max pooling, feature extraction, feature extraction, max pooling and max pooling. From this model, the output feature vectors were of less dimensions which helps in better processing speeds, but the resulting images differ from the query image. The ResNet 50 model which consists of 50 convolution layers is used, and the final fully connected output layer is removed. Using ResNet 50 was inspired by researching the literature surveys we had done.

ResNet 50 Model

Figure 3 denotes the ResNet50 model. ResNet 50 is a 50 layer deep convolutional neural network. It consists of 50 convolutional layers, 2 pooling layers (max and average) and 1 fully connected layer [10, 11]. The ResNet 50 architecture can be into blocks as follows:

Block 1: This block is the input block, where the image is given as input. It has a predefined shape of ($1 \times 224 \times 224 \times 3$), where 1 is the batch size, 3 is the channel width, 224×224 is the dimension of the input image. This is passed through the first convolutional layer which applies 64 convolutions on it with a kernel size of 7×7. The output of this block is of shape ($1 \times 112 \times 112 \times 64$). After passing through the first convolutional layer, the output is passed on to a max pooling layer (3×3). The output shape of this max pooling layer is ($1 \times 56 \times 56 \times 64$).

Block 2: These blocks start with a convolutional layer of 64 convolutions and a kernel size of 1×1; it is followed by another convolutional layer of 64 convolutions and a kernel size of 3×3. The remaining 2 convolutional layers have 256 convolutions with a kernel size of 1×1. The output shape of this block is ($1 \times 56 \times 56 \times 256$).

Block 3, 4: These blocks contain three convolutional layers; there are two layers of 64 convolutions with a kernel size of 1×1 and 3×3. The final layer is of 256 convolutions with a kernel size of 1×1. The output shape of these blocks is ($1 \times 56 \times 56 \times 256$).

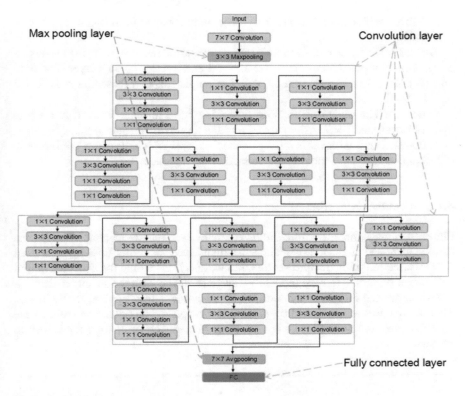

Fig. 3 ResNet50 model

Block 5: This block contains four convolutional layers; there are two layers of 128 convolutions with a kernel size of 1 × 1 and 3 × 3. The final two layers are of 512 convolutions with a kernel size of 1 × 1. The output shape of this block is (1 × 28 × 28 × 256).

Block 6, 7, 8: These blocks contain three convolutional layers; there are two layers of 128 convolutions with a kernel size of 1 × 1 and 3 × 3. The final layer is of 512 convolutions with a kernel size of 1 × 1. The output shape of these blocks is (1 × 28 × 28 × 512).

Block 9: Have 4 convolutional layers; there are two layers of 256 convolutions with a kernel size of 1 × 1 and 3 × 3. The final two layers are of 1024 convolutions with a kernel size of 1 × 1.The output shape of this block is (1 × 14 × 14 × 1024).

Block 10, 11, 12, 13, 14, 15: Have three convolutional layers; there are two layers of 256 convolutions with a kernel size of 1 × 1 and 3 × 3. The final layer is of 1024 convolutions with a kernel size of 1 × 1. The output shape of these blocks is (1 × 14 × 14 × 1024).

Block 16: This has four convolutional layers; there are two layers of 512 convolutions with a kernel size of 1 × 1 and 3 × 3. The final two layers are of 2048 convolutions with a kernel size of 1×1. The output shape of this block is $(1 \times 7 \times 7 \times 2048)$.

Block 17, 18: These blocks contain three convolutional layers; there are two layers of 512 convolutions with a kernel size of 1×1 and 3×3. The final layer is of 2048 convolutions with a kernel size of 1×1. The output shape of these blocks is $(1 \times 7 \times 7 \times 2048)$.

Block 19: The output of previous layers is passed through an average pooling layer of kernel size 7×7. After passing through this layer, the output shape becomes (1×2048). The final layer is a fully connected layer which classifies the image in different classes which could be predefined or user defined.

4.2.2 Feature Extraction

Figure 4 shows the procedure followed to extract feature vectors. Firstly, the ResNet50 model is loaded excluding the final layer, since the final layer is used for classification of the feature vectors into various classes, and the expected output are feature vectors. After the model the image is loaded with the specified shape, it is then converted into an array and further pre-processing is applied to the image. Then its features are extracted by passing it through the model, these features are then flattened and normalized to prevent any bias during comparing these feature vectors.

```
model = ResNet50(weights='imagenet', include_top=False, input_shape=(224, 224, 3))

def extract_features(img_path, model):
    input_shape = (224,224,3)

    img = image.load_img(img_path, target_size=(input_shape[0], input_shape[1]))
    img_array = image.img_to_array(img)

    expanded_img_array = np.expand_dims(img_array, axis=0)
    preprocessed_img = preprocess_input(expanded_img_array)

    features = model.predict(preprocessed_img)

    flattened_features = features.flatten()
    normalized_features = flattened_features/norm(flattened_features)

    return normalized_features
```

Fig. 4 Feature vector extraction

```
neighborsNearest = NearestNeighbors(n_neighbors = 5, algorithm='brute', metric='cosine'
neighbors = neighborsNearest.fit(feature_list)
distances, indices = neighbors.kneighbors([feature_list[0]])
```

Fig. 5 Comparing feature vectors

4.2.3 Comparing Feature Vectors

Figure 5 shows the calculation of distance between two feature vectors. On uploading an image, its feature vectors are extracted and compared with all the feature vectors in the database. This comparison happens by using nearest neighbour algorithm. Cosine distance is used to find the distance between two images. The nearest neighbour algorithm finds the cosine distance of query image feature vectors with other feature vectors in the database and returns nearest N feature vectors indexes according to the value of N specified. The most similar feature vector appears on the top.

4.2.4 Integrating the Image Search ML Model with Website

On the website, when a seller creates a product its feature vectors are extracted and stored in our database. Now, whenever a buyer wants to search with an image, he can click the image search button and upload the image which will be the query image to get the products that look the most similar to his uploaded image. Figure 6 represents

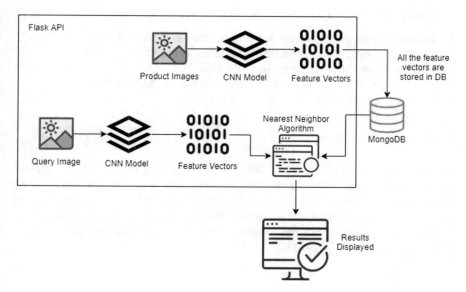

Fig. 6 Implementation of image search

the steps followed while performing image search. An ML model into a Flask App has been made which is deployed to a server to receive requests from the website and return data accordingly.

Seller Section
When the server starts, feature vectors of existing images are fetched from the database only once and stored in a list in flask application. When the seller creates a product, he/she is prompted to upload an image of the product. The feature vectors of newly created products are stored in the database and in the list. If the server crashes, the process of fetching vectors from the database is repeated again.

Buyer Section
On uploading an image on the homepage of the website, the features of this uploaded image are extracted. These features are then compared linearly with the feature vectors of previously contained images in the database. Comparison between feature vectors of the images is done using the nearest neighbour algorithm. It is the distance between the query image and the images contained in the dataset. The products with the shortest distance are displayed first in the results.

4.3 Price Prediction

4.3.1 Data Pre-Processing

The data pre-processing has been divided into the following steps:

- Encoding of the phone brands from categorical values to numbers
- Encoding of the product condition from string to numbers
- Removing unnecessary strings like 'GB' and 'MB'
- Converting ram and storage which contained data in 'MB' to 'GB'
- Removing unnecessary strings from product brands

4.3.2 Model Training

Figure 7 represents the flowchart followed while training the model. The dataset contained categorical values which were converted into numerical values using label encoding. Since the dataset contained largely varying ranges of values, standard scalar was used to normalize the features. This data was then trained with various models, and the accuracy of these models were compared with root mean squared error as the parameter (Table 1).

Fig. 7 Flowchart for training the model

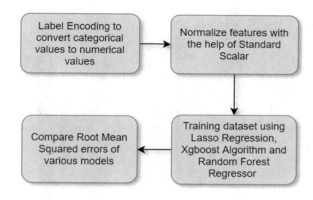

Table 1 Root mean squared errors of the models used

Models	Root mean square error
XGBoost	0.8
Random forest regressor	0.76
Lasso regression	0.6

4.3.3 Pseudocode

```
IMPORT models from scikit-learn
IMPORT dataset
PRE-PROCESSING to remove unnecessary characters and symbols, conversion
into a universal unit for storage and RAM.
LABEL-ENCODING to convert categorical values to numerical
STORE models in a variable named models
SPLIT dataset into training and testing categories
NORMALIZE data using StandardScalar
FOR model in models:
    TRAIN the model with the dataset
    PREDICT the output by using model.predict()
    CALCULATE and STORE Root Mean Squared errors
END FOR
COMPARE accuracies of all models
```

Root mean squared error in Lasso regression is the least. Hence, this algorithm was selected for creating the model.

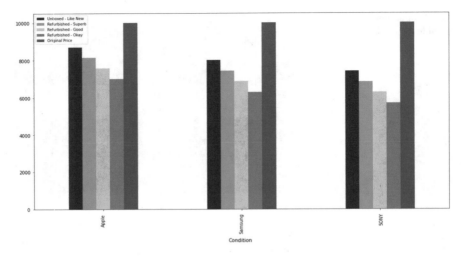

Fig. 8 Refurbished cost comparison with different brands

5 Results

5.1 Price Prediction

Figure 8 shows comparison of prices with refurbished costs of phones of varying brands and conditions. The condition 'Unboxed—Like New' depicts a new phone. 'Refurbished—Okay' depicts a phone that has been used for a long time and then refurbished. 'Original Price' is the current market price of a new phone with the same characteristics. The original cost is kept the same for all brands, i.e. Rs. 9999. Results shown in the figure suggest that refurbished cost for 'Apple' is the most while refurbished cost for 'SONY' is the least. Apple and Samsung are among the most bought refurbished phone brands. Since Apple provides better features and quality, the cost of refurbished phones of brand 'Apple' is higher. SONY is a relatively low supply and demand brand, and hence, it has the least cost of all the three brands.

Figure 9 represents the implementation of price prediction model on the website. A seller can predict a product using the form shown in Fig. 9. On entering specific features of a phone and clicking on 'Predict Price' button, the result fetched from ML model with the help of API's is displayed on the website as shown in Fig. 9.

5.2 Image Search

After extracting the feature vectors from the query image and comparing the feature vectors of 6000 images of phones in the dataset with the feature vector of the query

Fig. 9 Price prediction

image, the output displays images that look similar to the query image as shown in Fig. 10. The similarity is measured by calculating the distance of an image with the query image. This distance measured is the cosine distance and it is computed with nearest neighbour algorithm.

Figure 11 represents the distances of the resulting images in Fig. 10 when compared with the query image.

6 Conclusion

In this paper, an efficient solution for enhancing the searching of a product has been proposed. A universal solution for phone price prediction has been proposed. This feature will help the sellers to receive the predicted prices of a particular used phone, instead of searching various websites to check market price of similar phones. Also with image search, buyers need not know the exact product details before searching and can simply upload images and find similar products. These proposed solutions may help in enhancing the business in the e-commerce sector and will save a lot of time for the sellers as well as buyers.

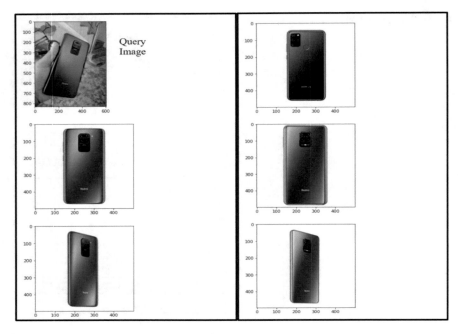

Fig. 10 Query image and image search results obtained

Fig. 11 Distances of
resulting images with the
query image

```
0.6164874
0.62891763
0.64027554
0.653972
0.6555522
```

References

1. Ministry of Commerce and Industry, Government of India, India Brand Equity Foundation, Indian E-commerce Industry Report, April 2021
2. Abhishek, C., Sandeep, C.: Study on recent trends in online shopping in India. Int. J. Sci. Eng. Res. **9**(2) (2018)
3. Komal, S.: A surge in e-commerce market in India after Covid-19 pandemic. Glob. J. Soc. Sci. (2020). ISSN – 2581-5830
4. Lo, S.-K., Hsieh, A.-Y., Chiu, Y.-P.: Why Expect lower prices online? Empirical examination in online and store-based retailers. Int. J. Electr. Commer. Stud. **5**(1), 27–38 (2014)
5. Gaillard, M., Egyed-Zsigmond, E.: Large scale reverse image search. Université de Lyon INSA Lyon, Inforsid (2017)
6. Mani, K.: Reverse image search—find similar images. Medium, 22 May 2017
7. Mohit, N.: Reverse image search with machine learning. Commerce Tools, 19 Aug 2019
8. Enes, G.: Predicting used car prices with Machine Learning techniques, Medium, 10 Jan 2018
9. Panwar A.A.: Used car price prediction using machine learning. Medium, 3 Aug 2018

10. Residual neural network, Youtube, uploaded by When Maths Meet Coding, 3rd September, 2020
11. Priya D.: Understanding and Coding a ResNet in Keras. Medium, Jan 4, 2019

Modified Geometric Mean Decomposition and Orthogonal Matching Pursuit Based Hybrid Precoding for Millimeter-Wave Massive MIMO Systems

V. Baranidharan, Dharini Subburajulu, S. Niveditha, V. S. Arun Prasad, Monalisa Sutar, S. Manoj Kumar, and R. Ramya

Abstract The hybrid precoding technique efficiently achieves the reduction of the number of chains in the radio frequency range namely the RF chains in mm-wave massive MIMO system. The method of combined hybrid precoding that is existent now is based on Singular Value Decomposition algorithm which gives rise to the problem of complication in the bit allocation in order to match the varying signal-to-noise ratio for each sub-channel. Thus, this problem of complexity in bit allocation could be easily sorted out with the proposed method of modified Geometric Mean Decomposition and orthogonal matching pursuit algorithm. This is achieved by manipulating and bringing up an identical signal-to-noise ratio enabling an easy and simplified bit allocation. But when the system is completely digital, the problem of increased number of RF chains sprouts out. Thus, the hybrid precoder which is a combination of analog and digital precoder comes into the picture. This pair of analog and digital precoder sufficiently brings out similar output as that of the unconstrained6fully8digital precoder. In order to achieve this initially, the analog precoder is designed and then fixed. This is then followed by the designing of digital precoder. Orthogonal Matching Pursuit algorithm is fetched to design the analog precoder while GMD is used for the design of digital precoder. The results after simulation of this algorithm are highly efficient and achieve a remarkably better performance than the existing methods.

Keywords Massive MIMO · Orthogonal matching Pursuit · Hybrid precoder · Singular value decomposition · Modified geometric mean decomposition

V. Baranidharan · D. Subburajulu (✉) · S. Niveditha · V. S. Arun Prasad · M. Sutar · S. Manoj Kumar · R. Ramya
Department of Electronics and Communication Engineering, Bannari Amman Institue of Technology, Sathy, India

V. Suma et al. (eds.), *Evolutionary Computing and Mobile Sustainable Networks*,
Lecture Notes on Data Engineering and Communications Technologies 116,
https://doi.org/10.1007/978-981-16-9605-3_58

1 Introduction

The current world needs the unveiling of 5G technology which is just a mile beyond to achieve since it spectacularly reduces the delay time from 70 ms in the current 4G to a minimal of less than an ms which could be achieved since the data rate is nearly 20 times higher than the existing methodology. This could be possible only with the deployment of massive MIMO (Multiple Input Multiple Output) systems. The MIMO is already in existence with an antenna array consisting of about 16 antenna in an array with 12 for transmitting and four for receiving approximately but this massive MIMO brings in the concept of mammoth array which would consist of more than 100 antennas at the single array. Since all the current used mobile phones operate in the radio frequency range the massive MIMO operates at a bit higher frequency with is extremely fast and a short-range around a frequency of 25–40 GHz due to which this wavelength is very small. This is the concept of mm-wave which is used to overcome the problem of interference. Since the frequency is very high the wavelength computes to around 10^{-3} and hence waves are called mm-wave.

This work aims to overcome this hindrance of complication in bit allocation happening in the SVD algorithm due to varying SNR and sub-channel gains by the implication of decomposition of channel matrix through the geometric mean algorithm which includes the process of decomposing into two semi unitary matrix and arbitrary one included in an upper triangular matrix which is unconstrained precoder solving the issue by making all the SNR identical. The simulation achieved shows remarkably better performance than the conventional completely digital precoder preceded with the SVD technique. The later sections of this work will describe the model of the system and the channel state information in the immediate chapter and the further chapters describing the decomposition of the existing technique and the hindrance and the component in the newly proposed work overcoming it in detail.

2 Related Works

More number of researches are carried out in recent years. Some of the recent proposed work is discussed in this section.

Qingfeng Ding et al. [1] have proposed a technique on hybrid precoding for massive MIMO which deals with different antenna arrays. They have used orthogonal matching pursuit in order to obtain hybrid precoding algorithm, they have done a detailed study on the four different types of antennas with respect to different parameters considered and tested it by implementing the hybrid precoding. The limitation of this work is that more bit allocation leads to complexity.

Huang et al. [2] have proposed a technique on deep learning-based massive MIMO, which looks at using the hybrid precoding in the deep neural network manner. In which they are capable of reducing the bit rate and enhancing the efficiency of the spectrum. The limitation of this work is that the results clearly show the degradation of the performance in deep learning as the batch size increases; too small batch size also does not seem to provide accurate results.

Wu et al. [3] have proposed a technique on downlink communication, which proposes a hybrid block Geometric Mean Decomposition with low complexity. In the sub-channel, they have applied successive interference cancellation at the receiver's end which also has identical SNR. They have introduced a new concept which is not been seen in other works that is bit loading. The limitation of this work is that the complexity of the system increases four times as they have used FLOP's, the operation that is done on complex values undergo multiplication followed by addition. There is no specified power allocation scheme being implemented, that could improve the efficiency and accuracy of the system.

Li et al. [4] is focused on the problem of hybrid precoding and combining design for MIMO system, which is based on orthogonal frequency division multiplexing operating or using mm-wave e frequency. This idea base uses an iterated scheme that was based on the codebook analog combiner and precoder, in the channels which are both forward and reverse. The limitation identified from this technique is that the efficiency or output gets better as the number of RF chains gets increased. This limitation identified in this is overcome in our work by coupling analog and digital.

Li et al. [5], have proposed the concept of joint measure matrix and estimation of channel, is used for the massive MIMO system. This also aims in exploitation of sparsity in a massive MIMO system in order to estimate the channel, based on the Toeplitz–structured measuring matrix and a CC-S which expands as the complete complementary sequence has been proposed. The limitation of this system is the limitations of length. This solution is calculated from length-restricted solutions. Many concepts like channel analysis based on angle domain and pilot pollution need to be realized.

Alemaishat et al. [6] proposed design with digital precoding unit followed by numerous RF chains and then analog precoding with the transmitting antenna succeeding it. Then the intermediate channel separating the receiving block preceded with the receiving antenna and analog combining units and RF chains and at the end with a digital synthesizer. The main drawback of the algorithm is time synchronization and eigenvalue calculation criteria, which are left unorganized and unsorted.

3 Proposed Work

3.1 System Model

In this section, the main idea of hybrid precoding for massive MIMO under the mm-wave channel has been explained. Hybrid precoding can be explained as the coupling of analog and digital precoder in order to match the performance as well as complexity of the system. The exploitation of the radio frequency in more than one

in number this technique is helpful in enabling mm-wave system to be advantages in case of patient multiplexing as well as beamforming gain. For consideration of a general hybrid precoding system with massive MIMO, with base station (BS) having A_T transmitting antennas and ($N_{streams}$) number of the data streams are independent with A_R as receiving antennas at the user end. Then the number of RF chains are considered to be $N_{T\text{-RF}}$ from the transmitting base station and $N_{R\text{-RF}}$ at the reception user end. The number of chains has to satisfy the conditions as follows:

$$N_{streams} \leq N_{T-RF} \leq A_T \text{ and } N_{streams} \leq N_{R-RF} \leq A_R$$

The signal that had been received at the user end can be realized as,

$$y = \sqrt{\rho} C^H HPs + C^H n \tag{1}$$

where ρ is the average received power, P is the hybrid precoder which is both analog (P_a) and digital (P_d) and C is the hybrid combiner which is also both analog (C_a) and digital (C_d) and S is the vector for source signal, H is the channel matrix and then n is the representation of noise signal. The distribution of the Additive white Gaussian noise is (0, $\sigma_2 I$), where the noise power is denoted or represented by σ_2. Since the realization of both precoder and combiner in analog is as phase shifters all their elements would have the same amplitude which would be equal to $1/\sqrt{t}$ and $1/\sqrt{r}$ for analog precoder and analog combiner, respectively.

3.2 Channel Model

The channel model is devised using the principle of Saleh—Valenzuela model for the massive MIMO mm-wave communications. This model has two discrete parts of channel matrix namely the Line of sight (LoS) and Non Line of Sight (NLoS) component. This also adds upto the point of reduction in the number of radio frequency chains in the hybrid precoding structure achieving an equally efficient and optimal structure. This model in brief would have,

$$H = \sqrt{(N_T * N_R)/L} (\text{LoS and NLoS}) \tag{2}$$

The number of antennas is comparatively higher than the rank of the channel matrix H, which exclaims the point that is the independent data streams beings exploited is very low which is due to the effect of very much reduced scat- tering happening in the mm-wave systems.

3.3 Completely Digital SVD- and GMD-Based Precoding

This section as titled would briefly deal with the fully digital precoder, then show some light in the existing algorithm namely Singular Value Decomposition (SVD) then add up the concept of GMD into hybrid precoding in detail.

Singular Value Decomposition involves the decomposition of the channel matrix into three main components which is given as,

$$H = U\Sigma V^H = U_1\Sigma_1 V_1^H + U_2\Sigma_2 V_2^H \tag{3}$$

U, V are two semi unitary matrices and another diagonal matrix with diagonals having largest singular values that have been in the arrangement of decreasing values. The output wave received is denoted by,

$$y = \sqrt{\rho}\Sigma_1 s + U_1^H n \tag{4}$$

where $Y = U$ and $Z = V$ and σ represent the channel gains. The value of gain would vary between the LoS and NLoS by 15 dB with the LoS component with a greater gain value than the counterpart. This is the major cause for the reported variation in the SNR. This will make the problem of high complexity in bit allocation since more careful coding and decoding have to be incurred. This problem could be overcome by the GMD by decomposing the channel matrix into semi unitary matrix upper triangular matrix and arbitrary matrix.

$$H = GRQ^H \tag{5}$$

The upper triangular matrix will be manipulated from the previous technique by having the geometric mean of the largest values in the singular diagonals and R represents diagonal matrix in order to simplify the equation. By diving the system.

into precoder (T) and combiner (M) the equation of received signal could be rewritten as,

$$y = \sqrt{\rho}R_1 s + G_1^H n \tag{6}$$

This will result in an equal gain for all the sub-channel. This would instantly reduce the bit allocation complexity by the SNR differing in the preceding method of SVD. But in this way of completely digital precoding, the problem of increased RF chains is yet to be resolved since this eventually led to increased consumption of energy. Thus, the concept of hybrid precoding comes into picture.

3.4 GMD-Based Hybrid Precoding Techniques

The conventional hybrid precoding [7, 8] is not advisable for effective bit allocation in the complicated situation of massive MIMO systems. To overcome these issue, we proposed a modified Geometric Mean Decomposition-based hybrid precoding technique which is based on the hybridization of both the analog and digital combiners and precoders. For taking unconstrained GMD—precoder techniques, we are going to use Q_1, Q_A, and Q_D for the given GMD precoding with digitizer. By using this GMD precoder technique, it should seek a pair of analog and digital type of precoders are sufficiently close to obtaining near-optimal value of the precoders. In this GMD-based hybrid decoder with an implementation procedure,

$$Q_1 = V_1 S_R; \ G_1 = U_1 S_L; \ R_1 = S_L \Sigma_1 S_R \tag{7}$$

where S_R and S_L are the unitary matrices which is always the product value of the mutation and given matrices. By recalling the objective functions, the hybrid decoders will form the analog and digital precoder which is based on the combination coefficients. By using the satisfies the constant modulus based A_t matrix which is cannot the matrix with L columns has a smaller value of L. By using this way of constraints, the analog precoder value-based design problems is given by the selecting the T matrix,

$$T = \arg \min \| Q_1 - A_t T Q_D \|_F \text{s.t.} \left\| \text{diag}\left(T T^H\right) \right\|_0 = N\_T^{RF} \tag{8}$$

Here, the T is selecting matrix by using the non-zero rows. The various number of smaller RF chains values is always smaller than the hybrid precoding matrix with number of paths values. T is the sparse aware algorithms, such this orthogonal pursuit algorithms (OMP) to solve the above equations. This equation will be fulfilled with digital precoder normalization of all the algorithms. By using the analog precoder, the Norm minimization problems are formed as,

$$Q_D = \arg \min \| Q_1 - Q_A Q_D \|_F \text{s.t.tr}\left(Q_A Q_D Q^H Q a^H\right) < N_s \tag{9}$$

This proposed modified GMD-based hybrid algorithm is based on the three important parts. The first part is analog precoder is constructed based on the OMP algorithms with steering vectors. The second part is about the digital precoder is constructed and its normalizations. The computation complexity of the digital precoder is used to compute the pseudo-inverse matrix of the Modified GMD precoding.

4 Simulation Results

The section explains the performance of the proposed modified the GMD-based precoding with all the other existing systems. In this mm-wave MIMO systems will be considered for the frequency at 28 GHz. Here, we consider the 256 various ULA Transmitting antenna elements with antenna spacing $\lambda/2$ is employed over the Base station. Similarly, we consider the 16 various elements of ULA which is also spacing over the $\lambda/2$ is employed over the user. Both the base station will have the RF chains as 4. In this mm-wave channel, there are only one LOS and 4 NLOS paths are used. For this channel, power normalization factor is always adjusted with the power distribution between the various LoS paths component and NLos Components. Here, we have considered the various channel modulation scheme, 16QAM modulation on based on the same complexity and it is compared with the performance of different precoding methods (Fig. 1).

Here, we consider the LoS and NLoS environment is considered for mm-wave propagations is shown in the given figure. We observed that the proposed modified GMD-based hybrid precoding is achieved by the better performance than the conventional SVD based precoding methods. This method will convert the mm-wave massive MIMO channels are converted into the several sub-channels that will be identical SNR values. This performance degradation in various sub-channels is always very low with other SVD precoding methods (Table 1).

These results show that the approximation performance is analysis of the various LoS and NLoS environment through MATLAB simulations. These results show that the proposed modified GMD-OMP based hybrid precoding systems give the better results than the other existing precoding schemes. This hybrid precoding approximation is achieved because of these hybrid precoders will disperse the channel powers

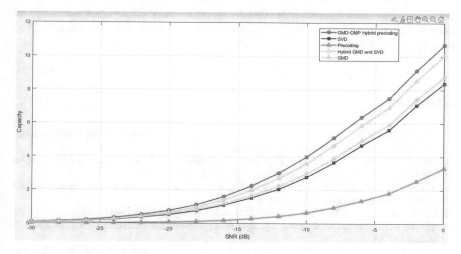

Fig. 1 Comparison of SNR versus capacity

Table 1 Statistical comparison of proposed modified GMD-OMP based hybrid precoding with all the other systems

Parameters	GMD-OMP hybrid precoding	SVD	Precoding	Hybrid GMD and SVD	GMD
Min	0.08924	0.0654	0.008129	0.07465	0.06948
Max	10.85	8.496	3.354	10.2	8.885
Mean	3.322	2.453	0.7488	3.044	2.598
Median	1.907	1.3	0.2406	1.672	1.4
Mode	0.08924	0.0654	0.008129	0.07465	0.06948
Standard deviation	3.523	2.718	1.029	3.305	2.852
Range	10.76	8.43	3.345	10.13	8.816

to the several paths by using GMD-OMP algorithms. It is observed that these prosed methods gain the sub-channel after precoding and combining are also very similar to the NLoS environments.

5 Conclusion

In this article, In order to avoid the complicated bit allocation, we have proposed a hybrid GMD-OMP based precoder for mm-wave MIMO channels. In this proposed scheme, the mm-wave MIMO channel is converted into several sub-channels with identical SNRs, which can be done with the help of GMD. Using GMD-OMP scheme, we obtain digital precoder and using the principle of basis pursuit we obtain analog precoder, which is done for decoupling the design of both digital and analog precoders. Simulation results show that the proposed work GMD-OMP based hybrid precoding technique can achieve better performance than the conventional SVD based hybrid precoding.

References

1. Ding, Q., Deng, Y., Gao, X., Liu, M.: Hybrid precoding for mmwave massive MIMO systems with different antenna arrays. IEEE Xplore (2019)
2. Huang, H., Song, Y., Yang, J., Gui, G., Adachi, F.: Deep learning-based millimeter-wave massive MIMO for hybrid precoding. IEEE Trans. Veh. Technol. **68**(3) (2019)
3. Wu, W., Liu, D.: Hybrid BD-GMD precoding for multiuser millimeter-wave massive MIMO systems. IEICE Trans. Commun (2018)
4. Li, M., Liu, W., Tian, T., Wang, Z., Liu, Q.: Iterative Hybrid precoder and Combiner Design for mmWave MIMO-OFDM Systems. Springer Science+Business Media, LLC, part of Springer Nature (2018)
5. Li, S., Su, B., Jin, L., Cai, M., Wu, H.: Joint measure matrix and channel estimation for millimeter-wave massive MIMO with hybrid precoding. EURASIP J. (2019)

6. Alemaishat, S., Saraereh, O.A., Khan, I., Affes, S.H., Li, X., Lee, J.W.: An efficient precoding scheme for millimeterwave massive MIMO systems. Electronics (2019)
7. Wang, Y., Zou, W.: Low complexity hybrid precoder design for millimeter wave MIMO systems. IEEE Commun. Lett. **23**(7) (2019)
8. Xie, T., Dai, L., Gao, X., Shakir, M.Z., Li, J.: Geometric mean decomposition based hybrid precoding for millimeter-wave massive MIMO. China Commun. (2018)

Design and Simulation of a Direct-PSK Based Telecommand Receiver for Small Satellite

N. S. Srihari, S. Amruth Kumar, and B. S. Premananda

Abstract Telecommand receiver is responsible for receiving the commands uplinked from the ground stations and transmits the demodulated data to satellite management unit which is responsible for executing them. The existing telecommand receivers contain RF front end having VCO, mixers, analog I/Q downconverters, LPF, and two ADCs. Modulation schemes in the existing telecommand receivers are PM-PSK and FM-PSK. The objective of the proposed receiver design is to have minimum RF involvement having one high-speed ADC and direct digitally demodulated output. Using Direct-PSK modulation high data rates of 128/256 kbps can be achieved which simplifies the design. The design and simulation is performed in MATLAB/Simulink and involves telecommand transmitter, ADC, BPF, DDC with decimator, Direct-PSK demodulator, FFT-based frequency estimator, and bit synchronizer. The bit error rate (BER) is 0 after 302 bits are received and demodulated successfully after 2.34 ms. The designed telecommand receiver thus works for an input gain of 0 dB to -40 dB. BER is observed at -40 dB. The band-limited noise is 10^{-15}. The input frequency sweep rate is 250 Hz for a range of frequencies 5 MHz $\pm$ 500 kHz.

Keywords BER calculation · Bit synchronizer · Digital down-conversion · Direct-PSK demodulator · FFT-based frequency estimator · TT&C

1 Introduction

A satellite's telemetry, tracking, and control (TT&C) subsystem establish a communication link between the satellite and ground stations (user or control segment). To achieve a successful mission, the TT&C subsystem must execute three specialized tasks: telemetry, tracking, and command and control [1]. The telecommand receiver,

N. S. Srihari · B. S. Premananda (✉)
Department of Electronics and Telecommunication Engineering, RV College of Engineering, Bengaluru, India
e-mail: premanandabs@rvce.edu.in

S. Amruth Kumar
Design Department, Centum Electronics Ltd., Bengaluru, India

© The Author(s), under exclusive license to Springer Nature Singapore Pte Ltd. 2022
V. Suma et al. (eds.), *Evolutionary Computing and Mobile Sustainable Networks*,
Lecture Notes on Data Engineering and Communications Technologies 116,
https://doi.org/10.1007/978-981-16-9605-3_59

which has been created and qualified to function on Low Earth Orbit (LEO) satellites, is in charge of receiving orders uplinked from ground stations and transmitting the demodulated data to the satellite management unit, which is in charge of carrying them out. Control orders with a time stamp are delivered to the satellite to perform tasks such as directing the spacecraft's movement or turning on and off its payloads. Telecommand receivers are currently state-of-the-art RF technologies that operate on S-Band frequencies. Small satellites, also known as smallsats or miniature satellites, are satellites with a mass of less than 500 kg (1100 lbs) [2]. New mission concepts have resulted from the advancement of small satellite technologies in domains such as microelectronics, cubesats, micro-propulsion, long-distance communications, and improved launch availability of cheap ridesharing. The telecommand data is modulated directly on the RF carrier in Direct-PSK modulation with Non-Return-to-Zero (NRZ) coding, completely suppressing it. If a direct modulation technique is adopted, significant data sideband power falls into the RF carrier phase lock loop (PLL) bandwidth. As a result, direct modulated PSK is more bandwidth efficient than subcarrier-based PSK. It also solves pulse code modulation's drawbacks, such as the peak of the data's frequency spectrum being present in the RF residual carrier's frequency, and performance degradation owing to 1's and 0's imbalances in the data stream. To segregate data streams, direct modulation systems with Consultative Committee for Space Data Systems (CCSDS) specified virtual channels should be utilized instead of subcarrier modulation schemes. The RF front end of telecommand receivers includes mixers, a Voltage Controlled Oscillator (VCO), and an LPF. PM-PSK and FM-PSK are two existing modulation methods. The objective of the work is to design and simulate a Direct-PSK-based telecommand receiver at 128/256 kbps data rate having one high-speed ADC for small satellite in MATLAB/Simulink.

The rest of the paper is organized as follows: Section II discusses literature review. Section III includes design and implementation details. Section IV discusses about design and implementation as well as analysis of results is done in MATLAB/Simulink. Finally, Section V presents the conclusions and future work.

2   Literature Review

Several researches have been conducted based on the design and simulation of telecommand receivers. Several literatures have been explored in order to conduct a thorough investigation and develop a methodology, or to improve on existing results. Every type of space mission requires a dependable space link between a spacecraft and its associated ground station(s). In today's world, TT&C systems are frequently built in compliance with the recommendations of the CCSDS or the European Cooperation for Space Standardization (ECSS) [3]. TT&C as well as transmission of data across different nodes is required for satellite or aircraft networks. To achieve high-precision measurement, traditional measurement often uses the frequency division duplex mode and a consistent measuring system [4].

Telemetry represents the satellite's operation and health status, telecommand represents the tasks that the mission control center (MCC) on the ground needs the satellite to achieve, and orbit data characterizes the satellite's space setting, location, and velocity, which directly determines the satellite's tasks. Furthermore, the TT&C system is in charge of downloading application statistics from the satellite to the appropriate application centers in particular space missions [5]. The TT&C system onboard the Space-based multi-band astronomical Variable Objects Monitor satellite is critical. It provides a handy approach to observe the Gamma Ray Burst and Target of Opportunity [6].

The loop filter, low-pass filter (LPF), phase detector, and VCO are the key components of the BPSK Costas loop. The BPSK signals are routed to two multipliers in the top branch, referred to as the in-phase channel or I-channel, and the bottom branch, referred to as the quadrature-phase channel or Q-channel [7]. DBS, television service, WLANs, GPS, and RF identification systems, which are either point-to-point or point-to-multipoint, are currently used as ground systems for satellite communication. Furthermore, because of the restricted spectrum available, the modulation technique used is critical for signal transmission accuracy. PSK such as BPSK, QPSK, and OQPSK are used for greater bit data rates [8].

Experiments indicate that varying ratios of the carrier frequency to the frequency of transmission of information bits result in proper demodulation of these signals [9]. Doppler rate and acceleration are created and tuned for spacecraft-to-spacecraft communications where Doppler shifts occur in acquisition and tracking systems. This might also be applied to space-to-ground links, albeit frequency sweeping is usually done at the ground station rather than on-board in these cases [10].

The conventional telecommand receivers contain RF front end having VCO, mixers, analog I/Q downconverters, LPF, and two ADCs. Modulation schemes in the existing telecommand receivers are PM-PSK and FM-PSK at 4 kbps to 64 kbps data rates. They also use subcarriers having PM of 32 kbps data rate. Hence, they are complex in design. Proposed receiver design has minimum RF involvement with one high-speed ADC and the output is direct digitally demodulated. Using Direct-PSK modulation high data rates of 128/256 kbps can be achieved which also simplifies the design. The proposed design reduces cost, space, weight, and operational complexity of the telecommand receiver.

3 Design and Implementation

Key features and specifications of the proposed Direct-PSK based telecommand receiver are as follows:

- The telecommand module has encoder complying with CCSDS standards.
- The module receives Direct-PSK modulated uplink telecommand at 2.1 GHz and 70 MHz intermediate frequency having 128/256 kbps Direct-PSK modulated telecommand data.

Table 1 General
characteristics of the
Direct-PSK Based
telecommand receiver

Parameters	Description
Telecommand modulation format	Direct-PSK at 70 MHz IF upconverted to 2.1 GHz
Data rate	128/256 kbps
Data modulation and encoding format	PCM (NRZ-L) format
Input power range	0 to -40 dBm
Carrier tracking range	± 250 Hz

Table 1 General characteristics of the Direct-PSK Based telecommand receiver

- The telecommand module consists of RF signal conditioning, high-speed space-grade ADC, digital downconverters, Direct-PSK demodulator, FFT-based frequency estimator, and bit synchronizer as per CCSDS standards.
- The module has carrier sweep provision of transmit 70 MHz signal with selectable sweep ranges and rates. Sweep hold and continuation and 70 MHz carrier ON/OFF provision.

Table 1 provides the general characteristics of the proposed telecommand receiver. Modulation format chosen is Direct-PSK since the telecommand data is modulated directly on the RF carrier with NRZ coding format and significant data sideband power falls into the RF carrier phase lock loop (PLL) bandwidth. Therefore, it is bandwidth efficient than subcarrier-based PSK. It also solves pulse code modulation's drawbacks such as performance degradation owing to 1's and 0's imbalances in the data stream. The high data rate of 128/256 kbps is chosen to enable faster reconfiguration of FPGA's LUT in the receiver. Input power range of 0 to -40 dBm validates the telecommand receiver's sensitivity and performance. The methodology of the proposed work is detailed in the following sections. The implementation of the telecommand transmitter and receiver chain as well as its analysis is performed in MATLAB/Simulink, which is a simulation tool for modeling and analyzing multidomain dynamical systems. Two main toolboxes used in the simulation are Communications toolbox and DSP System toolbox. MATLAB programming is used to obtain Kaiser Window filter coefficients for LPF and BPF. It is also used to obtain coefficients for loop filter present in the Direct-PSK Costas loop and bit synchronizer. Parabolic interpolation in FFT-based frequency estimator block is realized by writing a function in MATLAB and incorporating it in Simulink environment.

3.1 Telecommand Transmitter and Receiver

Direct-PSK modulated telecommand data is generated for transmission. The input is command data which has data rate of 128 kbps and is modulated with Direct-PSK on a 70 MHz IF carrier and upconverted to 2.1 GHz (S-band). Figure 1 depicts a general block diagram of the telecommand transmitter. It consists of a telecommand data

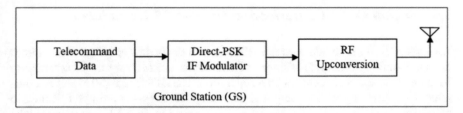

Fig. 1 General block diagram of the telecommand transmitter

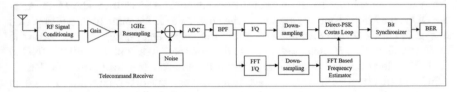

Fig. 2 Block diagram of the proposed direct-PSK based telecommand receiver

generator which can be simulated using a Bernoulli random data generator. It also has an IF signal generator over which the modulated data is superimposed prior to the upconversion for transmission. Figure 2 depicts the block diagram of the proposed Direct-PSK-based telecommand receiver.

The received telecommand data is demodulated after passing through a series of blocks such as RF signal conditioning block, 1 GHz resampler, ADC, BPF, I/Q downconverters, Direct-PSK demodulator, FFT-based frequency estimator, and bit synchronizer. The functions of each block are explained in the following sections.

3.2 Analog to Digital Converter and Band Pass Filter

Analog to digital conversion using ADC12D1620 provides 2.1 GHz aliased output at 100 MHz (5th Nyquist zone). The ADC gives data in DDR mode i.e., 8 samples of data are sent during every clock cycle (at 125 MHz). Sampling frequency is 1 GHz. The output of the ADC consists of signals at 100 ± 70 MHz since the intermediate frequency is 70 MHz. A band pass filter allows 30 MHz signal and rejects the 170 MHz component. An eight parallel bandpass filter (BPF) is designed with cut-off frequencies at around 29 MHz and 31 MHz to allow 30 MHz signal that contains the data of 128 kbps to pass through it. The coefficients of the BPF are obtained using Kaiser Window.

3.3 Digital Down Converter and Direct-PSK Demodulator

An 8-parallel digital down converter (DDC) with decimator is designed and simulated in Simulink. One cycle of sine and cosine signal (25 MHz w.r.t. 1 GHz sampling frequency) is stored in RAM for DDC operation. The DDC input (30 MHz) is multiplied by 25 MHz signal. The output of the multipliers is passed through FIR filter to eliminate higher frequency component and pass only 5 MHz ± 128 kbps signal at the output of DDC. The cutoff frequency of the FIR filter in DDC is 5.3 MHz. The output of the DDC provides an *I* and *Q* output each for the Direct-PSK Demodulator. When downsampling value is 16, the sampling rate reduces from 1 GHz to 62.5 MHz. After the DDC, the *I* and *Q* outputs are sent to the Direct-PSK Demodulator block depicted in Fig. 3, which corrects for any remaining frequency and phase offset between the input signal and the *IQ* detector's front end numerically controlled oscillator (NCO) [11]. The demodulated output comes from the block's In-phase arm. The Kaiser Window FIR filters with order 128 are designed for a cut-off equal to 1.5 times data rate of 128 kbps, i.e., 192 kbps. The sampling rate is 62.5 MHz. The output is sent to the bit synchronizer, which retrieves the clock and data.

The demodulated data is passed through an IIR LPF having a sharp cut-off to determine whether the Costas loop has successfully locked or not. Appropriate threshold is set after trial and error. Sweep control is given by the carrier lock detector to analyze the demodulated data coming out of the Costas loop for frequency shifts of ± 500 kHz due to Doppler shifts in the AWGN channel.

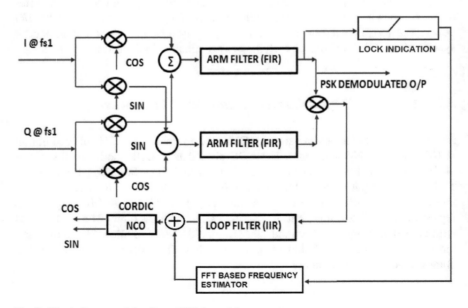

Fig. 3 Block diagram of the direct-PSK demodulator

3.4   FFT-Based Frequency Estimator

A Fast Fourier Transform (FFT) based frequency estimator is used to detect the 5 MHz offset. It continuously corrects large offsets and once the Direct-PSK modulator gets locked at exactly 5 MHz, it is removed. In PM receivers, the use of an FFT-based algorithm guarantees that the PLL works near the carrier frequency offset. This greatly decreases the chances of latching onto the sideband or causing a false lock [12]. When the FFT is performed on a non-periodic block of data, inaccuracies occur. Appropriate windowing functions must be used to resolve this issue. When windowing is not done correctly, problems in the FFT amplitude, frequency, or general form of the spectrum might occur. With rectangular or non-rectangular windowing, inter bin tone peak localization for isolated peaks may be done successfully.

Assume a band-limited compound signal as follows:

$$s(t) = \sum_{l} A_l \sin(2\pi f_l t) + A_{in} \sin(2\pi f_{in} t) \tag{1}$$

The sinusoidal element $\sin(t)$, which has frequency f_{in} to be calculated [13], has been uniformly sampled with frequency f_s. The spectral magnitude $S[k]$ of the sampled signal obtained via the Discrete Fourier Transform (DFT) technique is defined by:

$$S[k] = \left| \sum_{n=0}^{N-1} s[n] \exp\left(-j\frac{2\pi nk}{N}\right) \right| \tag{2}$$

where the signal sample sequence is $s[n] = s(nT_s)$, the sampling period is T_s, and the total number of samples is N. At frequencies that are integer multiples of Δ_f, the discrete spectrum is determined,

$$\Delta_f = \frac{f_s}{N} \tag{3}$$

If the local maximum of the discrete spectral magnitude $S[k]$ is at spectrum bin k_m corresponding to component $\sin(t)$, the frequency f_{in} can be calculated as:

$$f_{in} \cong f_m = k_m \Delta_f = k_m \frac{f_s}{N} = \frac{k_m}{N T_s} = \frac{k_m}{L} \tag{4}$$

where the input signal's sampling duration is $L = NT_s$.

When frequency f_{in} is calculated using (4), the frequency in the midst of two discrete spectrum bins has the biggest inaccuracy. This inaccuracy might be viewed as the FFT frequency measurement method's resolution. As we get closer to the highest frequency error, we get:

$$\varepsilon = \max \left| k_m \Delta_f - f_{in} \right| = \frac{1}{2} \Delta_f = \frac{f_s}{2N} = \frac{1}{2NT_s} = \frac{1}{2L} \tag{5}$$

Equation (5) proposes two techniques to improve the frequency calculation resolution of the FFT: expand the FFT input data points in size; reduce the frequency of sampling. Because the calculation period of an N-point FFT increases with $N \cdot \log_2 N$, one magnitude estimate resolution expansion will increase the calculation period by more than one magnitude estimate. A larger N indicates an extended acquisition period L. The least sample frequency is determined by the sampling theorem. Due to the limited slope of a low-pass anti-alias filter's frequency characteristics, its value is usually larger than the theoretical bound. Furthermore, for some applications, using a greater sampling frequency (oversampling) at the cost of a lower resolution of the spectrum may be preferable. As a result, this method leaves minimal room for improving frequency measurement resolution. The principle of FFT-based three-node interpolation scheme is discussed in [13].

If $S[k]$ denotes the discrete spectral magnitude of a signal with a sinusoidal element of frequency f_{in}, and k_m denotes the index of the largest bin of the discrete spectral magnitude peak. Assuming the bin has a local maximum in a particular range, index k_m can be found. Fitting the equation of a parabola where φ is the abscissa:

$$S_p(\varphi) = a(\varphi - \varphi_m)^2 + h \tag{6}$$

The abscissa of the interpolation maximum is found by interpolating nodes $S[k_m - 1]$, $S[k_m]$, and $S[k_m + 1]$ and calculating the abscissa of the interpolation maximum:

$$\varphi_{in} \cong \varphi_m = k_m + \Delta_m = k_m + \frac{S[k_m + 1] - S[k_m - 1]}{2(2S[k_m] - S[k_m + 1] - S[k_m - 1])} \tag{7}$$

Coefficient a in (6) is less than zero, which infers that $2S[k_m] > S[k_m + 1] + S[k_m - 1]$. Δ_m is the abscissa correction component.

3.5 Bit Synchronizer and BER Calculator

Bit synchronizer based on Gardner algorithm is used to recover the clock and synchronize the output telecommand data and clock [14]. The BER block calculates the bit errors in the stream of demodulated telecommand data by comparing the received telecommand data with the transmitted telecommand data. The BER for Direct-PSK is calculated by Eq. (8).

$$BER = \frac{1}{2} erfc \left(\sqrt{\frac{E_b}{N_0}} \right) \tag{8}$$

4   Simulation Results and Discussion

The simulation results of the telecommand receiver are discussed in this section.

4.1   Telecommand Transmitter

The Direct-PSK data transmitter with NRZ encoding is simulated using a Bernoulli binary random data generator at a data rate of 128 kbps. A switch is used to alternate the output between -1 and $+1$ when the data is 0 or 1, respectively. The IF frequency is 70 MHz and is upconverted to 2.1 GHz. The Direct-PSK modulated output at 70 MHz is obtained at an amplitude of 27 dBm and observed at 69.987 MHz in the spectrum analyzer in Simulink. The sampling rate is 21 GHz. The upconverted signal at 2.03 and 2.167 GHz with peak of 20.9599 dBm is observed on a spectrum analyzer. A BPF is designed to reject the 2.167 GHz RF signal and allow only 2.03 GHz signal. Output obtained after under-sampling the RF carrier of 2.1 GHz is observed at 1 GHz sampling frequency. Two peaks at 29.948 and 169.922 MHz with amplitude of 20.8255 dBm are observed on a spectrum analyzer. The sampling rate is 1 GHz.

4.2   Analog to Digital Converter and Band Pass Filter

ADC12D1620 uses dual-edge sampling to obtain the output. The frequency of the input is 2.1 GHz carrier having telecommand data and the sampling frequency is 1 GHz. The ADC uses dual-edge clock (DCLK) of 500 MHz. There are two internal ADCs that sample at rising edge In-phase and Quadrature-phase (I and Q) and falling edge In-phase and Quadrature-phase (I_d and Q_d), respectively. The samples are stored such that eight samples are captured and pushed out at a frequency of 125 MHz using double data rate (DDR) mode. The conversion from dual edge to single edge takes place to provide the final output in IDDR single edge mode. BPF is designed using the specifications given in Table 2.

Table 2 Specifications of the band pass filter

Parameters	Value
F_{pass1}	29 MHz
F_{pass2}	31 MHz
Sampling frequency	1 GHz
Window	Kaiser window
Order	16
Beta	0.5

The design equations for the 8-tap BPF are as follows. Two 8-tap BPFs are combined to obtain a single 16-tap BPF.

$$y_0 = x_0h_0 + z^{-1}(x_1h_7 + x_2h_6 + x_3h_5 + x_4h_4 + x_5h_3 + x_6h_2 + x_7h_1)$$
$$y_1 = x_0h_1 + x_1h_0 + z^{-1}(x_2h_7 + x_3h_6 + x_4h_5 + x_5h_4 + x_6h_3 + x_7h_2)$$
$$y_2 = x_0h_2 + x_1h_1 + x_2h_0 + z^{-1}(x_3h_7 + x_4h_6 + x_5h_5 + x_6h_4 + x_7h_3)$$
$$y_3 = x_0h_3 + x_1h_2 + x_2h_1 + x_3h_0 + z^{-1}(x_4h_7 + x_5h_6 + x_6h_5 + x_7h_4)$$
$$y_4 = x_0h_4 + x_1h_3 + x_2h_2 + x_3h_1 + x_4h_0 + z^{-1}(x_5h_7 + x_6h_6 + x_7h_5)$$
$$y_5 = x_0h_5 + x_1h_4 + x_2h_3 + x_3h_2 + x_4h_1 + x_5h_0 + z^{-1}(x_6h_7 + x_7h_6)$$
$$y_6 = x_0h_6 + x_1h_5 + x_2h_4 + x_3h_3 + x_4h_2 + x_5h_1 + x_6h_0 + z^{-1}(x_7h_7)$$
$$y_7 = x_0h_7 + x_1h_6 + x_2h_5 + x_3h_4 + x_4h_3 + x_5h_2 + x_6h_1 + x_7h_0 \qquad (9)$$

The BPF is realized using the design equations given above. The coefficients obtained for Kaiser Window are substituted in the structure. ADC12D1620 is interfaced with BPF to filter the 30 MHz output having the telecommand data. The output of 70 MHz IF upconverted to 2.03 GHz after filtering eliminates 2.167 GHz and allows 2.03 GHz at amplitude of 20.4979 dBm. The BPF is designed at cut-off frequencies of 29 and 31 MHz to allow the 30 MHz signal component and reject the spurious signals. The output of 30 MHz obtained at ADC12D1620 after band pass filtering eliminates 170 MHz and allows 30 MHz at an amplitude of 79.2508 dBm. ADC provides 30 MHz output at 1 GHz sampling frequency.

4.3 Digital Down Converter and Direct-PSK Demodulator

The telecommand transmitter, ADC12D1620, and BPF realized in previous steps are interfaced with the DDC and LPF to obtain the chain in Simulink as illustrated in Fig. 4. DDC is designed to obtain 5 MHz signal from 30 MHz signal. A decimator using downsampling value of 16 is used to lower the sampling rate from 1 GHz to 62.5 MHz. One cycle of sine and cosine signal (25 MHz w.r.t. 1 GHz sampling frequency) is stored in RAM for DDC operation. A 64-tap LPF is designed using the specifications given in Table 3. DDC output before LPF is 5.208 MHz with amplitude of 17.3818 dBm. DDC output after LPF is 5.208 MHz with amplitude of 17.0602 dBm, with the spurious signals being filtered out. The detailed structure of Direct-PSK Demodulator Costas loop in Simulink is illustrated in Fig. 5. Any remaining phase or frequency offset between the input signal and the *IQ* detector's NCO front end is compensated for by the Direct-PSK Demodulator block. The demodulated output is obtained either in the in-phase or the 180° out of phase arm.

The lock indication detector is designed to detect if the PLL has locked successfully or not and is used to enable or disable the FFT-based frequency estimator.

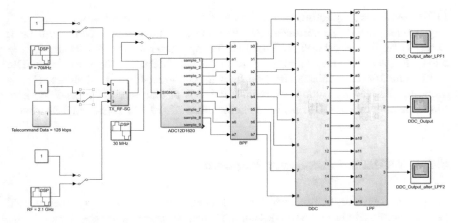

Fig. 4 Structure of telecommand receiver upto DDC with LPF in Simulink

Table 3 Specifications of the low-pass filter

Parameters	Value
F_{pass}	5.3 MHz
Sampling frequency	1 GHz
Window	Kaiser window
Order	64
Beta	5

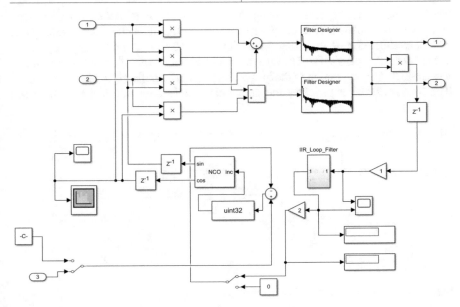

Fig. 5 Structure of direct-PSK demodulator costas loop in simulink

FFT-based frequency estimator block is needed if the output goes out of loop and large offsets need to be corrected. The lock indication detector is designed using a Direct-Form II Chebyshev Type I IIR LPF having sharp roll off of 80 dB and $F_{pass} = 5$ kHz, $F_{stop} = 10.5$ kHz, sampling frequency is 62.5 MHz. The threshold is 0.1. The coefficients for the feedback IIR loop filter in the Direct-PSK Demodulator Costas loop are obtained for a loop bandwidth of 1–3% of data rate of Direct-PSK data. Since the data rate is 128 kbps, loop bandwidth is 3% of 128 kbps, i.e., 3.84 kbps.

4.4 FFT-Based Frequency Estimator

FFT-based frequency estimator is based on Blackman windowing. Number of FFT points is 4096. Either parabolic or Gaussian interpolation can be used to obtain the index at which peak value of FFT is obtained. Blackman window provides main lobe width of 6 bin, highest sidelobe of −68.2 dB and sidelobe asymptotic fall-off of 6 dB/oct. Parabolic interpolation provides maximum error of 4.66 (% of Δ_f) and gain factor of 10.7. Gaussian interpolation provides maximum error of 0.578 (% of Δ_f) and gain factor of 86.5 [13].

4.5 Bit Synchronizer and BER Calculator

The bit synchronizer synchronizes the demodulated output with the recovered clock from the demodulated data. Gardner algorithm is the algorithm that is employed. The demodulated data is further downsampled by 32 to obtain a sampling rate of 1.953 MHz from 62.5 MHz. The bit synchronizer contains a timing error detector and a resampler. The simulation of the bit synchronizer in Simulink is illustrated in Fig. 6. The bit synchronizer provides the demodulated telecommand data and the clock

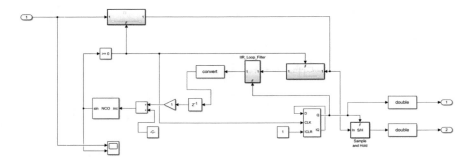

Fig. 6 Simulation of bit synchronizer in Simulink

extracted from it as outputs. By comparing the received and transmitted telecommand data, the BER block estimates the bit errors in the stream of demodulated telecommand data.

4.6 Analysis of the Demodulated Output

The demodulated telecommand data obtained is analyzed using the parameters: input frequency sweep, input amplitude, and band-limited noise. Sweep generator is designed to sweep frequencies of 5 MHz ± 500 kHz. The sweep rate is 250 Hz. Input amplitude is varied from −40 to 0 dB. The band-limited white noise is 10^{-15}. The Direct-PSK Demodulator Costas loop is tested with frequency sweep (Doppler) and noise with varying input amplitude. The results obtained from the model are found to be satisfactory. The demodulated telecommand data obtained at the output of Costas loop in Simulink is illustrated in Fig. 7. The pink-colored line infers the in-phase output, which is close to zero. The blue-colored line infers the 180 degrees out of phase demodulated output, used for retrieving telecommand data. The orange-colored line is the telecommand data transmitted from the telecommand transmitter (at 128 kbps) at the ground station.

The bit synchronizer retrieves the data and the clock (at 128 kbps) from the demodulated telecommand data collected at the output of Costas loop. The demodulated telecommand data latches onto the rising edge of the clock extracted from it. The bit synchronizer output in Simulink is illustrated in Fig. 8. The clock retrieved from the

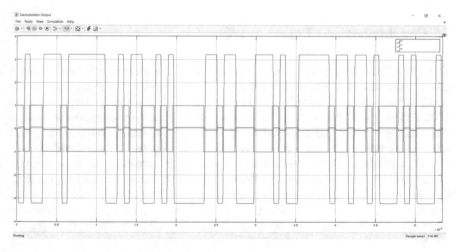

Fig. 7 Demodulated telecommand output in Simulink

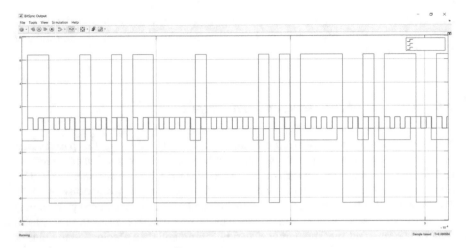

Fig. 8 Bit synchronizer output in Simulink

demodulated data is depicted on the pink-colored line. The demodulated telecommand data is represented by the blue-colored line. The orange-colored line represents telecommand data sent from the ground station's telecommand transmitter.

The BER output is obtained by comparing the transmitted telecommand data and the demodulated output obtained at the output of the bit synchronizer with respect to the clock extracted from the demodulated output.

5   Conclusions and Future Work

The Direct-PSK based telecommand receiver design has minimum RF involvement due to a high-speed ADC and direct digitally demodulated output from 2.1 GHz (S-band) to baseband at 128/256 kbps data rate. The receiver works for large data rates of 128/256 kbps with a simpler design compared to subcarrier-based demodulation methods. As a result, the proposed design reduces cost, space, weight, and operational complexity of the telecommand receiver. The proposed telecommand receiver demodulates the telecommand data for the satellite management unit to perform its operations. After 302 bits are received and demodulated correctly in 2.34 ms, the BER is 0. As a result, the developed telecommand receiver works well with input gains ranging from 0 to −40 dB. At −40 dB, BER is observed. The band-limited noise is in the range of 10^{-15}. For a frequency range of 5 MHz ± 500 kHz, the input frequency sweep rate is 250 Hz. The Direct-PSK Demodulator's loop bandwidth is 1% of the data rate or 1.28 kHz. As a result, the developed Direct-PSK-based telecommand receiver outperforms conventional subcarrier-based telecommand receivers in terms of acquisition range and tracking capability. Frame synchronizer can be designed to obtain the output without any phase shift w.r.t the input. Further improvements can be done to the design to achieve successful reception of telecommand data up to a frequency sweep rate of 32 kHz from the achieved sweep rate of 250 Hz. LDPC, BCH, and other coding techniques can be applied to the telecommand data. Telecommands are stored in OBC/BMU with GPS command time and are executed.

References

1. Nguyen, H.H., Nguyen, P.S.: Communication subsystems for satellite design, pp. 1–24. InTechOpen (2020)
2. Lappas, V., Kostopoulos, V.: A survey on small satellite technologies and space missions for geodetic applications, pp. 1–22. InTechOpen (2020)
3. Budroweit, J., Gärtner, T., Greif, F.: Design of a fully-integrated telemetry and telecommand unit for CCSDS spacecraft communication on a generic software-defined radio platform. In: 2020 IEEE Space Hardware and Radio Conference, pp. 13–15 (2020)
4. Xue, L., Li, X., Wu, W., Yang, Y.: Design of tracking, telemetry, command (TT&C) and data transmission integrated signal in TDD Mode. MDPI Remote Sens. 1–21 (2020)
5. Zhan, Y., Wan, P., Jiang, C., Pan, X., Chen, X., Guo, S.: Challenges and solutions for the satellite tracking, telemetry, and command system. IEEE Wirel. Commun. 27(6), 12–18 (2020)
6. Liu, Y., Yu, S., Hu, H., Li, Z., Dai, Y., Zhang, X.: Design and validation of SVOM on board TT&C system. In: 2020 IEEE International Instrumentation and Measurement Technology Conference, pp. 1–5
7. Shamla, B., Gayathri Devi, K.G.: Design and implementation of costas loop for BPSK demodulator. In: 2012 Annual IEEE India Conference, pp. 785–789
8. Singh, K., Nirmal, A.V.: Overview of modulation schemes selection in satellite based communication. ICTACT J. Commun. Technol. 11(03), 2203–2207 (2020)
9. Bondariev, A., Maksymiv, I., Altunin, S.: Simulation and experimental research of the enhanced BPSK and QPSK demodulator. In: 2020 IEEE 15th International Conference on Advanced Trends in Radioelectronics, Telecommunications and Computer Engineering, pp. 767–770 (2020)
10. Divsalar, D., Net, M.S., Cheung, K.M.: Adaptive sweeping carrier acquisition and tracking for dynamic links with high uplink Doppler. In: 2020 IEEE Aerospace Conference, pp. 1–14 (2020)
11. Sharma, A., Syed, A.H., Midhun, M., Raghavendra, M.R.: Realization of programmable BPSK demodulator-bit synchronizer using multirate processing. Int. J. Reconfig. Embed. Syst. 3(1), 18–24 (2014)
12. Nargund, A.A., Kashyap, C.A., Prashanth, D., Premananda, B.S., Sharma, A.: A novel method of wideband acquisition and anti-sideband lock in PM receivers using FFT. In: 2016 IEEE International Conference on Recent Trends in Electronics, Information and Communication Technology, pp. 2098–2101 (2016)
13. Gasior, M., Gonzalez, J.L.: Improving FFT frequency measurement resolution by parabolic and Gaussian interpolation. In: European Organization for Nuclear Research Organisation Europeenne Pour La Recherche Nucleaire Cern—AB Division, AB-Note-2004–021 BDI, pp. 1–15 (2004)
14. Zhang, Q., Gao, W., Zhao, H.: A new lock detection algorithm for Gardner's timing recovery. In: 2011 13th International Conference on Communication Technology, pp. 319–322

Insect-Inspired Advanced Visual System for AI-Based Aerial Drones

Vijay A. Kanade

Abstract The research proposes principles for developing advanced visual systems that assist AI-based aerial drones in navigation. The system mimics color vision mechanics as observed in insects and animals. It harnesses a greater level of optical perception as different wavelengths of light are used for navigation of UAVs. The proposed system is employed to program drones that allow them to pilot autonomously by using their 'eyes' (i.e., cameras).

Keywords Advanced visual system · Drones · Color perception · Thermoelectric (TE) nanostructures · Chip · GPS signals

1 Introduction

Drone popularity has gained steam in the past few years. With the increased adoption of drone technology for multiple applications, new use cases of drones continue to pop-up every day. This has further led to extensive research and development into the aerial tech machines.

With the technological advancement, change in manufacturing dynamics and exponential scaling of nanotech components, drone tech can only be expected to become an integral part of our daily lives in years to come. These ubiquitous machines use internal navigation systems in coordination with satellite systems and manual pilot control by a personnel. Drone automation still poses a challenge as a full-fledged navigation system seems a distant reality as of now.

Drones use sophisticated computer vision for navigation—technically referred to as simultaneous localization and mapping (SLAM). The method generates a map of the nearby geography to better comprehend the surroundings of a drone. However, the method is not entirely foolproof—implying, in situations when the surrounding objects are out of place, the drone is susceptible to crashing into them.

V. A. Kanade (✉)
Pune, India

© The Author(s), under exclusive license to Springer Nature Singapore Pte Ltd. 2022   865
V. Suma et al. (eds.), *Evolutionary Computing and Mobile Sustainable Networks*,
Lecture Notes on Data Engineering and Communications Technologies 116,
https://doi.org/10.1007/978-981-16-9605-3_60

On the other hand, commercial drones combine GPS with inertial sensors for effective navigation. Sensors include cameras or lidar. These measurement devices help in detecting obstacles in a drone's path. However, the speed of drone movement is directly proportional to accuracy of these sensors, size of the machines and even the count of obstacles.

As the accuracy of the sensors cannot be guaranteed, it has also been identified that with time, the inertial sensors gradually accumulate errors. Also, drones face challenges when GPS signals bounce off a number of obstacles. This leads to false positional information. In a crowded environment, the drone's forward movement and speed suffers considerably as there are plenty of obstacles (man-made structures, buildings) to navigate around for a typical drone. Although, there are sizable filters to tackle this problem, yet, it does not present a wholesome solution for all use cases.

Looking at the flaws of the current navigation system, there seems to be a long standing need to explore better visual systems for aerial drones.

2 Color Perception

The universe is colorful and each color communicates some useful information. Consider nature for example, a flower markets its nectar via colors, fruit color dictates its maturity stage (ripe) and light from a glowing object reveals its temperature.

Insects and animals are able to see colors. Insect vision plays a critical role in food hunting, avoiding obstacles, keeping the enemies away and attracting mates. Some insects also use their vision to navigate, guide and control flight around objects as they move around. Photoreception capability also allows the insect to predict approaching winter based on reduction in the duration of sunlight. Such useful data assists a cold-blooded animal to prepare well in advance for the onset of cold winter [1].

Humans on the other hand are known to have a 'trichromatic color vision.' This implies, all the colors perceivable by humans can be produced just by using primary colors, i.e., red, green and blue.

3 Mechanics of Vision

Humans rely heavily on color vision to understand the world around them. We possess the capability to interpret different wavelengths of light, rather than just understand the brightness of it.

We have light-sensing cells in the form of cones and rods, with each one performing a different function. Rods are sensitive to light, hence they are operational more so in dim light, for example, at night. But, rods are not capable of detecting colors. This is the reason we perceive black and white worlds in moonlight. Color perception comes from cones. We have three types of cones, each depicting primary colors: red (R), green (G) and blue (B). Each cone type is sensitive to a specific wavelength: R cones respond to 564 nm, G cones 534 nm and B cones 420 nm [2]. Wavelengths outside the visible band cannot be perceived by humans as our visual physiology is not equipped enough to detect it.

Similarly, different species, animals and insects have different types of light-sensing cells. Honeybees have three types of cells, however they also are known to possess cells that can sense UV light. Meanwhile, the common bluebottle butterfly, Graphium sarpedon, has 15 cell types and small white butterfly, Pieris rapae, has seven [3]. Insects, birds, animals and we humans perceive colors due to these specific light-sensing cells that reside in our respective visual apparatuses (Figs. 1 and 2).

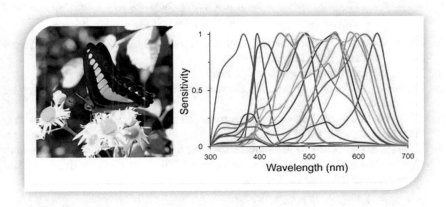

Fig. 1 Spectral sensitivity of 15 light-sensing cells for Graphium sarpedon [3]

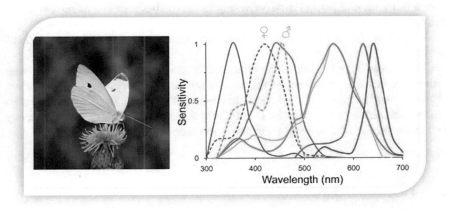

Fig. 2 Spectral sensitivity of 7 light-sensing cells for Pieris rapae [3]

4 Insect-Inspired Advanced Visual System for AI-Based Aerial Drones

In the proposed model, we intend to design a visual system in a way that allows the drones to see different wavelengths of light and pilot their flight autonomously, similar to insects or animals. For this purpose we embed a program code into drones that detect specific wavelengths and navigate the machine skillfully. This reduces the risk of GPS signals banging off objects, causing a problem in drone navigation. Also, the chances of drones banging onto obstacles occurring due to errors in inertial sensors are cut down.

To target the visual system, we program a code on a hardware chip and attach it to the drones. The chip is composed of different nanostructures which vary in their diameter, width and structure. The underlying tech works in the following way:

1. Nanostructures within the chip have different widths that absorb different wavelengths of light, i.e., colors [4, 5].
2. The chip captures incident light photons.
3. This excites the electrons within the nanostructures that are further detected by the detector.
4. The movement of light-excited electrons leads to generation of subtle electric current, i.e., a signal.
5. The generated electric current has a strength that corresponds to the light wavelength that is absorbed by the nanostructures. Thus, the signal is measured and quantified to navigate the drones.

The proposed technology uses the principles of thermoelectric to use light energy. The temperature differences (heat) between light photons of different energies is translated directly into electron voltage. This helps to distinguish between different wavelengths of light, i.e., colors within the visible spectrum and utilize them for drone navigation. Also, infrared wavelengths are distinguishable using the above tech (Fig. 3).

Note: The chip is positioned in a way that keeps it exposed to sunlight all the while

The above figure illustrates the cross-sectional view of the hardware chip with nanostructures that is placed on the drones for visual perception and navigation. According to the proposal, the nanostructures within the chip have different widths that absorb different wavelengths (λ_n) of light to generate different eVs. Following notations describe the relevant variables of the proposed hardware chip.

Notations

E_n photonic energy
h Planck's constant
c speed of light

Fig. 3 Cross-sectional view of hardware chip

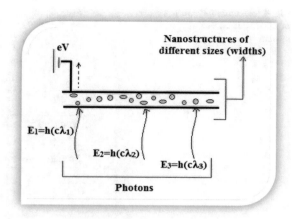

λ_n wavelengths of different colors.

eV electron voltage translated from different photonic energies.

Nanostructures of different widths = Nanostructures of different sizes.

4.1 Schematic Flowchart

The operation of the proposed visual system is illustrated in the below flowchart (Fig. 4).

Use Case

For example, consider a drone that is programmed to guide flight in blue light. As energy photons corresponding to the blue band and other colors hit the nanostructures, electric current of corresponding magnitude is generated within the chip. Here, the temperature gradient between light photons of different colors (or energies) is translated directly into electron voltage which aids in blue wavelength detection. On detection of the wavelength, the drone guides its flight. On the other hand, the drone does not respond to photons whose energies lie within the red or green wavelength. This leads to significant noise reduction for the drones as the visual system stays unaffected by the unwanted wavelengths of light.

4.2 Preliminary Research Results

The employed thermoelectric (TE) nanostructures are known to perform selective absorption upon illumination. They tend to create substantive temperature gradients that are enough to have quantifiable thermoelectric voltages (eV). According to the

Fig. 4 Schematic flowchart
of the advanced visual
system

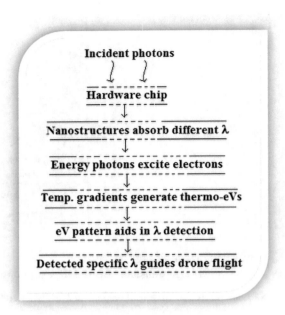

preliminary research, the nanostructures are explicitly tunable and help in precise wavelength detection.

The nanostructures are essentially TE thermocouple junctions. These are made up of known TE materials such as chromel/alumel and bismuth telluride/antimony telluride. They have a strong affinity for photon absorption. Upon optical stimulation, they generate substantial heat with variable thermal gradients. Research reveals, for a combination of bismuth telluride and antimony telluride structures, the TE voltages (eVs) around 850 μV are generated with 3.4 W/cm^2 incident illumination densities. Further, the responsivity of nanostructure was around 119 V/W for incident optical power [4]. Thus, it is thereby established that the selected nanostructures are suitable to generate TE voltages (eVs) on light illumination for specific wavelength detection.

Note: TE junctions generate the optical power (i.e., hot end), while the ends of the TE nanostructures are at ambient temperature (i.e., cold end).

Responsivity is determined with the aid of TEV which is directly proportional to the Seebeck coefficient (S), and the temperature difference ΔT, between cold and hot ends of the nanostructure. Thus, TEV $= S\Delta T$ [4].

At equilibrium, the absorbed power and dissipated power are balanced by taking following variables into account (Fig. 5).

$$\int_{V_{TE}} \frac{1}{2}\omega\varepsilon''|E|^2 dV = \int_{A_{air}} e\sigma(T^4 - T_0^4)dA - \int_{A_{slice}} \kappa\nabla T dA$$

$$+ \int_{A_{air}} h(T - T_0)dA + \int_{A_{interface}} h_c(T - T_{sub})dA,$$

Fig. 5 Steady state (absorbed vs. dissipitated power) for TE nanostructures [4]. where ω = incident light frequency, ε'' = the imaginary part of the TE nanostructure dielectric function, $|E|$ = electric field within the nanostructure, VTE = the illuminated volume of nanostructure, e = the nanostructure emissivity, A_{air} = the air exposed area of the nanostructure, σ = the Stefan-Boltzmann constant, T_0 = the initial temperature (cold/ambient end), κ = the thermal conductivity of nanostructure, A_{slice} = the cross-sectional area of nanostructure segregating optically illuminated and unilluminated areas, h = the heat transfer coefficient between the nanostructure and air, h_c = the conductance between the nanostructure and the substrate, T_{sub} = the substrate temperature near the nanostructure and $A_{interface}$ = the intersection area of the nanostructure and the substrate

5   Night Color Perception

For color perception during the night, we embed high performance sensors and image signal processors (ISP) in the hardware chip. This hardware also has an achromatic lens that allows the lens to stay in the color mode 24/7. The achromatic lens is a blend of convex and concave glass pieces that is assembled in a way to focus different light color wavelengths distinctly on a single plane. It reduces chromatic aberrations and operates even in minimal lighting conditions as that of 1 lx of ambient light [6].

Additionally, the target hardware also utilizes back-illuminated CMOS technology to have equal color distribution, thereby providing colorful vision in dark environments. Here, the back-illuminated section allows more light to enter through each pixel as it pushes the photosensitive surface in the outward direction for extra light exposure. Also, ISP further enhances the vision as it corrects the error caused by lens, or overall sensors, thereby producing high quality night color vision.

6   Conclusion

The research proposes a novel visual system that enables drones to navigate seamlessly by employing advanced color perception to guide their flight. The drones are retro-fitted with programmed chips that have nanostructures of various widths embedded within them. The research proposal reduces the navigational risk arising from conventional GPS signals and errors in inertial sensors. Thus, the paper gives a first-hand account of a unique navigation system that allows drones to see different wavelengths of light and use their visual system to move around in various geographies.

7   Future Work

In future, we intend to extend the research to harness different wavelengths (colors) of light within the visible spectrum for different applications. For example, as sunlight is a combination of all colors, we could use each wavelength for specific applications. Implying blue light could be used for drone navigation, red light could be used to determine the approach of winter, and fading orange/yellow shade could be used to predict the onset of rainfall.

Acknowledgements I would like to extend my sincere gratitude to Dr. A. S. Kanade for his relentless support during my research work.

Conflict of Interest   The authors declare that they have no conflict of interest.

References

1. Turpin, T.: Professor of Entomology. Purdue University, Insects See the Light (2012)
2. Land, M.F., Nilsson, D.-E.: Animal Eyes. Oxford University Press, Oxford (2002)
3. Chen, P.-J., Awata, H., Matsushita, A., Yang, E.-C., Arikawa, K.: Extreme spectral richness in the eye of the common bluebottle butterfly, Graphium sarpedon. Front. Ecol. Evol. **4**, 1–12 (2016). https://doi.org/10.3389/fevo.2016.00018
4. Mauser, K.W., Kim, S., Mitrovic, S., Fleischman, D., Pala, R., Schwab, K.C., Atwater, H.A.: Resonant thermoelectric nanophotonics. Nat. Nanotechnol. **12**(8):770–775 (2017). ISSN 1748-3387
5. Perkins, P.: Caltech, Nanostructures Detect Colors, June 28, 2017. https://www.caltech.edu/about/news/nanostructures-detect-colors-78861
6. Dahua technology, Seeing in the Dark for Around the Clock Surveillance. https://us.dahuasecurity.com/?page_id=51720

Comparison of Machine Learning Algorithms for Hate and Offensive Speech Detection

Mehant Kammakomati⊙, P. V. Tarun Kumar⊙, and K. Radhika

Abstract Hate speech is not uncommon and is likely practiced almost on every networking platform. In recent times, due to exponential increase in Internet users and events such as the unprecedented pandemic and lockdown, it showed increased usage of social platforms for communicating thoughts, opinions, and ideas. Hate speech has a strong impact on people's lives and is one of the reasons for suicidal events. There is certainly a strong need to make progress toward the mitigation of hate speech. Detection is the primary step to eradicate hate speech. In the following paper, the comparative analysis of different machine learning algorithms to detect hate speech was shown. Data from the Twitter social platform was considered. From the analysis, it was found that the long short-term memory method is a highly performant machine learning algorithm.

Keywords Classification · Hate speech · Machine learning · Natural language processing

1 Introduction

In recent years, the Internet has become an easy and vital medium for communication [1]. However, the Internet is under continuous exploitation to quickly spread hatred through speech, text, or other multimedia among the masses. Online hate speech can be understood as any form of communication that discriminates against a person or group of people based on their religion, ethnicity, nationality, race, color, descent, gender, or other identity factors. With densely connected social platforms,

M. Kammakomati (✉)
National Institute of Technology, Andhra Pradesh, Tadepalligudem, India
e-mail: 411843@student.nitandhra.ac.in

P. V. Tarun Kumar · K. Radhika
Chaitany Bharathi Institute of Technology, Gandipet, Telangana, India
e-mail: ugs18051_it.tarun@cbit.ac.in

K. Radhika
e-mail: kradhika_it@cbit.ac.in

© The Author(s), under exclusive license to Springer Nature Singapore Pte Ltd. 2022
V. Suma et al. (eds.), *Evolutionary Computing and Mobile Sustainable Networks*,
Lecture Notes on Data Engineering and Communications Technologies 116,
https://doi.org/10.1007/978-981-16-9605-3_61

these hate artifacts are spreading much faster and reaching a wider range. Hate speech may include posting hateful online criticisms that insult a person or a group of people. Online hate speech is one of the causes of the increase in suicidal cases.

This is quite a serious problem across the globe. Based on the statistics recorded in 2015 [2], the percentage of agreement toward making hate comments against minority groups is highest in the USA followed by Latin America, Europe, Africa, and Asia. Moreover, the number of active users of the Internet has increased tremendously in recent months owing to the unprecedented pandemic and lockdown. Hate speech is not uncommon, it is everywhere, and with the increase in Internet users, violent acts are also increasing on social networking platforms every day. Social networking platforms are widely used to share thoughts and ideas and communicate with a wide range of people. Social platform-owning organizations are spending large amounts of money and resources to legally handle and eradicate hate speech on their platforms.

The primary step in the prevention is to detect or identify hate speech on the platforms. Manual moderation is not helpful with the amounts of data that each social platform produces every day. The idea is to come up with a scalable, reliable, and privacy-preserving solution to scan through billions of information over the platform and tag hate-specific threads for further steps of mitigation. In this work, a comparative analysis of certain machine learning algorithms to detect hate speech was shown. The widely used social platform Twitter was chosen as a source of the data, and the dataset is taken from Kaggle [3].

2   Related Work

There have been various researches done in the sphere of sentiment analysis to classify hate speech with datasets from various sources.

Lin Jiang et al. [4] have done a comparative analysis where their best precision score obtained was 79% for long short-term memory (LSTM). The work did not include hyperparameter tuning and handling of missing values in the dataset. Bujar Raufi et al. [5] have proposed a system for identifying hate and offensive speech using artificial neural networks specifically on mobile platforms. Their focus was on presenting lightweight machine learning algorithm; however , the work ignores deeper language constructs. The research was also carried out for the detection of hate speech in different language communications. Ali et al. [6] presented work on identifying hate speech in Urdu tweets. The work was more focused on dataset preparation, and comparison study did not include deep learning algorithms. Similar work on Nepali tweets was done by Tripathi and Milan et al. [7], where they have used Naive Bayes, support vector machine, and LSTM algorithms. Putra et al. [8] have presented work on hate speech detection in Indonesian language. Luu et al. [9] work considers gate recurrent unit model instead of LSTM to save on training time. However, the compromise on model score for that decision was not mentioned. Their best $F1$-score obtained was 83%. Hajime Watanabe et al. [10] propose an approach

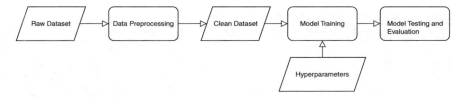

Fig. 1 Proposed work flow diagram

to detect hate speech from a dataset of 25020 tweets. The work considered random forest, SVM, and J48graft for comparison. The best model accuracy obtained from their proposed approach was 87.4%.

3 Proposed Work

The main objective of the proposed work was to perform a comparative analysis of different machine learning models for hate speech detection. For each of the machine learning models included in the analysis, follow the following steps: data collection, data preprocessing, train–test split, model preparation, model training, model testing, and calculating evaluation parameters (see Fig. 1).

3.1 Dataset Description

This dataset contains 25000 tweets that were taken from Kaggle [3], which was further divided into two subsets: the training dataset and the testing dataset. The training dataset contains 16,000 observations, and the testing dataset contains 8000 observations. The dataset consists of the following columns:

- Unnamed: A numeric value that describes the cumulative sum of named tweets in the top to bottom order through the dataset.
- Count: A numeric value that describes the number of flagged words.
- Hate_Speech: A numeric value that describes the total number of hate words in a tweet.
- Offensive_language: A numeric value that describes the degree of hate speech. This value ranges from 0 to 10.
- Neither: A numeric value that describes the number of words that are neither positive nor negative.
- Class: This column takes three values:

  - 0: The tweet contains words from which are considered hate speech.
  - 1: The tweet contains offensive words.
  - 2: The tweet does not contain any offensive or hate speech words.

Table 1 Statistical description of the dataset

	Unnamed: 0	Count	Hate_speech	Offensive_language	Neither	Class
Count	25,000	25,000	25,000	25,000	25,000	25,000
Mean	12500	3.243473	0.280515	2.413711	0.549247	1.110277
Std	7299.553863	0.883060	0.631851	1.399459	1.113299	0.462089
Min	0.000000	3.000000	0.000000	0.000000	0.000000	0.000000
25%	6372.500000	3.000000	0.000000	2.000000	0.000000	1.000000
50%	12703.000000	3.000000	0.000000	3.000000	0.000000	1.000000
75%	18995.500000	3.000000	0.000000	3.000000	0.000000	1.000000
Max	25296.000000	9.000000	7.000000	9.000000	9.000000	2.000000

Table 1 gives a statistical description of the dataset.

3.2 Data Preprocessing

Data preprocessing is the process of preparing the raw data that is taken directly from sources, and it is done by modifying it wherever needed and retaining just the required details. Incorrectly formatted data is generally not necessary, and it can hinder training performance and is highly prone to produce inaccurate results. Pattern matching was chosen to remove unnecessary characters from the dataset. A language-specific stop words dataset was obtained to clean the dataset by removing those words which do not contribute to the model training. Tuples were removed having missing values. Complete data was converted to lower case considering the fact that case of letters does not contribute to classification. Finally, word embeddings were obtained using Word2Vec technique, specifically skip-gram method was used.

3.3 Hyperparameter Tuning

Finding optimal hyperparameters is crucial for proper learning of the model parameters during the training phase. There are a couple of methods to find the hyperparameters to any model. The grid search method performs search exhaustively trying every combination of the provided hyperparameters before returning the optimal choice. The random search method involves picking a set of hyperparameters randomly, and this process is run multiple times as sufficient to arrive at an optimal choice. The Bayesian search method is a probabilistic approach to find optimal hyperparameters. Instead of searching in isolation, this approach chooses the set of hyperparameters based on the previous result.

The Bayesian search was used for each of the models mentioned in the paper to find the hyperparameters. It needs choosing a range for each of the hyperparameters, defining an objective function, choosing a surrogate method, model score, and the number of iterations. In this work, the objective function was the machine learning algorithm. Tree Parzen estimator was used as the surrogate method. Accuracy was chosen as the model score.

3.4 Decision Tree Classifier

Decision tree classification [11] is a supervised learning technique that is used for both classification and regression problems. It is a tree-structured classifier, where internal nodes represent the features of a dataset, leaf nodes represent the outcome, and branches represent the decision rules. Decision nodes are used to make any decision, and they have multiple branches, whereas leaf nodes are the output of those decisions and do not contain any further branches.

3.5 Random Forest

Random forest is a classifier that contains a cluster of decision trees on various subsets of the given dataset and takes the mean of all the outputs to improve the predictive accuracy of that dataset [12]. Instead of relying on a single decision tree, the random forest takes the predictions from each tree of the cluster and based on the majority votes on predictions predicts the final output.

3.6 AdaBoost

AdaBoost can be used for classification problems. Its central idea is to fit a set of weak classifiers (small decision trees) to repeatedly update versions of the data [13]. This way multiple weak classifiers are combined into the single strong classifier. To get the final prediction, the predictions from all of them are combined using a weighted majority vote (or sum).

3.7 Support Vector Clustering

Support vector machines in classification problems are referred to as support vector clustering. The central idea is to find a straight line or a hyperplane to separate data into classes [14]. In the case of multi-class classification, multiple classifiers are built, and each one trains data from two classes. This way one-vs-one classifiers are produced that are then converted into one-versus-many classifiers.

3.8   Long Short-Term Memory

A recurrent neural network (RNN) [15] is a sort of artificial neural network in which nodes are connected in a directed graph in a temporal order. The preceding step's output is used as input in the RNN's following phase. Long short-term memory (LSTM) classification is derived from RNN [16]. It deals with the problem of RNN long-term dependency, in which the RNN is unable to predict words stored in long-term memory but can make more accurate predictions using current data. RNN does not provide efficient performance as the gap length rises. By default, the LSTM can keep the information for a long time. It is used for processing, predicting, and classifying on the basis of time series data.

3.9   Training Phase

In this phase, each of the machine learning models was trained by providing the training data and required hyperparameters. The training data was 80% of the pre-processed dataset. All the models were trained in a sequential fashion to ensure a fair comparison.

3.10   Testing Phase

In this phase, each of the machine learning models was tested to obtain various evaluation parameters. Testing data was 20% of the preprocessed dataset.

4   Result Analysis

In this section, a set of evaluation parameters were calculated to analyze the performance of each machine learning algorithm. The evaluation parameters are precision, recall, F1-score, and accuracy. These parameters can be calculated on knowing the true positive (TP), true negative (TN), false positive (FP), and false negative (FN) that are often represented as a confusion matrix. Tables 2 and 3 show values of these evaluation parameters for each model.

Precision can be understood as the number of correctly predicted positive observations by the model to the total number of predicted positive observations by the model. It can be observed that all the models were nearly placed in terms of precision. However, LSTM slightly outperforms.

Precision formula for any given model.

Table 2 Precision, recall, and $F1$-score values

Algorithm	Precision (Micro)	Recall (Micro)	F1 (Micro)	Precision (Macro)	Recall (Macro)	F1 (Macro)	Precision (Weighted)	Recall (Weighted)	F1 (Weighted)
Decision Tree	0.88	0.88	0.88	0.70	0.70	0.70	0.87	0.87	0.87
Random Forest	0.89	0.89	0.89	0.74	0.69	0.71	0.88	0.88	0.88
AdaBoost	0.89	0.89	0.89	0.74	0.67	0.65	0.88	0.87	0.87
SVC	0.90	0.90	0.90	0.76	0.64	0.64	0.88	0.90	0.88
LSTM	0.92	0.92	0.92	0.80	0.80	0.78	0.91	0.92	0.92

Table 3 Accuracy values

Algorithm	Accuracy
Decision tree	0.8752
Random forest	0.8908
AdaBoost	0.8908
SVC	0.8950
LSTM	0.9201

$$\text{Precision} = \text{TP}/(\text{TP} + \text{FP}) \tag{1}$$

Recall can be defined as the number of correctly predicted positive observations to the total number of actually positive observations. It can be observed that all the models were nearly placed in terms of recall as well. Similar to precision, LSTM slightly outperforms the other algorithms.

Recall formula for any given model.

$$Recall = \text{TP}/(\text{TP} + \text{FN}) \tag{2}$$

$F1$-score is an evaluation parameter that is a function of both precision and recall, and it is used to strike a balance between them. It can be observed that all the models were nearly placed in terms of $F1$-score as well. Similar to precision and recall, LSTM was slightly high performant.

$F1$-score formula for any given model.

$$F1 = 2 * (\text{Precision} * \text{Recall})/(\text{Precision} + \text{Recall}) \tag{3}$$

Accuracy represents the number of correctly classified data instances over the total number of data instances. LSTM tops in terms of accuracy.

Accuracy formula for any given model.

$$\text{Accuracy} = (\text{TN} + \text{TP})/(\text{TN} + \text{FP} + \text{TP} + \text{FN}) \tag{4}$$

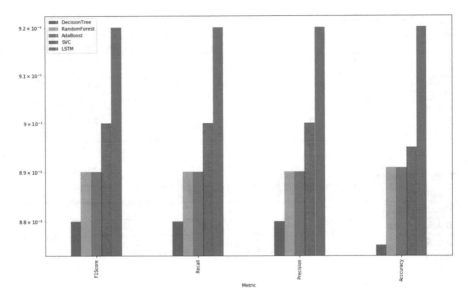

Fig. 2 Comparison of evaluation parameters

Figure 2 clearly shows how each algorithm is compared among each other.

5 Conclusion

In conclusion, the work delineates a comparative analysis of various machine learning algorithms to detect hate speech using a huge corpus of tweets as the dataset to train and test the models. LSTM method was highly performant among other machine learning algorithms that were considered. The work achieved a best accuracy score of 92%. In the future, it can be extended by utilizing a much larger dataset. The dataset can be from multimedia data that includes dynamic and static elements. It can involve extracting more features from the multimedia information associated with a post on a social media platform.

References

1. Persia, F., D'Auria, D.: A survey of online social networks: challenges and opportunities. In: 2017 IEEE International Conference on Information Reuse and Integration (IRI), pp. 614-620. https://doi.org/10.1109/IRI.2017.74

2. Janto-Petnehazi, I.: Grey areas: a cross-country comparative study of user-generated hate speech, incivility and journalistic responsibility for online comments. Ph.D. Thesis, University of Westminster, London (2017)
3. Samoshyn, A.: Hate speech and offensive language dataset (2020). https://www.kaggle.com/mrmorj/hate-speech-and-offensive-language-dataset. Accessed 7 Sept 2021
4. Jiang, L., Suzuki, Y.: Detecting hate speech from tweets for sentiment analysis. In: 6th International Conference on Systems and Informatics (ICSAI), pp 671-676 (2019). https://doi.org/10.1109/ICSAI48974.2019.9010578
5. Raufi, B., Xhaferri, I.: Application of machine learning techniques for hate speech detection in mobile applications. In: 2018 International Conference on Information Technologies (InfoTech), pp. 1-4 (2018) . https://doi.org/10.1109/InfoTech.2018.8510738
6. Ali, M.Z., Ehsan-Ul-Haq, S., Rauf, K., Javed, S. Hussain.: Improving hate speech detection of Urdu tweets using sentiment analysis. IEEE Access 9, 84296–84305 (2021). https://doi.org/10.1109/ACCESS.2021.3087827
7. Tripathi, M.: Sentiment analysis of Nepali COVID19 Tweets using NB, SVM AND LSTM. J. Artif. Intell. 3(03), 151–168 (2021)
8. Putra, I.G.M., Nurjanah, D.: Hate speech detection in Indonesian Language Instagram. In: 2020 International Conference on Advanced Computer Science and Information Systems (ICACSIS), pp. 413–420 (2020). https://doi.org/10.1109/ICACSIS51025.2020.9263084
9. Luu, S.T., Nguyen, H.P., Van Nguyen, K., Luu-Thuy Nguyen, N.: Comparison between traditional machine learning models and neural network models for Vietnamese hate speech detection. In: 2020 RIVF International Conference on Computing and Communication Technologies (RIVF), pp. 1–6 (2020). https://doi.org/10.1109/RIVF48685.2020.9140745
10. Watanabe, H., Bouazizi, M., Ohtsuki, T.: Hate speech on Twitter: a pragmatic approach to collect hateful and offensive expressions and perform hate speech detection. IEEE Access 6, 13825–13835 (2018). https://doi.org/10.1109/ACCESS.2018.2806394
11. Safavian, S.R., Landgrebe, D.: A survey of decision tree classifier methodology. IEEE Trans. Syst. Man Cybern. 21(3), 660–674 (1991). https://doi.org/10.1109/21.97458
12. Paul, A., Mukherjee, D.P., Das, P., Gangopadhyay, A., Chintha, A.R., Kundu, S.: Improved Random Forest for classification. IEEE Trans. Image Process. 4012-4024 (2018). https://doi.org/10.1109/TIP.2018.2834830
13. Zhang, Y., et al.: Research and application of AdaBoost algorithm based on SVM. In: 2019 IEEE 8th Joint International Information Technology and Artificial Intelligence Conference (ITAIC), pp. 662–666. https://doi.org/10.1109/ITAIC.2019.8785556
14. Mohan, L., Pant, J., Suyal, P., Kumar, A.: Support vector machine accuracy improvement with classification. In: 2020 12th International Conference on Computational Intelligence and Communication Networks (CICN), pp. 477-481 (2020). https://doi.org/10.1109/CICN49253.2020.9242572
15. Marhon, S., Cameron, C., Kremer, S.: Recurrent neural networks. In: Intelligent Systems Reference Library, pp. 29–65 (2013)
16. Skovajsová, L.: Long short-term memory description and its application in text processing. In: 2017 Communication and Information Technologies (KIT), pp. 1–4 (2017). https://doi.org/10.23919/KIT.2017.8109465

A Review on Preprocessing and Segmentation Techniques in Carotid Artery Ultrasound Images

K. V. Archana and R. Vanithamani

Abstract Epidemiological studies reveal that anatomical changes of carotid artery due to deposition of fatty lesions are effective signs of cardiovascular diseases. Ultrasound imaging modality provides the cross-sectional view of these arteries to identify the deposited plaques. Segmentation of common carotid artery (CCA) wall is important in determining the thickness of intima-media region of carotid artery. With the development of automated systems in medical field, a plethora of algorithms and strategies to analyze the carotid artery are proposed by various researchers over a period. This review focus on the techniques that have been developed for preprocessing and segmentation of various parts of the carotid artery from the longitudinal B-mode ultrasound images using both automated and semi-automated techniques. The review ends with a discussion about the techniques and future perspectives to make the computer-based carotid artery analysis more accurate and reliable.

Keywords Common Carotid Artery · Longitudinal ultrasound images · Despeckle · Segmentation

1 Introduction

Over the past decade, a rise of 26.6% of cardiovascular cases is noted. According to the experts of [1], the long-term effects of COVID-19 are likely to influence cardiovascular health, and the global burden of cardiovascular disease is expected to grow exponentially over the next few years. The pathological mechanism of cardiovascular

K. V. Archana (✉)
Department of Electronics and Communication Engineering, School of Engineering,
Avinashilingam Institute for Home Science and Higher Education for Women, Coimbatore, Tamil Nadu, India
e-mail: archana_ece@avinuty.ac.in

R. Vanithamani
Department of Biomedical Instrumentation Engineering, School of Engineering, Avinashilingam Institute for Home Science and Higher Education for Women, Coimbatore, Tamil Nadu, India
e-mail: vanithamani_bmie@avinuty.ac.in

© The Author(s), under exclusive license to Springer Nature Singapore Pte Ltd. 2022
V. Suma et al. (eds.), *Evolutionary Computing and Mobile Sustainable Networks*,
Lecture Notes on Data Engineering and Communications Technologies 116,
https://doi.org/10.1007/978-981-16-9605-3_62

diseases starts with deposition of lipoprotein cholesterol, thickening of vessel walls, increased degrees of vascularization, and formation of atherosclerotic plague on the vessel walls [2]. Figure 1a represents a longitudinal B-mode ultrasound image of the carotid artery, and Fig. 1b depicts a plague region in carotid artery. This plague can break off from the inner walls of the blood vessels and enter the circulation causing arterial obstruction. This stimulates thrombus or vasospasm in heart, brain, kidneys, and lower extremities [3–5].

Precise imaging of the common carotid artery is important for diagnosis as well as assessing the risk of various vascular diseases. There are various imaging modalities used for this purpose, namely ultrasound, invasive coronary angiography, and Magnetic Resonance Imaging (MRI). Angiography is a routine procedure to locate the plague or affected area, but it does not provide structural information [6]. Ultrasonography is the preferred method of vascular imaging due to its reliability, low cost, non-invasiveness, and better visualization of anatomical structures. Important quantitative information such as lumen area, thickness of the carotid walls, distribution and composition of plaques, and intima-media thickness are available in longitudinal ultrasound images [7]. Therefore, analysis of the arterial wall and plague deposits in coronary artery has significant clinical relevance for assessment and management of cardiovascular diseases. Conventionally, these details are measured from ultrasound images by trained personals in clinical setting which is highly user-dependent, time-consuming, and prone to errors. Due to the advent of automation, several computerized techniques have been developed to reduce the subjectivity and time of analysis while increasing the accuracy and efficiency. A large number of follow-up studies have been done on B-mode ultrasound images to determine the possibilities of carotid atherosclerosis or stenosis [3, 8, 9]. A cohort study in [10] reveals that diabetic patient along with renal dysfunction are more susceptible to carotid stenosis leading to mortality.

Since the ultrasound image acquisition procedure is completely manual, the parameter settings of the machine depend on the subjective judgment of the operator. Thus, the acquisition of good quality ultrasound images is complicated and requires

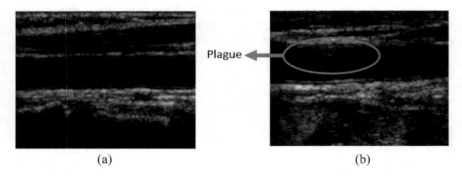

(a)   (b)

Fig. 1 Longitudinal B-mode ultrasound image of carotid artery: **a** healthy, **b** plague deposits on the arterial wall

high skilled personal. Regardless of the acquisition protocols, the echogenicity depends on tissue composition, insonation angle, tissue attenuation, and blood flow. This leads to local changes in intensity, contrast, and adds speckle pattern in ultrasound images. Hence, particular attention should be given to suppress the speckle pattern before segmentation. The intention of this article is to give an extensive appraisal on the analysis of carotid artery ultrasound images for disease diagnosis starting from image denoising for speckle reduction and segmentation approaches.

2 Preprocessing

Due to ultrasonic echoes from tissues, ultrasound images are often accompanied by multiplying speckle noise. This falsifies the fine details of the image, making it difficult for the computer system to analyze the information. Therefore, despeckling of ultrasound images is critical for better diagnosis of pathologies. Several speckle reduction techniques have been proposed, especially for ultrasound B-mode images. Table.1 shows some of the speckle reduction techniques considered in this study. Commonly available despeckling filters are mean, Lee, Kuan, Wiener, and Gaussian filters. Speckle reducing anisotropic diffusion (SRAD) filter developed by [11] is a benchmark filter for despeckling [12, 13]. Filtering in wavelet domain has also gained popularity among researchers [12, 14]. Linear scaling filter and local statistical filter that utilizes mean and variance are also applied to despeckle the image [15–17]. Whereas, authors of [13] developed an integrated toolbox in MATLAB that includes ten despeckling filters specially utilized for carotid artery ultrasound images. On contrary, multiplicative noise is converted into additive noise in [18] and applied wavelet decomposition. Then, three different filters, namely non-local means (NLM), vectorial total variation (VTV), and block matching and 3D filtering (BM3D) algorithms are applied. Whereas, utilized non-local variational model is used in [19] for despeckling of ultrasound images. Table 1 shows the comparative analysis of the despeckling algorithm that are implemented and the performance metrics used for analysis.

3 Segmentation

This section highlights the state-of-the art techniques that facilitate segmentation of carotid artery from longitudinal ultrasound images and are listed in Table 2. Researchers developed different methodologies such as edge detection methods, thresholding methods, contours or snakes, and classifier-based approaches.

Thresholding Simple thresholding methods like Otsu thresholding using wind-driven optimization, global thresholding followed by morphological operations are

Table 1 Despeckling techniques implemented for longitudinal B-mode carotid artery ultrasound images

Name, Year, and Ref. No.	Despeckling techniques	Features extracted for analysis	Advantages and limitations
Loizou et al. 2007 [20, 21]	Linear scaling mean variance	Mean and variance	• Image normalization prior to speckle reduction • Improved image quality
Izquierdo-Zaragoza et al. 2011 [22]	Low pass Gaussian filter, morphological filtering	–	• Clean the image highlighting only the relevant information
Petroudi et al. 2012 [15]	Speckle removal linear scaling filter	Local mean and variance	• Edge preservation is better • No comparison with other despeckling filters
Ramasamy and Jayanthi 2013 [23]	Gaussian filter	–	• The smoothing effect reduces the impact of noise in the image • Alters the position of the object boundaries for increased radius
Chaudhry et al. 2013 [24]	Median filter	–	• Smooth out noise and preserves image details in better way compared to average and bilateral filters
Loizou et al. 2014 [13]	(a) Linear filter, (b) Wiener linear filter, (c) linear filter, (d) Nonlinear filter, (e) Geometric filter, (f) Median filter, (g) Hybrid median filter, (h) Anisotropic diffusion filter, (i) Coherent nonlinear anisotropic diffusion filter, (j) SRAD	Statistical features, spatial gray-level dependence matrices, gray-level difference statistics, neighborhood gray tone difference matrix, neighborhood gray tone difference matrix, laws texture energy measures, fractal dimension texture analysis, Fourier power spectrum, shape parameters	• Developed a freeware despeckle filtering toolbox which also supports image normalization, delineation of structures, texture extraction • Filters can be applied over entire image as well as in ROI. Validated the performance for IVUS images • Filter parameters are made default. Cannot be configured by user which limits their applications over other medical images • Processing time of the proposed method is high

(continued)

Table 1 (continued)

Name, Year, and Ref. No.	Despeckling techniques	Features extracted for analysis	Advantages and limitations
Nagaraj et al. 2018 [14]	Wiener filtering in wavelet domain	–	• Smoothens the edges still preserving the edge information • Faster ROI extraction • Reduced execution time
Nagaraj et al. 2018 [25]	Optimized Bayesian non-local means filter	–	• Filter suppresses the multiplicative noise by utilizing optimally tuned parameters of grayscale images
Gupta et al. 2018 [17]	Local statistics mean variance filter	Mean, standard deviation, mean square error, quality index, peak signal-to-noise ratio, structural similarity index, correlation coefficient, edge preserving index	• Noise reduction without much loss of information • High similarity index
Latha and Dhanalakshmi 2019 [18]	Calculus approach: • Convert the multiplicative speckle noise into additive • Wavelet decomposition • Three different filters: (a) NLM, (b) VTV, and (c) block matching and BM3D	Peak signal-to-noise ratio, mean square error, mean absolute error, root mean square error, structural similarity, quality factor, correlation, and image enhancement factor	• Preserving edge features to a great extent and better denoising • Noise in high frequency content is removed by identifying the dependent components of the image
Febin and Jidesh 2021 [19]	Non-local variational model	Entropy, equivalent number of looks, natural image quality evaluator, global contrast factor, contrast-to-noise ratio	• Despeckling with contrast enhancement and illumination correction • Retaining fine texture details • Less execution time
Latha et al. 2021 [26]	Curvelet decomposition with smooth edge and sparsity	–	• Best denoised results and also reserves maximum of the necessary edge features • Smoothens low frequency noise and the oscillatory high frequency noise are filtered

Table 2 Segmentation techniques implemented for longitudinal B-mode carotid artery ultrasound images

Reference No.	No. of images	Technique	Performance metrics	Merits and demerits
[12]	–	Modified Otsu threshold	–	• Low computational complexity • Better edge detection • Algorithm is unstable for the images with changes in morphological structure of carotid artery
[25]	90	Otsu thresholding using wind-driven optimization	CC, CoV	• Better boundary estimation compared to model based and snake segmentation • No significant deviation between the automated and the ground truth data measurement
[23]	50	Global thresholding + Hole filling	–	• Excellent in quantifying the elasticity and stiffness of carotid artery • No classification made between health and unhealthy subjects
[16, 27]	20	Edge detection using Prewitt operator Edge detection—global threshold + Sobel operator, Hough transform	Error –	• Better speckle noise reduction • Effective detection of inner edges of artery wall • IMT measurement with less error • Implemented for minimum dataset
[28]	287	Watershed algorithm	RoC	• Automatic system for identification and discrimination of plaques and non-plaques • Reduced false detection • Minimum number of sample dataset is taken for analysis

(continued)

Table 2 (continued)

Reference No.	No. of images	Technique	Performance metrics	Merits and demerits
[15]	100	Active contours	MAD, HD, PD	• Fully automatic system providing lower absolute mean IMT Error • Aids in evaluation of early stages of atherosclerosis • Does not work appreciably for images with artery wall irregularities
[24]	250	Active contour	–	• Automatic snake initialization using image registration technique • Precise selection of control points for image registration • Plaque characterization is poor
[30]	84	Active oblongs	DI, JI, Se, Sp, Acc, L	• Segmentation accuracy is 97.5% • Segmentation and initialization of luminal wall are done subsequently • Not all performance metrics are compared with the existing techniques
[22]	–	B-spline active contour	Mean, max, min	• Automatic segmentation of arterial wall with contrast enhancement • B-spline provides smooth edges by removing the rough texture in ultrasound images • No automatic initialization of snakes • Validation with more number image set is required to be done

(continued)

Table 2 (continued)

Reference No.	No. of images	Technique	Performance metrics	Merits and demerits
[26]	361	Affinity propagation + DBSCAN, GVF snake model, PSO clustering	JI, DI, HI, RI Vol, Ck, C	• Computation is less time-consuming • Full automatic with no need to initiation of ROI • Suitable for differently oriented carotid images • Proposed algorithm occupies more memory and is computationally marginally costly
[34]	153	(i) $H\infty$ grayscale-derivative constraint snake algorithm; (ii) Kalman snake method, (iii) snake method; (iv) dynamic programming; (v) level set method using Chan-Vese energy functional	MAE	• Minimizes worst case error • Grayscale constrain will automatically correct the position of snake along the ROI • Algorithms segments are better even for images with blurred flocculent structure • The parameters for filter and grayscale contours are selected empirically • Method maintains a fixed shape for IM border which is not optimal when the IM border changes during the sequencing
[37]	30	Mixtures of Nakagami distributions and stochastic optimization	GoF	• Semi-automatic method of segmentation with minimal mean difference from the experts' views • Echogenicity of the tissue is verified well • High computational time

(continued)

Table 2 (continued)

Reference No.	No. of images	Technique	Performance metrics	Merits and demerits
[39]	200	Dynamic programming	Acc, Ct	• Algorithm can selectively detect the object boundary with desired patterns • Robustness and high accuracy in edge detection with minimal computational time • Limitation in plaque identification
[40]	47	Dynamic programming, smooth intensity thresholding surfaces and geometric snakes	–	• Accurate segmentation of arterial wall even for arteries with large, hypo-echogenic or irregular plaques • Detection of luminal wall is difficult with poor quality images
[41]	29	Classifiers (linear SVM, SVM + radial bias function, AdaBoost, random forest)	Se, Sp, DSC overlap, HD EArea, PPD, AEArea	• Auto-context model is improving the accuracy of segmentation making the method more robust and stable • Limited number of images • Not efficient in plaque characterization • High training time
[43]	90,000 image patches	• CNN (patch-based approach)	median, mean, SD, min, max, Sp, Se	Good correlation between estimated output and experts view Though training time is high, testing is achieved in near real-time limited dataset

utilized in [23, 25]. In [12], modified Otsu thresholding followed by morphological operations is used to segment upper and lower walls.

Edge detection methods Usually, the blood tissue interface at the walls of blood vessels gives rise to typical edge patterns. These clearly manifest as borders due to discontinuities in the intensity values and reduced edge reflections. Hence, edge detection methods like Prewitt and Sobel operators followed by Hough transform for segmentation are used [16, 27]. Whereas, gradient filter is applied to detect the edges in image followed by watershed algorithm to segment the lesion in ultrasound images [28]. A drawback of this method is over segmentation leading to false detections which is avoided by cluster analysis.

Snake or contours A variety of snake models are used by researchers in which some are fully automated and some need human intervention for contour initiation. Snake or contour models are edge-based models, wherein the initial contour attracts toward local maxima in the edge map, where the shape of the snake is governed by energy functions. In [15], the boundaries of blood vessels are identified by Chan-Vese level set algorithm which provides excellent initialization for segmentation. The intima-media complex is segmented using active contours as in [21]. Whereas, authors of [29] defined two masks as initial contours for Chan-Vese segmentation of carotid artery walls as well as its bifurcation. The obtained contours are smoothed by cubic spline interpolation and projecting contour points toward local regression line. Active oblongs are utilized in [30] to segment the carotid artery. Hough transform is used to automatically initialize the active oblong in the arterial region followed by pixel offset operations and growing the oblong. The growth of the oblong is optimized using gradient descent technique and Green's theorem. A similar approach using active oblong with five degrees of freedom is proposed by [31] and added a post-processing step (median filter, the canny edge operator, and the cubic curve fitting) to provide a smooth curved area. In [22], a semi-automatic approach using two frequency implemented B-spline snakes is implemented. In addition, a small gravity force is added to the upper contour, and a take-off force to the lower contour to make sure the contours does not remain motionless. Recently, [26] compared three methods to segment lumen area, namely combining affinity propagation and (DBSCAN) density-based spatial clustering of applications with noise, gradient vector flow (GVF) snake model, and particle swarm optimization clustering-based segmentation. The combination of affinity propagation and DBSCAN outperformed the other models with low computational power. Moreover, the outliers were removed by Z-score method. A limitation of contour-based segmentation is the initialization of contour or snakes. To overcome this problem [24, 32], proposed an intelligent algorithm that locates arterial wall areas to set the initial contour, thus making the system fully automated. Jin et al. [33] proposed fully automated algorithm for region identification, contour initialization, and segmentation of intima-lumen and media-adventia layers using general snake and GVF snake. Authors of [34] proposed H∞ grayscale-derivative constraint snake algorithm to segment intima-media borders and compared with the following models: Kalman snake method, snake method, dynamic programming, and level set method using Chan-Vese energy functional. The layers segmented by H∞

algorithm are precisely defined and robust to system error. The performance of four different snake models (Williams and Shah, Lai and Chin, Balloon, and GVF snake) with manual segmentation is compared [20]. Lai and Chin snake model depicted slightly better region of curve than the other models.

Nakagami models Nakagami distributions are considerably used in image processing applications as it provides information about the spatial arrangement and statistical distribution of ultrasound imaging data [35, 36]. An iterative method using mixtures of Nakagami distributions and stochastic optimization is applied in [37]. Moreover in [38], they applied Bayesian segmentation approach modeled by mixtures of three Nakagami distributions. Authors have stated that this kind of segmentation approach is semi-computerized and no longer sensitive to the degree of stenosis or calcification. In addition, these images are not preprocessed due to the fact that the application of filter affects the statistics of data.

Dynamic models Some authors proposed dynamic approach in segmentation. In [39], a combination of dual line detection with edge maps of two edge detectors and coupled snake model is used to maintain parallelism in the intima-media segmentation. Whereas in [40], best fit for cubic spline is searched to identify the adventitia layer. Dynamic programming is employed to segment the lumen boundary. Since dynamic programming causes irregularities in boundaries, a combination of smooth intensity thresholding and hybrid Chan-Vese model is applied as post-processing step.

Other approaches Simple iterative clustering algorithm was used in [41] to produce super pixels of ultrasound image and RealAdaBoost is utilized to produce classification map. A learning-based algorithm extracted the plague region from the classification map. They segmented the plague region using classifiers such as linear support vector machine (SVM), AdaBoost, random forest, and SVM + radial bias function by considering each pixel in the image as features belonging to either normal or plague area. Additionally, auto-context model is implemented with ten iterations. It was found that random forest is superior, and the context features help to stabilize the model. Whereas, [42] proposed multiclass framework using k-means classifier and proved its superior performance with manual tracings. Conversely, convolutional neural network is used to characterize the plague composition using patch-based approach [43]. In paper [44], an integrated graph model and Markov random fields is used to segment the plague region in coronary artery. Whereas, researchers in [45] suggested segmentation of coronary artery using clustering algorithms. They have utilized fuzzy c-means clustering, spatial fuzzy c-means, modified spatial fuzzy c-means, k-means clustering, and self-organizing maps. K-means and self-organizing maps yielded similar results. Furthermore, equal weightage problem has been overcome by modified spatial fuzzy c-means segmentation. Authors of [42] combined scale space paradigm with a boundary-based approach using level sets, while some used level set method without re-initialization with uniform length [46].

The performance metrics used for comparison of segmentation techniques are as follows: mean, max, min, error, relative error, standard deviation (SD), mean absolute distance (MAD), mean absolute error (MAE), Hausdorff distance (HD), point-to-point distance (PPD), polyline distance (PD), Jaccard index (JI), dice index (DI), Rand index (RI), variation of information (VoI), Cohens kappa (Ck), cophenet (C), sensitivity (Se), specificity (Sp), DSC overlap, EArea, AEArea, point-to-point distance (PPD), Daviea Bouldin index (DBI), partition coefficient (PC), classification entropy (CE), accuracy (Acc), localization (L), true-positive fraction (TPF), and false-positive fraction (FPF), true-negative fraction (TNF), false-negative fraction (FNF), similarity kappa index (KI), and the overlap index (OI), confidence interval (CI), correlation coefficient (CC), coefficient of variation (CoV), percent statistics (PS), goodness-of-fit (GoF), mean contour distance (MCD), computation time (Ct), region of curve (RoC), precision of merit (PoM).

4 Discussion and Conclusion

Several advancements in the field of ultrasonic imaging have been proposed; however, a number of factors hinder automated analysis and disease diagnosis. On analyzing the despeckling approaches, it is evident that some approaches suffer from degraded spatial resolution and increased system complexity. Some techniques face limitations as follows: the window size of filters affects the resultant image quality; hence, a fair choice of window size is necessary; inability to preserve the edges may lead to loss of information; post-processing becomes necessary in some cases.

Whereas, segmentation techniques also had some complications. Boundary-based methods are prone to errors as they are sensitive to gradient variations at the edges [42]. Some authors assumed the arterial structures as straight lines, which is not true in all cases. This has led to false estimation in some scenarios. It is observed that most authors are interested in utilizing contours or snakes for segmentation, but still, they possess some drawbacks such as vulnerable to discontinuities and false edges. Some integrated and dynamic approaches were proposed to overcome these drawbacks. After segmentation, most articles had calculated the characteristics of plague region or thickness of boundary walls which helps in further analysis.

A plethora of techniques has been discussed in this paper. Although better results have been reported, a direct comparison cannot be made due to the following reasons (i) image set used is not same; (ii) some required manual intervention which may introduce human specific errors in final outcome, (iii) different areas of carotid artery (far-end wall, near-end wall, plague, intima-media complex, lumen boundary) are segmented; (iv) some approaches are user-independent, and others were semi-automated. In this context, it is obvious that there is still room for improvement in segmenting carotid artery for real-time diagnosis. Some studies have disclosed that these pathologies may manifest either chronically or acutely in all arterial territories. Hence, a growing interest in vascular imaging and computational analysis helps to expand the diagnostic abilities. Articles like [47, 48] reveal smaller artery size is reported in India, as compared to other western regions of the world due to genetics and lifestyle. Some interpret smaller diameter of arteries in women than men. These conditions are not taken into account in any of the methods. Hence, future works in automated analysis of carotid artery should consider these anatomical differences for better results.

References

1. Information on https://www.news-medical.net/news/20210127/COVID-19-will-impact-cardio vascular-disease-deaths-for-years-to-come-warn-experts.aspx
2. Insull, W., Jr.: The pathology of atherosclerosis: plaque development and plaque responses to medical treatment. Am. J. Med. **122**, S3–S14 (2009)
3. Steinl, D.C., Kaufmann, B.A.: Ultrasound imaging for risk assessment in atherosclerosis. Int. J. Mol. Sci. **16**(5), 9749–9769 (2015)
4. Cattaneo, M., Wyttenbach, R., Corti, R., Staub, D., Gallino, A.: The growing field of imaging of atherosclerosis in peripheral arteries. Angiology **70**(1), 20–34 (2019)
5. Oikonomou, E., Latsios, G., Vogiatzi, G., Tousoulis, D.: Atherosclerotic Plaque. Coronary Artery Disease, Elsevier, pp. 31–41 (2018)
6. Qiu, W., et al.: An open system for intravascular ultrasound imaging. IEEE Trans. Ultrason. Ferroelectr. Freq. Control **59**(10), 2201–2209 (2012)
7. Sidhu, P.S.: Ultrasound of the carotid and vertebral arteries. Br. Med. Bull. **56**(2), 346–366 (2020)
8. Kastelein, J.J.P., de Groot, E.: Ultrasound imaging techniques for the evaluation of cardiovascular therapies. Eur. Heart J. **29**(7), 849–858 (2008)
9. Liguori, C., Paolillo, A., Pietrosanto, A.: An automatic measurement system for the evaluation of carotid intima-media thickness. IEEE Trans. Instrum. Measure. **50**(6), 1684–1691 (2001)
10. Koroleva, E.A., Khapaev, R.S.: The prevalence and risk factors of carotid artery stenosis in type 2 diabetic patients. In: 2020 Cognitive Sciences, Genomics and Bioinformatics (CSGB) (2020)
11. Yu, Y., Acton, S.T.: Speckle reducing anisotropic diffusion. IEEE Trans. Image Process. **11**, 1260–1270 (2002)
12. Nithya, A., Kayalvizhi, R.: Measurement of lower and upper IMT from ultrasound video frames. Biomed. Pharmacol. J. **8**, 355–364 (2015)
13. Loizou, C.P., Theofanous, C., Pantziaris, M., Kasparis, T.: Despeckle filtering software toolbox for ultrasound imaging of the common carotid artery. Comput. Methods Programs Biomed. **114**(1), 109–124 (2014)
14. Nagaraj, Y., Asha, C.S., Teja, H.S., Narasimhadhan, A.V.: Carotid wall segmentation in longitudinal ultrasound images using structured random forest. Comput. Electr. Eng. **69**, 753–767 (2018)
15. Petroudi, S., Loizou, C., Pantziaris, M., Pattichis, C.: Segmentation of the common carotid intima-media complex in ultrasound images using active contours. IEEE Trans. Biomed. Eng. **59**(11), 3060–3069 (2012)
16. Gupta, R., Pachauri, R., Singh, A.K.: Despeckle and segmentation of carotid artery for measurement of intima-media thickness. In: 2019 International Conference on Signal Processing and Communication (ICSC) (2019)
17. Gupta, R., Pachauri, R., Singh, A.: Linear despeckle approach for ultrasound carotid artery images. J. Intell. Fuzzy Syst. **35**(2), 1807–1816 (2018)
18. Latha, S., Dhanalakshmi, S.: Despeckling of carotid artery ultrasound images with a calculus approach. Curr. Med. Imag. **15**(4), 414–426 (2019)
19. Febin, I.P., Jidesh, P.: Despeckling and enhancement of ultrasound images using non-local variational framework. Vis. Comput. 1–14 (2021)

20. Loizou, C.P., Pattichis, C.S., Pantziaris, M., Nicolaides, A.: An integrated system for the segmentation of atherosclerotic carotid plaque. IEEE Trans. Inf. Technol. Biomed. **11**(6), 661–667 (2007)
21. Loizou, C.P., Pattichis, C.S., Pantziaris, M., Tyllis, T., Nicolaides, A.: Snakes based segmentation of the common carotid artery intima media. Med. Biol. Eng. Comput. **45**, 35–49 (2007)
22. Izquierdo-Zaragoza, J.L., Bastida-Jumilla, M.C., Verdu-Monedero, R., Morales-Sanchez, J., Berenguer-Vidal, R.: Segmentation of the carotid artery in ultrasound images using frequency-designed B-spline active contour. In: 2011 IEEE International Conference on Acoustics, Speech and Signal Processing (ICASSP) (2011)
23. Ramasamy, N., Jayanthi, K.B.: Automated lumen segmentation and estimation of numerical attributes of common carotid artery using longitudinal B-mode ultrasound images. In: 2013 IEEE Point-of-Care Healthcare Technologies (PHT) (2013)
24. Chaudhry, A., Hassan, M., Khan, A., Kim, J.Y.: Automatic active contour-based segmentation and classification of carotid artery ultrasound images. J. Digit. Imaging **26**(6), 1071–1081 (2013)
25. Nagaraj, Y., Madipalli, P., Rajan, J., Kumar, P.K., Narasimhadhan, A.V.: Segmentation of intima media complex from carotid ultrasound images using wind driven optimization technique. Biomed. Signal Process. Control **40**, 462–472 (2018)
26. Latha, S., Samiappan, D., Muthu, P., Kumar, R.: Fully automated integrated segmentation of carotid artery ultrasound images using DBSCAN and affinity propagation. J. Med. Biol. Eng. (2021)
27. Golemati, S., Stoitsis, J., Sifakis, E.G., Balkizas, T., Nikita, K.S.: Using the Hough transform to segment ultrasound images of longitudinal and transverse sections of the carotid artery. Ultrasound Med. Biol. **33**(12), 1918–1932 (2007)
28. Sottile, F., Marino, S., Bramanti, P., Bonanno, L.: Validating a computer-aided diagnosis system for identifying carotid atherosclerosis. In: 2013 6th International Congress on Image and Signal Processing (CISP) (2013)
29. Santos, A.M.F., Tavares, J.M.R.S., Sousa, L., Santos, R., Castro, P., Azevedo, E.: Automatic segmentation of the lumen of the carotid artery in ultrasound B-mode images. In: Medical Imaging 2013: Computer-Aided Diagnosis (2013)
30. Harish Kumar, J.R., Teotia, K., Raj, P.K., Andrade, J., Rajagopal, K.V., Sekhar Seelamantula, C.: Automatic segmentation of common carotid artery in longitudinal mode ultrasound images using active oblongs. In: ICASSP 2019—2019 IEEE International Conference on Acoustics, Speech and Signal Processing (ICASSP) (2019)
31. Dhupia, A., Harish Kumar, J.R., Andrade, J., Rajagopal, K.V.: Automatic segmentation of lumen intima layer in longitudinal mode ultrasound images. In: Annual International Conference of the IEEE Engineering in Medicine and Biology Society, pp. 2125–2128 (2020)
32. Chaudhry, A., Hassan, M., Khan, A., Kim, J.Y., Tuan, T.A.: Automatic segmentation and decision making of carotid artery ultrasound images. In: Advances in Intelligent Systems and Computing, pp. 185–196. Springer Berlin Heidelberg, Berlin (2013)
33. Jin, J., Ding, M., Yang, X.: Automatic detection of the intima-media layer in ultrasound common carotid artery image based on active contour model. In: 2011 International Conference on Intelligent Computation and Bio-Medical Instrumentation (2011)
34. Zhao, S., et al.: Robust segmentation of intima–media borders with different morphologies and dynamics during the cardiac cycle. IEEE J. Biomed. Health Inform. **22**(5), 1571–1582 (2018)
35. Chang, M., Varghese, B., Gunter, J., Lee, K.J., Hwang, D.H., Duddalwar, V.: Feasibility of Nakagami parametric image for texture analysis. In: 15th International Symposium on Medical Information Processing and Analysis (2020)
36. Tsui, P.H., Huang, C.C., Wang, S.H.: Use of Nakagami distribution and logarithmic compression in ultrasonic tissue characterization. J. Med. Biol. Eng. **26**, 69–73 (2006)
37. Destrempes, F., Meunier, J., Giroux, M.-F., Soulez, G., Cloutier, G.: Segmentation in ultrasonic B-mode images of healthy carotid arteries using mixtures of Nakagami distributions and stochastic optimization. IEEE Trans. Med. Imaging **28**(2), 215–229 (2009)

38. Destrempes, F., Meunier, J., Giroux, M.-F., Soulez, G., Cloutier, G.: Segmentation of plaques in sequences of ultrasonic B-mode images of carotid arteries based on motion estimation and a Bayesian model. IEEE Trans. Biomed. Eng. **58**(8), 2202–2211 (2011)
39. Zhou, Y., Cheng, X., Xu, X., Song, E.: Dynamic programming in parallel boundary detection with application to ultrasound intima-media segmentation. Med. Image Anal. **17**(8), 892–906 (2013)
40. Rocha, R., Campilho, A., Silva, J., Azevedo, E., Santos, R.: Segmentation of the carotid intima-media region in B-mode ultrasound images. Image Vis. Comput. **28**(4), 614–625 (2010)
41. Qian, C., Yang, X.: An integrated method for atherosclerotic carotid plaque segmentation in ultrasound image. Comput. Methods Programs Biomed. **153**, 19–32 (2018)
42. Araki, T., et al.: Two automated techniques for carotid lumen diameter measurement: Regional versus boundary approaches. J. Med. Syst. **40**(7), 182 (2016)
43. Lekadir, K., et al.: A convolutional neural network for automatic characterization of plaque composition in carotid ultrasound. IEEE J. Biomed. Health Inform. **21**, 48–55 (2017)
44. Gastounioti, A., Sotiras, A., Nikita, K.S., Paragios, N.: Graph-based motion-driven segmentation of the carotid atherosclerotique plaque in 2D ultrasound sequences. In: Lecture Notes in Computer Science, pp. 551–559. Springer International Publishing, Cham (2015)
45. Hassan, M., Chaudhry, A., Khan, A., Kim, J.Y.: Carotid artery image segmentation using modified spatial fuzzy C-means and ensemble clustering. Comput. Methods Programs Biomed. **108**(3), 1261–1276 (2012)
46. Sumathi, K., Mahesh, V., Ramakrishnan, S.: Analysis of intima media thickness in ultrasound carotid artery images using level set segmentation without re-initialization. In: 2014 International Conference on Informatics, Electronics and Vision (ICIEV) (2014)
47. Lip, G.Y., Rathore, V.S., Katira, R., Watson, R.D., Singh, S.P.: Do Indo-Asians have smaller coronary arteries? Postgrad. Med. J. **75**, 463–466 (1999)
48. Reddy, S., et al.: Coronary artery size in North Indian population—Intravascular ultrasound-based study. Indian Heart J. **71**(5), 412–417 (2019)

Performance of UDP in Comparison with TCP in Vehicular Communication Networks

B. Seetha Ramanjaneyulu, K. Annapurna, and Y. Ravi Sekhar

Abstract If the physical medium is error-prone, the congestion control mechanisms implemented by using transmission control protocol (TCP) might generate some unforeseen challenges. Packets lost due to the bit errors of wireless channel may provide the impression of a crowded network, forcing the sender to reduce transmission rates. As a result, throughput will be reduced. Several mechanisms are proposed in literature in the last two decades to address this problem of TCP. In the context of vehicular networks, maintaining the transport connection between the end users is considered as an additional challenge. In such scenarios, if the application can be designed for user datagram protocol (UDP) by incorporating the essential retransmission policies at application layer level, where a better throughput can be achieved. This paper includes a study of it for the use of railway signaling systems.

Keywords Bit error rate · Bandwidth-delay product · Window size · Packet retransmission · Timeouts · Roundtrip delay · Transmission rate · Throughput

1 Introduction

In data communication networks, TCP and UDP are the two popular protocols of transport layer that are meant for reliable and quick data transfers, respectively. While TCP protocol has the feature of retransmitting the missed packets with the help of acknowledgments, UDP protocol does not have this capability. 'Sequence Number' and 'Acknowledgment Number' fields of TCP header assist in implementing this packet retransmission feature. Hence, if an application requires the transmission of data without missing any packets, the TCP protocol is needed in such cases. TCP protocol ensures to have proper reception of all the packets of the file, even if the physical channel through which the packets travel is erroneous and non-perfect. During the early years of data communications, using the cables of those days, bit error rates (BER) used to be very high in the order of 10^{-4} or more so that one in

B. Seetha Ramanjaneyulu (✉) · K. Annapurna · Y. Ravi Sekhar
Vignan's Foundation for Science Technology and Research (VFSTR), Vadlamudi, Guntur, India

© The Author(s), under exclusive license to Springer Nature Singapore Pte Ltd. 2022
V. Suma et al. (eds.), *Evolutionary Computing and Mobile Sustainable Networks*,
Lecture Notes on Data Engineering and Communications Technologies 116,
https://doi.org/10.1007/978-981-16-9605-3_63

10,000 bits or more might be received with errors. The error detection and correction mechanisms came to the rescue there to detect and correct the bit errors at the receiver if possible, or just to notify the transmitter with negative acknowledgment to indicate that bit errors have occurred. Sometimes, if the packet does not reach the destination at all, and gets dropped at one of the intermediate routers in the network, the receiving computer may not be able to send the packet error notification to the transmitter. In such cases, the transmitter will wait for some duration of time (called 'timeout interval') for getting the acknowledgment from the receiver; after the lapse of which, it retransmits the same packet one more time. This kind of packet retransmission can also happen even in the case of successful reception of packet also at receiver, if the positive acknowledgment that is sent back to the transmitter gets missed in the network transit. So, based on these acknowledgment notifications and timeout intervals, packet retransmissions take place at the transmitting system to ensure that all the packets of the file are delivered correctly to the receiver [1, 2]. During the later three decades, fiber optic cables and copper cables of high quality are introduced for data communication networks with which the bit error rates have come down drastically to the order of 10^{-9} or lesser.

The other popular transport protocol UDP, which is of unreliable type and it does not support any acknowledgments and retransmission. Yet, UDP protocol is also useful for important applications like real-time voice or multimedia transmission in the networks. The main purpose of using UDP is that there is no use of getting the retransmission of a missed voice syllable or a multimedia frame of real time at a later instance of time because it cannot be replayed after a few milliseconds when some other voice syllable or multimedia frame that belongs to later time instance is being played to the user. As a result, rather than retransmitting packets, correcting bit errors will be beneficial in this case. Even if all bit errors cannot be corrected, the speech and multimedia frames can be sent to the application layer for presenting it to the user although with some acceptable errors and quality deterioration. As a result, as long as bit error rates are maintained within acceptable limits, the UDP protocol may be used in real-time transmission applications. However, if bit mistakes are too high, the quality of received music and video degrades [3, 4].

In the context of wireless networks, bit error rates are much higher when compared to those of present day wired networks. They can be as high as 10^{-4} which is comparable to the BER values of yesteryears' cable-based transmission. Presence of interference from other users, attenuation, and multipath fading are the main reasons of these higher values. During earlier years, on-the-ground microwave and satellite-based communications were the main types of wireless communication that were in use, in addition to the wireless broadcasting systems. Over the years, several other wireless data communication systems like WLAN and cellular networks of 1–5G have been introduced. In this context, discussion of bit error rates and the use of TCP and UDP protocols on these wireless links have gained significant importance [1, 2, 5]. In this paper, the effect of bit errors on the throughput of wireless networks is studied. The network considered here is meant for the application of railway signaling system that makes use of wireless communications for its signaling and control purposes. The network considered here is a typical wide area network (WAN) with possible wired and wireless links between the routers, and the presence of wireless interfaces at the host systems on either side, to connect them with the WAN. GSM-R system that is based on the second generation GSM cellular system, LTE-R that is based on the fourth generation LTE system and the WLAN or WPAN-based systems that use the ISM bands are the typical wireless interfaces considered in this application [5–9].

In general, throughput will be reduced when bit errors take place. This can be fairly observed while using UDP as the transport protocol because UDP offers a plain calculation of it without the baggage of acknowledgments and retransmissions. However, when TCP is used as the transport protocol, it results in further reduction of throughput. It is due to the congestion control mechanisms used with TCP protocol. The main phenomenon implemented in congestion control mechanisms is that the sender will reduce its transmission rate drastically, whenever it notices the acknowledgment timeouts of its previously transmitted packets. The rationale for implementing the reduced transmission rate phenomenon is that the sender assumes that timeouts have occurred due to network congestion somewhere in the WAN, and hence, the sending system should reduce the transmission rate. This rationale is perfect as long as the timeouts are due to the network congestion. However, if the timeouts are due to the bit errors in the wireless channel, reducing the transmission rate will not help to improve the situation. Instead, it will reduce the throughput. In such cases, it would be desirable to continue with the same rate of transmission [10, 11]. Simulation study of this phenomenon is carried out in this article, in the context of railway signaling application and comparing the performance of TCP with UDP in such scenarios. Section 2 of the article discusses the error control and flow control mechanisms of TCP along with the retransmission policies of it. Bandwidth-delay product and the related analysis about congestion window size and throughput calculations are also included in it. Various congestion control schemes that are there for wired networks are discussed in Sect. 3. Section 4 is about the options of wireless communication in railway signaling systems. Various network scenarios that are considered in this work and their simulation results are discussed in Sect. 5. Section 6 discusses the modifications proposed in literature to TCP to deal with the problems of wireless. The option of using UDP is also discussed there. Section 7 concludes the paper.

2 Error Control, Flow Control, and Bandwidth-Delay Product

In TCP, every transmitted packet is acknowledged. It may be acknowledged individually or confirmed of its receipt through 'cumulative acknowledgment' mechanism where an acknowledgment of a later transmitted packet indicates the receipt of all the packets transmitted up to that packet. While transmitting a packet, a timer is set with a retransmission timeout (RTO) duration. If the acknowledgment of that packet does not come back within this timeout period, retransmission of the same packet will take place. After receiving this acknowledgment, that packet is removed from the transmitting buffer, and another new packet is included into this buffer. It works as per the sliding window mechanism, where the acknowledged packets are removed from the leading side of the slider, and new packets are included for transmission, on the trialing side of the slider. So, instead of waiting for an immediate acknowledgment to its transmitted packet, the sender keeps on sending the packets that are there in the slider or buffer. As the acknowledgments of the previously sent packets arrive, new packets are added to the buffer [4, 12]. If no new acknowledgments are received, it stops adding the packets to buffer. Instead, it would be doing the retransmission of those packets for which the RTOs have elapsed. The permitted size of this buffer is the minimum of the following two parameters. First one is 'congestion window' size, and the other one is the 'buffer space' available at receiver. Congestion window size varies as per the congestion scenario prevailing in the network. Usually, the congestion window starts at the size of one segment (packet), and keeps on incrementing it for every successful reception of an acknowledgment. Different TCP variants use slightly different approaches on fixing the 'congestion window' size. These are discussed in Sect. 5.

In addition to sending back the acknowledgments, the receiver also sends the information about its limitations of receiving the packets. The typical limitation is the receiver buffer size. This information is also used by the transmitter while fixing the slider size. The transmissions as well as retransmissions happen at a transmission rate set for that link. The congestion control mechanisms included in TCP also play a role here. If an acknowledgment is not returned within the timeout period, it is understood as congestion somewhere in the network, and the transmitting system reduces the transmission rate drastically, with an objective of pumping lesser number of packets into the network and thereby easing the network. Some of the popular congestion control algorithms of TCP are TAHO, RENO, and new RENO. Each of them has their own specific mechanisms that are slightly different from other. Slow start, fast recover are some of the words used to describe these mechanisms [1, 11, 13]. The TCP protocol has maximum window size of 65,535 bytes. That means after sending these many bytes it must get acknowledgments of the first few transmitted bytes to be able to transmit next few bytes. But, on communication links like satellites where the roundtrip delay is very high, after sending these 65,535 bytes, the transmitter has to sit idle for much of its time waiting for the acknowledgments. This is known as the effect of bandwidth-delay product. If the window size is equal to the product of link bandwidth and roundtrip delay of the link, the problem of waiting would have been avoided. So, the solution for this is to increase the window size to higher value. With mechanisms like 'window scaling' option, it can be increased up to 2^{30} bytes, due to which idle sitting time of the transmitter can be avoided. It results in improved link efficiency [2, 11, 13].

These kind of delays play a major role in the context of satellite-based data networks, where the typical roundtrip delay in a communication that involves one satellite link is 250 ms. That means the transmitter can get back acknowledgment for its successful transmission after a minimum time lapse of 250 ms only. Network of these delayed links is modeled in the Sect. 5.

3   Congestion Control Mechanism of TCP

TCP has introduced various options for controlling the congestion in network. It has the provisions of avoiding the congestion and as well as easing the already congested network. These are offered through mechanisms like slow start, fast retransmit, and fast recovery. Old Tahoe, Tahoe, Reno, New Reno, and BIC are the popular variants that make use of these mechanisms. Slow start is the common feature in all of them, where the size of congestion window increments by one for every successful receipt of acknowledgments. The incrementing mechanism continues up to a threshold, after which it slows down. Still, it grows up to a point where packet timeout takes place. Whenever packet timeout occurs, the window size is brought down to 'one' and continues to grow again, as described earlier. The threshold up to which it can increment now is fixed as half of the size at which packet error has occurred in the previous pass. This process is known as 'Old Tahoe' mechanism.

'Tahoe' is the next method in which 'fast retransmit' mechanism was added. Using this, a packet is retransmitted immediately after receiving three duplicate acknowledgments of the previous packet. The advantage here is that the transmitter is not waiting for the RTO to lapse. Instead, it is retransmitting the packet, whenever it sees duplicate acknowledgment of the previous packet, which means the packet has not reached the receiver.

The next improvement with 'Reno' method includes 'fast recovery' option, with which the congestion window does not restart at 'one' after the packet failure, but starts at half of the congestion window, at which packet failure occurred. New Reno method is a further improvement where the congestion window is not reduced to half for successive packet failures. It is based on the fact that due to burst error types, successive packets may see the failure and hence should not halve the window size for every such packet failure. BIC method is based on fixing the window size to the mid value of the two values where packet error has occurred and not occurred. This mid value keeps on changing based on the congestion scenario of the network from time-to-time [2–4, 13]. These five variants of TCP are considered in the models that are used in this study.

4   Role of Wireless Networks in Railway Signaling

For many years, signaling systems in railways were based on visual indications passed to the cabin driver, through various color lamps fixed on the lamp posts erected on the sides of the rail tracks. Due to their limitations like the poor visibility during rain, fog and snow conditions and also due to the possibility of missing visual indications at higher speeds of the trains, alternate methods were needed. The early methods in these systems were based on balise units that were installed on the tracks, which pass the signal information to the cabin unit, when a receiving unit that is fixed below the cabinet unit travels over the balise unit. Radio communication-based systems were introduced in the subsequent developments. Yet, many railway systems in the world are still dependent on visual color light indications that are based on visual indications observed from the trackside fixtures [14–16].

Whether it is a visual indication or electronically transferred indication that appears on the monitor inside the driven cabin, the purpose of signaling is to convey some vital information to the train driver, to make him act upon that. The main indications are to 'Proceed', 'Stop', or 'Go Slow' that are brought through Green, Red, and Yellow color indications. In addition to this basic information, there are several other kinds of visual indications that are brought to driver's notice, for proper operation of the train and efficient utilization of the tracks. Block sectioning is one important aspect that is regulated through signaling system, where only one train is permitted in each block of the rails that spans over a few kilometers. In conventional signaling systems, this block sectioning was of 'fixed block' of few kilometers that starts at some geo-location and ends at another location. Track circuits and axle counters are fixed at the beginning and end of each such 'fixed blocks'. These devices are used to detect the entry and exit of a train on that fixed block. The color lights that display the track status are fed from these devices such that a train is allowed onto the track after the previously entered train leaves that block [15, 17–19].

Another kind of block is 'moving block' where there is no fixed start and end points of a block, but some safe distance are maintained between the successive trains. This is possible by measuring the location of the trains on continuous basis and passing this information to the control rooms. Then, this information is shared to all the trains that are on that track so that safe distance is maintained between the trains, and advising on the speeds to be followed at various times. This helps in better utilization of the tracks and lesser wear and tear on the rails and train wheels, by avoiding the need of frequent acceleration and braking. GSM-R was proposed and implemented for radio communication between the cabins and control rooms. LTE-R is the later version implemented in some railway systems. WLAN technologies are widely used in CBTC systems that are popular on limited distance railway networks like the metros. Message transfers that occur in real time are critical in all of these implementations [7, 14, 16]. As a result, the selection of a suitable transport layer protocol is also critical, as investigated in this work.

5   Effect of Congestion Control on Network Throughput

Assessing the suitability of UDP for wireless deployments in applications like railway signaling, in comparison with TCP is the main objective of the work here. To study the phenomena related to the effect of bit error rates and congestion control algorithms of TCP on the throughput values, the following network scenarios are considered. At

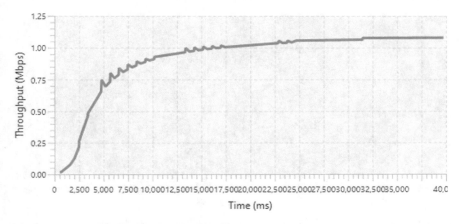

Fig. 1 Throughput of the system for the roundtrip delay of 250 ms

first, a network that consists of a source and a destination that are connected through a router is considered. A roundtrip delay of 250 ms, which is the typical roundtrip delay for a multi-hop communication link, is considered. The bit error rates of the links are considered to be zero at first. A link capacity of 10 Mbps is considered. The network scenario is simulated using Netsim simulator. The throughput for it is observed as 1 Mbps, which is shown in Fig. 1. It is close to the theoretically calculated value.

Simulation is carried out for different versions of TCP, namely Old Tahoe, Tahoe, Reno, New Reno, and BIC. Throughput is observed to be almost same for all these versions. Even though link speed is 10 Mbps, the resultant throughput obtained here is only 1 Mbps because of the higher bandwidth-delay product value and the constraint of maximum window size of 65,535. Hence, in the next model of simulation, window scaling is incorporated, with which the window size can grow up to 2^{30} bytes, by which the throughput is increased up to 10 Mbps gradually, as shown in Fig. 2. Then, UDP is considered in the place of TCP to repeat the above simulations. It is found that it offers the full 10 Mbps throughput from the start of transmission because bandwidth-delay product does not affect this. It is shown in Fig. 3.

Then, roundtrip delay is decreased to 10 ms, which is the typical value on local area networks or metropolitan area network. Then, the simulations are carried out for bit error rates of 'zero' and 10^{-5}. For the case of 'zero' bit error rate, the throughput is found to be close to 10 Mbps, as shown in Fig. 4. It is found to be almost same for all the versions of TCP. These values are unchanged after including the 'Window Scaling' option. However, for the case of 10^{-5} BER value, the throughput is found to be in the range of 3 Mbps, as shown in Fig. 5. Again, the throughput values are found to be unchanged with window scaling option also.

By analyzing these results, the following observations can be made. While the bandwidth-delay product value was large for the case of 250 ms roundtrip delay, it was the smaller window size due to which the lesser throughput has occurred. In such

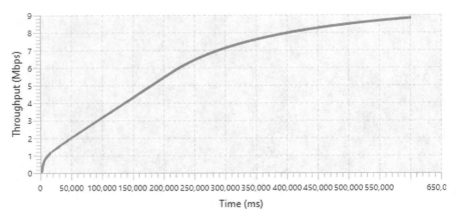

Fig. 2 Throughput of the system, with window scaling (roundtrip delay is 250 ms)

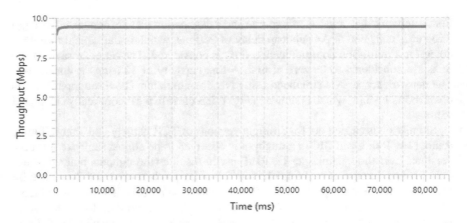

Fig. 3 Throughput of the system for the case of UDP (roundtrip delay is 250 ms)

cases, by implementing the 'window scaling' option, window sizes were increased to match the bandwidth-delay product, and hence, the throughputs also got increased. Later, when the bandwidth-delay product value was small for the case of 10 ms roundtrip delay, the existing 'window size' was sufficient, and so, the throughput did not improve any further, by implementing the 'window scaling' option also. Now, when the experiment is repeated with UDP protocol in the place of TCP, a throughput of 6 Mbps is recorded, which double when compared with 3 Mbps is observed with TCP.

It can be understood here that, when TCP was used, the packet failures that occurred due to the bit errors on the links are assumed as the packet failures due to network congestion, and so, appropriate congestion control mechanisms were introduced. The main mechanism there is to reduce the transmission data rate of the

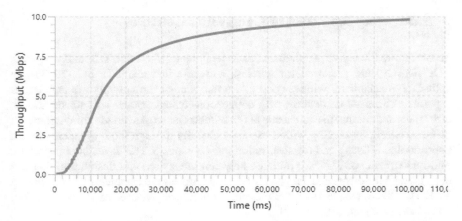

Fig. 4 Throughput of the system for the roundtrip delay of 10 ms (for the case of 'zero' BER)

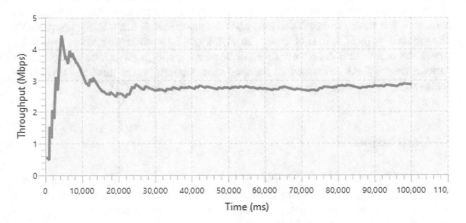

Fig. 5 Throughput of the system in the presence of non-zero BER (without window scaling option)

sender. But, due to this reduction of data rate, throughput has reduced drastically, by half in this experiment.

So, increasing the window size also did not improve the throughput because the earlier window size itself was sufficient. Hence, adaptive window mechanism is not required and not useful also in such cases. The higher throughput of UDP here is due to the non-incorporation of congestion control due to which constant data rate is possible.

6 Mitigating the Effect of Congestion Control

To overcome the above problems, various techniques like Split-TCP and multipath TCP (MTCP) are proposed for wireless networks that use TCP [11, 20–23]. In split-TCP mechanism, wireless part and wired part of the network are handled as separate entities while dealing with congestion-related issues. This makes the bit error rates and retransmission needs of the wireless networks hidden from wired part of the network, by dealing with those errors locally through the implementation of appropriate buffering and retransmission mechanisms. MTCP is a technique where multiple streams of TCP are initiated between the source and destination that can cope up with the errors and delays that occur in wireless environment [22–25]. Snoop TCP, TCP Veno, etc., are some of the other techniques proposed in literature to deal with these problems [20, 26–29].

Alternately, if the application layer is equipped with handling the packet errors by implementing appropriate acknowledgments, UDP also can be used. In the case of Ad-hoc networks and sensor network applications where cross-layer designs are common, and prior connection-based packet transmissions are not always possible like the ones with TCP, in such cases also, UDP can be a preferred choice, especially where the end-to-end delays are more or window scaling may not be an option for various reasons. As discussed in the previous section, better throughputs can be achieved by using UDP wherever it is possible, in the place of TCP.

7 Conclusion

Though TCP protocol offers reliable transfer of data through its flow control, packet acknowledgments, and retransmission mechanisms, its congestion control policies may sometimes pose problems in wireless network environments of high bit error rates. The problems add up if the wireless environments are mobile. In such cases, if the application layer can include some additional mechanisms for reliable packet deliveries, the usage of UDP can also be explored for better throughput achievements. The application scenario of railway signaling is considered here for studying this phenomenon. For the environments of moderate bit error rates, a throughput improvement of double the value is observed. Exploring it further to assess its suitability in terms of meeting the real-time deadlines of the packet deliveries needed for such applications will offer further insight.

References

1. Tian, Y., Xu, K., Ansari, N.: TCP in wireless environments: problems and solutions. IEEE Commun. Mag. **43**(3), S27–S32 (2005)

2. Dias, M.C., Caldeira, L.A., Perkusich, A.: Traditional TCP congestion control algorithms evaluation in wired-cum-wireless networks. In: 2015 International Wireless Communications and Mobile Computing Conference (IWCMC), pp. 805–810. IEEE Press (2015)
3. Panda, M., Vu, H.L., Mandjes, M., Pokhrel, S.R.: Performance analysis of TCP NewReno over a cellular last-mile: buffer and channel losses. IEEE Trans. Mob. Comput. **14**(8), 1629–1643 (2015)
4. Pokhrel, S.R., Panda, M., Vu, H.L., Mandjes, M.: TCP performance over Wi-Fi: joint impact of buffer and channel losses. IEEE Trans. Mob. Comput. **15**(5), 1279–1291 (2016)
5. Farooq, J., Bro, L., Karstensen, R.T., Soler, J.: Performance evaluation of a multi-radio, multi-hop Ad-hoc radio communication network for communications-based train control (CBTC). IEEE Trans. Veh. Technol. **67**(1), 56–71 (2018)
6. Ai, B., et al.: Challenges toward wireless communications for high-speed railway. IEEE Trans. Intell. Transp. Syst. **15**(5), 2143–2158 (2014)
7. Zayas, A.D., Garcia Perez, C.A., Gomez, P.M.: Third-generation partnership project standards: for delivery of critical communications for railways. IEEE Veh. Technol. Mag. **9**(2), 58–68 (2014)
8. Farooq, J., Soler, J.: Radio communication for communications-based train control (CBTC): a tutorial and survey. IEEE Commun. Surv. Tutorials **19**(3), 1377–1402 (2017)
9. Sniady, A., Soler, J.: LTE for railways: impact on performance of ETCS railway signaling. IEEE Veh. Technol. Mag. **9**(2), 69–77 (2014)
10. Mammadov, A., Abbasov, B.: A review of protocols related to enhancement of TCP performance in wireless and WLAN networks. In: IEEE 8th International Conference on Application of Information and Communication Technologies (AICT), pp. 1–4 (2014)
11. Liu, J., Han, Z., Li, W.: Performance analysis of TCP new Reno over satellite DVB-RCS2 random access links. IEEE Trans. Wireless Commun. **19**(1), 435–446 (2020)
12. Lin, F., Li, X., Li, W.: Research on TCP protocol in wireless network and network simulation. In: 4th International Conference on Wireless Communications, Networking and Mobile Computing, pp. 1–4 (2008)
13. Waghmare, S., Nikose, P., Parab, A., Bhosale, S.J.: Comparative analysis of different TCP variants in a wireless environment. In: 3rd International Conference on Electronics Computer Technology, pp. 158–162 (2011)
14. Macucci, M., Pascoli, S.D., Marconcini, P., Tellini, B.: Derailment detection and data collection in freight trains, based on a wireless sensor network. IEEE Trans. Instrum. Measure. **65**(9), 1977–1987 (2016)
15. Lopez, I., Aguado, M., Pinedo, C.: A step up in European rail traffic management systems: a seamless fail recovery scheme. IEEE Veh. Technol. Mag. **11**(2), 52–59 (2016)
16. Allotta, B., D'Adamio, P., Faralli, D., Papini, S., Pugi, L.: An innovative method of train integrity monitoring through wireless sensor network. In: 2015 IEEE International Instrumentation and Measurement Technology Conference, pp. 278–283. Pisa (2016)
17. Del Signore, E., et al.: On the suitability of public mobile networks for supporting train control/management systems. In: IEEE Wireless Communications and Networking Conference, pp. 3302–3307 (2014)
18. Kim, J., Choi, S.W., Song, Y.S., Yoon, Y.K., Kim, Y.K.: Automatic train control over LTE: design and performance evaluation. IEEE Commun. Mag. **53**(10), 102–109 (2015)
19. Guan, K., Zhong, Z., Ai, B., Kürner, T.: Propagation measurements and analysis for train stations of high-speed railway at 930 MHz. IEEE Trans. Veh. Technol. **63**(8), 3499–3516 (2014)
20. Xiao, S., Gao, Z., Gao, F., Wang, N., Zhao, R.: An enhanced TCP Veno over wireless local area networks. In: 5th International Conference on Wireless Communications, Networking and Mobile Computing, pp. 1–4 (2009)
21. Kim, B.H., Calin, D.: On the split-TCP performance over real 4G LTE and 3G wireless networks. IEEE Commun. Mag. **55**(4), 124–131 (2017)
22. Hurtig, P., Grinnemo, K., Brunstrom, A., Ferlin, S., Alay, Ö., Kuhn, N.: Low-latency scheduling in MPTCP. IEEE/ACM Trans. Network. **27**(1), 302–315 (2019)

23. Palash, M.R., Chen, K., Khan, I.: Bandwidth-need driven energy efficiency improvement of MPTCP users in wireless networks. IEEE Trans. Green Commun. Network. **3**(2), 343–355 (2019)
24. Xu, J., Ai, B., Chen, L., Pei, L., Li, Y., Nazaruddin, Y.Y.: When high-speed railway networks meet multipath TCP: supporting dependable communications. IEEE Wireless Commun. Lett. **9**(2), 202–205 (2020)
25. Polese, M., Jana, R., Zorzi, M.: TCP and MP-TCP in 5G mmWave networks. IEEE Internet Comput. **21**(5), 12–19 (2017)
26. Saedi, T., El-Ocla, H.: Performance analysis of TCP CERL in wireless networks with random loss. In: 2019 IEEE Canadian Conference of Electrical and Computer Engineering, pp. 1–4 (2019)
27. Beshay, J.D., Taghavi Nasrabadi, A., Prakash, R., Francini, A.: Link-coupled TCP for 5G networks. In: IEEE/ACM 25th International Symposium on Quality of Service, pp. 1–6 (2017)
28. Chawhan, M.D., Kapur, A.R.: TCP performance enhancement using ECN and snoop protocol for Wi-Fi network. In: Second International Conference on Computer and Network Technology, pp. 186–190 (2010)
29. Ha, L., Fang, L., Bi, Y., Liu, W.: A TCP performance enhancement scheme in infrastructure based vehicular networks. China Commun. **12**(6), 73–84 (2015)

Semi-Supervised Self-Training Approach for Web Robots Activity Detection in Weblog

Rikhi Ram Jagat, Dilip Singh Sisodia, and Pradeep Singh

Abstract Due to the significant added value of web servers, they are vulnerable to attacks, so web security has received a lot of attention. Web server logging systems that record each user request performed by users have become an important data analysis object in web security. Traditionally, system experts' analyses log data manually using keyword searches and regular expressions. However, the amount of log data and attack types makes routine detection ineffective. Machine learning-based supervised and unsupervised detection approaches have been employed extensively during the last decade to improve traditional detection methods. The proposed semi-supervised STBOOST web robot detection system uses self-training with XGBoost as its base classifier. Experimental data are taken from the open-source data repository, the NASA 95 dataset, and e-commerce site access logs. In both datasets, self-training XGBoost outperforms XGBoost and can detect anonymous web robots using unlabeled data.

Keywords Web Robot · Web attacks · Semi-supervised learning · Self-training · XGBoost · Weblog files

1 Introduction

Web applications provide a variety of services and support social infrastructure. Every year, new vulnerabilities are discovered, and new attacks are devised [1]. Web robots are computer programs that travel around the internet and autonomously collect data

R. R. Jagat (✉) · D. S. Sisodia · P. Singh
Department of Computer Science and Engineering, National Institute of Technology Raipur, Raipur 492010, India
e-mail: rrjagat.phd2019.cse@nitrr.ac.in

D. S. Sisodia
e-mail: dssisodia.cs@nitrr.ac.in

P. Singh
e-mail: psingh.cs@nitrr.ac.in

© The Author(s), under exclusive license to Springer Nature Singapore Pte Ltd. 2022
V. Suma et al. (eds.), *Evolutionary Computing and Mobile Sustainable Networks*,
Lecture Notes on Data Engineering and Communications Technologies 116,
https://doi.org/10.1007/978-981-16-9605-3_64

Table 1 Good bot versus bad bot versus human traffic from 2014 to 2020 [4]

	2014	2015	2016	2017	2018	2019	2020
Bad bots (%)	22.8	18.8	19.9	21.8	20.4	24.1	25.6
Good bots (%)	36.3	27.0	18.8	20.4	17.5	13.1	15.2
Humans (%)	40.9	54.4	61.3	57.8	62.1	62.8	59.2

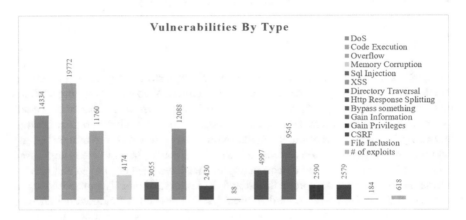

Fig. 1 Vulnerability by type from last six years [5]

from websites, known as spiders, crawlers, hikers, and harvesters [2]. The web robot is as ancient as the internet. Over the years, there has been an enormous increase in the number of web robots and remarkable technological advances [3]. In 2020, web bots accounted for nearly half (40.8%) of total web traffic, with 25.6% being bad and 15.2% being good bots (see Table 1) [4]. Web bots produce most web traffic on vulnerable web applications. Figure 1 shows the new vulnerabilities discovered from 2014 to 2020 [5]. SQL injection and remote code execution attacks cause data spillage or website defacement.

Malicious bots can jeopardize an online application's security, privacy, and non-malicious bots distort data, affecting measurement reliability and decision-making [6]. Large websites with unique content, such as e-commerce sites, online news-papers, digital libraries, blogs, and e-government portals, attract bots the most. For example, hostile and non-malicious robots can falsify metrics and ratings, tainting their validity [7]. Social bots also help spread rumors and false information [8]. Detecting and restricting the activities of web robots are crucial tasks in the battle for a safe web, and they should not be overlooked. IoT devices are also being used extensively for automation. Some intruders attack these devices, and some inconsistencies are created. Which makes several IoT applications are prone to delay that needs to be addressed [9].

Malicious web robots are classified by their skills and behaviors. Unfair techniques are used to obtain things and services that are limited in availability [10].

Impersonators are capable of spamming and account registration, as well as ad fraud and last-minute product bidding [11]. The Open Online Application Security Project (OWASP) has created a taxonomy of automated online application threads [12].

Bots that are not malicious have also been categorized in the same way. Its crawlers and feed fetchers collect data from websites and deliver it to mobile or web applications. At the same time, its monitoring bots assist developers in monitoring the performance and health of their websites [13]. Fetchers of RSS feeds make for about 12% of overall traffic, and mobile applications of Facebook accounting for 4.4% of total traffic [4].

According to prior web bot detection research, there are four basic categories of web bot detection that have been used: Syntactic log analysis uses text processing technologies. Analytical learning uses machine learning to extract properties from web server logs; traffic patterns are compared to human and robot attributes to identify robots in real-time. For web robot identification, current research has focused on analytical learning techniques, which outperform other approaches since they depend on algorithms or procedures that an appropriately developed web robot can bypass [5]. Supervised learning methods require a lot of labeled data to train the model; it may be biased to the majority class in the test dataset. While unsupervised learning need not require any labeled information but it takes a lot of time to compute each instance, and it gives less accuracy as compared to supervised learning.

In this research, employ a semi-supervised based method called self-training to train a supervised base classifier and compare the results. The assumption is that utilizing less labeled data and more unlabeled data for semi-supervised self-training is superior to using the entire dataset for supervised learning. As for the remainder of the paper, it is organized as follows: Sect. 2 describes the related work. Section 3 provides an overview of the system, followed by Sects. 4 and 5, which illustrate the details of design, implementation, and evaluation. Section 6 concludes the paper and provides a glimpse into the future.

2   Related Work

Several supervised analytical learning techniques based on various learning algorithms and features have been developed in the past. Tan and Kumar [14] trained a model utilizing 25 distinct characteristics collected from each user session using decision trees (C4.5 method). Included in the feature vector were percentages for different categories of content (HTML, images, multimedia, etc.), as well as information about time (total_time, average_time, etc.) and HTTP request types (GET method, POST method, HEAD method, and others methods) and other features like (host IP address, host user-agent, etc.). Stassopoulou and Dikaiakos [15] labeled the training data with a semi-automatic technique based on heuristics and classified the sessions with a Bayesian approach. Bomhardt et al. [16] utilized neural networks to analyze their data, and they took into account features such as the total number of response bytes and the response codes percentage. Stevanovic et al. [17] used a variety of classifiers were employed (C4.5, SVM, Bayesian Network, RIPPER, Nave Bayesian, k-nearest, and Neural Networks). In addition, they proposed two unique features that took into consideration the session's requests page depth as well as the HTTP request sequentiality. Doran et al. [13] have finally developed a unique technique for the real-time detection of web robots. Their method is based on the visitors' request patterns and the Markov chain model of the first-order discrete-time. Smys and Wang [18] proposed a Naive Bayes and entropy-based system that classify the humans and chatbots from a commercial chat network which is surveyed in their study. The proposed system takes a long time due to the need for a large number of messages. They found that it is faster to use the Naive Bayes Classifier than the Entropy Classifier since the amount of messages required is smaller.

Contrary to the above-supervised methods, Mittal et al. [19] standard validity metrics and indices are used to assess threshold-based algorithms and K-means clustering using synthetic data. For standard validity metrics and validity indices, the experiments revealed that the threshold-based method outperforms the K-means clustering algorithm. The adaptive K-means clustering technique is used in a suggested clustering approach called sample-based hierarchical adaptive K-means (SHAKM) proposed by Liao et al. [20] to find the correct number of clusters and create an imbalanced tree for large-scale video retrieval. A study conducted by Stevanovic et al. [21] utilized unsupervised neural networks to detect humans, whereas Zabihi et al. [22] DBSCAN clustering technique was employed with only four distinct characteristics to assess malicious and non-malicious web robot activity. Cho and Cha [23] developed a Bayesian estimate methodology SAD (Session Anomaly Detection). They used the Whisker software to simulate the web attacks. The SAD system can be detecting previously unknown web attacks.

3   System Overview

Figure 2 depicts the proposed system architecture of STBoost. The whole system is divided into three phases in this paper: (i) data preparation, which includes row log data cleaning, session identification, feature extraction, and session labeling tasks. Cleaning unwanted data from raw log files, such as incomplete entries, and extracting the desired features from cleaned log data to create tabulated data, then creating session using some analytical learning applied to feature extracted data, and labeling it as 1 for malicious activity session and 0 for a normal session. (ii) STBoost, in this phase of the system, utilized the data prepared in the previous section and used XGBoost as the base classifier for the self-training algorithm and created a model and trained them until self-training conditions were not satisfied. (iii) Empirical study, this phase study the different performance measures to check the created model performance.

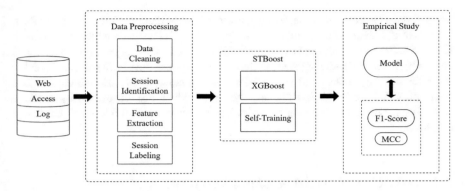

Fig. 2 The framework of web robot activity detection system

4 System Design and Implementation

This part of the study further explains the proposed system structure. It starts with a description of web server logs and then moves on to preprocess raw log data. It includes Data Cleaning, Session Identification, Feature Extraction, and Session Labeling. Then introducing the base classification model and self-training approaches. Weblog files, or web access log files, are the subject of this study.

4.1 Web Server Logs

Whenever a website's users make requests for resources, the webserver records these requests in a simple plain text file; this file is called a weblog file [24]. Agent log files, error log files, access log files, and referrer log files are the four categories of server logs [25]. For web servers such as Apache, the Common Log Format (CLF) is the standard log format [25, 26]. Each HTTP request is logged in the CLF log file. In each line, tokens are separated by spaces. A hyphen (-) represents a token that has no value. There are nine fields in a log entry shown in Fig. 3: [Ip Address] of the

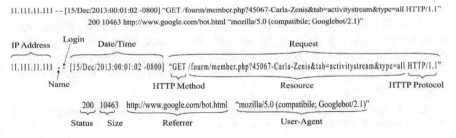

Fig. 3 Example of an entry in a weblog file with each field details

users, [Name] is the user name, [Login] is the HTTP-authentication of the login user, [Date/Time] is request date and time, [Request] is HTTP request which contains the HTTP Method, the resource URL, and HTTP-protocol; [Status] is a server response code which has three-digit, [Size] is the server response size in bytes, [Referrer] is the referring page URL, and [User-Agent] is the name of browser signature. The client provides the Request, Referrer, and agent, which are untrustworthy in terms of analysis. The remaining fields are the web server-generated fields and are hence reliable.

4.2 Data Cleaning and Session Identification

Raw log data is read line-by-line, and any unnecessary or incomplete entries are removed. When a server is experiencing issues, certain unnecessary entries appear in the access log file; for web robot detection, these lines are useless. Apart from removing unwanted log lines in these steps, create a tabulated form of each field, as shown in Fig. 3 with session identification when it came to identifying sessions, analytic learning methods used for session identification. Sessions may be determined using an analytical learning technique, which involves breaking down the clickstream into sessions. The effects on the performance of time threshold have been researched in the past. In which, threshold of 10 min [27], threshold of 30 min [14], and a threshold that are dynamically adaptable ranging from 30 to 60 min [28] have all been investigated. Session time thresholds of 30 min were used in this study to create sessions for each client.

4.3 Feature Extraction

It is a crucial stage in feature extraction from identifying the session, which is based on the diversity of information contained in weblog entries. Such as host's IP address that made the server request, the date and time, an indication of when and where the Request was made, the resource that was requested for, which HTTP Method (e.g., GET, POST, HEAD, DELETE, PUT, PATCH) used for generating Request, the server response code or status (200, 300, 404, 500, etc.) for current request return back to the client. The returned object's size in bytes, the HTTP request header referer, which is the website that connects to the resource which is requested, and the string of User-Agent, which identifies the browser or application of client's which make Request. A typical entry from a server access log (Fig. 3), Session-specific attributes include the following:

- Total Requests: This is the total amount of requests that have been made.
- Total Page: Total HTML page request in a session.

- Total Session Time: The amount of time in seconds elapsed between the first Request and last Request in the log file entries.
- Average Time: The amount of time that elapses between two successive requests, measured in seconds.
- Repeated Requests: Requests that have been made more than once, their percentage. Repeating requests using the same HTTP Method where it was used for earlier requests.
- Standard deviation of time: The time between two successive requests with a standard deviation.
- HTTP Methods: There are four features, each of which contains the proportion of the Request linked with one of the HTTP methods: GET, POST, HEAD, and others (PUT, PATCH, DELETE).
- HTTP Response: which include the Successful response code with (2xx), Redirection response code with (3xx), Client Errors response code with (4xx), and Server Errors response code with (4xx) are four characteristics that each include the percentage of requests associated with one of the HTTP response code.
- Robots.txt: The standard for excluding robots.
- Requests for Specific Types. The number of requests of a particular category is divided by the total number. Depending on the application, this feature is available.

4.4 Session Labeling

In session labeling, each unit data is assigned a label indicating whether the session is deemed a web robot. Session labeling should be done by a security expert manually. This study is automatically conducted in three steps. Using the useragentstring.com API, label each session in the first step. As an input, this API accepts a user-agent. It returns one of the categories listed of agents: Browser, Offline Browser, Mobile Browser, Crawler, Console, Validator, Feed Fetcher, Cloud Client, Link Checker, etc. It was determined that robots were all sessions whose user-agent was Link Checker, Feed Fetcher, and Crawler.

The second stage makes use of two sets of regular expressions that are designed to match the user-agent string of known bots. This includes the project COUNTER's "robot/spider" list of user agents, which outlines a set of guidelines for publishers and librarians to collect and online resource consumption statistics was reported in a uniform and reliable manner. Finally, Matomo's open-source web analytics software uses a constantly updated list. Robots were defined as sessions that match one of the regular expressions using a string of user agents.

4.5 XGBoost Classification Model

XGBoost, a boosting technique, is a good implementation of Chen and Guestrin's Gradient Boosting Framework [29]. Equation (1) illustrates how the XGBoost ensemble of classification and regression trees (CART) combines homogenous weak learners to produce a stronger learner.

$$\hat{Y} = \sum_{n=1}^{N} f_n(X) \tag{1}$$

In this case, $\hat{Y}$ is the predicted class label for the given problem, and the characteristics variables are X, i.e., TotalRequest, TotalTime, AverageTime, StDevTime. RepeatedRequest, GET, POST, HEAD, etc.; Number of trees are represented by N; The direction in which the residual of the $(n-1)$th tree declines is represented by f_n, which is a CART tree.

XGBoost outperforms the gradient boosting decision tree (GBDT) in three key ways. Firstly, regularization is the requirement term added to the objective function, which reduces model complexity and prevents overfitting from taking place. Because of this, the objective function of XGBoost consists of both a loss function and a regularization function. Equation (2) describes the objective function for optimizing the nth iteration.

$$O^n = \sum_{i=1}^{m} l\left(y_i, \hat{Y}_i^n\right) + \sum_{j=1}^{n} \Omega(f_j) \tag{2}$$

The number of samples is denoted by the letter m. The ith sample's actual value is y_i. $\hat{Y}_i^n$ is the predicted value of the ith sample in the nth iteration. $l\left(y_i, \hat{Y}_i^n\right)$ is the differentiable loss function that quantifies the difference between the predicted $\hat{Y}_i^n$ and the target y_i, and $\Omega(f_j)$ denotes the regularization term, which is composed of the number of nodes and the weight of each node. Secondly, the loss function is expanded using a second-order Taylor expansion to obtain a more precise estimate of the model error. Thirdly, the block storage structure allows for parallel tree creation, considerably increasing executive speed. Due to the mentioned benefits, XGBoost has demonstrated cutting-edge performance in classification and regression.

4.6 Semi-Supervised Self-Training Method

The function to map an input to output is trained using the input–output sample pairs (x, y), which are obtained via the supervised learning approaches. The input sample is labeled x, and the output sample is labeled y. Although powerful supervised learning algorithms have been created, they need a lot of labeled data to train. There are two types of algorithms used in supervised learning: classification and regression. Decision trees, support vector machines, neural networks, and other classification algorithms are common. Unsupervised learning is a sort of machine learning technique that uncovers the data's hidden structure. It exclusively uses data samples that have not been labeled, i.e., unlabeled data. Cluster analysis, density estimation, dimensionality reduction, and other unsupervised learning approaches are common.

Semi-supervised learning is a sort of machine learning that falls somewhere between supervised and unsupervised methods. It is possible to use both labeled and unlabeled data for training purposes in semi-supervised learning. The main reason for developing a semi-supervised learning method is because of the fact that labeling data is typically expensive, and the vast majority of data samples in real systems are unlabeled. Due to the lack of labeled data in this circumstance, supervised learning approaches are unable to produce good results. When dealing with unlabeled data, semi-supervised learning algorithms provide an efficient solution by combining this knowledge with a small number of labeled training samples in order to accomplish the supervised learning task.

Standard Self-Training is implemented in accordance with the algorithm described in the paper [30, 31]. An iterative procedure is required to train a Standard Self-Training base classifier. Initially, the base classifier is trained on labeled examples. The unlabeled samples are then utilized to label them, and the most confident predictions are added to the labeled set. This new labeled set will then be used to retrain the classifier. The procedure of training and labeling is repeated until all training samples have been labeled or the maximum number of iterations has been reached.

The algorithm can be summarized as follows: Notation:

C—used as a base classifier
D—initial dataset
L—training samples with labeled, $L \subseteq D$
U—training samples without labeled, $U \subseteq D$

---

1. Train C on L
2. Predict labels on U using C, and then select the most confident predictions from the results.
3. Remove the most confident predictions from U and place them in L alongside the labels that have already been predicted.
4. Repeat steps 1–4 until U is completely exhausted or the maximum number of iterations has been achieved

---

The final C can then be used to categorize data that has not yet been seen.

5　Experiment and Results

This section involves the empirical study of the proposed system. It starts with a dataset description of used weblogs, then experiment environment setup, and finally discusses the result of the experiment.

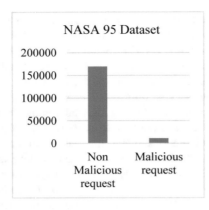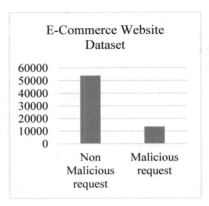

Fig. 4 Class distribution of web access log datasets

5.1 Dataset

The web access log data is used from two different sources: the NASA 95 weblog dataset and an e-commerce website web access log data. NASA log dataset has collected from 1st August to 31st August 1995 and consist of 1,569,898 requests; from there, we extract the 26 essential features, with 180,602 sessions where 11,282 sessions are malicious (robot sessions), and 169,320 sessions are non-malicious session (Browsers). The E-commerce dataset collects from 1st March to 31st March 2018 and consists of 4,091,155 requests. After preprocessing, 30 features are extracted from raw log data, where 26 features are common to the first one and four features are belong to e-commerce characteristics and generate 67,352 sessions where 13,494 sessions are malicious (robot sessions), and 53,858 sessions are non-malicious (Browsers) sessions. Their class distribution is shown in Fig. 4. The dataset where split into 70% of unlabeled data and 30% labeled. From 30% of labeled data, it used five-fold cross-validation for training the initial model. Then consumed 70% of unlabeled data in the self-training stage for model training.

5.2 Evaluation Metrics

This study compares the performance of a web robot activity detection system to a comprehensive supervised approach based on a machine learning algorithm, namely XGBoost, and the self-training XGBoost algorithm.

The classification performance is evaluated using the following evaluation metrics. The accuracy, precision, recall, $f1$-score, and Matthews correlation coefficient (MCC) are defined as follows, while Table 2 illustrates the confusion matrix.

Table 2 Confusion matrix

		Actual label	
		Malicious	Non-malicious
Predicted label	Malicious	num_{mm}	num_{mn}
	Non-malicious	num_{nm}	num_{nn}

$$\text{Accuracy} = \frac{(\text{num}_{mm} + \text{num}_{nn})}{(\text{num}_{mm} + \text{num}_{mn} + \text{num}_{nm} + \text{num}_{nn})} \tag{3}$$

$$\text{Precision} = \frac{\text{num}_{mm}}{(\text{num}_{mm} + \text{num}_{mn})} \tag{4}$$

$$\text{Recall} = \frac{\text{num}_{mm}}{(\text{num}_{mm} + \text{num}_{nm})} \tag{5}$$

$$f1 - \text{score} = \frac{2 \times \text{num}_{mm}}{(2 \times \text{num}_{mm} + \text{num}_{mn} + \text{num}_{nm})} \tag{6}$$

$$\text{MCC} = \frac{(\text{num}_{mm} \times \text{num}_{nn}) - (\text{num}_{mm} \times \text{num}_{nm})}{\sqrt{\begin{array}{c}(\text{num}_{mm} \times \text{num}_{mn}) \times (\text{num}_{mm} + \text{num}_{nm}) \\ \times (\text{num}_{nn} + \text{num}_{mn}) \times (\text{num}_{nn} + \text{num}_{nm})\end{array}}} \tag{7}$$

5.3 Environment for Experiment

The effectiveness and scalability of web robot activity detection systems were investigated in this study by running the entire process on a workstation equipped with a Xeon 10 cores at 3.30 GHz CPU, 32 GB RAM, 1 TB hard drive, and Windows environment.

5.4 Results

This study compared the performance of the proposed STBoot self-learning XGBoost web robot activity detector with ensemble XGBoost algorithms for web robot activity detection. This paper's classification algorithm makes use of scikit-learn, a widely-used machine learning package, as well as XGBoost API [32], which are both available on the internet. There are many parameters in XGBoost, which are manually adjusted during training to achieve the best performance results.

As can be seen from Table 3, the accuracy when only the XGBoost algorithm is used is merely 96.08% in NASA 95 dataset and 96.44% in the E-Commerce dataset. In comparison, the number can jump to 99.60% and 97.84% when using

Table 3 Comparison of test performance of XGBoost and self-training XGBoost in NASA 95 and e-commerce dataset

Dataset	Model	Accuracy	Precision	Sensitivity (recall)	f1-score	MCC
NASA 95 dataset	XGBoost	96.08	85.95	44.14	58.33	59.97
	Self-training XGBoost	**99.60**	**99.50**	**94.40**	**96.88**	**96.71**
E-commerce dataset	XGBoost	96.44	92.69	89.21	90.92	88.73
	Self-training XGBoost	**97.84**	**94.18**	**95.05**	**94.61**	**93.26**

Bold in the table show the good result obtained from the experiment

the self-training XGBoost based on the previous result. As Fig. 4 shows that class distribution in the data set is imbalanced. We consider the f1-score and MCC score for our model performance. Table 3 clearly show that without self-training, f1-score and MCC are 58.33% and 59.97% for NASA 95 dataset and 90.94% and 88.73% for the E-commerce dataset, which improved significantly when applying self-training XGBoost method (STBoost) as 96.88% and 96.71% for NASA dataset and 94.61% and 93.26% for E-commerce dataset. The result shows that using unlabeled data to train the model improves the performance of the detection model.

6   Conclusion and Future Work

Using machine learning, this study describes a technique implemented to detect web robot activity in weblog files. The analytical methods based on some heuristics are used to preprocess the raw log data and extract important features from log data. This processed data is then used for training the proposed self-training XGBoost (STBoost) algorithm to build a model which can then be able to detect web robot activity from log sessions. The data belongs to a well-known data repository from both the old and new time periods 1995 and 2018. Compared with XGBoost with proposed self-training XGBoost, self-training XGBoost improves the performance results. The limitation of the proposed method is that if there is an error in the label data being used, it will lead to mislabeling even at the time of self-training, which may reduce the performance of the final model.

References

1. Sisodia, D.S., Verma, N.: Framework for pre-processing and feature extraction from weblogs for identification of HTTP flood request attacks. In: 2018 International Conference on Advanced Computation and Telecommunication (ICACAT), pp. 8–11 (2018)

2. Udapure, T.V., Kale, R.D., Dharmik, R.C.: Study of web crawler and its different types. IOSR J. Comput. Eng. **16**(1), 01–05 (2014)
3. Chen, H., He, H., Starr, A.: An overview of web robots detection techniques. In: International Conference on Cyber Security and Protection of Digital Services (Cyber Security) (2020)
4. Imperva: Bad Bot Report 2021, p. 28 (2021)
5. CVE Details: Vulnerabilities by types 2021. [Online]. Available: https://www.cvedetails.com/vulnerabilities-by-types.php
6. Doran, D., Gokhale, S.S.: An integrated method for real time and offline web robot detection. Expert Syst. **33**(6), 592–606 (2016)
7. Greene, J.W.: Web robot detection in scholarly open access institutional repositories. Libr. Hi Tech **34**(3), 500–520 (2016)
8. Ferrara, E., Varol, O., Davis, C., Menczer, F., Flammini, A.: The rise of social bots (2016)
9. Shakya, D.S., Smys, S.: Anomalies detection in fog computing architectures using deep learning. J. Trends Comput. Sci. Smart Technol. **2**(1), 46–55 (2020)
10. Lee, J., Cha, S., Lee, D., Lee, H.: Classification of web robots: an empirical study based on over one billion requests. Comput. Secur. **28**(8), 795–802 (2009)
11. Wang, B., Zheng, Y., Lou, W., Hou, Y.T.: DDoS attack protection in the era of cloud computing and software-defined networking. Comput. Networks **81**, 308–319 (2015)
12. Watson, C., Zaw, T.: OWASP Automated Threat Handbook (2018)
13. Doran, D., Morillo, K., Gokhale, S.S.: A comparison of web robot and human requests. In: Proceedings of 2013 IEEE/ACM International Conference on Advances in Social Network Analysis and Mining, ASONAM, pp. 1374–1380 (2013)
14. Tan, P.N., Kumar, V.: Discovery of web robot sessions based on their navigational patterns. Data Min. Knowl. Discov. **6**(1), 9–35 (2002)
15. Stassopoulou, A., Dikaiakos, M.D.: Web robot detection: a probabilistic reasoning approach. Comput. Networks **53**(3), 265–278 (2009)
16. Bomhardt, C., Gaul, W., Schmidt-Thieme, L.: Web robot detection—preprocessing web logfiles for robot detection. Stud. Classif. Data Anal. Knowl. Organ. (211289), 113–124 (2005)
17. Stevanovic, D., An, A., Vlajic, N.: Feature evaluation for web crawler detection with data mining techniques. Expert Syst. Appl. **39**(10), 8707–8717 (2012)
18. Smys, S., Wang, H.: Naive Bayes and entropy based analysis and classification of humans and chat bots. J. ISMAC **3**(1), 40–49 (2021)
19. Mittal, M., Sharma, R.K., Singh, V.P.: Validation of k-means and threshold based clustering method. Int. J. Adv. Technol. **5**(2), 153–160 (2014)
20. Liao, K., Liu, G., Xiao, L., Liu, C.: A sample-based hierarchical adaptive K-means clustering method for large-scale video retrieval. Knowl. Based Syst. **49**, 123–133 (2013)
21. Stevanovic, D., Vlajic, N., An, A.: Detection of malicious and non-malicious website visitors using unsupervised neural network learning. Appl. Soft Comput. J. **13**(1), 698–708 (2013)
22. Zabihi, M., Jahan, M.V., Hamidzadeh, J.: A density based clustering approach for web robot detection. In: Proceedings of 4th International eConference on Computer and Knowledge Engineering (ICCKE 2014), pp. 23–28 (2014)
23. Cho, S., Cha, S.: SAD: web session anomaly detection based on parameter estimation. Comput. Secur. **23**(4), 312–319 (2004)
24. Salama, S.E., Marie, M.I., El-Fangary, L.M., Helmy, Y.K.: Web server logs preprocessing for web intrusion detection. Comput. Inf. Sci. **4**(4), 123–133 (2011)
25. Joshila Grace, L.K., Maheswari, V., Nagamalai, D.: Analysis of web logs and web user in web mining. Int. J. Netw. Secur. Appl. **3**(1), 99–110 (2011)
26. Castellano, G., Fanelli, A.M., Torsello, M.A.: Log data preparation for mining web usage patterns. Computing 371–378 (2007)
27. Alnoamany, Y., Weigle, M.C., Nelson, M.L.: Access patterns for robots and humans in web archives. In: Proceedings of the 13th ACM/IEEE-CS Joint Conference on Digital Libraries, pp. 339–348 (2013)
28. Stassopoulou, A., Dikaiakos, M.D.: A probabilistic reasoning approach for discovering web crawler sessions. Lecture Notes in Computer Science (including Subseries Lecture Notes in Artificial Intelligent Lecture Notes Bioinformatics), vol. 4505, pp. 265–272. LNCS (2007)

29. Agarwal, A.K., Wadhwa, S., Chandra, S.: XGBoost a scalable tree boosting system. J. Assoc. Phys. India **42**(8), 665 (2016)
30. Yarowsky, D.: Unsupervised word sense disambiguation rivaling supervised methods. 189–196 (1995)
31. Triguero, I., García, S., Herrera, F.: Self-labeled techniques for semi-supervised learning: taxonomy, software and empirical study. Knowl. Inf. Syst. **42**(2), 245–284 (2015)
32. Pedregosa, F., et al.: Scikit-learn: machine learning in Python. J. Mach. Learn. Res. **12**, 2825–2830 (2011)

Analysis of Data Aggregation and Clustering Protocol in Wireless Sensor Networks Using Machine Learning

P. William, Abhishek Badholia, Vijayant Verma, Anurag Sharma, and Apurv Verma

Abstract Wireless Sensor Networks (WSNs) monitor dynamic environments that change rapidly over time. As the Wireless Sensor Networks are resource-constrained, energy-efficient data transmission is required by considering various applications. A Strong Clustering Algorithm and Data Aggregation (SCADA) is the name of the study. A new routing protocol was developed utilizing machine learning in order to accomplish scalability and QoS optimization with minimal cost (SCADA-ML). Cluster-Head selection and data aggregation are focused on when applying machine learning techniques to WSNs of different sizes. The neural network machine learning approach known as Artificial Neural Networks is used to boost overall CH yield and cluster formation, to help optimize CH yield and cluster formation (ANN). A given sensor node characteristic, such as residual energy, distance from the base station, and allotted bandwidth, is utilized to instruct the ANN architecture which CH is the best for the given cluster. For this reason, the second contribution explains how Machine Learning is used to reduce cluster energy usage by performing effective data aggregation on the CH nodes of each cluster (ICA). The results must be replicated on groups of databases that all have comparable data. The algorithm is computationally more efficient than previous data aggregation techniques and eliminates extraneous data by use of differential entropy. The SCADA-ML method beats machine learning-based solutions for grouping and data aggregation in controlled testing.

Keywords Artificial neural network · Cluster-head selection · Machine learning · Clustering · Data transmission · Data aggregation

1 Introduction

Sensors on sensor nodes are part of the Wireless Sensor Network (WSN). As a sensor node is very tiny, uses little power, and is inexpensive, power consumption is of critical importance and influences the life expectancy of the network. As a result,

P. William (✉) · A. Badholia · V. Verma · A. Sharma · A. Verma
Department of Computer Science and Engineering, School of Engineering and Information Technology, MATS University, Raipur, India

© The Author(s), under exclusive license to Springer Nature Singapore Pte Ltd. 2022   925
V. Suma et al. (eds.), *Evolutionary Computing and Mobile Sustainable Networks*,
Lecture Notes on Data Engineering and Communications Technologies 116,
https://doi.org/10.1007/978-981-16-9605-3_65

the sensor node consumes more energy throughout the activities of data collection, processing, transmission, receiving, and forwarding There are a lot of factors that affect the energy efficiency of a WSN. These include the WSN's architecture, topology design, routing protocol, MAC protocol, and data aggregation techniques [1]. WSNs have been referred to be "energy efficient" and "network lifetime."

Energy Efficiency: Extending the WSN procedure as far as feasible can help save energy. Topology control protocol is energy-efficient, but when it is feasible to increase the network lifespan of WSNs, that is considered an additional benefit. Each sensor node's energy consumption should be reduced as they provide identical purposes.

Network Lifetime: The network lifetime, which is equal to the number of data collecting rounds or the overall life in minutes until the first sensor node in a WSN dies, is called the "Network Lifetime."—WSN applications maintain that every sensor node should operate in the network, with lifespan defined as the time it takes for one node to die.

In this area, WSNs are now confronting a significant research challenge. For these reasons, it has been argued that saving energy is critical: A battery with a capacity of 200 mAh is good for a period of many months after deployment. If the WSN has been constructed and deployed in an inaccessible area, extending the battery life of the network is a need.

Network layer protocols such as routing techniques have proven critical to WSNs because of the high degree of activity and energy consumption that is associated with them. A number of routing methods have been developed to increase the energy efficiency of WSNs [2, 3]. In terms of energy efficiency, cluster-based routing protocols outperformed other methods. Other than cluster formation, a number of approaches for optimum CH selection and cluster formation have also been proposed, and they are all NP-hard. Using clustering techniques, the study of data aggregation has also been researched. Data duplication occurs at the CH node, necessitating the collection and merger of duplicate data in order to help limit the number of transmissions and energy consumption. Using the CH node, the data was combined into an aggregate.

It has been challenging to provide estimates that are helpful across a broad variety of applications. So, WSN architects must deal with data collection, information quality that cannot be swayed and limitations such as hub grouping and energy-efficient direction as well as event planning and deficiency detection. To create artificial intelligence (computer-based intelligence), machine learning (ML) was originally suggested in the late 1950s [4]. Later, its core developed into a more powerful and computationally viable set of calculations. As many industries have seen advances in AI, applications like bioinformatics and voice recognition have used AI systems to perform relapse detection and thickness estimation, as well as spam identification and computer vision. A significant variety of issues connected with WSNs may be categorized as progress or exhibiting concerns since WSNs were first introduced. This is done by sifting through huge amounts of data and utilizing design recognition, factual and scientific procedures to discover important new linkages, instances, and patterns that are frequently previously opaque. Information sharing, i.e., the recognition of new wonders, may be helpful, but it may also improve our understanding of old wonders. Finally, artificial intelligence systems may help to build choice guidance devices, and they may also motivate researchers to investigate sensor data collected via wireless sensor networks. We believe that machine learning can be used to solve WSN clustering (i.e., optimal CH selection) and data aggregation problems. Different researchers utilized machine learning methods in a variety of ways, with a range of applications and preferences. For future researchers who want to build on the present findings, these differences and assumptions will be the biggest barrier. General frameworks for WSN machine learning were needed to support WSN machine learning machine learning.

To make this possible, the aim is to implement general machine learning-based WSN algorithms that can balance energy economy and performance while accommodating the different end-user applications. Section 2 briefly covers machine learning-based clustering and data aggregation techniques. Section 3 describes the protocol's design. The experimental findings are given in Sect. 4. Section 5 contains a conclusion and recommendations for further study.

2 Related Works

In this section, an overview of current research on machine learning-based clustering and data aggregation for wireless sensor networks will be given (WSNs).

A. ML-Based Clustering

Authors [5–18] make available a diverse range of machine learning-based methods for optimum clustering and CH selection. He also offers a real IoT-organized design for comparing the proposed convention to widely utilized ones in [5] which proposes a productive cross breed vitality conscientious grouping correspondence convention for the registration of green IoT systems. Determining how to best use the hub's vitality material extends both the system's life and its ability to send packages to the base station. On the basis of a weighted race probability, Hy-IoT selected a Cluster-head that took into account the location's heterogeneity dimension When the inventor returned to [6], he created Internet of Things (IoT) gadgets that are particularly helpful for a wide range of purposes. In order to guarantee the WSN-based IoT system's dependable data transmission under these conditions, the system chooses an IoT gadget group leader from among the numerous IoT gadgets in the system. Using Fuzzy C-Means (FCM) bunching count, they were able to determine the most productive group head. The eight most notable IoT-related problems and their primary concern is the subject of vitality proficiency, are all ideas for potential solutions. One of the attractive aspects of the Internet of Things (IoT) is its capacity to integrate sensors and actuators into the natural environment, as well as the ability to exchange data across several phases to provide a unified, complete picture.

Protocols for IoT applications are addressed in [8–14]. Furthermore, there are proposals for IoT-enabled WSNs. They developed the Energy-Efficient Content-Based Routing (EECBR) standard for WSNs in order to help them use less energy. The importance of implementing green IoT is discussed in [9, 10]. An awesome morning In order for IoT to be extensively embraced, various energy-efficient communications and improved steering systems must be addressed. In [11] is shown a novel technique for the conveyance of open information from diverse cloud and heterogeneous assets using an assessed open detection framework. At the heart of the project is a chain of free-market activity created by the open data contained in mobile phones. In an IoT and Smart City setting, Device-to-Device (D2D) transfers were presented by the inventor in [12]. So, you get a better return on your investment and less energy use. A single device combines data from a few adjacent gadgets and sends it to the cell station, instead of each gadget sending information individually. Using standard WSN protocols to provide device-to-device communication in the IoT was shown in [13]. Using the fluffy c-implies bunching approach, a method for estimating the information total in WSNs was presented in recent research [14]. They used a c-implies method to get similitude data that would allow them to do so. Fluffy C methods were used to assign sensor data categories based on the data's proximity to related information. The abnormality was discovered while working on an aid degree.

A routing system that utilizes machine learning was suggested in [15] and implemented as an energy-efficient routing method. Energy efficiency may be accomplished via the application of reinforcement learning. The first step is to use an innovative technique for clustering the network, and the next step is to use a linked graph to set up the network. The Q-value parameter of reinforcement learning is then used to transfer data. The article also discusses a new machine learning-based sensor node validation method. Validation sensors in the domain may be implemented using spectral clustering. The simplest method for grouping sensors based on their location was employed. The author of [17] has recently proposed MLProph, a new routing protocol for OppNets that incorporates machine learning (ML) techniques. In this case, the results were estimates provided by using machine learning methods like neural networks and decision trees. Predictability, the ability to be predicable, was heavily relied on while training the machine learning model. Factors such as node popularity, the amount of energy required to utilize it, and its mobility were all taken into consideration. Presented a distributed cluster organization approach that does not need a centralized processor (CPU). Distributed eigenvector computing along with distributed K-means clustering is used in this method. Laplacian eigenvector computation is done via distributed power iteration.

B. **ML-Based Data Aggregation**

In [19], a new technique for calculating the Center at Nearest (CNS) was presented. A central aggregator receives the data from each center that has identified an event as quickly as feasible. The aggregator is the closest center to the sink when an event is recognized (under bounces). Further proposed characteristic foundation management was facilitated by the Directed Diffusion calculation in [20]. It is possible to painstakingly complete information when these two routes cross at a transitional center point.

It was in [21] that the Greedy Incremental Tree (GIT) technique was introduced. For example, an excellent way to develop essentiality is produced by the GIT calculation which eagerly combines different sources onto the created method.

Using fictional neural networks, Authors [22–24] proposes a technique for remote sensor frameworks to aggregate information. According to its creators, the Information Combination Tree was intended to decrease pack stream and regenerate leaf centers rapidly. By giving [25] an arrangement of collaborative information collecting and an innovation in coordination, as well as a technique for network premise guidance and aggregator selection, the author was able to balance low essentiality distribution, and low idleness while still meeting quality standards. Investigating the need for communication between a combination center and each sensor.

For remote sensor frameworks, the author of [26] presented a target tracking instrument based on quantized improvements and Kalman filtering. Data acquired via the transfer hub may be merged at once by adding a delay interval, decreasing energy usage. Assuring the information's quality.

In [27], the author suggests various QoS (quality of service) measures for the length of time spent on information. To guarantee that above-QoS values are maintained even when subtle differences exist, this method has also been thoroughly tested. For similar reasons as tree-based approaches, bunch-based designs have many levels of structure. As a consequence of these methods, hubs have been segmented into groups.

3 Design of Proposed ML Protocol

A suggested SCADA-ML protocol for WSNs is discussed in this part, with a focus on machine learning-based grouping and data aggregation. Using machine learning techniques, SCADA-ML provided broader approaches for prolonging the network's lifetime and improving QoS performance. Among the two contributions to the SCADA-ML routing protocol are:

ML-based Optimum CH Selection (ANN): To build the optimal CH selection issue for each group, a neural network architecture (input layer, hidden layer, and yield layer) is utilized. Contributions to the ANN are produced for each sensor hub's remaining energy, good routes to the BS, and data transmission allocation. Because the hidden layer is capable of selecting CH in the yield layer, we can see how beneficial it is to utilize ANN's flexible learning to choose CH in the yield layer. Data is exchanged between and among groups once clustering has been completed.

Efficient Data Aggregation Using Machine Learning: Because of the inherent possibility for duplicated information being broadcast at the CH hub, the second commitment is focused on utilizing Independent Component Analysis, a Machine

Learning technique, to efficiently aggregate data from each bunch's CH hub to minimize energy usage (ICA). To gather similar data, use data aggregation across sizes
that have comparable data. ICA reduces the amount of redundant data by using a
different standard of entropy [28, 29].

A. **ML-based Data Transmission and Clustering**
     Our machine learning-based clustering cost measure is based on residual
energy and distance to the BS node, as opposed to the one we developed for
route design, which included just distance. As a result, the use of dense regions
to find CHs leads to overpopulation in heavily inhabited areas [30]. Conversely,
it is uncommon to find CHs in less populated areas. Shorter network life results
from using more costly sensors in these circumstances. ANN-based method is
used to determine the CHs for each cluster. Figure 1 illustrates the CH election
using an ANN-based architecture.

Our research includes the ANN because of its capacity to manage a range of
problems owing to its better learning capabilities, with the goal of improving the CH
election process on a periodic basis using it. Other machine learning techniques are
not as flexible as the ANN. It is also not as quick or as versatile as other approaches.
As shown in Fig. 1, the two-layer feed-forward neural network is used to select the
best CH node from a collection of sensor nodes.

B. **ML-based Data Aggregation**
     As an added bonus, the SCADA-ML routing protocol will benefit from this
addition. Data aggregation results in a decreased number of transfers, further
optimizing energy efficiency and quality of service (QoS) performance.

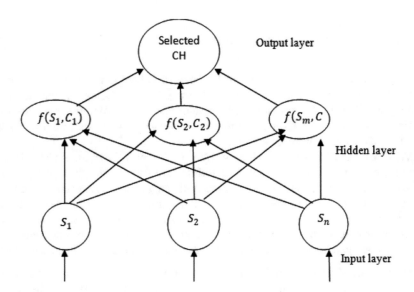

Fig. 1 ANN base CH selection

For data aggregation, we used a machine learning-based approach. In this research, clustering is done using an ANN-based approach after the nodes have been installed. Data correlation is also used during the data aggregation phase to determine the data similarity between sensor nodes. The preset value may be used to assess if two data items are in the same cluster [31, 32]. It is assumed that two data items are part of the same cluster if the correlation value reaches a certain threshold value. To avoid information loss, data aggregation is performed to only comparable data clusters. Without using data aggregation, data clusters with disparate data conduct data transfer activities.

The Independent Component Analysis (ICA) method is introduced in this study for cluster-based data aggregation and is used to cluster comparable data. The ICA that the CH node applies. The ICA approach outperforms other techniques such as PCA and compression. It utilizes the idea of differential entropy to reduce mutual information [33, 34]. The sink node receives aggregated data from comparable clusters. Computation and energy consumption may be reduced as a consequence of the decreased number of aggregating processes. Figure 2 shows the ICA-based data aggregation used by the SCADA-ML protocol.

4 Performance and Analysis

The simulation results for the SCADA-ML protocol are split into two sections: this part provides the results for the protocol, and the next section describes how the simulation results are generated. Section A provides the findings of a comparison research, in which machine learning-based clustering methods were used to compare to more contemporary methodologies. In Sect. B, we study the current machine learning-based data aggregation methods and compare our methodology to it. The performance of a network is measured by the average throughput, latency, PDR, energy consumption, and lifespan.

A. ML-based Clustering Protocols Evaluation

According to the researchers, "SCADA-ML is compared to existing machine learning-based clustering protocols [35, 36].

It is examined in this part how well the proposed SCADA-ML (which does not include machine learning-based data aggregation) performs in comparison to other machine learning-based clustering methods. From Figs. 3 and 4, it can be seen that as the number of sensor nodes utilized in the experiment increases, the average throughput and PDR performance rise. As a consequence of these findings, it is clear that increasing density results in improved throughput and PDR performance for all machine learning-based clustering methods. Due to the novel clustering and data transmission methods presented, the SCADA-ML protocol outperformed previous approaches in terms of throughput and PDR. Along with optimum CH selection, SCADA-ML places a premium on effective data transmission while taking energy and distance limitations into account. Among the various machine learning-based

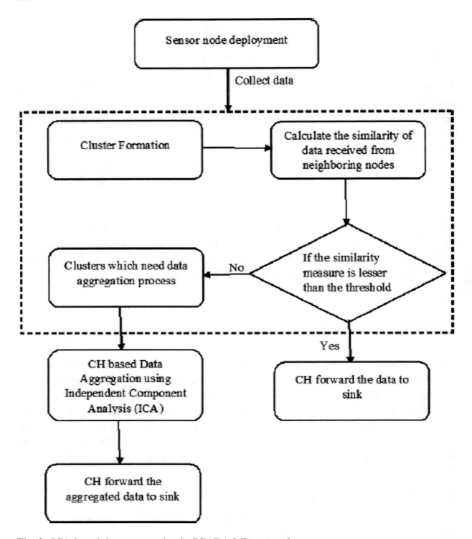

Fig. 2 ICA-based data aggregation in SCADA-ML protocol

clustering techniques, the recently developed K-means algorithm-based clustering outperforms Q-learning.

This figure shows the results of the computation for the average end to end latency by using machine learning-based clustering methods. The time it takes to establish new connections as well as the growing amount of data being transmitted causes latency to increase substantially as the number of sensor nodes increases. When compared to existing methods, the SCADA-ML protocol exhibits a reduction in

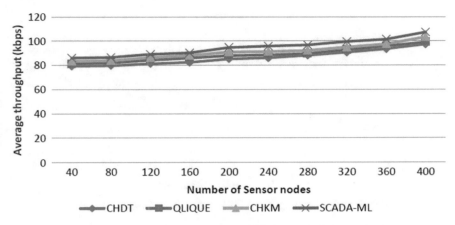

Fig. 3 Average throughput analysis versus density of ML-based clustering methods

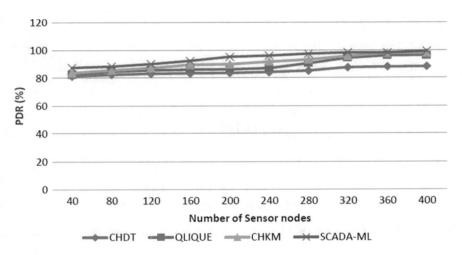

Fig. 4 PDR analysis versus density of ML-based clustering methods

latency since data transmission is accomplished via the use of a strong connection optimization mechanism in combination with optimal CH selection. Another interesting result from these experiments is that the QLIQUE protocol has a lower delay performance than the CHKM protocol, which is due to the fact that Q-learning operations take less time than K-means operations (Fig. 5).

Figures 6 and 7 illustrate the average energy usage and network lifespan for machine learning-based clustering techniques. On the other hand, as energy use grows, the network lifetime declines. The efficiency findings from SCADA-ML show that compared to earlier protocols, it was better in regards to energy efficiency. This is due to the energy-aware cost function and the excellent CH selection that

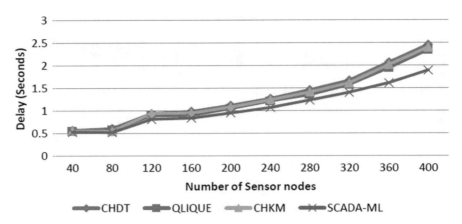

Fig. 5 Average end to end delay analysis versus density of ML-based clustering methods

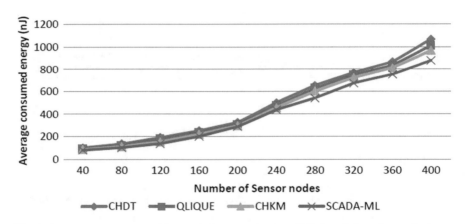

Fig. 6 Average energy consumption analysis versus density of ML-based clustering methods

was made, where energy efficiency was prioritized. No other machine learning-based clustering approaches take into account energy-efficient data transfer while designing the system.

B.   Data Aggregation Protocols Evaluation based on ML

Self-organizing maps (CODA) and principal component analysis (DAPCA) are evaluated with SCADA-ML data aggregation to find their relative value.

Machine learning-based aggregation methods, as well as clustering computations, are used in the final SCADA-ML model reviewed in this part of the article. CODA and DAPCA, two machine learning-based aggregation techniques that have been around for a while, are compared with SCADA-ML. Each aggregation technique has a different number of sensor hubs, as shown in Figs. 8 and 9. In all of the techniques examined, throughput and PDR exhibits rise as thickness increases. If

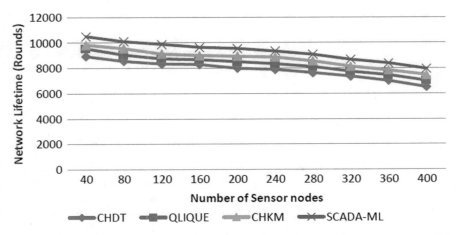

Fig. 7 Network lifetime analysis versus density of ML-based clustering methods

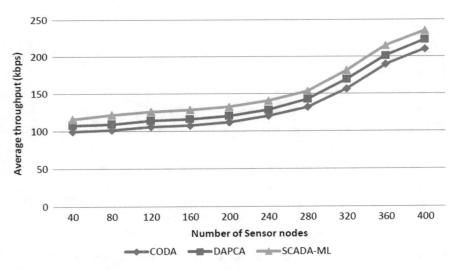

Fig. 8 Average throughput analysis versus density

you compare the SCADA-ML convention to other techniques, you will find that it increases throughput and PDR. Why? Unlike previous technologies, because of its knowledge of reducing transmissions, and the ANN-based method for optimum CH selection, which ensures stable clusters in the network. The enhanced SCADA-ML displays are enhanced by the efficient data transmission made possible by connection cost minimization (Fig. 10).

According to Fig. 10, a variety of machine learning-based aggregation techniques have an average delay. Since it takes so long to establish connections and transmit so many data packets, the delay performance declines with an increase of sensor nodes.

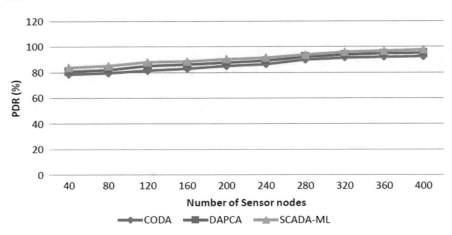

Fig. 9 PDR analysis versus density of ML-based aggregation methods

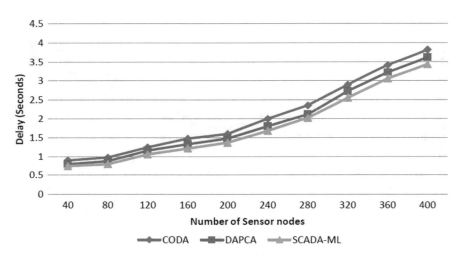

Fig. 10 Average end to end delay analysis versus density of ML-based aggregation methods

Due to the ICA technique employed in the SCADA-ML protocol, SCADA-delay ML's performance is better than CODA and DAPCA.

Figuring out each WSN's energy usage using all the machine-learning-based aggregation techniques is shown in Fig. 11. Several factors contributed to this improvement, including an energy-aware connection cost function, optimum CH selection with energy as a major component, and efficient data aggregation algorithms. SCADA-ML outperformed previous protocols in terms of energy efficiency.

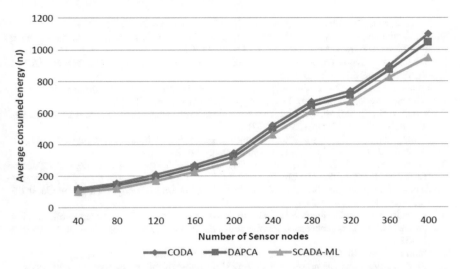

Fig. 11 Average energy consumption analysis versus density of ML-based aggregation methods

5 Future Work and Conclusions

In this paper, the design of the SCADA-ML protocol is given and explained in detail in detail. As part of the SCADA-ML process of optimal CH selection and cluster creation, the ANN technique was effectively utilized. Sensor node features such as residual energy and geographical distance are used to compute the neuron weights. Using a link-cost-based optimal route selection technique, data transfers between and within clusters are carried out to ensure the dependability and QoS performance of WSNs. An ICA-based data aggregation approach was created in SCADA-ML in order to reduce energy consumption, and it performs better than current machine learning data aggregation techniques. Comparing the simulation results for different densities and data rates, we can see that machine learning-based clustering methods have a much higher SCADA-scalability and reliability. Throughput, latency, PDR, and energy efficiency are all improved using SCADA-ML.

References

1. Cagalj, M., Hubaux, J.-P., Enz, C.C.: Energy-efficient broadcasting in all-wireless networks. Wireless Netw. **11**(1/2), 177–188 (2005)
2. Chen, Y.P., Wang, D., Zhang, J.: Variable-base tacit-communication: a new energy efficient communication scheme for sensor networks. In: Proceedings of the First International Conference in Integrated Internet Ad Hoc and Sensor Networks, InterSense 2006, Nice, France, 30–31 May (2006)
3. Chen, Y.P., Liestman, A.L., & Liu, J., Energy-efficient data aggregation hierarchy for wireless sensor networks. In: Proceedings of 2nd International Conference on Quality of Service in

Heterogeneous Wired/Wireless Networks (QShine '05), Orlando (2005)

4. Sundararaman, B., Buy, U., Kshemkalyani, A.D.: Clock synchronization for wireless sensor networks: a survey. J. Ad Hoc Netw. **3**, 281–323 (2005)

5. Sadek, R.A.: Hybrid energy aware clustered protocol for IoT heterogeneous network (2018). https://doi.org/10.1016/j.fcij.2018.02.003

6. Reddy, P.K., Babu, R.: An Evolutionary Secure Energy Efficient Routing Protocol in Internet of Things. Vellore Institute of Technology University, Vellore (2017)

7. Vellanki, M., Kandukuri, S.P.R., Razaque, A.: Node level energy efficiency protocol for Internet of Things. J. Theor. Comput. Sci. (2017)

8. Samia Allaou, C.: Energy-efficient content-based routing in Internet of Things. J. Comput. Commun. **3**, 9–20 (2015). Published Online December 2015 in SciRes

9. Shaikh, F.K., Zeadally, S., Exposito, E.: Enabling technologies for green Internet of Things. IEEE Syst. J. **11**(2), 983–994 (2017)

10. Zhu, C., Leung, V.C.M., Shu, L., E.C.-H. Ngai, Green internet of things for smart world. IEEE Access **3**, 2151–2162 (2015)

11. Al-Fagih, A.E., Al-Turjman, F.M., Alsalih, W.M., Hassanein, H.S.: A priced public sensing framework for heterogeneous IoT architectures. IEEE Trans. Emerg. Top. Comput. **1**, 133–147 (2013)

12. Orsino, A., Araniti, G., Militano, L., Alonso-Zarate, J., Molinaro, A., Iera, A.: Energy efficient IoT data collection in smart cities exploiting D2D communications. Sensors **16**, 836 (2016)

13. Bello, O., Zeadally, S.: Intelligent device-to-device communication in the Internet of things. IEEE Syst. J. **10**, 1172–1182 (2016)

14. Wan, R., Xiong, N., Hu, Q., Wang, H., Shang, J.: Similarity-aware data aggregation using fuzzy c-means approach for wireless sensor networks. EURASIP J. Wireless Commun. Netw. **2019**, 59 (2019)

15. Kiani, F., Amiri, E., Zamani, M., Khodadadi, T., Abdul Manaf, A.: Efficient intelligent energy routing protocol in wireless sensor networks. Int. J. Distrib. Sens. Netw. (2015)

16. Abdo, M.T.N., Pawar, V.P.: Machine learning approach for sensors validation and clustering. In: International Conference on Emerging Research in Electronics, Computer Science and Technology (2015)

17. Sharma, D.K., Dhurandher, S.K., Woungang, I., Srivastava, R.K., Mohananey, A., Rodrigues, J.J.P.C.: A machine learning-based protocol for efficient routing in opportunistic networks. IEEE Syst. J. **12**(3) (2018)

18. Muniraju, G., Zhang, S., Tepedelenlioğlu, C., Banavar, M.K.: Location based distributed spectral clustering for wireless sensor networks. In: Sensor Signal Processing for Defence Conference (SSPD) (2017)

19. Krishnamachari, B., Estrin, D., Wicker, S.B.: The impact of data aggregation in wireless sensor networks. In: Proceedings of 22nd Int'l Conference on Distributed Computing Systems (ICDCSW 02), pp. 575–578 (2002)

20. Intanagonwiwat, C., Govindan, R., Estrin, D., Heidemann, J., Silva, F.: Directed diffusion for wireless sensor networking. IEEE/ACM Trans. Netw. **11**(1), 2–16 (2003)

21. Intanagonwiwat, C., Estrin, D., Govindan, R., Heidemann, J.: Impact of network density on data aggregation in wireless sensor networks. In: Proceedings of 22nd Int'l Conference on Distributed Computing Systems, pp. 457–458 (2002)

22. Nakamura, E.F., de Oliveira, H.A.B.F., Pontello, L.F., Loureiro, A.A.F.: On demand role assignment for event-detection in sensor networks. In: Proceedings of IEEE 11th Symposium on Computers and Communications (ISCC '06), pp. 941–947 (2006)

23. Madden, S., Franklin, M.J., Hellerstein, J.M., Hong, W.; Tag: a tiny aggregation service for Ad-Hoc sensor networks. In: ACM SIGOPS Operating Systems Review, vol. 36, no. SI, pp. 131–146 (2002)

24. Sun, L.Y., Huang, X.X., Cai, W.: Data aggregation of wireless sensor networks using artificial neural networks. Chin. J. Sens. Actuat. **24**(1), 122–127 (2011)

25. Aikaraki, J.N., Uimustafa, R., Kamal, A.E.: Data aggregation and routing in wireless sensor networks: optimal and heuristic algorithms. Comput. Netw. **53**(7), 945–960 (2009)

26. Xu, J., Li, J.X., Xu, S.: Data fusion for target tracking in wireless sensor networks using quantized innovations and Kalman filtering. Sci. China Inf. Sci. Ed. **55**(3), 530–544 (2012)
27. Li, H., Yu, H.Y.: Research on data aggregation supporting QoS in wireless sensor networks. Appl. Res. Comput. **25**(1), 64–67 (2008)
28. Chen, J.I.Z.: Optimal multipath conveyance with improved survivability for WSN's in challenging location. J. ISMAC **2**(02), 73–82 (2020)
29. Bhalaji, N.: Cluster formation using fuzzy logic in wireless sensor networks. IRO J. Sustain. Wireless Syst. **3**(1), 31–39 (2021)
30. Raj, J.S.: Machine learning based resourceful clustering with load optimization for wireless sensor networks. J. Ubiquit. Comput. Commun. Technol. (UCCT) **2**(01), 29–38 (2020)
31. Polastre, J., Szewczyk, R., Culler, D.: Telos: enabling ultra-low power wireless research. In Proceedings of International Symposium on Information Processing in Sensor Networks, pp. 364–369 (2005)
32. Ahmed, G., Khan, N.M., Khalid, Z., Ramer, R.: Cluster head selection using decision trees for Wireless Sensor Networks. In: IEEE International Conference on Intelligent Sensors, Sensor Networks and Information Processing (2008)
33. Lee, S., Chung, T.: Data Aggregation for Wireless Sensor Networks Using Self-organizing Map. Springer, Berlin (2005)
34. Morell, A., Correa, A., Barceló, M., Vicario, J.L.: Data aggregation and principal component analysis in WSNs. IEEE Trans. Wireless Commun. **15**(6), 3908–3919 (2016)
35. Forster, A., Murphy, A.L.: CLIQUE: role-free clustering with Q-learning for wireless sensor networks. In: 29th IEEE International Conference on Distributed Computing Systems (2009)
36. Muniraju, G., Zhang, S., Tepedelenlioglu, C., Banavar, M.K.: Location based distributed spectral clustering for wireless sensor networks. In: IEEE Sensor Signal Processing for Defence Conference (SSPD) (2017)

Improved Reranking Approach for Person Re-identification System

C. Jayavarthini and C. Malathy

Abstract Safety is ensured in all the public places with the help of surveillance cameras. It can be used to find any person involved in crime or lost person. Human operators do this job by enabling correspondence between images of same person captured across diffrent cameras. Automation of the same is called as person re-identification (PRID). One of the major challenges of the PRID process is to determine the representation of images which discriminates the person identity irrespective of the view-points, poses, illumination variations, and occlusions. With this motivation, a new deep learning-based inception network (DLIN) with ranking is proposed for person re-identification (PRID) in surveillance videos. The DLIN technique mainly uses Adadelta with Inception-v4 model as feature extraction technique in which the hyperparameters of the Inception-v4 model are optimally tuned by the use of Adadelta technique. The features are extracted from the probe and gallery images. Euclidean distance-based similarity measurement and expanded neighborhood distance reranking (ENDRR) is employed for determining the similarity and ordering the output of the PRID process along with Mahalanobis distance. A wide range of simu-lations are carried out using CUHK01 benchmark dataset and the experimental outcomes ensured the goodness of the DLIN technique over the other existing techniques.

Keywords Person reidentification · Deep learning · Similarity measurement · Surveillance · Reranking process

C. Jayavarthini (✉) · C. Malathy
Department of Computer Science and Engineering, SRM Institute of Science and Technology, Chengalpattu, Tamilnadu, India
e-mail: jayavarc@srmist.edu.in

C. Malathy
e-mail: malathyc@srmist.edu.in

1   Introduction

Due to the familiarity of the surveillance camera, security observation models are widely used in almost all public areas like airport, hospital, educational institutions, government offices, etc. These systems include several camera linked to an operation center [1]. The images captured from cameras are shown to human operators to examine and carry out several processes namely, detection, recognition, and tracking. It is highly difficult since increase in number of camera results in an increase in number of persons appear in the camera. Hence, person re-identification (PRID) becomes a hot research topic in recent days [2]. Figure 1 illustrates the person being monitored by network of surveillance cameras fixed in the building [3]. The numbers are the 'id' assigned to the human present in the building. Dotted lines with arrow marks show the direction of human movement. Same person is captutred by multiple cameras. If a person is captured by camera 1 first time, a new id is assigned to the person. When the same person appears in camera 2, same id assigned by camera 1 should be re-assigned to the person instead of generating a new one. This process is called as person re-identification.

The person re-identification (PRID) model will receive the image or video as input to match the people with many images or videos. The general process invovled in the PRID is shown in Fig. 2. Owing to the poor quality of the surveillance videos, the conventional recognition approaches like biometrics or iris are not feasible to attain precise outcomes. Besides, the differences in viewpoint and illumination across dissimilar cameras causes appearance variations. It might result in misidentification. In addition, occlusions and background cluttering are also the issues faced in the PRID process. A conventional PRID technique includes two major modules namely,

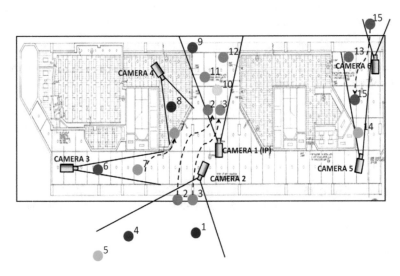

Fig. 1   Building monitored by network of cameras

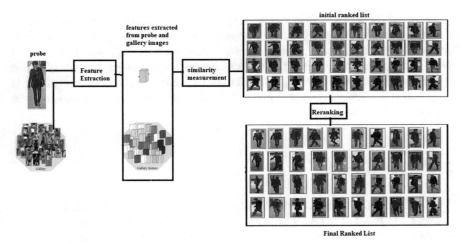

Fig. 2 Process invovled in reranking in person re-identification system

feature extraction and image matching [4]. The existing PRID techniques generally make use of several feature extraction techniques. The SDALF combines 3 types of low level features whereas semantic features are incorporated to the low level features [5]. Other methods include learning of proper distance measures. With the available PRID techniques, a set of probe images of every person is handled individually. For a probe image, the distances from the gallery images are computed. The available PRID techniques commonly develop an effective feature representation or learns a distance measure.

The features employed for re-identification are mostly the versions of color histogram, local binary pattern (LBP) or gabor feature [6, 7]. Few of the models like ELF, SADALF, LOMO utilizes the features which are especially developed for handling the appearance variation. Some of the important measures are Mahalanobis metrics, euclidean metrics. Due to the advent of deep learning (DL) approaches for computer vision applications, several CNN models are presented.

Dissimilar kinds of features and metric learning methods from a label attributes view are summarized [8]. Advanced methods are combined for data enhancement and feature extraction. Besides, the simulation analysis also take place on the metric learning approaches. An unsupervised ranking optimization technique depending upon the discriminant context information analysis was devised [9]. The presented method refined a provided initial ranking through the elimination of the visual ambiguity commonly related to the first rank. It is accomplished by the investigation of the content and context data. The concept of RANkinG Ensembles (RANGE) was discovered [10]. It learnt the data from the ranking list. Particularly, for an off-the-self deep PRID feature representation approach, a pet-probe ranking list is designed and use them for learning the inter-ranking ensemble representation. For mitigating the effect of inevitably false positive gallery, intra ranking ensemble depiction is defined [11, 12].

From the study of existing systems, it is inferred that there is a need to develop an automated system to identify the same pesron across multiple cameras irrespective of change in lighting, background, occlusion and viewpoint. Hence, a new deep learning layer with inception network (DLIN) is proposed to find robust features of probe (person to be searched) across gallery images. Expanded neighborhood reranking is proposed and calculate the similarity and finally rank the matched results to solve person re-identification problem.

2   Proposed Work

A new PRID model is developed using DLIN technique and diagrammatically expressed in Fig. 3. The DLIN technique utilizes Adadelta with Inception-v4 model as feature extraction technique. Then, Euclidean distance-based similarity measurement is used to calculate the similarity score between the probe and gallery images. Initial list of matching images are obtained. For improving the matches, expanded neighborhood distance reranking (ENDRR) with Mahalanobis distance is employed for ordering the output of the PRID process.

2.1   Feature Extraction

At the initial stage, the Inception-v4 model is employed for the extraction of the features from the input images [13]. The old inception version works good in training different blocks. Every repetitive blocks are divided to an equal amount of subnetworks using the overall memory. But, the inception network is simply tuned to represent various changes that are executed based on the amount of filters in distinct layers. These layers do not control the quality of fully trained network. For optimally selecting the training rate, layer size should be optimally set for reaching an efficient tradeoff among processing and different subnetworks. Figure 4 shows the network representation of Inception-v4. On the other hand, in TensorFlow, innovative inception methods are denoted without some replica partitions. Modern memory is applied for optimizing the back propagation by enabling tensors that are critical for gradient processing.

For the residual form of inception network, the lower inception blocks are attainable on standard inception. Each inception block has a filter expansion layer to increase the dimensionality of filtering bank prior to summation. This is done to match with the input depth. It is mandatory to reduce dimension request to inception blocks. Amongst the different developments of the inception techniques, the time step of Inception-v4 becomes significantly slow, due to several layers. The additional variations between the non-remaining and remaining inception models are that batch normalization (BN) is employed on the convention layer, but not on the peak value of the residual synopses. It could be expected to the unique exploitation

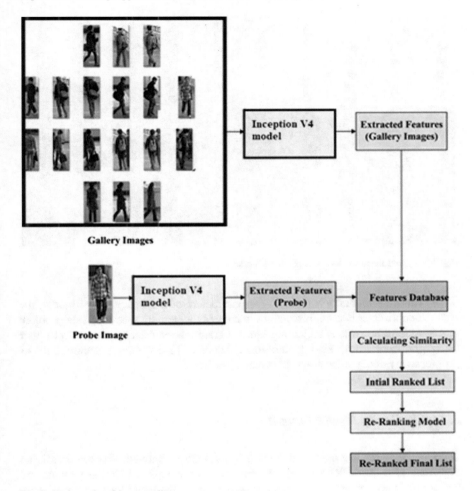

Fig. 3 Proposed architecture diagram

of BN is beneficial, but still the design of BN in TensorFlow requires huge memory and it becomes significant for minimalizing the amount of layers and BN is applied. It could be investigational that if the filter amount go beyond thousands, the residual versions start to give ambiguity and the network "dies" before training, demonstrating that the last layer previous to the average pooling creates 0's on distinct amount of iterations. It is not removed by reducing the rate of learning or having an additional BN layer. It could be denoted that the minimizing remaining previous to attaching to the earlier activation layer is stable during the training. Generally, some of the scaling factor exists in the range of [0.1–0.3] to scale the remaining previous layer to attach to the gathered layer activation.

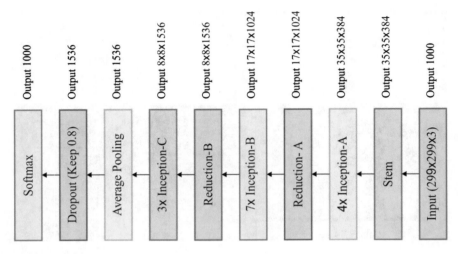

Fig. 4 Layered framework of Inception-v4 model

In order to effectually tune the hyperparameters of the Inception-v4 model, the Adadelta optimizer is used. It manages to change the reducing learning rate problem in Adagrad framework by collecting squared gradients over fixed-size window rather using gradients from existing iterations. Adadelta [14] executes accumulator as progressively decaying average of squared gradients.

2.2 Similarity Measurement

Once the features are extracted, measuring similarity between the features is an important step for a PRID system. Similarity score is high for similar persons and less for different images. Euclidean distance is one of the way to measure similarity between probe features and gallery features [7]. The Euclidean distance between $x_i = [x_{i1}, x_{i2}, ..., x_{in}]$ and $x_j = [x_{j1}, x_{j2}, ..., x_{jn}]$ can be calculated as

$$d(x_i, x_j) = \sqrt{\sum_{k=1}^{n} (x_{ik} - x_{jk})^2}. \tag{1}$$

whereas x_i and x_j are the features extracted between probe and gallery.

2.3   *Reranking Process*

The main objective of reranking the matched results is to get more relevant images in the top rank. Reranking procedure is done by ENDRR technique in which two level neighborhoods are selected in the initial ranking list. Generally, the probe q and N gallery images $G = \{g_i \mid i = 1, 2, \ldots, N\}$, and real distances from 2 images q and g_i can be determined using Euclidean distance, but it does not consider the vector orientation [15] and therefore Cosine distance is employed for the initial ranking process as given below:

$$d(q, g_i) = \frac{x_q x_{g_i}}{\|x_q\| \|x_{g_i}\|} \tag{2}$$

In Eq. (2), x_q and $x(g_i)$ are the gained features of probe q and gallery image g_i. Then, the final distance is re-ranked using the Mahalanobis distance measure, which defines the detection of person from gallery. Consider a pair of feature vector as x_i and x_j from databases X, and the distance can be determined as follows,

$$d_m(x_i, x_j) = \sqrt{(x_i - x_j)^T M (x_i - x_j)}, \tag{3}$$

where $M = \Sigma_I^{-1} - \Sigma_E^{-1}$, Σ_I^{-1} denotes the covariance matrix of interior differences, and Σ_E^{-1} means the covariance matrix of exterior variances.

3   Performance Validation

The proposed DLIN technique is simulated using the CUHK01 dataset [16]. This dataset is available in public. It includes a set of 3884 persons taken from 971 identities correspondingly. Every identity has two instances per camera view. Few sample images from the CUHK01 dataset is shown in Fig. 5. Implemenation is carried out using GPU system with python language. The probe image is highlighted in yellow color. Correct images are highlighted in green color. Incorrect images are highlighted in red color. Screenshot of the results before and after applying DLIN technique can be seen in Figs. 6 and 7, respectively.

Table 1 and Fig. 8 showcases the comparative analysis of the proposed DLIN technique with other techniques under varying ranks. Ranked accuracy metric is used for comparison. Rank 1 accuracy is calculated as the percentage of number of times the top most matched image is the same as ground truth image. Rank 5 accuracy is calculated as as the percentage of number of times the matched image is the same as ground truth image and appear in the top 5 of matched results. Rank 10 accuracy is calculated as as the percentage of number of times the matched image is the same as ground truth image and appear in the top 10 of matched results. Rank 20 accuracy is calculated as as the percentage of number of times the matched image is the

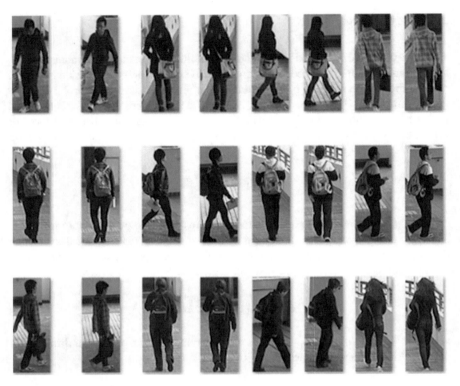

Fig. 5 Sample test images

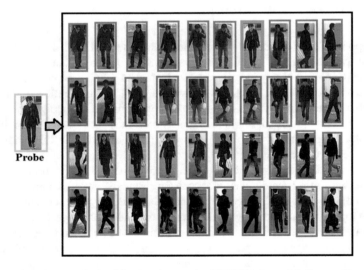

Fig. 6 Resultant images obtained for the given probe (CUHK01 dataset) before ranking

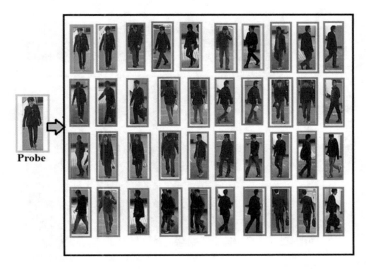

Fig. 7 Resultant images obtained (CUHK01 dataset) after applying DLIN ranking algorithm

Table 1 Results analysis of DLIN with existing techniques [18]

Methods	Rank 1 (%)	Rank 5 (%)	Rank 10 (%)	Rank 20 (%)
Proposed DLIN	89.74	90.95	92.75	94.86
MCC [19]	12.00	33.66	47.96	67.00
ITM [20]	21.67	41.80	55.12	71.31
Adaboost [21]	22.79	44.41	57.16	70.55
LMNN [22]	23.70	45.42	57.32	70.92
Xing's [23]	23.18	45.24	56.90	70.46
L1-Norm [24]	26.73	49.04	60.32	72.07
Bhat	24.76	45.35	56.12	69.31
PRDC [25]	32.60	54.55	65.89	78.30
CDPI	32.41	55.19	67.70	81.54

same as ground truth image and appear in the top 20 of matched results. Among the other compared methods, the proposed DLIN technique has accomplished maximum recognition accuracy under all the ranks. MCC used quadratic Gaussian metric to group similar images to same class and dissimilar images to different class. It worked good in reducing the feature space of the original image. But failed to retrieve the most perfect match. ITM is good in handling different contraints in metric learning in a fast and scalable manner. Its performance is good compared to MCC. Ensemble of localized features were generated and Adaboost was used to select best features to give better results even when there is change in viewpoint and pose. LMNN showed further improvement in KNN classification so that similar images are grouped closely

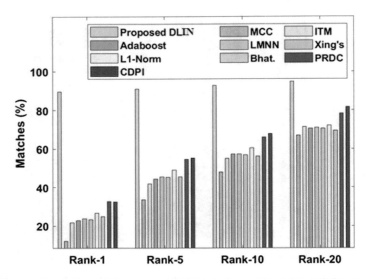

Fig. 8 Comparative analysis of the proposed DLIN technique with existing techniques

and dissimilar images are separated with huge distance. Xing's improvised the results by conentrating on k-means clustering method to increase the distance between dissimilar images and decrease the distance between similar images. L1 norm showed better results by generating occurence matrix based on shape and appearance of the pedestrians. PRDC improved the results by calculating the probability of relative distance between appearance changes of a person [17]. These state of art methods did not use deep learning to solve PRID problem. Using deep learning with inception network, the ability to identify the persons across different camera is improved and the same is shown graphically.

4 Conclusion

The process of enabling correspondence between same person captured by different cameras with the help of computer is increasing research interest due to its application in realtime surveillance. Its really challenging due to change in camera location, clibration, poor illumination, background changes, etc. A new PRID model is developed using deep learning with inception network (DLIN). The proposed model encompasses three stages of operations namely, DLIN feature extraction to extract robust discriminant features of probe and gallery images, Euclidean distance to measure the similarity between probe and gallery images and ENDRR model with Mahalanobis distance to reranking the results. A comprehensive set of simulation analysis is performed on the benchmark CUHK01 dataset. Ranked acuuracy metric is used for performance validation. The propsed model achieved more accuracy under rank1,rank 5, rank 10 and rank 20. The experimental outcomes ensured the goodness of the DLIN technique over other existing techniques.

References

1. Bedagkar-Gala, A., Shah, S.K.: A survey of approaches and trends in person re-identification. Image Vis. Comput. **32**(4), 270–286 (2014)
2. Leng, Q., Hu, R., Liang, C., Wang, Y., Chen, J.: Person re-identification with content and context re-ranking. Multimedia Tools Appl. **74**(17), 6989–7014 (2015)
3. Prosser, B.J., Zheng, W.S., Gong, S., Xiang, T., Mary, Q.: Person re-identification by support vector ranking. In: BMVC, vol. 2, No. 5, p. 6 (2010)
4. Chen, Y., Duffner, S., Baskurt, A., Stoian, A., Dufour, J.Y.: Similarity learning with listwise ranking for person re-identification. In: 2018 25th IEEE International Conference on Image Processing (ICIP), pp. 843–847. IEEE (2018)
5. Nguyen, N.B., Nguyen, V.H., Ngo, T.D., Nguyen, K.M.: Person re-identification with mutual re-ranking. Vietnam J. Comput. Sci. **4**(4), 233–244 (2017)
6. Liu, C., Loy, C.C., Gong, S., Wang, G.: Pop: Person re-identification post-rank optimisation. In: Proceedings of the IEEE International Conference on Computer Vision, pp. 441–448 (2013)
7. Zhong, Z., Zheng, L., Cao, D., Li, S.: Re-ranking person re-identification with k-reciprocal encoding. In: Proceedings of the IEEE Conference on Computer Vision and Pattern Recognition, pp. 1318–1327 (2017)
8. Wu, W., Tao, D., Li, H., Yang, Z., Cheng, J.: Deep features for person re-identification on metric learning. Pattern Recogn. **110**, 107424 (2021)
9. Garcia, J., Martinel, N., Micheloni, C., Gardel, A.: Person re-identification ranking optimisation by discriminant context information analysis. In: Proceedings of the IEEE International Conference on Computer Vision, pp. 1305–1313 (2015)
10. Wu, G., Zhu, X., Gong, S.: Person re-identification by ranking ensemble representations. In: 2019 IEEE International Conference on Image Processing (ICIP), pp. 2259–2263. IEEE (2019)
11. Bindhu, V., Ranganathan, G.: Hyperspectral image processing in internet of things model using clustering algorithm. J. ISMAC **3**(02), 163–175 (2021)
12. Sungheetha, A., Sharma, R.: 3D image processing using machine learning based input processing for man-machine interaction. J. Innov. Image Process. (JIIP) **3**(01), 1–6 (2021)
13. Szegedy, C., Ioffe, S., Vanhoucke, V., Alemi, A.A.: Inception-v4, inception-ResNet and the impact of residual connections on learning. In: Thirty-first AAAI Conference on Artificial Intelligence (2017)
14. Zeiler, M.D.: Adadelta: an adaptive learning rate method (2012). arXiv preprint arXiv:1212. 5701
15. Lv, J., Li, Z., Nai, K., Chen, Y., Yuan, J.: Person re-identification with expanded neighborhoods distance re-ranking. Image Vis. Comput. **95**, 103875 (2020)
16. https://www.ee.cuhk.edu.hk/~xgwang/CUHK_identification.html
17. Yuan, M., Yin, D., Ding, J., Luo, Y., Zhou, Z., Zhu, C., Zhang, R.: A framework with updateable joint images re-ranking for person re-identification (2018). arXiv preprint arXiv:1803.02983
18. Hu, Y., Yi, D., Liao, S., Lei, Z., Li, S.Z.: Cross dataset person re-identification. In: Asian Conference on Computer Vision, pp. 650–664. Springer (2014)
19. A. Globerson and S. T. Roweis. Metric learning by collapsing classes. In NIPS, pages 451–458, 2005.
20. Davis, J.V., Kulis, B., Jain, P., Sra, S., Dhillon, I.S.: Information-theoretic metric learning. In: ICML, pp. 209–216 (2007)
21. Weinberger, K.Q., Blitzer, J., Saul, L.K.: Distance metric learning for large margin nearest neighbor classification. In: NIPS, pp. 1473–1480 (2005)

22. Gray, D., Tao, H.: Viewpoint invariant pedestrian recognition with an ensemble of localized features. In: ECCV, pp. 262–275 (2008)
23. Xing, E.P., Jordan, M.I., Russell, S., Ng, A.Y.: Distance metric learning with application to clustering with side-information. In: NIPS, pp. 505–512 (2002)
24. Wang, X., Doretto, G., Sebastian, T., Rittscher, J., Tu, P.: Shape and appearance context modeling. In: ICCV, pp. 1–8 (2007)
25. Zheng, W.-S., Gong, S., Xiang, T.: Person re-identification by probabilistic relative distance comparison. In: CVPR, pp. 649–656 (2011)

An Insight into Deep Learning Methods for Pulmonary Medical Imaging

Rachna Sethi and Monica Mehrotra

Abstract Artificial neural networks (ANNs) modeled on the human brain are proving to be very close to the human brain for analysis of medical images. This paper explores the work done in the field of lung diseases using deep artificial neural networks. The survey criteria were medical imaging tasks, architectures, types of images, and learning methods used. Broad categories of tasks in medical imaging analysis are the diagnosis, segmentation of pathological features, and quantification of the severity of the disease. Convolutional neural networks (CNNs) have shown the potential of providing a solution for each of these phases. The work in the area of segmentation and quantification is in the early stages. To resolve the problem of domain shift, research has started in domain adaptation-based architectures with semi-supervised and unsupervised learning. Adversarial networks have shown improved performance for data augmentation and segmentation. Thus, an automated system that is capable of generating synthetic images from sparsely annotated lung image datasets, and detecting and assessing the pulmonary disease using those synthetic and actual images seems to be possible in near future.

Keywords Pulmonary disease · Convolutional neural networks · Deep neural network · COVID-19 · Transfer learning · Adversarial networks

1 Introduction

Building intelligent machines are the main priority of artificial intelligence (AI). Intelligent machines are the ones that are capable of performing complex tasks of the human brain like speech recognition, image analysis, and decision-making. Artificial neural networks (ANNs), the learning algorithms modeled on the human brain have shown the significant capability of bringing human-like intelligence in machines.

R. Sethi (✉) · M. Mehrotra
Department of Computer Science, Jamia Millia Islamia, New Delhi, Delhi, India

R. Sethi
Sri Guru Gobind Singh College of Commerce, University of Delhi, New Delhi, India

© The Author(s), under exclusive license to Springer Nature Singapore Pte Ltd. 2022　　953
V. Suma et al. (eds.), *Evolutionary Computing and Mobile Sustainable Networks*,
Lecture Notes on Data Engineering and Communications Technologies 116,
https://doi.org/10.1007/978-981-16-9605-3_67

Like biological neurons, these algorithms process input data through a machine neuron and predict the output. A deep ANN is a multiple-layered network of machine neurons that are dynamic and learn through several iterations of execution on the data. These techniques have proved very helpful in understanding complex functions like image and speech recognition, visual object recognition, and object detection.

The last few years have seen tremendous growth in medical imaging analysis using deep ANN. The reasoning for this is the capability of deep learning approaches to depict complex medical images easily. This paper explores the work done in the field of lung diseases using deep ANN.

2 Overview of Medical Imaging

Medical imaging analysis is a branch of medical science that identifies, treats, and regulates diseases using images of infected organs. There are many different techniques to capture images of human body parts. X-ray technology is used in conventional X-ray, computed tomography (CT), and mammography. Nuclear medicine imaging uses radiopharmaceuticals to see the biological processes occurring in the cells of organisms. MRI and ultrasound are based on the concept of magnetic fields [23]. The features and location of the infectious region assist in the diagnosis of the disease and the precise coordinates of the lesions provide a detailed measure of the disease. The severity and progression of life-threatening diseases can easily be determined and the follow-ups and recoveries can also be traced through these images. Medical imaging analysis, therefore, plays a very important role in medical science.

An automated system of medical imaging analysis can be very beneficial as it will reduce the load on expert radiologists and physicians thereby reducing the time and effort required for the treatment of diseases. The system will be called efficient only if it is capable of performing the following tasks [40]:

(i)   **Segmentation**: Outlining the organ, a pathological characteristic, or an infected region.
(ii)   **Diagnosis**: Classification of an image, e.g., labeling the chest X-ray as either normal or infected with COVID-19.
(iii)   **Quantification**: Risk assessment of the disease. The exact location of lesions is identified to determine the severity and progression.
(iv)   **Optimization**: Enhancement of image resolution and formulation of synthetic images from the input image.

3 Significance of Medical Imaging for Lung Diseases

Lung diseases are abnormalities appearing as areas of increased or decreased density in the lungs. In most lung diseases abnormality appears as increased density or opacity. Depending upon the pattern of opacities lung diseases are identified. Some of the lung diseases are discussed below:

- **Interstitial lung disease (ILD)**: Around two hundred chronic lung disorders with opacities in the form of lung tissue inflammation are collectively known as ILD. It leads to scarring that is commonly referred to as pulmonary fibrosis [2].
- **Chronic obstructive pulmonary disease (COPD)**: Opacities in COPD occur in airways with varying patterns and magnitude. The common COPD diseases are emphysematous lung tissue damage, gross/functional airway disease. Emphysema is a category of COPD characterized by the widening of air spaces and the loss of alveolar walls [39].
- **Pneumonia**: Pneumonia is a very common lung disease in which opacities appear in air sacs. The cause of these opacities may be bacteria, viruses, or fungi.
- **Pleural Effusion**: Pleural effusion has opacity in the form of fluid-filled in the area between the chest and the lung.
- **Lung Cancer**: Lung cancer is the unchecked development of rare or unusual cells in the main part of the lungs near the air sac due to DNA mutations. These pathological growths are considered tumors or nodules that interfere with the normal operations of the lungs.
- **Atelectasis**: A portion of the lung collapses in atelectasis due to the reduction in the volume of air in the alveoli.

4   Image Analysis with Convolutional Neural Network

Deep neural networks follow the biological neuron system and hence can easily be used to perform complex functions of the human brain like image recognition. CNN has shown remarkable performance in the analysis of all kinds of images [1]. CNNs are deep feed-forward ANN capable of identifying images through extraction and labeling of features. The whole concept of CNN is based on the convolutional layer [40]. The convolutional layer consists of multiple sets of neurons or filters that slide over the input volume covering a specific area of the input volume. In this slide, a linear convolution operation is applied and results are stored in an activation map or feature map. Activation layers are generally followed by pooling layers that combine small subgroups of values to sub-sample the previous layer. This helps in reducing the computation and controls overfitting. Classification is done by the fully connected (FC) dense layers at the end of the network.

Learning in CNN is through labeled data therefore it falls under the category of supervised learning algorithms. But supervised CNN is very expensive to train as they require a high volume of labeled data, high-end computation resources, and time for training. For each specific task, there is a need to build a specific model. For better efficiency some other learning methods are also being used to train CNN models as discussed below:

- **Semi-Supervised Learning**

  Semi-supervised learning is used in situations where very few instances are labeled and the remaining instances are unlabeled. From labeled data, supervised loss is calculated. Unsupervised loss computation may be purely based on unclassified data or may be based on both types of data. These two losses are aggregated and then optimised.

- **Transfer Learning**

  Transfer learning for CNN-based image classification is a two-step process. The first step is the training of a model on a large amount of labeled data to acquire knowledge. In second step this base model is fine-tuned to make them more relevant for specific tasks. Training of this model takes much lesser time as compared to a model built from scratch. Also, they require a small learning rate during fine-tuning as compared to the one used in their initial training. Based on this concept many variants of CNN architecture have been developed in recent years that can be used as generic models. These generic models have been trained using a voluminous database of images called ImageNet consisting of accurate and diverse images from day-to-day life [11]. The most commonly used pre-trained CNN models are VGG (variants are VGG16 and VGG19), GoogLeNet/InceptionV3, Residual network (variants are ResNet50, ResNet101, etc.), Xception and MobileNet.

- **Domain Adaptation**

  Domain adaptation is a learning method in which one or more source domains are used for model training, but testing is conducted on a completely different but related target domain. Source domain has labeled data. The target domain data may be fully labeled, partially labeled, or fully unlabeled. One-step domain adaptation has a single domain for training while in multi-step domain adaptation, the model traverses through many source domains for learning [45].

  Domain adaptation concept works very well with deep neural networks as they are capable of giving more transferable representations. A combination of domain adaptation and deep neural network is termed deep domain adaptation. For image analysis, the base neural network for deep domain adaptation is CNN. Techniques used in one-step deep domain adaptation are as follows:

  (i) **Divergence-based Domain Adaptation**: It reduces the divergence between two domains. The most commonly used divergence tests are maximum mean discrepancy, similarity alignment, and contrast domain discrepancy.

  (ii) **Adversarial-based Domain Adaptation**: This is a data augmentation technique in which synthetic adversarial data is generated and the model is trained on real images as well as on synthetic images so that model becomes efficient to label all variations of images in the target domain. This technique can be implemented either through generative adversarial networks (GAN) or domain confusion loss. The concept of GAN came into existence in 2014 [17]. These networks have two deep learning models. The first one known as generator produces synthetic instances and the second one called discriminator works as a classifier, identifying whether a particular instance is actual or synthetic. The domain confusion loss approach works like the discriminator aligning the source and target domain as in the domain-adversarial neural networks.

(iii) **Reconstruction-based Domain Adaptation**: In this auxiliary, reconstruction is used to represent both domains. Deep reconstruction classification network and cycle GAN use this concept.

5 Discussion

Research in pulmonary medical image analysis using deep learning is in a very young stage. Around 50% of publications were in 2020 only followed by 23% in 2019 and 17% in 2018 (Fig. 1). The COVID-19 pandemic brought a sudden increase in research in this area in 2020. Chest X-rays and CT scans have come out to be significant diagnostic tools for COVID-19. 50% of papers focused on COVID-19 in 2020. But the disease that has been the focus of the research in this area is lung cancer because of its life-threatening nature. It is not only important to detect this deadly disease in the early stages but also the visual and quantified assessment of risk helps in treatment. It was also observed that most of the analysis has been done using datasets consisting of either X-rays or CT scans (Fig. 2). X-rays have been used in 44% of publications and CT scans have been used in 52% of publications (Fig. 2). Only a few types of research have used MRI-based datasets for lung cancer. Diseases like pneumonia, tuberculosis, and COVID-19 can be reliably detected using

Fig. 1 Year-wise publications distribution

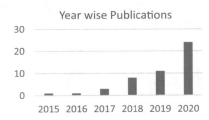

Fig. 2 Image wise publications distribution

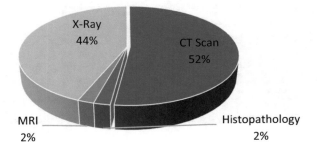

X-rays (Fig. 3). But lung cancer, COPD, ILD require analysis of tissue patterns for segmentation and quantification of disease. CT scans are better options as they

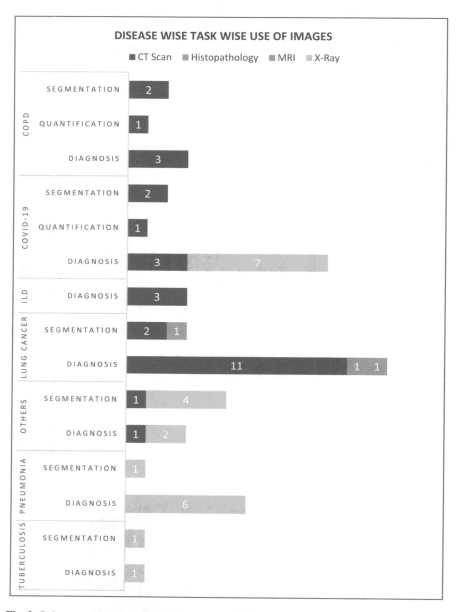

Fig. 3 Pulmonary disease and task wise image type distribution

provide a 3-D view of the lungs and infected areas or opacities can be observed easily.

If medical analysis tasks are considered then the diagnosis is the task that has been covered well (Fig. 3). Research in the segmentation and quantification of pulmonary diseases is still in the early stages. Lung segmentation and quantification are crucial for complete analysis and follow-up of the disease. Besides detection, infected region segmentation helps in assessing the risk and severity of the disease. Several frameworks have been developed for segmenting the different body organs. For segmentation of lung regions, most of the frameworks proposed are based on U-Net and region-based CNN (R-CNN) architecture.

ConvNet for the lung regions segmentation is a modified U-Net-based architecture [39]. VB-Net, a modified 3-D CNN, had two distinct paths for extracting global image features and then integrating those features [38]. Inf-Net, a semi-supervised model to classify COVID-19 infection, modified base CNN by using a decoder to combine the features from high-level layers [13]. Jaiswal et al. classified and segmented chest X-rays through a modified faster region-based convolutional network [22]. In another framework for classification and segmentation by Han et al., feature extraction, classification, and segmentation of CT images were done using two pre-trained CNN architectures in combination with SVM and modified mask R-CNN [19].

Table 1 lists all the references that have proposed CNN-based architecture using supervised learning. The CNN model for ILD tissue pattern classification implemented by Anthimopoulos et al. had five convolutional layers, LeakyReLU activations, a pooling layer, and three dense layers. Classification into seven labels was carried by the last dense layer [2]. CheXNet trained on the ChestX-ray14 public dataset labeled for fourteen types of lung diseases had 121-layers [34]. Abiyev and Ma'aitah did a comparison of three types of networks—(i) CNN with supervised learning, (ii) backpropagation neural networks (BPNNs) with supervised learning, and (iii) competitive neural networks (CpNNs) with unsupervised learning [1].

Varalakshmi et al. created different CNN architectures by changing the hyperparameters for pneumonia classification of the chest X-rays [42]. Bondfale and Bhagwat implemented a CNN framework for multi-label classification of ILD tissue patterns such as normal, ground glass, honeycombing, etc. [6]. FissureNet—a supervised learning framework for COPD disease used a coarse-to-fine strategy for segmenting a thin structure of the image pixels of CT Scans [16]. 3-D CNN architecture for lung nodules classification using CT images of lungs has been proposed by Cengil and Cinar [7], Ozdemir et al. [30]. Du et al. constructed a deep CNN model using the Bayesian optimization algorithm to identify COPD [12]. Another CNN model used a variety of data augmentation algorithms to create a large number of pneumonia datasets for classification [41]. COVID-Net model was trained and tested on a COVIDx dataset consisting of chest X-rays from five publicly available data repositories [44]. The multi-scene deep learning framework (MSDLF), a modified four-channel model used CT scans to segment and classify four-stage cancer nodules [50].

Table 1 List of references that used CNN with supervised learning

Reference	Year	Disease	Type of image	Task	Features
[2]	2016	ILD	CT scan	Classification	Five convolutional layers, LeakyReLU activations, average pooling layer of the size of feature maps, and in the last three FCN layers
[34]	2017	Pneumonia	X-ray	Diagnosis	CheXNet—a dense CNN with 121 convolutional layers
[6]	2018	ILD	CT scan	Diagnosis	Tailored CNN
[42]	2018	Pneumonia	X-ray	Diagnosis	Multiple CNN architectures differing by hyperparameters
[7]	2018	Lung cancer	CT scan	Diagnosis	3-D CNN
[39]	2018	Lung cancer	CT scan	Segmentation	U-Net-based segmentation
[21]	2018	Others	X-Ray	Segmentation	U-Net-based segmentation
[22]	2019	Pneumonia	X-ray	Segmentation and Diagnosis	Region-based CNN
[41]	2019	Pneumonia	X-ray	Diagnosis	Tailored CNN with data augmentation algorithms
[4]	2019	Lung cancer	CT Scan	Diagnosis	Tailored CNN
[16]	2019	COPD	CT Scan	Lobar segmentation	FissureNet developed using two convolutional neural networks through coarse-to-fine cascading
[44]	2020	COVID-19	X-Ray	Diagnosis	Tailored CNN
[19]	2020	COPD	CT Scan	Classification and segmentation	Region-based CNN
[13]	2020	Others	CT Scan	Segmentation	CNN with edge attention module
[49]	2020	Lung cancer	CT Scan	Segmentation and diagnosis	K-Means for segmentation and tailored CNN for classification
[5]	2020	Lung cancer	CT Scan	Diagnosis	Tailored CNN
[30]	2020	Lung cancer	CT Scan	Diagnosis	3-D cascaded CNN

(continued)

Table 1 (continued)

Reference	Year	Disease	Type of image	Task	Features
[12]	2020	COPD	CT Scan	Diagnosis	CNN model optimized it using Bayesian optimization algorithm
[50]	2020	Lung cancer	CT Scan	Diagnosis	Multi-scene deep learning framework (MSDLF) based on CNN
[38]	2020	COVID-19	CT Scan	Segmentation and quantification	3-D CNN in a bottle-neck structure called VB-Net
[18]	2020	COVID-19	CT Scan	Segmentation and diagnosis	U-Net-based segmentation

Table 2 lists all the references that have worked with pre-trained CNN architectures using transfer learning. Many researchers have used pre-trained CNN in the preprocessing stage for feature extraction and then used support vector machine (SVM), K-nearest neighbor (KNN), linear discriminant analysis (LDA), regression, or decision tree for classification. Ming et al. applied transfer learning from AlexNet and VGG16 for feature extraction. AlexNet had eight layers while VGG16 had sixteen weight layers with 3X3 filters [28]. Han et al. used transfer learning from Xception and VGG16 networks on CT images for feature extraction [19]. Ozkaya et al. also tried feature extraction using three pre-trained architectures namely, VGG16, GoogleNet, and ResNet50. The resulting feature sets from these models were combined to obtain higher dimensional characteristics and then rated for training the SVM classifier [31]. Coudray et al. used inception V3 to diagnose lung cancer from histopathology slides with remarkably high accuracy [9]. Varshani et al. extracted features using DenseNet 169, DenseNet 121, ResNet 50, Xception, VGG16, and VGG 19 [43] (Table 3).

Some researches indicate that end-to-end solution for the automated diagnosis of lung diseases is possible with pre-trained architectures using supervised transfer learning. Rahman et al. experimented with supervised transfer learning with nine different models for the classification as TB and non-TB X-ray. They also did segmentation using two variations of U-Net models and used segmented images along with the original for training [33]. Apostolpoulos and Mpesiana applied transfer learning from six pre-trained CNN architectures for the diagnosis of COVID-19 [3]. Shuai Wang et al. fine-tuned the inception model to diagnose COVID-19 from CT scans [47]. Another end-to-end solution for the diagnosis and quantification of emphysema was proposed by Negadhar et al. They modified AlexNet architecture to get multi-label learning. Their model first performs segmentation and then does classification corresponding to different patterns. For quantification, each image is graded as per the size of the patch [29].

Table 2 List of references that used transfer learning with pre-trained CNN architectures

Reference	Year	Disease	Type of image	Task	Features
[9]	2018	Lung cancer	Histopathology	Diagnosis and mutation prediction	Pre-trained inception V3 architecture
[28]	2018	Others	CT Scan	Diagnosis	AlexNet and VGG16 with SVM
[43]	2019	Pneumonia	X-Ray	Diagnosis	DenseNet 169, Densenet 121, ResNet 50, Xception, VGG16 and VGG19 for feature extraction and SVM, KNN, and random forest for classification
[25]	2019	Lung cancer	MRI	Diagnosis and segmentation	Pre-trained CNN architectures used for feature extraction and R-CNN used for classification
[20]	2019	Lung cancer	CT scan	Diagnosis	Used a 3-D CNN pre-trained architecture (on sorts-1 M dataset) and fine-tuned it for lung nodule CT dataset
[35]	2019	Lung cancer	CT scan	Diagnosis and quantification	a lightweight multi-section CNN architecture based on pre-trained MobileNet
[29]	2019	COPD	CT Scan	Diagnosis and quantification	Alexnet with multi-label classification
[3]	2020	COVID-19	X-ray	Diagnosis	Experimented with six pre-trained CNN architectures
[47]	2020	COVID-19	CT scan	Diagnosis	Pre-trained inception V3 architecture
[26]	2020	Pneumonia	X-ray	Diagnosis	CNN framework with 49 convolutions and 2 dense layers that combined residual thought and dilated convolution
[14]	2020	COVID-19	X-ray	Diagnosis	ResNet-based architecture
[31]	2020	COVID-19	CT Scan	Diagnosis	VGG16, GoogleNet, ResNet50 and SVM

(continued)

Table 2 (continued)

Reference	Year	Disease	Type of image	Task	Features
[32]	2020	COVID-19	X-Ray	Diagnosis	Inception for feature extraction, resampling algorithms with multi-class and hierarchical classifiers
[37]	2020	COVID-19	X-Ray	Diagnosis	Pre-trained CNN architectures—Inception, MobileNet, ResNet50, and Xception
[33]	2020	Tuberculosis	X-Ray	Diagnosis, segmentation and visualisation	ResNet18, ResNet50, ResNet101, ChexNet, InceptionV3, Vgg19, DenseNet201, SqueezeNet, and MobileNet

Hussein et al. fine-tuned a pre-trained 3-D CNN for lung nodule CT dataset using multi-task learning supporting supervised and unsupervised learning [20]. Liang and Zheng's model for the detection of childhood pneumonia used residual thought and dilated convolutions [26]. A lightweight multi-section CNN representing the nodules compactly for differentiation and prediction used transfer learning from MobileNet [35]. Christodoulidis et al. used experimented with pre-training from multiple sources. Using multi-task learning, six copies of the proposed framework were pre-trained separately using six different publicly available texture databases. All these models were then fine-tuned independently and in the end combined merged knowledge of these models was used to identify patterns [8].

Dai et al. used adversarial training for lung and heart segmentation in chest X-ray images [10]. In DL-CRC framework, data augmentation was performed by generating artificial images of infected lungs using a GAN. This synthetic data along with actual data was fed into customized CNN model for diagnosis of COVID-19 [36]. Madani et al. applied unsupervised GAN for increasing the input volume and then applied semi-supervised GAN for classification [27]. Another domain adaptation-based framework used semi-supervised as well as unsupervised learning for aligning data across various domains. Chest X-ray-14 was used as a source domain and the target domain contained images collected from various public sources [52].

Adversarial networks have also been used for the classification of pulmonary nodules. Kuang et al. suggested a multi-discriminator, unsupervised learning-based GAN paired with an encoder [24]. A semi-supervised adversarial classification network (SSAC) used both labeled and unlabeled data [48]. Wang et al. applied Wasserstein generative adversarial network (WGAN) for resolving data imbalance and then used simple CNNs for classification [46]. Gaál et al. tried lung segmentation of lungs in chest X-rays using adversarial methods [15].

Table 3 List of references that used CNN with domain adaptation

Reference	Year	Disease	Type of image	Task	Features
[10]	2017	Others	X-ray	Segmentation	Used FCN and the adversarial critic network for segmenting the lung fields and the heart
[27]	2018	Others	X-ray	Diagnosis	Unsupervised Gan for augmentation of chest X-rays images and then semi-supervised GAN for classification
[46]	2019	Lung cancer	CT scan	Diagnosis	WGAN to resolve the issue of data imbalance and then fine-grained classification using normal CNN
[48]	2019	Lung cancer	CT Scan	Diagnosis	A semi-supervised adversarial classification model (SSAC) for both labeled and unlabeled data
[36]	2020	COVID-19	X-ray	Diagnosis	DL-CRC framework—data augmentation by generating synthetic images using (GAN), both actual and synthetic images are fed into customized CNN model for classification
[52]	2020	COVID-19	X-ray	Diagnosis	The proposed model can align the data distributions across different domains
[24]	2020	Lung cancer	CT scan	Diagnosis	GAN-based multi-discriminator unsupervised learning model using encoder
[15]	2020	Others	X-ray	Segmentation	U-Net-based adversarial architecture
[51]	2020	Others	X-Ray	Segmentation	Unsupervised GAN

6 Conclusion

The researches indicate that accurate and efficient deep learning-based automated systems can be developed for pulmonary medical image analysis. CNNs have the potential of providing an end-to-end solution in this area. Annotated medical image datasets are a prerequisite for the training of CNN models. But most of the publicly available datasets are consisting of chest X-rays and CT scans only. Research has also shown that CNN architectures pre-trained on large image datasets can be used on small datasets of lung images with transfer learning and fine-tuning. But these proposed solutions might face performance issues across datasets from different domains. Large differences in the distribution, texture, and scale of pathogens across domains can contribute to false negative outcomes.

The recent COVID-19 pandemic has opened up another challenge. Though the volume of COVID-19 patients was very high it increased the workload on radiologists, hence only small datasets of labeled and annotated images are publicly available. Pre-trained CNN architectures with transfer learning have proved to provide complete end-to-end solutions successfully for these small datasets. However, likely, these models working efficiently on small datasets may not produce the recorded success on a different and larger dataset. Data augmentation approaches using adversarial networks are proving to be efficient in solving data imbalance issues. Semi-supervised learning needs to be explored more to avoid reliance on experts. It was observed that the preferred choices for diagnosis of lung diseases out of all the different types of lung medical images are chest X-rays and CT scans. CT scans score over X-rays for analysis of these diseases as they have better resolution, and provide a 3-D view of the lungs. But the models trained on CT scans and MRI are of no use in countries with weak medical infrastructure as availability of these datasets will be scarce. Also, the chest X-ray exposes only a smaller dosage of radiation and is relatively more affordable. Hence more research is required in the field of study of lung tissue patterns using X-rays.

References

1. Abiyev, R.H., Ma'aitah, M.K.S.: Deep convolutional neural networks for chest diseases detection. J. Healthc. Eng. **2018**, 1–11 (2018). https://doi.org/10.1155/2018/4168538
2. Anthimopoulos, M., et al.: Lung pattern classification for interstitial lung diseases using a deep convolutional neural network. IEEE Trans. Med. Imaging. **35**(5), 1207–1216 (2016). https://doi.org/10.1109/TMI.2016.2535865
3. Apostolopoulos, I.D., Mpesiana, T.A.: Covid-19: automatic detection from X-ray images utilizing transfer learning with convolutional neural networks. Phys. Eng. Sci. Med. **43**(2), 635–640 (2020). https://doi.org/10.1007/s13246-020-00865-4
4. Ardila, D., et al.: End-to-end lung cancer screening with three-dimensional deep learning on low-dose chest computed tomography. Nat. Med. **25**(6), 954–961 (2019). https://doi.org/10.1038/s41591-019-0447-x
5. Arslan, A.K., et al.: An intelligent system for the classification of lung cancer based on deep learning strategy. In: 2019 International Artificial Intelligence and Data Processing Symposium (IDAP). pp. 1–4. IEEE, Malatya (2019). https://doi.org/10.1109/IDAP.2019.8875896
6. Bondfale, N., Bhagwat, D.S.: Convolutional neural network for categorization of lung tissue patterns in interstitial lung diseases. In: 2018 Second international conference on inventive communication and computational technologies (ICICCT), pp. 1150–1154. IEEE, Coimbatore (2018). https://doi.org/10.1109/ICICCT.2018.8473302
7. Cengil, E., Cinar, A.: A deep learning based approach to lung cancer identification. In: 2018 International Conference on Artificial Intelligence and Data Processing (IDAP). pp. 1–5. IEEE, Malatya (2018). https://doi.org/10.1109/IDAP.2018.8620723
8. Christodoulidis, S., et al.: Multisource transfer learning with convolutional neural networks for lung pattern analysis. IEEE J. Biomed. Health Inform. **21**(1), 76–84 (2017). https://doi.org/10.1109/JBHI.2016.2636929

9. Coudray, N., et al.: Classification and mutation prediction from non–small cell lung cancer histopathology images using deep learning. Nat. Med. **24**(10), 1559–1567 (2018). https://doi.org/10.1038/s41591-018-0177-5

10. Dai, W., et al.: SCAN: Structure correcting adversarial network for organ segmentation in chest X-rays. ArXiv170308770 Cs (2017)

11. Deng, J. et al.: ImageNet: A large-scale hierarchical image database. In: 2009 IEEE Conference on Computer Vision and Pattern Recognition, pp. 248–255. IEEE, Miami (2009). https://doi.org/10.1109/CVPR.2009.5206848

12. Du, R., et al.: Identification of COPD from multi-view snapshots of 3D lung airway tree via deep CNN. IEEE Access **8**, 38907–38919 (2020). https://doi.org/10.1109/ACCESS.2020.2974617

13. Fan, D.-P., et al.: Inf-Net: automatic COVID-19 lung infection segmentation from CT images. IEEE Trans. Med. Imaging **39**(8), 2626–2637 (2020). https://doi.org/10.1109/TMI.2020.2996645

14. Farooq, M., Hafeez, A.: COVID-ResNet: a deep learning framework for screening of COVID19 from Radiographs. ArXiv200314395 Cs Eess (2020)

15. Gaál, G., et al.: Attention U-net based adversarial architectures for chest X-ray lung segmentation. ArXiv200310304 Cs Eess (2020)

16. Gerard, S.E., et al.: FissureNet: a deep learning approach for pulmonary fissure detection in CT images. IEEE Trans. Med. Imaging **38**(1), 156–166 (2019). https://doi.org/10.1109/TMI.2018.2858202

17. Goodfellow, I.J., et al.: Generative adversarial networks. ArXiv14062661 Cs Stat (2014)

18. Gozes, O. et al.: Rapid AI development cycle for the coronavirus (COVID-19) pandemic: initial results for automated detection & patient monitoring using deep learning CT image analysis. ArXiv200305037 Cs Eess (2020).

19. Han, T., et al.: Internet of medical things—based on deep learning techniques for segmentation of lung and stroke regions in CT scans. IEEE Access **8**, 71117–71135 (2020). https://doi.org/10.1109/ACCESS.2020.2987932

20. Hussein, S., et al.: Lung and pancreatic tumor characterization in the deep learning era: novel supervised and unsupervised learning approaches. IEEE Trans. Med. Imaging **38**(8), 1777–1787 (2019). https://doi.org/10.1109/TMI.2019.2894349

21. Islam, J., Zhang, Y.: Towards robust lung segmentation in chest radiographs with deep learning. ArXiv181112638 Cs (2018)

22. Jaiswal, A.K., et al.: Identifying pneumonia in chest X-rays: a deep learning approach. Measurement **145**, 511–518 (2019). https://doi.org/10.1016/j.measurement.2019.05.076

23. Ker, J. et al.: Deep learning applications in medical image analysis. IEEE Access **6**, 15 (2018)

24. Kuang, Y., et al.: Unsupervised multi-discriminator generative adversarial network for lung nodule malignancy classification. IEEE Access **8**, 77725–77734 (2020). https://doi.org/10.1109/ACCESS.2020.2987961

25. Li, Y., et al.: Lung nodule detection with deep learning in 3D thoracic MR images. IEEE Access **7**, 37822–37832 (2019). https://doi.org/10.1109/ACCESS.2019.2905574

26. Liang, G., Zheng, L.: A transfer learning method with deep residual network for pediatric pneumonia diagnosis. Comput. Methods Programs Biomed. **187**, 104964 (2020). https://doi.org/10.1016/j.cmpb.2019.06.023

27. Madani, A. et al.: Semi-supervised learning with generative adversarial networks for chest X-ray classification with ability of data domain adaptation. In: 2018 IEEE 15th International Symposium on Biomedical Imaging (ISBI 2018), pp. 1038–1042. IEEE, Washington, DC (2018). https://doi.org/10.1109/ISBI.2018.8363749

28. Ming, J.T.C., et al.: Lung disease classification using different deep learning architectures and principal component analysis. In: 2018 2nd International Conference on BioSignal Analysis, Processing and Systems (ICBAPS), pp. 187–190. IEEE, Kuching (2018). https://doi.org/10.1109/ICBAPS.2018.8527385

29. Negahdar, M., et al.: An end-to-end deep learning pipeline for emphysema quantification using multi-label learning. In: 2019 41st Annual International Conference of the IEEE Engineering in Medicine and Biology Society (EMBC), pp. 929–932. IEEE, Berlin (2019). https://doi.org/10.1109/EMBC.2019.8857392

30. Ozdemir, O., et al.: A 3D probabilistic deep learning system for detection and diagnosis of lung cancer using low-dose CT scans. IEEE Trans. Med. Imaging. **39**(5), 1419–1429 (2020). https://doi.org/10.1109/TMI.2019.2947595

31. Ozkaya, U., et al.: Coronavirus (COVID-19) classification using deep features fusion and ranking technique. ArXiv200403698 Cs Eess (2020)

32. Pereira, R.M., et al.: COVID-19 identification in chest X-ray images on flat and hierarchical classification scenarios. Comput. Methods Programs Biomed. **194**, 105532 (2020). https://doi.org/10.1016/j.cmpb.2020.105532

33. Rahman, T., et al.: Reliable tuberculosis detection using chest X-ray with deep learning, Segmentation and Visualization. IEEE Access **8**, 191586–191601 (2020). https://doi.org/10.1109/ACCESS.2020.3031384

34. Rajpurkar, P., et al.: CheXNet: radiologist-level pneumonia detection on chest X-rays with deep learning. ArXiv171105225 Cs Stat (2017)

35. Sahu, P., et al.: A lightweight multi-section CNN for lung nodule classification and malignancy estimation. IEEE J. Biomed. Health Inform. **23**(3), 960–968 (2019). https://doi.org/10.1109/JBHI.2018.2879834

36. Sakib, S., et al.: DL-CRC: deep learning-based chest radiograph classification for COVID-19 detection: a novel approach. IEEE Access **8**, 171575–171589 (2020). https://doi.org/10.1109/ACCESS.2020.3025010

37. Sethi, R., et al.: Deep learning based diagnosis recommendation for COVID-19 using chest X-rays images. In: 2020 Second International Conference on Inventive Research in Computing Applications (ICIRCA), pp. 1–4. IEEE, Coimbatore (2020). https://doi.org/10.1109/ICIRCA48905.2020.9183278

38. Shan, F., et al.: Lung infection quantification of COVID-19 in CT images with deep learning. ArXiv200304655 Cs Eess Q-Bio (2020)

39. Shaziya, H., et al.: Automatic lung segmentation on thoracic CT scans using U-Net convolutional network. In: 2018 International Conference on Communication and Signal Processing (ICCSP), pp. 0643–0647. IEEE, Chennai (2018). https://doi.org/10.1109/ICCSP.2018.8524484

40. Soffer, S., et al.: Convolutional neural networks for radiologic images: a radiologist's guide. Radiology **290**(3), 590–606 (2019). https://doi.org/10.1148/radiol.2018180547

41. Stephen, O., et al.: An efficient deep learning approach to pneumonia classification in healthcare. J. Healthc. Eng. **2019**, 1–7 (2019). https://doi.org/10.1155/2019/4180949

42. Varalakshmi, P., et al.: Diminishing fall-out and miss-rate in the classification of lung diseases using deep learning techniques. In: 2018 Tenth International Conference on Advanced Computing (ICoAC), pp. 373–376. IEEE, Chennai (2018). https://doi.org/10.1109/ICoAC44903.2018.8939073

43. Varshni, D. et al.: Pneumonia detection using CNN based feature extraction. In: 2019 IEEE International Conference on Electrical, Computer and Communication Technologies (ICECCT), pp. 1–7. IEEE, Coimbatore (2019). https://doi.org/10.1109/ICECCT.2019.8869364

44. Wang, L., Wong, A.: COVID-Net: a tailored deep convolutional neural network design for detection of COVID-19 cases from chest X-ray images. ArXiv200309871 Cs Eess (2020)

45. Wang, M., Deng, W.: Deep visual domain adaptation: a survey. ArXiv180203601 Cs (2018)

46. Wang, Q. et al.: WGAN-based synthetic minority over-sampling technique: improving semantic fine-grained classification for lung nodules in CT images. IEEE Access **7**, 14 (2019)

47. Wang, S., et al.: A deep learning algorithm using CT images to screen for Corona Virus Disease (COVID-19). Infect. Dis. (except HIV/AIDS) (2020). https://doi.org/10.1101/2020.02.14.20023028

48. Xie, Y., et al.: Semi-supervised adversarial model for benign–malignant lung nodule classification on chest CT. Med. Image Anal. **57**, 237–248 (2019). https://doi.org/10.1016/j.media.2019.07.004

49. Yu, H. et al.: Deep learning assisted predict of lung cancer on computed tomography images using the adaptive hierarchical heuristic mathematical model. IEEE Access **8**, 11 (2020)
50. Zhang, Q., Kong, X.: Design of automatic lung nodule detection system based on multi-scene deep learning framework. IEEE Access **8**, 10 (2020)
51. Zhang, Y., et al.: Unsupervised X-ray image segmentation with task driven generative adversarial networks. Med. Image Anal. **62**, 101664 (2020). https://doi.org/10.1016/j.media.2020.101664
52. Zhou, J., et al.: SODA: detecting covid-19 in chest X-rays with semi-supervised open set domain adaptation. ArXiv200511003 Cs Eess (2020)

Traffic Prediction Using Machine Learning

H. R. Deekshetha, A. V. Shreyas Madhav, and Amit Kumar Tyagi

Abstract The paper deals with traffic prediction that can be done in intelligent transportation systems which involve the prediction between the previous year's dataset and the recent year's data which ultimately provides the accuracy and mean square error. This prediction will be helpful for the people who are in need to check the immediate traffic state. The traffic data is predicated on a basis of 1 h time gap. Live statistics of the traffic is analyzed from this prediction. So this will be easier to analyze when the user is on driving too. The system compares the data of all roads and determines the most populated roads of the city. I propose the regression model in order to predict the traffic using machine learning by importing Sklearn, Keras, and TensorFlow libraries.

Keywords Traffic · Regression · Intelligent transport system (ITS) · Machine learning · Prediction

1 Introduction

Machine learning (ML) is one of the most important and popular emerging branches these days as it is a part of artificial intelligence (AI). In recent times, machine learning becomes an essential and upcoming research area for transportation engineering, especially in traffic prediction. Traffic congestion affects the country's economy directly or indirectly by its means. Traffic congestion also takes people's valuable time, cost of fuel every single day. As traffic congestion is a major problem for all classes in society, there has to be a small-scale traffic prediction for the people's sake of living their lives without frustration or tension. For ensuring the country's economic growth, the road user's ease is required in the first place. This is possible

H. R. Deekshetha · A. V. Shreyas Madhav · A. K. Tyagi (✉)
School of Computer Science Engineering, Vellore Institute of Technology, Chennai, Tamilnadu, India

A. K. Tyagi
Centre for Advanced Data Science, Vellore Institute of Technology, Chennai, Tamilnadu, India

only when the traffic flow is smooth. To deal with this, traffic prediction is needed so that we can estimate or predict the future traffic to some extent.

In addition to the country's economy, pollution can also be reduced. The government is also investing in the intelligent transportation system (ITS) to solve these issues. The plot of this research paper is to find different machine learning algorithms and speculating the models by utilizing python3. The goal of traffic flow prediction is to predict the traffic to the users as soon as possible. Nowadays, the traffic becomes really hectic and this cannot be determined by the people when they are on roads. So, this research can be helpful to predict traffic. Machine learning is usually done using anaconda software but in this paper, I have used the python program using command prompt window which is much easier than the usual way of predicting the data [1].

The constructs of this paper consist of ten major sections. These are:

- Abstract, Introduction, Purpose of Traffic Prediction, Problem Statement, Related Work, Overview, Methodology, Software Implementation and Conclusion with Future work

1.1 Purpose of Statement

Many reports of the traffic data are of actual time but it is not favorable and accessible to many users as we need to have prior decision in which route we need to travel. For example, during working days, we need to have daily traffic information or at times we need hourly traffic information but then the traffic congestion occurs; for solving this issue, the user needs to have actual time traffic prediction. Many factors are responsible for the traffic congestion. This can be predicted by taking two datasets: one with the past year and one with the recent year's dataset. If traffic is so heavy, then the traffic can be predicted by referring to the same time in the past year's dataset and analyzing how congested the traffic would be. With the increasing cost of the fuel, the traffic congestion changes drastically. The goal of this prediction is to provide real-time gridlock and snarl-up information. The traffic on the city becomes complex and are out of control these days, so such kind of systems are not sufficient for prediction. Therefore, research on traffic flow prediction plays a major role in ITS [2].

1.2 Problem Statement

To overcome the problem of traffic congestion, the traffic prediction using machine learning which contains regression model and libraries like pandas, os, numpy, matplotlib.pyplot are used to predict the traffic. This has to be implemented so that the traffic congestion is controlled and can be accessed easily. Users can collect the traffic information of the traffic flow and can also check the congestion flow from the start of the day till the end of the day with the time span of one hour data. In this way, users can know the weather conditions of the roads that they would probably opt to take. This also tells the accuracy of the traffic by comparing their mean square errors of the past year's data and the recent year's data. Users can also know how many vehicles are traveling on average by the traffic prediction.

2 Related Works

A person developed the LSTM-based prediction models by using machine learning approaches, which involve structure designing or network training designing and prediction. The next goal is to deal with prediction errors that may occur during the prediction process with deep learning methods. This method is applied to big data that has been collected from the performance measurement system. Then the experiments show that the LSTM model performed well.

Vlahogianni implemented a feeding device knowledge with an artificial neutral network (ANN), but some researchers implemented a completely different spatial associated temporal holdup to the setup. [3–6]

Some researchers worked on the traffic prediction using the neural networks and hybrid by various techniques. Traffic patterns have been identified by grouping them, and the traffic flow has been identified by the neural networks. On other hand, studies conduct two kinds of prediction models:

Models which are not available in online are accessed through on historical information that can be trained when important changes appear in the system as like changes and updating the whole framework,

Online models are measured by historical data, and the system gets updated using usual transport condition achieved through v2v/v21 communication. [7, 8]

Many researchers say that LSTM is more capable compared to shallow machine learning models. Machine learning approaches big data for the prediction problems. With the lots of the data, the real-time traffic flow can be predicted. LSTM is based on the deep learning. RNN may be used to communicate the prior information to the present situation [9–11].

3 Overview

In traffic congestion forecasting, there are data collection and prediction model. The methodology has to be done correctly so that there won't be any flaws while predicting. After data collection, the vital role is the data processing which is to train and test the datasets that are taken as the input. After processing the data, the validation of the model is done by using necessary models. Figure 1 highlights the outline of traffic prediction using machine learning.

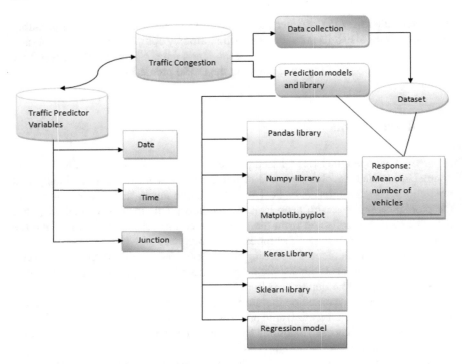

Fig. 1 Outline of the traffic prediction in this paper

4 Methodology

Many researchers have been used various discussed approaches. This paper contains the technique of predicting the traffic using regression model using various libraries like Pandas, Numpy, OS, Matplotlib.pyplot, Keras, and Sklearn.

4.1 Dataset

Traffic congestion is rising a lot these days and factors like expanding urban populations, uncoordinated traffic signal timing, and a lack of real-time data. The effect of the traffic congestion is very huge these days. Data collected in this paper are from the Kaggle website for the implementations of machine learning algorithms using python3 to show outputs in the traffic prediction. Two datasets are collected in which one is the 2015s traffic data which comprises of date, time, and number of vehicles, junction and the rest one is the 2017s traffic data with the same details so as to compare easily without any misconception. The unwanted data has been deleted by preprocessing the data aggregated from 1 to 24 h time interval to calculate traffic flow prediction with each 1 h interval [12].

4.2 Regression Model

Regressor model analysis could even be a mathematical technique for resolving the connection in the middle of one dependent (criterion) variable and one or more independent (predictor) variables. The evaluation yields a foretold value for the benchmark resulting from a sum of scalar vectors of the predictors. The accuracy is measured by computing mean square error. Thus obtaining the expected error from the observed value and also truth value which is equivalent to the standard deviation deployed within the statistical methods [13, 14]. Figure 2 shows the regression model for the traffic prediction.

JupyterLab is a browser-based communal development. JupyterLab is a limber and which can construct and exhibit the user interface to support a far-flung of metadata

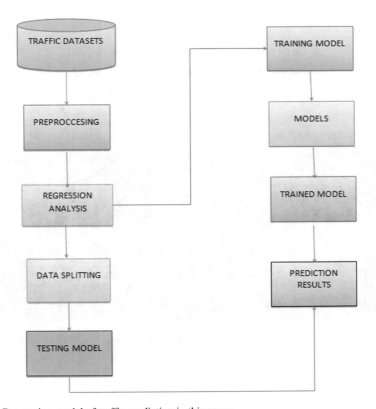

Fig. 2 Regression model of traffic prediction in this paper

in machine learning [15]. Python3 is the status quo environment where the code is implemented in Jupyter notebook. This can be accessed/installed using command prompt. This is done in order to get the access from the local drive. So the Jupyter notebook is installed through command prompt and then a local host is created. The file is accessed though this host and the prediction are done using various libraries and models in the python environment.

5 Software Implementation

5.1 Simulation

The command prompt is the local host in this paper to initialize the jupyter notebook (Fig. 3).

The local host contains the nbextenisons which we modify to our convenience (Figs. 4, 5 and 6).

The below mentioned are the libraries used for the prediction of traffic (Tables 1, 2, 3, 4, 5, 6 and 7).

$$import\ pandas\ as\ pd\ \#\ library\ imported$$
$$import\ numpy\ as\ np\ \#\ library\ imported$$
$$import\ os\ \qquad \#\ library\ imported$$

Fig. 3 Initializing the jupyter notebook through command prompt

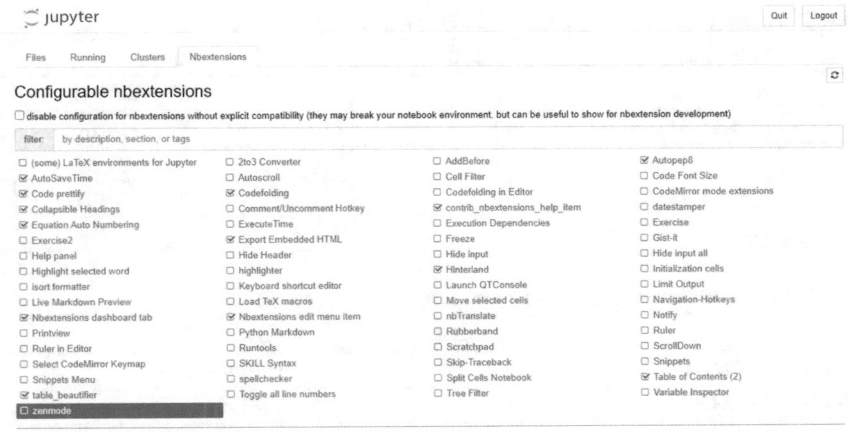

Fig. 4 Necessary nbextenisons that is needed for the prediction to take place

Fig. 5 Page of the jupyter notebook

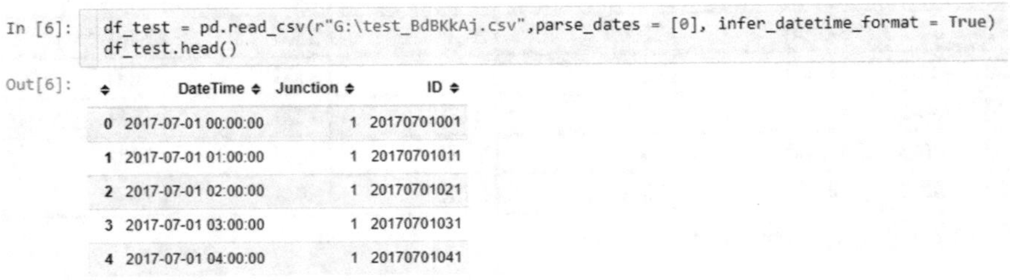

```
In [6]:   df_test = pd.read_csv(r"G:\test_BdBKkAj.csv",parse_dates = [0], infer_datetime_format = True)
          df_test.head()
```

Out[6]:

	DateTime ⬍	Junction ⬍	ID ⬍
0	2017-07-01 00:00:00	1	20170701001
1	2017-07-01 01:00:00	1	20170701011
2	2017-07-01 02:00:00	1	20170701021
3	2017-07-01 03:00:00	1	20170701031
4	2017-07-01 04:00:00	1	20170701041

Fig. 6 Result after running the particular command in the prediction process will be displayed under out command

Table 1 Training model of the junction 1 for the prediction that is taken [16]

	DateTime	Junction	Vehicles	ID
0	2015-11-01 00:00:00	1	15	20151101001
1	2015-11-01 01:00:00	1	13	20151101011
2	2015-11-01 02:00:00	1	10	20151101021
3	2015-11-01 03:00:00	1	7	20151101031
4	2015-11-01 04:00:00	1	9	20151101041

Table 2 Testing model of the junction 1 for the prediction that is taken

DateTime	Junction	ID
2017-07-01 00:00:00	1	20170701001
2017-07-01 01:00:00	1	20170701011
2017-07-01 02:00:00	1	20170701021
2017-07-01 03:00:00	1	20170701031
2017-07-01 04:00:00	1	20170701041

time_targets = df_tmp.groupby([level_values(0)] + [pd. Grouper(freq = '1 M', level = −1)])
['Vehicles'].sum()
train = df_train.pivot(index = 'DateTime',columns = 'Junction', values = 'Vehicles')
train = train.fillna(0)

Table 3 Pivot table of both the training and testing model without NaN values

Junction	1	2	3	4
DateTime				
2015–11-01 00:00:00	15.0	6.0	9.0	0.0
2015–11-01 01:00:00	13.0	6.0	7.0	0.0
2015–11-01 02:00:00	10.0	5.0	5.0	0.0
2015–11-01 03:00:00	7.0	6.0	1.0	0.0
2015–11-01 04:00:00	9.0	7.0	2.0	0.0
…	…	…	…	…
2017–06-30 19:00:00	105.0	34.0	33.0	11.0
2017–06-30 20:00:00	96.0	35.0	31.0	30.0
2017–06-30 21:00:00	90.0	31.0	28.0	16.0
2017–06-30 22:00:00	84.0	29.0	26.0	22.0
2017–06-30 23:00:00	78.0	27.0	39.0	12.0

Xy_train =gen_lag_features(train)

Table 4 Both the training and testing models of concatenated values

Junction 1 (H-1) DateTime	Junction 2 (H-1)	Junction 3 (H-1)	Junction 4 (H-1)	Junction 1 (H)	Junction 2 (H)	Junction 3 (H)	Junction 4 (H)
2015-11-01 01:00:00	15.0	6.0	9.0	0.0	13.0	6.0	7.0
2015-11-01 02:00:00	13.0	6.0	7.0	0.0	10.0	5.0	5.0
2015-11-01 03:00:00	10.0	5.0	5.0	0.0	7.0	6.0	1.0
2015-11-01 04:00:00	7.0	6.0	1.0	0.0	9.0	7.0	2.0
2015-11-01 05:00:00	9.0	7.0	2.0	0.0	6.0	2.0	2.0
...	...	...	...	...	...	...	...
2017-06-30 19:00:00	95.0	34.0	38.0	17.0	105.0	34.0	33.0
2017-06-30 20:00:00	105.0	34.0	33.0	11.0	96.0	35.0	31.0
2017-06-30 21:00:00	96.0	35.0	31.0	30.0	90.0	31.0	28.0
2017-06-30 22:00:00	90.0	31.0	28.0	16.0	84.0	29.0	26.0
2017-06-30 23:00:00	84.0	29.0	26.0	22.0	78.0	27.0	39.0

Xy_train[Xy_train.columns] = scaler.fit_transform(Xy_train[Xy_train.columns])

5.2 *Results from the Simulation*

The results of the traffic are as follows which is by the matplotlib library [18] (Figs. 7, 8, 9 and 10).

In the last, researchers are suggested to go through various uses of machine learning in many useful applications like churn prediction, health care, agriculture, and transportation in [19–34].

Table 5 Training model's results where the scaling of the training data is simulated

Junction 1 (H-1) DateTime	Junction 2 (H-1)	Junction 3 (H-1)	Junction 4(H-1)	Junction 1 (H)	Junction 2 (H)	Junction 3 (H)	Junction 4 (H)
2015-11-01 01:00:00	0.066225	0.106383	0.044693	0.000000	0.052980	0.106383	0.033520
2015-11-01 02:00:00	0.052980	0.106383	0.033520	0.000000	0.033113	0.085106	0.022346
2015-11-01 03:00:00	0.033113	0.085106	0.022346	0.000000	0.013245	0.106383	0.000000
2015-11-01 04:00:00	0.013245	0.106383	0.000000	0.000000	0.026490	0.127660	0.005587
2015-11-01 05:00:00	0.026490	0.127660	0.005587	0.000000	0.006623	0.021277	0.005587
…	…	…	…	…	…	…	…
2017-06-30 22:00:00	0.562914	0.638298	0.150838	0.444444	0.523179	0.595745	0.139665
2017-06-30 23:00:00	0.523179	0.595745	0.139665	0.611111	0.483444	0.553191	0.212291

X_train = Xy_train[Xy_train.index < '2017–04-01'].iloc [:,0:4]

Table 6 Training model's first 4 rows and columns and the testing model's first 4 rows and columns [17]

Junction 1 (H-1) DateTime	Junction 2 (H-1)	Junction 3 (H-1)	Junction 4 (H-1)
2015-11-01 01:00:00	0.066225	0.106383	0.044693
2015-11-01 02:00:00	0.052980	0.106383	0.033520
2015-11-01 03:00:00	0.033113	0.085106	0.022346
2015-11-01 04:00:00	0.013245	0.106383	0.000000
2015-11-01 05:00:00	0.026490	0.127660	0.005587
…	…	…	…
2017-03-31 19:00:00	0.476821	0.574468	0.178771
2017-03-31 20:00:00	0.496689	0.531915	0.156425
2017-03-31 21:00:00	0.483444	0.638298	0.156425
2017-03-31 22:00:00	0.403974	0.574468	0.150838
2017-03-31 23:00:00	0.423841	0.553191	0.162011

y_train = Xy_train[Xy_train.index < '2017–04-01'].iloc [:,4:]
y_train
me = pd.concat([d1,d2],axis = 1, join = 'outer')

Table 7 The combining of the two datasets

	DateTime	Junction	Vehicles	ID	DateTime	Junction	ID
0	2015-11-01 00:00:00	1	15	20151101001	2017–07-01 00:00:00	1	20170701001
1	2015-11-01 01:00:00	1	13	20151101011	2017–07-01 01:00:00	1	20170701011
2	2015-11-01 02:00:00	1	10	20151101021	2017–07-01 02:00:00	1	20170701021
3	2015-11-01 03:00:00	1	7	20,51101031	2017–07-01 03:00:00	1	20170701031
4	2015-11-01 04:00:00	1	9	20151101041	2017–07-01 04:00:00	1	20170701041
...	...	...	...	...	...	...	...
11,806	2017-03-06 22:00:00	1	81	20170306221	2017–10-31 22:00:00	4	20171031224
11,807	2017-03-06 23:00:00	1	72	20170306231	2017–10-31 23:00:00	4	20,171,031,234

Fig. 7 Traffic prediction of Junction 1 from the datasets

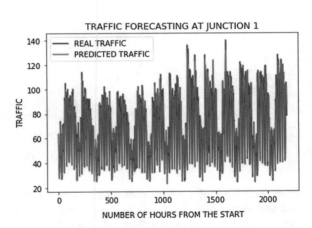

6 Conclusion

In the system, it has been concluded that we develop the traffic flow prediction system by using a machine learning algorithm. By using regression model, the prediction is done. The public gets the benefits such as the current situation of the traffic flow, and they can also check what will be the flow of traffic on the right after one hour of the situation and they can also know how the roads are as they can know to mean of the vehicles passing, a particular junction that is 4 here. The weather conditions have been changing from years to years. The cost of fuel is also playing a major role in the transportation system. Many people are not able to afford the vehicle because of the fuel cost. So, there can be many variations in the traffic data. There is one

Fig. 8 Traffic prediction of
Junction 2 from the datasets

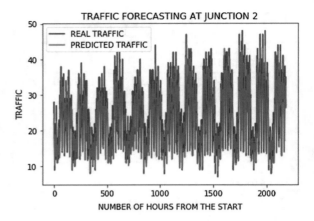

Fig. 9 Traffic prediction of
Junction 3 from the datasets

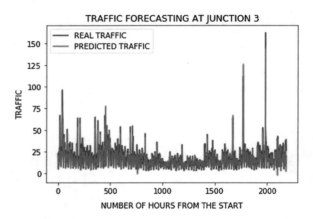

more scenario where people prefer going on their own vehicle without carpooling, and this also matters in the traffic congestion. So, this prediction can help judging the traffic flow by comparing them with these 2 years datasets. The forecasting or the prediction can help people or the users in judging the road traffic easier beforehand and even they can decide which way to go using their navigator and also this will prediction will be also helpful.

7 Future Work

In the future, the system is often further improved using more factors that affect traffic management using other methods like deep learning, artificial neural network, and even big data. The users can then use this technique to seek out which route would be easiest to achieve on destination. The system can help in suggesting the users

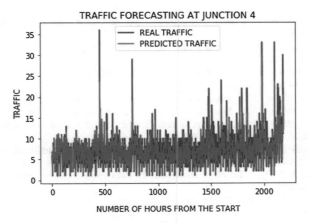

Fig. 10 Traffic prediction of Junction 4 from the datasets

with their choice of search, and also, it can help to find the simplest choice where traffic isn't in any crowded environment. Many forecasting methods have already been applied in road traffic jam forecasting. While there's more scope to create the congestion prediction more precise, there are more methods that give precise and accurate results from the prediction. Also, during this period, the employment of the increased available traffic data by applying the newly developed forecasting models can improve the prediction accuracy. These days, traffic prediction is extremely necessary for pretty much every a part of the state and also worldwide. So, this method of prediction would be helpful in predicting the traffic before and beforehand. For better congestion prediction, the grade and accuracy are prominent in traffic prediction. Within the future, the expectation is going to be the estimation of established order accuracy prediction with much easier and user-friendly methods so people would find the prediction model useful and that they won't be wasting their time and energy to predict the information. There will be some more accessibility like weather outlook, GPS that's the road and accident-prone areas will be highlighted in order that people wouldn't prefer using the paths which aren't safe and simultaneously they'll predict the traffic. This will be done by deep learning, big data, and artificial neural networks [35].

References

1. Azzouni, A., Pujolle, G.: A long short-term memory recurrent neural network framework for network traffic matrix prediction. Comput. Sci. **3**(6), 18–27 (2017)
2. Ratanaparadorn, A., Meeampol, S., Siripachana, T., Anussornnitisarn, P.: Identification Traffic Flow Prediction Parameters, pp. 19–21. International Conference. Department of Industrial Engineering, Kasetsart University, Thailand, Zadar, Croatia (2013)
3. Vlahogianni, E.I., Karlaftis, M.G., Golias, J.C.: Optimized and meta-optimized neural networks

for short term traffic flow prediction: a genetic approach. Transp. Res. Part C Emerg. Technol. **13**(3), 211–234 (2005)

4. Jia, Y., Wu, J., Xu, M.: Traffic flow prediction with rainfall impact using a deep learning method. J. Adv. Transp. (2017)
5. Kunde, F., Hartenstein, A., Pieper, S., Sauer, P.: Traffic prediction using a Deep Learning paradigm. CEUR-WS.org (2017)
6. Loumiotis, I.: Road traffic prediction using artificial neural networks (2018)
7. Alfar, A., Talebpour, A., Mahmassani, H.S.: Machine Learning Approach to Short-Term Traffic Congestion Prediction in a Connected Environment. National Academy of Sciences: Transportation Research Board (2018)
8. Jiang, X., Adeli, H.: Dynamic wavelet neural network model for traffic flow forecasting. J. Transp. Eng. **131**(10), 771–779 (2005)
9. Kong, F., Li, J., Jiang, B., Zhang, T., Song, H.: Big data-driven machine learning-enabled traffic flow prediction (2018)
10. Bao, G., Zeng, Z., Shen, Y.: Region stability analysis and tracking control of memristive recurrent neural network. Neural Netw. **5**(1), 74–89 (2017)
11. https://www.scitepress.org/Papers/2016/58957/pdf/index.html
12. https://www.kaggle.com/fedesoriano/traffic-prediction-dataset
13. https://www.ncbi.nlm.nih.gov/pmc/articles/PMC2845248/
14. https://www.hindawi.com/journals/jat/2021/8878011/
15. https://jupyter.org/
16. https://www.geeksforgeeks.org/formatting-dates-in-python/
17. https://www.shanelynn.ie/pandas-iloc-loc-select-rows-and-columns-dataframe/
18. https://matplotlib.org/2.0.2/api/pyplot_api.html
19. Malik, S., Mire, A., Tyagi, A.K., Arora, V.: A novel feature extractor based on the modified approach of histogram of oriented gradient. In: Gervasi, O., et al. (eds) Computational Science and Its Applications—ICCSA 2020. ICCSA 2020. Lecture Notes in Computer Science, vol 12254. Springer, Cham (2020). https://doi.org/10.1007/978-3-030-58817-5_54
20. Gudeti, B., Mishra, S., Malik, S., Fernandez, T.F., Tyagi, A.K., Kumari, S.: A novel approach to predict chronic kidney disease using machine learning algorithms. In: 2020 4th International Conference on Electronics, Communication and Aerospace Technology (ICECA), Coimbatore, 2020, pp. 1630–1635. https://doi.org/10.1109/ICECA49313.2020.9297392
21. Rekha, G., Krishna Reddy, V., Tyagi, A.K.: KDOS—Kernel density based over sampling—a solution to skewed class distribution. J. Inf. Assur. Secur. (JIAS) **15**(2), 44–52, 9p
22. Ambildhuke, G.M., Rekha, G., Tyagi, A.K.: Performance analysis of undersampling approaches for solving customer churn prediction. In: Goyal, D., Gupta, A.K., Piuri, V., Ganzha, M., Paprzycki, M. (eds) Proceedings of the Second International Conference on Information Management and Machine Intelligence. Lecture Notes in Networks and Systems, vol. 166. Springer, Singapore (2021). https://doi.org/10.1007/978-981-15-9689-6_37
23. Nair, M.M., Kumari, S., Tyagi, A.K., Sravanthi, K.: Deep Learning for Medical Image Recognition: Open Issues and a Way to Forward. In: Goyal, D., Gupta, A.K., Piuri, V., Ganzha, M., Paprzycki, M. (eds) Proceedings of the Second International Conference on Information Management and Machine Intelligence. Lecture Notes in Networks and Systems, vol 166. Springer, Singapore (2021). https://doi.org/10.1007/978-981-15-9689-6_38
24. Tyagi, A.K., Nair, M.M., Deep learning for clinical and health informatics. In: Computational Analysis and Deep Learning for Medical Care: Principles, Methods, and Applications, 28 July 2021. https://doi.org/10.1002/9781119785750.ch5
25. Kanuru, L., Tyagi, A.K., Aswathy, S.U., Fernandez, T.F., Sreenath, N., Mishra, S.; Prediction of pesticides and fertilizers using machine learning and Internet of Things. In: 2021 International Conference on Computer Communication and Informatics (ICCCI), pp. 1–6 (2021). https://doi.org/10.1109/ICCCI50826.2021.9402536
26. Rekha, G., Krishna Reddy, V., Tyagi, A.K.: An Earth mover's distance-based undersampling approach for handling class-imbalanced data. Int. J. Intell. Inf. Database Syst. **13**(2/3/4)

27. Pramod, A., Naicker, H.S., Tyagi, A.K.: Machine learning and deep learning: open issues and future research directions for next ten years. In: Computational Analysis and Understanding of Deep Learning for Medical Care: Principles, Methods, and Applications. Wiley Scrivener (2020)

28. Tyagi, A.K., Rekha, G.: Machine learning with Big Data (March 20, 2019). In: Proceedings of International Conference on Sustainable Computing in Science, Technology and Management (SUSCOM). Amity University Rajasthan, Jaipur, India, 26–28 Feb 2019

29. Tyagi, A.K., Chahal, P.: Artificial intelligence and machine learning algorithms. In: Challenges and Applications for Implementing Machine Learning in Computer Vision. IGI Global (2020). https://doi.org/10.4018/978-1-7998-0182-5.ch008

30. Kumari, S., Vani, V., Malik, S., Tyagi, A.K., Reddy, S.: Analysis of text mining tools in disease prediction. In: Abraham, A., Hanne, T., Castillo, O., Gandhi, N., Nogueira Rios, T., Hong, T.P. (eds) Hybrid Intelligent Systems. HIS 2020. Advances in Intelligent Systems and Computing, vol. 1375. Springer, Cham (2021). https://doi.org/10.1007/978-3-030-73050-5_55

31. Majumdar, S., Subhani, M.M., Roullier, B., Anjum, A., Zhu, R.: Congestion prediction for smart sustainable cities using IoT and machine learning approaches (2020). https://doi.org/10.1016/j.scs.2020.102500

32. Jamal, A., Zahid, M., Rahman, M.T., Al-Ahmadi, H.M., Almoshaogeh, M., Farooq, D., Ahmad, M.: Injury severity prediction of traffic crashes with ensemble machine learning techniques: a comparative study. Int. J. Inj. Contr. Saf. Promot. (2021). https://doi.org/10.1080/17457300.2021.1928233

33. Ahmed, A.A., Pradhan, B., Chakraborty, S., Alamri, A.: Developing vehicular traffic noise prediction model through ensemble machine learning algorithms with GIS. Saudi Society for Geosciences (2021). https://doi.org/10.1007/s12517-021-08114-y

34. Behnood, A., Golafshani, E.M.: Predicting the dynamic modulus of asphalt mixture using machine learning techniques: an application of multi biogeography-based programming (2021). https://doi.org/10.1016/j.conbuildmat.2020.120983

35. https://www.catalyzex.com/s/Traffic%20Prediction

Comparative Analysis of Boosting Algorithms Over MNIST Handwritten Digit Dataset

Soumadittya Ghosh

Abstract Several machine learning algorithms have been adopted for detection of handwritten digits from the MNIST dataset. However, most commonly implemented algorithms are convolutional neural network (CNN), for the extraction of features and support vector machine (SVM), artificial neural network (ANN) and random forest, for classification. In this paper, boosting algorithms like extreme gradient boost (XGBoost), advanced gradient boost (AdaBoost), and gradient boosting have been used to detect handwritten digits from MNIST dataset. To compare the performance of the algorithms confusion metrics, recall, F1 score, and precision were used. From the results, it was found that AdaBoost outperformed the XGBoost and gradient boosting (GB) algorithms with 96.86% accuracy.

Keywords Handwritten character recognition · MNIST · XGBoost · AdaBoost · Gradient boosting

1 Introduction

MNIST stands for modified National Institute of Standards and Technology which is a subset of a bigger dataset called NIST dataset. It was developed by LeCun et al. in 1999 [1]. MNIST contains 70,000 handwritten digits (0–9); out of which, the training set constitutes of 60,000 images, and the test set constitutes of 10,000 images. All the images are of dimension 28×28. It is the most widely used elementary dataset for computer vision because of its simplicity which enables the researchers to reduce the complexity of work at early stages. The purpose of using this dataset is to make computers detect handwritten digits by itself. Though it is not a very complicated dataset, it still contains some of the digits which are quite difficult to understand

S. Ghosh (✉)
School of Computer Science and Engineering, Vellore Institute of Technology, Vellore, India

Fig. 1 Sample data from MNIST dataset

even by humans which makes it a good choice for testing the behavior of different classification algorithms. Some of the sample images from the dataset have been shown in Fig. 1.

Lots of research have been done on MNIST handwritten dataset for detecting handwritten digits using several machine learning algorithms; out of which, most common are convolutional neural network for feature extraction and support vector machine, artificial neural network, and random forest for classification. Vijaykumar [2] developed a machine learning-based multimodal biometric feature recognition model using SVM as a classifier and CNN as a feature extractor to overcome the limitations of unimodal approaches. In another study, Dhaya [3] also used CNN and SVM to detect relevant changes between different temporal images. Using their own dataset and SVM classifier, Shruti and Patel [4] proposed a new model for handwritten word detection where several parameters for feature extraction such as density, long run, and structural features were considered. In 2016, Ramraj et al. [5] came up with a comparative study between two boosting algorithms, namely XGBoost and gradient boosting over several datasets in order to know the difference in terms of accuracy and execution time. They observed that XGBoost did not perform better in terms of accuracy every time, but it managed to achieve higher execution speed most of the time. Ghosh and Maghari [6] compared three neural network-based approaches,

CNN, deep neural network (DNN), and deep belief network (DBN) in terms of accuracy, execution time, etc. Out of all the three algorithms, DNN showed the highest accuracy of 98.08% and performed the best in terms of execution speed. Accuracy of DBN and CNN was close to each other. Using decision tree classifier Korac et al. [7] worked on classification of handwritten digits in MNIST dataset. The performance of the model was found satisfactory with accuracy of 85%. It was also observed that as the dataset is handwritten and comprises of a collection of data from 250 different writers, the developed model was getting confused in several digits such as predicting digit 9 as digit 3. Dixit et al. [8] compared the accuracy of support vector machines (SVM), multilayer perceptron (MLP), and convolution neural network for handwritten digit recognition using MNIST dataset. Using CNN model, they achieved the maximum accuracy of 98.72%; however, the execution time for the same was the highest. Again in another study, Rahman et al. [9] compared different boosting algorithms like XGBoost, light gradient boosting machine (LGBM), gradient boosting, cat boosting (CB), and AdaBoost for the purpose of classification over data of day-to-day activities collected using smartphone sensor. The whole process was performed to see which of the abovementioned algorithms can perform the best in terms of accuracy to decrease the physical inactivity patterns in order to improve the overall lifestyle and health of people. It was found that these algorithms have performed decently with approximately 90% of accuracy. Joseph and his coworkers [10] used XG Boost and gradient boosting models in conjunction with CNN as a feature extraction tool to analyze the NIST special database and achieved an accuracy of 97.18% for handwritten digit recognition. In another work, researchers [11] have compared five different machine learning algorithms, such as k-nearest neighbor (KNN) classifier, random forest, neural network, decision trees, and bagging with gradient boost over MNIST handwritten dataset for the detection of handwritten digits and to compare the performance of the above algorithms. It has been found that the neural network and KNN classifier have performed better than all the other algorithms with 96.8% and 96.7% accuracy, respectively, the KNN being 10 times faster in terms of processing speed. Zhang et al. [12] came with a new kind of classification algorithm by blending two different algorithms support vector machine (SVM) and AdaBoost, in which AdaBoost was used to combine several weak learners (SVM algorithm was used as a weak learner) to develop a strong learner. It was found that the resultant algorithm had performed better in terms of accuracy and speed on datasets of different sizes. Recently Jiao et al. [13] used CNN for feature extraction and XGBoost for classification. The hyperparameters of the overall architecture were optimized by APSO which helped in increasing the diversity of particle population and prevented the algorithm from getting struck in local optima. In 2021, Lai et al. [14] carried out a comparative analysis of AdaBoost, XGBoost, and logistic regression on several unbalanced datasets. The results showed that XGBoost performed the best on the datasets which were highly imbalanced, but the performance has shown inconsistency with the increase in size of the dataset with the same minority class percentage, whereas logistic regression worked the best on the datasets which were not having severe imbalance, while AdaBoost outperformed the other two when it was tested on comparatively more balanced, and larger dataset and accuracy increased with the increase in size of the dataset. In another study [15], XGBoost, gradient boosting, and random forest algorithms were compared on several datasets. Gradient boosting performed the best out of the three and also suggested that XGBoost will not necessarily perform the best in every case. However, information on the comparison of different boosting algorithms is meager in open literature. Therefore, in the present investigation, a comparative study has been done among AdaBoost, XGBoost, and gradient boosting algorithms on the basis of accuracy, precision, F1 score, and recall over the MNIST dataset.

2 Methodology

Boosting algorithms develop n number of trees. It works in such a way that after the first tree is formed the improperly classified records are prioritized and passed on to the next tree. This process continues until and until the algorithm arrives at the nth tree.

AdaBoost stands for adaptive boosting. It is a boosting algorithm. In this, the number of leaf node created is same as that of the different classes for prediction. The leaves in this case are known as stumps. The objective of this algorithm is to make the weak learners a strong learner [16, 17]. First, it initializes the weight for the first model and then does the prediction. After the first iteration, the incorrectly classified points will be given higher weights (means giving more priority) and will be fitted to the next in the next iteration. The iterations will go on till it reaches a specified number. The mean of all the weighted outputs will be calculated and according to mean. At last, the final model should be able to predict all the points correctly.

The basic working principle of AdaBoost and gradient boost [18] is similar as both of them try to make the weak learner a strong learner by making the predictions better of the very first tree created using the subsequent tree. In case of gradient boosting, the weak learner remains the same, i.e., a decision tree. The difference between AdaBoost and gradient boost is that AdaBoost uses high weight data points, while gradient boost uses gradient descent to minimize the loss function which has been used. The loss function used varies with the nature of the problem. So, trees are added in such a way that it minimizes the loss.

XGBoost is a decision tree-based machine learning algorithm. It is an extension of gradient boosting [19, 20]. Some of the advantages provided by XGBoost are as follows:

- Regularization—it has been designed in such a way that it automatically takes care of the regularization and prevents overfitting of the model.
- Cache awareness—it can store some of the intermediate data in cache for faster access and hence reduces time.

- Out of core computation—this comes to play when there is a huge dataset, and the dataset cannot be fitted into the memory. So, the data are compressed and stored to hard drive.
- Use of block structure—with the help of block structure, it can achieve parallel processing. Block structure is used to store data in a proper way in order to use the data layout for the subsequent iterations instead of computing it again.

MNIST dataset has been imported from Keras (part of Tensorflow 2.6.0) library in the form of training set and test set. Initially, the input data of the training set were in the form of a NumPy array of shape (60,000, 28, 28), which indicates an array containing 60,000 images of height and width both as 28 pixels. The test set input was similar but with 10,000 records only. The input data were reshaped as (60,000, 784) and (10,000, 784) for training and test set, respectively. The output data of both the training and test sets are a one dimensional array of length 60,000 and 10,000, respectively, and has been left intact.

After importing and flattening the images, the data were fitted into three different boosting algorithms, namely AdaBoost, gradient boosting, and XGBoost for creating the models with the default parameters. The AdaBoost and gradient boosting classifiers were imported from Scikit—learn library in Python, whereas a standalone library was used for XGBoost. After training the models, four different performance measures, namely accuracy, precision, F1 score, and recall, have been used on the test data to evaluate the performance of different models. The whole workflow has been pictorially represented in Fig. 2.

The performance of all the three models has been pictorially represented using confusion matrix and horizontal bar plot using visualization libraries like seaborn and matplotlib. Confusion matrix is a technique to summarize the performance of a classification model on a set of test data. It shows how a classification model gets confused while making predictions by summarizing the number of correct and incorrect predictions with respect to every single class [21, 22].

Accuracy = Number of correct predictions/Total number of predictions.
Precision = True positive/Actual results.
Recall = True positive/Predicted results.
F1 Score = 2 * ((Precision * Recall)/(Precision + Recall)).
(True positive is the case when a model predicts a positive class correctly.)

3 Results and Discussion

Figure 3 shows the confusion matrix for the results achieved using AdaBoost algorithm. The confusion matrix indicates that the AdaBoost algorithm has predicted the classes correctly for the maximum number of time with an accuracy of 96.86%. Similarly, Figs. 4 and 5 show the performance of gradient boosting and XGBoost, respectively.

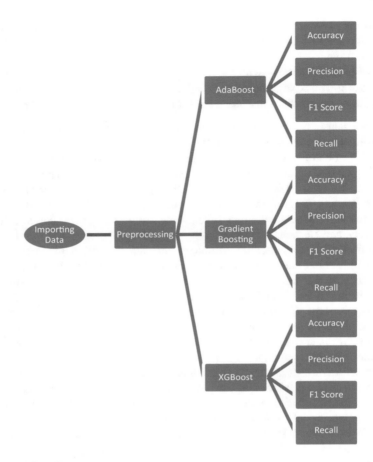

Fig. 2 Workflow diagram

It can be seen that after AdaBoost, the gradient boosting algorithm has performed the best with an accuracy of 94.59% followed by XGBoost with 93.68%. Similar observations have been reported earlier [5, 15] on several datasets. It has been found that all the three algorithms (AdaBoost, gradient boosting, and XGBoost) can predict digit 1 most efficiently as shown in Figs. 3, 4 and 5. From these figures, it can also be seen that the corresponding diagonal block for digit 1 belongs to the highest range of scale (above 1000).

Apart from accuracy, three other considered performance measures (precision, F1 score, and recall) have been shown in Figs. 6, 7 and 8.

From Figs. 6, 7 and 8, it can be observed that among the three tested algorithms AdaBoost has shown the best performance in terms of F1 score (96.84%), precision (96.84%), and recall (96.85%). The results are in agreement with the observations of

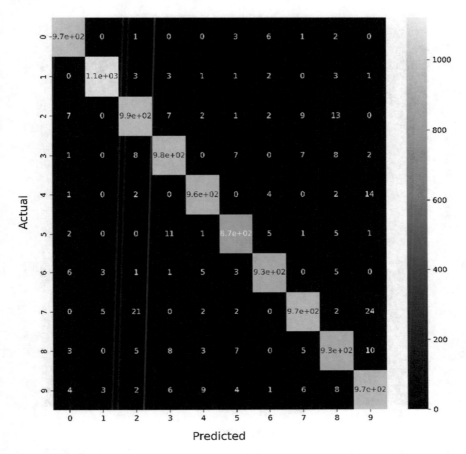

Fig. 3 Confusion matrix for AdaBoost

Lai et al. [14] as MNIST handwritten digit dataset is also quite balanced and contains large number of records.

4 Conclusions

Three different boosting algorithms have been compared, namely AdaBoost, gradient boosting, and XGBoost over the MNIST handwritten digit dataset in order to know how these boosting algorithms perform in case of an image classification problem. Though ANN along with feature extractor like CNN is a popular choice for image classification in today's time, but in the present work, it has been found that the three considered boosting algorithms have also performed quite well in predicting the classes correctly as all the algorithms have gained accuracy of more than 93%

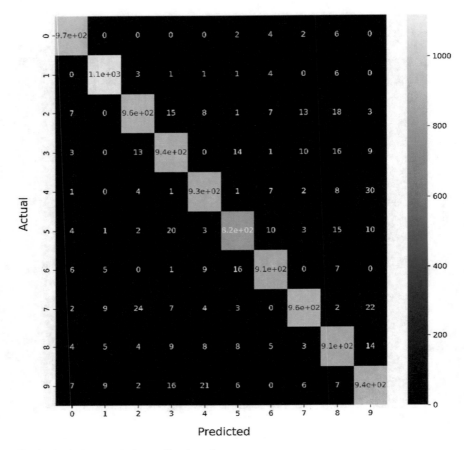

Fig. 4 Confusion matrix for gradient boosting

with default parameters. For comparing the three algorithms, performance measures like accuracy, precision, F1 score, and recall have been considered. Among the three algorithms, AdaBoost has performed the best in terms of all the four performance measures for MNIST dataset which is quite balanced and large. After the AdaBoost algorithm, the gradient boosting algorithm has achieved the second spot, followed by the XGBoost algorithm. However, further studies may be done on a wide range of classification algorithms coupled with different CNN architectures to examine their performance in recognizing handwritten digits more accurately.

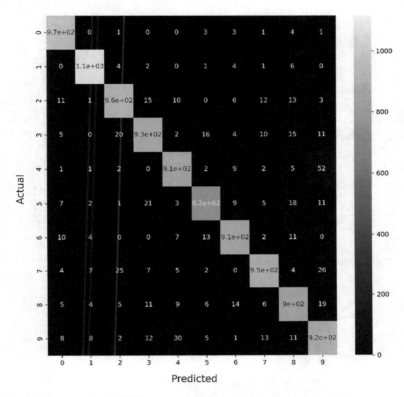

Fig. 5 Confusion matrix for XGBoost

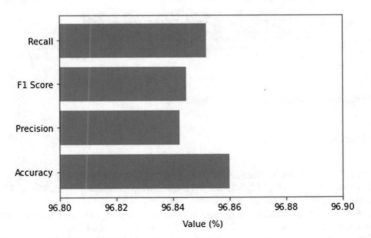

Fig. 6 Performance analysis of AdaBoost algorithm

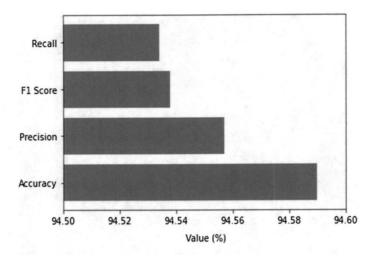

Fig. 7 Performance analysis of gradient boosting algorithm

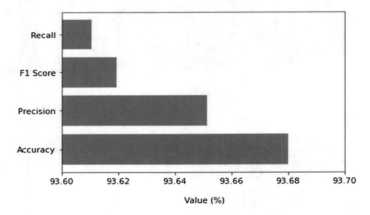

Fig. 8 Performance analysis of XGBoost algorithm

References

1. MNIST Dataset of Handwritten Digits. http://www.pymvpa.org/datadb/mnist.html
2. Vijaykumar, T.: Synthesis of Palm Print in feature fusion techniques for multimodal biometric recognition system online signature. J. Innov. Image Process. **3**(2) (2021)
3. Dhaya, R.: Hybrid machine learning approach to detect the changes in SAR images for salvation of spectral constriction problem. J. Innov. Image Process. **3**(2) (2021)
4. Shruti, A., Patel, M.S.: Offline handwritten word recognition using multiple features with SVM classifier for holistic approach. Int. J. Innov. Res. Comput. Commun. Eng. **3**(6) (2015)
5. Ramraj, S., Uzir, N., Sunil, R., Banerjee, S.: Experimenting XGBoost algorithm for prediction and classification of different datasets. Int. J. Control Theory Appl. **9**(40) (2016)

6. Ghosh, M.M.A., Maghari, A.Y.: A comparative study on handwriting digit recognition using neural networks. In: IEEE International Conference on Promising Electronic Technologies (2017). https://doi.org/10.1109/ICPET.2017.20

7. Korac, D., Jukic, S., Hadzimehanovic, M.: Handwritting digit recognition using decision tree classifiers. J. Nat. Sci. Eng. **2** (2020). https://doi.org/10.14706/JONSAE2020214

8. Dixit, R., Kushwah, R., Pashine, S.: Handwritten digit recognition using machine and deep learning algorithms. Int. J. Comput. Appl. **176**(42) (2020)

9. Rahman, S., Irfan, M., Raza, M., Ghori, K.M., Yaqoob, S., Awais, M.: Performance analysis of boosting classifiers in recognizing activities of daily living. Int. J. Environ. Res. Publ. Health **17** (2020). https://doi.org/10.3390/ijerph17031082

10. Joseph, J.S., Lakshmi, C., Uday, K.P., Parthiban: An efficient offline hand written character recognition using CNN and Xgboost. Int. J. Innov. Technol. Explor. Eng. **8**(6) (2019)

11. Chen, S., Almamlook, R., Gu, Y., Wells, L.: Offline handwritten digit recognition using machine learning. In: Proceedings of the International Conference on Industrial Engineering and Operations Management, Washington DC, USA, 27–29 Sept 2018

12. Zhang, Y., Ni, M., Zhang, C., Liang, S., Fang, S., Li, R., Tan, Z.: Research and application of AdaBoost algorithm Based on SVM. In: IEEE 8th Joint International Information Technology and Artificial Intelligence Conference (2019)

13. Jiao, W., Hao, X., Qyn, C.: The image classification method with CNN – XGBoost model based on adaptive particle swarm optimization. Information **12**(156) (2021). https://doi.org/10.3390/info12040156. https://www.mdpi.com/journal/information

14. Lai, S.B.S., Shahri, N.H.N.B.M., Mohamad, M.B., Rahman, H.A.B.A., Rambli, A.B.: Comparing the performance of AdaBoost, XGBoost, and logistic regression for imbalanced data. Math. Stat. **9**(3) (2021). 10.13.189/ms.2021.090320

15. A Comparative Analysis of XGBoost. https://arxiv.org/pdf/1911.01914.pdf

16. The Ultimate Guide to AdaBoost Algorithm|What is AdaBoost Algorithm? https://www.mygreatlearning.com/blog/adaboost-algorithm/

17. The Intuition and Working Behind AdaBoost. https://medium.com/analytics-vidhya/intuition-and-working-behind-adaboost-1c559f8d7a73

18. Understanding Gradient Boosting Machines. https://towardsdatascience.com/understanding-gradient-boosting-machines-9be756fe76ab

19. An End-to-End Guide to Understand the Math Behind XGBoost. https://www.analyticsvidhya.com/blog/2018/09/an-end-to-end-guide-to-understand-the-math-behind-xgboost/

20. A Gentle Introduction to XGBoost for Applied Machine Learning. https://machinelearningmastery.com/gentle-introduction-xgboost-applied-machine-learning/

21. Simple guide to confusion matrix terminology. https://www.dataschool.io/simple-guide-to-confusion-matrix-terminology/

22. What is a Confusion Matrix in Machine Learning. https://machinelearningmastery.com/confusion-matrix-machine-learning/

DetecSec: A Framework to Detect and Mitigate ARP Cache Poisoning Attacks

Debadyuti Bhattacharya, N. Sri Hari Karthick, Prem Suresh, and N. Bhalaji

Abstract As the use of computers in professional and educational places in non-technical fields has increased over the years, so have the number of people with malicious intent. This has given rise to a number of people who use scripts they find on the Internet to attack other computers on their networks for fun or with other malicious intent and are often known as "script kiddies." Most solutions in the market for such problems are either extremely bulky or have to be paid for and often require command line mastery and sufficient knowledge on network administration. In this paper, a framework is proposed to help prevent one of the most common types of attack known as man-in-the-middle attack which is often done to either sniff or steal personal data. It can also be used to perform denial-of-service attacks by dropping the packets as the man-in-the-middle. The proposed framework will detect any rogue user in the network attempting to impersonate another user for such activities and will report them to whoever the administrator is, without requiring much technical knowledge. The framework also has built-in security measures to prevent attacks on the framework itself and to prevent having a single point of failure.

Keywords Man-in-the-middle · ARP-poisoning · Spoofing attack · Local area network scan · Denial-of-service

1 Introduction

The address resolution protocol (ARP) is a stateless protocol used to map Internet protocol (IP) addresses to their respective media access control (MAC) addresses. In other words, it is a mechanism to discover the link layer address of a given device in

D. Bhattacharya (✉) · N. Sri Hari Karthick · P. Suresh · N. Bhalaji
SSN College of Engineering, Chennai, India
e-mail: debadyutibhattacharya16021@it.ssn.edu.in

N. Bhalaji
e-mail: bhalajin@ssn.edu.in

V. Suma et al. (eds.), *Evolutionary Computing and Mobile Sustainable Networks*,
Lecture Notes on Data Engineering and Communications Technologies 116,
https://doi.org/10.1007/978-981-16-9605-3_70

a network when the Internet layer address is known. The ARP is cached in an entry known as the ARP table in each individual node in the network.

Generally, the ARP is a *request-response* protocol whose messages are encapsulated by a link layer protocol and these messages are further used to update the ARP cache entries in the network. However, instead of following through this *request-response* protocol, it can also be used as a simple *announcement* protocol. In this scenario, a specific device announces its IP and MAC addresses through the announcement packet and the hosts that receive this packet, update their local cached entries of the ARP tables. Since, this *announcement* method is not authenticated or verified in any way, the cached ARP entries will be updated with the message sent even if it is false or misleading [1, 2].

This lack of verification during an ARP announcement leads to a common attack known as ARP cache poisoning. In this attack, a node in a network sends an ARP announcement to a particular host with an incorrect IP address mapped to its MAC address. This host upon receiving the message, updates its ARP table and therefore, in a sense, has its ARP cache "*poisoned*" with an incorrect IP-MAC mapping [3, 4].

Now, with a host with a "*poisoned*" ARP cache, it may incorrectly identify the attacker in this scenario as the device with the announced IP. This sort of ARP cache poisoning attack most commonly further leads to two types of attacks, man-in-the-middle (MITM) attacks, and denial-of-service (DoS) attacks [5, 6].

DoS attack. The DoS attack, similar to the MITM attack, can form a relay between the communicating nodes and can choose to drop packets or make a resource unavailable to the node requesting a service. This, therefore, results in a DoS attack and can cause temporary or indefinite disruption of services.

MITM attack. The MITM attack involves a variety of scenarios where the attacker forms a secret relay between the two communicating nodes and can either eavesdrop and learn sensitive information or alter the messages being relayed without either of the communicating nodes realizing. This is possible because the nodes believe that they are talking to each other over a private connection while they are actually communicating with the attacker. Even with the most basic MITM attacks, the attacker can eavesdrop on unsecure communications (over HTTP) of victim devices and can see the entire traffic in plain text, which can be anything from passwords to banking information and hence can have a devastating effect if not identified. Furthermore, an MITM attack can be escalated to a DNS spoofing attack which can be used to lead the victims into imposter Websites and collect sensitive information and other credentials.

In this paper, we propose a framework that collects the ARP cache from all the nodes in a system and sends them to a node running a server in the local network on a periodic basis (user-defined or randomly generated period), detects a poisoned ARP cache by comparing the tables received from all the participating nodes, and mitigates the attack by launching a deauthentication attack against the attacker and forcefully removing him from the network and notifying the registered network administrator. The proposed framework is lightweight, does not require over the top security mechanisms, and can run as a system service with minimal dependencies.

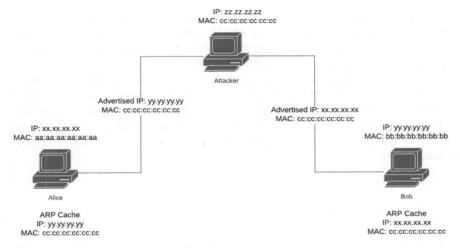

Fig. 1 Depiction of a man-in-the-middle attack

In Fig. 1, we see that Alice, Bob, and the attacker are in the same network and their original IP and MAC addresses are depicted, where Alice and Bob are the communicating hosts. It is clearly observed that the attacker poisons Alice's cache to impersonate Bob and simultaneously poisons Bob's cache to impersonate Alice and thereby successfully launching an MITM attack intercepting all traffic between Alice and Bob and can even view unencrypted communication between the two in plain text.

2 Literature Review

Bruschi et al. in their paper discuss the usage of a secure version of the ARP, wherein, the structure of the ARP packet is revamped to include headers for supporting signature and encryption using digital signature algorithm, while maintaining backwards compatibility with non-secure version [7].

Abad et al. in their paper review various tools such as packet monitoring and IDS software to detect ARP cache poisoning and discuss ideal scenarios under which the attacks can be detected and prevented [8].

Nam et al. in their paper propose a non-centralized voting-based conflict resolution mechanism for detecting the attacker in the local area network while maintaining a long-term IP/MAC mapping table [9].

Kumar et al. in their paper discuss an alternate centralized server-based solution for detection and prevention of ARP cache poisoning through an ARP central server and discuss the possible scenarios of attacks of MITM involving the central server and detection mechanisms [10].

Arote et al. propose a solution which involves the central server sending trap Internet control messaging protocol (ICMP) ping packet, analyze the response in terms of ICMP reply, and successfully detect the attacker with four alternate algorithms based on the method of use [11].

Our proposed solution tries to provide a feasible, efficient, and hybrid server solution without the necessity of modification of existing standards (stack, packet format, etc.). This, therefore, maintains compatibility, providing a plug and play software solution at the application level with zero modification of lower-level internet functionality, while maintaining robustness in detecting MITM attacks with high probability [8].

3   Proposed Work

The solution proposed in this paper is a framework that will effectively detect and mitigate ARP cache poisoning attacks through a lightweight framework.

The client-side service makes system calls and recovers the ARP cache of that particular host. This ARP table is then formatted and sent to the server for further processing in intervals.

The server-side service runs a server that receives the ARP cache sent by the clients and also uses its own, to create a map with the IP address as key and the corresponding MAC address received as value. This map persists and is only updated, and along with this, the count of each of the records is stored, that is, if n nodes send a message stating that IP xx.xx.xx.xx is bound to the MAC address aa:aa:aa:aa:aa:aa, the corresponding count for this map entry would be n. This way, if there is a node with its ARP cache poisoned, then the server would maintain a record of two MACs aa:aa:aa:aa:aa:aa and bb:bb:bb:bb:bb:bb both mapped to the same address xx.xx.xx.xx, but with different counts n and 1. The count is used to check how many systems' ARP table corroborates the IP-MAC mapping of either record. Now, the server would know who the attacker is and could launch the mitigation mechanism which kicks the attacker from the network by flooding it with *deauth* packets. The working of the proposed system is shown in Fig. 2.

The fact that the client sends the requests in an asynchronous manner helps with the detection, as, if the attacker flooded every other system with its ARP announcement, the server would check the new entry against the existing map and its own ARP cache [12–14].

Initially, the client sends an "init" message to the server which uses RSA encryption due to its fixed size where it shares the time interval between subsequent messages after performing a key exchange. The subsequent messages are encrypted with AES—256 in counter mode. The AES key is further encrypted by the RSA public key. This is used to encrypt the communication in order to maintain a lower computational cost of encryption while maintaining robust security.

An edge case arises when the attacker spoofs the MAC address of the victim along with its IP. This is where the timestamp of the message comes into picture,

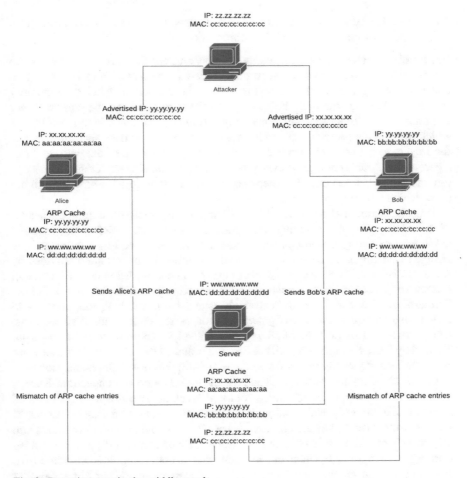

Fig. 2 Detecting man-in-the-middle attack

where the server compares the timestamps of successive messages from the hosts to measure the time interval between them. Every host has a set time interval that is randomly generated when the process is first run and this is then compared to the interval between successive messages' timestamps. If this interval is lesser than the set interval that was set, a warning is generated.

The proposed framework's client–server architecture supports two modes of operation, *persistent centralized server,* and *randomly selected server.*

Persistent centralized server. In this mode of operation, a particular node in the network is pre-emptively nominated as a server and runs the service. This can be nominated as the centralized server by the network administrator but it has a few drawbacks. The attacker may try to physically tamper with the device and this creates a single point of failure for the entire framework.

Randomly selected server. In this mode of operation, the current server has the ability to randomly assign a new server after a certain interval of time. Every system that runs the service has the capability to become the server. After the randomly selected node sends the periodic message to the current server, the current server responds with a JSON message containing a Boolean variable which is set to true which triggers the server service in the selected node. The current server then waits for a cycle to get over and sends the other nodes the location of the new server as a part of the JSON message. It then sends its map to the new server and then stops running the server service. This keeps happening and therefore the attacker will not know which system is the server.

Bruschi, Rosti, et al. approach this problem by completely revamping the header of the ARP packet by including encryption and signature, which is not required in the case of our approach and hence incurs no performance loss nor require any change to existing industry standards. Another solution to this problem provided by Abad, Christina L. et al. is to use packet monitoring and intrusion detection system software both require heavy resources to run as well as pose a single point of failure. Furthermore, this could become extremely expensive in the long run, contrary to which the proposed solution is a lightweight process and since it runs in all the hosts, it does not have a single point of failure. Alternatively, a voting-based solution is put forward by Nam, Seung Yeob et al. but it assumes that all the members of the network are connected physically by a wire whereas our solution just requires the processes to be run and does not pose any requirement on the mode of connection. Kumar, Sumit et al. investigate a centralized solution which again poses a single point of failure, unlike our proposed solution. Solutions which require sending extra packets such as Arote, Prerna et al. are also avoided as we just populate the information from the pre-existing ARP table of each host. Finally, Shakya et al. [15] propose a deep learning-based approach, which due its nature can be both computationally intensive and require a large amount of training to be effective in real-world performance either of which is not required by the proposed solution.

4 Experimentation and Result

The language used for implementing the client and server programs was GoLang due to its built-in support of concurrency. The client program is designed to recover the system's ARP cache and then send it to the server as a JSON string. This messaging is done over the Gin framework. The central server at the time receives the messages from all the clients and creates a consolidated map [16].

Upon receiving messages, the server first checks if the entry already exists in the map. If it does not exist, the entry is inserted into the table and if it does, the entry and the new message are compared. If the IP-MAC mapping satisfies the existing entry,

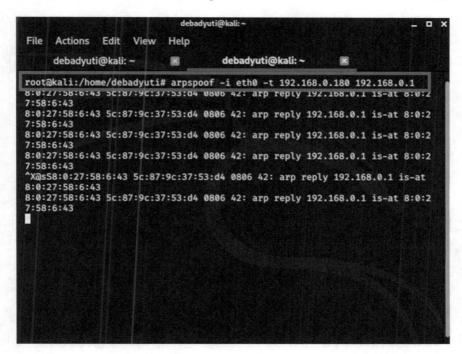

Fig. 3 Simulation of attack using Arpspoof

the operations continue as usual. However, if there is a conflict in the mappings, the network admin is alerted about suspicious behavior. A Kali Linux virtual machine (VM) is set up with the host running the centralized server and the VM using the Arpspoof framework to become the MITM. Figure 3 shows an instance of such an attack where the attacker becomes the MITM between hosts 192.168.0.180 and 192.168.0.1.

The VM is now the MITM between the target (host) and the gateway (router). However, by default the MITM does not forward the IP packets and therefore this needs to be separately done as shown in Fig. 4.

When the attack is performed, the ARP cache can be seen changing on the machines. This is then sent to the server that detects the inconsistency in the ARP caches. This detection then causes the server to send an alert to the network administrator and performs a deauthentication attack if the available hardware permits.

The persisting table in the centralized server at that time is stored in a PostgreSQL database along with the timestamp and the time interval of ARP cache messages. Another Node.js server runs on the same system that uses the data in the database to create visual indicators of the network in a locally hosted Website that the network administrator can access from any other system in the same network. This provides a workable and intuitive user interface that the network administrator can use to

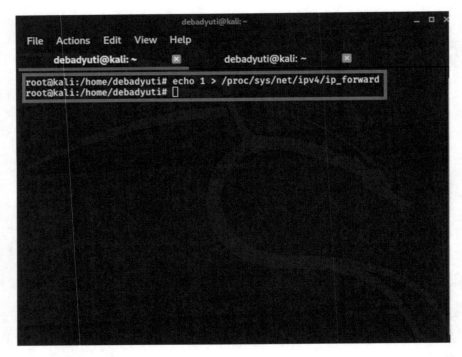

Fig. 4 Forwarding packets as the man-in-the-middle

resolve conflicts, view existing alerts and does not require command line arguments to the server [17].

When there is no alerts or warning, the home page of the framework will look as shown in figure and the network administrator can monitor the consolidated ARP cache as well as shown in Figs. 5 and 6.

Conversely, when there is an attack detected in the network, the UI reflects as shown in Fig. 7 where it is implied that there is a discrepancy in the consolidated ARP cache. Furthermore, in the logs shown in Fig. 8, the IP of the gateway (192.168.0.1) has two different MAC addresses corresponding to it, which are highlighted in red.

5 Conclusion

The proposed solution is, therefore, able to detect an' MITM attack successfully and is able to successfully isolate the attacker and the victim. Furthermore, the user interface is quite simple and efficiently reports problems when they arise and is able to highlight the entries in the cache causing the problem.

The framework also uses robust security mechanisms including RSA for initializing the random time intervals that prevent other devices from sending client

Fig. 5 Admin interface on normal case

ID	MAC Address	IP Address
1	01:00:5e:00:00:02	224.0.0.2
2	44:00:49:ba:20:4a	192.168.0.151
3	fc:03:9f:ec:59:f6	192.168.0.166
4	0c:d2:b5:73:e1:6c	192.168.0.2
5	01:00:5e:7f:ff:fa	239.255.255.250
6	b0:4e:26:03:09:c0	192.168.0.1
7	40:4e:36:88:12:40	192.168.0.144
8	c0:ee:fb:4d:d9:1e	192.168.0.150
9	d8:e0:e1:94:38:6f	192.168.0.161
10	01:00:5e:00:00:16	224.0.0.22
11	01:00:5e:00:00:fb	224.0.0.251
12	f4:f5:24:5e:4d:b8	192.168.0.140
13	40:4e:36:87:83:a3	192.168.0.141
14	5c:87:9c:37:53:d4	192.168.0.180
15	b8:27:eb:90:58:59	192.168.0.188
16	01:00:5e:00:00:fc	224.0.0.252

Fig. 6 No attacks hence no suspicious addresses have been highlighted

messages and uses AES encryption for all subsequent messages. The degree of randomness and the multilayer encryption provided, both provide a secure experience to the user.

The framework is built using GoLang and the nature of messages served are also asynchronous in nature. The database used by the framework is also open-source and does not require any licensing to use and can therefore be used by anyone.

Therefore, this allows the use of lightweight clients and servers which are platform independent, communicating with each other to detect discrepancies in the network and allow the network administrator to take appropriate action immediately without consuming an enormous amount of user resources.

Fig. 7 Admin interface after detecting an attack

ID	MAC Address	IP Address
1	b0:4e:26:03:04:c0	192.168.0.1
2	44:00:49:ba:20:4a	192.168.0.151
3	01:00:5e:00:00:02	224.0.0.2
4	01:00:5e:00:00:fc	224.0.0.252
5	40:4e:36:87:83:a3	192.168.0.141
6	f4:f5:24:5e:4d:b8	192.168.0.140
7	40:4e:36:88:12:40	192.168.0.144
8	c0:ee:fb:4d:d9:1e	192.168.0.150
9	d8:a0:e1:94:38:6f	192.168.0.161
10	fc:03:9f:ec:99:fb	192.168.0.166
11	b8:27:eb:90:58:59	192.168.0.188
12	01:00:5e:00:00:16	224.0.0.22
13	01:00:5e:7f:ff:fa	239.255.255.250
14	5c:87:9c:37:53:64	192.168.0.180
15	08:00:27:58:06:43	192.168.0.189
16	0c:d2:b5:73:e1:6c	192.168.0.2
17	01:00:5e:00:00:fb	224.0.0.251
18	08:00:27:58:06:43	192.168.0.1

Fig. 8 Suspicious hosts highlighted

Although the solution proposed is decentralized in nature, it can be expanded to distributed environments by replicating the processes to each of the networks separately as if they are individual networks reporting MITM attacks within, these logs can then be forwarded to a single admin console for analyzing the state of the distributed network.

References

1. Plummer, D.C.: Ethernet address resolution protocol: or converting network protocol addresses to 48 bit Ethernet address for transmission on Ethernet hardware. RFC **826**, 1–10 (1982)
2. Finlayson, R., et al.: A reverse address resolution protocol (1984)

3. Whalen, S.: An introduction to ARP spoofing (2001). Node99 [Online Document], Apr 2001
4. Fleck, B., Dimov, J.: Wireless access points and ARP poisoning. Online document (2001)
5. Hwang, H., Jung, G., Sohn, K., Park, S.: A study on MITM (Man in the Middle) vulnerability in wireless network using 802.1X and EAP. In: 2008 International Conference on Information Science and Security (ICISS 2008), Seoul, 2008, pp. 164–170. https://doi.org/10.1109/ICISS.2008.10.
6. Herzberg, A., Shulman, H.: Stealth-MITM DoS attacks on secure channels (2009). arXiv preprint arXiv:0910.3511
7. Bruschi, D., Ornaghi, A., Rosti, E.: S-ARP: a secure address resolution protocol. In: 19th Annual Computer Security Applications Conference, 2003. Proceedings. IEEE (2003)
8. Abad, C.L., Bonilla, R.I.: An analysis on the schemes for detecting and preventing ARP cache poisoning attacks. In: 27th International Conference on Distributed Computing Systems Workshops (ICDCSW'07). IEEE (2007)
9. Nam, S.Y., Kim, D., Kim, J.: Enhanced ARP: preventing ARP poisoning-based man-in-the-middle attacks. IEEE Commun. Lett. **14**(2), 187–189 (2010)
10. Kumar, S., Tapaswi, S.: A centralized detection and prevention technique against ARP poisoning. In: Proceedings Title: 2012 International Conference on Cyber Security, Cyber Warfare and Digital Forensic (CyberSec). IEEE (2012)
11. Arote, P., Arya, K.V.: Detection and prevention against ARP poisoning attack using modified ICMP and voting. In: 2015 International Conference on Computational Intelligence and Networks. IEEE (2015)
12. Sun, J.H., Tu, G.-H.: Address resolution protocol (ARP) cache management methods and devices. U.S. Patent Application No. 11/552678
13. Abdel Salam, A.M., Elkilani, W.S., Amin, K.M.: An automated approach for preventing ARP spoofing attack using static ARP entries. Proc. IEEE (IJACSA) Int. J. Adv. Comput. Sci. Appl. **5**(1) (2014)
14. Lootah, W., Enck, W., McDaniel, P.: TARP: ticket-based address resolution protocol. Comput. Netw. **51**(15), 4322–4337 (2007)
15. Shakya, S., Pulchowk, L.N., Smys, S.: Anomalies detection in fog computing architectures using deep learning. J. Trends Comput. Sci. Smart Technol. **2020**(1), 46–55 (2020)
16. Gin Web Framework https://github.com/gin-gonic/gin.git
17. Postgres database https://github.com/postgres/postgres.git

Design and Performance Analysis of Multiported Memory Module Using LVT and XOR Approaches on FPGA Platform

S. Druva Kumar and M. Roopa

Abstract Most of the FPGAs support dual-port RAMs, and it is challenging to implement MPM on FPGA. In this manuscript, the MPM module using Live Value Table (LVT) and XOR approaches are designed to improve the performance metrics. The 2W4R and 3W4R examples are considered to create the MPM using LVT, and XOR approaches. The MPM (2W4R and 3W4R) modules are synthesized on Virtex-7 FPGA and discuss the performance parameters like chip area (Slices and LUTs), BRAM usage, frequency, and efficiency using both LVT and XOR approaches. The XOR-based MPM module provides better performance than LVT-based MPM module. The 2W4R and 3W4R modules using LVT and XOR approaches are compared individually with existing similar MPM modules on the same FPGAs with better improvement hardware constraints.

Keywords 2W4R · 2W4R · FPGA · LVT · Multiported memory · XOR

1 Introduction

The Multiported Memory (MPM) provides parallel access to the shared memory architecture and commonly uses microprocessors. The performance of the shared memory is effectively used in typical applications like communication, Digital signal processing, and Network system. The shared memory module has a simple conflict mechanism. So, a flexible and efficient MPM module is required to manage conflicts [1]. Typically, most FPGAs use dual-ported BRAM and build the MPM module; the designer has to configure the available Logic elements (LE) and BRAMs on FPGA [2]. The FPGA-based BRAMs provide better speed requirements than conventional approaches and are also adopted to improve the Instruction level parallelism (ILP)

S. Druva Kumar (✉) · M. Roopa
Electronics and Communication Enggineering, Dayananda Sagar College of Engineering, Bangalore, Karnataka, India
e-mail: druva-ece@dayanandasagar.edu

M. Roopa
e-mail: roopa-ece@dayanandasagar.edu

© The Author(s), under exclusive license to Springer Nature Singapore Pte Ltd. 2022 1009
V. Suma et al. (eds.), *Evolutionary Computing and Mobile Sustainable Networks*,
Lecture Notes on Data Engineering and Communications Technologies 116,
https://doi.org/10.1007/978-981-16-9605-3_71

and access multiple-register files to construct MPM [3, 4]. The high-performance microprocessor architectures are built using multiported shared memory, allowing parallel data access for both write and read ports [5, 6].

There are many multi-porting approaches available to construct MPM modules. (1) MPM module using only Logic elements—in this approach, as memory depth increases, the FPGA's chip area rapidly increases, and operating frequency decreases significantly. (2) Replication approach: This is one of the best approaches to increase the read ports, but it supports one write port access for all the read ports. (3) Banking approach: in this approach, the memory banks support both write and read ports; Users can increase additional write and read ports by creating different memory banks. (4) Multi-pumping approach: The approach needs additional multiplexers and registers to hold the write and read data temporarily; clock synchronization is the prime factor in this approach while reading the data from memory ports [7–10].

Multiported memory (MPM) using logic elements (LE) are one of the conventional approaches to implement on any FPGA. This type of module includes 'n' writes and read ports, memory locations, along with many decoders and multiplexers. As the memory depth increases, the memory module's complexity increases, which affects the chip area—in contrast, decoders and multiplexers limit the maximum operating frequency. So, one of the best ways to implement the multiport memory is using BRAM on FPGA, which offers the best operating frequency and moderate chip area. So, implementing the MPMs with high operating frequency and less chip area on FPGA is challenging.

On the Virtex-7 FPGA, this manuscript uses LVT and XOR-based techniques to construct efficient multiported memory modules. The performance measurements are realized using the 2W4R and 3W4R memory modules. The XOR techniques used in the 2W4R and 3W4R memory modules provide better performance benefits than LVT-based alternatives. The manuscript organizes as follows: Sect. 2 describes the existing MPM architectures using different approaches. Section 3 provides 2W4R and 3W4R memory module architecture and its detailed description. The results and discussion of the design work are analyzed in Sect. 4 with comparison. Section 5 provides the overall work summary and performance improvement with future suggestion.

2   Related Works

This section reviews the existing multiport memory (MPM) module design using different approaches and application scenarios. Laforest et al. [11] present the multiported memory modules using LVT and XOR methods on Xilinx and Altera FPGA. The construction of memory modules with more than two ports, using either BRAMs or logic elements. So, this work demonstrates using different multiport memory modules like 2W4R, 4W8R, and 8W16R using LVT and XOR methods. Lin et al. [12] analyze the multiported memory with BRAM efficient architecture on FPGA. The designed work reduces the BRAM 37.5% for the 1W2R module than the replication method.

Similarly, around 25% BRAM reduction in 2W1R than LVT approach from 2K to 8K memory depths. The designed work reduces the BRAM by approximately 30 and 37.5% in the 2W4R module for 8K memory depth than existing XOR,8I, and LVT approaches. Muddebihal et al. [13] present the MPM module with write conflict by optimizing the FPGA area. The MPM module uses LVT and pumping process and realizes the BRAM and Logical elements for different 256 memory depths. Kawashima et al. [14] adopted the register write-back port prediction method for banked register files (BRF). This work improves the performance degradation, which is affected by write bank conflict and estimates the performance metrics like latency and power.

Abdelhadi et al. [15] present the Multiported SRAM-based memory module with modular switch architecture. The work analyzes the invalidation-based LVT approach for the MPM module by providing high performance with less BRAM utilization than the conventional LVT-based approach. Lai et al. [16] present the MPM with an efficient design on FPGA, which includes a hierarchical based structure for 1W2R/4R memory module. The design work is compared with the existing XOR and LVT approach, reducing BRAMs utilization around 53%, 69%, respectively. Shahrouzi et al. [17] present the Embedded MPM module for future FPGA with better efficient performance. The MPM contains simple dual-port BRAMs to realize the read and write ports, counter, and decision-making unit to simplify the architecture. The work recognizes the BRAM and Slices utilization for different memory depths compared with existing LVT, and XOR approaches with better improvement in resource utilization. Lai et al. [18] present the Algorithmic based MPM module using hierarchical banking architecture on FPGA. The design work uses hierarchical banking architecture to improve the BRAM usage. The design work of the 4W2R module is compared with the existing memory approach by reducing 62% of BRAMs for 32K depth.

Ullah et al. [19] present the Ternary content-addressable memory (TCAM) based MPM with multi-pumping enable features on FPGA. This work reduces the BRAM usage and improves the storage limitations. This work improves 2.85 times better performance/memory than the existing TCAM based MPM module. Manivannan et al. [20] describe the pulsed latch-based 2W4R multiport register file module. The work improves the area and power by incorporating the pulsed latch approach in the multiport register file module. The design work is limited with 8-bit data width and implemented on 180 nm technology. Humenniy et al. [21] present the shared access memory module for data protection and transmission system. The shared access memory is designed based on Galois field codes, providing a parallel data process for MPM users to perform both write and read operations simultaneously. Shahrouzi et al. [22] present the MPM module future FPGAs using an Optimized counter-based mechanism. The work improves the existing drawbacks like extra logic and routing requirements and gives importance to parallelism to enhance the FPGA performance.

Kwan et al. [23] describes the lossy MPM module, which provides high bandwidth, and can access parallel to a single address through multiple read ports. Chen et al. [24] present the Algorithm based MPM using a written scheme, which provides better performance by utilizing the Remapping table. The hash write controller is used to handle the write conflict and reduce the latency time. Jain et al. [25] present the single-port memory module using better coding schemes. The coding access is used in memory banks, best-case and worst-case analysis of MPM. The work analyzes the write latency and works read performance against existing approaches with improvements. Shahrouzi et al. [26] present the bi-directional MPM for future FPGA with optimization in resources. The design has bi-direction R/W ports, which contain an arbitrary number. The work reduces the design complexity and improves the routing efficiency by removing the delay paths. Zhang et al. [27] present an XOR-based MPM module using an FPGA parallel hash table approach. The design work provides high throughput by incorporating hash tables in memory modules. The hash table architecture is customized to multiple read and write ports per entry. The work analyzes the throughput and supported quires with existing approaches.

3   Multiported Memory (MPM) Modules

The multiported memory (MPM) using logic elements (LE) on FPGA drains the maximum frequency and consumes more chip area, which affects the memory module's overall performance. So, in this section, efficient MPMs are designed using the Live Value Table (LVT) and XOR approach to overcome the drawbacks of the conventional approach. These LVT and XOR approaches increase the write ports and integrate them with the replication approach for reading port increment. In this work, two-write-four read (2W4R) and three-write-four read (3W4R) modules are designed and discussed using both LVT and XOR approach.

3.1   LVT-Based Approach

LVT-based approach uses the BRAM unit in the implementation of MPM. This approach uses banked modules and reads the corresponding data based on which bank holds the most recent write value. The LVT supports a more significant number of write ports by concerning BRAMs. The LVT approach has many banking modules based on design requirements and reads the data based on the most written bank. The name itself suggests which bank module holds current live value for each memory location. Figure 1 illustrates the example of a 2W4R memory module using the LVT approach. The 2W4R module contains two write ports, and four read ports. It is decomposed into two main memory modules for W_{data0} and W_{data1}. The first set includes four memory banks (M_{0_0}, M_{0_1}, M_{0_2}, M_{0_3}) and the second set contain other four memory banks (M_{1_0}, M_{1_1}, M_{1_2}, M_{1_3}). The LVT module receives two Write addresses (W_{addr0} and W_{addr1}), four read addresses (R_{addr0}, R_{addr1}, R_{addr2}, R_{addr3}) as input, and four updated 1-bit read values (r_0, r_1, r_2, and r_3).

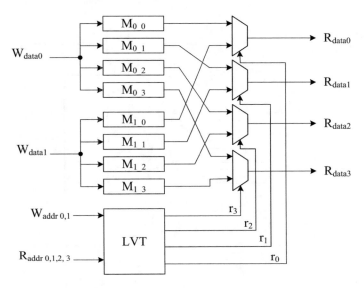

Fig. 1 2W4R memory module using LVT approach

First, define the Write address (W_{addr0} and W_{addr1}) and write data (W_{data0} and W_{data1}); the write data are stored in corresponding memory banks based on the write address. The LVT provides the related four read select lines (acts as line value) to the four multiplexors to read the recent read data from the memory banks. The LVT holds either the '0' or '1' value. The depth of the LVT is the same as the memory depth of the BRAMs (or bank). So, each LVT entry is defined for each memory location of the bank. The LVT also acts as a memory module, which contains the same memory depth and holds either '0' or '1' on each location. The LVT values are either '0' or '1' stored in LVT memory based on write addresses. So read the LVT values from LVT memory based defined read addresses. The LVT provides four values (r_0, r_1, r_2, and r_3) and is used in corresponding four multiplexors to read the most updated write values. LVT controls each read port of the corresponding multiplexor output. The four multiplexors provide four outputs: final read outputs (Rdata0, Rdata1, Rdata2, Rdata3) of the 2W4R memory module.

Similarly, an example of a 3W4R memory module is represented in Fig. 2. It contains three write data, three write addresses, four read addresses as inputs, and four as outputs. The operation of the 2W4R module is the same as the 3W4R module, with few changes in LVT. Once the write data and addresses are defined, the write data is stored in the corresponding memory bank. The LVT contains '0', '1', '2' values in the table. The LVT also acts as a memory module with the same memory depth of the bank and holds three values on each memory location. The LVT values are stored in LVT memory based on write addresses. So read those values from LVT memory based on reading addresses. The multiplexor reads the most recent BRAM bank value using LVT, which acts as the final 3W4R output.

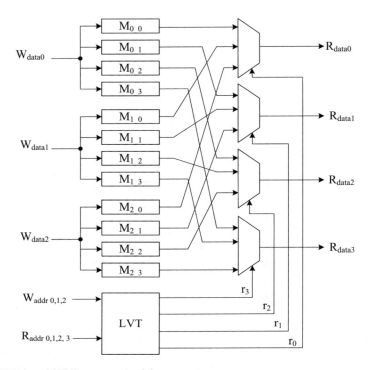

Fig. 2 LVT-based 3W4R memory module

The LVT uses a write address in BRAM to update the corresponding location with a number of writes ports for a write operation. These write ports are used to find the BRAM memory bank's written data. The LVT uses the defined read addresses to fetch the corresponding memory bank numbers to find the read port values in a read operation. The LVT-based MPM modules need more logic elements as memory depth increases and critical paths using both multiplexor output and LVT. So, there is a need for an alternate MPM approach to overcome these drawbacks.

3.2 XOR-Based Approach

The MPM using the XOR approach overcomes the LVT-based approach's drawbacks like area overhead and critical path. The XOR approach is one of the approaches to increase the write ports. The system aims to read the data from BRAM, which are computed with XOR values in each read port. To perform XOR operation in MPM, a few of the properties as to follows as given below:

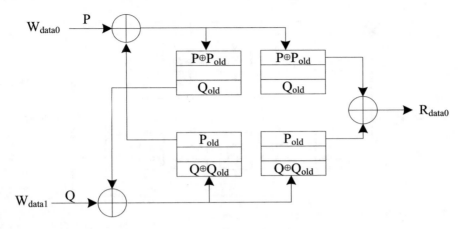

Fig. 3 An example of 2W1R memory module using XOR approach

Property-1: $P \oplus 0 = P$;
Property-2: $Q \oplus Q = 0$;
Property-3: $P \oplus Q \oplus Q = P$.

The property-3 is constructed using the first two properties. Perform the XOR operation between P and Q values and later XOR these results with Q value to recover the original P-value. The same XOR concept is applied to the MPM module to read the updated write value. An example of a 2W1R memory module using the XOR approach is represented in Fig. 3. Each write port data value ('P' or 'Q') is copied to its BRAM, so it can hold the same values after reading the memory location using XOR operation.

The write port (W_{data0}) data 'P' from the upper location is XOR with an old value (P_{old}) from the exact memory location of the write port (W_{data1}) bank. Similarly, The Write port (W_{data1}) data 'Q' from the lower location is XOR with an old value (Q_{old}) from the exact memory location of the Write port (W_{data0}) bank. The read port will read data by performing the XOR operation of ($P \oplus P_{old}$) with the old value of 'P' (P_{old}) from the exact memory location of upper and lower memory banks. So, the read port reads (R_{data0}) the most recent updated data by computing ($P \oplus P_{old}$) $\oplus P_{old}$ and retains the original value of 'P'.

In this work, 2W4R and 3W4R memory modules are designed using the XOR approach. Figure 4 illustrates the 2W4R memory module using the XOR approach. The 2W4R module contains two write ports (W_{data0} and W_{data1}), and four read ports (R_{data0}, R_{data1}, R_{data2}, R_{data3}). The same procedure of 1W2R using the XOR approach is deployed to this module also. The Write port (W_{data0}) from the memory bank (M_{w0_w1}) is XOR with old contents of a memory bank (M_{w1_w0}) of the same memory location, and the results are copied to all the upper memory banks in a row. Similarly, The Write port (W_{data1}) from the memory bank (M_{w1_w0}) is XOR with old contents of the memory bank (M_{w0_w1}) of the same memory location, and the results are

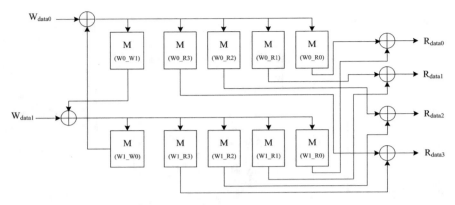

Fig. 4 2W4R memory module using XOR approach

copied to all the lower memory banks in a row. The four read ports read the data by performing XOR operation from the corresponding memory bank (Upper and lower) of the same location column-wise to generate the final read port outputs.

The XOR-based 3W4R memory module is represented in Fig. 5. Additionally, one row and column of memory banks are increased in the 3W4R memory module. For 3-writes and 4-reads, the XOR approach requires $3 * (3 - 1) + 4 = 18$ BRAMs. In general, for p-writes and q-reads, the XOR approach requires $p * (p - 1) + q$ BRAMs.

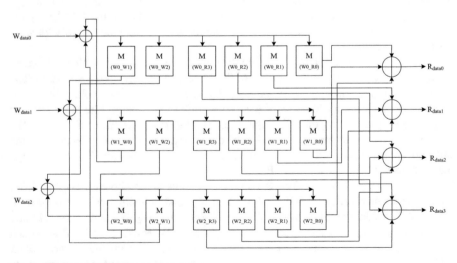

Fig. 5 OR-based 3W4R memory module

4   Results and Discussions

This section analyzes the results and discussion of the multiport memory designs concerning FPGA. The multiport memory modules using LVT and XOR approaches are synthesized and implemented on Virtex-7 FPGA. The synthesized results include Slices, LUTs, Maximum operating frequency (Fmax), and BRAM utilization on FPGA for the corresponding designs. The proposed techniques are compared with existing similar approaches with better hardware constraints improvements. In this work, The Virtex-7 FPGA is considered for synthesis and implementation, which includes the device of XC7V585T with FFG1761 Package. The Virtex-7 FPGA provides high performance with optimization features for any standard design. The Virtex-7 FPGA contains 91K Slices, 582K Logic cells, 728.4K CLB Flip-flops, BRAM of 795, distributed RAM of 6938 Kb, and 1260 of DSP slices [28]. In this work, the integration of write and read port-based 2W4R and 3W4R memory modules are designed using LVT, and XOR approaches. The different resources analysis for 2W4R memory modules using LVT and XOR approach is represented in Fig. 6.

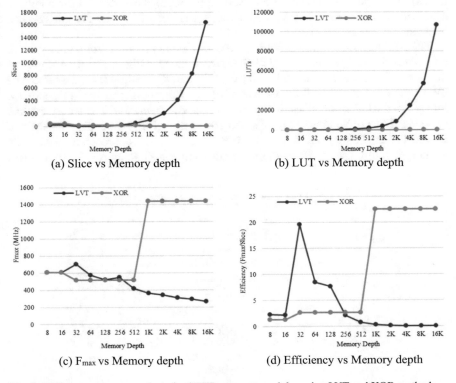

(a) Slice vs Memory depth

(b) LUT vs Memory depth

(c) F$_{max}$ vs Memory depth

(d) Efficiency vs Memory depth

Fig. 6 Different resources analysis for 2W4R memory modules using LVT and XOR methods

The 2W4R memory module synthesized for different memory depth ranges from 8 to 16K on Virtex-7 FPGA. Figure 6a depicts the slices versus. memory depth. The LVT and XOR approaches are used in the 2W4R memory module to create 16.3K and 64 slices, respectively. Figure 6b shows the LUTs of the 2W4R memory module versus. memory depth. It employs 107K and 192 LUTs, respectively, using the LVT and XOR approaches. Figure 6c depicts the Fmax (MHz) versus memory depth of the 2W4R memory module. The LVT and XOR approaches operate the 2W4R module at 267.3 MHz and 1443 MHz, respectively. Figure 6d depicts the hardware efficiency of the 2W4R memory module as a function of memory depth. The hardware efficiency of the 2W4R module is 0.0164 Fmax/slice for LVT and 22.55 Fmax/slice for the XOR method.

The 2W4R memory module using XOR approaches utilizes fewer chip area resources (Slices and LUTs), provides better operating frequency and hardware efficiency than the 2W4R memory module using the LVT approach. The Slices and LUTs utilization exponentially increase as the memory depth increases in the LVT approach. In contrast, the slices and LUT's utilization downwards linearly as the memory depth increases in the XOR approach. Because these resources (slices and LUTs) are replaced with BRAMs and XOR approaches, it avoids the write conflicts, even with multiple reads. In the LVT approach, write conflicts arise even more when increasing the read port sizes.

The BRAM utilization for 2W4R and 3W4R memory modules for memory depth range from 512 to 16K on Virtex-7 FPGA is represented in Fig. 7. The BRAM utilization of the 2W4R memory module using the LVT and XOR approach is the same. The 2W4R memory module utilizes BRAM of 64 and 128 for 8K and 16K, respectively, for LVT and XOR approaches. Using the LVT approach, the 3W4R memory module utilizes BRAM of 96 and 192 for 8K and 16K, respectively. Similarly, BRAM's of 89 and 177 are being used for 8K and 16K using the XOR approach on the 3W4R memory module. The 3W4R memory module using the XOR approach provides less

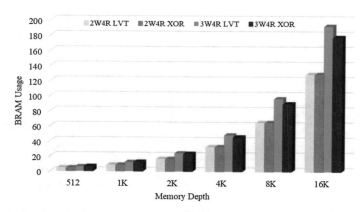

Fig. 7 BRAM Utilization for 2W4R and 3W4R memory modules on Virtex-7 FPGA

Table 1 Performance comparison of different 2W4R memory approaches on Virtex-7 FPGA

Design approaches	BRAM		Fmax (MHz)		Slices (%)	
	8K	16K	8K	16K	8K	16K
LVT-based design [16]	64	128	149	116	16.7	32.1
LVT-proposed	64	128	293	268	1.12	2.24
XOR-based design [16]	80	160	270	260	0.8	0.4
XOR-proposed	64	128	1442	1422	0.008	0.006

BRAM overhead of 7.29% and 7.85% for 8K and 16K memory depths, respectively than the 3W4R memory module using the LVT approach.

The performance comparison of LVT and XOR-based 2W4R memory modules with an existing similar approach [16] on Virtex-7 is tabulated in Table 1. The LVT-based 2W4R memory module utilizes 2.24% Slices, 128 BRAMs, and works at 268 MHz frequency for 16K memory depth.

The existing LVT-based 2W4R memory module [16] uses 32.1% Slices, 128 BRAMs, and operates at 116 MHz frequency for 16K memory depth. In contrast, The XOR-based 2W4R memory module utilizes 0.006% Slices, 128 BRAMs, and works at 1422 MHz frequency for 16K memory depth. The existing XOR-based 2W4R memory module [16] utilizes 0.4% Slices, 160 BRAMs, and operates at 260 MHz frequency for 16K memory depth. The proposed LVT and XOR-based 2W4R memory modules use fewer slices, BRAM s and work at high operating frequencies than existing LVT and XOR-based 2W4R memory modules [16].

Table 2 compares the performance of LVT and XOR-based 3W4R memory modules with an existing similar technique [16] on the Virtex-7. The LVT-based 3W4R memory module has 4.49% Slices, 192 BRAMs, and operates at 268 MHz for a memory depth of 16K. The existing LVT-based 2W4R memory module [16] uses 66.4% Slices, 192 BRAMs, and operates at 86 MHz for a memory depth of 16K. The XOR-based 3W4R memory module, on the other hand, uses 0.018% Slices, 177 BRAMs, and operates at 1422 MHz for a 16K memory depth. The existing XOR-based 3W4R memory module [16] uses 0.9% Slices, 288 BRAMs, and operates at a frequency of 199 MHz for a memory depth of 16K. Compared to existing LVT and XOR-based 3W4R memory modules, the proposed LVT and XOR-based 3W4R memory modules employ fewer slices and BRAMs and operate at higher frequencies [16].

Table 2 Performance comparison of different 3W4R memory approaches on Virtex-7 FPGA

Design approaches	BRAM		Frequency (MHz)		Slices (%)	
	8 K	16 K	8 K	16 K	8 K	16 K
LVT-based design [16]	96	192	111	86	35.4	66.4
LVT-proposed	96	192	293	268	2.25	4.49
XOR-based design [16]	144	288	220	199	0.6	0.9
XOR-proposed	89	177	1442	1422	0.013	0.018

5 Conclusions

In this manuscript, the multiported memory modules like 2W4R and 3W4R are designed using LVT and XOR approaches on Virtex-7 FPGA. The LVT and XOR strategy improves write port access while integrating with the replication mechanism to improve read port access. On the Virtex-7 FPGA, the 2W4R and 3W4R memory modules with varied memory levels are generated and implemented. At 16K memory depth, the 3W4R module employing the XOR technique improves BRAM utilization by 7.85% over the LVT approach. The 2W4R and 3W4R modules using LVT and XOR approaches are compared individually with existing methods on Virtex-7, improving hardware constraints like Slices, BRAM, and frequency. In the future, a hierarchical banking approach is adapted to MPM using LVT and XOR approaches to reduce the BRAM usage and enhance the operating frequency on FPGA.

References

1. Zuo, W., Zuo, Q., Li, J.: An intelligent multiport memory. In: 2008 International Symposium on Intelligent Information Technology Application Workshops, pp. 251–254. IEEE (2008)
2. Trang, H.: Design choices of multi-ported memory for FPGA. Special. Inf. Commun. Technol. edition (JICT), Le Quy Don University, **2** (2013)
3. Malazgirt, G.A., Yantir, H.E., Yurdakul, A., Niar, S.: Application-specific multiport memory customization in FPGAs. In: 2014 24th International Conf on Field Programmable Logic and Applications (FPL), pp. 1–4. IEEE (2014)
4. Townsend, K.R., Attia, O.G., Jones, P.H., Zambreno, J: A scalable unsegmented multiport memory for FPGA-based systems. Int. J. Reconfigurable Comput (2015)
5. Dighe, S., Pagare, R.A.: FPGA based data transfer using the multiport memory controller. In: Int. Conf. for Convergence for Technol. 2014, pp. 1–3. IEEE (2014)
6. Strollo, E., Trifiletti, A.: A shared memory, parameterized and configurable in FPGA, for use in multiprocessor systems. In: 2016 MIXDES-23rd Int. Conf. Mixed Design of Integrated Circuits and Systems, pp. 443–447. IEEE (2016)
7. LaForest, C.E., Gregory Steffan, J.: Efficient multi-ported memories for FPGAs. In: Proceedings of the 18th Annual ACM/SIGDA İnternational Symposium on Field Programmable Gate Arrays, pp. 41–50 (2010)
8. LaForest, C.E., Liu, M.G., Rapati, E.R., Steffan, J.G.: Multi-ported memories for FPGAs via XOR. In: Proceedings of the ACM/SIGDA İnternational Symposium on Field Programmable Gate Arrays, pp. 209–218 (2012)
9. Yantir, H.E., Bayar, S., Yurdakul, A.: Efficient implementations of multi-pumped multiport register files in FPGAs. In: 2013 Euromicro Conference on Digital System Design, pp. 185–192. IEEE (2013)
10. Abdelhadi, A.M.S., Lemieux, G.G.F.: Modular multi-ported SRAM-based memories. In: Proceedings of the 2014 ACM/SIGDA İnternational Symposium on Field-Programmable Gate Arrays, pp. 35–44 (2014)
11. Laforest, C., Li, Z., O'rourke, T., Liu, M., Gregory Steffan, J.: Composing multi-ported memories on FPGAs. ACM Trans Reconfigurable Technol Syst (TRETS) **7**(3), 1–23 (2014)
12. Lin, J-L., Lai, B-C.C.: BRAM efficient multi-ported memory on FPGA. In: VLSI Design, Automation and Test (VLSI-DAT), pp. 1–4. IEEE (2015)
13. Muddebihal, A.A., M., Purdy, C.: Design and implementation of area efficient multi-ported memories with write conflict resolution. In: 2015 IEEE 58th International Midwest Symposium on Circuits and Systems (MWSCAS), pp. 1–4. IEEE (2015)

14. Kawashima, H., Sasaki, T., Fukazawa, Y., Kondo, T.: Register port prediction for a banked register file. In: 2015 Third International Symposium on Computing and Networking (CANDAR), pp. 551–555. IEEE (2015)
15. Abdelhadi, A.M.S., Lemieux, G.G.F.: Modular switched multiported SRAM-based memories. ACM. Trans. Reconfigurable. Technol. Syst. (TRETS) 9(3), 1–26 (2016)
16. Lai, B-C.C., Lin, J-L.: Efficient designs of multiported memory on FPGA. IEEE Trans Very Large-Scale Integr (VLSI) Syst. 25(1), 139–150 (2016)
17. Shahrouzi, S.N., Perera, D.G., An efficient embedded multi-ported memory architecture for next-generation FPGAs. In: 2017 IEEE 28th Int. Conf. on Application-specific Systems, Architectures and Processors (ASAP), pp. 83–90. IEEE (2017)
18. Lai, B-C.C., Huang, K-H.: An efficient hierarchical banking structure for algorithmic multi-ported memory on FPGA. IEEE Trans Very Large-Scale Integr (VLSI) Syst. 25(10), 2776–2788 (2017)
19. Ullah, I., Ullah, Z., Lee, J.A.: Efficient TCAM design based on multipumping-enabled multiported SRAM on FPGA. IEEE Access 6, 19940–19947 (2018)
20. Manivannan, T.S., Meena S.: A 4-read 2-write multiport register file design using pulsed-latches. In: 2018 Second International Conference on Electronics, Communication and Aerospace Technology (ICECA), pp. 262–267. IEEE (2018)
21. Humenniy, P., Volynskyy, O., Albanskiy, I., Voronych, A.: Designing a shared access memory and its application in data transmission and protection systems. In: 2018 14th International Conference on Advanced Trends in Radioelecrtronics, Telecommunications and Computer Engineering (TCSET), pp. 143–147. IEEE (2018)
22. Shahrouzi, S.N., Darshika G.P.: Optimized counter-based multi-ported memory architectures for next-generation FPGAs. In: 2018 31st IEEE International System-on-Chip Conference (SOCC), pp. 106–111. IEEE (2018)
23. Kwan, B.P., Chow, G.C.T., Todman, T., Luk, W., Xu, W.: Lossy multiport memory. In: 2018 Int. Conf. on Field-Programmable Technology (FPT), pp. 250–253. IEEE (2018).
24. Chen, B-Y., Chen, B-E., Lai, B-C.: Efficient write scheme for algorithm-based multi-ported memory. In: 2019 International Symposium on VLSI Design, Automation and Test (VLSI-DAT), pp. 1–4. IEEE (2019)
25. 'Jain, H., Edwards, M., Elenberg, E. R., Rawat, A. S., Vishwanath, S.: Achieving Multiport Memory Performance on Single-Port Memory with Coding Techniques. In: 2020 3rd International Conference on Information and Computer Technologies (ICICT), pp. 366–375. IEEE (2020)
26. Shahrouzi, S. N., Alkamil, A., Perer, D. G.: Towards composing optimized bi-directional multi-ported memories for next-generation FPGAs. IEEE Access 8, pp. 91531–91545 (2020)
27. Zhang, R., Wijeratne, S., Yang, Y., Kuppannagari, S. R., Prasanna, V. K.: A high throughput parallel hash table on fpga using xor-based memory. In: 2020 IEEE High-Performance Extreme Computing Conf. (HPEC), pp. 1–7. IEEE (2020).
28. Xilinx 7-series-product-selection-guide, XMP101. 8 (2021)

Low-Cost Smart Cart with Nutritional Information

M. Florance Mary and B. Hemakumar

Abstract In this work, an architecture and solution for the acquisition of products and navigation in grocery stores using a smart cart are proposed. The cart also can provide nutritional information of certain products. This SMART cart was named as billing, navigation, and nutrition (BNN) cart highlighting the advantages of improved quality of services provided by retailers to augment the consumer value. Thus, BNN cart provides smart billing of products, easy navigation inside the shop for identification of products, and nutrient information display, which will be a more wanted choice in the upcoming days and thereby providing a great customer experience.

Keywords IoT · Smart shopping · Navigation · Nutrition · Smart billing

1 Introduction

Shopping is the process where individuals get their everyday necessities running from usual commodities, garments, groceries, and so forth. In modern day, growth of shopping malls has tremendously increased. Different problems faced by people buying commodities from shopping malls are experimented and experienced. The top three problems were identified by interviewing consumers from different shopping malls around Puducherry, a small town in Southern India. The following malls in Puducherry were taken for study, viz. Pothys, Providence Mall, and different fruits shops. The top three problems customers experienced were.

1. Time consumption for billing because of the queue.
2. Navigation inside the store to identify different items is not available.
3. Nutrient information of food products is not available.

M. Florance Mary (✉) · B. Hemakumar
Department of Electronics and Instrumentation, Puducherry Technological University, Puducherry, India
e-mail: florancemary@pec.edu

B. Hemakumar
e-mail: hemakumarb@pec.edu

© The Author(s), under exclusive license to Springer Nature Singapore Pte Ltd. 2022   1023
V. Suma et al. (eds.), *Evolutionary Computing and Mobile Sustainable Networks*,
Lecture Notes on Data Engineering and Communications Technologies 116,
https://doi.org/10.1007/978-981-16-9605-3_72

Conventionally billing is done by barcode scanning and for paying the price of the commodities: individuals use cash, cards and digital wallets. For the sake of navigation inside the shop, the shop personal and name boards help individuals to identify the products. This is quite cumbersome and quite misleading at times. For nutrient information on the foods, there are only standard android and IOS applications which display them for standard products only.

Hence to overcome these shortcomings, a novel low-cost SMART cart is proposed which has features like automatic billing, navigation inside the shop, and nutrient display of foods using image processing technique.

2   Literature Review

In order to identify objects, Kalyani Dawkhar et al. [1] proposed usage of radio waves to transfer data from an electronic tag. Further, a reader was employed for tracking. Thereby each product has the individual card based on RFID representing the product name.

Identifying the typical problems faced by supermarket visiting customers, P. Chandrasekar and T. Sangeetha [2] have proposed a central automated billing system. The proposed system has access to the product database and while purchasing, it calculates the total amount. Further, customer will obtain their billing information by their cart identification number.

Shreyas J Jagtapet et al. [3] have proposed a technique for employing a pocket PC as a shopping mall navigator, to help the users to find their interested shops efficiently and effectively, thereby creating an awareness in usage of smart mobile devices.

Automation in all fields has been made to exist with the help of Internet of things. Abul Bashar [4] has proposed one such system using IoT for automation. Aashis Kumar et al. [5] have also proposed an Internet of things for navigation support for visually impaired. They have used Android platform in the calculation of the shortest distance between source and location by employing ultrasonic sensors. These sensors were attached to the person's knees. When sensor detects an obstacle, it triggers a signal and accordingly the mobile application describes the kind of obstacle to the person.

Real-time nutritional value of food detection was developed by Raza Yunus et al. [6] using deep learning techniques. These techniques help in accurate identification of food by using image analysis. Attributes and ingredients are predicted by deriving semantically correlated words from a database, collected over the Internet.

Another interesting research study by Jun Zhou, et al. [7] focused on calorie estimation from food pictures. Their system can estimate calories in 20 pictures of food chosen in 2000 individuals, thereby proposing a framework for healthy diet.

Many researchers have proposed the use of RFID as an excellent tool for automatic billing system. But still now the researchers have employed methods like cash and cards for checkout purposes. Further, IoT-based billing and navigation will be useful in reducing time and increasing the efficiency of customer experience. Further, nutrient information display will be added value to the consumers who are health conscious. Hence, in this work a novel low-cost SMART shopping cart (BNN) is proposed, which mixes the technologies of automatic billing, navigation, and food nutrition information.

3 Architecture of BNN Cart

The main building blocks of BNN cart include (1) billing module, (2) navigation module, and (3) nutrient detection module. All the three modules are integrated by IoT to enable automation and control. The block diagram of the BNN cart is shown in Fig. 1.

3.1 Billing Module

The main part of the billing module is the RFID Tag. The RFID Tag was placed on the product. The trolley employs a RFID reader interfaced with Raspberry Pi board and Wi-Fi module. The electronics in the trolley system communicates with the server via IoT. EM18 RFID module was used here. It is a 125 kHz RFID reader with serial output and a range of 812 cm. Upon placement of products in the trolley, its code will be detected and automatically its price will be stored in the memory. When the number of unit products purchased increases, the costs will be automatically appended, thereby facilitating billing. A provision for product name and it is cost to be displayed on local LCD display which is also provided. Once the person crosses,

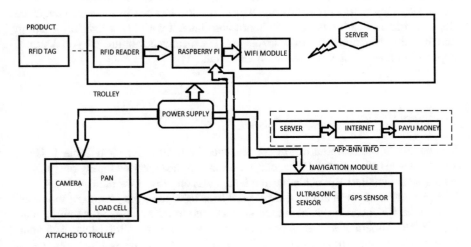

Fig. 1 Architecture of BNN cart

the billing area of the shop the amount will be automatically debited from the digital wallet connected to their account via IoT. So customers need not wait in the billing counter, thereby reducing the time to stand in the queue.

3.2 Navigation Module

Navigation module consists of Raspberry Pi board and GPS sensors. The floor map of the entire shop is fed as the base location for navigation. The places where different items are placed will be shown to the customer. Using a GPS sensor, the navigation module identifies the different items placed inside the shop. The trolley will be able to show the path to navigate through the entire base location like a map. The Pi board communicates with the GPS sensor and obtains the coordinates and the direction to traverse the base location. Based on that information, the users can move the trolley in respective direction.

3.3 Nutrient Detection Module

Nutrient detection module includes of a camera with a pan. The pan is inbuilt with a load cell. The pan with load cell sensor calculates the weight of the object. The camera takes snapshot of the object placed in the pan. This image gets transferred to the microcontroller Pi. The microcontroller initially preprocesses the image and further extract shape features, texture features, and color of the object placed in the pan. The extracted features will then be compared with the database stored, and based on the extent of pattern match, the object identification is done using TensorFlow logic. At a time, only one type of product can be placed in the pan. Once the object identification is done, Pi board correlates it with the weight information and the calories of nutrient will be displayed. The display is facilitated by an Android application interacting with the microcontroller.

4 Hardware

The details of the main hardware components are used in listed in Table 1. Here, Rpi Zero WH (wireless Wi-Fi, header) comes under the extended family of Rpi Zero which is the smallest Rpi available and hence was employed for the BNN cart.

Rpi 5 MP camera board module is used here to capture the picture of the items kept in the BNN cart. Pi camera module was used to take pictures for training and testing. The camera was initially programmed to take pictures after a delay of 5 s and save it. The specifications such as angle, brightness, and other settings were set to capture a perfect snapshot without any distortion. A python code was developed to

Table 1 List of main hardware components

S. No.	Name of the Component	Specification
1	RFID Reader—EM18	125 kHz
2	Raspberry Pi Board—Rpi Zero WH	512 MB RAM with Wi-Fi
3	Rpi camera board module	5MP
4	Robocraze digital load cell weight sensor	Weight sensor 10 kg 24 BIT precision ADC
5	Global positioning system (GPS) NEO-6MV2	50 Channels—GPS L1 frequency —SBAS; Maximum navigation update rate: 5 Hz

ensure that the camera module is activated to take pictures when the object is placed in the BNN cart. The overall cost of setting up the BNN cart was 5500 rupees.

5 IoT Communication

The transmitted data from the trolley section will be received through server to the billing system which is a PC in our case and the product details with the price are displayed onto the screen using free software for IoT. 000webhost.com is a free web hosting service that was used in this work. It is quite easy to generate shopping cart using this website. It supports PHP and MySQL language. It also provides unlimited space and bandwidth for the user. The calculated data was stored in the server. 000webhost.com supports to develop the database and take proper corrective action with immediate alertness. It is a free web service to develop IoT-based online carts. PHP code was written to retrieve the data from the ESP module and store it in the database using GET request. The stored value was read back using JQUERY and displayed in webpage using HTML language. The outputs were observed in two sections: one in the trolley section and the other in the PC section. The products with RFID tags attached were placed in the trolley provided with local LCD display as shown in Fig. 2.

On sensing the RFID tag, the item name of that particular tag (Here-ITEM1) and the cost of it (Here-Total) will be displayed in the locally mounted LCD display which is fixed to the trolley for the local indication. The same was also transmitted

Fig. 2 Local LCD display in the trolley

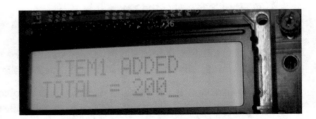

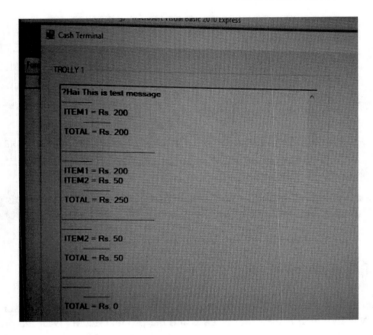

Fig. 3 Output snapshot of the customer billing

by Wi-Fi module inbuilt with Raspberry Pi board in the trolley side to the server and its cost is displayed in the PC screen as shown in Fig. 3.

The nutrient detection module uses TensorFlow algorithm for object identification. Nutrition detection using image processing was implemented only for selected fruits and vegetables. 50 different images for each fruit type were taken. Fruits taken up for the study include apples, oranges, and guava. Totally, 150 images were taken for the study. Out of which 100 images were used for training and remaining 50 were used for testing. The steps employed in object identification were listed below:

1. Apple image has been trained as an object, and it has been detected.
2. An open-source algorithm, TensorFlow, has been used for object detection
3. About 100 images / fruit have been used to train the model library
4. Object (apple) in each image has been labeled using labeling software, which is saved as xml file
5. This xml file is converted to csv file
6. Csv file contains the image name and coordinates of the object that is present in that image, which is labeled before
7. Using this csv file, record file is created to train the model
8. After training, this trained file is directly used in the program.

Image comparison is the method used to detect object in the real-time images. Object identification with its weight was used to calculate the nutritional information from the database. Figure 4 shows the identification of the object (viz. apple) with 58% accuracy.

Fig. 4 Object identification after training

Figure 5 shows the Android application developed to display the nutritional information. The application has provision to show the information for the captured image. In this way, the nutritional information display will be useful to make it a smart cart for shopping.

6 Testing

For testing the efficiency of the method, a grocery and fruit shop in the locality of Kalapet (part of Puducherry) was chosen. The shop was normally crowded in the evenings around 7 pm. This proposed billing module and nutrition detection module in the trolley were installed there. RFID tags were placed on the object. Some of the customers were used to opt for our proposed method. Initially, people hesitated. Later around 4 volunteers in local shops were ready to use this proposed IoT-based billing system. Initially, their wallet numbers were enrolled by us into the PC. The users found the product quite interesting, and they gave good feedback. A summary of their feedback is listed below:

a. Our billing system was limited only to digital wallets.
b. All the customer wallet numbers have to be enrolled before they start shopping.
c. Displaying Nutritional information was very useful.

Fig. 5 Android application nutritional information

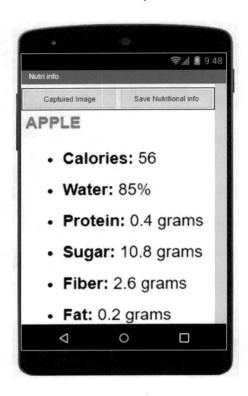

7 Conclusion

A low-cost smart shopping trolley to enhance the customer experience in malls is proposed in this paper. Quality billing module with printer alone costs around Rs 6000 in the market. Our proposed system has additional nutrition detection and navigation modules, and the cost comes around Rs 5500. Here, Raspberry Pi board was used for controlling the total process. The BNN cart works with three modules: one for billing with RFID and IoT; the second for navigation inside mall; and third module for nutrition detection for fruits and vegetables. TensorFlow is used for object detection and was operated using python code. Proposed BNN cart will be able to revolutionize the future of shopping malls by enabling automatic billing, easy navigation, and nutrition information display. Thus, BNN cart serves in enhancing the customer experience to a greater extent.

References

1. Dawkhar, K., Dhomase, S., Mahabaleshwarkar, S.: Electronic shopping cart for effective shopping based on RFID. Int. J. of Innovative Res. Electri. Electron. Instrum. Control Eng. 3(1), 84–86 (2015)
2. Chandrasekar, P., Sangeetha, T.: Smart shopping cart with automatic billing system through RFID and Zigbee. IEEE International Conference on Information Communication and Embedded Systems (ICICES). IEEE Xplore, (2014)
3. . Jagtap, S.J., Khodke, S., Korde, V., Deokule, A., Baviskar, C.: Pocket mall navigator: bridging digital and traditional shopping experience. Int. Res. J. Eng. Technol. 5(4), 409–412 (2018)
4. Bashar, A.: Agricultural machine automation using iot through android. J. Electri. Eng. Autom. 1(2), 83–92 (2019).
5. Kumar, A, Jain, M., Saxena, R., Jain, V., Siddharth. : IOT-Based navigation for visually impaired using Arduino and android. Advances in communication, devices and networking: proceedings of ICCDN, 2018, Lecture Notes in Electrical Engineering 537, Springer (2019)
6. Yunus, R., Arif, O., Afzal, H., Amjad, M.F., Abbas, H., Bokhari, H.N., Haider, S.T., Zafar, N., Nawaz, R.: A framework to estimate the nutritional value of food in real time using deep learning techniques. Spec. Sect. Mob. Multimedia. Healthc. IEEE Access, Vol. 7(2643–2652) (2019)
7. Zhou, J., Bell, D., Nusrat, S., Hingle, M., Surdeanu, M., Kobourov, S.: Calorie estimation from pictures of food: crowdsourcing study. Interact. J. Medical Res. 7(2), (2018)

ML-Based Comparative Analysis of Interconnect RC Estimation in Progressive Stacked Circuits

M. Parvathi and Anwar Bhasha Pattan

Abstract From the MOS R–C models, wire resistance and parallel plate capacitance are proportional to length of the wire. Hence it is essential to calculate parasitic components of interconnect with accuracy. The existing empirical approaches involve Maxwell equations with repeated delay models that result in time-consuming process for complex SoCs. A novel approach is proposed using ML algorithms for estimation of interconnect R, C in progressive stacked SRAM circuits designed from 6T to 10T. Using linear regression, for micron meter technologies, the prediction accuracies for both R and C observed are 61 and 73.8% under pre-layout simulations, 76 and 80% under post-layout simulations, respectively. Similarly, for nanometer technologies the prediction accuracies for both R and C observed are 76 and 81% in the case of pre-layout modeling and 63 and 80% in the case of post-layout modeling, respectively. Further, it is observed the accuracy has reached 100% using k means clustering.

Keywords Interconnect RC · 6T to 10T SRAM · ML algorithms · Linear regression · Prediction accuracy

1 Introduction

The switching speed is the key indicator of circuit's dynamic performance, hence it must be predicted and optimized at the initial design phase of any digital block. The traditional method of estimating the dynamic speed of a digital system is based on the assumption that the output drive elements are mainly either capacitive or resistive. Comparatively few delay models of logic gates use capacitive load at their output node, hence, the estimation of dynamic behavior of the circuit is made easy. The contemporary delay estimation models are developed based on the type of gate load, assuming purely capacitive, and are as: (i) transistor internal parasitic capacitances,

M. Parvathi (✉) · A. B. Pattan
Deptartment of ECE, BVRIT HYDERABAD College of Engineering for Women, Hyderabad, Telangana, India
e-mail: parvathi.m@bvrithyderabad.edu.in

© The Author(s), under exclusive license to Springer Nature Singapore Pte Ltd. 2022
V. Suma et al. (eds.), *Evolutionary Computing and Mobile Sustainable Networks*,
Lecture Notes on Data Engineering and Communications Technologies 116,
https://doi.org/10.1007/978-981-16-9605-3_73

(ii) capacitances due to interconnect lines, and (iii) input capacitances due to fan-out gates. Among these three components, the capacitance due to load conditions imposed by the interconnection lines causes serious issue in the performance phase. Net delay or wire delay or Interconnect delay is due to the finite resistance and capacitance of the net, as shown in Eq. (1).

$$\text{Wire delay} = f(\text{Rnet, Cnet} + \text{Cp.in}) \tag{1}$$

Several delay models are existing as of today, some of which can provide more accurate result, however, they take more runtime to do the calculations and those which are fast provides less accurate value of delay. Most popular delay models are Wire Load models of type L, T, Pi, and Elmore Delay model. However, these models result in poor accuracy and bad option for an RC line. Moreover, delay models cannot be mixed within a single library. Ideally, physical wire will not persist in the design, so the Wire/Net delay cannot be calculated. Issues with interconnect lines are in the form of thickness, as the number of layers increases the interconnect layer thickness increases, which leads to the use of higher dielectric layer with high permittivity. This causes the inter wire capacitance to rise.

ICs are manufactured with different electronic components in the form of layers of different materials each with 5 μm to 0.1 mm thick on the semiconductor surface. Each material follows a specific shape, size, and location and must be accurately formed for proper fabrication. Each layer will be formed using corresponding mask layer with a specific geometrical shape. The ICs may have different layers like diffusion, polysilicon and various metal layers, perhaps 30 or more layers may be formed in a final chip with every single layer separated by necessary interconnect layer. In the olden days of IC fabrication, aluminum was the most commonly used material for interconnect. But the advanced fabrication procedures aluminum-based alloys containing silicon, copper, or both in order to reduce the delay that is caused by interconnect layers.

Due to the advantage of scaling nowadays, the technology has gone down which in turn increases the density of the standard cells with increase in interconnections as well. In fact, these interconnections are made of metal wires, which means number of metal wires increases. Each metal layer is different with its dimensions. The metal layers are given names like metal1, metal2, etc. based on the order of their fabrication style. For example, for 0.12 μm technology, lower level metal layer is labeled as metal1 and upper level metal layer is labeled as metal6. Usually, the lower metal layers are thinner than the upper metal layers and proportionately increase their dimension as the level increases. Silicon di-oxide (SiO2) is the material used for isolation between each layer. The material used as junction between diffusion and metal is generally known as "Contact", and is used in general for poly to metal junction, "Via" junction, poly2 to metal1, similarly between metal2 to metal3, and so on.

Generally, the upper metal layer will be thinly populated that increases the spacing between interconnects and hence leads to form sidewall capacitance between parallel adjacent lines. This further leads to cause interconnect delay, performance degradations will be triggered as well. Interconnect delay is also called net/wire delay or extrinsic delay i.e., the time gap between the signal application to the net and device reception. The main reason behind this delay is R and C that exhibit by the net line.

Now a day's technology made the interconnect delay dominant and has become the most important limiting factor in the IC performance, rather than concern with the switching speeds. The developments in deep-submicron process geometries have enabled the companies to build compact, efficient, and cost-effective devices. It is observed that in the present-day technology that the interconnect delay is predominant by 75 percent than that of total path delay. As a result, the interconnect technology also scaled down proportionately with the technology factor. Thus, the scaling of the interconnect width leads to decrease in cross-sectional area and overall resistance perhaps would increase. At the same time, scaling of interconnect wires increases the sidewall capacitance.

In the advanced technology processes as the wire width reduces the impact of wire resistance has become critical than the previous technology nodes. Hence, the chip implementations in this new era of technologies beyond 28 nm have become complex and challenging. Larger wire delay affects the transition time that results in poor timing in clock tree, further leading to poor performance in physical implementation. The proposed method shows how to estimate wire capacitance and resistance of chosen length and further that helps in delay estimation and design optimization as well to achieve best performance during the physical implementation flow. The wire layers on the layout have resistance; capacitance and inductance will introduce interconnect parasitic effects in real scenario. If the wiring complexity increases the parasitic effect will be more and degrade the performance. The important parameters that may get affected with parasitic RC are delay, energy consumption, power, noise and reliability of IC, etc. In order to estimate the design performance, it is required to identify the accurate delay model for the interconnect parasitic RC. Section 2 includes a review on existing delay models; Sect. 3 will discuss the proposed method of estimation of parasitic values for chosen length using ML algorithms. Section 4 will give details on result analysis. Section 5 will brief on conclusions.

2 Review on Related Work

The latest lowest technology with complex circuit layouts shows much effect on circuit performance in terms of changing power dissipation and delay characteristics. Currently very few EDA tools coming up with solutions to the RC delay calculation with high precision in the process of RC extraction from full chip layouts. The extraction can be done on selected area, or a single net. The parasitic R,Cs can be extracted in terms of distributed or lumped values based on type of tool that is used for extraction.

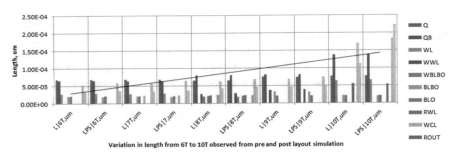

Fig. 1 Impact of interconnect length on post and prior simulations

Some of the tools provide physically accurate solutions in resulting exact values that are mapped with real-world parasitic values. However, these tools rely on electromagnetic parameters for calculation using Maxwell's equations. This increases the mathematical complexity while striving for high accurate parameters. Hence it is required to use some assumptions in the process of extracting parasitic values using approximate solutions with pattern matching techniques. The previous work [1, 2] has suggested estimation of interconnect parasitics after the floor plan, placement, and routing steps. However, the same method limits its use to analog circuits. The research toward generation of automatic layouts for analog circuits has come up with [3, 4]. Also, the tools are not fully automated and mature yet to apply in all the possible layout extraction environments. It will be difficult for the schematic designers if required accuracy level is not met in the extraction process and redesign phases would become time-consuming, also cost-effective. Hence estimation of both capacitance and resistance is equally important for both analog and digital circuits, as well as the process technology, scales down [5].

The impact of interconnect lengths during pre and post-layout phase are observed as shown in Fig. 1, whereas the impact of length over R and C observed is shown in Figs. 2 and 3, using 6T to 10T SRAM designs, extracted from pre and post-layout simulations. "L" refers to the length extracted in pre-layout simulation and similarly "LPS" refers to values extracted in post-layout simulations.

These models are developed in Microwind tool, in which DSCH is used for circuit simulation and Microwind tool is used for layout simulation and CIF extraction as well. Traditionally methods like distributed RLC networks were used to reduce π model. From this π model, the effective capacitances were determined [3]. Majority of the implementations use the mathematical models [4, 5] through which new algorithms have been formulated for the effective capacitance calculation in very deep submicron (VDSM) technologies, [6–8].

Most of the present EDA tools provide the facility for extraction of interconnect parasitic R and C and further the extracted RCs will be used in classifying them into different categories. Parasitics will be categorized into three types based on their association with the internal components on a typical semiconductor die. Parasitics if they are associated with the semiconductor devices on the die are called Front-end of the Line (FEOL) parasitics, if they are associated with contacts then they are

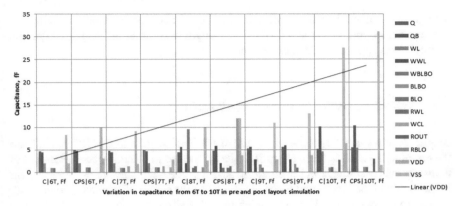

Fig. 2 Impact of interconnect length on capacitance as observed in post and prior simulations

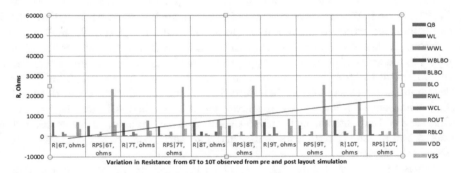

Fig. 3 Impact of interconnect length on resistance as observed in post and prior simulations

called as Middle-end of the Line (MEOL), and if they associate with the interconnect layers then are called as Back-end of the Line (BEOL). Parasitic extraction is a bit challenging if the design gets larger and the process geometries are smaller as well. The primary parameter that gives utmost impact on RCs is length of the wire. In addition to that the large run time while achieving desirable accuracy and are in trade-off always observed in various computational methods.

3 Proposed Method of ML-Based Interconnect Parasitic R&C Estimation

In the proposed method of RC estimation using machine learning techniques the issues that are mentioned in Sect. 2 are addressed. The machine learning techniques are deployed in the process of estimation of RCs from layouts that are obtained in both pre and post-layout simulations. The simulation data obtained from layouts is used to train the models and further predict the value of interconnect parasitic RC. It is equally important to the use of ML paradigm in the analog mix signal domain to emphasize ease of parasitic estimation.

3.1 SRAM Circuit Design Considerations

In order to introduce the proposed method of ML-based estimation of RC, considered the circuits like SRAM from 6T to 10T, the advantage in selecting such circuit is to maintain the consistency in the circuit behavior while having progressive increment in the layers in the layout that helps in estimating the essentiality of RC growth with respect to increased metal layers. Traditional SRAM 6T is shown in Fig. 4. In order to improve the read and write stability, the original circuitry of 6T is modified using one or two transistors for separating the read and write operations and as a result, 7T, 8T, 9T, and 10T have been evolved. The corresponding from 6T to 10T SRAM cell layouts are observed at 120 nm technology node in Figs. 5, 6, 7, 8, and 9.

From pre and post layouts one can observe that the addition of metal or diffusion layers in the areas of devices and supply rails. That leads to an increase in parasitic RC values in the post-layout simulation. The need from 6T to 10T SRAM cell development is observed in the literature [9–12], and is mainly to avoid right and read disturbances in the cell by enhancing the cell stability. All these layouts are simulated and extracted using CIF file in Microwind environment for the creation of dataset that includes length, capacitance and resistance observed at each individual node of the design.

Tables 1 and 2 show the relation among the parameters and corresponding design parameters that are considered under the design. Absolute permittivity ε_0 is taken as

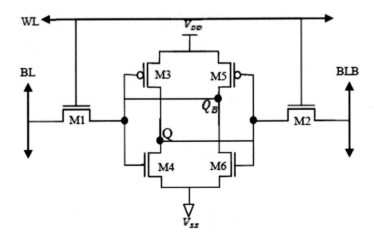

Fig. 4 Traditional 6T SRAM cell

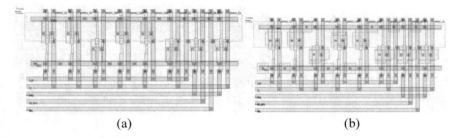

Fig. 5 **a** SRAM 6T Pre layout **b** post layout

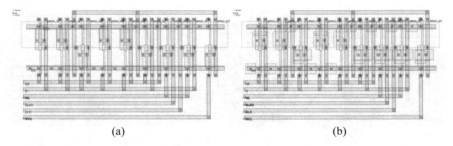

Fig. 6 **a** SRAM 7T circuit pre layout **b** post layout

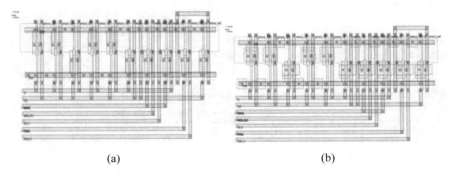

Fig. 7 **a** SRAM 8T circuit pre layout **b** post layout

8.854×10^{-12} F/m and relative permittivity $\varepsilon_r = k\varepsilon_0$, where k is small to be chosen for lowering the leakage currents. In our work, assumed the process parameters are constant for chosen technologies.

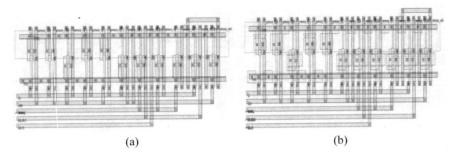

(a)   (b)

Fig. 8 **a** SRAM 9T circuit pre layout **b** post layout

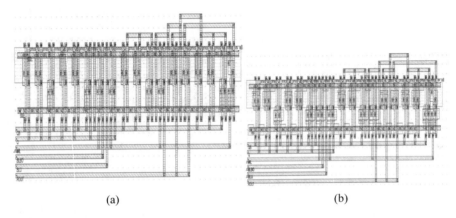

(a)   (b)

Fig. 9 **a** SRAM 10T circuit pre layout **b** post layout

Table 1 Relation among the parameters in Voltage and Current expressions

Mode	Condition	Expression for the current Ids
Cut-off	$V_{gs} < 0$	$I_{ds} = 0$
Linear	$V_{ds} < V_{gs} - V_t$	$I_{ds} = \text{UO} \frac{\varepsilon_0 \varepsilon_r}{\text{TOX}} \cdot \frac{W}{L} \left((V_{gs} - vt).V_{ds} - \frac{(V_{ds})^2}{2} \right)$
Saturated	$V_{ds} > V_{gs} - V_t$	$I_{ds} = \text{UO} \frac{\varepsilon_0 \varepsilon_r}{\text{TOX}} \cdot \frac{W}{L} (V_{gs} - vt)^2$

3.2 ML Algorithms for Modeling SRAM R, and C

Each SRAM circuit is extracted to its corresponding layout that consists of several nets, devices, and interconnects.

Each layout is different from others in terms of number of layers, interconnects, and routing as shown in Tables 3, 4 and 5. In order to make use of ML algorithms, the pre and post-layout simulation data is collected and used as CSV file. Pre and post-layout data CSV files are shown in Tables 6 and 7.

Table 2 Typical values used for parameters in the model of 6T to 10T SRAM

Mos level1 parameters

Paramter	Definition	Typical value 45 nm NMOS	PMOS
VTO	Threshold voltage	0.18 V	−0.15 V
U0	Carrier mobility	0.016 m²/V-s	0.012 m²/V-s
TOXE	Equivalent gate oxide thickness	3.5 nm	3.5 nm
PHI	Surface potential at strong inversion	0.15 V	0.15 V
Gamma	Bulk threshold parameter	$0.4 \text{ V}^{0.5}$	$0.4 \text{ V}^{0.5}$
W	MOS channel width	80 nm minimum	80 nm minimum
L	MOS channel length	40 nm minimum	40 nm minimum

Table 3 Variation of length from 6T SRAM to 10T SRAM of pre and post layout simulations

Node/CKT	L|6T,um	LPS|6T,um	L|7T,um	LPS|7,um	L|8T,um	LPS|8T,um	L|9T,um	LPS|9T,um	L|10T,um	LPS|10T,um
Q	6.90E-05	6.80E-05	6.90E-05	6.80E-05	6.60E-05	6.50E-05	7.60E-05	7.50E-05	7.70E-05	7.70E-05
QB	6.50E-05	6.50E-05	6.40E-05	6.40E-05	8.00E-05	8.00E-05	8.20E-05	8.20E-05	1.38E-04	1.38E-04
WL	2.80E-05	2.90E-05	2.80E-05	2.90E-05	NA	NA	NA	NA	6.40E-05	6.60E-05
WWL	NA	NA	NA	NA	2.80E-05	2.90E-05	3.80E-05	3.90E-05	NA	NA
WBLBO	NA	NA	NA	NA	2.00E-05	1.90E-05	NA	NA	NA	NA
BLBO	2.00E-05	1.90E-05	2.00E-05	1.90E-05	NA	NA	3.30E-05	3.20E-05	2.10E-05	2.00E-05
BLO	2.10E-05	2.10E-05	2.10E-05	2.10E-05	2.10E-05	2.10E-05	2.10E-05	2.10E-05	2.20E-05	2.20E-05
RWL	NA	NA	NA	NA	2.30E-05	2.30E-05	NA	NA	NA	NA
WCL	NA	NA	2.30E-05	2.30E-05	NA	NA	NA	NA	NA	NA
ROUT	NA	NA	NA	NA	NA	NA	NA	NA	5.50E-05	5.40E-05
RBLO	NA	NA	NA	NA	2.40E-05	2.30E-05	NA	NA	NA	NA
VDD	5.20E-05	5.80E-05	5.70E-05	6.40E-05	6.30E-05	6.90E-05	6.80E-05	7.50E-05	1.70E-04	1.83E-04
VSS	3.50E-05	3.60E-05	3.50E-05	3.70E-05	4.40E-05	4.60E-05	4.70E-05	4.90E-05	1.13E-04	2.23E-04

Table 4 Variation of capacitance in from 6T SRAM to 10T SRAM of pre and post-layout simulations

Node/CKT	C|6T, Ff	CPS|6T, Ff	C|7T, Ff	CPS|7T, Ff	C|8T, Ff	CPS|8T, Ff	C|9T, Ff	CPS|9T, Ff	C|10T, Ff	CPS|10T, Ff
Q	4.70E-15	4.99E-15	4.70E-15	5.01E-15	4.51E-15	4.82E-15	5.34E-15	5.63E-15	5.16E-15	5.47E-15
QB	4.43E-15	4.77E-15	4.46E-15	4.72E-15	5.55E-15	5.80E-15	5.60E-15	5.86E-15	1.02E-14	1.04E-14
WL	1.98E-15	1.96E-15	1.98E-15	1.96E-15	NA	NA	NA	NA	4.61E-15	5.43E-15
WWL	NA	NA	NA	NA	1.98E-15	1.96E-15	2.82E-15	2.78E-15	NA	NA
WBLBO	NA	NA	NA	NA	9.50E-16	1.03E-15	NA	NA	NA	NA
BLBO	9.50E-16	1.03E-15	9.50E-16	1.03E-15	NA	NA	1.67E-15	1.83E-15	1.02E-15	1.10E-15
BLO	1.00E-15	1.04E-15	1.00E-15	1.04E-15	1.00E-15	1.04E-15	1.00E-15	1.04E-15	1.04E-15	1.08E-15
RWL	NA	NA	NA	NA	1.33E-15	1.32E-15	NA	NA	NA	NA
WCL	NA	NA	1.33E-15	1.32E-15	NA	NA	NA	NA	NA	NA
ROUT	NA	NA	NA	NA	NA	NA	NA	NA	2.74E-15	2.99E-15
RBLO	NA	NA	NA	NA	1.13E-15	1.21E-14	NA	NA	NA	NA
VDD	8.25E-15	9.93E-15	9.12E-15	1.11E-14	9.88E-15	1.22E-14	1.08E-14	1.33E-14	2.76E-14	3.11E-14
VSS	2.04E-15	3.05E-15	1.85E-15	2.82E-15	2.63E-15	3.67E-15	2.76E-15	3.80E-15	6.39E-15	1.57E-15

Initially, the values of R, and C from pre layout are observed taking length as parameter and is shwon in Fig. 10. Linear regression algorithm is chosen optimum as the curves are fillting to linear relation using both resistance and capacitance parameters.

Similarly, the same procedure is repeated for post-layout simulation data. The extracted post-layout paramters are collected in another CSV file as shown in Table

Table 5 Variation of resistance in from 6T SRAM to 10T SRAM of pre and post-layout simulations

Node/CKT	R|6T, ohms	RPS|6T, ohms	R|7T, ohms	RPS|7T, ohms	R|8T, ohms	RPS|8T, ohms	R|9T, ohms	RPS|9T, ohms	R|10T, ohms	RPS|10T, ohms
Q	7240	2415	7427	2977	7469	3019	7678	3019	7381	2931
QB	6673	5005	6557	4889	6744	5076	6798	5076	7461	5793
WL	369	369	369	369	NA	NA	NA	NA	677	677
WWL	NA	NA	NA	NA	369	369	553	369	NA	NA
WBLBO	NA	NA	187	187	2094	282	NA	NA	NA	NA
BLBO	2094	282	2094	282	NA	NA	4188	563	2095	283
BLO	1158	2127	1158	2127	1158	2127	1158	2127	1158	2127
RWL	NA	NA	NA	NA	187	187	NA	NA	NA	NA
WCL	NA	NA	NA	NA	NA	NA	NA	NA	NA	NA
ROUT	NA	NA	NA	NA	NA	NA	NA	NA	4784	2146
RBLO	NA	NA	NA	NA	2096	284	NA	NA	NA	NA
VDD	7064	23510	7660	24610	8031	24981	8401	25351	16904	54927
VSS	3609	5547	2468	3437	4763	7669	4768	7674	9777	35023

7. The values of R, and C from post layout are observed in Fig. 11. Further simulations were carried out using pre-layout data with 80% for training and 20% for testing, followed the same for post layout too. In this process, observed the predictions of capacitance and resistance individually for a chosen length in micronmeter technology. This is done using R and L as independent and C as dependent parameter in first case, similarly C and L as independent and R as dependent parameter in second case.

3.3 Multivariate Linear Regression Algorithm

Using Multivariate linear regression algorithm, the prediction accuracies are observed as 61 and 73.8% under pre layout, 76 and 80% under post-layout simulations for capacitances and resistances, respectively using micrometer technology. The predicted capacitance and resistance values under post-layout simulation are compared with the existing post-layout trained data for validity and are shown in Fig. 12. The two variations are comparable to say that the predictions are closer to practical validity.

The same process is extended to both pre and post-layout simulations using nanometer technology, in which the model is trained and tested as well, whereas the nanometer technology is taken by scaled down the 120μm node to 0.9 nm node. Corresponding simulations are shown in Figs. 13 and 14. The prediction accuracy is observed as 76 and 81% for both capacitance and resistance, respectively in the case of pre-layout modeling. Similarly, 63 and 80% of accuracy are observed for both capacitance and resistance, respectively in the case of post layout modeling. This work is extended further by modeling to predict R and C under nanometer technology using both pre and post-layout micron meter technology parameters as trained data. The results are calculated in terms of Absolute Error and Maximum Absolute Error as shown in Table 8.

- **R^2 error $= (1\text{-diff}^2)/(\text{actual-mean})$** is a statistic measurement to say the goodness of fit of a model.
- If **$R^2 = 1$(one)** indicates that the regression predictions perfectly fit the data.

Table 6 CSV file for pre-layout

L	C	R
6.80E−08	4.99E−15	2415
6.50E−08	4.77E−15	5005
2.90E−08	1.96E−15	369
1.90E−08	1.03E−15	282
2.10E−08	1.04E−15	2127
5.80E−08	9.93E−15	23,510
3.60E−08	3.05E−15	5547
6.80E−08	5.01E−15	2977
6.40E−08	4.72E−15	4889
2.90E−08	1.96E−15	369
1.90E−08	1.03E−15	282
2.10E−08	1.04E−15	2127
2.30E−08	1.32E−15	187
6.40E−08	1.11E−14	24,610
3.70E−08	2.82E−15	3437
6.50E−08	4.82E−15	3019
8.00E−08	5.80E−15	5076
2.90E−08	1.96E−15	369
1.90E−08	1.03E−15	282
2.10E−08	1.04E−15	2127
2.30E−08	1.32E−15	187
2.30E−08	1.21E−14	284
6.90E−08	1.22E−14	24,981
4.60E−08	3.67E−15	7669
7.50E−08	5.63E−15	3228
8.20E−08	5.86E−15	5130
3.90E−08	2.78E−15	553
3.20E−08	1.83E−15	563
2.10E−08	1.04E−15	2127
7.50E−08	1.33E−14	25,351
4.90E−08	3.80E−15	7674
7.70E−08	5.47E−15	2931
1.38E−09	1.04E−14	5793
6.60E−08	5.43E−15	677
2.00E−08	1.10E−15	283
2.20E−08	1.08E−15	2127
5.40E−08	2.99E−15	2146
1.83E−07	3.11E−14	54927
2.23E−07	1.57E−15	35023

Table 7 CSV file for post layout

L	C	R
6.90E−05	4.70E−15	7240
6.50E−05	4.43E−15	6673
2.80E−05	1.98E−15	369
2.00E−05	9.50E−16	2094
2.10E−05	1.00E−15	1158
5.20E−05	8.25E−15	7064
3.50E−05	2.04E−15	3609
6.90E−05	4.70E−15	7427
6.40E−05	4.46E−15	6557
2.80E−05	1.98E−15	369
2.00E−05	9.50E−16	2094
2.10E−05	1.00E−15	1158
2.30E−05	1.33E−15	187
5.70E−05	9.12E−15	7660
3.50E−05	1.85E−15	2468
6.60E−05	4.51E−15	7469
8.00E−05	5.55E−15	6744
2.80E−05	1.98E−15	369
2.00E−05	9.50E−16	2094
2.10E−05	1.00E−15	1158
2.30E−05	1.33E−15	187
2.40E−05	1.13E−15	2096
6.30E−05	9.88E−15	8031
4.40E−05	2.63E−15	4763
7.60E−05	5.34E−15	7678
8.20E−05	5.60E−15	6798
3.80E−05	2.82E−15	553
3.30E−05	1.67E−15	4188
2.10E−05	1.00E−15	1158
6.80E−05	1.08E−14	8401
4.70E−05	2.76E−15	4768
7.70E−05	5.16E−15	7381
1.38E−04	1.02E−14	7461
6.40E−05	4.61E−15	677
2.10E−05	1.02E−15	2095
2.20E−05	1.04E−15	1158
5.50E−05	2.74E−15	4784
1.70E−04	2.76E−14	16,904

(continued)

Table 7 (continued)

L	C	R
1.13E−04	6.39E−15	9777

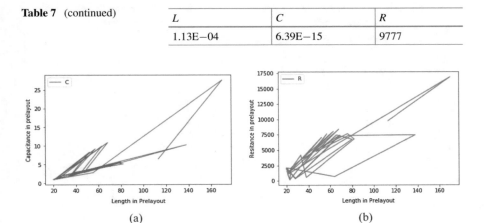

Fig. 10 Pre simulation layout capacitance and resistance variation over chosen lengths using micronmeter technology

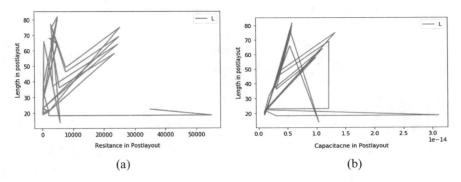

Fig. 11 Post simulation layout capacitance and resistance variation over chosen lengths using micronmeter technology

3.4   K Means Clustering Algorithm

For the proposed RC estimation method using pre-layout CSV file, k is selected as 4 and corresponding resistance clusters formed are shown in Fig. 15. Corresponding accuracy score observed using k-means algorithm observed is 100 percent in estimating the R or C values against to other two chosen parameters.

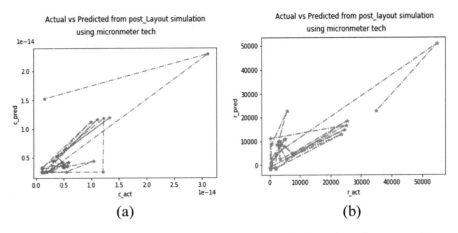

Fig. 12 Post simulation layout capacitance and resistance predictions using micrometer tech

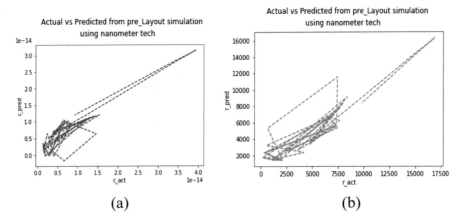

Fig. 13 Pre-layout simulation capacitance and resistance predictions using nanometer tech

4 Result Analysis

By considering the circuit of SRAM from 6T to 10T using 120μm and 0.9 nm, using linear regression model the capacitance and resistances are estimated for chosen length. In this process, the accuracy level is improved from micron meter to nanometer technology prediction model and is observed as 76 and 81% for both capacitance and resistance, respectively in the case of pre-layout modeling. Similarly, 63 and 80% of accuracy are observed for both capacitance and resistance, respectively in the case of post-layout modeling. Further, the model is applied with various test and train data chosen for different technologies and observed absolute average error and maximum absolute error. Finally, the CSV data set is applied with k-means

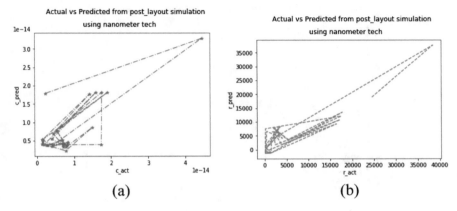

Fig. 14 Post-layout simulation capacitance and resistance predictions using nanometer tech

Table 8 Accuracy and error calculations for chosen models

S. No	Model		% Accuray	Absolute error	Maximum absolute error
	Train_data	Test_data			
1	Pre layout_μm	Post layout_μm	52.31	62.426	11.646
2	Pre layout_μm	Post layout_nm	13.6	131.56	113.67
3	Pre layout_μm	Pre layout_nm	−9.9	129	71
4	Pre layout_μm	Post layout_nm	35	120	81
5	Post layout_μm	Post layout_nm	78	80	38
6	Post layout_μm	Pre layout_nm	−15.25	72.6	12.85

clustering model and observed 100% accuracy in estimating the predicted R or C against the chosen other two parameters. Compared with the existing models [13], in the proposed method the accuracy and error predictions are better, as the data is considered from a progressive layered architecture built from 6T to 10T rather than using individual circuits. Prediction accuracies can be enhanced when dataset is considered maximum, rather than using small datasets obtained from individual circuits.

5 Conclusions

In this work, we have projected a method of interconnect RC estimation using machine learning algorithms for the chosen length, using layout schematic information. This method unanimously can be used for any sort of designs that belongs to

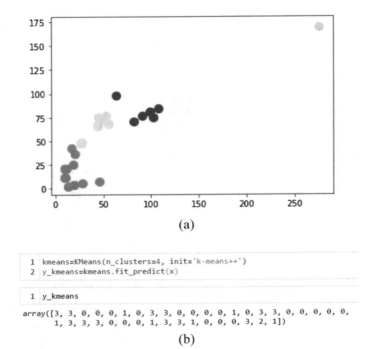

(a)

```
1  kmeans=KMeans(n_clusters=4, init='k-means++')
2  y_kmeans=kmeans.fit_predict(x)
```

```
1  y_kmeans
```

```
array([3, 3, 0, 0, 0, 1, 0, 3, 3, 0, 0, 0, 0, 1, 0, 3, 3, 0, 0, 0, 0, 0,
       1, 3, 3, 3, 0, 0, 0, 1, 3, 3, 1, 0, 0, 0, 3, 2, 1])
```

(b)

Fig. 15 Clusters of resistances using K-means clustering

either digital or analog or mixed. The advantage of the proposed method is, simple in extraction of R and C using any existing industry-supported EDA tools using to any technology node. The model with experimental results proved significant improvement in prediction of R and C with better accuracies compared to existing methods and is proved by k-means. Further, this work can be extended by making the model more empirical and realistic by involving a few more processing parameters to achieve greater accuracies even at deeper nanometer technologies.

References

1. Foo, H.Y., Leong, K.W.C., Mohd-Mokhtar, R.: Density aware interconnect parasitic estimation for mixed signal design. In: Proceedings: IEEE International Conference on Circuits and Systems (ICCAS), 258–262 (2012)
2. Shah, S., Nunez, A.: Pre-layout parasitic estimation in interconnects. In: 48th Midwest Symposium on Circuits and Systems, 1442–1445 (2005)
3. Martins, R., Lourenço, N., Canelas, A., Povoa, R. Horta N.:AIDA: robust layout-aware synthesis of analog ICs including sizing and layout generation, In: International Conference on Synthesis, Modeling, Analysis and Simulation Methods and Applications to Circuit Design (SMACD), 1–4 (2015)

4. Chang, E., et al., BAG2: a process-portable framework for generator based AMS circuit design, In: IEEE Custom Integrated Circuits Conference (CICC), 1–8 (2018)
5. The International Roadmap for Devices and Systems, IEEE (2017) [Online], Available: https://irds.ieee.org/images/files/pdf/2017/2017IRDS_MM.pdf
6. Barboza, E.C., Shukla, N., Chen, Y., Hu, J.: Machine learning-based pre-routing timing prediction with reduced pessimism, In: 56th ACM/IEEE Design Automation Conference (DAC), 1–6 (2019)
7. Abu-Mostafa, Y.S., Magdon-Ismail, M., Lin, H.-T.: Learning from Data, vol. 4. AMLBook, New York, NY, USA (2012)
8. Odabasioglu, A., Celik, Pileggi, M.L.T., Pileggi, L.T., Pileggi, L.T.: PRIMA: passive reduced-order interconnect macro modeling algorithm. In: IEEE/ACM International Conference on Computer-Aided Design, pp. 58–65 (1997)
9. Takeda, K., Hagihara, Y., Aimoto, Y., Nomura, M., Nakazawa, Y., Ishii, T., Kobatake, H.: A read-static-noise-margin-free SRAM cell for low-VDD and high-speed applications. IEEE J. Solid-State Circuits **41**(1), 113–121 (2005)
10. Kulkarni, J.P., Goel, A., Ndai, P., Roy, K.: A read-disturb-free, differential sensing 1r/1w port, 8t Bitcell Array. IEEE Trans. Very Large Scale Integrat (Vlsi) Syst. **19**(9) (2011)
11. Birla, S., Singh, R.K., Pattnaik, M.: Static noise margin analysis of various sram topologies. IACSIT J. Eng. Technol. **3**(3), 304 (2011)
12. Glocker, E., Schmitt-Landsiedel, D., Drapatz, S.: Countermeasures against NBTI degradation on 6T-SRAM cells. Adv. Radio Sci. **9**, 255–261 (2011)
13. Shook, B., Bhansali, P., et.al.: MLParest: machine learning based parasitic estimation for custom circuit design. In: IEEE International Conference on Block Chain and Crypto Currency, 2–6 May 2020, doi: 978-1-7281-1085-1/20/$31.00 © IEEE (2020)

PAPR Reduction in SDR-Based OFDM System

L. Chandini and A. Rajagopal

Abstract Orthogonal frequency division multiplexing (OFDM) is a promising new technology as it manages multicarrier modulation. Due to the high spectral bandwidth efficiency and robustness against multipath fading and intersymbol interference (ISI), the OFDM has been accepted for several high data rate wireless communication systems. But the OFDM system's major disadvantage is its high peak-to-average power ratio (PAPR). Hence, the execution of PAPR reduction methods in the OFDM system assists in defeating the high PAPR problem. Software-defined radio (SDR) is a radio platform that is flexible and makes wireless networks highly adaptive. Hence, in this paper, the performance of the selective mapping (SLM) and clipping and filtering PAPR reduction techniques are analysed for SDR-based OFDM systems using LabVIEW (Laboratory Virtual Instrument Engineering Workbench). The proposed work not only provides a simpler method for PAPR reduction but also scalable for a large value of data in the form of bits.

Keywords Clipping and filtering technique · Selective mapping (SLM) technique · Software defined radio (SDR) · Orthogonal frequency division multiplexing (OFDM) · Peak-to-average power ratio (PAPR) · Laboratory virtual instrument engineering workbench (LabVIEW)

1 Introduction

Most communication systems depend on the OFDM multicarrier modulation technique since it is robust against intersymbol interference, multipath fading, delay spread, and scarcity of bandwidth. But high PAPR is a serious issue in OFDM systems. Because it results in high out-of-band radiation, in-band distortions [1, 2], intersymbol interference etc. which lowers the OFDM system's performance. To lessen the high PAPR in the OFDM system, two frequently utilized PAPR reduction schemes are selective mapping and clipping and filtering. The key idea in selected

L. Chandini (✉) · A. Rajagopal
Department of Electronics and Communication, Dayananda Sagar College of Engineering, Bengaluru, India

© The Author(s), under exclusive license to Springer Nature Singapore Pte Ltd. 2022   1051
V. Suma et al. (eds.), *Evolutionary Computing and Mobile Sustainable Networks*,
Lecture Notes on Data Engineering and Communications Technologies 116,
https://doi.org/10.1007/978-981-16-9605-3_74

mapping is the development of several sequences which carry the same information. This is accomplished by modulating the data block with different phase sequences generated. The PAPR value of every sequence is determined, and the sequence with the least PAPR is chosen for broadcast [3]. In the clipping and filtering technique, the signal's peaks that surpass the reference threshold are clipped, while the peaks that are less than or equal to the threshold level will not be changed. The gain introduced by clipping, in a multicarrier system, can be used to increase the range and coverage and can also provide significant power savings [4].

SDR is a vital equipment for wireless communication because it can interact with various radios. Also, it offers a flexible platform for the operation of the latest technologies [5]. In SDR technology, radio functionalities are accomplished by running software modules on a generic hardware platform. It delivers an inexpensive and effective solution for building multi-range, multi-mode, and multi-functional wireless devices. LabVIEW and USRP (Universal Software Radio Peripheral) are easy-to-use software defined radio platforms. Today's computers may contain very fast processors and high-speed interfaces, so we can leverage these capabilities for the SDR by executing them on a computer rapidly, using LabVIEW.

LabVIEW communication is the hardware-aware design environment for prototyping communications systems that streamlines prototyping by offering a single, cohesive environment that enables both the processor and FPGA (Field Programmable Gate Array) programming. It supports graphical approaches, instrument control, data acquisition, test automation, and embedded system design, as well as shortening the design process. LabVIEW is mainly used for interaction with an SDR transceiver (USRP 2930) to observe the working of the project in real-time. It is a highly productive development environment for creating custom applications that interact with real-world data or signals. In this paper, the implementation of an SDR-based OFDM scheme by using LabVIEW software is demonstrated. The entire OFDM system and PAPR reduction techniques are designed in LabVIEW software with the help of different signal processing blocks and communication blocks.

2 Related Works

In this section, some of the previous and recent research on OFDM, PAPR reduction techniques and SDR are discussed.

Mitola in [6] described that software radio is a meta-level arrangement for combining the digital signal processing (DSP) primitives into communications systems functions. In [7], Jiang and Wu have analysed the different techniques for reducing PAPR in the OFDM system, all of which have the ability to provide significant PAPR reduction at the expense of transmit signal power increase, computational complexity increase, bit error rate (BER) performance degradation, and so on.

Using a programmable fixed-point DSP, Yeh and Ingerson in [8] proposed a software reconfigurable OFDM system for wireless communication. Janjić et al. [9] worked on the implementation of a secondary cognitive link based on an OFDM system that was accomplished by utilizing USRP N210 kit platforms. But all the processing in their work was done over MATLAB, so the speed of the transmission and reception was very slow. In [10], Bhosle and Ahmed described the advantages of USRP and LabVIEW for improving the operation of OFDM signals. Ann and Jose in [11] implemented different PAPR reduction techniques at the OFDM transmitter and provided their comparison based on CCDF and BER performance of the system. Sravanti and Vasantha in [12] evaluated the OFDM system with assorted precoding approaches in order to reduce the high PAPR. Hafsa Iqbal and Khan presented the implementation of the selective mapping PAPR reduction approach in OFDM on the SDR platform using MATLAB and Xilinx in [13].

When the techniques for PAPR reduction are employed in the SDR technology, as Gordillo et al. showed in [14], range and coverage can be extended by proper power utilization. In [15], El Bahi et al. designed an SDR-based prototype to demonstrate two spectrum-sensing approaches for detecting the presence of OFDM primary user's signals. The evaluation of the performance of BER for a massive multi-input multi-output (MMIMO) system with a spatial time shift keying (STSK) scheme over three-dimensions (3-D) $\alpha-\lambda-\mu$ fading model is proposed by Chen and Joy in [16]. Agarwal et al. in [17] gave the readers a correct education about the PAPR problem in OFDM systems and also gave a comparison of the results for clipping and filtering and SLM PAPR reduction techniques. Lalwani et al. proposed the use of reinforcement learning for security of an LDPC (low-density parity check) coded cognitive radio in [18], where the LDPC decoder and software portion of the cognitive radio are done using LabVIEW software.

2.1 Analysis of Different PAPR Reduction Techniques

In the literature, several PAPR reduction techniques have been proposed. Thus, by using different techniques, it is possible to minimize the high PAPR.

The PAPR reduction technique should be chosen with awareness according to various system requirements. Before employing PAPR reduction techniques in a digital communication system, there are a number of issues to be considered. These issues include power increase in transmit signal, PAPR reduction capacity, loss in data rate, BER increase at the receiver, and so on. A comparisons between some of the PAPR reduction techniques are depicted below [7, 19] (Table 1).

3 Orthogonal Frequency Division Multiplexing (OFDM)

OFDM is a mechanism for delivering parallel data by employing a huge number of subcarriers spaced at harmonic frequencies so that the carriers are orthogonal to one another [20].

Table 1 Comparison of PAPR reduction techniquess

PAPR reduction techniques	Parameters		
	Decrease distortion	Power increases	Defeat data rate
Clipping and filtering	No	No	No
Selective mapping	Yes	No	Yes
Partial transmit sequence (PTS)	Yes	No	Yes
Coding	Yes	No	Yes
Tone reservation	Yes	Yes	Yes
Tone injection	Yes	Yes	No

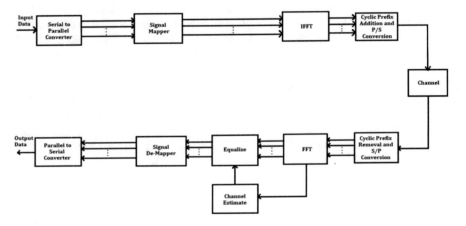

Fig. 1 OFDM system model

Figure 1 depicts the OFDM system in block diagram form. In order to achieve strong spectral performance, the OFDM technology relies on orthogonality among its subcarriers. The chosen frequencies for multiple carrier communication systems must be orthogonal to each other. When subcarrier frequencies are orthogonal to one another, crosstalk interference between subcarriers is reduced, resulting in higher spectrum utilization. The main idea of the OFDM approach is to split a large stream of accessible data into numerous data streams with lower rates. After that, numerous orthogonal (independent) subcarriers are used to broadcast these lower-rate streams simultaneously. As a result, the symbol's time duration grows, while the time dispersion caused by multipath delay spread decreases. Hence, OFDM has been used as a most important modulation technique in several communication protocols.

In an OFDM system, incoming serial data is converted into parallel at the serial-to-parallel converter, then the parallel data is modulated into complex-valued symbols in the signal mapper. After that, using an IFFT operation, the modulated data is transformed from the frequency to the time domain. Subsequently, a cyclic prefix is added to the symbols to overcome ISI, and then the data is converted back from parallel to serial. Finally, the receiver performs a reverse operation to restore the original data. At the receiver, equalizer is used to reverse the distortion incurred by a signal transmitted through a channel by estimating the channel at all subcarriers.

4   Peak-to-Average Power Ratio (PAPR)

PAPR is the ratio of the signal's maximum power to its average power. It is expressed in decibel as

$$\mathrm{PAPR(dB)} = 10 \log_{10} \left(\frac{P_{\mathrm{peak}}}{P_{\mathrm{average}}} \right)$$

$$= 10 \log_{10} \frac{\max[|x_{(n)}|^2]}{E[|x_n|^2]} \tag{1}$$

where P_{peak} denotes the maximum output power. P_{average} is the average output power. x_n is the OFDM signal received after applying IFFT to modulated symbols. E denotes the expected value [21]. High PAPR is caused when a number of subcarriers which are modulated independently are added up coherently.

To ensure linear amplification, high-power amplifiers (HPA) employed in the RF communication systems are forced to have very large back-off by high PAPR. This leads to a decrease in the efficiency of the amplifier. And if an amplifier works with nonlinear characteristics, unwanted distortion will be caused. Also, the orthogonality between the carriers is destroyed by high PAPR, which causes high adjacent channel interference. Also, this high PAPR reduces the battery life of mobile devices. Hence, PAPR reduction is essential for the implementation of OFDM systems. The selective mapping technique and the clipping and filtering technique are the most often used PAPR reduction techniques while designing the OFDM system, which are described below.

4.1   Clipping and Filtering Technique

Figure 2 shows the signal distortion technique called clipping and filtering and its implementation in the OFDM system is given in Fig. 3. In this technique, when the incoming OFDM signal's amplitude exceeds the defined level of threshold, then the amplitude is clipped to the level of threshold, and the signal is delivered without changing its phase and amplitude when the amplitude is less than or equal to the threshold level.

Let X_K be the input data block which is in the frequency domain where $K = 1$, 2,..., N. To capture the signal peaks, X_K is padded with $(L-1)$ N zeros between $K = N/2-1$ and $K = N/2$ samples. Then the signal in frequency domain is subjected to

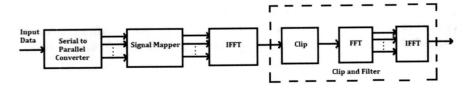

Fig. 2 Block diagram of clipping and filtering technique

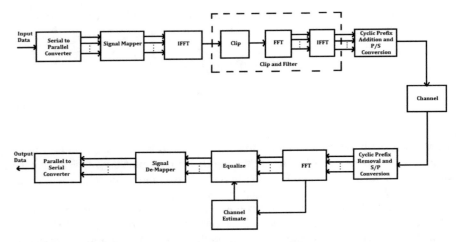

Fig. 3 Clipping and filtering technique implementation in OFDM system

IFFT to produce the oversampled discrete signal in time domain. It can be expressed as

$$x_n = \frac{1}{\sqrt{LN}} \sum_{K=0}^{LN-1} X_K e^{j\frac{2\pi Kn}{LN}}, \quad 0 \leq n \leq (LN-1) \tag{2}$$

In the above equation, L represents the oversampling factor. The level for clipping is denoted by 'A' which is then applied to the sampled signal.

The clipped signal, $y(x_n)$ is written as

$$y(x_n) = x_n, \quad \text{if } |x_n| \leq A$$
$$= Ae^{j\phi}(x_n), \quad \text{if} |x_n| > A \tag{3}$$

The altered PAPR after the clipping process can be calculated by

$$PAPR = \frac{A^2}{E\{|y(x_{(n)})|^2\}} \tag{4}$$

Amplitude clipping causes unwanted distortion in the OFDM signal, which results in error performance degradation and a decrease in spectral efficiency. In this PAPR reduction method, two FFT operations are used to carry out the filtering process. The clipped OFDM signal is changed from time to frequency domain by the first FFT operation and out of band components are eliminated by the second IFFT operation.

4.2 Selective Mapping Technique

The SLM technique is a signal scrambling approach, which is shown in Fig. 4. Figure 5 gives its implementation in the OFDM system. At the transmission side, this approach generates a group of various data blocks whose data content is similar to the data of the original data block. And then the data block which gives the least PAPR is chosen for transmission.

At first, modulation of N subcarriers which are independent of each other with the various pseudorandom patterns generates M sequences with similar information. The pseudorandom sequences can be written as

$$R_m = \left[R_{m,0}, R_{m,1}, \ldots, R_{m,N-1} \right]^{\mathrm{T}}, \quad m = 1, 2, \ldots, M \tag{5}$$

In the above equation, $R_{m,n} = e^{j\phi_{m,n}}$ denotes the rotation factor and $\phi_{m,n}$ is evenly distributed from 0 to 2π. The branch m's symbols can be expressed as

$$S_m = \left[X_0 R_{m,0}, X_1 R_{m,1}, \ldots, X_{N-1} R_{m,N-1} \right] \tag{6}$$

After that, M frequency domain data blocks are transmitted into the IFFT block, which converts them into the time domain. Then, the PAPR values of the data blocks are calculated individually, and finally, the one that has the lowest PAPR value is

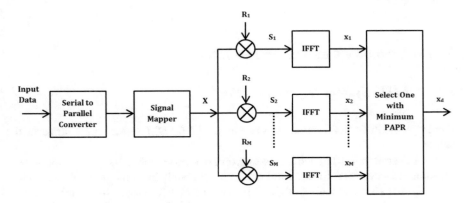

Fig. 4 Block diagram of selective mapping technique

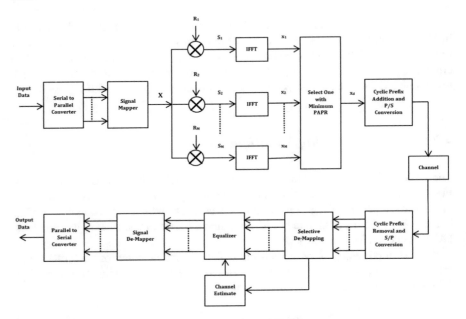

Fig. 5 Selective mapping technique implementation in OFDM system

transmitted. Its expression is given as

$$x_d = \arg \min_{1 \leq m \leq M} [\text{PAPR}(x_m)] \tag{7}$$

The argmin(.) confirms that the argument of its value is decreased. The receiver must recognize the sequence having the least PAPR in order to properly demodulate the received signal. Hence, as side information, the receiver will need to receive the entire sequence of branch number m [21].

5 Performance Analysis and Results

5.1 OFDM System Implementation

The implementation and simulation results of the OFDM system are summarized below.

The transmitter blocks such as bits-to-word, word-to-symbol mapping, symbol-to-fft, addition of cyclic prefix and windowing are separately implemented and converted into sub virtual instruments (Sub VIs) in LabVIEW. Afterwards, these Sub VIs are combined to implement the OFDM transmitter, which generates the OFDM signal that is then transmitted through the channel.

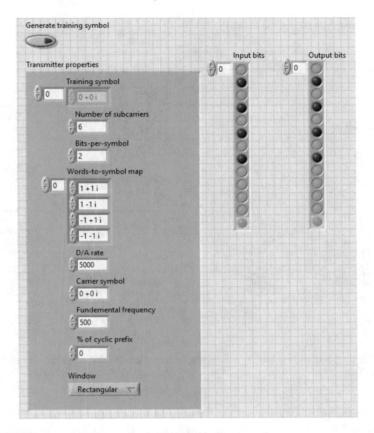

Fig. 6 OFDM system implementation in LabVIEW—Front panel

Similar to the transmitter side, the receiver blocks such as frequency synchronization, symbol-to-fft, equalizer, symbol-to-word, word-to-bit are separately implemented and converted into Sub VIs to implement the OFDM receiver which demodulates the OFDM signal. The front panel of the OFDM system implementation in LabVIEW is given in Fig. 6. This verifies that the input bit stream sent at the transmitter is received at the receiver side without any flipping, which confirms that the implemented generalized OFDM system is working. Parameters configured in LabVIEW are given in Table 2.

5.2 Implementation of Clipping and Filtering Technique

At first, the clipping and filtering technique is implemented as subVI and then interfaced with the OFDM transmitter as shown in Fig. 7. The time domain output of the

Table 2 Parameters configured in LabVIEW

Parameter	Value
Number of subcarriers	6
Bits-per-symbol	2
Modulation	QPSK
D/A rate	5000 samples/s
Fundamental frequency	500 Hz

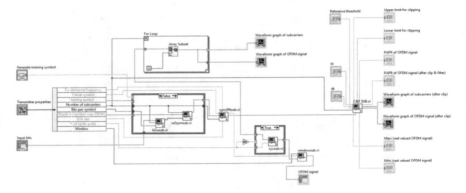

Fig. 7 Clipping and filtering technique implementation at the OFDM transmitter in LabVIEW—block diagram

symbol-to-fft subVI (IFFT block) in the OFDM transmitter is taken as input to the clipping and filtering subVI, and its waveform graph is plotted using the for loop and array subset. The for loop function executes its subdiagrams *n* times, where n is the value connected to the count (*N*) terminal. Array subset returns a portion of the array starting at the index.

In the clipping and filtering subVI, the time domain output of symbol-to-fft is clustered with t0 and dt, which is given as input to the PAPR functional block. t0 specifies the starting time of the time domain output. dt specifies the time interval between data points in the time domain output. The default value of t0 and dt is '0' and '1', respectively. The output terminal of the PAPR functional block is connected to the numerical indicator to display the PAPR value of the time domain OFDM data sequence without clipping and filtering. Then, the real and imaginary part of the output of symbol-to-fft are extracted by using complex to real/imaginary functional block. Then the real part of the OFDM signal is given as input to an array max and min functional block to get the maximum and minimum value of the real-valued OFDM signal.

Afterwards, the reference threshold is defined, which varies from 0.1 to 0.9. The reference threshold value is multiplied with both the maximum and minimum values of the real-valued OFDM signal to obtain the upper limit and lower limit for clipping, respectively. Then the real part of the OFDM signal is given as input to a clipping functional block which clips the input to within the bounds specified by upper limit and lower limit. For example, consider the maximum and minimum value of the real-valued OFDM signal as 100 and −100%. Multiplication of reference threshold 0.1 with these values gives 10 and −10% as the upper limit and lower limit for clipping, respectively. Hence, the amplitude of the OFDM signal above 10% of the OFDM signal and below −10% of the OFDM signal are clipped.

Then the real-valued clipped OFDM signal and previously extracted imaginary part of the OFDM signal are given as input to a real/imaginary to complex functional block to obtain the complex OFDM signal, and its waveform graph is plotted to show the clipped amplitude. Then it is passed through the two FFT operations for filtering. The output obtained after filtering is then clustered with t0 and dt and sent into the PAPR functional block which outputs the PAPR value of the clipped and filtered OFDM signal. Front panel result of the clipping and filtering technique implementation at the OFDM transmitter for the input data bits 101010101010 with a reference threshold of 0.5 is shown in Fig. 8, and Table 3 gives the results of this technique for different threshold values.

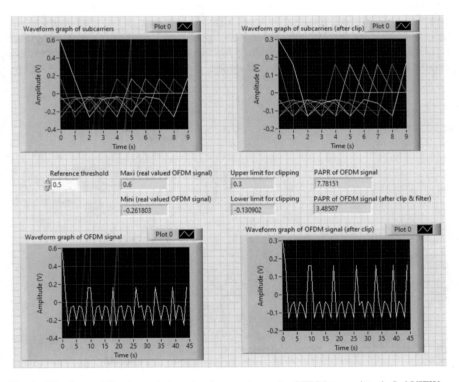

Fig. 8 Clipping and filtering technique implementation at the OFDM transmitter in LabVIEW—front panel

Table 3 Results for the clipping and filtering technique in LabVIEW for the input bits 101010101010	Reference threshold	PAPR (dB)	PAPR (dB) [Clip and filter]
	0.1	7.78151	3.21364
	0.2	7.78151	3.51303
	0.3	7.78151	3.6972
	0.4	7.78151	3.62022
	0.5	7.78151	3.48507
	0.6	7.78151	3.29703
	0.7	7.78151	3.06285
	0.8	7.78151	2.78999
	0.9	7.78151	2.48606

5.3 Implementation of Selective Mapping Technique

The selective mapping technique is implemented as subVI and interfaced with the OFDM transmitter as shown in Fig. 9. The parallel symbol output of the word-to-sym subVI (signal mapper) in the OFDM transmitter is taken as input to the selective mapping subVI. In the Selective mapping subVI, the entire parallel symbol output is wired to the first inbuilt for loop function of LabVIEW, which performs two function. First, it multiplies the parallel symbol output X with five different imaginary

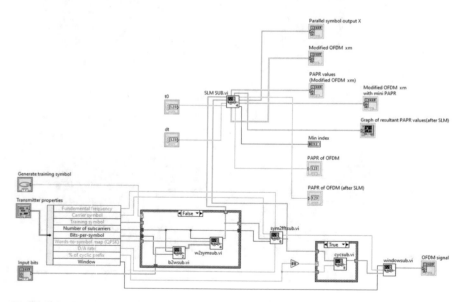

Fig. 9 Selective mapping technique implementation at the OFDM transmitter in LabVIEW—block diagram

pseudorandom sequences R_1, R_2, R_3, R_4 and R_5, which outputs the OFDM sequences S_m, and then it converts the modified OFDM sequences S_m from the frequency domain to the time domain using the IFFT functional block. Next, PAPR values of the time domain-modified OFDM sequences x_m are calculated by using the index array and PAPR functional block of LabVIEW in the second for loop. The index array returns the elements or subarrays of an n-dimension array at index.

The resultant PAPR values of the OFDM sequences x_m are given as input to the array max and min functional block to get the minimum index value. Then the modified OFDM x_m and minimum index value are given as inputs to another index array which outputs the OFDM data sequence with the least PAPR. Further, the obtained output is connected to the array indicator to display the selected OFDM data sequence with the least PAPR and is also given as input to the PAPR functional block by clustering with t0 and dt, which outputs the PAPR value of the selective mapped OFDM sequence separately. Also, the PAPR value of the OFDM sequence without selective mapping is calculated. The front panel result of the selective mapping technique implementation at the OFDM transmitter for the input data bits 101010101010 is shown in Fig. 10. A comparison between the PAPR values of the above two techniques is given in Table 4.

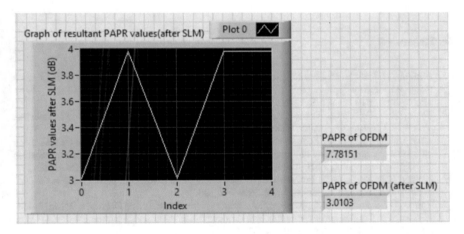

Fig. 10 Selective mapping technique implementation at the OFDM transmitter in LabVIEW—front panel

Table 4 Results for the clipping and filtering and selective mapping technique in LabVIEW for the input bits 101010101010	PAPR estimation (dB)	Value
	PAPR	7.78151
	PAPR after clip and filter with reference threshold 0.5	3.48507
	PAPR after selective mapping	3.0103

6 Conclusion

For high-speed data broadcasting, OFDM is the most widely utilized multicarrier communication technique. But the major disadvantage of the OFDM system is the high PAPR value. This paper demonstrates the detailed description of the two different PAPR reduction techniques such as selective mapping and clipping and filtering and also provides their successful implementation in the SDR-based OFDM system using LabVIEW. Clipping and filtering technique and SLM technique achieved a considerable PAPR reduction of 4.29644 dB for reference threshold 0.5 and 4.77121 dB, respectively, in the OFDM signal compared to the OFDM signal without clipping and filtering and SLM. Among them, the PAPR reduction efficiency of the clipping and filtering technique is higher than that of the selective mapping technique. The clipping and filtering technique results in signal distortion due to amplitude clipping, which causes degradation in the performance of bit error rate, which decreases the spectral efficiency, but this technique does not defeats the data rate, whereas the selective mapping technique defeats the data rate but decreases the signal distortion. Hence, these techniques need to be chosen based on the communication system requirements.

References

1. Gautam, P., Lohani, P., Mishra, B.: Peak-to-average power ratio reduction in OFDM system using amplitude clipping. In: IEEE Region 10 Conference (TENCON). 1101–1104 (2016). https://doi.org/10.1109/TENCON.2016.7848179
2. Sandoval, F., Poitau, G., Gagnon, F.: Hybrid peak-to-average power ratio reduction techniques: review and performance comparison. IEEE Access. **5**, 27145–27161 (2017)
3. Reddy, D.S.P, Sudha, V., Sriramkumar, D.: Low complexity PAPR reduction in OFDM using both selective mapping and clipping methods. In: International Conference on Communication and Signal Processing. 1201–1204(2014). https://doi.org/10.1109/ICCSP.2014.6950042
4. Baxley, R.J., Zhou, G.T.: Power savings analysis of peak-to-average power ratio in OFDM. IEEE Trans. Consum. Electron. **50**(3), 792–798(2004)
5. Rohde, U.L., Poddar, A.K., Marius, S.A.: Next generation radios: SDR and SDN.In: IEEE-APS Topical Conference on Antennas and Propagation in Wireless Communications (APWC). 296-299(2017). https://doi.org/10.1109/APWC.2017.8062305
6. Mitola, J: Software radios-survey, critical evaluation and future directions. In: [Proceedings] NTC-92: National Telesystems Conference, Washington, DC, USA, 13/15–13/23 (1992). https://doi.org/10.1109/NTC.1992.267870
7. Jiang, T., Wu, Y.: An overview: peak-to-average power ratio reduction techniques for OFDM signals. In IEEE Trans. Broadcast. **54**(2), 257–268 (2008). https://doi.org/10.1109/TBC.2008.915770
8. Yeh, H., Ingerson, P. Software-defined radio for OFDM transceivers. IEEE Int. Syst. Conf. 261–266(2010). https://doi.org/10.1109/SYSTEMS.2010.5482434
9. Janjić, M., Brković, M., Erić, M.: Development of OFDM based secondary link: some experimental results on USRP N210 platform. In: 21st Telecommunications Forum Telfor (TELFOR). 216–219(2013). https://doi.org/10.1109/TELFOR.2013.6716211

10. Bhosle, A.S. Ahmed, Z.: Modern tools and techniques for OFDM development and PAPR reduction. In: International Conference on Electrical, Electronics, and Optimization Techniques (ICEEOT). Chennai. 290–292(2016) https://doi.org/10.1109/ICEEOT.2016.7755205
11. Ann, P.P. Jose, R.: Comparison of PAPR reduction techniques in OFDM systems. In: International Conference on Communication and Electronics Systems (ICCES). Coimbatore, India. 1–5(2016). https://doi.org/10.1109/CESYS.2016.7889995
12. Sravanti, T. Vasantha, N.: Precoding PTS scheme for PAPR reduction in OFDM. In: International Conference on Innovations in Electrical, Electronics, Instrumentation and Media Technology (ICEEIMT). Coimbatore. 250–254(2017). https://doi.org/10.1109/ICIEEIMT.2017.8116844
13. Iqbal, H. Khan, S.A.: Selective mapping: implementation of PAPR reduction technique in OFDM on SDR platform. In: 24th International Conference on Automation and Computing (ICAC).1–6(2018). https://doi.org/10.23919/IConAC.2018.8749039
14. Gordillo, B., Sandoval, F., Ludeña-González, P., Rohoden, K.: Increase the range and coverage on OFDM system using PAPR reduction by clipping on SDR. In: IEEE Third Ecuador Technical Chapters Meeting (ETCM). Cuenca, Ecuador.1–6(2018). https://doi.org/10.1109/ETCM.2018.8580261
15. El Bahi, F.Z., Ghennioui, H. Zouak, M.: Real-time spectrum sensing of multiple OFDM signals using low cost SDR based prototype for cognitive radio. In: 15th International Wireless Communications & Mobile Computing Conference (IWCMC). Tangier, Morocco. 2074–2079(2019) https://doi.org/10.1109/IWCMC.2019.8766788
16. Chen, J.I.Z.: The evaluation of performance for a mass—MIMO system with the STSK scheme over 3-d α–λ–μ fading channel. In: IRO J. Sustain. Wireless Syst. $\mathbf{1}(1)$, 1–19(2019)
17. Agarwal, K., Jadon, J.S., Chahande, M., Pant, M. Rana, A.: Mitigation the issue of peak to average power ratio in OFDM setups. In: 8th International Conference on Reliability. Infocom Technologies and Optimization (Trends and Future Directions) (ICRITO). Noida, India. 1258–1261(2020). https://doi.org/10.1109/ICRITO48877.2020.9197860
18. Lalwani, P., Anantharaman, R.: Reinforcement learning for a LDPC coded cognitive radio. In: 3rd International Conference on Innovative Data Communication Technologies and Application. 20–21 August 2021, 830–842, ISSN: 2367-4512
19. Sudha, V, Balan, S., Sriram Kumar, D.: Performance analysis of PAPR reduction in OFDM system with distortion and distortion less methods. In: International Conference on Computer Communication and Informatics. 1–4(2014). https://doi.org/10.1109/ICCCI.2014.6921809
20. Wu, Y., Zou, W.Y.: Orthogonal frequency division multiplexing: a multi-carrier modulation scheme. In IEEE Trans. Consum. Electron. $\mathbf{41}(3)$, 392–399 (1995). https://doi.org/10.1109/30.468055
21. Munni, T.N., Hossam-E-Haider, M.: Performance analysis of peak to average power ratio (PAPR) reduction techniques in OFDM system for different modulation schemes. In: 4th International Conference on Electrical Engineering and Information & Communication Technology (iCEEiCT). 605–610 (2018). https://doi.org/10.1109/CEEICT.2018.8628163

Printed in the United States
by Baker & Taylor Publisher Services